REMOTE SENSING
FOR THE EARTH SCIENCES

Manual of Remote Sensing
Third Edition

Edited by
Robert A. Ryerson

Volume 1

Earth Observing Platforms & Sensors
Edited by

Stanley A. Morain and Amelia M. Budge
(Available only as a CD-ROM)

Volume 2

Principles and Applications of Imaging Radar
Edited by

Floyd M. Henderson and Anthony J. Lewis

Volume 3

Remote Sensing for the Earth Sciences

Edited by
Andrew N. Rencz

REMOTE SENSING
FOR THE
EARTH SCIENCES

edited by

Andrew N. Rencz
Geological Survey of Canada

Manual of Remote Sensing

Third Edition, Volume 3
Robert A. Ryerson, editor

Published in Cooperation with the
American Society for Photogrammetry and Remote Sensing

John Wiley & Sons, Inc.
New York • Chichester • Weinheim • Brisbane • Singapore • Toronto

Library of Congress Cataloging-in-Publication Data:
(Revised for vol. 3)

Manual of remote sensing / Robert A. Ryerson editor-in-chief;
    published in co-operation with the American Society for
    Photogrammetry and Remote Sensing.—3rd ed.
        p.   cm.
    Includes bibliographical references and index.
    Contents: —v.3. Remote Sensing for the Earth
Sciences / edited by Andrew N. Rencz.
    ISBN 0471-29405-5 (alk. paper)
    1. Remote sensing. 2. Radar.   1. Ryerson, Robert A.
II. American Society for Photogrammetry and Remote Sensing.
G70.6.M36   1998
621.36'78—dc21   98-78865

Contents

Preface

Over 15 years has lapsed since the second edition of the *Manual of Remote Sensing* was published in 1983. It seemed timely that a new edition be compiled that updates developments and serves as a source of information between those time periods. Accordingly this manual, *Remote Sensing for the Earth Sciences*, was conceived. This volume, along with several others will cover specific disciplines that are designed to highlight topics that have occurred since publication of the 1983 edition of the *Manual of Remote Sensing*.

In 1983, digital imagery for remote sensing was typically acquired from orbital sensors with relatively coarse spatial and spectral bands. Data from airborne sensors provided more detailed information, but this was generally acquired for research purposes. Software and hardware for data analysis were expensive. The second edition of the *Manual* reflects developments to that time and remains a very useful source of information. However, much of the promise anticipated at that time has been realized, and there are significantly new directions in remote sensing Concepts such as hyperspectral, interferometry, and unmixing models are accepted. There are new sources of data. As noted in Chapter 11, the series of thematic mapper sensors have fulfilled expectations, and new earth resource satellites from France, India, Canada, and Japan were launched. Orbital imaging radar systems became operational. There is new spectral analysis software for a variety of applications. Similarly, other fields of research, now aligned with remote sensing, such as GIS and visualization, have advanced. Another far-reaching change has been the evolution of computer technology, which has led to the development of affordable and user-friendly software for desktop use. The user of the technology is faced with most of the same issues as in 1983 plus a new set of opportunities and problems. This book is meant to serve as a source for insight into some of these issues within the context of remote sensing applications in the earth sciences. It is no longer possible to contain all this information in one book, as earlier manuals did, and even within the field of earth sciences there are certain topics that cannot be covered.

In the tradition of the *Manual of Remote Sensing,* this book is meant to serve as a fundamental source of information for students and professionals in the field of remote sensing and for people in related fields. To this end the 13 chapters are provided by an international team of scientists who are active and well known in the remote sensing field. The book is divided into four sections. The first section concentrates on the theory of rock, mineral, soil, and vegetation spectra. This is mainly within the visible and infrared parts of the electromagnetic spectrum (0.4 to 12.5 μm). The second section is devoted to procedures used in information extraction and techniques used in the visual display of data, particularly in the integration of various geospatial data. The third section provides applications of remote sensing, including mineral exploration, hydrocarbon exploration, stratigraphic mapping, and a chapter on planetary geology. The final section provides details on sensors (particularly airborne and new satellites) relevant to the earth sciences, including radar, visible, infrared, and geophysical sensors and examples of the data in case studies. Although there are four separate sections, the sections are highly interdependent. For example, new applications build on the availability of new data and analytical techniques developed in response to ground-based studies relating "geofeatures" to a spectral response.

It is hoped that information in this book will serve as a useful source of information for the years to come and that some of the promises noted will be fulfilled in the next edition of the *Manual of Remote Sensing*.

Andrew Rencz

Contributors

Elsa A. Abbott, *Jet Propulsion Laboratory, California Institute of Technology, Mail Stop 183-501, 4800 Oak Grove Drive, Pasadena, CA 91109.*
Elsa A. Abbott is a geologist in the Earth and Space Sciences Division of the Jet Propulsion Laboratory. She received the B.Sc. degree in geology from California State University at Los Angeles in 1965. Her current research is concentrated on the use of thermal infrared data to study geological problems.

James F. Bell III, *Department of Astronomy, Cornell University, 424 Space Sciences Building, Ithaca, NY 14853*; jimbo@marswatch.tn.cornell.edu.
James F. Bell III is a planetary scientist and an assistant professor in the Department of Astronomy at Cornell University. He received his Ph.D. in geology and geophysics in 1992 from the University of Hawaii, focusing on the composition and mineralogy of the Martian surface. His current research involves telescopic and spacecraft observations of Mars, the Moon, outer solar system satellites, asteroids, and comets. He is a member of the science teams for the Near Earth Asteroid Rendezvous, Comet Nucleus Tour, Mars-98 Orbital Imager, and Mars-01 Rover spacecraft investigations.

Eyal Ben-Dor, *Department of Geography, Tel Aviv University, P.O. Box 39040, Tel Aviv, Israel 69978*; bendor@ccsg.tau.ac.il.
Eyal Ben-Dor received his Ph.D. in soil sciences from the Hebrew University of Jerusalem in 1991. He was a visiting scientist at the Center for Study of the Earth from Space (CSES) at the University of Colorado. He is serving as a full-time senior lecturer and senior researcher in the Department of Geography at Tel-Aviv University, in two fields: soil science and remote sensing. His current studies focus on quantitative mapping of soils, water, and urban areas from hyperspectral data, and developing low-cost remote sensing capabilities.

John Berry, *Shell E&P Technology Co., P.O. Box 481, Houston, TX 77001-0481*; berry@shellus.com.
John Berry obtained a B.A. in geology from the University of Pennsylvania and an M.Sc. in geology and oceanography from Columbia University. He began his geological career with Anglo American Corporation on the Zambain Copperbelt, where he became interested in remote sensing techniques as a possible means of mapping

the bedrock geology beneath the thick tropical residual soils without the necessity for laborious digging of pits to the C horizon. After returning to the United States he taught in North Carolina and then joined Earth Satellite Corporation, where he was involved in remote sensing projects on all five continents. He then joined Shell to pursue the use of remote sensing techniques for gold exploration, later serving at the research laboratory and with Pecten International.

John Broome, *Geological Survey of Canada, 615 Booth Street, Ottawa, Ontario, Canada K1A 0E8*; jbroome@gis.nrcan.gc.ca.
John Broome graduated form Queen's University in 1975 with a B.Sc. in geology. He has worked at Sander Geophysics Ltd. (Ottawa) from 1979 to 1983 collecting and interpreting airborne geophysical data. He joined the Geological Survey of Canada in 1983 and conducted research on the display and interpretation of potential field and other geophysical data. In 1989 he obtained an M.Sc. in geology from Carleton University in Ottawa. Since 1994 Mr. Broome has been Head of the Geoscience Integration Section at the Geological Survey of Canada.

Bruce A. Campbell, *National Air and Space Museum, MRC 315, Smithsonian Institution, Washington, DC 20560*; campbell@ceps.nasm.edu.
Bruce A. Campbell is a geophysicist and chair of the Center for Earth and Planetary Studies at the Smithsonian Institution. He received his Ph.D. in geology and geophysics in 1991 from the University of Hawaii. His current research includes radar studies of the Earth, Moon, Mars, and Venus, with an emphasis on understanding volcanic processes on each planet. He also pursues field studies of volcanic landforms on the Earth and analysis of their radar scattering properties.

Roger N. Clark, *U.S. Geological Survey, MS 964, Box 25046, Federal Center, Denver, CO 80225-0046*; Rclark@speclab.cr.usgs.gov.
Roger N. Clark is a research scientist at the U.S. Geological Survey in Denver, Colorado. He studies the surface mineralogy of the Earth, planets, satellites, asteroids, and comets in our solar system using spectroscopy and imaging spectroscopy. Dr. Clark obtained his Ph.D. from the Massachusetts Institute of Technology in 1980 in the field of planetary science. At the USGS, Dr. Clark heads several projects, ranging from surface compositional studies of the planets to mineral spectroscopy and the development of spectral libraries (from 0.2 to 150 µm), environmental assessments, and ecosystem studies of the Earth using imaging spectroscopy. Dr. Clark works on several NASA spacecraft teams, including those dealing with the Visual/Infrared Mapping Spectrometer (VIMS) Team on the Cassini mission to Saturn, the Thermal Emission Spectrometer (TES) on Mars Global Surveyor, and the Galileo Near Infrared Mapping Spectrometer (NIMS) team currently orbiting Jupiter.

Marc A. D'Iorio, *Canada Centre for Remote Sensing, 588 Booth Street, Ottawa, Ontario, Canada K1A 0Y7*; marc.diorio@ccrs.nrcan.gc.ca.
Marc A. D'Iorio is a scientist at the Canada Centre for Remote Sensing. He received his B.Sc. in solid earth geophysics at McGill University and his Ph.D. in geostatistics at the University of Ottawa in 1988. His principal field of research is applications of radar remote sensing in geology. Dr. D'Iorio is head of industrial cooperation at CCRS and an adjunct professor in remote sensing at the Université de Montréal.

Gerrit F. Epema, *Department of Environmental Sciences, Laboratory of Soil Science and Geology and Laboratory of Geoinformation Processing and Remote Sensing,*

Wageningen Agricultural University, P.O. Box 37, 6700 AA, Wageningen, The Netherlands; Gerrit.Epema@bodlan.beng.wau.nl.
Gerrit F. Epema obtained his M.Sc. degree from Utrecht University in 1980. He received his Ph.D. degree at Wageningen Agricultural University, focusing on spectral reflectance measurements in the Tunisian desert. Since 1982 he has been involved in teaching and research in the field of remote sensing. The research projects in remote sensing range from spectroscopy for soils to application of remote sensing in combination with GIS for sustainable land use. Most of the work has been done in southern Europe and various countries in Africa. Since 1994 he has been course director of an international M.Sc. course in geoinformation for rural applications.

Yves M. Govaerts, *Space Applications Institute, EC Joint Research Center, I-21020 Ispra (VA), Italy.*
Yves M. Govaerts received his Ph.D. degree in physics from the Université Catholique de Louvain in Louvain-la-Neuve, Belgium, in 1995. In 1990, he joined the Royal Meteorological Institute of Belgium as a research scientist. He worked for the Space Applications Institute, European Community Joint Research Center in Ispra, Italy from 1993 to 1996. In 1997 he joined Eumetsat. His principal research fields are radiative transfer modeling and the development of retrieval algorithms for remote sensing applications. He worked on the design of vegetation indices for the MERIS instrument on the ESA Envisat platform.

Cindy Grove, *Jet Propulsion Laboratory, California Institute of Technology, Mail Stop 183-501, 4800 Oak Grove Drive, Pasadena, CA 91109*; C.Grove@jpl.nasa.gov.
Cindy Grove is a member of the technical staff for the Terrestrial Science Group at the Jet Propulsion Laboratory. She received her B.Sc. in geology in 1982 from the California State University–Long Beach. Her current research involves developing a spectral library of minerals in the visible/near-infrared and infrared regions of the electromagnetic spectrum.

Jeff R. Harris, *Geological Survey of Canada, 615 Booth Street, Ottawa, Ontario, Canada K1A 0E8*; jharris@pf.emr.ca.
Jeff R. Harris is a geologist with the Geological Survey of Canada. He holds B.Sc. and M.A. degrees in Earth Science and is presently working on a Ph.D. in geology at Ottawa University. His current research involves developing GIS and remote sensing applications for mineral exploration, geochemistry processing and visualization, and lithological and structural mapping.

Simon J. Hook, *Jet Propulsion Laboratory, California Institute of Technology, Mail Stop 183-501, 4800 Oak Grove Drive, Pasadena, CA 91109*; simon.j.hook@jpl.nasa.gov.
Simon J. Hook holds a B.Sc. in geology from the University of Durham, England, a M.Sc. in geology from the University of Alberta, Canada, and a Ph.D. in geology from the University of Durham, England. After completing his Ph.D. he undertook a two-year National Research Council resident research associateship at the Jet Propulsion Laboratory, Pasadena, California. He is now employed as the U.S. Project Scientist for the advanced spaceborne thermal emission reflectance radiometer (ASTER) at the Jet Propulsion Laboratory. His current research is focused on the use of thermal infrared data for geological and climatological studies.

James R. Irons, *Biospheric Sciences Branch, NASA Goddard Space Flight Center, Greenbelt, MD 20771;* Jim_irons@gsfc.nasa.gov.
James R. Irons received his B.Sc. degree in environmental resources management in 1976 and a M.Sc. degree in agronomy in 1979 from Pennsylvania State University. He received his Ph.D. degree in agronomy in 1993 from the University of Maryland. He has worked as an earth scientist in the Biospheric Sciences Branch, NASA Goddard Space Flight Center, since 1979. His research interests include the process of radiation scattering by soils and the application of multiangle observation to the characterization of forest canopies. He currently serves as Goddard's deputy Landsat project scientist.

Stephane Jacquemoud, *Laboratoire Environnement et Développement, Université Paris 7, 75215 Paris Cedex 05, France.*
Stephane Jacquemoud received the diploma of Agricultural Engineer in 1988 from the Ecole Nationale Supérieure Agronomique de Montpellier, France, a DEA in 1989, and completed a thèse de troisième cycle in 1992 on Méthodes Physiques en Télédétection from the Université Paris. He was a visiting postdoctoral fellow at the European community Joint Research Centre (JRC) in Ispra, Italy (1992–1993), the Université Blaise Pascal in Clermont-Ferrand, France (1993–1994), and the University of California–Davis (1994–1995). He is currently assistant professor of physics at the Université Paris.

Anne B. Kahle, *Jet Propulsion Laboratory, California Institute of Technology, Mail Stop 183-501, 4800 Oak Grove Drive, Pasadena, CA 91109.*
Anne B. Kahle holds a B.Sc. in physics and a M.Sc. in geophysics from the University of Alaska, and a Ph.D. in atmospheric science from the University of California at Los Angeles. Currently she is a senior research scientist and the U.S. ASTER science team leader at the Jet Propulsion Laboratory.

Fred A. Kruse, *Analytical Imaging and Geophysics LLC, 4450 Arapahoe Avenue, Suite 100, Boulder, CO 80303; phone: 303-499-9471; fax: 303-665-6090;* kruse@aigllc.com.
Fred A. Kruse has been involved in multispectral, hyperspectral, and SAR remote sensing research and applications for over 17 years in positions with the U.S. Geological Survey, the University of Colorado, and private industry. He is currently senior scientist, Analytical Imaging and Geophysics, Boulder, Colorado, and also serves as an adjunct assistant professor with the Department of Geological Sciences, University of Colorado–Boulder. Dr. Kruse holds a B.S. degree in geology from the University of Massachusetts–Amherst, and M.S. and Ph.D. degrees in geology from the Colorado School of Mines, Golden, Colorado. His primary scientific interests are in the development of analysis and visualization techniques for imaging spectrometer (hyperspectral) data, and the use of knowledge-based artificial intelligence (AI) techniques to identify and map Earth-surface materials. Dr. Kruse is also cofounder of BSC, the company that developed the remote sensing analysis software package ENVI.

Harold Lang, *Jet Propulsion Laboratory, California Institute of Technology, Mail Stop 183-501, 4800 Oak Grove Drive, Pasadena, CA 91109;* Harold.R.Lang@jpl.nasa.gov or Harold.R.Lang@fax.jpl.nasa.gov.
Harold Lang is a research scientist at the Jet Propulsion Laboratory, California In-

stitute of Technology. After working as a petroleum geologist for the state of California and teaching geology at California State University–Long Beach, he joined JPL in 1980 as the principal investigator for oil and gas test site studies for the Joint NASA/Geosat Test Case Project. Following that, he was principal investigator for NASA-funded tectonostratigraphic studies in the Wind River Basin, Wyoming, and now in the Guerrero-Morelos Basin, Mexico. Since 1991, he has also been team associate responsible for the digital topography experiment for the ASTER instrument, which will fly on the first Eos platform. From 1991 to 1993, he was manager of JPL's 100-member Geology and Planetology Section. He holds degrees in geology from California State University Long Beach (B.S. and M.S.) and the University of Calgary (Ph.D.).

John F. Mustard, *Department of Geological Sciences, Brown University, Box 1846, Providence, RI 02912*; John_Mustard@brown.edu.
John F. Mustard is an assistant professor in the Department of Geological Sciences at Brown University He received a B.S. in geology from the University of British Columbia and a Ph.D. from Brown University in 1990. He has worked in the field of remote sensing for over a decade and has had experience in using multispectral and hyperspectral imaging of earth and planetary surfaces. This effort is supported by laboratory spectroscopic studies and analytical modeling to developed advanced information extraction techniques. His primary interests are in the application of remotely sensed data in the understanding of processes that shape planetary surfaces, and in the dynamics of earth system processes as revealed by multitemporal data sets.

Frank D. Palluconi, *Jet Propulsion Laboratory, California Institute of Technology, Mail Stop 183-501, 4800 Oak Grove Drive, Pasadena, CA 91109.*
Frank D. Palluconi is a scientist in the earth science element at the California Institute of Technology's Jet Propulsion Laboratory. He received a B.Sc. degree in physics from the Michigan Technological University in 1961 and a M.Sc. degree in physics from Pennsylvania State University in 1963. His scientific expertise and interests include thermal infrared broadband and spectral measurements, including interpretation of remote measurements, correction for atmospheric emission and attenuation, and instrument calibration.

Jeffrey J. Plaut, *Jet Propulsion Laboratory, California Institute of Technology, Mail Stop 183-501, 4800 Oak Grove Drive, Pasadena, CA 91109*; plaut@jpl.nasa.gov.
Jeffrey J. Plaut is a research scientist in the Geophysics and Planetary Geology group at the Jet Propulsion Laboratory, California Institute of Technology. He received his Ph.D. in earth and planetary sciences from Washington University in St. Louis, Missouri. He served as deputy project scientist on the Magellan radar mapping mission to Venus, and has ongoing research projects that utilize the Magellan data set. He participated in the spaceborne imaging radar-C/synthetic aperture radar (SIR-C/X-SAR) missions that flew aboard the Space Shuttle in 1994 as experiment scientist, geology experiment representative, and real-time science coordinator. He is the study scientist for planetary radar sounder missions (Mars and Europa) at JPL.

Gary L. Prost, *Golf Canada Resources, 1 Northwest Center, 1700 Lincoln, Suite 5000, Denver, CO 80203-4525; Gary_Prost@Golf.ca*
Gary L. Prost obtained a B.Sc. in geology from Northern Arizona University and a M.Sc. and Ph.D. in geology from the Colorado School of Mines. Over the past 18 years, Dr. Prost has worked for the U.S. Geological Survey, an independent oil and

mineral company, and a major oil company, serving four and a half years as the supervisor of the remote sensing group at Amoco Production Company in Houston. His remote sensing experience includes both mineral and oil exploration in the United States and over 30 foreign countries.

Andrew N. Rencz, *Geological Survey of Canada, 601 Booth Street, Ottawa, Ontario Canada K1A 0E8*; arencz@gsc.nrcan.gc.ca.
Andrew N. Rencz is a research scientist at the Geological Survey of Canada. He received his Ph.D. from the University of New Brunswick, M.Sc. from McGill University, and B.Sc. from the University of Alberta. He has held research positions at the Canada Centre for Remote Sensing and the Canadian International Development Agency. He is an adjunct professor in the Geography Department at Carleton University. His interests center on the extraction and visualization of remote sensing data, particularly for mineral exploration and environmental studies.

Benoit Rivard, *Department of Earth and Atmospheric Sciences, University of Alberta, Edmonton, Alberta, Canada T6G 2E3*; benoit.rivard@ualberta.ca.
Benoit Rivard is an associate professor at the Department of Earth and Atmospheric Sciences of the University of Alberta. He received his Ph.D. in earth and planetary sciences in 1990 from Washington University in St. Louis, focusing on the effect of rock coatings on the reflectance characteristics of rocks. His research interests have included the use of ground and airborne hyperspectral remote sensing for lithologic and structural mapping, the infrared characterization of terrestrial materials, and the development of methodologies for precise measurement of emissivity. More recently, his primary focus has been in the analysis of radar data for geological applications.

Mark S. Robinson, *Northwestern University, Department of Geological Sciences, 1847 Sheridan Road, Evanston, IL 60208*; robinson@earth.nwu.edu.
Mark S. Robinson is a planetary scientist in the Department of Geological Sciences at Northwestern University. He received his Ph.D. in geology and geophysics in 1993 from the University of Hawaii, focusing on volcanic processes on the Moon and Mars. His current research involves spacecraft observations of Mars, the Moon, Mercury, and asteroids and he is a member of the science team for the Near Earth Asteroid Rendezvous mission.

Robert A. Ryerson, *Kim Geomatics Corporation, Box 1125, Manotick, Ontario Canada K4M 1A9*; Bryerson@kim-geomatics.com.
Robert A. Ryerson, who is the editor-in-chief of the *Manual of Remote Sensing,* received his Ph.D. from the University of Waterloo, the first Canadian Ph.D. in remote sensing. He was with the Canada Centre for Remote Sensing (1973–1996) and has recently formed the consulting firm Kim Geomatics Corp. His research interests include agricultural and land use applications of satellite imagery, and he served the United Nations Economic and Social Commission for Asia and the Pacific and many other agencies working in Asia and Africa. Dr. Ryerson received a Canada 125 Medal for outstanding service to the nation, and he was the 1995 recipient of the Alan Gordon Memorial award from the American Society of Photogrammetry and Remote Sensing for outstanding work in the field.

Charles Sabine, *Geopix, 1280 Bodega Drive, Sparks, Nevada 89436* drpixel@ *untercomm.com.*

Charles Sabine is an independent consultant specializing in remote sensing and GIS applications to mineral exploration and mining. Dr. Sabine has worked with applications of remotely sensed data to mineral resource assessment and exploration problems for over 20 years in positions with the U.S. Bureau of Land Management, U.S. Bureau of Mines, University of Nevada and DRI. He holds a Ph.D. in geology from the University of Nevada, an M.S. from the University of California, Riverside, and a B.S. from Beloit College, Wisconsin. His current research examines remote sensing applications to mine safety problems.

Milton O. Smith, *Department of Geological Sciences, University of Washington, Seattle, WA 98195.*
Milton O. Smith received his B.S. in forestry and natural resources at the University of California–Berkeley in 1970 and his Ph.D. in forestry from the University of Washington in Seattle in 1983. He has held a research faculty position in the Department of Geological Sciences at the University of Washington since 1983. He has served as a member of various advisory committees for NASA, is a co-investigator on the Amazon Earth Observing System Interdisciplinary Science team, and a participant on the ASTER and SIR-C science teams. His interests are in the development of multispectral image analysis, resulting in extendible methods for detection of forest canopy and soil processes and landscape composition.

Jessica M. Sunshine, *Science Applications International Corp., 4501 Daly Drive, Suite 400, Chantilly, VA 20151*; sunshinej@saic.com.
Jessica M. Sunshine is a senior staff scientist at Science Applications International Corporation (SAIC) with responsibilities in the analysis of a variety of spectral imaging data sets of the Earth, Moon, Mars, and asteroids. Her principle duties are analysis and scientific support to the HYDICE (Hyperspectral Digital Imagery Collection Experiment) program. She is also principle investigator in NASA's Planetary Geology and Geophysics Program. She received her Ph.D. in geological sciences from Brown University in 1993.

Susan L. Ustin, *Department of Land, Air, and Water Resources, University of California, Davis, CA 95616.*
Susan L. Ustin received B.S. and M.A. degrees in biology in 1974 and 1978 from California State University, Hayward, California, and a Ph.D. degree in botany in 1983 from the University of California–Davis in the area of plant physiological ecology. She held a research faculty position at the University of California–Davis until she was appointed assistant professor in the Department of Land, Air, and Water resources in 1991. She has been active in NASA's imaging spectrometry program and on the Earth observing system HIRIS team and atmosphere–biosphere interactions interdisciplinary science team and SIR-C shuttle missions. Her environmental research includes both basic and applied ecological and ecophysiological research and development of new remote sensing methodologies for analysis of ecological processes. She has worked in many different ecological communities, including boreal forests, semiarid shrublands, temperate conifer forests, grasslands, and agricultural systems.

Michel M. Verstraete, *Applications Space Institute, EC Joint Research Center, I-21020 Ispra (VA), Italy.*

Michel M. Verstraete received the License en Physique in 1974 from the Université Catholique de Louvain in Louvain-la-Neuve, Belgium, the License Spéciale in Géophysique in 1976 from the Université Libre de Bruxelles, Belgium, and both the M.Sc. degree in meteorology in 1978 and the D.Sc. in atmospheric sciences in 1985 from the Massachusetts Institute of Technology, Cambridge. He was with the World Meteorological Organization from 1979 through 1981 in Geneva, Switzerland and Nairobi, Kenya, at the National Center for Atmospheric Research from 1982 through 1989 in Boulder, Colorado. He taught at the University of Michigan in 1989–1990 in Ann Arbor, Michigan, and is currently employed at the Space Applications Institute, European Community Joint Research Center in Ispra, Italy. He is a member of various scientific advisory committees of the European Space Agency (e.g., MERIS) and co-investigator on the NASA EOS MISR science team. His initial work on topics such as the modeling of atmosphere–biosphere interactions and desertification led him to his current interest in the quantitative exploitation of satellite remote sensing data for the detection and characterization of terrestrial surface properties.

David W. Viljoen, *Geological Survey of Canada, 615 Booth Street, Ottawa, Ontario Canada K1A 0E8.*
David W. Viljoen is a GIS specialist at the Geological Survey of Canada. He received his B.Sc. and M.A. degrees from Carleton University in Ottawa. His interests center on customizing and utilizing GIS to integrate and visualize geoscientific data for geological applications.

REMOTE SENSING
FOR THE EARTH SCIENCES

Spectral Characteristics

Chapter

1

Spectroscopy of Rocks and Minerals, and Principles of Spectroscopy

Roger N. Clark

U.S. Geological Survey
Denver, Colorado

1.1 INTRODUCTION

1.1.1 About This Chapter

Spectroscopy is the study of light as a function of wavelength that has been emitted, reflected, or scattered from a solid, liquid, or gas. In this chapter I discuss primarily the spectroscopy of minerals, but the principles apply to any material. No single chapter can cover this topic adequately, and one could argue, not even a single book. Thus, in some ways, this chapter may fall short of some reader's expectations. This chapter constitutes an overview of what is already known (some of which may be covered better in other reviews) and some of the practical lessons of spectroscopy, some of which have been in use by spectroscopists as common knowledge but have not necessarily been published previously in detail. See Farmer (1974), Adams (1975), Hunt (1977, 1982), Clark and Roush (1984), Clark et al. (1990a), Gaffey et al. (1993), Salisbury (1993), and references in those papers for more details.

1.1.2 Absorption and Scattering

As photons enter a mineral, some are reflected from grain surfaces, some pass through the grain, and some are absorbed. Those photons that are reflected from grain sur-

Remote Sensing for the Earth Sciences: Manual of Remote Sensing, 3 ed., Vol. 3, edited by Andrew N. Rencz.
ISBN: 0471-29405-5 © 1999 John Wiley & Sons, Inc.

faces or refracted through a particle are said to be *scattered*. Scattered photons may encounter another grain or be scattered away from the surface so they may be detected and measured. Photons may also originate from a surface, a process called *emission*. All natural surfaces emit photons when they are above absolute zero. Emitted photons are subject to the same physical laws of reflection, refraction, and absorption to which incident photons are bound.

Photons are absorbed in minerals by several processes. The variety of absorption processes and their wavelength dependence allow us to derive information about the chemistry of a mineral from its reflected or emitted light. The human eye is a crude reflectance spectrometer: We can look at a surface and see color. Our eyes and brain are processing the wavelength-dependent scattering of visible-light photons to reveal something about what we are observing, such as the red color of hematite or the green color of olivine. A modern spectrometer, however, can measure finer details over a broader wavelength range and with greater precision. Thus a spectrometer can measure absorptions due to more processes than can be seen with the eye.

1.1.3 Spectroscopy Terms

There are four general parameters that describe the capability of a spectrometer: (1) spectral range, (2) spectral bandwidth, (3) spectral sampling, and (4) signal/noise ratio. Spectral range is important to cover enough diagnostic spectral absorptions to solve a desired problem. There are general spectral ranges that are in common use, each to first order controlled by detector technology: (1) ultraviolet (UV): 0.001 to 0.4 μm, (2) visible: 0.4 to 0.7 μm, (3) near-infrared (NIR): 0.7 to 3.0 μm, (4) mid-infrared (MIR): 3.0 to 30 μm, and (5) far infrared (FIR): 30 μm to 1 mm (see, e.g., the *Photonics Design and Applications Handbook*, 1996 and the *Handbook of Chemistry and Physics*, any recent year). The approximate wavelength range 0.4 to 1.0 μm is sometimes referred to in the remote sensing literature as the VNIR (for visible/near-infrared: 0.4 to 1.0 μm), and the range 1.0 to 2.5 μm is sometimes referred to as the SWIR (shortwave-infrared). It should be noted that these terms are not recognized standard terms in fields other than remote sensing, and because the NIR in VNIR conflicts with the accepted NIR range, the VNIR and SWIR terms probably should be avoided. The mid-infrared covers thermally emitted energy, which for the Earth starts at about 2.5 to 3 μm, peaking near 10 μm, decreasing beyond the peak, with a shape controlled by gray-body emission.

Spectral bandwidth is the width of an individual spectral channel in the spectrometer. The narrower the spectral bandwidth, the narrower the absorption feature the spectrometer will measure accurately, if enough adjacent spectral samples are obtained. Some systems have a few broad channels, not contiguously spaced, and thus are not considered spectrometers (Figure 1.1a), Examples include the *Landsat* thematic mapper (TM) system and the moderate resolution imaging spectroradiometer (MODIS), which cannot resolve narrow absorption features. Others, such as the NASA JPL airborne visual/infrared imaging spectrometer (AVIRIS) system, have many narrow bandwidths, contiguously spaced (Figure 1.1b). Figure 1.1 shows spectra for the mineral alunite that could be obtained by some broadband and spectrometer systems. Note the loss in subtle spectral detail in the lower-resolution systems compared to the laboratory spectrum. Bandwidths and sampling greater than 25 nm rapidly lose the ability to resolve important mineral absorption features. All the

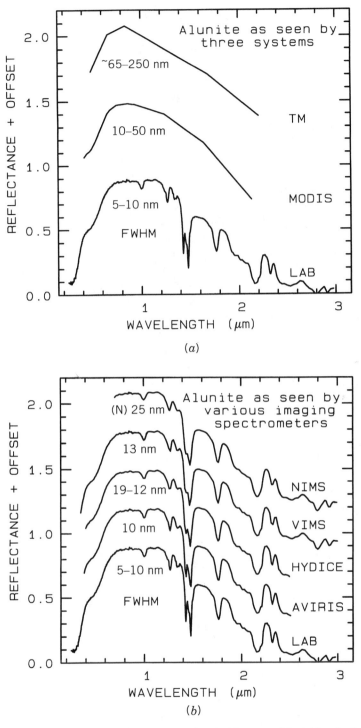

Figure 1.1 Spectra of the mineral alunite shown as measured in the laboratory and for (*a*) broadband remote sensing instruments and (*b*) some imaging spectrometers (see the text). The FWHM is the full width at half maximum, defined in Figure 1.2. The alunite is sample HS295.3B from the USGS spectral library (Clark et al., 1993b). *Note:* The NIMS and VIMS systems measure to 5 μm. Each spectrum is offset upward (*a*) 0.6 unit and (*b*) 0.3 unit from the one below it, for clarity.

spectra in Figure 1.1*b* are sampled at half-Nyquist (critical sampling) except the near-infrared mapping spectrometer (NIMS), which is at Nyquist sampling. Note, however, that the fine details of the absorption features are lost at about the 25-nm bandpass of NIMS. For example, the shoulder in the 2.2-μm absorption band is lost at 25-nm bandpass. The visual and infrared mapping spectrometer (VIMS) and NIMS systems measure out to 5 μm, thus can see absorption bands not obtainable by the other systems.

The shape of the bandpass profile is also important. Ideally, each spectrometer channel rejects all light except that from within a given narrow-wavelength range, but occasionally, due to optical effects too complex to discuss in detail here, light may leak in from out of the bandpass (e.g., scattering within the optical system, or inadequate blocking filters). The most common bandpass in spectrometers is a Gaussian profile. While specific spectrometer designs may have well-defined theoretical bandpass profiles, aberrations in the optical system usually smears the profile closer to a Gaussian shape. The width of the bandpass is usually defined as the width in wavelength at the 50% response level of the function, as shown in Figure 1.2, called the *full width at half maximum* (FWHM).

Spectral sampling is the distance in wavelength between the spectral bandpass profiles for each channel in the spectrometer as a function of wavelength. Spectral sampling is often confused with bandpass, with the two lumped together and called *resolution*. Information theory tells us that to resolve two spectral features, we must have two samples. Further, in order not to introduce sampling bias, the samples must be close enough together to measure the peak and valley locations. The Nyquist theorem states that the maximum information is obtained by sampling at one-half the FWHM. Spectrometer design, however, sometimes dictates a different sampling, and many modern spectrometers in use (e.g., AVIRIS, VIMS) sample at half-Nyquist, a sampling interval approximately equal to the FWHM. Note that the AVIRIS system has a bandpass of about 0.01 μm (10 nm), a sampling of about 0.01 μm, and thus

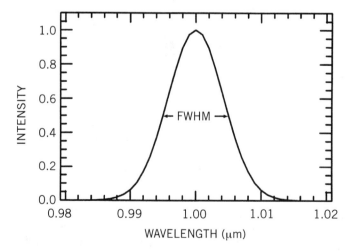

Figure 1.2 Gaussian profile with a full width at half maximum (FWHM) of 10 nm. This profile is typical of spectrometers such as AVIRIS, which has 224 such profiles spaced at about 10 nm.

a spectral resolution of about 0.02 μm (20 nm). The NIMS system in Figure 1.1 can sample at Nyquist (shown), half-Nyquist, and lower.

Finally, a spectrometer must measure the spectrum with enough precision to record details in the spectrum. The signal/noise ratio (S/N) required to solve a particular problem will depend on the strength of the spectral features under study. S/N is dependent on the detector sensitivity, spectral bandwidth, and intensity of the light reflected or emitted from the surface being measured. A few spectral features are quite strong and a S/N value of only about 10 will be adequate to identify them; whereas others are weak, and a S/N value of several hundred (and higher) is often needed (Swayze et al., submitted).

1.1.4 Imaging Spectroscopy

Today, spectrometers are in use in the laboratory, in the field, in aircraft (looking both down at the Earth and up into space), and on satellites. Reflectance and emittance spectroscopy of natural surfaces are sensitive to specific chemical bonds in materials, whether solid, liquid, or gas. Spectroscopy has the advantage of being sensitive to both crystalline and amorphous materials, unlike some diagnostic methods, such as x-ray diffraction. Spectroscopy's other main advantage is that it can be used up close (e.g., in the laboratory) to far away (e.g., to look down on the Earth or up at other planets). Spectroscopy's historical disadvantage is that it is too sensitive to small changes in the chemistry and/or structure of a material. The variations in material composition often cause shifts in the position and shape of absorption bands in the spectrum. Thus, with the vast variety of chemistry typically encountered in the real world, spectral signatures can be quite complex and sometimes unintelligible. However, that is now changing, with increased knowledge of the natural variation in spectral features and the causes of the shifts, so that the previous disadvantage is turning into a huge advantage, allowing us to probe ever more detail about the chemistry of our natural environment.

With the advance in computer and detector technology, the new field of imaging spectroscopy is developing (Goetz et al., 1985; Green et al., 1990; Vane et al., 1993; Chapter 5; Chapter 11; and references therein). Imaging spectroscopy is a new technique for obtaining a spectrum in each position of a large array of spatial positions so that any one spectral wavelength can be used to make a recognizable image. The image might be of a rock in the laboratory, a field study site from an aircraft, or an entire planet from a spacecraft or Earth-based telescope. By analyzing the spectral features, and thus specific chemical bonds in materials, one can map where those bonds occur, and thus map materials. Such mapping is best done by spectral feature analysis.

Imaging spectroscopy has many names in the remote sensing community, including *imaging spectrometry, hyperspectral,* and *ultraspectral imaging. Spectroscopy* is the study of electromagnetic radiation. *Spectrometry* is derived from *spectrophotometry,* the measure of photons as a function of wavelength, a term used for years in astronomy. However, spectrometry is used increasingly to indicate the measurement of nonlight quantities, such as in mass spectrometry (e.g., Ball, 1995). *Hyper* means excessive, but no imaging spectrometer in use can be considered hyperspectral—after all, a couple of hundred channels pales in comparison to a truly high-resolution spectrometer with millions of channels. *Ultraspectral* is beyond hyperspectral, a lofty goal that we have not yet reached. Terms such as *laboratory spectrometer, spectroscopist, reflectance spectroscopy,* and *thermal emission spectroscopy* are in common use. One rarely, if ever, sees the converse: *spectrometrist, reflectance spectrometry,* and so on. So it seems prudent to keep the terminology consistent with *imaging spectroscopy.*

This chapter provides an introduction to the factors affecting the spectra of natural materials, including scattering and absorption, and the causes of absorption features. We also discuss doing quantitative estimates of mixtures and show sample spectra of minerals and other common materials that might be encountered in the natural world.

1.1.5 Atmospheric Transmittance: Windows for Remote Sensing

Any effort to measure the spectral properties of a material through a planetary atmosphere must consider where the atmosphere absorbs. For example, the Earth's atmospheric transmittance is shown in Figure 1.3. The drop toward the ultraviolet is due to scattering and strong ozone absorption at wavelengths short of 0.35 μm. Ozone also displays an absorption at 9.6 μm. Oxygen absorbs at 0.76 μm in a narrow feature. CO_2 absorbs at 2.01 and 2.06, with a weak doublet near 1.6 μm. Water causes most of the rest of the absorption throughout the spectrum and hides additional (weaker) absorptions from other gases. The mid-infrared spectrum in Figure 1.3*b* shows the effect of doubling CO_2, which in this case is small compared to the absorption due to water. Although we will see that the spectral region near 1.4 and 3 μm can be diagnostic of OH-bearing minerals, we cannot usually use these wavelengths when remotely measuring spectra through the Earth's atmosphere (it has been done from high-elevation observatories during dry weather conditions). However, these spectral regions can be used in the laboratory where the atmospheric path lengths are thousands of times smaller or when measuring spectra of other planets from orbiting spacecraft.

1.2   REFLECTION AND ABSORPTION PROCESSES

1.2.1 Reflection and Absorption

When a stream of photons encounter a medium with a change in the index of refraction, some are reflected and some are refracted into the medium. It is beyond this chapter to review all the physical laws of reflection and refraction; a good optics or physics book can do that (e.g., Hecht, 1987). However, the basics of reflection should be understood. All materials have a complex index of refraction:

$$m = n - jK \qquad (1.1)$$

where m is the complex index of refraction, n is the real part of the index, $j = (-1)^{1/2}$, and K is the imaginary part of the index of refraction, sometimes called the extinction coefficient.

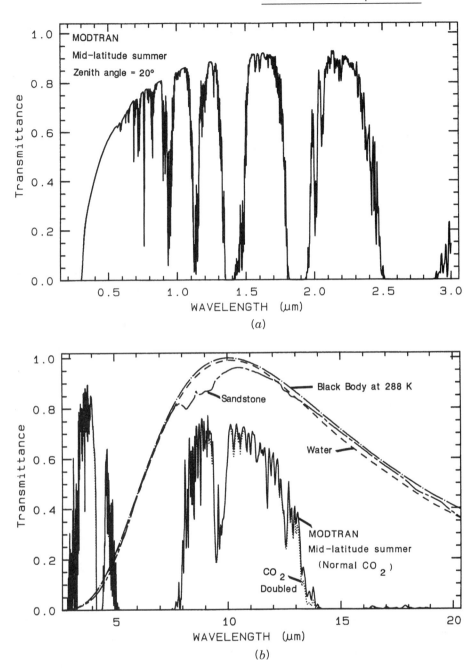

Figure 1.3   (*a*) Modtran (Berk et al., 1989) modeled atmospheric transmittance, visible to near-infrared. Most of the absorptions are due to water. Oxygen occurs at 0.76 μm, carbon dioxide at 2.0 and 2.06 μm. (*b*) Atmospheric transmittance, mid-infrared, is compared to scaled gray-body spectra. Most of the absorption is due to water. Carbon dioxide has a strong 15-μm band, and the dotted line shows the increased absorption due to doubling CO_2. Also shown is the blackbody emission at 288 K and the gray-body emission from water and a sandstone scaled to fit on this transmittance scale. The water and sandstone curves were computed from reflectance data using $1 -$ reflectance $\times$ a blackbody at 288 K.

When photons enter an absorbing medium, they are absorbed according to *Beer's law:*

$$I = I_o e^{-kx} \qquad (1.2)$$

where I is the observed intensity, I_o is the original light intensity, k an absorption coefficient, and x the distance traveled through the medium.

The absorption coefficient is related to the complex index of refraction by the equation:

$$k = \frac{4\pi K}{\lambda} \qquad (1.3)$$

where λ is the wavelength of light. A sample index of refraction, n, and extinction coefficient, K, are shown in Figure 1.4a for quartz. The reflection of light, R, normally incident onto a plane surface is described by the *Fresnel equation:*

$$R = \frac{(n - 1)^2 + K^2}{(n + 1)^2 + K^2} \qquad (1.4)$$

At angles other than normal, the reflectance is a complex trigonometric function involving the polarization direction of the incident beam and is left to the reader to study in standard optics or physics textbooks. The reflection from quartz grains as measured on a laboratory spectrometer is shown in Figure 1.4b. While the spectrum is of a particulate surface, first surface reflection dominates all wavelengths and so is similar to the spectrum of a slab of quartz.

The absorption coefficient is traditionally expressed in units of cm^{-1} and x in cm. Equations (1.1) to (1.4) hold for a single wavelength. At other wavelengths, the absorption coefficient and index of refraction are different, and the reflected intensity observed varies. The absorption coefficient as a function of wavelength is a fundamental parameter describing the interaction of photons with a material. So is the index of refraction, but it generally varies less than the absorption coefficient as a function of wavelength, especially at visible and near-infrared wavelengths. At fundamental absorption bands, both n and K vary strongly with wavelength, as shown in Figure 1.4a, although K still varies over more orders of magnitude than n does.

The complex index of refraction in Figure 1.4a shows important properties of materials. As one moves to longer wavelengths (left to right in Figure 1.4a), the index of refraction decreases to a minimum just before a sharp rise (e.g., at 8.5 and 12.6 μm in Figure 1.4a). The minimum is often near or even below $n = 1$. The wavelength where $n = 1$, called the *Christensen frequency*, usually results in a minimum in reflected light because of the small (to zero) difference in the index of refraction compared to the surrounding medium (e.g., air or vacuum). The location of the observed reflectance minimum is also controlled by the extinction coefficient according to equation (1.4). Note that the Christensen frequency sometimes occurs at a wavelength shorter than the maximum in the extinction coefficient (e.g., Figure 1.4a). This maximum is called the *reststrahlen band*: the location of fundamental vibrational stretching modes in the near and mid-infrared. The combination of n and K

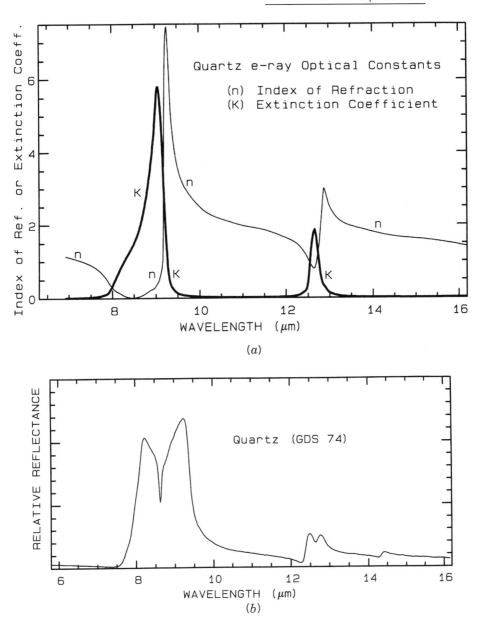

Figure 1.4   (*a*) Optical constants of quartz, SiO_2 (from Spitzer and Kleinman, 1960); (*b*) relative reflectance of powdered quartz.

at these wavelengths often results in high reflectance. See Hapke (1993) for more details.

1.2.2 Emittance

At mid-infrared wavelengths, materials normally receive thermally emitted photons. In the laboratory one can shine enough light on a sample to ignore emitted photons and measure reflectance, but that cannot be done in typical remote sensing situations. Measuring energy emitted in the laboratory is not easy because all materials emit energy unless cooled to very low temperatures. Trying to measure thermal emission at room temperatures would be like trying to take a picture with a camera with transparent walls and light bulbs turned on inside the camera! However, *Kirchhoff's law* (e.g., Nicodemus, 1965) states that

$$E = 1 - R \qquad (1.5)$$

where E is emissivity. Several studies have been conducted to show that the rule generally holds (see, e.g., Salisbury, 1993, and references therein). Although some discrepancies have been found, they may be due to the difficulty of measuring emittance or due to temperature gradients in the samples (e.g., Henderson et al., 1996, and references therein). Considering that and the fact that one rarely measures all the light reflected or emitted (usually a directional measurement is made), the law is basically true except in the most rigorous studies where absolute levels and band strengths are critical to the science. In practical terms, small changes in grain size result in spectral changes that are usually larger than the discrepancies in the law.

1.2.3 Summary

The complex interaction of light with matter involves reflection and refraction from index of refraction boundaries, a process we call scattering, and absorption by the medium as light passes through the medium. The amount of scattering versus absorption controls the amount of photons we receive from a surface.

1.3 CAUSES OF ABSORPTION

What causes absorption bands in the spectra of materials? There are general processes: electronic and vibrational. Burns (1993) examines the details of electronic processes, and Farmer (1974) covers vibrational processes. These two books are significant works which provide the fundamentals as well as practical information. Shorter introductions to the causes of absorption bands in minerals are given by Hunt (1977, 1982) and Gaffey et al. (1993) for the visible and near-infrared.

1.3.1 Electronic Processes

Isolated atoms and ions have discrete energy states. Absorption of photons of a specific wavelength causes a change from one energy state to a higher-energy state. Emission of a photon occurs as a result of a change in an energy state to a lower-energy state. When a photon is absorbed, it is usually not emitted at the same wavelength. For example, it can cause heating of the material, resulting in gray-body emission at longer wavelengths.

In a solid, electrons may be shared between individual atoms. The energy level of shared electrons may become smeared over a range of values called *energy bands*.

However, bound electrons will still have quantized energy states (see, e.g., Burns, 1970, 1993).

1.3.1.1 CRYSTAL FIELD EFFECTS.

The most common electronic process revealed in the spectra of minerals is due to unfilled electron shells of transition elements (Ni, Cr, Co, Fe, etc.). Iron is the most common transition element in minerals. For all transition elements, d orbitals have identical energies in an isolated ion, but the energy levels split when the atom is located in a crystal field (see, e.g., Burns, 1970, 1993). This splitting of the orbital energy states enables an electron to be moved from a lower level into a higher level by absorption of a photon having an energy matching the energy difference between the states. The energy levels are determined by the valence state of the atom (e.g., Fe^{2+}, Fe^{3+}), its coordination number, and the symmetry of the site it occupies. The levels are also influenced by the type of ligands formed, the extent of distortion of the site, and the value of the metal–ligand interatomic distance (e.g., Burns, 1993). The crystal field varies with crystal structure from mineral to mineral; thus the amount of splitting varies and the same ion (e.g., Fe^{2+}) produces obviously different absorptions, making specific mineral identification possible from spectroscopy (Figures 1.5 to 1.7).

Example Fe^{2+} absorptions are shown in Figure 1.5a (olivines) and Figure 1.6a (pyroxenes). Note the shift in band position and shape between the different compositions. Sample Fe^{3+} absorptions are shown in goethite (FeOOH) and hematite (Fe_2O_3) in Figure 1.7a. Compositional changes also shift vibrational absorptions, discussed below, and as seen in Figures 1.5b, 1.6b, and 1.7b. The compositional shifts of the electronic absorptions have been studied by Adams (1974, 1975) and Cloutis and Gaffey (1991) for pyroxenes and are shown in Figure 1.6c, and by King and Ridley (1987) for olivines.

The unfilled shells of rare earth ions involve deep-lying electrons that are well shielded from surrounding crystal fields, so the energy levels remain largely unchanged. Thus absorption bands due to rare earth elements are not diagnostic of mineralogy but of the presence of the ions in the mineral (Figure 1.8).

1.3.1.2 CHARGE TRANSFER ABSORPTIONS.

Absorption bands can also be caused by charge transfers, interelement transitions where the absorption of a photon causes an electron to move between ions or between ions and ligands. The transition can also occur between the same metal in different valence states, such as between Fe^{2+} and Fe^{3+}. In general, absorption bands caused by charge transfers are diagnostic of mineralogy. Their strengths are typically hundreds to thousands of times stronger than those of crystal field transitions. The band centers usually occur in the ultraviolet, with the wings of the absorption extending into the visible. Charge transfer absorptions are the main cause of the red color of iron oxides and hydroxides (Figure 1.7a). Morris et al. (1985) studied the details of submicron iron oxides, where it was found that the absorption bands decrease rapidly in intensity. This occurs because the increased surface/volume ratio at small grain size results in a greater proportion of grain boundaries where crystal field effects are different, resulting in lower magnetic coupling and reduced absorption strength. Other iron oxides probably show similar effects. Reflectance spectra

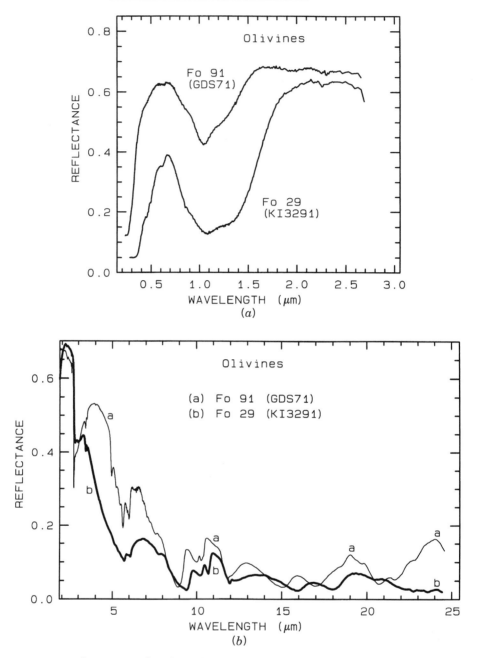

Figure 1.5 (*a*) Reflectance spectra of two olivines, showing the change in band position and shape with composition. The 1-μm absorption band is due to a crystal field absorption of Fe^{2+}. "Fo" stands for forsterite (Mg_2SiO_4) in the forsterite–fayalite ($Fe_2^{2+}SiO_4$) olivine solid solution series. The Fo 29 sample (KI3291 from King and Ridley, 1987) has an FeO content of 53.65%, while the Fo 91 sample (GDS 71; labeled Twin Sisters Peak in King and Ridley, 1987) has an FeO content of 7.93%. The mean grain size is 30 and 25 μm respectively. The 1-μm band position varies from about 1.08 μm at Fo 10 to 1.05 μm at Fo 90 (King and Ridley, 1987). (*b*) Same as for part (*a*) but for mid-infrared wavelengths. Note the shifts in the spectral features due to the change in composition. See the text for a discussion of vibrational absorption bands.

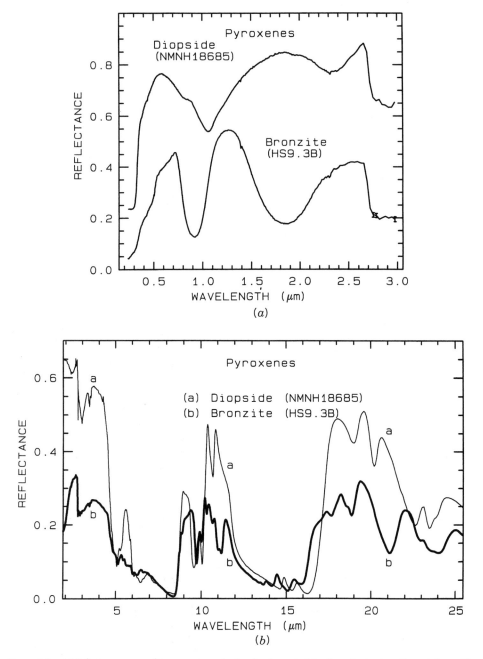

Figure 1.6 (*a*) Reflectance spectra of two pyroxenes, showing the change in Fe^{2+} absorption band position and shape with composition (from Clark et al., 1993b). Diopside, sample NMNH18685, is $CaMgSi_2O_6$, but some Fe^{2+} substitutes for Mg. Bronzite, sample HS9.3B, is $(Mg,Fe)SiO_3$ with mostly Mg. The 1-μm versus the 2-μm band position of a pyroxene describes the pyroxene composition [part (*c*)]. The diopside spectrum is offset 0.2 unit upward. (*b*) Same as for part (*a*) but for mid-infrared wavelengths. Note the shifts in the spectral features due to the change in composition.

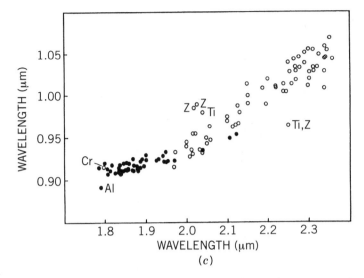

Figure 1.6 (c) Pyroxene 1-μm versus 2-μm absorption band position as a function of composition, as adapted from Adams (1974) by Cloutis and Gaffey (1991). Open circles have greater than 11% wollastonite (Wo), and solid symbols less than 11% Wo. Samples with zoned or exolved phases are marked by "Z." Other samples not following the "normal" trend include those with greater than 1% TiO_2 (Ti), greater than 1% Cr_2O_3 (Cr), or greater than 4% Al_2O_3. (From Cloutis and Gaffey, 1991.)

of iron oxides have such strong absorption bands that the shape changes significantly with grain size. This is discussed later in this chapter and is illustrated in Figure 1.29. Small shifts in absorption band position are also observed due to substitution of other elements, such as aluminum for iron in hematite (e.g., Morris et al., 1985, and references therein), but more work needs to be done to fully understand the effects.

1.3.1.3   CONDUCTION BANDS.

In some minerals there are two energy levels in which electrons may reside: a higher level called the *conduction band*, where electrons move freely throughout the lattice, and a lower-energy region called the *valence band*, where electrons are attached to individual atoms. The difference between the energy levels is called the *band gap*. The band gap is typically small or nonexistent in metals and is very large in dielectrics. In semiconductors, the band gap corresponds to the energy of visible to near-infrared wavelength photons, and the spectrum in these cases is approximately a step function. The yellow color of sulfur is caused by such a band gap. The minerals cinnabar (HgS) and sulfur (S) have spectra showing the band gap in the visible (Figure 1.9).

1.3.1.4   COLOR CENTERS.

A few minerals show color due to absorption by *color centers*. A color center is caused by irradiation (e.g., by solar ultraviolet radiation) of an imperfect crystal. Crystals in nature have lattice defects that disturb the periodicity of the crystal. For example, defects might be caused by impurities. These defects can produce discrete energy levels and electrons can become bound to them. The movement of an electron into the defect requires photon energy. The yellow, purple, and blue colors of fluorite are caused by color centers. See Hunt (1977) and references therein for more details.

More detailed discussions of electronic processes can be found in review papers

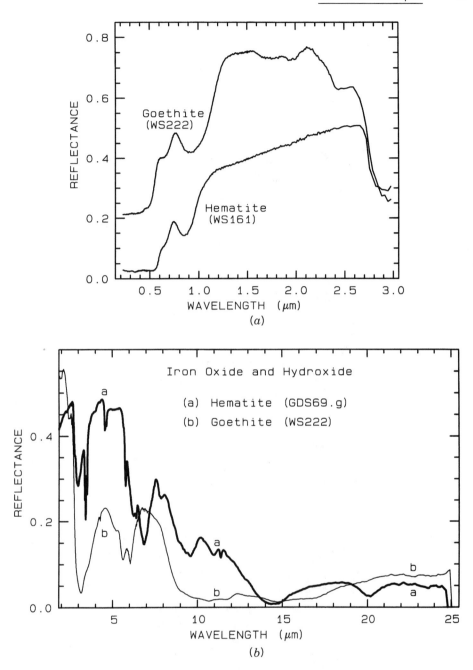

Figure 1.7 (*a*) Reflectance spectra of the iron oxide hematite (Fe₂O₃) and iron hydroxide goethite (FeOOH) (from Clark et al., 1993b). The intense charge transfer band in the ultraviolet (<0.4 μm) is "saturated" in reflectance, so only first surface (specular) reflection is seen in these spectra. The 0.9-and 0.86-μm absorption features are due to Laporte-forbidden transitions (e.g., Morris et al., 1985; Sherman, 1990; and references therein). The absorption at 2.7 to 3 μm is due to trace water in the samples, and in the case of goethite, the OH. The goethite spectrum is offset upward 0.2 unit. (*b*) Same as for part (*a*) but for mid-infrared wavelengths.

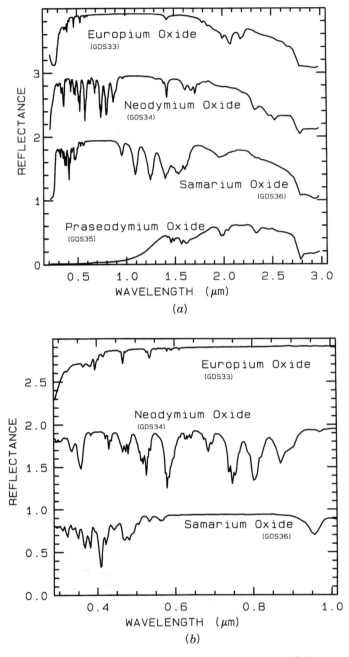

Figure 1.8 (*a*) Reflectance spectra of rare earth oxides. These absorptions are due to crystal field transitions involving deep-lying electrons of the rare earth element and do not shift when the rare earth ion is in another mineral. Each spectrum is offset by 1.0 unit, for clarity. (Spectra from Clark et al., 1993b.) (*b*) Same as for part (*a*) except showing absorptions in the visible region. Spectra are offset 1.0 unit for clarity. Spectral resolution is about 1 nm, critically sampled.

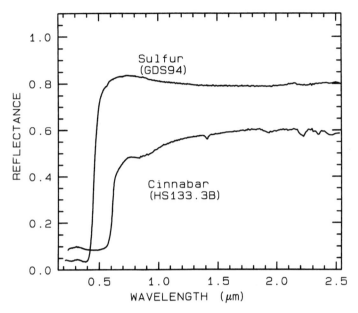

Figure 1.9 Reflectance spectra of sulfur, S, and cinnabar, HgS, showing conduction bands in the visible. (From Clark et al., 1993b.)

by Hunt (1977) and Gaffey et al. (1993) and in a book by Burns (1993). A summary diagram of the causes of absorption bands is shown in Figure 1.10.

1.3.2 Vibrational Processes

The bonds in a molecule or crystal lattice are like springs with attached weights: the entire system can vibrate. The frequency of vibration depends on the strength of each spring (the bond in a molecule) and their masses (the mass of each element in a molecule). For a molecule with N atoms, there are $3N - 6$ normal modes of vibrations called *fundamentals*. Each vibration can also occur at roughly multiples of the original fundamental frequency. The additional vibrations are called *overtones* when they involve multiples of a single fundamental mode, and *combinations* when they involve different modes of vibrations.

A vibrational absorption will be seen in the infrared spectrum only if the molecule responsible shows a dipole moment (it is said to be infrared active). A symmetric molecule such as N_2 is not normally infrared active unless it is distorted (e.g., when under high pressure). Vibrations from two or more modes can occur at the same frequency, and because they cannot be distinguished, are said to be degenerate. An isolated molecule with degenerate modes may show the modes at slightly different frequencies in a crystal because of the nonsymmetric influences of the crystal field.

A free molecule can rotate and move translationally, but even in a solid, partial rotation and slight translation can occur. These motions are called *lattice modes* and

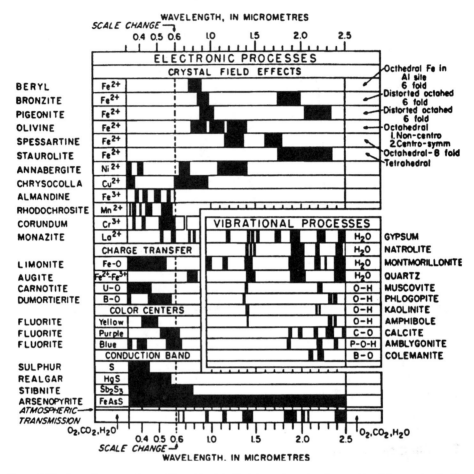

Figure 1.10   Spectral signature diagram. The widths of the black bars indicate the relative widths of absorption bands. (From Hunt, 1977.)

typically occur at very low energies (longer mid-infrared wavelengths), beyond about 20 μm.

Traditionally, the frequencies of fundamental vibrations are labeled with the Greek letter v and a subscript (Herzberg, 1945). If a molecule has vibration fundamentals v_1, v_2, v_3, it can have overtones at approximately $2v_1$, $3v_1$, $2v_2$ and combinations at approximately $v_1 + v_2$, $v_2 + v_3$, $v_1 + v_2 + v_3$, and so on. These examples used summations of modes, but subtractions are also possible (e.g. $v_1 + v_3 - v_2$). Each higher overtone or combination is typically 30 to 100 times weaker than the last. Consequently, the spectrum of a mineral can be quite complex. In reflectance spectroscopy, these weak absorptions can be measured easily and diagnostic information routinely gained from second and third overtones and combinations (e.g., Figures 1.5b, 1.6b, 1.7b, and 1.11 to 1.13).

Lattice modes are sometimes denoted by v_T and v_R and also couple with other

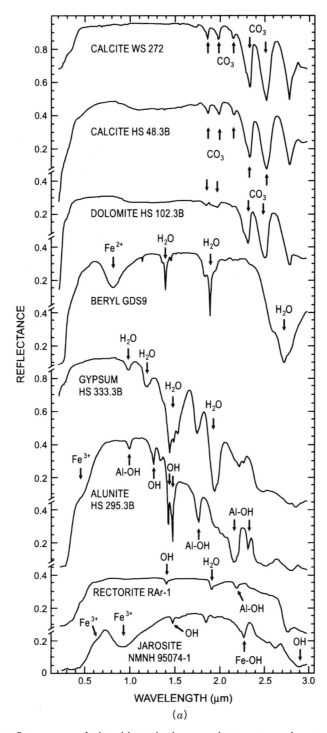

Figure 1.11 (*a*) Reflectance spectra of calcite, dolomite, beryl, gypsum, alunite, rectorite, and jarosite, showing vibrational bands due to OH, CO_3, and H_2O;

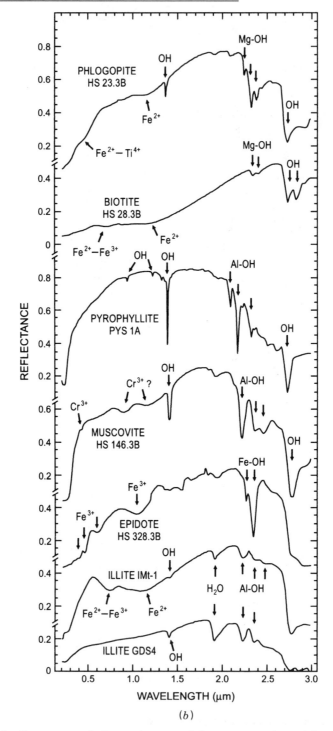

Figure 1.11 (*b*) reflectance spectra of phlogopite, biotite, pyrophyllite, muscovite, epidote, and illite showing vibrational bands due to OH and H_2O;

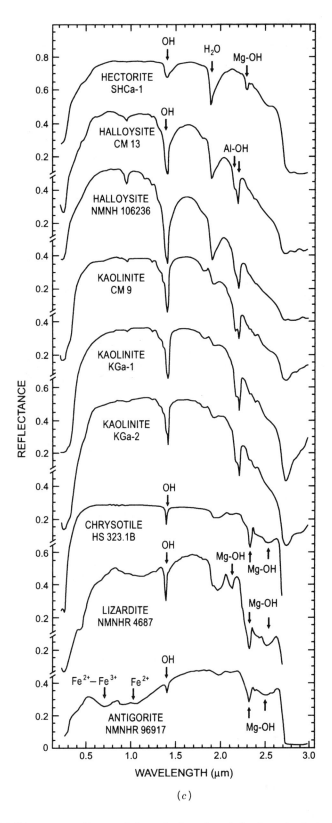

Figure 1.11 (*c*) Reflectance spectra of hectorite, halloysite, kaolinite, chrysotile, lizardite, and antigorite showing vibrational bands due to OH. (From Clark et al., 1990a.) Figure 1.19 shows an expanded view of the 1.4-μm region for chrysotile, lizardite, and antigorite.

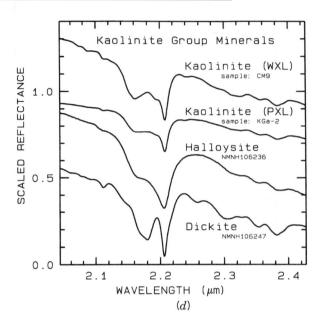

Figure 1.11 (*d*) Subtle spectral differences in the kaolinite group minerals near 2.2 μm. Kaolinite CM9 is well crystallized (WXL), while KGa-2 is poorly crystallized (PXL). Spectral bandwidth is 1.9 nm and sampling is 0.95 nm. Each spectrum was scaled to 0.7 at 2.1 μm, then offset up or down so that the curves do not overlap. Original reflectances were between 0.5 and 0.8.

fundamentals, resulting in the finer structure seen in some spectra. The causes of vibrational absorptions in mid-infrared spectra are summarized in Figure 1.14.

Mid-infrared reflectance spectra of quartz are shown in Figure 1.4*b*. The strong 9-μm Si–O–Si asymmetric stretch fundamental is obvious from the reflection maximum. The O–Si–O bending mode occurs near 25 μm and is the second strongest absorption. The absorption between 12 and 13 μm is the Si–O–Si symmetric stretch fundamental.

Olivine spectra in the mid-infrared are shown in Figure 1.5*b*. When Mg is present, a strong absorption appears near 23 μm, perhaps seen in the Fo 91 spectrum. The Si–O–Si asymmetric stretch fundamental occurs near 11 μm, and a weaker symmetric absorption occurs near 12 μm. The absorptions shift with composition as shown in Figure 1.5*b* and discussed in more detail in Farmer (1974, pp. 288–290).

Pyroxene mid-infrared spectra are shown in Figure 1.6*b*. The Si–O fundamentals are at similar to those of other silicates, as indicated in Figure 1.14. Grain size effects are discussed in Section 1.6.2 and illustrated in Figure 1.23.

Iron oxide and iron hydroxide mid-infrared spectra are shown in Figure 1.7*b*. Because iron is more massive than silicon, Fe–O fundamentals will be at longer wavelengths than Si–O stretching modes. Hematite, Fe_2O_3, has three strong stretching modes between 16 and 30 μm. Because iron oxides and hydroxides tend to be fine grained, typically less than the wavelength of mid-infrared photons, and because of the strong absorption in the mid-infrared, iron oxides tend to be dark in reflectance, showing few features beyond about 12 μm. The hematite in Figure 1.7*b* has a small amount of water, as evidenced by the 3-μm absorption, and a moderate amount of organics, as seen by the C–H stretch fundamental near 3.4 μm. The goethite, FeOOH, having hydroxyl, has a strong 3-μm absorption. The olivines (Figure 1.5) and py-

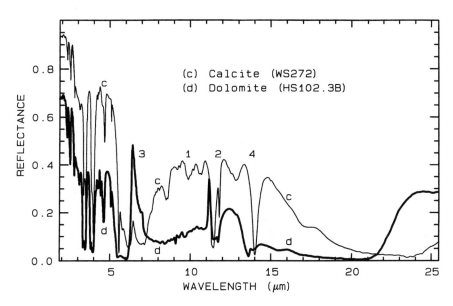

Figure 1.12   Comparison of calcite ($CaCO_3$) and dolomite ($CaMg(CO_3)_2$) spectra in the mid-infrared, showing small band shifts due to the change in composition between the two minerals. The level change (calcite higher in reflectance than dolomite) is because the calcite has a smaller grain size. The numbers indicate the fundamental stretching positions of ν_1, ν_2, ν_3, and ν_4. The ν_1 stretch is infrared innactive but may be weakly present in carbonates. The ν_3 fundamental is so strong that only a reflection peak is seen in these spectra.

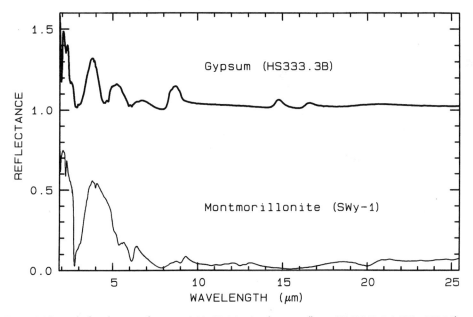

Figure 1.13   Mid-infrared spectra of gypsum, $CaSO_4 \cdot 2H_2O$ (top) and montmorillonite, $(Al,Mg)_8(Si_4O_{10})_3(OH)_{10} \cdot 12H_2O$ (bottom). The gypsum curve is offset upward 1.0 unit, for clarity. Both samples have very low reflectance because of the water content of the samples. Water is a strong infrared absorber. The montmorillonite also has a small grain size, which also tends to produce low mid-infrared reflectance because of the strong absorption in the mid-infrared.

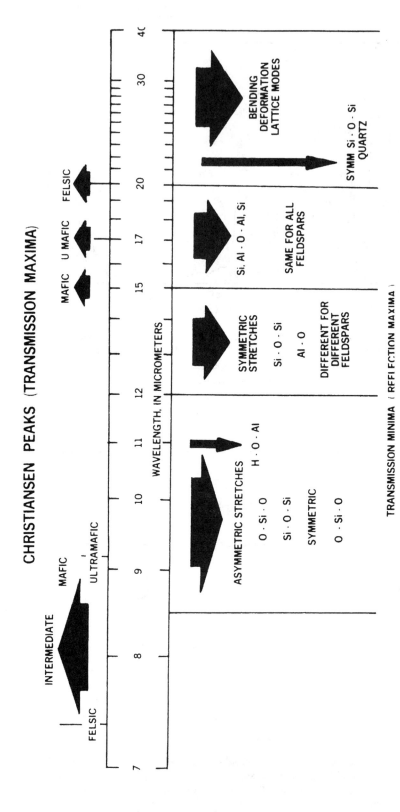

Figure 1.14 Locations and causes of absorptions in mid-infrared spectra of silicates. (From Hunt, 1982.)

roxenes (Figure 1.6) also show small amounts of water in the sample, as shown by the 3-μm absorptions in their spectra.

1.3.2.1 WATER AND HYDROXYL.

Water and OH (hydroxyl) produce particularly diagnostic absorptions in minerals. The water molecule (H_2O) has $N = 3$, so there are $3N - 6 = 3$ fundamental vibrations. In the isolated molecule (vapor phase) they occur at 2.738 μm (v_1, symmetric OH stretch), 6.270 μm (v_2, H–O–H bend), and 2.663 μm (v_3, asymmetric OH stretch). In liquid water, the frequencies shift due to hydrogen bonding: $v_1 = 3.106$μm, $v_2 = 6.079$ μm, and $v_3 = 2.903$ μm.

The overtones of water are seen in reflectance spectra of H_2O-bearing minerals (Figure 1.11). The first overtones of the OH stretches occur at about 1.4 μm and the combinations of the H–O–H bend with the OH stretches are found near 1.9 μm. Thus a mineral whose spectrum has a 1.9-μm absorption band contains water (e.g., hectorite and halloysite in Figure 1.11c), but a spectrum that has a 1.4-μm band but no 1.9-μm band indicates that only hydroxyl is present (e.g., kaolinite in Figure 1.11c has only a small amount of water because of the weak 1.9-μm absorption but a large amount of OH). The hydroxyl ion has only one stretching mode, and its wavelength position is dependent on the ion to which it is attached. In spectra of OH-bearing minerals, the absorption is typically near 2.7 to 2.8 μm but can occur anywhere in the range from about 2.67 to 3.45 μm (see, e.g., Clark et al., 1990, and references therein). The OH commonly occurs in multiple crystallographic sites of a specific mineral and is typically attached to metal ions. Thus there may be more than one OH feature. The metal–OH bend occurs near 10 μm (usually superimposed on the stronger Si–O fundamental in silicates). The combination metal–OH bend plus OH stretch occurs near 2.2 to 2.3 μm and is very diagnostic of mineralogy (see, e.g., Clark et al., 1990a, and references therein).

1.3.2.2 CARBONATES.

Carbonates also show diagnostic vibrational absorption bands (Figures 1.11a and 1.12). The observed absorptions are due to the planar CO_3^{2-} ion. There are four vibrational modes in the free CO_3^{2-} ion: the symmetric stretch, v_1: 1063 cm^{-1} (9.407 μm); the out-of-plane bend, v_2: 879 cm^{-1} (11.4 μm); the asymmetric stretch, v_3: 1415 cm^{-1} (7.067 μm); and the in-plane bend, v_4: 680 cm^{-1} (14.7 μm). The v_1 band is not infrared active in minerals. There are actually six modes in the CO_3^{2-} ion, but two are degenerate with the v_3 and v_4 modes. In carbonate minerals, the v_3 and v_4 bands often appear as a doublet. The doubling has been explained in terms of the lifting of the degeneracy (see, e.g., White, 1974) due to mineral structure and anion site.

Combination and overtone bands of the CO_3 fundamentals occur in the near-infrared. The two strongest are $v_1 + 2v_3$, at 2.50 to 2.55 μm (4000 to 3900 cm^{-1}), and $3v_3$ at 2.30 to 2.35 μm (4350 to 4250 cm^{-1}; e.g., Figure 1.11a). Three weaker bands occur near 2.12 to 2.16 μm ($v_1 + 2v_3 + v_4$ or $3v_1 + 2v_4$; 4720 to 4630 cm^{-1}), 1.97 to 2.00 μm ($2v_1 + 2v_3$; 5080 to 5000 cm^{-1}), and 1.85 to 1.87 μm ($v_1 + 3v_3$; 5400 to 5350 cm^{-1}; e.g., Figure 1.11a) (e.g., Hunt and Salisbury, 1971). The band positions in carbonates vary with composition (Hunt and Salisbury, 1971; Gaffey, 1986, Gaffey et al., 1993). An example of such a band shift is seen in Figures 1.11a, 1.12, and in more detail in Figure 1.15, showing the shift in absorption position from calcite to dolomite.

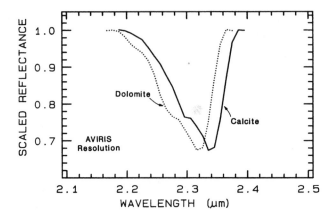

Figure 1.15 Comparison of calcite and dolomite continuum-removed features. The dolomite absorption occurs at a shorter wavelength than the calcite absorption.

1.3.2.3   OTHER MINERALS.

Phosphates, borates, arsenates, and vanadates also have diagnostic vibrational spectra. Space precludes inclusion of spectra here. See Hunt et al. (1972), and Clark et al. (1993b) for visual to near-infrared spectra. In general, the primary absorptions (e.g., P–O stretch) occurs at mid-infrared wavelengths. However, many of these minerals contain OH or H_2O and have absorptions in the near-infrared.

In the mid-infrared, minerals with H_2O, or those that are fine grained, like clays, have very low reflectance and show only weak spectral structure (e.g., Figure 1.13). Therefore, in emittance, spectral features will also be weak and thus difficult to detect. Grain size effects are discussed further below.

Typical spectra of minerals with vibrational bands are shown in Figures 1.4*b*, 1.5*b*, 1.6*b*, 1.7*b*, and 1.11. See Hunt and Salisbury (1970, 1971), Hunt et al. (1971a,b, 1972, 1973), Farmer (1974), Hunt (1979), Gaffey (1986, 1987), King and Clark (1989a), Clark et al. (1990a), Swayze and Clark (1990), Mustard (1992), Gaffey et al., 1993, and Salisbury (1993), and for more details. A summary of absorption band positions and causes is shown in Figures 1.10 and 1.14.

1.4   SPECTRA OF MISCELLANEOUS MINERALS AND MATERIALS

1.4.1   Organics

Organic materials are found all over the Earth and in the solar system. Organics can be important compounds in some environmental problems. The C–H stretch fundamental occurs near 3.4 μm (e.g., Figure 1.16*a*), the first overtone is near 1.7 μm, and a combination band is near 2.3 μm (Figure 1.16*b*). The combinations near 2.3 μm can sometimes be confused with OH and carbonate absorptions in minerals (e.g., Figure 1.11), especially at low spectral resolution. Further discussion and references appear in Chapter 3.

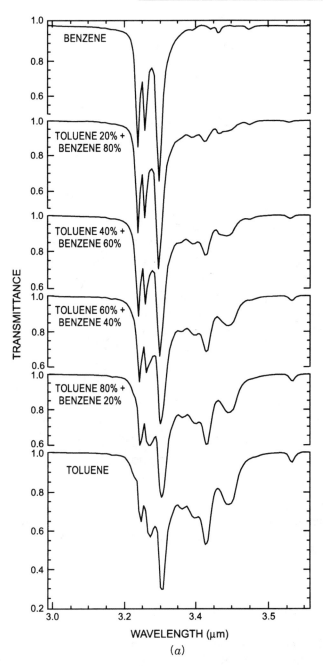

Figure 1.16   (*a*) Transmittance spectra of organics and mixtures showing the complex absorptions in the CH-stretch fundamental spectral region;

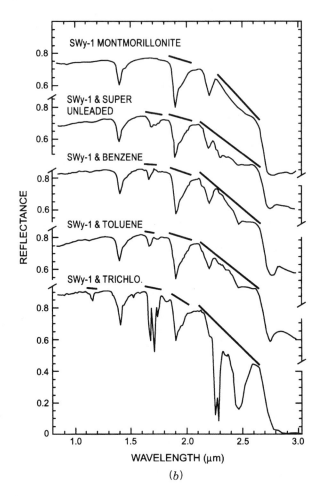

(*b*)

Figure 1.16   (*b*) reflectance spectra montmorillonite and montmorillonite mixed with super-unleaded gasoline, benzene, toluene, and trichlorethylene. Montmorillonite has an absorption feature at 2.2 μm, whereas the organics have a CH combination band near 2.3 μm. The first overtone of the CH stretch can be seen at 1.7 μm, and the second overtone near 1.15 μm. (From King and Clark, 1989b.)

1.4.2 Ices

Just as water in minerals shows diagnostic absorption bands, ice (crystalline H_2O), which is formally a mineral, also shows strong absorption bands. In the planetary literature it is referred to as water ice so as not to confuse it with other ices. Spectra of solid H_2O, CO_2, and CH_4 are shown in Figure 1.17. The spectral features in Figure 1.17 are all due to vibrational combinations and overtones, whose fundamentals have previously been discussed in general. Note that the H_2O spectra show broad absorptions compared to the others. The reason is that while ice is normally a hexagonal structure, the hydrogen bonds are orientationally disordered (e.g., Hobbs, 1974), and the disorder broadens the absorptions. There are many ices in the solar system (see, e.g., reviews by Cruikshank et al., 1985 and Clark et al., 1986,).

Ice, being ubiquitous in the solar system is found mixed with other minerals, on

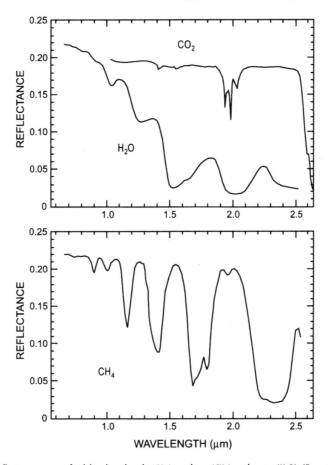

Figure 1.17   Reflectance spectra of solid carbon dioxide (CO_2), methane (CH_4), and water (H_2O). (From Clark et al., 1986.)

the Earth, as well as elsewhere in the solar system (e.g., Clark et al., 1986). The spectral properties of ice and ice–mineral mixtures have been studied by Clark (1981a,b), Clark and Lucey (1984), Lucey and Clark (1985), and references therein.

1.4.3 Vegetation

Spectra of vegetation come in two general forms: green and wet (photosynthetic) and dry nonphotosynthetic, but there is a seemingly continuous range between these two end members. The spectra of these two forms are compared to a soil spectrum in Figure 1.18. Because all plants are made of the same basic components, their spectra appear generally similar. However, in the spectral analysis section we will see methods for distinguishing subtle spectral details. The near-infrared spectra of green vegetation are dominated by liquid-water vibrational absorptions. The water bands are shifted to slightly shorter wavelengths than in liquid water, due to hydrogen bonding. The absorption in the visible is due to chlorophyll and is discussed in more detail in

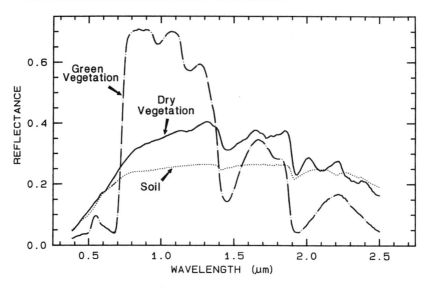

Figure 1.18 Reflectance spectra of photosynthetic (green) vegetation, nonphotosynthetic (dry) vegetation, and a soil. The green vegetation has absorptions short of 1 μm due to chlorophyll. Those at wavelengths greater than 0.9 μm are dominated by liquid water. The dry vegetation shows absorptions dominated by cellulose, but also lignin and nitrogen. These absorptions must also be present in the green vegetation but can be detected only weakly in the presence of the stronger water bands. The soil spectrum shows a weak signature at 2.2 μm due to montmorillonite.

Chapter 4. The dry nonphotosynthetic vegetation spectrum shows absorptions due to cellulose, lignin, and nitrogen. Some of these absorptions can be confused with mineral absorptions unless a careful spectral analysis is done.

1.5 SENSITIVITY OF ABSORPTION BANDS TO CRYSTAL STRUCTURE AND CHEMISTRY

Reflectance spectroscopy shows a wealth of information about mineralogy. Why, then, is spectroscopy not used more widely? In many cases spectroscopy is very sensitive to subtle changes in crystal structure or chemistry. This has resulted in confusion in the past over cause and effect. More recently, this sensitivity has been recognized as a powerful means of studying the structure and composition of minerals. Additional problems occur with reflectance spectra due to scattering and are discussed below.

Because spectroscopy is sensitive to so many processes, the spectra can be very complex and there is still much to learn. However, it is because of this sensitivity that spectroscopy has great potential as a diagnostic tool. In fact, for some materials, spectroscopy is an excellent tool not only for detecting certain chemistries, but also at abundance levels unmatched by other tools. For example, each layer of a layered silicate absorbs radiation almost independently from its neighbors. The absorption of photons does not depend on the longer-range crystallographic order as is required to give distinctive x-ray diffraction patterns. Thus, many processes (e.g., clay dehydroxylation) are detectable with spectroscopy before other methods (see, e.g., Farmer, 1974, p. 355). Spectroscopy is more sensitive to the presence of clays, iron oxides, iron hydroxides, quartz, and other minerals with strong absorption bands at levels significantly lower than other methods, such as x-ray diffraction (e.g., Farmer, 1974, p. 355). Next, a few examples of the possibilities are shown.

1.5.1 Pyroxenes

The iron bands near 1 and 2 µm shift with pyroxene composition as shown in Figure 1.6*a* and *c*. This series has been calibrated by Adams (1975), Cloutis et al. (1986), and Cloutis and Gaffey (1991). The olivine 1-µm band also shifts with composition (Figure 1.5*a*), although more subtly than with pyroxenes, and the shift has been calibrated by King and Ridley (1987). Note also the shifts, seen as positions of absorption minima and reflectance maxima, in the mid-infrared with different compositions in Figures 1.5*b* and 1.6*b*.

1.5.2 OH

The sharper OH-related absorption bands allow smaller band shifts to be measured. These bands can be so sensitive that it is possible to distinguish between the isochemical end members of the Mg-rich serpentine group, chrysotile, antigorite, and lizardite (King and Clark, 1989a; Figure 1.19). The Fe:Fe + Mg ratio can be estimated from reflectance spectra of minerals with brucitelike structure (Clark et al., 1990a, Mustard, 1992; Figure 1.19). Mustard (1992) calibrated changes in the 1.4- and 2.3-µm absorptions in the tremolite–actinolite solid solution series; sample spectra of the 1.4-µm absorptions are shown in Figure 1.19.

The structure of the 2.2-µm Al–OH band has been shown to be diagnostic of disorder in kaolinite–dickite mixtures (Crowley and Vergo, 1988) and the degree of kaolinite crystallinity (Clark et al., 1990a), which is illustrated in Figure 1.11*c* and *d*.

The strong and sharp OH features have proven particularly diagnostic of clay mineralogy, perhaps better than with x-ray diffraction (XRD) analysis (like any method, spectroscopy has advantages in some areas, and XRD in others). For example, it appears easy to distinguish kaolinite from halloysite with spectroscopy (e.g., Clark et al., 1990a), as shown in Figures 1.11*c* and *d*. Montmorillonite is easily distinguished from illite or muscovite (e.g., Clark et al., 1990a), whereas XRD analysis combines them into the general term *smectites*.

1.5.3 Al in Muscovite

More recently, subtle shifts have been found in muscovite series with aluminum composition (e.g., Post and Noble, 1993; Duke, 1994; Swayze, 1997). As elements substitute for aluminum in the crystal structure, the crystal becomes slightly distorted relative to no substitutions. This causes slight changes in Al–O–H bond lengths and thus shifts absorption band position. Sample shifts with composition are shown in Figure 1.20. In this case the shift of the 2.2-µm absorption appears continuous with

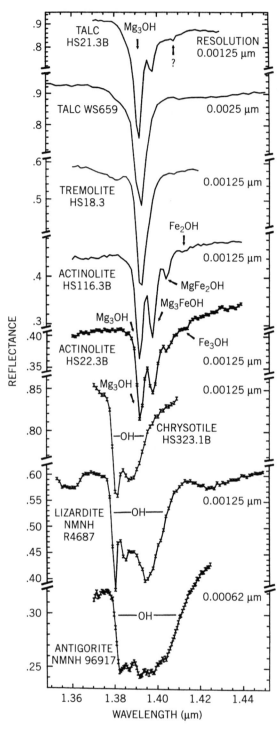

Figure 1.19 High-spectral-resolution reflectance spectra of the first overtone of OH in talc, tremolite, actinolite, crysotile, lizardite, and antigorite. The three sharp absorption bands in talc, tremolite, and actinolite are caused by Mg and Fe ions associated with the hydroxyls, causing small band shifts. The Fe:Fe + Mg ratio can be estimated. In chrysotile, lizardite, and antigorite, the absorptions change with small structural differences, even though the composition is constant. (From Clark et al., 1990a.)

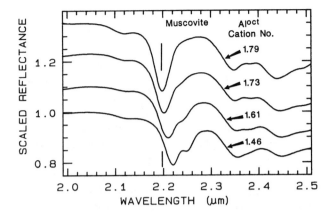

Figure 1.20   Reflectance spectra of muscovite, showing band shifts due to changing aluminum composition. (From Swayze, 1997.)

composition, compared to the growth of specific absorptions as in the tremolite–actinolite series shown in Figure 1.19. It is likely that all muscovite-group minerals show similar behavior, and illites may also.

1.5.4 Discussion

Reflectance spectroscopy can be used without sample preparation, and it is nondestructive. This makes mapping of exposed minerals from aircraft possible, including detailed clay mineralogy (e.g., Clark et al., 1993a). Visual and near-infrared spectroscopy, on the other hand, is insensitive to some minerals that do not have absorptions in this wavelength region. For example, quartz has no diagnostic spectral features in the visible and near-infrared; in fact, it is used as optical components in many telescopes and prisms. Quartz must be detected at its fundamental Si–O stretching region near 10 μm, as shown in Figure 1.4*b*.

Now that we have explored the causes of absorption features in minerals, we explore how those features get modified.

1.6   SCATTERING PROCESS

Scattering is the process that makes reflectance spectroscopy possible. Photons enter a surface, are scattered one or more times, and while some are absorbed, others are scattered from the surface, so we may see and detect them. Scattering can also be thought of as scrambling information. The information is made more complex, and because scattering is a nonlinear process, recovery of quantitative information is more difficult.

Consider the simple Beer's law in equation (1.2). In transmission, light passes through a slab of material. There is little or no scattering (none in the ideal case; but there are always internal reflections from the surfaces of the medium). Analysis is relatively simple. Reflectance of a particulate surface, however, is much more complex and the optical path of photons is a random walk. At each grain the photons encounter, a certain percentage are absorbed. If the grain is bright, like a quartz grain at visible wavelengths, most photons are scattered and the random walk process can go on for hundreds of encounters. If the grains are dark, like magnetite, the majority of photons will be absorbed at each encounter and essentially all photons will be absorbed in only a few encounters. The random walk process, scattering, and the mean depth of photon penetration are discussed in Clark and Roush (1984).

The random walk process of photons scattering in a particulate surface also enhances weak features not normally seen in transmittance, further increasing reflectance spectroscopy as a diagnostic tool. Consider two absorption bands of different strengths, such as a fundamental and an overtone. The stronger absorption will penetrate less deeply into the surface, encountering fewer grains because the photons are absorbed. At the wavelengths of the weaker absorption, fewer photons are absorbed with each encounter with a grain, so the random walk process goes further, increasing the average photon path length. The greater path length will result in more absorption, thus strengthening the weak absorption in a reflectance spectrum.

1.6.1 Mixtures

The real world (and for that matter, the universe) is a complex mixture of materials, at just about any scale at which we view it. In general, there are four types of mixtures:

1. *Linear mixture.* The materials in the field of view are optically separated, so there is no multiple scattering between components. The combined signal is simply the sum of the fractional area times the spectrum of each component. This is also called *areal mixture.*

2. *Intimate mixture.* An intimate mixture occurs when different materials are in intimate contact in a scattering surface, such as the mineral grains in a soil or rock. Depending on the optical properties of each component, the resulting signal is a highly nonlinear combination of the end-member spectra.

3. *Coatings.* Coatings occur when one material coats another. Each coating is a scattering–transmitting layer whose optical thickness varies with material properties and wavelength.

4. *Molecular mixtures.* Molecular mixtures occur on a molecular level, such as two liquids or a liquid and a solid mixed together. Examples: water adsorbed onto a mineral; gasoline spilled onto a soil. The close contact of the mixture components can cause band shifts in the adsorbate, such as the interlayer water in montmorillonite or the water in plants.

An example mixture comparison is shown in Figure 1.21 for alunite and jarosite. Note in the intimate mixture how the jarosite dominates in the region 0.4 to 1.3 μm. The reason is because in an intimate mixture, the darker material dominates because photons are absorbed when they encounter a dark grain. In the areal mixture, the brighter material dominates.

In a mixture of light and dark grains (e.g., quartz and magnetite) the photons have

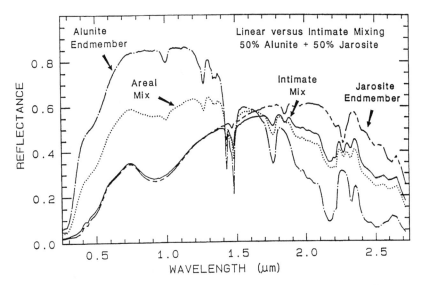

Figure 1.21   Reflectance spectra of alunite, jarosite, and mixtures of the two. Two mixtures types are shown: intimate and areal. In the intimate mixture, the darker of the two spectral components tends to dominate, and in the areal mixture, the brighter component dominates. The areal mixture is a strictly linear combination and was computed from the end members, whereas the intimate mixture is nonlinear and the spectrum of a physical mixture was measured in the laboratory. The jarosite dominates the wavelength region 0.3 to 1.4 μm in the intimate mixture because of the strong absorption in jarosite at those wavelengths and because the jarosite is finer grained than the alunite and tends to coat the larger alunite grains.

such a high probability of encountering a dark grain that a few percent of dark grains can drastically reduce the reflectance, much more than their weight fraction. The effect is illustrated in Figure 1.22, with spectra of samples having various proportions of charcoal grains mixed with montmorillonite.

1.6.2 Grain Size Effects

The amount of light scattered and absorbed by a grain is dependent on grain size (e.g., Clark and Roush, 1984; Hapke, 1993; Figures 1.23 and 1.24). A larger grain has a greater internal path where photons may be absorbed according to Beer's law. It is the reflection from the surfaces and internal imperfections that control scattering. In a smaller grain there are proportionally more surface reflections than internal photon path lengths, or in other words, the surface/volume ratio is a function of grain size. If multiple scattering dominates, as is usually the case in the visible and near-infrared, the reflectance decreases as the grain size increases, as shown in the pyroxene visible to near-infrared spectra in Figure 1.23*a*. However, in the mid-infrared, where absorption coefficients are much higher and the index of refraction varies strongly at the Christensen frequencies, the first surface reflection is a larger or even dominant component of the scattered signal. In these cases the grain size effects are much more complex, even reversing trends commonly seen at shorter wavelengths (e.g., Figure 1.23*b*).

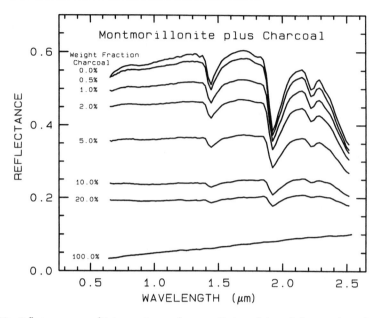

Figure 1.22   Reflectance spectra of intimate mixtures of montmorillonite and charcoal illustrates the nonlinear aspect of reflectance spectra of mixtures. The darkest substance dominates at a given wavelength. (From Clark, 1983.)

1.6.3 Continuum and Band Depth

Absorptions in a spectrum have two components: continuum and individual features. The continuum is the background absorption onto which other absorption features are superimposed (see, e.g., Clark and Roush, 1984). It may be due to the wing of a larger absorption feature. For example, in the pyroxene spectra in Figure 1.23a, the weak feature at 2.3 μm is due to a trace amount of tremolite in the sample and the absorption is superimposed on the broader 2-μm pyroxene band. The broader pyroxene absorption is the continuum to the narrow 2.3-μm feature. The pyroxene 1.0-μm band is superimposed on the wing of a stronger absorption centered in the ultraviolet.

The depth of an absorption band, D, is usually defined relative to the continuum, R_c:

$$D = 1 - \frac{R_b}{R_c} \tag{1.6}$$

where R_b is the reflectance at the band bottom and R_c is the reflectance of the continuum at the same wavelength as R_b (Clark and Roush, 1984).

The depth of an absorption is related to the abundance of the absorber and the grain size of the material. Consider a particulate surface with two minerals, one whose spectrum has an absorption band. As the abundance of the second mineral is increased, the band depth, D, of the absorption in the first mineral will decrease.

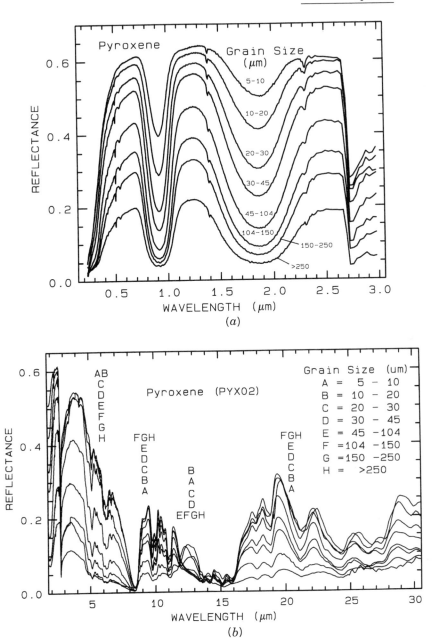

Figure 1.23 (*a*) Reflectance spectra of pyroxene as a function of grain size. As the grain size becomes larger, more light is absorbed and the reflectance drops. (From Clark et al., 1993b.) (*b*) Same series as in part (*a*), but for the mid-infrared. The position of letter identifiers indicates the relative position of the spectra at the various wavelengths. Note the reversal in the trends at some wavelengths and not others. Grain size effects on the shapes of spectral features in the mid-infrared can be quite large.

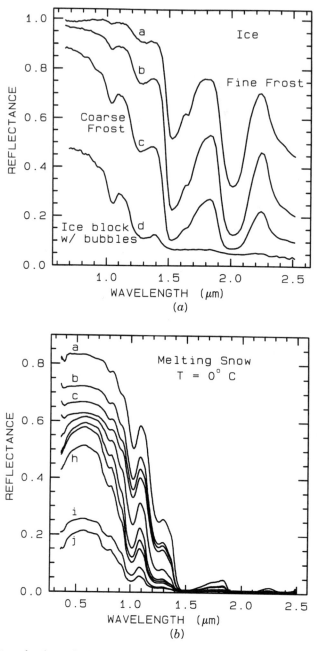

Figure 1.24 (*a*) Near-infrared spectral reflectance of a fine-grained (about 50 μm) water frost (curve a), medium-grained (about 200 μm) frost (curve b), coarse-grained (400 to 2000 μm) frost (curve c), and an ice block containing abundant microbubbles (curve d). The larger the effective grain size, the greater the mean photon path that photons travel in the ice, and the deeper the absorptions become. Curve d is very low in reflectance because of the large path length in ice. The ice temperatures for these spectra are 112 to 140 K. (From Clark et al., 1986.) (*b*) Series of reflectance spectra of melting snow. Curve a is at 0°C and has only a small amount of liquid water, whereas the lowest spectrum (curve j) is of a puddle of about 3 cm of water on top of the snow. Note in the top spectrum that there is no 1.65-μm band as in the ice spectra in part (*a*) because of the higher temperature. The 1.65-μm feature is temperature dependent and decreases in strength with increasing temperature (see Clark, 1981a, and references therein). Note the increasing absorption at about 0.75 μm and in the short side of the 1-μm ice band as more liquid water forms. The liquid water becomes spectrally detectable at about spectrum e, when the UV absorption increases. (Spectra from Clark et al., submitted).

Next consider the visual and near-infrared reflectance spectrum of a pure powdered mineral. As the grain size is increased from a small value, the absorption-band depth, D, will first increase, reach a maximum, and then decrease. This can be seen with the pyroxene spectra in Figure 1.23a and more so in the ice spectra in Figure 1.24. If the particle size were made larger and larger, the reflectance spectrum would eventually consist only of first surface reflection, as at most wavelengths beyond 1.45 μm in the ice spectra in Figure 1.24. The reflectance can never go to zero because of this reflection unless the index of refraction of the material is 1.0. These concepts, called *band saturation*, are explored further by Clark and Lucey (1984) and Lucey and Clark (1985).

A sloping continuum causes an apparent shift in the reflectance minimum, as shown in Figure 1.25. Continua can be thought of as an additive effect of optical constants, but in reflectance spectra, scattering and Beer's law make the effects non-linearly multiplicative (see Clark and Roush, 1984, for more details). So the continuum should be removed by division whether you are working in reflectance or emittance. The continuum should be removed by subtraction only when working with absorption coefficients. In a spectrum with a sloping continuum, correction removes

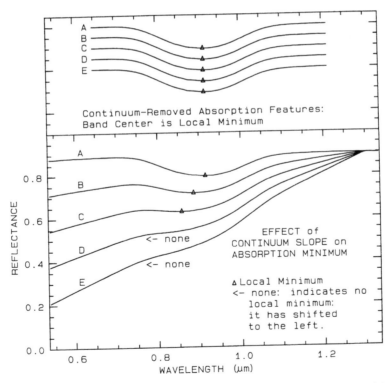

Figure 1.25 Continua and continuum removal. In the lower set of curves, the local minimum in the curve shifts to shorter wavelengths with increasing slope. Removal of the continuum by division isolates the spectral features so they may be compared (top). The top set of curves are offset for clarity. In the continuum-removed spectra, we can see that there is no real shift in the absorption-band center.

the effect of shifts in the local reflectance minimum (Figure 1.25). Note your perception of spectrum E versus A in Figure 1.25. The spectral features do not appear to be the same, but if you remove the continuum, it is obvious that they are the same (Figure 1.25, top).

1.6.4 Continuum-Removed Spectral Feature Comparison

The continuum-removal process isolates spectral features and puts them on a level playing field so they may be intercompared. Continuum removal and feature comparison is the key to successful spectral identification. For example, compare the spectra of calcite ($CaCO_3$) and dolomite ($CaMg(CO_3)_2$) in Figure 1.11a. If we isolate the spectral features, remove the continuum, and scale the band depth (or band area) to be equal, we can see subtle band shifts and shapes (Figure 1.15). Now compare a harder case: halloysite and kaolinite (Figure 1.11c). You might note that halloysite has a different absorption feature at 1.9 μm. However, if you were obtaining the spectrum through the Earth's atmosphere, you would have virtually no data in that wavelength region because atmospheric water absorbs too much of the signal. The diagnostic feature is the 2.2-μm band. The continua-removed 2.2-μm features for halloysite and kaolinite are shown in Figure 1.26, where we can see significant differences between the spectra of the two minerals.

One of the most challenging spectral features to distinguish between are those in spectra of various plant species. Figure 1.27a shows four plant spectra (the spectra are offset for clarity). The overall shapes are quite similar. If we remove the continuum according to Figure 1.27b, we see the detailed chlorophyll absorption spectral variations for these as well as other plants in Figure 1.27c. Shape-matching algorithms, such as that presented in Clark et al. (1990b), can distinguish between these spectra and compare them accurately to spectral libraries (see, e.g., Clark et al., 1993b; Chapter 5).

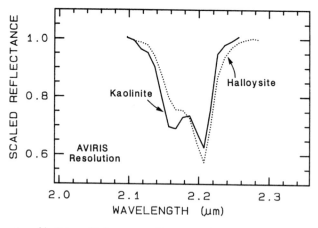

Figure 1.26 Comparison of kaolinite and halloysite spectral features. Both mineral spectra have the same band position at 2.2 μm. However, the kaolinite spectrum shows a stronger feature at 2.16 μm than in the halloysite spectrum.

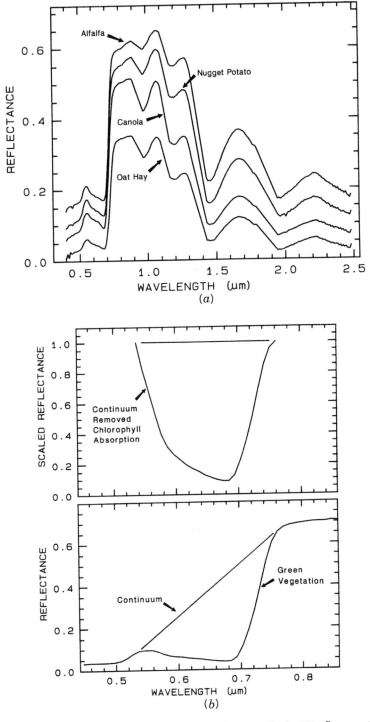

Figure 1.27 (*a*) Reflectance spectra for four types of vegetation. Each curve is offset by 0.05 reflectance unit from the one below. (From Clark et al., 1995, submitted.) (*b*) Continuum-removal example for a chlorophyll absorption in vegetation.

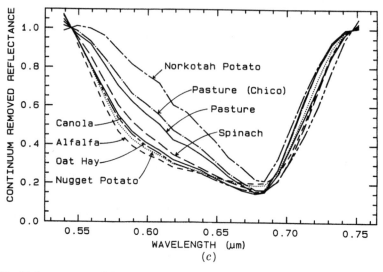

Figure 1.27 (c) Continuum-removed chlorophyll absorptions for eight vegetation types [including the four from part (a)], showing that the continuum-removed features can show subtle spectral differences. (From Clark et al., 1995, submitted.)

1.6.5 Other Spectral Variability and Rules

1.6.5.1 VIEWING GEOMETRY.

We have seen tremendous variation in the spectral properties of minerals and materials in general, due to composition, grain size, and mixture types. So far, viewing geometry has not been discussed. Viewing geometry, including the angle of incidence, angle of reflection, and the phase angle [the angle between the incident light and observer (the angle of reflection)] all affect the intensity of light received. Varying the viewing geometry results in changes in shadowing and the proportions of first surface to multiple scattering (e.g., Nelson, 1986; Mustard and Pieters, 1989; Hapke, 1993), which can affect band depths a small amount except in rare cases (e.g., extreme specular reflection off a mirror or lake surface). Whereas measuring precise light levels are important for such things as radiation balance studies, they are less important in spectral analysis. The following illustrates why.

First, your eye is a crude spectrometer, able to distinguish the spectral properties of materials in a limited wavelength range by the way we interpret color. Pick up any colored object around you. Change the orientation of the local normal on the surface of the object with respect to the angle of incident light, the angle at which you observe it (called the *emission* or *scattering angle*), and the angle between the incident and scattered light (the *phase angle*). As you do this, note any color changes. Unless you chose an unusual material (e.g., a diffraction grating or very shiny object), you will see no significant color changes. Plant leaves appear green from any angle, a pile of hematite appears red from any angle. This tells you that the spectral features do not change much with viewing geometry. Your eye–brain combination normalizes intensity variations so that you see the same color, regardless of the actual brightness of the object (the amount of light falling on the surface). The continuum removal does a similar but more sophisticated normalization. The band depth, shape, and position are basically constant with viewing geometry. Band depth will change only with the proportion of specular reflection added to the reflected light. For surfaces (and at wavelengths) where multiple scattering dominates, that change in band depth is minimized.

1.6.5.2   RATIOING SPECTRA.

Ratioing two spectra with spectral features can cause spurious features in the ratio (e.g., Clark and King, 1987). However, this ratio can be used to advantage. Consider two spectra, with an absorption edge, such as conduction bands in cinnabar (Figure 1.9), sulfur, or the chlorophyll-absorption edge in plants at 0.7 μm. If one spectrum is shifted relative to the other and then the two ratioed, the resulting ratio has a residual feature that looks like either an absorption or emission feature (depending on the direction of the shift and which spectrum is the numerator and which is the denominator). A sample residual caused by such shifts is shown in Figure 1.28. This effect has recently been used to determine subtle shifts in the chlorophyll absorption edge in plants (Clark et al., 1995, 1996). The intensity of the feature in the reflectance

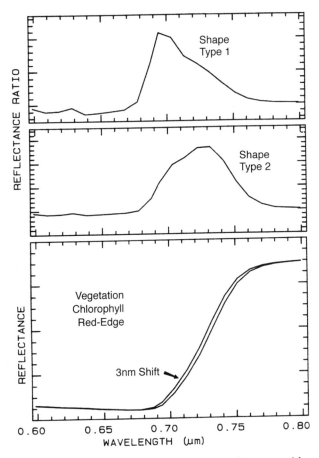

Figure 1.28   The ratio of two spectra, one slightly shifted from the other, results in a spectral feature. The two shapes in this example are from different chlorophyll bands. Shifts as small as 0.01 times the full width at half maximum of the spectrometer system produce 1% features in strong chlorophyll absorptions. The shape, type 1 profile, is from the shifted spectra shown at the bottom of the figure, while the shape, type 2, is from shallower chlorophyll absorptions. (From Clark et al., submitted.)

ratio is proportional to the amount of shift between the two spectra, and the shape of the ratio does not change if the shifts are small. Clark et al. (1995, 1998) showed that chlorophyll-edge shifts as small as 0.01 times the bandwidth of the spectrometer can be detected. Thus for the AVIRIS system with 10-nm bandpass and sampling, shifts of 0.1 nm can readily be detected. Such sensitivity indicates that the wavelength stability of a spectrometer must be very good for such analyses.

1.6.5.3 IRON OXIDE, HYDROXIDE, AND SULFATE COMPLEXITY.

Iron oxides, hydroxides, and sulfates are a special case for remote sensing because they are so ubiquitous. Further, because of the strength of the iron absorptions in the ultraviolet to about 1 μm, at least one if not all are saturated in reflectance. Several hematite reflectance spectra at different grain sizes are shown in Figure 1.29a. Nanocrystalline hematite (Morris et al., 1985) has such a fine grain size, ≤25 nm, that the grain surface boundary modifies the electronic transitions, changing and weakening them. The iron absorption at 0.9 μm is reduced in depth, the 0.65-μm band is absent, and the ultraviolet absorption is weak. Absorptions in transmittance, as in the thin-film case, are $\sqrt{2}$ times narrower in width (see Clark and Roush, 1984). Larger grain sizes show increased saturation of the 0.9-μm absorption, broadening and shifting the apparent reflectance minimum to longer wavelengths. The 0.9-μm absorption also shifts position with elements substituted for iron (see, e.g., Morris et al., 1985, and references therein). Continuum removal and scaling the hematite absorption to similar depth show the wide variety of band shapes and positions that can be found in nature (Figure 1.29b).

There are a whole suite of iron oxides, iron hydroxides, iron sulfates, and so on, some only now being discovered, and many amorphous phases, all with similar electronic absorption bands in the visible and near-infrared. A few examples are shown in Figure 1.30a. Note that hematite has a narrower absorption at a slightly shorter wavelength than goethite. However, a coarse-grained hematite has a broader absorption, approaching the position and width of a fine-grained goethite (or a thin-film goethite).

Jarosite has a narrow absorption near 0.43 μm, but it sometimes appears weak because of the saturated UV absorption. Jarosite, an iron sulfate, has a diagnostic absorption at 2.27 μm due to a combination OH stretch and Fe–OH bend. However, this feature is weaker than the electronic absorptions in the visible and is often masked by clay or alunite (jarosite often occurs in hydrothermal deposits with alunite). The features near 1.475 μm and 1.8 μm are OH related and are commonly seen in sulfate spectra.

Ferrihydrite is an amorphous iron oxide, and its spectrum appears very similar to the orange precipitate, an amorphous iron hydroxide, obtained downstream from the Summitville, Colorado mine (King et al., 1995). However, if we remove the continuum and compare the positions and shapes of the bands (Figure 1.30b), we see that they are indeed different.

As discussed above, there are many iron-bearing minerals and amorphous materials with similar but distinct absorption bands. How many can be distinguished with reflectance spectroscopy is still a matter of research. An even harder question to answer is how many can be distinguished and not be confused with mixtures of other iron-bearing materials. Detailed spectral analysis, including continuum removal to isolate absorption features, can certainly improve the success of distinguishing them.

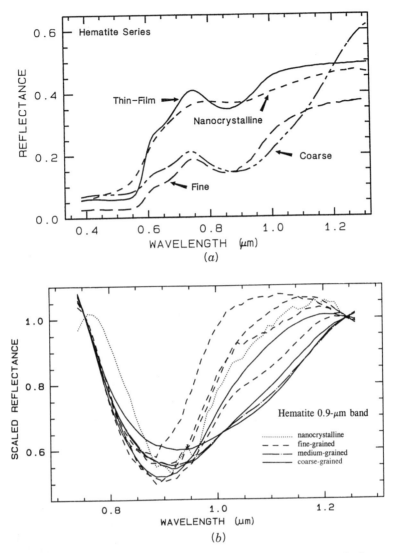

Figure 1.29 (a) Reflectance spectra of different grain sizes of hematite; (b) continuum-removed reflectance spectra of different grain sizes of hematite.

Iron oxides, hydroxides, and sulfates are additional cases where spectroscopy detects at very low levels because of the strong absorption bands in the visible and ultraviolet. In nature, there appear to be many amorphous iron oxides, hydroxides, and so on, with equally intense absorptions. Thus spectroscopy cannot only detect them at levels below other methods (e.g., x-ray diffraction), but in the case of amorphous materials can detect them when other methods are not sensitive to their presence when they are major fractions of the sample!

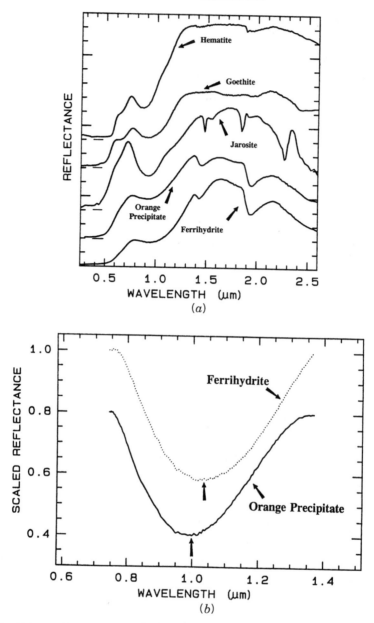

Figure 1.30 (*a*) Spectra of iron oxide, iron hydroxide, and iron sulfate spectra; (*b*) comparison of the 1-μm continuum-removed absorption feature of ferrihydrite and an amorphous iron hydroxide [the orange precipitate from part (*a*)]. The spectra are offset for clarity. Note the shift in the band center between the two spectra despite the similarities in part (*a*).

1.7 QUANTITATIVE UNDERSTANDING: RADIATIVE TRANSFER THEORY

There have been many attempts over the years to quantify the scattering process. Kubelka–Munk theory was one of the first and still finds uses today (e.g., Wendlandt and Hecht, 1966; Clark and Roush, 1984). A method growing in popularity in some industries is log (1/R), where R is reflectance, but this is a less robust attempt at quantifying the scattering process than the decades-old Kubelka–Munk theory (Clark and Roush, 1984). The latter method is usually combined with computing the derivatives of the log (1/R) spectra and doing a correlation analysis to find particular trends. This has popularly become known as *near-infrared reflectance analysis* (NIRA). Either method has its uses in controlled situations, but there is a modern, more effective alternative. The limitations of these older methods are due to a poor representation of the scattering process and are discussed in Clark and Roush (1984) and Hapke (1993).

Fortunately, in the early 1980s three independent investigations (Hapke, 1981; Goguen, 1981; Lumme and Bowell, 1981) provided reasonable solutions to the complex radiative transfer problem as applied to particulate surfaces. These theories provide for nonisotropic scattering of light from particles, shadowing between particles, and first surface reflection from grain surfaces, important processes not considered in earlier theories. One theory, that of Hapke (1981, 1993), also provides for mixtures, and because of its relative simplicity compared to the other two, has become the dominant theory used in the planetary and to some degree the terrestrial remote sensing communities.

1.7.1 Hapke Theory

From the optical constants of a mineral, the reflectance (called the *radiance factor* by Hapke) can be computed from Hapke's (1981) eq. 36:

$$r'(\bar{w}, \lambda, \mu, g) = \frac{\bar{w}}{4\pi} \frac{\mu}{\mu + \mu_0} \{[1 + B(g)]P(g) + H(\mu)H(\mu_0) - 1\} \qquad (1.7)$$

where r' is the reflectance at wavelength λ, $\bar{w}$ the average single scattering albedo, μ the cosine of the angle of emitted light, g the phase angle, μ_0 the cosine of the angle of incident light, $B(g)$ a backscatter function, $P(g)$ the average single-particle phase function, and H the Chandrasekhar (1960) H-function for isotropic scatterers. When $r' > 0.9$ Hapke's approximation of the H-function shows considerable error and equation (1.7) deviates from measurements (Hapke, 1981). Because of this deviation, a table interpolation subroutine using "exact" values from Chandrasekhar (1960) can be used. The table interpolation is faster computationally than the Hapke approximation, as well as being more accurate.

The single-scattering albedo is the probability that a photon survives an interaction with a single particle, which includes Fresnel reflection, absorption, scattering, and diffraction due to the presence of an individual grain. Hapke (1981) developed the theory further by deriving a relation between the single-scattering albedo, the complex index of refraction, the grain size, and a scattering parameter to describe scattering centers within nonperfect grains. The single-scattering albedo of a grain can be found from his eq. 24:

$$w = S_E + \frac{(1 - S_E)(1 + S_I)\{r_1 + \exp[-2(k(k + s))^{1/2}d/3]\}}{1 - r_1 S_I + (r_1 - S_I)\exp[-2(k(k + s))^{1/2}d/3]} \tag{1.8}$$

where S_E and S_I are the external and internal scattering coefficients, respectively, which can be computed from the complex index of refraction (Hapke, 1981, eq. 21), s is a scattering coefficient, d is the particle diameter, k is the absorption coefficient (note that Hapke uses α instead of k here), and (from Hapke's eq. 23)

$$\begin{aligned} r_1 &= \frac{1 - [k/(k + s)]^{1/2}}{1 + [k/(k + s)]^{1/2}} \\ &= \frac{1 - [kd/(kd + sd)]^{1/2}}{1 + [kd/(kd + sd)]^{1/2}} \end{aligned} \tag{1.9}$$

In a monomineralic surface, $w = \bar{w}$. For a multimineralic surface, $\bar{w}$ can be computed from eq. 17 of Hapke (1981):

$$\bar{w} = \frac{\sum_i \dfrac{M_i w_i}{\rho_i d_i}}{\sum_i \dfrac{M_i}{\rho_i d_i)}} \tag{1.10}$$

where i refers to the ith component, M_i is the mass fraction, w_i the single-scattering albedo of the ith component, ρ_i the density of the material, and d_i the mean grain diameter.

With the Hapke (1981, 1993) reflectance theory and the optical constants of minerals, reflectance spectra of pure minerals at a single grain size, spectra of a pure mineral with a grain size distribution, and mineral mixtures with varying grain size components can all be computed. Clark and Roush (1984) also showed that a reflectance spectrum can be inverted to determine quantitative information on the abundances and grain sizes of each component. The inversion of reflectance to quantitative abundance has been tested in laboratory mixtures (e.g., Johnson et al., 1983, 1992; Clark, 1983; Mustard and Pieters, 1987a, 1989; Shipman and Adams, 1987; Sunshine and Pieters, 1990, 1991; Sunshine et al., 1990; Gaffey et al., 1993; and references therein). Some quantitative inversion attempts have been undertaken with imaging spectroscopy data (e.g., Mustard and Pieters, 1987b; Adams et al., 1993; Li et al., 1996; and references therein).

1.8 SPECTRAL LIBRARIES

The spectra presented in this paper are available on the World Wide Web site http://speclab.cr.usgs.gov, as are spectra from the Clark et al. (1993b) USGS digital spectral library. Other spectral libraries include the mid-infrared work of Salisbury et al. (1991). A recently available spectral library Web site is the NASA ASTER site (http://asterweb.jpl.nasa.gov), managed by Simon Hook, Jet Propulsion Laboratory. This site includes the Salisbury et al. (1991) library and additions since the original publication. As spectral libraries are currently a focus of activity, it is probably best to search the Internet and check with the authors cited in this chapter for the latest information on what is available.

A word of caution concerning spectral libraries and spectra obtained from other sources in general: Wavelength errors are common except from data obtained on interferometers. This author and colleagues at the USGS have evaluated many spectrometers and other spectral libraries and have found many to have significant wavelength shifts. Other specific libraries and spectrometers are not mentioned here because some may have wavelength shifts and must each be validated. One mineral with a stable absorption feature is a well-crystallized kaolinite which has a sharp absorption at 2.2086 ± 0.0003 μm and is commonly found in visible and near-infrared libraries. When obtaining spectral library data, confirm that wavelength positions of known features are measured at the correct positions. Absorptions due to rare earth oxides are often used as wavelength standards in the visible. Mid-infrared systems can be checked by interferometer measurements, which is now probably the most common spectrometer in use for this wavelength region.

Also be cautious of spurious spectral features from incomplete reduction to true reflectance. All measurements are made relative to a "white" standard. However, these standards also have spectral features. For example, the common visible and near-infrared standards, Halon and Spectralon and derivatives, have significant spectral features in the 2.14-μm region and beyond (see, e.g., Clark et al., 1990a) which must be corrected properly. Mid-infrared standards are more difficult, due primarily to the wide wavelength range usually covered. Nash (1986) reviewed some common mid-infrared reflectance standards.

1.9 CONCLUSIONS AND DISCUSSION

Reflectance spectroscopy is a rapidly growing science that can be used to derive significant information about mineralogy with little or no sample preparation. It may be used in applications when other methods would be too time consuming or require destruction of precious samples. For example, imaging spectrometers are already acquiring millions of spatially gridded spectra over an area from which mineralogical maps are being made. It is possible to set up real-time monitoring of processes using spectroscopy, such as monitoring the mineralogy of drill cores at the drilling site. Research is still needed to better understand the subtle changes in absorption features before reflectance spectroscopy will reach its full potential. Good spectral databases documenting all the absorption features are also needed before reflectance spectroscopy can be as widely used as XRD. Spectral databases are now becoming available (e.g., Clark et al., 1993b) and research continues on the spectral properties of minerals, but it will probably take about a decade before general software tools are available to allow reflectance spectroscopy to challenge other analytical methods in the commercial marketplace. For certain classes of minerals, however, spectroscopy is already an excellent tool. Among these classes are clay mineralogy, OH-bearing minerals, iron oxides and hydroxides, carbonates, sulfates, olivines, and pyroxenes.

Space limits the contents of any review article covering such a broad topic. Other review articles are Adams (1975), Hunt (1977, 1982), Gaffey et al. (1993), Salisbury

(1993), and Clark (1995). The Hunt (1982) article in particular presents more spectra, both visible–near-infrared and mid-infrared, than most other works and seems to be an overlooked but important work.

ACKNOWLEDGMENTS

This work was supported by NASA interagency agreement W15805. Thanks goes to reviewers John Mustard, Gregg Swayze, and Eric Livo, whose comments improved the manuscript substantially.

References

Adams, J. B., 1974. Visible and near-infrared diffuse reflectance spectra of pyroxenes and applied to remote sensing of solid objects in the solar system, *J. Geophys. Res.*, 79, 4829–4836.

Adams, J. B., 1975. Interpretation of visible and near-infrared diffuse reflectance spectra of pyroxenes and other rock-forming minerals, in *Infrared and Raman Spectroscopy of Lunar and Terrestrial Minerals*, C. Karr, ed., Academic Press, San Diego, Calif., pp. 94–116.

Adams, J. B., M. O. Smith, and A. R. Gillespie, 1993. Imaging spectroscopy: interpretation based on spectral mixture analysis, in *Remote Geochemical Analysis: Elemental and Mineralogical Composition*, C. M. Pieters and P. A. J. Englert, eds., Cambridge University Press, Cambridge, pp. 145–166.

Ball, D. W., 1995. Defining terms, *Spectroscopy*, 10, 16–18.

Berk, A., L. S. Bernstein, and D. C. Robertson, 1989. *MODTRAN: A Moderate Resolution Model for LOWTRAN 7, Final Report*, AFGL-TR-0122, Air Force Geophysics Laboratory, Hanscomb AFB, Mass., 42 pp.

Burns, R., 1970. *Mineralogical Applications of Crystal Field Theory*, Cambridge University Press, Cambridge, 224 pp.

Burns, R., 1993. *Mineralogical Applications of Crystal Field Theory*, 2nd ed., Cambridge University Press, Cambridge, 551 pp.

Chandrasekhar, S., 1960. *Radiative Transfer*, Dover, Mineola, N.Y. 393 pp.

Clark, R. N., 1981a. Water frost and ice: the near-infrared spectral reflectance 0.65–2.5 μm, *J. Geophys. Res.*, 86, 3087–3096.

Clark, R. N., 1981b. The spectral reflectance of water–mineral mixtures at low temperatures, *J. Geophys. Res.*, 86, 3074–3086.

Clark, R. N., 1983. Spectral properties of mixtures of montmorillonite and dark carbon grains: implications for remote sensing minerals containing chemically and physically adsorbed water, *J. Geophys. Res.*, 88, 10635–10644.

Clark, R. N., 1995. Reflectance spectra, in *AGU Handbook of Physical Constants*, American Geophysical Union, Washington, D.C., 12 pp.

Clark, R. N., and T. V. V. King, 1987. Causes of spurious features in spectral reflectance data, in *Proceedings of the 3rd Airborne Imaging Spectrometer Data Analysis Workshop*, JPL Publ. 87-30, Jet Propulsion Laboratory, California Institute of Technology, Pasadena, Calif., pp. 132–137.

Clark, R. N., and P. G. Lucey, 1984. Spectral properties of ice–particulate mixtures and implications for remote sensing: I. Intimate mixtures, *J. Geophys. Res.*, 89, 6341–6348.

Clark, R. N., and T. L. Roush, 1984. Reflectance spectroscopy: quantitative analysis techniques for remote sensing applications, *J. Geophys. Res.*, 89, 6329–6340.

Clark, R. N., F. P. Fanale, and M. J. Gaffey, 1986. Surface composition of satellites, in *Natural Satellites*, J. Burns and M. S. Matthews, eds., University of Arizona Press, Tucson, Ariz., pp. 437–491.

Clark, R. N., T. V. V. King, M. Klejwa, G. Swayze, and N. Vergo, 1990a. High spectral resolution reflectance spectroscopy of minerals, *J. Geophys. Res.*, 95, 12653–12680.

Clark, R. N., A. J. Gallagher, and G. A. Swayze, 1990b. Material absorption band depth mapping of imaging spectrometer data using a complete band shape least-squares fit with library reference spectra, in *Proceedings of the 2nd Airborne Visible/Infrared Imaging Spectrometer (AVIRIS) Workshop*, JPL Publ. 90–54, Jet Propulsion Laboratory, California Institute of Technology, Pasadena, Calif., pp. 176–186.

Clark, R. N., G. A. Swayze, and A. Gallagher, 1993a. *Mapping Minerals with Imaging Spectroscopy*, USGS Bull. 2039, U.S. Geological Survey, Office of Mineral Resources, Washington, D.C., pp. 141–150.

Clark, R. N., G. A. Swayze, A. Gallagher, T. V. V. King, and W. M. Calvin, 1993b. *The U.S. Geological Survey, Digital Spectral Library, Version 1: 0.2 to 3.0 μm*, USGS Open File Rep. 93–592, U.S. Geological Survey, Washington, D.C., 1326 pp.

Clark, R. N., T. V. V. King, C. Ager, and G. A. Swayze, 1995. Initial vegetation species and senescence/stress mapping in the San Luis Valley, Colorado using imaging spectrometer data, in *Proceedings of the Summitville Forum '95*, H. H. Posey, J. A. Pendelton, and D. Van Zyl, eds., CGS Spec. Publ. 38, Colorado Geological Survey, Colo., pp. 64–69.

Clark, R. N., T. V. V. King, C. Ager, and G. A. Swayze, submitted. Vegetation species and stress indicator mapping in the San Luis Valley, Colorado using imaging spectrometer data, *Remote Sensing Environ.* 1998.

Cloutis, E. A., and M. J. Gaffey, 1991. Pyroxene spectroscopy revisited: spectral–compositional correlations and relationships to geotherometry, *J. Geophys. Res.*, 96, 22809–22826.

Cloutis, E. A., M. J. Gaffey, T. L. Jackowski, and K. L. Reed, 1986. Calibrations of phase abundance, composition, and particle size distribution of olivine–orthopyroxene mixtures from reflectance spectra, *J. Geophys. Res.*, 91, 11641–11653.

Crowley, J. K., and N. Vergo, 1988. Near-infrared reflectance spectra of mixtures of kaolin group minerals: use in clay studies, *Clays Clay Miner.*, 36, 310–316.

Cruikshank, D. P., R. H. Brown, and R. N. Clark, 1985. Methane ice on Triton and Pluto, in *Ices in the Solar System*, J. Klinger et al., eds., D. Reidel, Dordrecht, The Netherlands, pp. 817–827.

Duke, E. F., 1994. Near infrared spectra of muscovite, Tschermak substitution, and metamorphic reaction progress: implications for remote sensing, *Geology*, 22, 621–624.

Farmer, V. C., ed., 1974. *The Infra-red Spectra of Minerals*, Mineralogical Society, London, 539 pp.

Gaffey, S. J., 1986. Spectral reflectance of carbonate minerals in the visible and near infrared (0.35–2.55 μm): calcite, aragonite and dolomite, *Am. Mineral.*, 71, 151–162.

Gaffey, S. J., 1987. Spectral reflectance of carbonate minerals in the visible and near infrared (0.35–2.55 μm): anhydrous carbonate minerals, *J. Geophys. Res.*, 92, 1429–1440.

Gaffey, S. J., L. A. McFadden, D. Nash, and C. M. Pieters, 1993. Ultraviolet, visible, and near-infrared reflectance spectroscopy: laboratory spectra of geologic materials, in *Remote Geochemical Analysis: Elemental and Mineralogical Composition*, C. M. Pieters, and P. A. J. Englert, eds., Cambridge University Press, Cambridge, pp. 43–78.

Goetz, A. F. H, G. Vane, J. E. Solomon, and B. N. Rock, 1985. Imaging spectrometry for earth remote sensing, *Science*, 228, 1147–1153.

Goguen, J. D., 1981. A theoretical and experimental investigation of the photometric functions of particulate surfaces, Ph.D. thesis, Cornell University, Ithaca, N.Y.

Green, R. O., J. E. Conel, V. Carrere, C. J. Bruegge, J. S. Margolis, M. Rast, and G. Hoover, 1990. Determination of the in-flight spectral and radiometric characteristics of the airborne visible/infrared imaging spectrometer (AVIRIS), in *Proceedings of the 2nd Airborne Visible/Infrared Imaging Spectrometer (AVIRIS) Workshop*, JPL Publ. 90–54, Jet Propulsion Laboratory, California Institute of Technology, Pasadena, Calif., pp. 15–22.

Hapke, B., 1981. Bidirectional reflectance spectroscopy: 1. Theory, *J. Geophys. Res.*, 86, 3039–3054.

Hapke, B., 1993. *Introduction to the Theory of Reflectance and Emittance Spectroscopy*, Cambridge University Press, New York.

Hecht, E., 1987. *Optics*, Addison-Wesley, Reading Mass., 676 pp.

Henderson, B. G., P. G. Lucey, and B. M. Jakosky, 1996. New laboratory measurements of mid-IR emission spectra of simulated planetary surfaces, *J. Geophys. Res.*, 101, 14969–14975.

Herzberg, G., 1945. *Molecular Spectra and Molecular Structure, Vol. 2, Infrared and Raman Spectra of Polyatomic Molecules*, Van Nostrand Reinhold, New York, 632 pp.

Hobbs, P. V., 1974. *Ice Physics*, Clarendon Press, Oxford, 837 pp.

Hunt, G. R., 1977. Spectral signatures of particulate minerals, in the visible and near-infrared, *Geophysics*, 42, 501–513.

Hunt, G. R., 1979. Near-infrared (1.3–2.4 μm) spectra of alteration minerals: potential for use in remote sensing, *Geophysics*, 44, 1974–1986.

Hunt, G. R., 1982. Spectroscopic properties of rocks and minerals, in *Handbook of Physical Properties of Rocks*, Vol. 1, R. S. Carmichael, ed., CRC Press, Boca Raton, Fla., pp. 295–385.

Hunt, G. R., and J. W. Salisbury, 1970. Visible and near infrared spectra of minerals and rocks: I. Silicate minerals, *Mod. Geol.*, 1, 283–300.

Hunt, G. R., and J. W. Salisbury, 1971. Visible and near infrared spectra of minerals and rocks: II. Carbonates, *Mod. Geol.*, 2, 23–30.

Hunt, G. R., J. W. Salisbury, and C. J. Lenhoff, 1971a. Visible and near infrared spectra of minerals and rocks: III. Oxides and hydroxides, *Mod. Geol.*, 2, 195–205.

Hunt, G. R., J. W. Salisbury, and C. J. Lenhoff, 1971b. Visible and near infrared spectra of minerals and rocks: IV. Sulphides and sulphates, *Mod. Geol.*, 3, 1–14.

Hunt, G. R., J. W. Salisbury, and C. J. Lenhoff, 1972. Visible and near infrared spectra of minerals and rocks: V. Halides, arsenates, vanadates, and borates, *Mod. Geol.*, 3, 121–132.

Hunt, G. R., J. W. Salisbury, and C. J. Lenhoff, 1973. Visible and near infrared spectra of minerals and rocks: VI. Additional silicates, *Mod. Geol.*, 4, 85–106.

Johnson, P., M. Smith, and S. Taylor-George, 1983. A semi-empirical method for analysis of the reflectance spectra of binary mineral mixtures, *J. Geophys. Res.*, 88, 3557–3561.

Johnson, P. M., M. O. Smith, and J. B. Adams, 1992. Simple algorithms for remote determination of mineral abundances and particle sizes from reflectance spectra, *J. Geophys. Res.*, 97, 2649–2657.

King, T. V. V., and R. N. Clark, 1989a. Spectral characteristics of serpentines and chlorites using high resolution reflectance spectroscopy, *J. Geophys. Res.*, 94, 13997–14008a.

King, T. V. V., and R. N. Clark, 1989b. Reflectance spectroscopy (0.2 to 20μm) as an analytical method for the detection of organics in soils; in *Proceedings of the First International Symposium: Field Screening Methods for Hazardous Waste Site Investigations*, U.S. Environmental Protection Agency, Washington, D.C., pp. 485–488.

King, T. V. V., and W. I. Ridley, 1987. Relation of the spectroscopic reflectance of olivine to mineral chemistry and some remote sensing implications, *J. Geophys. Res.*, 92, 11457–11469.

King, T. V. V., R. N. Clark, C. Ager, and G. A. Swayze, 1995. Remote mineral mapping using AVIRIS data at Summitville, Colorado and the adjacent San Juan Mountains, in *Proceedings of the Summitville Forum '95*, H. H. Posey, J. A. Pendelton, and D. Van Zyl eds., CGS Spec. Publ. 38, Colorado Geological Survey, Boulder, Colo., pp. 59–63.

Li, L., J. F. Mustard, and G. He, 1996. Mixing across simple mare–highland contacts: new insights from *Clementine* UV–VIS data of the Grimaldi basin, in *Lunar and Planetary Science XXVII*, Lunar and Planetary Institute, Houston, Texas, pp. 751–752.

Lucey, P. G., and R. N. Clark, 1985. Spectral properties of water ice and contaminants, in *Ices in the Solar System*, J. Klinger et al., eds., D. Reidel, Dordrecht, The Netherlands, pp. 155–168.

Lumme, K., and Bowell, E., 1981. Radiative transfer in the surfaces of atmosphereless bodies: I. Theory, *Astron. J.*, 86, 1694–1704.

Morris, R. V., H. V. Lauer, C. A. Lawson, E. K. Gibson, Jr., G. A. Nace, and C. Stewart, 1985. Spectral and other physiochemical properties of submicron powders of hematite (α-Fe_2O_3), maghemite (γ-Fe_2O_3), maghemite (Fe_3O_4), goethite (α-FeOOH), and lepidochrosite (γ-FeOOH), *J. Geophys. Res.*, 90, 3126–3144.

Mustard, J. F., 1992. Chemical composition of actinolite from reflectance spectra, *Am. Mineral.*, 77, 345–358.

Mustard, J. F., and C. M. Pieters, 1987a. Quantitative abundance estimates from bidirectional reflectance measurements, in *Proceedings of the 17th Lunar and Planetary Science Conference, J. Geophys. Res.*, 92, E617–E626.

Mustard, J. F., and C. M. Pieters, 1987b. Abundance and distribution of serpentinized ultramafic microbreccia in Moses Rock dike: quantitative application of mapping spectrometer data, *J. Geophys. Res.*, 92, 10376–10390.

Mustard, J. F., and C. M. Pieters, 1989. Photometric phase functions of common geologic minerals and applications to quantitative analysis of mineral mixture reflectance spectra, *J. Geophys. Res.*, 94, 13619–13634.

Nash, D. B., 1986. Mid-infrared reflectance spectra (2.3–22 µm) of sulfur, gold, FBr, Mgo and Halon, *Appl. Opt.*, 25, 2427–2433.

Nelson, M. L., 1986. Application of radiative transfer theory to the spectra of mixtures of minerals with anisotropic phase functions, Master's thesis, University of Hawaii, Honolulu, Hawaii, 71 pp.

Nicodemus, F. E., 1965. Directional reflectance and emissivity of an opaque surface, *Appl. Opt.*, 4, 767–773.

Post, J. L., and P. N. Noble, 1993. The near-infrared combination band frequencies of dioctahedral smectites, micas, and illites, *Clays Clay Miner.*, 41, 639–644.

Salisbury, J. W., 1993. Mid-infrared spectroscopy: laboratory data, in *Remote Geochemical Analysis: Elemental and Mineralogical Composition*, C. M. Pieters and P. A. J. Englert, eds., Cambridge University Press, Cambridge, pp. 79–98.

Salsibury, J. W., L. S. Walter, N. Vergo, and D. M. D'Aria, 1991. *Infrared (2.1–25 µm) Spectra of Minerals*, Johns Hopkins University Press, Baltimore, 267 pp.

Sherman, D. M., 1990. Crystal chemistry, electronic structures and spectra of Fe sites in clay minerals, in *Spectroscopic Characterization of Minerals and Their Surfaces*, L. M. Coyne, S. W. S. McKeever, and D. F. Drake, eds., American Chemical Society, Washington, D.C., pp. 284–309.

Shipman, H., and J. B. Adams, 1987. Detectability of minerals on desert alluvial fans using reflectance spectra, *J. Geophys. Res.*, 92, 10391–10402.

Spitzer, W. G., and D. A. Kleinman, 1960. Infrared lattice bands of quartz, *Phys. Rev.*, 121, 1324–1335.

Sunshine, J. M., and C. M. Pieters, 1990. Extraction of compositional information from olivine reflectance spectra: new capability for lunar exploration (abstract), in *Lunar and Planetary Science XXI*, Lunar and Planetary Institute, Houston, Texas, pp. 962–963.

Sunshine, J. M., and C. M. Pieters, 1991. Identification of modal abundances in spectra of natural and laboratory pyroxene mixtures: a key component for remote analysis of lunar basalts (abstract), in *Lunar and Planetary Science XXII*, Lunar and Planetary Institute, Houston, Texas, pp. 1361–1362.

Sunshine, J. M., and C. M. Pieters, and S. R. Pratt, 1990. Deconvolution of mineral absorption bands: an improved approach, *J. Geophys. Res.*, 95, 6955–6966.

Swayze, G. A., 1997. The hydrothermal and structural history of the Cuprite mining district, southwestern Nevada: an integrated geological and geophysical approach, Ph.D. dissertation, University of Colorado at Boulder, Boulder, Colo., 430 pp.

Swayze, G. A., and R. N. Clark, 1990. Infrared spectra and crystal chemistry of scapolites: implications for Martian mineralogy, *J. Geophys. Res.*, 95, 14481–14495.

Swayze, G. A., R. N. Clark, A. F. H Goetz, N. S. Gorelick, and T. G. Chrien, submitted. Spectral identification of surface materials using imaging spectrometer data: evaluating the effects of detector sampling, bandpass, and signal to noise ratio using the U.S.G.S. Tetracorder algorithm: Parts I and II, *J. Geophys. Res.*

Vane, G., J. E. Duval, and J. B. Wellman, 1993. Imaging spectroscopy of the Earth and other solar system bodies, in *Remote Geochemical Analysis: Elemental and Mineralogical Composition*, C. M. Pieters and P. A. J. Englert, eds., Cambridge University Press, Cambridge, pp. 121–143.

Wendlandt, W. W., and H. G. Hecht, 1966. *Reflectance Spectroscopy*, Interscience, New York, 298 pp.

White, W. B., 1974. The carbonate minerals, in *The Infrared Spectra of Minerals*, V. C. Farmer, ed., Mineralogical Society, London, pp. 227–284.

Use of Multispectral Thermal Infrared Data in Geological Studies

Simon J. Hook, Elsa A. Abbott, C. Grove, A. B. Kahle, and F. Palluconi

California Institute of Technology
Pasadena, CA 91109

2.1 INTRODUCTION

Mineral spectra exhibit diagnostic features at various wavelengths which provide a means for their remote discrimination and identification. These features are produced by electronic or vibrational–rotational processes resulting from the interaction of electromagnetic energy with the atoms and molecules which comprise the minerals that make up a rock. The different processes require different amounts of energy to proceed, and therefore are manifest in different wavelength regions. Electronic processes require the most energy and result in spectral features in the visible and near-infrared wavelength regions. Fundamental vibrational processes require less energy, and evidence for them occurs beyond 2.5 μm. Between 0.5 and 2.5 μm there is an overlap of features due to electronic processes and the excitation of overtone and combination-tone vibrations (Hunt, 1980).

Iron-, hydroxyl-, water-, sulfate-, and carbonate-bearing minerals display spectral features in the wavelength region 0.4 to 2.5 μm (Figure 2.1). By contrast, silicate minerals such as quartz, which dominate most crustal rocks, exhibit spectral features

Remote Sensing for the Earth Sciences: Manual of Remote Sensing, 3 ed., Vol. 3, edited by Andrew N. Rencz.
ISBN: 0471-29405-5 © 1999 John Wiley & Sons, Inc.

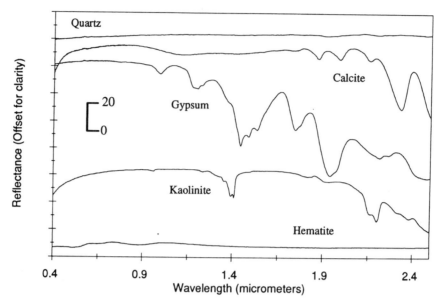

Figure 2.1 Hemispherical reflectance spectra of a variety of minerals from 0.4 to 2.5 μm. Minerals include hematite (oxide), kaolinite (phyllosilicate), gypsum (sulfate), calcite (carbonate), and quartz (silicate). Size fraction 125 to 500 μm.

in the thermal infrared and can be detected in the atmospheric window between 8 and 12 μm (Figure 2.2). Therefore, data from the visible through shortwave infrared complement data from the thermal infrared, the former providing information on the alteration products and the latter providing information on the composition of the rocks.

To date, the majority of research efforts have focused on the visible through short-wave infrared. There are three reasons for this focus: (1) global coverage and availability of multispectral visible through shortwave-infrared data, in particular the *Landsat* thematic mapper, (2) well-understood data sets, and (3) commercial off-the-shelf (COTS) software for ingesting and analyzing the data. This will change in 1999 with the launch of the advanced spaceborne thermal emission reflectance radiometer (ASTER), which will provide multispectral, high-spatial-resolution thermal infrared data globally. In addition, several new airborne instruments are planned with multispectral and hyperspectal thermal infrared capability, and thus it is likely that thermal infrared data will be used much more widely in the near future.

This chapter is divided into two parts related to thermal infrared imagery. The first part deals with the theory, instrumentation, and extraction of spectral information. This includes descriptions of field, airborne, and spaceborne instruments and the techniques used to analyze data from the field and airborne instruments. These techniques are generally applicable to field, airborne, and spaceborne multispectral thermal infrared instruments. The second part presents case histories that utilize the techniques described in the first part to analyze and interpret multispectral thermal infrared data in geological studies.

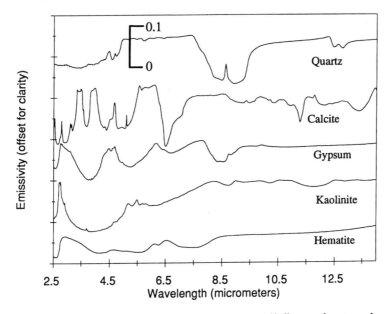

Figure 2.2 Hemispherical reflectance spectra, converted to emissivity using Kirchhoff's Law, of a variety of minerals from 2.5 to 14 μm. Minerals include hematite (oxide), kaolinite (phyllosilicate), gypsum (sulfate), calcite (carbonate), and quartz (silicate). Size fraction 125 to 500 μm. (Spectra provided by J. Salisbury, Johns Hopkins University.)

2.2 THEORY AND INSTRUMENTATION

2.2.1 Thermal Emission

Heat energy is the kinetic energy of the random motion of particles of matter. This random motion results in particle collisions that cause changes in the electron orbital or vibrational and rotational motions of the molecular or atomic particles. The change from higher to lower energy states of motion results in the emission of electromagnetic radiation (EMR), and heat energy is changed to radiant energy (see Suits, 1983).

A material that transforms heat energy to radiant energy at the maximum rate possible is termed a *blackbody*. Planck derived a formula for the spectral exitance, M, of a blackbody, given as

$$M_\lambda = \frac{C_1}{\lambda^5 \left[\exp\left(\dfrac{C_2}{\lambda T}\right) - 1 \right]} \tag{2.1}$$

where M_λ is the blackbody spectral exitance (W m^{-3}), λ the wavelength (m), T the absolute temperature (K), C_1 the first radiation constant = 3.74151 × 10^{-16} (W·m^2), and C_2 the second radiation constant = 0.0143879 (m·K).

The spectral exitance of a blackbody at a given wavelength increases with increas-

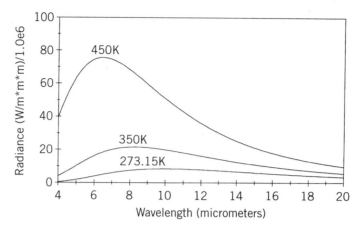

Figure 2.3   Spectral distribution of energy radiated by blackbodies at various temperatures in the range 4 to 20 μm. Note the shift to shorter wavelengths of radiance maximum with increasing temperature.

ing temperature. The wavelength of maximum spectral exitance at a given temperature can be obtained using the *Wien displacement law*:

$$\lambda_m = \frac{C}{T} \tag{2.2}$$

where $C = 2.898 \times 10^{-3}$ m·K. Figure 2.3 shows the spectral distribution of energy with wavelength for blackbodies at various temperatures and illustrates the shift to shorter wavelengths of the peak in spectral exitance with temperature.

A true blackbody surface is a Lambertian radiator (see Suits, 1983) and the relation between spectral radiance and spectral exitance is

$$L_\lambda = \frac{M_\lambda}{\pi} \qquad \text{W m}^{-3} \text{ Sr}^{-1} \tag{2.3}$$

2.2.1.1   SPECTRAL EMISSIVITY.

Materials are not perfect blackbodies, but instead emit radiation in accordance with their own characteristics. The ability of a material to emit radiation can be expressed as the ratio of the spectral radiance of a material to that of a blackbody at the same temperature. This ratio is termed the *spectral emissivity*:

$$\varepsilon_\lambda = \frac{L_\lambda(\text{material})}{L_\lambda(\text{blackbody})} \tag{2.4}$$

If the material is in thermal equilibrium with its surroundings, Kirchhoff's law states that the spectral absorptance is equal to the spectral emissivity. This important relationship can be used to determine the emissivity of a material by measuring its hemispherical reflectance (R) and assuming that $\varepsilon_\lambda = 1 - R$.

The spectral emissivity of most natural materials is between 0.9 and 1.0 in the region 8 to 12 μm; however, some common materials, such as quartz and many metals, have much lower emissivities in this region. (Quartz can have an emissivity as low as 0.55 at certain wavelengths within this region.)

2.2.1.2 ANGULAR VARIATIONS OF SPECTRAL EMISSIVITY.

The spectral emissivity of a surface does vary with view angle, and studies have been undertaken to examine the spectral bidirectional reflectance distribution (BRDF), which describes the variation in spectral emissivity at all inicident and reflected angles. Early studies focused on a couple of materials, sand and soil (Labed and Stoll, 1991). Recently, the BRDF of a larger number of materials has been examined (Becker et al., 1985; Snyder et al., 1997). The recent study by Snyder examined the effects of a change in the single-scattering albedo with wavelength. They constructed a spectral infrared bidirectional and reflectance and emissivity (SIBRE) instrument, which used a heated ceramic plate and thermal infrared source and measured the reflected radiance with a Fourier transform spectrometer. The instrument was mounted on a pointing system that allowed a wide range of geometrical combinations over the sample hemisphere to be examined. Snyder et al. (1997) report that the change in emissivity between 10 and 53° is small (<1%) for all materials examined (organic soil, sand, silt, vegetation and soil, gravel) except sand. The greater sensitivity of sand was also noted by Becker et al. (1985). For bare sand, the change in emissivity with angle ranged up to 4% in the region 8 to 10 μm and was approximately 2% in regions 3 to 5 μm and 10 to 14 μm.

2.2.1.3 EMISSIVITY SPECTRA OF MINERALS AND ROCKS.

Early work by Lyon (1965) and Hunt and Salisbury (1974, 1975, 1976) provided the physical basis for interpreting rock spectra from the wavelength region 7 to 14 μm. This work has been continued by Salisbury, who recently published a book on thermal infrared mineral spectra and several papers on the spectra of natural and manufactured materials (Salisbury et al., 1988, 1991, 1994a, b Salisbury and D'Aria, 1992). The most intense absorption features in the spectra of all silicates occurs near 10 μm, and this region of the spectrum is referred to as the Si–O stretching region or reststrahlen band. Figure 2.4 illustrates the spectral features of a variety of silicate minerals in the region 8 to 14 μm. The emissivity minimum occurs at relatively short wavelengths (8.5 μm) for framework silicates (quartz, feldspar) and progressively longer wavelengths for silicates having sheet, chain, and isolated SiO_4 tetrahedra (Hunt, 1980) (Figure 2.4). The only other feature in silicate mineral spectra between 8.5 and 12 μm results from the H–O–Al bond and occurs near 11 μm, it is characteristic of aluminum-bearing clay minerals (Hunt, 1980).

Other nonsilicate molecular units also give rise to spectral features in the thermal infrared. These include carbonates, sulfates, phosphates, oxides, and hydroxides, which typically occur in sedimentary and metamorphic rocks. Figure 2.5 illustrates the spectral features of a variety of carbonate minerals, characterized by a sharp feature around 11.2 μm. This feature moves to slightly longer wavelengths as the atomic weight of the cation increases (Liese, 1975; Lyon and Green, 1975; van der Marel and Beutelspacher, 1976), providing a very useful property for remote identification.

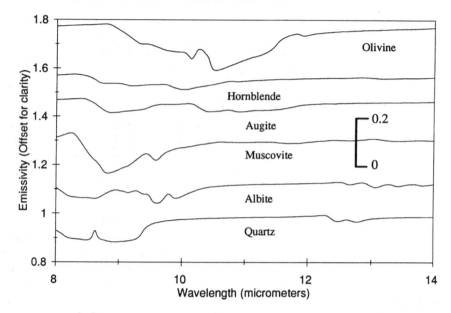

Figure 2.4 Biconical reflectance spectra, converted to emissivity using Kirchhoff's law, of silicate minerals from 8 to 14 μm. Minerals include quartz (tectosilicate), albite (tectosilicate), muscovite (phyllosilicate), augite (inosilicate, single chain), hornblende (inosilicate, double chain), and olivine (nesosilicate, single SiO$_4$ structure). Size fraction 125 to 500 μm. (Spectra provided by J. Salisbury, Johns Hopkins University.)

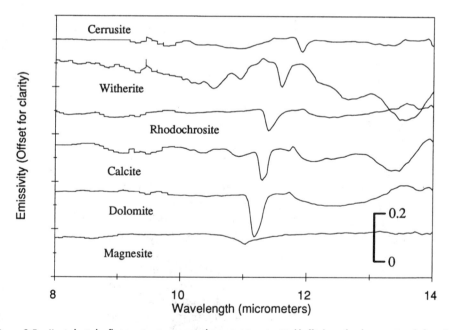

Figure 2.5 Hemispherical reflectance spectra, converted to emissivity using Kirchhoff's law, of carbonate minerals from 8 to 14 μm. Note the shift in the feature near 11 μm to longer wavelengths as the atomic weight of the cation increases. Minerals include magnesite (MgCO$_3$), dolomite [Ca(Mg,Fe)(CO$_3$)$_2$], calcite (CaCO$_3$), rhodochrosite (MnCO$_3$, witherite (BaCO$_3$), cerrusite (PbCO$_3$). Size fraction 125 to 500 μm. (Spectra provided by JPL.)

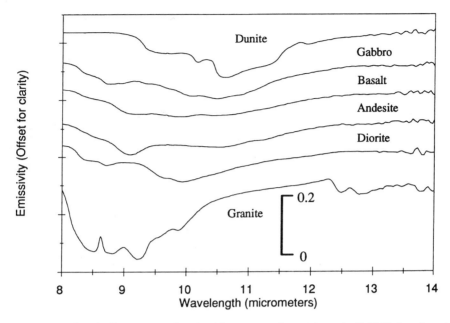

Figure 2.6 Hemispherical reflectance spectra from 8 to 14 μm, converted to emissivity using Kirchhoff's law, of a variety of rocks ranging from ultramafic to acidic in composition. Rocks include dunite, gabbro, basalt, andesite, diorite, and granite in order of increasing acidity. Size fraction 125 to 500 μm. (Spectra provided by J. Salisbury, Johns Hopkins University.)

Rocks are collections of minerals, and their spectra are composites of the individual spectra of the constituent minerals. Since the minerals that make up acidic rocks have emissivity minima at shorter wavelengths than those of mafic rocks, as the material becomes more mafic, the minimum of the reststrahlen band shifts to longer wavelengths (Figure 2.6).

2.2.1.4 ATMOSPHERIC WINDOWS.

The gases of the atmosphere selectively scatter, absorb, and emit electromagnetic radiation. The wavelength intervals where radiation is absorbed are referred to as *absorption bands* and the intervening regions of high-energy transmittance as *atmospheric transmission bands* or *windows*. Figure 2.7 shows the transmittance through a 100-km-thick terrestrial atmosphere. The atmospheric transmittance is high between 8 and 12 μm in the thermal infrared. Fortunately, this coincides with the wavelength of maximum emission (9.7 μm) for a blackbody at the average temperature of the earth, 290 K (Sabins, 1978).

2.2.2 Current Instrumentation

Several instruments are now available for the remote acquisition of surface information in the region 8 to 12 μm. If instruments with a single broad channel between 8 and 12 μm are excluded, the majority of instruments are either airborne or spaceborne with a few field instruments. The types of instruments range from interferom-

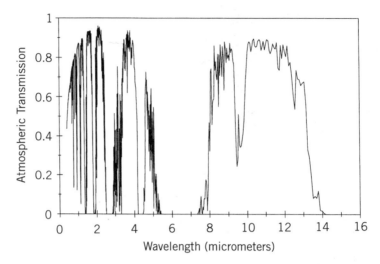

Figure 2.7   Atmospheric transmission through a 100-km vertical path through the 1976 U.S. standard atmosphere calculated with MODTRAN 3. (From Berk et al., 1989.)

eters to diffraction grating spectrometers to lasers. Interferometers and lasers tend to be used for nonimaging applications, whereas diffraction grating spectrometers tend to be used for imaging applications.

Rather than present detailed information on all known imaging and nonimaging systems, the majority of systems are reviewed briefly and then examples of field, airborne, and spaceborne instruments are presented in detail.

One of the early field instruments was the JPL portable field emission spectrometer, a filter wheel spectrometer that operated between 5 and 15 μm with a spectral resolution of about 2% of the wavelength (Hoover and Kahle, 1987). This instrument was large and heavy and was superseded by other, lighter grating spectrometers such as the THIRSPEC (Rivard et al., 1994) and interferometers such as the μFTIR (Korb et al., 1996; Hook and Kahle, 1996). The THIRSPEC is a thermal infrared grating spectrometer developed by Barringer Research Inc. for field studies in the earth sciences. The design is based on a reflection grating and a 60-element HgCdTe detector array. The useful spectral range of the instrument covers 7.9 to 11.3 μm with a Nyquist-limited resolution of 0.16 μm. The instrument averages over a 12° field of view and compares the exitance of the target to that of an internal blackbody at ambient temperature. The noise equivalent temperature is approximately 0.06 K over the useful spectral range. The weight of the instrument, including battery packs, is approximately 30 kg. The μFTIR is described in detail in Section 2.2.2.1 as an example of a field instrument.

There have been many more airborne imaging instruments, which are typically diffraction grating spectrometers. These include the airborne ASTER simulator (AAS), the airborne hyperspectral scanner (AHS), the airborne terrestrial applications sensor (ATLAS), the MkII airborne multispectral scanner (AMSS), the moderate resolution imaging spectrometer MODIS airborne simulator (MAS), the multispectral infrared and visible imaging spectrometer (MIVIS), and the thermal infrared multispectral scanner (TIMS).

The AAS was built by the Geophysical Environmental Research (GER) Corporation for the Japan Resources Observation System Organization (JAROS). The AAS has 24 channels, and 17 of these are in the thermal infrared. A comparison of the preliminary results from AAS with the TIMS is given in Kannari et al. (1992). The AHS is the commercial version of the MAS, which was developed initially under the NASA Small Business Innovative Research Program (SBIR). Both instruments were built by Daedalus Enterprises. The AHS has 48 channels between 0.4 and 13 μm, a 2.5-mrad instantaneous field of view (IFOV) and a digitized FOV of 86 degrees. The MAS has 50 channels between 0.5 and 14 μm. A detailed description of the MAS is given in King et al. (1996).

The ATLAS instrument was developed by Stennis Space Center. The instrument has 15 channels, an IFOV of 2 mrad, and a TFOV of 73 degrees. Six of the 15 channels are located in the thermal infrared between 8 and 12 μm. The AMSS was developed by Geoscan Pty Ltd., an Australian company. The MkII has 46 spectral channels available for recording up to 24 channels by selection. The channels cover the region 0.4 to 12 μm with six channels between 8 and 12 μm. The MIVIS instrument was recently developed by Daedalus Enterprises. MIVIS has 102 channels between 0.4 and 13 μm, a 2-mrad IFOV, and a TFOV of 71 degrees. Ten of the channels are located between 8 and 13 μm. Results from some of the early MIVIS experiments are give in Bianchi et al. (1996). The TIMS instrument was also developed by Daedalus Enterprises and is described in detail as an example of an airborne instrument in Section 2.2.2.

In addition to the passive airborne imaging instruments there are also some profiling instruments, including active lasers and passive interferometers. The lasers include the laser absorption spectrometer (LAS) and the Australian CO_2 laser. The LAS was developed for remote measurement of atmospheric gases (Shumate et al., 1981, 1982). The system uses two CO_2 infrared laser transmitter–receiver systems mounted in a small aircraft to measure the two-way transmittance between the aircraft and the ground. Heterodyne detection techniques are used to measure the laser radiation backscattered from the ground and distinguish it from the thermal background. The two laser wavelengths were 9.2 and 10.2 μm. Kahle et al. (1984) demonstrated that it was possible to distinguish a wide range of lithologies in Death Valley, California, using this instrument. Similar work on using lasers for geological mapping was under way in Australia around the same time, and this work is ongoing. The Australian Commonwealth Scientific and Industrial Research Organization (CSIRO) has developed a rapidly tuned CO_2 laser which tunes through approximately 90 discrete wavelengths, or laser, lines, between 9 and 11.2 μm in bursts lasting about 2 ms and repeated every 24 ms, with peak powers on each line of about 100 W. Using direct detection the instrument provides acceptable signal/noise ratios when operated from 500 m and a pixel size of about 2 m in diameter (Whitbourn et al., 1990). In 1993 the NASA TIMS was shipped to Australia and flown on a joint campaign with a CSIRO CO_2 laser. Some preliminary results from that study are given in Cudahay et al. (1994). In contrast to the high-spatial-resolution (2-m) laser measurements, most airborne profiling interferometers have coarse spatial resolution (1 km) and limited applicability to geological problems, although the high-spectral-resolution data they provide are useful for atmospherically correcting the broadband image data. An example of an airborne interferometer is the high resolution interferometer sounder (HIS) (Smith et al., 1988).

Currently, there are no high spatial and spectral *spaceborne* thermal infrared imagers, although there are a couple of instruments with 1-km resolution and two channels in the thermal infrared. These are the advanced very high resolution radiometer (AVHRR) and the along-track scanning radiometer (ATSR). There are no plans to increase the resolution of these instruments in the near future. The current *Landsat* thematic mapper has a single channel in the thermal infrared with 120-m spatial resolution. The next *Landsat* (*Landsat 7*) will have a single 60-m thermal infrared channel; however, current plans call for this channel to be removed on subsequent versions of *Landsat*. In 1998 the Earth Observation System (*EOS*) platform will be launched, which will include the advanced spaceborne thermal emission reflectance radiometer (ASTER), which is discussed below, and the moderate resolution imaging spectrometer (MODIS). ASTER has five channels between 8 and 12 μm with 90-m spatial resolution, MODIS has four channels between 8 and 12.5 μm with 1-km resolution.

Examples of field, airborne, and spaceborne instruments will now be reviewed in detail. Many of the techniques utilized with these instruments can be used with the other field and airborne instruments reviewed briefly above. The field instrument discussed is the micro FTIR built by Designs and Prototypes (Hook and Kahle, 1996). The airborne instrument is the thermal infrared multispectral scanner built by Daedalus Enterprises (Palluconi and Meeks, 1985), and the spaceborne instrument is the advanced spaceborne thermal emission reflectance radiometer built by the Japanese government (Fujisada and Ono, 1991).

2.2.2.1 MICRO FTIR.

The following description of the micro FTIR (μFTIR) is taken largely from Hook and Kahle (1996) and Korb et al. (1996).

A. Introduction. The micro FTIR (μFTIR) is a lightweight, rugged, moderate-spectral-resolution interferometer built by Designs and Prototypes based on a set of specifications provided by the Jet Propulsion Laboratory and J. W. Salisbury (Johns Hopkins University). The instrument permits the acquisition of infrared spectra of natural surfaces between 2 and 5 μm and 5 and 14 μm. The instrument has a spectral resolution of about six wavenumbers, weighs 16 kg, including batteries and computer, and can be operated easily by two people in the field. Laboratory analysis indicates that the instrument is spectrally calibrated to better than one wavenumber (over the region 7 to 9 μm evaluated) and the radiometric accuracy is < 0.5 K if the radiances from the blackbodies used for calibration bracket the radiance from the sample. Examples of field spectra are provided with case histories in Section 2.3.

B. Instrument Description. The μFTIR consists of two main components: a tripod-mounted optical head and a system unit set in an aluminum briefcase (Figure 2.8). The optical head weighs 4.1 kg and includes the interferometer, detector–dewar assembly, and the optics for measuring and observing the surface. The interferometer consists of a refractively scanned cavity driven by a mass-balanced torque motor. The refractively scanned crystals of the interferometer are made of potassium chloride (KCl) and sealed under nitrogen pressure. There is a single cooled detector assembly consisting of indium–antimonide (InSb) for the region 2 to 5 μm and mercury–cadmium–telluride (HgCdTe) for the region 5 to 14 μm. The detector assembly can be cooled using liquid nitrogen or a Sterling cycle cooler. The viewing optics are

Figure 2.8   Jet Propulsion Laboratory μFTIR in Death Valley, California during a 1994 field campaign. The gray box on top of the tripod is the optical head. The briefcase includes the computer and the system unit.

made from zinc–selenide and the detector lens focal length is 2.5 cm. The image of the surface may be alternately projected onto the detector or onto an aiming eyepiece. The optical head is tripod mounted; if the entrance aperture is 1 m above the surface, the imaged area should be 7.6 cm in diameter. Data acquired by the instrument are radiometrically calibrated with the aid of a small temperature-controlled blackbody that mounts on the front of the fore optics. Calibration requires measurement of the blackbody at a minimum of two temperatures to determine the gain and offset. Spectral (wavelength) calibration is achieved using a diode laser.

The system unit, excluding the laptop computer, weights 6.4 kg and is powered by a single rechargeable 12-V gel cell that weighs 2.5 kg. The laptop computer is powered by rechargeable nickel–cadmium batteries and weighs 3.2 kg. Designs and Prototypes includes a MS-DOS software package for the acquisition and analysis of the data. The software allows a set of measurements to be reduced to emissivity in near real time.

C. Data Reduction. There are several steps involved in reducing the sample spectra to emissivity. First, the sample spectrum must be reduced to radiance using the black-bodies. Second, a region within the calibrated spectrum is assumed to have a known emissivity, and this region is used to determine the temperature of the sample and the sample spectrum is then ratioed with a blackbody at the sample temperature to obtain the apparent emissivity. Third, the apparent emissivity spectrum is adjusted to compensate for the effect of downwelling radiation. The downwelling radiation is obtained by measuring the energy reflected from a gold standard. Each of these steps is described in more detail below.

Step 1: Radiometric Calibration. Radiometric calibration is the conversion of the instrument digital numbers (DNs) to radiance. It is achieved by measuring the black-body at two known temperatures, one above and the other below the sample temperature. If the output numbers (D_λ) are linearly related to the input radiance (R_λ), then for a given wavelength;

$$R_\lambda = a + bD_\lambda \tag{2.5}$$

where a and b are constants that relate radiance to data number. If the radiance for the cold blackbody (R_c) is given by

$$R_c = P(\lambda, T_c) \tag{2.6}$$

and the radiance from the hot blackbody (R_h) by

$$R_h = P(\lambda, T_h) \tag{2.7}$$

where $P(\lambda, T)$ is the Planck blackbody radiance [Equation (2.3)], then given R_c and R_h, one can solve for a and b using equation (2.5) with the following result:

$$a = \frac{R_h D_c - R_c D_h}{D_c - D_h}$$
$$\tag{2.8}$$
$$b = \frac{R_c - R_h}{D_c - D_h}$$

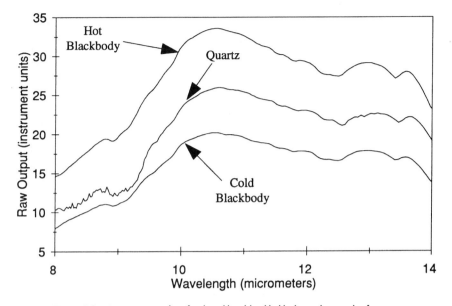

Figure 2.9   Raw output numbers for the cold and hot blackbodies and a sample of quartz.

Figure 2.9 shows the raw output numbers for cold and hot blackbodies and a sample of quartz. Note the double peak in the raw spectra that results from "doping" the HgCdTe detector to increase its sensitivity at longer wavelengths. Figure 2.10 shows the calibrated quartz spectrum. The main quartz doublet around 8.5 μm is clearly apparent, as is the weaker doublet around 12.5 μm. The main doublet is strongly affected by atmospheric absorption and is very "noisy" in appearance.

Step 2: Derivation of Apparent Emissivity. Once the spectrum is calibrated to radiance the next step is to obtain the apparent emissivity spectrum by ratioing the radiance spectrum with the radiance from a blackbody at the same temperature as the sample. This is obtained by assuming that the emissivity is constant in a given wavelength range in the sample spectrum and inverting the Planck equation for each wavelength in that range. Generally, the emissivity is assumed constant in the region 7.7 to 7.8 μm, where silicate minerals typically have their Christiansen frequency and the emissivity is very close to 1.0. The apparent emissivity (which is simply the ratio of the measured radiance to a blackbody at the same temperature as the surface) is shown in Figure 2.11. The spectrum still appears noisy, particularly the main quartz doublet, due to atmospheric effects.

Step 3: Derivation of True Emissivity. The radiance at the instrument (L_λ) is given by

$$L_\lambda = [\varepsilon_\lambda L_{\text{bb}\lambda}(T) + (1 - \varepsilon_\lambda)L_{\text{sky}\lambda}]\, \tau_{A\lambda} + L_{V\lambda} \qquad (2.9)$$

where ε_λ is the surface emissivity at wavelength λ, $L_{\text{bb}\lambda}(T)$ the spectral radiance from a blackbody at surface temperature (T), $L_{\text{sky}\lambda}$ the spectral radiance incident upon the

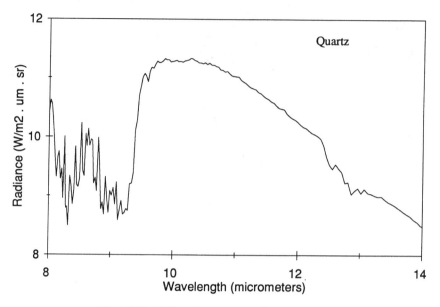

Figure 2.10   Calibrated quartz spectrum shown in Figure 2.9.

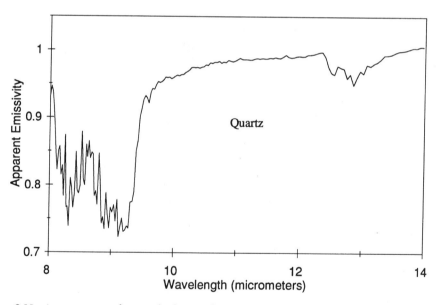

Figure 2.11   Apparent emissivity for a sample of quartz. The apparent emissivity is the ratio of the sample radiance to the radiance of a blackbody at the same temperature. The blackbody temperature is obtained by assuming that the emissivity is equal to 1.0 in some wavelength range in the measured radiance spectrum, then inverting Planck's equation at that wavelength to obtain the temperature. In this spectrum the emissivity was assumed equal to 1.0 in the wavelength range 7.7 to 7.8 μm.

surface from the atmosphere, $\tau_{A\lambda}$ the spectral atmospheric transmission, and $L_{\vee\lambda}$ the spectral radiance from atmospheric emission and scattering that reaches the sensor.

Since we are close to the surface, the spectral transmission and spectral radiance from the atmospheric emission and scattering terms can be ignored. If we then assume that the gold standard has an emissivity equal to zero, the radiance measured from the gold standard is equal to the spectral radiance incident on the surface from the atmosphere ($L_{sky\lambda}$). In reality, the emissivity of the gold standard may be slightly greater than zero and the radiance from the standard can be adjusted accordingly. The emissivity is derived by subtracting $L_{sky\lambda}$ (obtained by measuring the gold standard) from the measured surface radiance:

$$\varepsilon_\lambda = \frac{L_\lambda - L_{sky\lambda}}{L_{bb\lambda}(T) - L_{sky\lambda}} \tag{2.10}$$

To apply equation (2.10) it is necessary to know the temperature of the surface obtained during the calculation of the apparent emissivity. The true emissivity, utilizing equation (2.10) to remove the reflected atmospheric downwelling radiance, is shown in Figure 2.12 [field (corrected gold)]. The atmospheric lines within the dou-

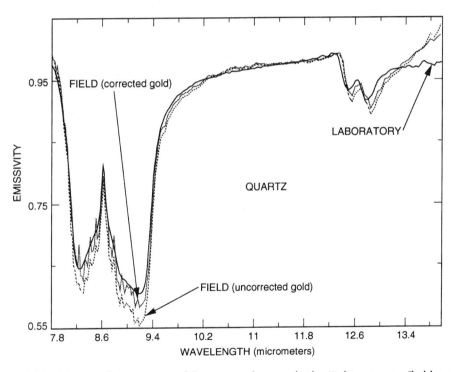

Figure 2.12 Laboratory reflectance spectrum of the quartz sample measured with a Nicolet spectrometer. The laboratory spectrum has been converted to emissivity using Kirchhoff's law ($1 - \varepsilon = R$). Emissivity spectrum of the quartz sample measured with the μFTIR. The temperature of the sample was obtained using the method outlined with Figure 2.11 and calibrated data processed using equation (2.10). Note the suppression of atmospheric lines within—and the increased depth of—the quartz doublet compared to Figure 2.11. Emissivity spectrum described above with the correction for the difference in reflectance of the laboratory gold standard and the field gold standard (field-corrected gold).

blet are now suppressed, and in addition, the main doublet is deeper due to the removal of the "filling in" effect of the reflected sky radiance, which adds radiance from the sky in areas where the emissivity is less than 1.0 [equation (2.9)]. The amount of radiance added increases as the emissivity decreases. Fortunately, the feature is not completely "filled in" since the sky is much colder than the surface and therefore less radiance is added.

An example of the size of the correction for the difference in emissivity between the laboratory gold standard and the field gold standard is shown in Figure 2.12. This does not include the additional correction for any radiation emitted by the gold standard. The depth of the main quartz doublet in the field spectrum with correction for the nonzero emissivity of the gold is greater, indicating that if this is not taken into account the sky correction will overcompensate for the reflected sky radiation. Figure 2.12 also shows a spectrum of the same quartz sample measured using a Nicolet Laboratory Interferometer with a hemispherical reflectance attachment. The laboratory reflectance spectrum was converted to emissivity using Kirchhoff's law ($1 - \varepsilon = R$). This comparison also indicates that the spectral calibration of the field instrument is poor at longer wavelengths around the position of the second quartz doublet.

2.2.2.2 THERMAL INFRARED MULTISPECTRAL SCANNER

A. Introduction. The thermal infrared multispectral scanner (TIMS) was built by Daedalus Enterprises for NASA in 1982, and some early results are described in Kahle and Goetz (1983). TIMS data have been acquired from the NASA Learjet at the Stennis Space Center since 1982, the NASA C-130 at Ames Research Center since 1985, and the NASA ER-2 at Ames since 1990. Numerous studies have now demonstrated the utility of data from TIMS in a wide range of applications, including urban monitoring, hydrology, ecology, and geology. TIMS data are analyzed either qualitatively or quantitatively and valuable information can be obtained with either approach. Examples of each approach are presented with case histories in Section 2.3.

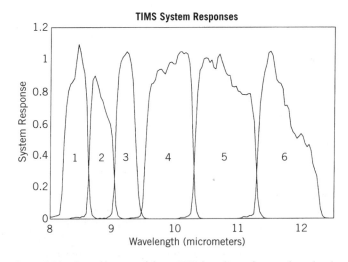

Figure 2.13   Spectral response of the six TIMS channels as a function of wavelength.

B. Instrument Description. TIMS has six channels between 8 and 12 μm, an instantaneous field of view (IFOV) of 2.5 mrad, and a total field of view of 76.6° (Palluconi and Meeks, 1985). The system response of each channel varies with wavelength and is checked in the laboratory approximately annually (Figure 2.13). Knowledge of the system response is required to calibrate the data accurately, especially for an instrument like TIMS, which has considerable variation in response with wavelength for the longer-wavelength channels.

The optical path of the energy incident on the TIMS is shown in Figure 2.14. Cross-track scanning is achieved through rotation of a 45° flat scan mirror that reflects incident energy to the parabolic primary mirror (M1). The energy is then directed into the spectrometer section of the instrument by a flat secondary mirror (M2). At the entrance to the spectrometer section and focal plane of the primary mirror is an 0.8-mm field stop aperture which defines the instruments 2.5-mrad

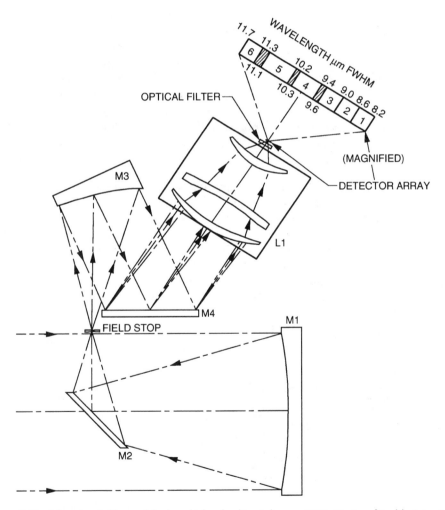

Figure 2.14 Schematic optical layout of the thermal infrared multispectral scanner (TIMS). (Courtesy of Daedalus Enterprises, Inc.)

IFOV. The energy that passes through the field stop is collimated by an off-axis parabolic mirror (M3) and directed to a replica grating (M4), which produces the spectral dispersion of the energy onto a germanium triplet lens (L1), which focuses the energy onto the detector array.

The in-flight radiometric calibration of TIMS is accomplished with two blackbodies whose temperatures are set to bracket the anticipated ground temperatures. One of the blackbodies is observed by the TIMS scanner at the start and one at the stop of each scan line. The blackbody temperatures and responses can then be used to derive a set of coefficients to convert the instrument digital numbers to radiance at the sensor with the method described for the field instrument.

(1) Calibration. To use data for a multispectral thermal infrared imager for quantitative analysis it is necessary to understand its spatial, spectral, and radiometric performance. Ideally, all these parameters should be monitored in flight. In the case of TIMS, in-flight radiometric calibration is accomplished with two blackbodies whose temperatures are set to bracket the anticipated ground temperatures. One of the blackbodies is observed by the TIMS scanner at the start and one at the stop of each scan line. The blackbody temperatures and responses can then be used to derive a set of coefficients to convert the instrument digital numbers to radiance at the sensor with the method described for the field instrument. The spatial and spectral calibration are not measured in flight; however, errors in the spectral calibration can be measured using atmospheric bands (Hook and Okada, 1996). These calibration parameters are also measured in the laboratory. The following description of how these parameters were determined for the TIMS instrument in the laboratory is taken from Realmuto et al. (1995).

(a) Spatial Calibration. The spatial calibration of TIMS was determined by mounting it 0.3 m above the surface of an optical bench. A mirror was then positioned beneath the instrument to reflect the output of a collimator into the instrument. A NiCr hot wire was heated to incandescence and positioned next to the input of the collimator such that the image of the wire was in the focal plane of the collimator. The hot wire was translated in both the horizontal and vertical directions, relative to the collimator. The movement of the hot wire across the TIMS field stop mapped the field stop into the image data. Profiles of these data in the horizontal and vertical directions resemble Gaussian functions whose full width at half maximum are a measure of the IFOV of TIMS. The derived IFOVs are given in Table 2.1.

TABLE 2.1   Spatial Calibration Results for the TIMS Instrument

TIMS Channel	Horizontal IFOV (mrad)	Vertical IFOV (mrad)
1	2.65 ± 0.096	2.58 ± 0.093
2	2.66 ± 0.094	2.64 ± 0.093
3	2.62 ± 0.095	2.50 ± 0.093
4	2.61 ± 0.090	2.71 ± 0.095
5	2.54 ± 0.096	2.69 ± 0.095
6	2.49 ± 0.105	2.56 ± 0.094

Source: Realmuto et al. (1995).

TABLE 2.2   Spectral Calibration Results for the TIMS Instrument

TIMS Channel	FWHM (μm)	Peak Position (μm)
1	0.374 ± 0.011	8.402 ± 0.050
2	0.358 ± 0.013	8.766 ± 0.052
3	0.379 ± 0.016	9.212 ± 0.055
4	0.690 ± 0.021	10.012 ± 0.060
5	0.888 ± 0.028	10.630 ± 0.063
6	0.659 ± 0.035	11.512 ± 0.068

Source: Realmuto et al. (1995).

(b) Spectral Calibration. The spectral calibration was determined by substituting a monochrometer and infrared glower for the hot wire. The monochrometer was positioned at the input port of the collimator such that the monochrometer slit was in the focal plane of the collimator. The monochrometer was scanned between 7 and 13 μm. Profiles of the TIMS data taken perpendicular to the scan direction produced spectral response profiles. These results are summarized in Table 2.2.

(c) Radiometric Calibration. The radometric calibration of the TIMS was determined by placing an extended area blackbody beneath the TIMS, which fills the TIMS aperture completely. The radiometric precision or NEΔT is the deviation in the TIMS temperature estimates over the uniform blackbody surface. The NEΔT values ranged between 0.08 and 0.16 for a range of temperatures between 20 and 45°C. The radiometric accuracy can be determined by comparing the blackbody set-point temperature with the temperature recovered from the TIMS data. In all cases the recovered temperatures were within 0.8°C of the blackbody temperature.

C. Data Reduction. There are two approaches to data reduction: qualitative and quantitative. In qualitative interpretations a decorrelation stretch is widely used with multispectral thermal infrared data, and this technique is described below. In quantitative interpretations a variety of techniques are used, but they tend to involve three steps: calibration, atmospheric correction, and separation into temperature and emissivity. Each of these steps is described with examples of techniques used with each step.

(1) Qualitative reduction. Multispectral images are typically prepared for photographic interpretation by selecting three of the image channels and displaying them in red, green, and blue as a color additive picture. Usually, the multispectral data do not occupy the full data range, and a contrast stretch is applied to each of the three input images prior to display. This approach works well provided that the spectral data are not highly correlated. If the data are strongly correlated, the image will appear nearly monochromatic (gray) both before and after the stretch. This is the case with multispectral thermal infrared data since the temperature information, which is highly correlated between channels, dominates the emissivity information, which provides the color contrast. One technique that has been successful in overcoming this problem and produces colorful displays from multispectral thermal infrared images is the decorrelation stretch developed by Soha and Schwartz in 1978. This technique has been widely used in geological studies with TIMS data (e.g., Kahle

and Goetz, 1983; Gillespie et al., 1984; Watson, 1985; Lang et al., 1987; Lahren et al., 1988; Abrams et al., 1991; Hook et al., 1992; Sabine et al., 1994).

The technique is based on a principal components transformation of the original data. Often, the transformed images are stretched, then displayed, but such a display results in image colors that are uncorrelated, with the subdued colors in the original image making interpretation difficult. Decorrelation stretching differs in that after contrast enhancement the statistically independant decorrelated images are stretched, then retransformed to the original coordinates for display, so that in general there is little distortion of the perceived hues due to the enhancement (Gillespie et al., 1986).

A decorrelation stretch of thermal infrared data enhances the emissivity information relative to the temperature information, and the product can be interpretated by assuming that most variations in color are due to emissivity differences, whereas differences in brightness are due to temperature differences. Several examples of decorrelation stretched images are provided in the case histories in Section 2.3. It should be noted that the resultant image is a function of the scene statistics used for the principal components transformation, and the scene statistics can be gathered from all or only part of the scene being studied. If the scene statistics are gathered from part of the scene that does not include a particular cover type, that cover type may be poorly enhanced in the resultant imagery. If the scene statistics are gathered from a single cover type, that cover type will be strongly enhanced, revealing any noise in the data, and other areas in the image will be saturated. Once an image has been decorrelation stretched, further processing such as spectral matching with a library of emissivity or reflectance spectra is no longer possible, making it difficult to explain the physical reason for the difference in color between two units in the decorrelation stretched imagery. As a result, such a product is qualitative; however, the colors in the decorrelation stretched images will appear similar for different scenes provided that there is a reasonable variety of materials in the scene and the method is very useful for discriminating different cover types.

(2) Quantitative reduction. Quantitive analysis typically uses either the emissivity or temperature information, which must be extracted from the raw TIMS data. Several steps are necessary to extract this information. They include calibration, atmospheric correction, and temperature/emissivity separation. Each of these steps is discussed in more detail below.

Step 1: Calibration. The calibration from instrument digital numbers (DNs) has already been described for the μFTIR, and the method used with TIMS data is very similar except that one sample of blackbody information is recorded for each scan line and could be applied for each scan line. However, experience has found that the blackbody temperatures change little and several of the blackbody digital number values (typically, 75 scan lines) can be averaged to minimize the effects of noise in the single sample of blackbody data provided with each scan line.

Step 2: Atmospheric Correction. After calibration the radiance at the sensor can be described by equation (2.9). To remove the atmospheric effects to extract the radiance at surface, it is necessary to correct for the atmospheric transmission, the radiance from atmospheric emission, and scattering that reaches the sensor and the radiance incident upon the surface from the atmosphere and reflected into the sensor as 1-R, all of which vary with wavelength. The values for these parameters can be obtained with a radiative transfer model (e.g., MODTRAN; Berk et al., 1989) fed by external sources such as climatology or atmospheric sounders. Alternatively, these parameters can either be derived from the data or at least corrected for using the data itself (e.g., the split-window technique).

MODTRAN derives atmospheric transmission and emission values used with the atmospheric correction based on an input atmospheric profile, which may be obtained from the default profiles in MODTRAN, or the profile may be modified or replaced with local atmospheric data. Figure 2.15 is an example of the data obtained

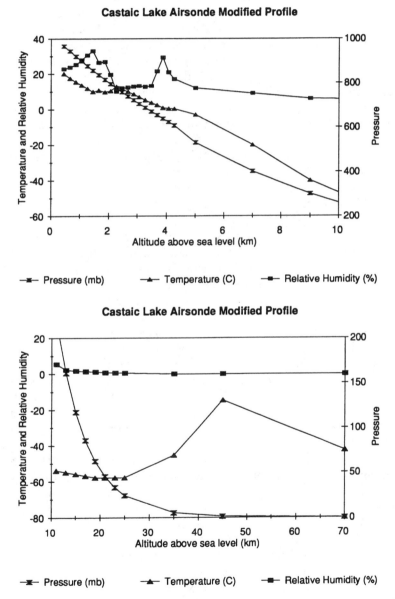

Figure 2.15 Atmospheric profile used for input to MODTRAN. Values below 4.5 km obtained from an airsonde. Values above 4.5 km obtained from the default midlatitude summer profile in MODTRAN.

with a radiosonde (airsonde) coupled with data from a default model atmosphere to provide a complete profile for MODTRAN. An airsonde is a instrument package attached to a balloon that transmits the properties of the atmosphere for different heights as the balloon rises. The airsonde provides atmospheric profiles of relative humidity, temperature, and pressure (Figure 2.15). The relative humidity is derived from the wet and dry bulb thermometers on the airsonde. The wet bulb froze at 4.5 km, and no useful data were provided by the airsonde above 4.5 km (Figure 2.15). Atmospheric data for the remainder of the profile were obtained from the default midlatitude summer model of MODTRAN.

An alternative approach to deriving the atmospheric parameters from external sources is to use the imagery itself to correct the atmospheric effects without deriving the physical values for the various atmospheric inputs. The most common example of this approach is the split-window technique, which is discussed in the next section since it is used to recover the land surface temperature. Recently, several other approaches have been developed that either identify random gray pixels (pixels whose emissivity is constant) within the scene and use these to determine the atmospheric correction parameters (H. Tonooka, 1996) or alternatively, utilize the overlap between two flight lines, which provides two look angles at the same pixel to derive the atmospheric correction parameters (Watson, 1996).

Once the data have been corrected to surface-leaving radiance, further processing is required to derive the spectral emissivity and surface kinetic temperature, and since both of these are (in general) unknown, at least one further assumption must be made for them to be determined separately.

Step 3: Extraction of Temperature and Emissivity. If it is assumed that a pixel is at one temperature and the radiance is measured from the surface in N spectral channels, there will always be $N + 1$ unknowns: N emissivities (one per channel) and a single unknown surface temperature. The system of equations described by a set of radiance measurements in N channels is thus underdetermined, and additional information is needed to extract the emissivity or temperature information. Both deterministic and nondeterministic approaches have been utilized to extract the temperature and emissivity information. The former approach is typically used to estimate temperatures from scenes with a known composition: for example, oceans, snowfields and glaciers, and closed-canopy forests. However, deterministic solutions require that the atmospheric parameters in equation (2.9) be measured directly and the measured radiance corrected for them, and this is not always possible. Over oceans this approach has been modified to use data from two channels or windows in the thermal infrared to compensate for atmospheric effects while solving for temperature (e.g., Prabhakara et al. 1974; McMillan and Crosby, 1984; Barton, 1985). The majority of ocean temperature studies utilize data from the advanced very high resolution radiometer (AVHRR), which has two channels in the thermal infrared, channel 4 (10.3 to 11.3 μm) and channel 5 (11.5 to 12.5 μm), with a spatial resolution of 1.1 km. Several authors have examined extending the split-window technique to land surfaces (e.g., Price, 1984; Becker, 1987; Wan and Dozier, 1989; Becker and Li, 1990; Vidal, 1991; Kerr et al., 1992; Ottle and Stoll, 1993; Prata, 1994). Initially, most authors concluded that large errors would arise using the split-window approach over land surfaces, due to unknown emissivity differences. Wan and Dozier (1989) examined the impact of errors due to atmospheric correction versus those due to an uncertainty in emissivity and concluded that the error associated with a

0.01 uncertainty in spectral emissivity was larger than that due to atmospheric correction. Since then several authors have modified their approaches and Wan and Li (1997) have developed a generalized split-window algorithm which they claim will permit the recovery of land surface temperature to 1 K. This problem does not arise in recovering ocean surface temperatures since the emissivity of water is well known (Masuda et al., 1988) unless the water is turbid (Wen-Yao et al., 1987).

Most geological studies are less concerned with temperature and more concerned with emissivity variations, since these relate to the composition of the surface and provide a means for remote geologic mapping. Most of the nondeterministic methods are designed to extract relative or absolute emissivity and require independant atmospheric correction. The methods which extract relative emissivity information that relates to the spectral shape but not the absolute emissivity include temperature-independent spectral indices (TISI) (Becker and Li, 1990), thermal log residuals and alpha residuals (Hook et al., 1992), and spectral emissivity ratios (Watson, 1987; Watson et al., 1990). The methods that extract absolute emissivity include the day–night method (Watson, 1992), model emissivity (Lyon, 1965; Kahle et al., 1980), normalized emissivity (Gillespie, 1986; Realmuto, 1990), alpha emissivity (Kealy and Gabell, 1990; Kealy and Hook, 1993), and min-max emissivity (Matsunaga, 1994).

The methods that allow the absolute emissivity to be determined can be summarized as follows. The day–night method increases the number of unknowns by 1 but doubles the number of measurements, making the problem overdetermined. In practice, however, the approach is sensitive to measurement noise and requires pixel-perfect registration between the day and night images. The model emissivity assumes that the emissivity is known at one wavelength, whereas the normalized emissivity assumes a maximum emissivity value at an unspecified wavelength. The last two methods, alpha emissivity and min-max emissivity, utilize an empirical relationship between either the mean and standard deviation or mean and min-max difference. A final method is being developed by the ASTER team which combines the attactive aspects of spectral ratios, normalized emissivity, alpha emissivity, and min-max emissivity termed the temperature–emissivity separation algorithm or TES (Gillespie et al., 1998).

Two of the techniques identified above that have been widely applied are described in detail below: the normalized emissivity calculation (Gillespie, 1986; Realmuto, 1990) and alpha residual calculation (Kealy and Gabell, 1990; Hook et al., 1992; Kealy and Hook, 1993). Examples of the application of these techniques are given in the case histories in Section 2.3. Each method makes slightly different assumptions in order to solve the underdetermined nature of the temperature–emissivity separation problem.

Normalized emissivity method.   This method involves calculating the surface temperature in each channel using a constant emissivity value from the surface radiance:

$$T_j = \frac{C_2}{\lambda_j \ln\left[\dfrac{(\varepsilon_j C_1)}{L_j \lambda_j^5 \pi)} + 1\right]} \tag{2.11}$$

where the subscript j refers to the channel.

The highest of the set of temperatures is considered to be the correct temperature of the pixel. This temperature is used with equation (2.4) to calculate emissivity values for the remaining channels. The emissivity to be assigned to the chosen channel is often 0.96, which represents a reasonable average of likely values for exposed geologic surfaces. A recent application of this technique to mapping basalt flows in Hawaii is given in Realmuto et al. (1992).

Alpha emissivity method. The alpha-derived emissivity method was developed by Kealy and Gabell (1990). This approach utilizes Wien's approximation of the Planck function. Wien's approximation neglects the -1 term contained in the Planck function (equation (2.1), making it possible to linearize the approximation with logarithms. For temperatures of 300 K and a wavelength of 10 m, Wien's approximation results in errors of up to 1% (Siegal and Howard, 1982). Taking natural logs of the surface radiance using Wien's approximation and multiplying by λ_j in order to separate the λ and T terms, we obtain

$$\lambda_j \ln L_j = \lambda_j \ln \varepsilon_j + \lambda_j \ln C_1 - \lambda_j 5 \ln \lambda_j - \lambda_j \ln \pi - \frac{C_2}{T} \tag{2.12}$$

Calculating the mean of the equation set for $j = 1, N$, where N is the number of channels, subtracting the mean from equation (2.12), and putting the elements containing λ on the left-hand side, six similar equations are created each without a T term. A new parameter, α_j, is then defined:

$$\lambda_j \ln \varepsilon_j - \frac{1}{N} \sum_{j=1}^{N} \lambda_j \ln \varepsilon_j = \alpha_j = \lambda_j \ln L_j - \frac{1}{N} \sum_{j=1}^{N} \lambda_j \ln L_j + K_j \tag{2.13}$$

where K_j contains only terms which do not include the measured radiances, L, and hence may be calculated from the constants.

The set of N values of α_j can be calculated using the right-hand side of equation (2.13) by using the ground radiance data. These α_j are temperature independent (containing no T term). Using laboratory-derived emissivity spectra, with the left-hand side of equation (2.13) it is possible to calculate α_j. Thus direct comparison of ground α spectra to laboratory α spectra is possible.

Figure 2.16 shows a TIMS-equivalent ε_j spectrum and its corresponding α_j spectrum. The main difference between ε_j and α_j spectra is not in shape but in the absolute values: α_j spectra will have a mean of zero whatever ε_j values are used to derive them. This manipulation, while producing temperature-independent spectra, has not solved the underdetermined equation set. The element of information missing has been confined to the mean value. An α_j spectrum provides information only on the shape of an ε_j curve. Figure 2.17 highlights this characteristic. The ε_j spectra, being of similar shape but varying in mean value, will all produce the same α_j spectrum.

An extension of the aforementioned method allows ε to be calculated. The left-hand side of equation (2.13) may be rearranged such that

$$\varepsilon_j = \exp\left[\frac{\alpha_j + \frac{1}{N} \sum_{j=1}^{N} \lambda_j \ln \varepsilon_j}{\lambda_j}\right] \tag{2.14}$$

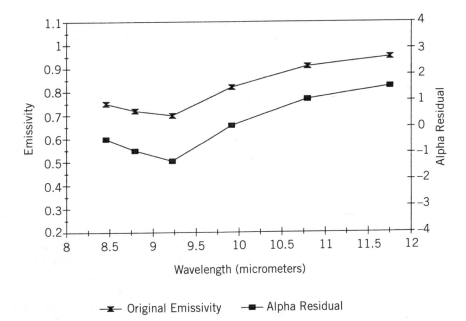

Figure 2.16 TIMS equivalent emissivity spectrum and its corresponding alpha residual spectrum.

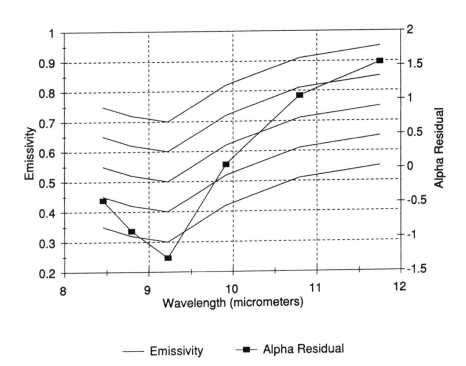

Figure 2.17 Suite of emissivity spectra that all produce the same alpha residual spectrum.

However, this is not solvable since the value of the mean in equation (2.14) is unknown. The mean may be estimated by considering the behavior of common thermal infrared spectra. Observation of emissivity spectra shows that in general, emissivity spectra with high mean values exhibit little variation (e.g., vegetation spectra), while those spectra that exhibit a greater variation have lower means. Spectra from a wide variety of igneous and sedimentary rocks and soils (Salisbury et al., 1988; Salisbury and D'Aria, 1992) were convolved to TIMS equivalent spectra, and the variance of λ_j and the mean in equation (2.13) were calculated. The mean and variance were used because the variance can be obtained directly from the instrument data, and the mean is the desired quantity in equation (2.13). Both these parameters are nearly independent of temperature.

Figure 2.18 is a plot of these parameters for spectra from a variety of materials resampled to the TIMS bandpasses, which can be fitted by a curve. Therefore, by measuring the variance of α_j derived from the instrument data, a mean may be estimated and used with equation (2.14) to produce an ε spectrum. Once the emissivity is found, equation (2.11), ignoring the +1 to put it in the Wien form, can be used to obtain the temperature. There will be no variation in the temperature calculated whichever band is used because the calculated spectrum is one of a suite of spectra that can be calculated from a single temperature, provided that the Wein form of the equation is used. Naturally, there is some variation in the best-fit equation, depending on the data set used. Where pixels do not have mean/variance characteristics consistent with the mean/variance curve used, the subsequent average value of the emissivity curve will be under- or overestimated, as will the temperature.

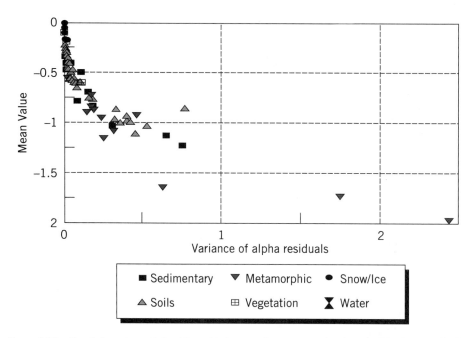

Figure 2.18 Plot of the variance of the alpha residuals versus their mean-wavelength-weighted log emissivity for data resampled to TIMS bandpasses.

2.2.2.3 ADVANCED SPACEBORNE THERMAL EMISSION REFLECTANCE RADIOMETER (ASTER).

The following description of the ASTER instrument was taken largely from Kahle et al. (1991).

A. Introduction. The advanced spaceborne thermal emission and reflectance radiometer (ASTER) is being built by the Japanese government and scheduled to fly in Earth orbit on the first (A) platform of NASA's Earth Observing System (*EOS-A*), which will be launched in 1998. ASTER is a high-spatial-resolution multispectral imager. The instrument will have three bands in the visible and near-infrared (VNIR), spectral range (0.5 to 1.0 μm) with 15-m resolution, six bands in the shortwave infrared (SWIR), spectral range (1.0 to 2.5 μm) with 30-m spatial resolution, and five bands in the thermal infrared (TIR), spectral range (8 to 12 μm) with 90-m resolution (Fujisada and Ono, 1991). An additional backward-viewing telescope with a single band in the near-infrared with 15-m spatial resolution will provide the capability, when combined with the nadir viewing band, for same-orbit stereo data. The swath width of an image will be 60 km. Companion cross-track pointing out to 136 km will allow viewing of any spot on Earth at least once every 16 days.

B. Instrument Description. ASTER continues the trend in high-spatial-resolution surface imaging begun by the *Landsat* thematic mapper and by *SPOT*. ASTER will increase the number of bands (14 versus the seven of the thematic mapper and four of *SPOT*) available simultaneously. It will increase the number of SWIR bands available, provide same-orbit stereo using a backward and a nadir telescope assembly, and for the first time, provide multispectral thermal emission capability in the window region, 8 to 12 μm on a global basis.

The VNIR subsystem is built on the technology from the optical sensor MESSR flown on the Japanese Marine Observation Satellites (MOS) *MOS-1* and *MOS-1b* and the OPS sensor to be on the Japanese Earth Resources Satellite 1 (*JERS-1*). The SWIR subsystem is the successor to the short-wavelength/infrared radiometer on *JERS-1*. The TIR subsystem is a new development.

Functional testing and calibration of the flight ASTER is not complete at the time of writing. Derived quantities such as signal/noise ratios, which depend on a systems analysis and a detailed definition of optics, filters, detectors, and other items, will not be final until the instrument is complete. Performance parameters such as NE delta rho and NE delta T, given in the following description, should be viewed only as a guide to expected performance.

Figure 2.19 displays a line drawing of the ASTER configuration. The VNIR and SWIR subsystems use line-array detectors in a pushbroom mode. For these subsystems, rotation of the entire telescope assembly is used to provide a cross-track pointing capability of ±136 km for the extremes of the swath. The TIR subsystem uses whiskbroom scanning with the scan mirror used for both the scanning involved in image generation and to provide the cross-track pointing function. The cross-track pointing capability in all telescopes is sufficient to allow imaging of any point on Earth over the 16-day repeat cycle of the *EOS-AM* platform orbit.

The overall performance requirements for the ASTER instrument are given in Table 2.3. Although the VNIR, SWIR, and TIR subsystems are separate and independent with respect to optics, filters, detectors, detector cooling, preamplifiers, and inflight calibration, the other system functions such as power conditioning, image

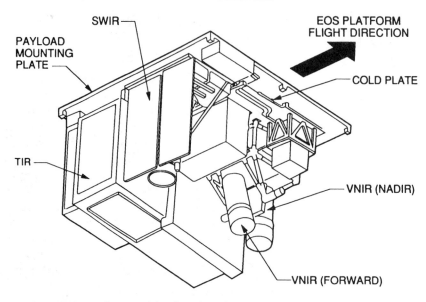

Figure 2.19 Configuration of the advanced spaceborne thermal emission reflectance radiometer.

data processing and packetization, command and telemetry handling, and safe mode capability are performed by common subsystems.

(1) Vnir design. The VNIR subsystem consists of two independent telescope assemblies to minimize image distortion in the backward-and nadir-looking telescopes (Takahashi et al., 1991). The detectors for each of the bands are listed in Table 2.4. On-board calibration of the two VNIR telescopes is accomplished with either of two independent calibration devices for each telescope (Ono and Sakuma, 1991). The radiation source is a halogen lamp. A diverging beam from the lamp filament is input to a portion of the first optical element (Schmidt corrector) of the telescope subsystem filling part of the aperture. The detector elements are uniformly irradiated by this beam.

The system signal/noise ratio is defined by specifying the NE delta rho to be ≤ 0.5% referenced to a diffuse target with a 70% albedo at the equator during equinox. The absolute radiometric accuracy is to be ±4% or better. The VNIR subsystem produces by far the highest data rate of the three ASTER imaging subsystems. With all four bands operating (three nadir and one forward) the data rate, including image data, supplemental information, and subsystem engineering data, is 62 Mbps.

(2) Swir design. The SWIR subsystem uses a single aspheric refracting telescope (Akasaka et al., 1991). The detector in each of the six bands (Table 2.4) is a platinum silicide–silicon (PtSi–Si) Schottky barrier linear array cooled to 80 K. Cooling is provided by a split Stirling cycle cryocooler (Kawada and Fujisada, 1991). A calibration device similar to that used for the VNIR subsystem is used for in-flight calibration. The NE delta rho will vary from 0.5 to 1.3% across the bands from short to long wavelength. The absolute radiometric accuracy is to be ±4% or better. The combined data rate for all six SWIR bands, including supplementary telemetry and engineering telemetry, is 23 Mbps.

TABLE 2.3   Overall Performance Requirements of the ASTER Instrument

Spatial resolution	15m (bands 1–3, VNIR)
	30m (bands 4–9, SWIR)
	90m (bands 10–14, TIR)
Stereo base/height ratio	0.6 (band 3)
Swath width	60 km (all bands)
Total edge-to-edge coverage capability in cross-track direction	± 136 km (all bands)
Cross-track pointing capability (array center)	± 8.54 (all bands)
Signal quantization	8 bits (bands 1–9)
	12 bits (bands 10–14)
Modulation transfer function (MTF) at Nyquist frequency	0.25 (all bands)
Peak data rate	89.2 Mbit/s (all bands, including stereo)
Mission life	5 years
Mass	352 kg
Maximum power	650 W
Physical size	$1.6 \times 1.6 \times 0.9$ m^3

(3) Tir design. The TIR subsystem uses a Newtonian catadioptric system with an aspheric primary mirror and lenses for aberration correction (Kitamura et al., 1991). Unlike the VNIR and SWIR telescopes, the telescope of the TIR subsystem is fixed, with pointing and scanning done by a mirror. Each band uses 10 mercury–cadmium–telluride (HgCdTe) detectors (Table 2.4).

As with the SWIR subsystem, the TIR subsystem will use a mechanical split Stirling cycle cooler. The scanning mirror functions for both scanning and pointing. For calibration, the scanning mirror rotates 180° from the nadir position to view an internal blackbody that can be heated. For the TIR subsystem, the signal/noise ratio

TABLE 2.4   Spectral Bandpasses of the ASTER Instrument

Subsystem	Band	Spectral Range (μm)
VNIR	1	0.52–0.60
	2	0.63–0.69
	3	0.76–0.86[a]
SWIR	4	1.6–1.7
	5	2.145–2.185
	6	2.185–2.225
	7	2.235–2.285
	8	2.295–2.365
	9	2.36–2.43
TIR	10	8.125–8.475
	11	8.475–8.825
	12	8.925–9.275
	13	10.25–10.95
	14	10.95–11.65

[a] This band is replicated in a backward-looking telescope for stereo.

TABLE 2.5   Summary of ASTER Products

Reformatted DN and ancillary data
Radiance at sensor with bands within and between telescopes coregistered
Brightness temperature at sensor
Decorrelation stretch of level 1b data
Surface radiance
Surface temperature
Surface emissivity
Surface reflectance
Digital elevation models
Polar cloud mask

can be expressed in terms of an NEδT. The requirement is that the NEδT be less than 0.3 K for all bands with a design goal of less than 0.2 K. The signal reference for NEδT is a blackbody at 300 K. The accuracy requirements on the TIR subsystem are given for each of several brightness temperature ranges as follows: 200 to 240 K, 3 K; 240 to 270 K, 2 K; 270 to 340 K, 1 K; and 340 to 370 K, 2 K.

The total data rate for the TIR subsystem, including supplementary telemetry and engineering telemetry, is 4.2 Mbps. Because the TIR subsystem can return useful data both day and night, the duty cycle for this subsystem has been set at 16%. The cryocooler, like that of the SWIR subsystem, will operate with a 100% duty cycle.

C. Data Reduction Algorithms and Data Products. The data obtained by the AS-TER instrument will be processed using a production set of algorithms in order to derive standard products and an experimental set of nonstandard data products. Four of the standard products—surface radiance, surface temperature, surface emissivity and surface reflectance—are considered to be of fundamental importance to the EOS program, with broad applicability to numerous EOS investigations and other applications. The full set of ASTER products are given in Table 2.5.

The algorithm used to derive surface radiance from the thermal infrared data is similar to that used for quantitative analysis of the TIMS data. The algorithm used to extract emissivity and temperature is a hybrid algorithm that uses the attractive features of the two algorithms described with the quantitative processing of the TIMS data (Gillespie *et al.*, 1998). The algorithms have been peer reviewed and are available through the NASA EOS Science Office Home Page (http://eospso.gsfc.nasa.gov) and the ASTER Home Page (http://asterweb.jpl.nasa.gov).

The accurate derivation of these four fundamental products involves atmospheric correction, which will require the input of supplemental information derived from existing sources and from the moderate resolution imaging spectrometer (MODIS) and multiangle imaging spectroradiometer (MISR) instruments onboard the *EOS-AM1* platform. An example of such data is that of local elevation data together with atmospheric data from MODIS/MISR.

2.3   CASE HISTORIES

Multispectral thermal infrared images have been used in geological studies for several decades. The early workers demonstrated that it was possible to discriminate silicate rocks using instruments with two channels in the thermal infrared (Vincent, 1972; Vincent et al., 1972). Kahle et al. (1980) published the first work that utilized data from an instrument with more than two channels in the thermal infrared to map alteration in the East Tintic mining district in central Utah. The study used data from a Bendix scanner, which was replaced by the thermal infrared multispectral scanner (TIMS) built by Daedalus Enterprises. Some early results from the analysis of TIMS data acquired over Death Valley, California were published by Kahle and Goetz (1983). Since that time TIMS data have been widely utilized for geologic studies in a variety of terrains: for example, alluvial fans (Gillespie et al., 1984), lava flows (Abrams et al., 1991), mineral exploration (Watson et al., 1990; Hook et al., 1992), sedimentary basins (Lang et al., 1987), igneous rocks (Lahren et al., 1988; Sabine et al., 1994), structurally complex metamorphosed terranes (Hook et al., 1994), and alkalic rock complexes (Watson et al., 1996).

In this section we provide examples on the use of the instruments and techniques described in Section 2.2 in geological remote sensing studies. Data from two sites are presented: Death Valley, California and Hawaii. Each case study follows the general format of introduction, geology, data processing, and interpretation. Detailed descriptions of the data processing techniques were given in Section 2.2. The Death Valley case history is taken from work by Gillespie et al. (1984) and Crowley and Hook (1996). The Hawaiian case history is taken from work by Kahle et al. (1988, 1995), Crisp et al. (1990), and Realmuto et al. (1992).

2.3.1 Case History: Death Valley

Death Valley has been used in remote sensing studies for many decades. Thermal infrared multispectral scanner (TIMS) data were first acquired from Death Valley shortly after the instrument was completed in 1984. Since then several more TIMS data sets have been acquired and used in a variety of studies. Two studies are presented here which illustrate the use of multispectral thermal infrared data for alluvial fan mapping and evaporite mineral mapping.

2.3.1.1 GEOLOGY.

Death Valley is a deep, narrow north/south-trending graben in the Basin and Range Province in southeastern California. It is partly filled with saline lake deposits and is bounded by great fault blocks. The largest and highest of these forms the Panamint Range along the west side of the valley and it rises in a distance of about 12 miles to more than 3353 m above the salt pan, which is at its lowest at Badwater, 86 m below sea level. The climate of the area is generally hot and the vegetation is sparse.

The Panamint Range is structurally dominated by a series of thrust faults known as the Amargosa thrust system, where younger rocks have been thrust westward over older rock in a series of blocks. The rocks exposed are dominantly Precambrian through Paleozoic in age but include some Tertiary volcanic and granitic intrusive rocks.

The Precambrian rocks have been organized into three major groups: (1) metamorphic rocks of the crystalline basement complex; (2) younger Pahrump Series, only slightly metamorphosed mostly clastic sedimentary rocks; and (3) even younger, mostly clastic rocks, but also including considerable dolomite and some limestone.

Much of the Black Mountain Range along the southeastern valley margin, as well as portions of the Panamint Range to the west, is composed of Precambrian schist and gneiss (Figure 2.20). A thick sequence of Palaeozoic sedimentary rocks, consisting mostly of carbonates, is exposed on the eastern slope of the Panamint Range. At the north end of the Black Mountains is the Tertiary Artist Drive Formation, composed mainly of felsic volcanic rocks and interbedded clastic sedimentary rocks. Unconformably overlying the Artist Drive Formation is the Furnace Creek Formation of Miocene and Pliocene (?) age. The Furnace Creek Formation is a mixture of con-

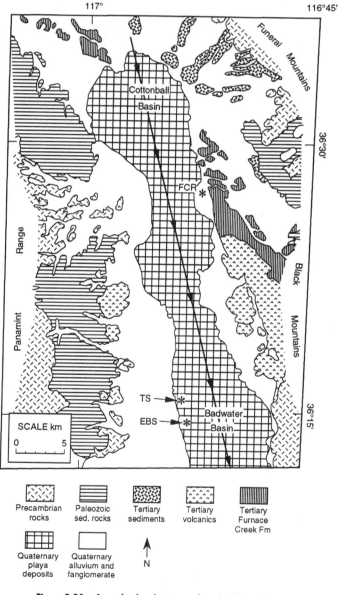

Figure 2.20 Generalized geologic map of Death Valley, California.

glomerate, saliferous playa sediments, and interbedded basalts and tuffs (McAllister, 1970, 1976). This formation hosts large borate mineral deposits that have been exploited since the early twentieth century (Barker and Wilson, 1975; Smith, 1985). The Furnace Creek Formation is unconformably overlain by the Funeral Formation of Pliocene and Pleistocene (?) age (McAllister, 1970). The Funeral Formation consists mainly of poorly stratified conglomerate, sandy mudstone, and basaltic flows. There are two granitic intrusions that appear to be related to the Sierra Nevada batholith.

Alluvial fans are very prominent on the lower slopes of Death Valley. On the west (Panamint) side of the valley they are large and mature, whereas on the east (Black Mountain) side of the valley they are immature and considerably smaller. This is due to both the greater height and thereby greater amount of discharged debris from the Panamint side and to the effects of the eastward tilt of the Panamint–Death Valley fault block. Hunt and Mabey (1966) mapped four Quaternary alluvial fan units based on relative weathering and geomorphic characteristics but not on lithologic composition. The three younger fan units occur within the study area. Of these, the youngest (Q4) comprises active channels containing silt, sand, and gravel reworked from the older fan deposits. The intermediate unit (Q3) consists of similar gravels in inactive channels. These are moderately coated by desert varnish (Hooke, 1972). The oldest unit (Q2) is typified by heavily varnished pebbles making up well-formed desert pavement. Hooke (1972) subdivided Q2 into three age units (as he did unit Q3).

The floor of Death Valley is covered by a large salt pan and contains surface crusts of evaporite minerals ranging from a few millimeters to more than 1 m in thickness. Hunt and Mabey (1966) noted that the salt pan displayed a zoned arrangement of evaporite minerals, including a central choride zone, surrounded by fringing zones of sulfate and carbonate salts. This zonation is due to the solubility of the salts, the more soluble minerals generally forming toward the center of the salt pan. Beneath the surficial deposits alternating layers of salt-and clay-rich sediment have been found to depths of more than 300 m.

The salt pan is divided into three areas: the northern Cottonball Basin, the central Middle Basin, and the southern Badwater Basin. The Cottonball Basin receives most of its inflow as groundwater, a significant fraction of which flows through the Furnace Creek and Funeral Formations. This inflow has greater Na/Ca values than inflow to other parts of the pan and is locally higher in bicarbonate (Hunt et al., 1966). Conversely, Badwater Basin receives groundwater inflow derived from the Palaeozoic sediments (mainly carbonates) and metamorphic rocks from the Panamint and Black Mountains, respectively. Badwater Basin also receives sporadic input from the Amargosa River which enters from the south. The Badwater Basin inflow sources generally have lower Na/Ca values and are less bicarbonate-rich than the inflow waters of Cottonball Basin (Hunt et al., 1966).

2.3.1.2 DATA PROCESSING.

The two studies used different approaches to extract geological information from the TIMS data and each approach is reviewed below:

A. Alluvial Fan Mapping. The TIMS data used in the original alluvial fan study were acquired on August 27, 1982. The TIMS data presented here were acquired on

June 12, 1995. The 1995 imagery is of slightly higher quality and covers nearly the same area. There appears to have been little change in the alluvial fans in the intervening decade. The 1995 imagery was acquired from an altitude of about 9 km, resulting in a pixel size of 20 × 20 m. The data were calibrated, then processed with the decorrelation stretch using channels 5, 3, and 1 displayed as red, green, and blue, respectively (Figure 2.21; see the color insert). The image shows units that are differentiable by texture and color. Textural differences related to topography allow differentiation of bedrock from alluvial fans in most cases. Color differences are related to composition. In this rendition quartzites appear deep red, carbonate rocks appear green, basalts and andesites appear blue, and rhyolites appear purple, as do most of the shales. The alluvial fans are represented by the same wide range of colors as the bedrock units they are derived from but are modified by the rock's weathering characteristics and tendency to hold desert varnish.

B. Evaporite Mineral Mapping. The TIMS data used in the evaporite mineral mapping study were acquired from an altitude of about 9 km on April 6, 1994. The swath width for this flight was about 11 km, and each pixel corresponds to a ground spot of about 20 × 20 m. The TIMS data were calibrated and atmospherically corrected using the MODTRAN radiative transfer code. No direct atmospheric measurements were available and the default midlatitude winter model was used with MODTRAN. Following the initial atmospheric correction, the brightness temperature of two areas, one of rough salt and another covered by vegetation, were examined. Both areas were expected to show little emissivity contrast at the spectral resolution of the TIMS, and therefore the brightness temperatures of the two targets should be constant in the six TIMS channels, although the brightness temperatures would vary between targets. The results indicated that the brightness temperatures for channel 1 were higher than the other channels, and the amount of water vapor in the default profile was reduced until there was good agreement between the temperature values in all six TIMS channels for the two types of calibration targets. This required reducing the amount of water in the original profile by 70%; it should be noted that there was very little water vapor in the profile to start with, so the effect of this reduction was fairly minor.

After the atmospheric correction, the temperature and emissivity information was extracted from the ground radiance values with the alpha residual technique (Hook et al., 1992, 1994; Kealy and Hook, 1993). The alpha residual spectra have shapes similar to emissivity spectra, although the mean value of each spectrum is zero. Initially, the alpha residual images were studied using an unsupervised K means classification program to identify spectrally distinct materials within the salt pan (Tou and Gonalez, 1974). A digital mask was applied to the alpha residual data to limit the unsupervised classification to the salt pan. The classification was able to discriminate many of the units in the salt pan, but further subdivision based on a priori knowledge was necessary since some important distinctions were not made (e.g., the distinction between areas of gypsum–anhydrite and thenardite).

Based on the modified classification, 11 alpha residual spectra representing different playa surface materials were identified. These spectra were used as input to a second classification procedure that calculated the vector angle (in six-channel space) between each spectrum and each unknown spectrum in the TIMS data (Kruse et al.,

1993). Smaller vector angle values signify closer spectral similarity and 11 thesholded images were produced, one for each class. These images were combined as overlays on a single base image with each spectral class displayed in a different color.

2.3.1.3 INTERPRETATION.

The decorrelation stretched image (Figure 2.21) in conjunction with the geological map of Hunt and Mabey (1966) was used by Gillespie et al. (1984) to produce the interpretative map (Figure 2.22; see the color insert) of the ages and provenance of the alluvial fans. This map was field checked, although this did not include field emissivity measurements, due to the lack of instrumentation in the early 1980s.

Spectra representing each of the classes derived from the alpha residual data were used to produce the interpretative evaporite mineral map in Figures 2.24 and 2.26 by Crowley and Hook (1996). These spectra were compared to field spectra acquired with the µFTIR and processed to emissivity using the methods outlined for that instrument in Section 2.2.

A. Alluvial Fan Mapping. Hunt and Mabey (1966) mapped the alluvial gravels based on relative age. Gillespie et al. (1984) used the decorrelation stretched imagery (Figure 2.21) to further subdivide the gravels according to composition and provenance (Figure 2.22). They recognized six suites of fan gravels, distinguished by their lithological assemblages. Suite I is dominated by shales with lesser amounts of dolomite (e.g., the fan gravels of Tucki Wash). Suite II has more quartzite and less shale (e.g., the fan gravels below Trail Canyon and Blackwater Wash). Suite III is lithologically intermediate between suites I and II and consists of quartzite and shale with little dolomite (e.g., the fan gravels between Tucki Wash and Blackwater Wash). Suite IV is dominated by carbonate clasts and found below the smaller drainages that cut the carbonate rocks only at the range front. Suite V consists of volcanic rocks and occurs beneath the large exposures of Tertiary volcanic rocks. Suite VI consists of fans made up of undifferentiated mixtures of lithologies.

The colors of the alluvial fans in the imagery are controlled primarily by the source of the fan material. If the source of the fan material is a single rock type, the fan will appear the same color as the rock type, and in some cases the contact between the fan debris and bedrock cannot be distinguished. The susceptibility of the rocks to erosion plays a major role in determining the lithologies present in the fan gravels. For example, the Trail Canyon gravels of easily eroded quartzite and shale, which crop out west of the study area, are transported through 4.5 km of resistant dolomite bedrock before deposition on the alluvial fan. This is apparent since the color of the fan gravels above and below the dolomite are virtually the same, showing that the admixture of dolomite is minor.

Differential erosion of gravels within a fan results in compositional changes over time which allow the color of the fan in the image to be used as a basis for relative age discrimination. For example, the gravels in the active channels of Tucki Wash are dominantly fragments of fissile shale. Surfaces of the older alluvial deposits (Q2) are a desert pavement of dolomite and quartzite; exposed shale has been reduced to fine grains and removed. This would have the effect of removing purple and could account for the pink color of this Q2. Compositional changes also occur in some carbonate fans as they weather, such as dissolving carbonate out of a gritty dolomite, thus concentrating quartzite and other clastic sediments. Dissolved carbonate may

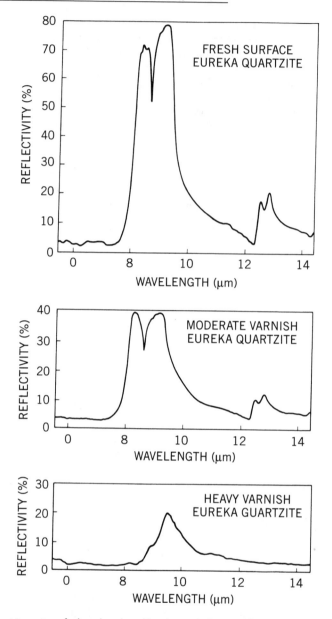

Figure 2.23 Quartzite spectra—fresh, moderately, and heavily varnished (top, middle, and bottom, respectively)—measured with the JPL Laboratory Nicolet spectrometer.

be precipitated as caliche near inactive surfaces of fans. Such caliche has been widely exposed by erosion and deflation of some Q2 surfaces (Hooke, 1972).

Many of the gravels found throughout the area, especially the quartzite and other clastic sedimentary rocks, are heavily coated with desert varnish. Laboratory reflectance spectra of varnished quartzite (Figure 2.23) indicate that the emissivity minimum is smaller and occurs at longer wavelengths as the amount of varnish increases than for unvarnished quartzite (Christensen and Harrison, 1993). In the decorrelation stretch image this should result in a decrease in the intensity of the red color of the quartzite and perhaps shift toward blue or purple.

B. Evaporite Mineral Mapping. Figure 2.24 (see the color insert) is a color image showing the distribution of the spectral classes identified in Cottonball Basin. Three of the original 11 classes were combined and displayed as a single class; therefore, the image has nine color patterns. The color overlays are superimposed on a pseudotemperature image created from the mean in the alpha residual calculation (see Section 2.2), in which dark areas are cool and bright areas warm. Areas not classified in the image did not have sufficient spectral contrast to fit in any of the classes.

Figure 2.25 shows laboratory and field spectral curves for the seven sample sites representing the main spectral classes in Figure 2.24 as well as the corresponding alpha residual spectrum taken from the TIMS data. Spectrum A on Figure 2.25 is extracted from an area shown as yellow on Figure 2.24, representing areas with thenardite (Na sulfate). The µFTIR spectrum for sample A has a distinctive spectral feature characteristic of thenardite which translates into a emissivity minimum in TIMS channels 1 and 2. The distribution of thenardite around the northern and eastern sides of Cottonball Basin probably reflects the inflow of Na-rich ground and surface waters, derived at least in part from the weathering of the Furnace Creek Formation rocks.

The orange areas on Figure 2.24 correspond to the silty halite-rich crusts without much thenardite. They have weak emissivity minima in TIMS channels 3 and 4 due to fine-grained feldspar and clay minerals (Figure 2.25, spectrum B). The red areas on Figure 2.24 correspond to gypsum–anhydrite and Ca sulfate. There are only a few small areas of gypsum–anhydrite in Figure 2.24; however, minor quantities of this mineral are present in field samples from many of the areas, together with fine-grained quartz, feldspar, and carbonate. Gypsum–anhydrite has a distinctive emissivity spectrum (Figure 2.25, spectrum C).

The dark blue areas on Figure 2.24 correspond to areas of illitic–clay and/or muscovite that has been reworked from the Furnace Creek Formation and deposited on the Furnace Creek alluvial fan. These materials are characterized by a minimum in TIMS channels 3 and 4 (Figure 2.25, spectrum D). The green areas on Figure 2.24 correspond to areas of quartz-rich fan gravel overlying silty efflorescent crusts in the northern portion of the image. The quartz is derived primarily from the adjacent Panamint Range. The green and orange/yellow classification patterns therefore show a distinction between the mudflat environment of primarily detrital sedimentation and the silty halite–thernardite environment of primarily chemical sedimentation, respectively (Figure 2.25, spectrum E).

The cyan areas on Figure 2.24 correspond to halite-rich rough salt which exhibits low-contrast spectra due to cavity-radiator behavior (Figure 2.25, spectrum F). The light green areas on Figure 2.24 correspond to halite-rich crusts that also contain the Na–Ca sulfate mineral glauberite. Figure 2.25, spectrum G illustrates the spectral behavior of these areas. Unlike the spectral curves already discussed, the laboratory spectrum of the glauberite-bearing crusts does not bear a close resemblance to alpha residual spectra for the very well defined image pattern (Figure 2.24). This discrepancy has been attributed to a change in the character of the area between the time

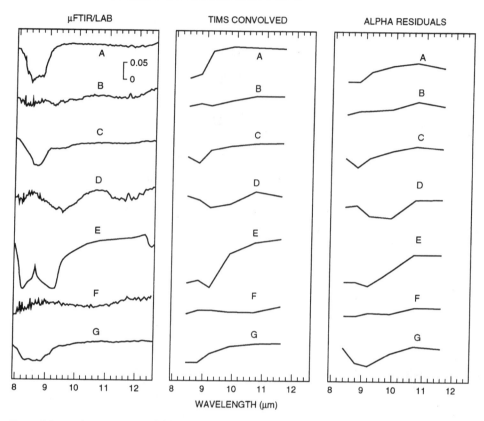

Figure 2.25   (Left) Emissivity spectra of field samples representing various image spectral classes. Spectra are offset on the vertical axis for clarity; bracket indicates an emissivity interval of 0.05 except for spectrum E, where the indicated interval is 0.15. Y intercept values are 0.98, 0.96, 0.90, 0.92, 0.97, and 0.95 (top to bottom). The majority of spectra were recorded in the field using the μFTIR field emission spectrometer. In the cases of spectra C, E, and G, laboratory directional-hemispherical reflectance spectra of field samples were converted to emissivity using Kirchhoff's law ($\varepsilon = 1 - R$). (Middle) Spectra from the left column after convolution to the six TIMS filter functions. (Right) Alpha residual spectra for single pixels extracted from the TIMS data. Areas used for the extraction are labeled A to G on Figure 2.24, with the exception of area B, which is not shown. Surficial units represented by these spectra are: A, thenardite-rich crusts in saline facies of sulfate zone; B, silty halite, smooth facies, and carbonate zone, silty facies; C, gypsum crusts; D, illite/muscovite-rich alluvial deposits; E, quartz-rich fan gravels and mudflats; F, massive halite and silty halite, rough facies; G, mixed silicate and evaporite mineral crusts on floodplains. (Unit description modified from Hunt et al., 1966.)

of the acquisition of the airborne data and the field work. This area is in the low-lying mudflat areas in the center of the salt pan that are generally the last areas to desiccate completely following a significant rainfall or inflow event.

Figure 2.26 is a geologic map of the Recent saline deposits in Cottonball Basin. There are several similarities between Figures 2.24 and 2.26. Areas mapped as the saline facies (Figure 2.26, unit cs) are indicated by the yellow and orange areas on Figure 2.24. Orange pixels also map some areas of silty halite (e.g., Figure 2.26, unit sh, in the west central part of the Cottonball Basin) that are transitional in salt content between the saline facies of the sulfate and carbonate zones (Figure 2.26, unit cs) and other unclassified silty halite areas. Several areas of rough and eroded salt (Figure

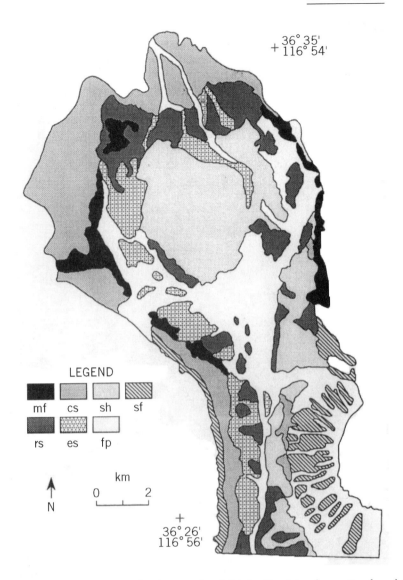

Figure 2.26   Geological map of Recent saline deposits in Cottonball Basin. Abbreviation of map units: mf, marsh facies; cs, saline facies of sulfate and carbonate zones; sh, silty halite, smooth facies; sf, sand facies of carbonate zone; rs, silty halite and massive halite, rough facies; es, eroded halite; fp, floodplain deposits. (Modified from Hunt et al., 1966.)

2.26 units rs and es) are distinguished by cyan pixels on Figure 2.24. The marsh facies (Figure 2.26, unit mf) is indicated on the image (Figure 2.24) by the darker gray-to-black areas within the salt pan. These areas are relatively cool, due to spring discharge and the evaporation of moisture from the surficial sediments.

2.3.1.4   SUMMARY.

The Death Valley case history illustrates two different approaches to data processing, qualitative and quantitative. It also shows how multispectral thermal infrared data can be used to address two different mapping problems, alluvial fan mapping and evaporite mineral mapping. The techniques used are equally applicable to calibrated multispectral thermal infrared data from other instruments provided that there is adequate spatial resolution.

2.3.2 Case History: Hawaii

Hawaiian basaltic lavas exhibit a variety of thermal infrared spectral signatures that typically do not correspond to differences at visible wavelengths. These spectral differences are correlated with differences in roughness and chemical and mechanical changes, beginning during initial cooling of the lava and continuing over a period of hundreds or thousands of years. These changes can be used to estimate relative ages of individual basalt flows.

The study area used in this case history is near the center of the island of Hawaii at an elevation of approximately 2200 on the north slope of Mauna Loa in an area known as the saddle. The area is generally fairly devoid of much vegetation, and even lichen does not grow quickly. The average annual precipitation is 40 in. or less. The basalts in the study area range in age from older than 8000 years to the most recent flow in 1935 (J. P. Lockwood, personal communication, 1988).

2.3.2.1 GEOLOGY.

Each of the basalt flows of the study area have been studied in the field to establish a correspondence between TIMS image color and visual estimates of weathering for the lavas. Pahoehoe and aa have identical chemical and mineralogical compositions within individual lava flows and differ in their genesis only by their physical conditions of emplacement (Peterson and Tilling, 1980).

The aa examined in the study is generally quite rough and similar in appearance from flow to flow. Incipient translucent coats, visible under magnification, appear on aa more than a few decades old (e.g., the 1843 flows). The youngest prehistoric aa flows (0.2 to 0.5 ka) are lighter in color than the black fresh aa. The principal effect of weathering appears to be the growth of a thin reddish or tan rind. Under the rind the basalt is black and fresh-looking. Farr and Adams (1984) report that this coat consists of alternating clear and red layers of silica and hydrous or amorphous iron oxides. In older aa the rind was thicker and the lava more pervasively weathered. Presumably, much of the basalt has been replaced by hydrous iron oxides (limonite or palagonite) and claylike minerals (Macdonald, 1971; Lipman, 1980).

The pahoehoe examined is everywhere characterized by a layer of surface glass commonly up to 1 cm in thickness, which grades downward into more crystalline basalt. Locally, this layer is topped by a smooth, thin (ca. 50 µm) glassy chill coat that is variously discontinuous or spalled off. Near fissures and old fumaroles the glass is commonly devitrified and has a waxy luster. These variations in the condition of the glassy crust have been seen even in pahoehoe that has just cooled. They thus represent a range in the initial state of the surface.

The two younger pahoehoe flows in the study area (1935 and 1843) are distinguishable only with difficulty in the field, primarily by their degree of luster. They are indistinguishable in air photos. These and older pahoehoe flows also support the same translucent accretionary rinds seen on the aa flows. Similarly, the oldest prehistoric flows exhibit the same ocher weathering as the aa flows.

2.3.2.2 DATA PROCESSING.
The TIMS data were processed with the decorrelation stretch algorithm.

2.3.2.3 DATA INTERPRETATION.
Figure 2.27 (see the color insert) shows a decorrelation-stretched image of the study area, with bands 5, 3, and 1 displayed as red, green, and blue, respectively. Despite the chemical and petrologic similarity of the unweathered basalt, it shows a wide range of differences in color both within and among the numerous individual lava flows. Field checking of this and other Hawaiian TIMS images and comparison with geological maps (Holcomb, 1987; J. P. Lockwood, personal communication, 1988). Lockwood and Lipman (1987) reveal systematic relationships between the TIMS colors and the type of basalt and its degree of weathering and hence age. Pahoehoe and aa flows are consistently separable in the images where there is little or no vegetation. Single basalt flows of either type may show some image color differences even immediately after eruption; however, the greatest differences appear to be related to age. Figure 2.28 is an index map for Figure 2.27 showing flow outlines and type of flow and flow ages.

Freshly broken and unweathered basaltic cinders and crushed basalt exposed in quarries consistently appear cyan or light blue-green, and this color shifts to dark brown or orange with increasing age. The oldest flows are heavily vegetated and appear dark green. In the false-color images these are not always easy to distinguish from young, largely unvegetated aa. However, they may be separated by temperature. In Figure 2.27, lightly vegetated aa flows ranged in surface temperature from about 35 to 43°C; heavily vegetated flows were about 29°C.

In this image and in TIMS images of other areas, very young pahoeoe generally appears blue in the false-color images. However, locally, it may be magenta or pink. These variations in image color can be seen even in year-old flows that have not had time to weather. They thus appear to represent a range in the initial state or condition of the surface of the lava.

There is a pronounced and systematic color change with increasing age of pahoehoe. The TIMS color shifts from dominantly blue to purple and magenta (compare, e.g., the 1935 and 1880 flows in Figure 2.27). This range of colors mimics the range for the different initial states but includes intermediate purple. Increasingly, older flows show colors not observed for young flows: red (1843) and orange (0.2 to 1.5 ka), mixed orange and green (1.5 to 4 ka), and ultimately light green (4 to 8 ka). The oldest lavas (>8 ka) are forested and appear dark green. Heavily vegetated aa and pahoehoe are probably indistinguishable from each other. Some areas that are difficult to map in the field, such as the boundary between the 1843 (red in Figure 2.27) and the 1935 (blue) pahoehoe flows, are very distinct in the TIMS image.

The thermal infrared spectral characteristics of basalts are controlled by (1) macroscopic properties, such as the amount and size of vesiculation; (2) differences related to the Si–O bonding, which are related to the silica content and degree of ordering of the material; (3) chemical alteration of the surfaces from acidic gases; and (4) coatings and weathering products added with time. The effect of roughness

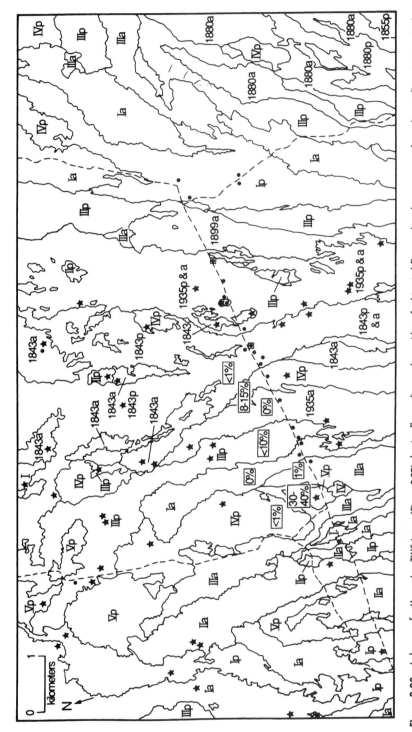

Figure 2.28 Index map for Mauna Loa TIMS image (Figure 2.27) showing flow outlines and ages (dates for historical flows; radiocarbon age groups for prehistoric flows: I, 0.2 to 0.5 ka; II, 0.5 to 1.5 ka; III, 1.5 to 4 ka; IV, 4 to 8 ka; V, 8 ka); "a" is aa; "p" is pahoehoe. Circles mark location for samples in Figure 2.30. Numbers in boxes give the vegetation cover measured in the field for selected flows. From Kahle et al. (1988)

and vesiculation can be illustrated by looking at the spectral character of aa. Radiation emitted by the aa is often partially trapped by the roughness of the surface, which tends to act like a large number of blackbody cavities. However, the spectral signature, while reduced, has features very much like those of similar age pahoehoe. Figure 2.29 shows lab spectra of aa and pahoehoe of the same age, which have been normalized to the same scale.

Crisp et al. (1990) investigated the relationship between changes in spectral emittance features and mineralogy of Hawaiian basalts. They found that basalt collected while still hot from Puu Oo and measured after cooling, the glassy interiors of older basalts, and glass made from older basalts which were fused in an oven and quenched, all exhibited a broad spectral feature peaked between 10.3 and 10.5 µm. This single feature is indicative of a strong degree of disorder (Simon and McMahon, 1953; Brawer and White, 1975; Dowty, 1987).

As the Hawaiian rocks age and weather, the spectral character of their outer surface changes. The first change evident in the spectra is that the single broad feature splits into two features which Kahle et al. (1988) and Crisp et al. (1990) call "B" at 9.2 to 9.5 µm and "C" at 10.5 to 10.8 µm (Figure 2.30). After just a few days or weeks of exposure to the elements (rain, atmosphere, and acidic volcanic fumes) the B and C features were evident in some but not all of the exposed features. These features were commonly found on the rapidly cooled top surfaces of flows but also appeared in the interiors when rocks were broken after emplacement and cooling, and these interiors were then exposed to the Hawaiian environment for a sufficient time. In samples that were only a week old, the C feature was always stronger than the B feature.

With increasing age the overall trend is for the B feature to strengthen relative to the C feature. A simple explanation for the evolution of these two features is that the structure of the metastable glass is becoming more ordered with time. Immediately after its rapid quenching in air, the glass is strongly disordered. With time this

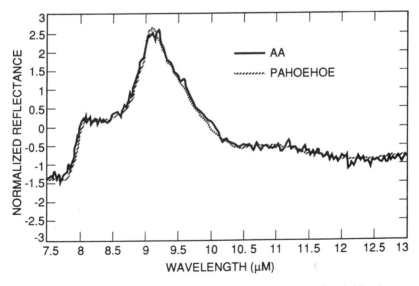

Figure 2.29 Laboratory reflectance spectra of aa and pahoehoe of the same age (0.2 to 0.5 ka) which have been normalized.

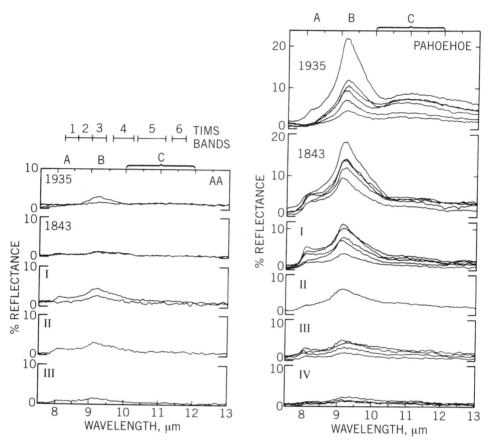

Figure 2.30   Laboratory reflectance spectra of Mauna Loa lavas collected in the area shown in Figures 2.27 and 2.28. Letters A, B, and C refer to spectral features discussed in the text.

unstable configuration breaks down as the silica tetrahedra become organized into silica tetrahedral sheetlike (B feature) and chainlike (C feature) units (e.g., Brawer and White, 1975). With more time the sheetlike units become the preferred mode. As aging continues, the height of the C feature decreases until it is undetectable.

A feature at about 8.1 or 8.2 μm which Kahle et al. (1988) called "A" appears soon after the B feature becomes stronger than the C feature. It is commonly a shoulder on the side of the B feature and is very common on samples greater than 50 years old. Kahle et al. (1988) proposed that the A feature results from the addition of a silica-rich coating. Farr and Adams (1984) describe the development of this coating as the accumulation and leaching of windblown tephra and dust. A scanning electron microscope (SEM) and thin-section investigation (Crisp et al., 1990) confirm an association with the A feature and a silica-rich rind that appears to be the addition of material to the basalt surface rather than a leaching process. As the lava ages further, the spectral contrast decreases slowly until, after a few hundred years, most of the units appear spectrally flat in the infrared.

Another influence on the TIR spectral characteristics of basalts is alteration associated with fumarolic activity. Near active vents and areas undergoing degassing, hydrous, silica-rich areas form, accompanied by the deposition of sulfate salts (Macdonald, 1994; Realmuto et al., 1992), which are thought to be caused by acid leaching of cations by aerosols and resulting enrichment of silica. The spectral characteristics of this material resemble those produced by the secondary silica coatings described by Crisp et al. (1990). Anderson (1950) demonstrated that very brief (45 s) exposure to strong hydrochloric acid at 50°C caused features to appear in the infrared spectrum of fresh barium silicate glass. Realmuto et al. (1992) demonstrated correlations between this finding and spectra of samples collected on the Kupaianaha flow field at Kilauea. It is important to realize that the top 5 to 10 µm of a material completely dominiates the emissivity spectrum.

2.3.2.4 SUMMARY.

The Hawaiian case history illustrates how multispectral thermal infrared data can be used to map Hawaiian lavas of different ages which are often indistinguishable to the human eye. Again the techniques utilized with the TIMS data can be applied to similar data from other multispectral thermal infrared imagers.

2.4 OVERALL SUMMARY AND CONCLUSIONS

In the preceding sections we have described some of the instrumentation and techniques available for geologic remote sensing in the thermal infrared region. This includes an operational field and airborne instrument and a planned spaceborne instrument. The techniques described to process data from the operational instruments can be applied equally to similar instruments and provide a good grounding in the methodology to adopt when processing multispectral thermal infrared data. One of the operational instruments, the micro FTIR, has become available only recently, and several others should come online in the next few years. The sheer number of new instruments that will become available in the next few years is amazing and reflects the gradual recognition of the value of thermal infrared data for a wide variety of land studies, including geology. Of all these new developments, perhaps the most exciting are the high spectral-and spatial-resolution thermal infrared spaceborne imagers such as ASTER, since these will provide data for the entire globe that will be accessible to all.

References

Abrams, M. J., E. A., Abbott, and A. B. Kahle, 1991. Combined use of visible, reflected infrared and thermal infrared images for mapping Hawaiian lava flows, *J. Geophys. Res.*, 96, 475–484.

Akasaka, A., M. Ono, Y. Sakurai, and B. Hayashida, 1991. Short wave infrared (SWIR) subsystem design status of ASTER, *Proc. SPIE*, 1490.

Anderson, S., 1950. Investigation of structure of glasses by their infrared reflection spectra, *J. Am. Ceram. Soc.*, 33, 45–51.

Barker, J. M., and J. L. Wilson, 1975. Borate deposits in the Death Valley region, in *Guidebook: Las Vegas to Death Valley and Return*, NBMG Rep. 26, Nevada Bururreau of Mines and Geology, Reno, New., pp. 23–32.

Barton, I. J., 1985. Transmission model and ground truth investigation of satellite derived sea surface temperatures, *J. Clim. Appl. Meterol.*, 24, 508–516.

Becker, F., 1987. The impact of spectral emissivity on the measurement of land surface temperatures from a satellite, *Int. J. Remote Sensing*, 8(10), 1509–1522.

Becker, F., and Z. L. Li, 1990. Temperature-independent spectral indices in thermal bands, *Remote Sensing Environ.*, 32, 17–33.

Becker, F., P. Ramanantsizehena, and M. P. Stoll, 1985. Angular variation of the bidirectional reflectance of bare soils in the thermal infrared band, *Appl. Opt.*, 24(3), 365–375.

Berk, A., L. S. Bernstein, and D. C. Robertson, 1989. *MODTRAN: A Moderate Resolution Model for LOWTRAN 7*, Tech. Rep. GL-TR-89-0122, Geophysics Laboratory, Bedford, Mass.

Bianchi, R., R. M. Cavalli, L. Fiumi, C. M. Marino, S. Pignatti, and G. Pizzaferri, 1996. The 1994–1995 CNR LARA Project airborne hyperspectral campaigns, in *Proceedings of the 11th Thematic Conference on Applied Geologic Remote Sensing*, Las Vegas, Nev., Feb. 27–29.

Brawer, S. A., and W. B. White, 1975. Raman spectroscopic investigation of the structure of silicate glasses: I. The binary alkali silicates, *J. Chem. Phys.*, 63, 2421–2432.

Christensen, P. R., and S. T. Harrison, 1993. Thermal-infrared emission spectroscopy of natural surfaces: application to desert varnish coatings on rocks, *J. Geophys. Res.*, 98, 19819–19834.

Crisp, J. A., A. B. Kahle, and E. A. Abbott, 1990. Thermal infrared spectral character of Hawaiian basaltic glasses, *J. Geophys. Res.*, 95, 21657–21669.

Crowley, J. K., and S. J. Hook, 1996. Mapping playa evaporite minerals and associated sediments in Death Valley, California, with multispectral thermal infrared images, *J. Geophys. Res.*, 101, 643–660.

Cudahy, T. J., P. M. Connor, P. Hausknecht, S. J. Hook, J. F. Huntington, A. B. Kahle, R. N. Phillips, and L. B. Whitbourn, 1994. Airborne CO_2 laser spectrometer and TIMS TIR data for mineral mapping in Australia in *Proceedings of the 7th Australasian Remote Sensing Conference*, Melbourne, Victoria, Australia, Mar. 1–4, pp. 918–924.

Dowty, E., 1987. Vibrational interactions of tetrahedra in silicate glasses and crystals: I. Calculations on ideal silicate–aluminate–germanate structural units, *Phys. Chem. Miner.*, 14, 80–93.

Farr, T. G., and J. B. Adams, 1984. Rock coatings in Hawaii, *Geol. Soc. Am. Bull.*, 95, 1077–1083.

Fujisada, H., and M. Ono, 1991. Overview of ASTER design concept, *Proc. SPIE*, 1490.

Gillespie, A. R., 1986. Lithologic mapping of silicate rocks using TIMS in *Proceedings of the TIMS Data User's Workshop*, JPL Publ. 86–38, Jet Propulsion Laboratory, California Institute of Technology, Pasadena, Calif., pp. 29–44.

Gillespie, A. R., A. B. Kahle, and F. D. Palluconi, 1984. Mapping alluvial fans in Death Valley, California using multichannel thermal infrared images, *Geophys. Res. Lett.*, 11, 1153–1156.

Gillespie, A. R., A. B. Kahle, and R. E. Walker, 1986. Color enhancement of highly correlated images: I, Decorrelation and HSI contrast stretches, *Remote Sensing Environ.*, 20, 209–235.

Gillespie, A., Rokugawa, S., Matsunaga, T., Cothern, J. S., Hook, S. and A. B. Kahle, 1998. Temperature and Emissivity Separation Algorithm for Advanced spaceborne thermal emission and reflection radiometer (ASTER) Images. Trans Geosci Remote Sensing., 36 pp. 1113–1126.

Holcomb, R. T., 1987. Eruptive history and long-term behavior of Kilauea volcano, in *Volcanism in Hawaii*, Vol. 1, R. W. Decker, T. L. Wright, and P. H. Stauffer, eds., USGS Prof. Pap. 1350, U.S. Geological Survey, Washington, D.C., pp. 261–350.

Hook, S. J., and A. B. Kahle, 1996. The micro Fourier transform interferometer (μFTIR): a new field spectrometer for acquisition of infrared data of natural surfaces, *Remote Sensing Environ.*, 56, 172–181.

Hook, S. J., and K. Okada, 1996. Inflight wavelength correction of thermal infrared multispectral scanner (TIMS) data acquired from the ER-2, *IEEE Trans. Geosci. Remote Sensing*, 34, 179–188.

Hook, S. J., A. R. Gabell, A. A. Green, and P. S. Kealy, 1992. A comparison of techniques for extracting emissivity information from thermal infrared data for geologic studies, *Remote Sensing Environ.*, 42, 123–135.

Hook, S. J., K. E. Karlstrom, C. F. Miller, and K. J. W. McCaffrey, 1994. Mapping the Piute Mountains, California, with thermal infrared multispectral scanner (TIMS) images, *J. Geophys. Res.*, 99, 15605–15622.

Hooke, R. L., 1972. Geomorphic evidence for Late-Wisconsin and Holocene tectonic deformation, Death Valley, California, *Geol. Soc. Am. Bull.*,83, 2073–2098.

Hoover, G., and A. B. Kahle, 1987. A thermal emission spectrometer for field use, *Photogramm. Eng. Remote Sensing*, 53, 627–632.

Hunt, C. B., and D. R. Mabey, 1966. *Stratigraphy and Structure, Death Valley, Cal-*

ifornia, USGS Prof. Pap. 494-A, U.S. Geological Survey, Washington, D.C., 163 pp.

Hunt, G. R., 1980. Electromagnetic radiation: the communication link in remote sensing, in *Remote Sensing in Geology*, B. S. Siegal and A. R. Gillespie, eds., Wiley, New York, pp. 5–45.

Hunt, G. R., and J. W. Salisbury, 1974. *Mid-infrared Spectral Behaviour of Igneous Rocks*, Environ. Res. Pap. 496-AFCRL-TR-74-0625, Air Force Cambridge Research Laboratory, Hanson Air Force Base, Mass., 142 pp.

Hunt, G. R., and J. W. Salisbury, 1975. *Mid-infrared Spectral Behaviour of Sedimentary Rocks*, Environ. Res. Pap. 520-AFCRL-TR-75-0256, Air Force Cambridge Research Laboratory, Hanson Air Force Base, Mass., 49 pp.

Hunt, G. R., and J. W. Salisbury, 1976. *Mid-infrared Spectral Behaviour of Metamorphic Rocks*, Environ. Res. Pap. 543-AFCRL-TR-76-0003, Air Force Cambridge Research Laboratory, Hanson Air Force Base, Mass., 67 pp.

Hunt, C. B., T. W. Robinson, W. A. Bowles, and A. L. Washburn, 1966. *Hydrologic Basin, Death Valley, California*, USGS Prof. Pap. 494-B, U.S. Geological Survey, Washington, D.C., 138 pp.

Kahle, A. B., and A. F. H. Goetz, 1983. Mineralogic information from a new airborne thermal infrared multispectral scanner, *Science*, 222, 24–27.

Kahle, A. B., D. P. Madura, and J. M. Soha, 1980. Middle infrared multispectral aircraft scanner data analysis for geological applications, *Appl. Opt.*, 19, 2279–2290.

Kahle, A. B., M. S. Shumate, and D. B. Nash, 1984. Active airborne infrared laser system for identification of surface rock and minerals, *Geophys. Res. Lett.*, 11, 1149–1152.

Kahle, A. B., A. R. Gillespie, E. A. Abbott, M. J. Abrams, R. E. Walker, G. Hoover, and J. P. Lockwood, 1988. Relative dating of Hawaiian lava flows using multispectral thermal infrared images: a new tool for geologic mapping of young volcanic terranes, *J. Geophys. Res.*, 93, 15239–15251.

Kahle, A. B., Pallucani, F. D., Hook, S. J., Realmuto, V. J. and G. Bothwell, 1991. The advanced spaceborne thermal emission and reflectance radiometer (ASTER). International Journal of Imaging Systems and Technology, 3, 144–156.

Kahle, A. B., M. J. Abrams, E. A. Abbott, P. J. Mouginis-Mark, and V. J. Realmuto, 1995. *Remote Sensing of Mauna Loa*, AGU Geophys. Monogr. 92, American Geophysical Union, Washington, D.C., pp. 145–170.

Kannari, Y., F. Mills, and H. Watanabe, 1992. Comparison of preliminary results from the airborne ASTER simulator (AAS) with TIMS data, in *Proceedings of the 3rd Airborne Geoscience Workshop*, JPL Publ. 92–14, Jet Propulsion Laboratory, California Institute of Technology, Pasadena, Calif., pp. 13–15.

Kawada, M., and H. Fujisada, 1991. Mechanical cooler development program for ASTER, *Proc. SPIE*, 1490.

Kealy, P. S., and A. R. Gabell, 1990. Estimation of emissivity and temperature using alpha coefficients in *Proceedings of the 2nd TIMS Workshop*, JPL Publ. 90–55, Jet Propulsion Laboratory, California Institute of Technology, Pasadena Calif.

Kealy, P. S., and S. J. Hook, 1993. Separating temperature and emissivity in thermal infrared multispectral scanner data: implications for recovering land surface temperatures, *IEEE Trans. Geosci. Remote Sensing*, 31, 1155–1164.

Kerr, Y. H., J. P. Lagouarade, and J. Imbernon, 1992. Accurate land surface temperature retrieval from AVHRR data with use of an improved split window algorithm, *Remote Sensing Environ*, 41, (2–3), 197–209.

King, M. D., W. P. Menzel, P. S. Grant, J. S. Myers, G. T. Arnold, S. E. Platnick, L. E. Gumley, S. Tsay, C. C. Moeller, K. S. Brown, F. G. Osterwisch, and M. C. Peck, 1996 Multiwavelength scanning spectrometer for airborne remote sensing of cloud, aerosol, water vapor and surface properties, *J. Atmos. Ocean. Technol.*, 13, 777–794.

Kitamura, S., H. Ohmae, and Y. Aoki, 1991. Thermal infrared (TIR) subsystem design status of ASTER, *Proc. SPIE*, 1490.

Korb, A. R., P. Dybwad, W. Wadsworth, and J. W. Salisbury, 1996. Portable FTIR spectroradiometer for field measurements of radiance and emissivity, *Appl. Opt.*, 35, 1679–1692.

Kruse, F. A., A. B. Lefkoff, J. W. Boardman, K. B. Heidebrecht, A. T. Spairo, P. J. Barloon, and A. F. H. Goetz, 1993. The spectral image processing system (SIPS): interactive visualization and analysis of imaging spectrometer data, *Remote Sensing Environ.*, 44, 145–163.

Labed, J., and M. Stoll, 1991. Angular variation of land surface spectral emissivity in the thermal infrared: laboratory investigations on bare soils, *Int. J. Remote Sensing*, 12, 2299–2310.

Lahren, M. M., R. A. Schweickert, and J. V. Taranik, 1988. Analysis of the northern Sierra accreted terrain, California, with airborne thermal infrared multispectral scanner data, *Geology*, 16, 525–528.

Lang, H. R., S. L. Adams, J. E. Conel, B. A. McGuffie, E. D. Paylor, and R. E. Walker, 1987. Multispectral remote sensing as stratigraphic and structural tool, Wind River Basin and Big Horn Basin areas, Wyoming, *Am. Assoc. Pet. Geol. Bull.*, 71, 389–403.

Liese, H. C., 1975. Selected terrestrial minerals and their infrared absorption spectral data (4000–300 cm^{-1}); in *Infrared and Raman Spectroscopy of Lunar and Terrestrial Minerals*, C. Karr, ed., Academic Press, San Diego, Calif., pp. 197–229.

Lipman, P. W., 1980. The southwest rift zone of Mauna Loa: implications for structural evolution of Hawaiian volcanoes, *Am. J. Sci.*, 280A, 752–776.

Lockwood, J. P., and P. W. Lipman, 1987. Holocene eruptive history of Mauna Loa volcano, in *Volcanism in Hawaii*, Vol. 1, R. W. Decker, T. L. Write, and P. H. Stauffer, eds., USGS Prof. Pap. 1350, U.S. Geological Survey, Washington, D.C., pp. 509–535.

Lyon, R. J. P., 1965. Analysis of rocks by spectral infrared emission (8 to 25 microns), *Econ. Geol.*, 60, 715–736.

Lyon, R. J. P., and A. A. Green, 1975. Reflectance and emittance of terrain in the mid-infrared (6–25 μm) region, in *Infrared and Raman Spectroscopy of Lunar and Terrestrial Minerals*, C. Karr, ed., Academic Press, San Diego, Calif., pp. 165–195.

Macdonald, G. A., 1971. *Geologic Map of the Mauna Loa Quadrangle*, USGS Map GQ-897, U.S. Geological Survey, Washington, D.C.

Macdonald, G. A., 1994. Solfataric alteration of rocks at Kilauea volcano, *Am. J. Sci.*, 242 pp.

Masuda, K., T. Takashima, and Y. Takayama, 1988. Emissivity of pure and sea waters for the model sea surface in the infrared window region, *Remote Sens. Environ.*, 24, 313–329.

Matsunaga, T., 1994. A temperature–emissivity separation method using an empirical relationship between the mean, the maximum and the minimum of the thermal infrared emissivity spectrum, *J. Remote Sensing Soc. Jpn.*, 14 230–241 (Japanese with English abstract).

McAllister, J. F., 1970. *Geology of the Furnace Creek Borate Area, Death Valley, Inyo County, California*, Map Sheet 14, California Division of Mines and Geology, San Francisco, 9 pp.

McAllister, J. F., 1976. *Columnar Section of the Main Part of the Furnace Creek Formation of Pliocene (Clarendonian and Hemphillian) Age Across Twenty Mule Team Canyon, Furnace Creek Borate Area, Death Valley, California*, USGS Open File Rep. 76–261, U.S. Geological Survey, Washington, D.C., 1 p.

McMillan, L. M., and D. S. Crosby, 1984. Theory and validation of the multiple window sea surface temperature technique, *J. Geophys. Res.*, 89 (C3), 3655–3661.

Ono, A., and F. Sakuma, 1991, ASTER calibration concept, *Proc. SPIE*, 1490.

Ottle, C., and M. P. Stoll, 1993. Effect of atmospheric absorption and surface emissivity on the determination of land temperature from infrared satellite data, *Int. J. Remote Sensing*, 14 (10), 2025–2037.

Palluconi, F. D., and G. R. Meeks, 1985. *Thermal Infrared Multispectral Scanner (TIMS): An Investigator's Guide to TIMS Data*, JPL Publ. 85–32, Jet Propulsion Laboratory, California Institute of Technology, Pasadena, Calif.

Peterson, D. W., and R. I. Tilling, 1980. Transition of basaltic lava from pahoehoe to aa, Kilauea volcano, Hawaii: field observations and key factors, *J. Volcanol. Geotherm. Res.*, 7, 271–293.

Prabhakara, C., G. Dalu, and V. G. Kunde, 1974. Estimation of sea surface temperature from remote sensing in the 11-to 13-μm window region, *J. Geophys. Res.*, 79 (33), 5039–5044.

Prata, A. J., 1994. Land surface temperature derived from the advanced very high resolution radiometer and the along-track scanning radiometer: 2. Experimental results and validation of AVHRR algorithms, *J. Geophys. Res.*, 79, 5039–5044.

Price, J. C., 1984. Land surface temperature measurements from the split window channels of the NOAA 7 advanced very high resolution radiometer, *J. Geophys. Res.*, 89, 7231–7237.

Realmuto, V. J., 1990. Separating the effects of temperature and emissivity: emissivity spectrum normalization, in *Proceedings of the 2nd TIMS Workshop*, JPL Publ. 90–55, Jet Propulsion Laboratory, California Institute of Technology, Pasadena, Calif.

Realmuto, V. J., K. Hon, A. B. Kahle, E. A. Abbott, and D. C. Pieri, 1992. Multispectral thermal infrared mapping of the 1 October 1988 Kupaianaha flow field, Kilauea volcano, Hawaii. *Bull. Volcanol.*, 55, 33–44.

Realmuto, V. J., P. Hajek, M. P. Sinha, and T. G. Chrien, 1995. The 1994 laboratory calibration of TIMS, in *Summaries of the 5th Annual JPL Airborne Earth Science Workshop*. JPL Publ. 95-1, Vol. 2, Jet Propulsion Laboratory, California Institute of Technology, Pasadena, Calif., pp. 25–28.

Rivard, B., P. Thomas, D. Pollex, A. Hollinger, J. Miller, and R. Dick, 1994. A field portable thermal infrared spectrometer (THIRSPEC), *IEEE Trans. Geosci. Remote Sensing*, 32, 307–314.

Sabine, C., V. J. Realmuto, and J. V. Taranik, 1994. Semiquantitative measurement of granitoid composition from thermal infrared multispectral scanner (TIMS) data, Desolation Wilderness, northern Sierra Nevada, California, *J. Geophys. Res.*, 99, 4261–4271.

Sabins, F., 1978. *Remote Sensing Principles and Interpretation*, W. H. Freeman, New York.

Salisbury, J. W., and D. D'Aria, 1992. Emissivity of terrestrial materials in the 8–14 μm atmospheric window, *Remote Sensing Environ.*, 42, 83–106.

Salisbury, J. W., L. S. Walter, and D. D'Aria, 1988. *Thermal Infrared (2.5 to 13.5 μm) Spectra of Igneous Rocks*, USGS Open File Rep. 88–686, U.S. Geological Survey, Washington, D.C.

Salisbury, J. W., L. S. Walter, N. Vergo, and D. M. D'Aria, 1991. *Infrared (2.1–25 μm) Spectra of Minerals*, Johns Hopkins University Press, Baltimore, 267 pp.

Salisbury, J. W., A. Wald, and D. M. D'Aria, 1994a. Thermal infrared remote sensing and Kirchhoff's law: I. Laboratory measurements, *J. Geophys. Res.*, 99, 11897–11911.

Salisbury, J. W., D. M. D'Aria, and A. Wald, 1994b. Measurements of thermal infrared spectral reflectance of frost, snow and ice, *J. Geophys. Res.*, 99, 24235–24240.

Shumate, M. S., R. T. Menzies, W. B. Grant, and D. S. McDougal, 1981. Laser absorption spectrometer: remote measurement of tropospheric ozone, *Appl. Opt.*, 20, 545–553.

Shumate, M. S., S. Lundquist, U. Persson, and S. T. Eng, 1982. Differential reflectance of natural and man-made materials at CO_2 laser wavelengths, *Appl. Opt.*, 21, 2386–2389.

Siegal, R., and J. R. Howell, 1982. *Thermal Radiation Heat Transfer*, 2nd ed. Hemisphere, New York.

Simon, I., and H. O. McMahon, 1953. Study of some binary silicate glasses by means of reflection in infrared, *J. Am. Ceram. Soc.*, 36, 160–164.

Smith, G. I., 1985. Borate deposits in the United States: dissimilar in form, similar in geologic setting, in *Borates: Economic Geology and Production*, J. M. Barker and S. J. Lefond, eds., American Institute of Mining, Metallurgical and Petroleum Engineers, New York, pp. 37–51.

Smith, W. L., H. M. Woolf, H. B. Howell, H. L. Huang, and H. E. Revercomb, 1988. High resolution interferometer sounder: the retrieval of atmospheric temperature and water vapor profiles, in *Proceedings of the 3rd Conference on Satellite Meteorology and Oceanography*, American Meterological Society, Boston.

Snyder, W. C., Z. Wan, Y. Zhang, and Yue-Zhong Feng, 1997. Thermal infrared (3–14 μm) bidirectional reflectance measurements of sands and soils, *Remote Sensing Environ.*, 60, 101–109.

Soha, J. M., and A. A. Schwartz, 1978. Multispectral histogram normalization contrast enhancement, in *Proceedings of the 5th Canadian Symposium on Remote Sensing*, Victoria, British Columbia, Canada, pp. 86–93.

Suits, G. H., 1983. *The Nature of Electromagnetic Radiation: Manual of Remote Sensing* 2nd ed., American Society of Photogrammetry, Falls Church, Va.

Takahashi, F., M. Hiramatsu, F. Watanabe, Y. Narimatsu, and R. Nagura, 1991. Visible and near infrared (VNIR) subsystem and common signal processor (CSP) design status of ASTER, *Proc. SPIE*, 1490.

Tonooka, H., 1996. Simultaneous determination of atmospheric correction parameters, LST and spectral emissivity from TIR multispectral data over land, presented at the International Land Surface Workshop, University of California at Santa Barbara, Sept. 17–19.

Tou, J. T., and R. C. Gonalez, 1974. Pattern recognition principles, *Appl. Math. Comput.*, 7, 94–104.

van der Marel, H. W., and H. Beutelspacher, 1976. *Atlas of Infrared Spectroscopy of Clay Minerals and Their Admixtures*, Elsevier, New York, 376 pp.

Vidal, A., 1991. Atmospheric and emissivity correction of land surface temperatures measured from satellite using ground measurements or satellite data, *Int. J. Remote Sensing*, 12(12), 2449–2460.

Vincent, R. K., 1972. Rock-type discrimination from ratioed infrared scanner images of Pisgah Crater, California, *Science*, 175, 986–988.

Vincent, R. K., F. Thomson, and K. Watson, 1972. Recognition of exposed quartz sand and sandstone by two-channel infrared imagery, *J. Geophys. Res.*, 77, 2473–2477.

Wan, Z., and J. Dozier, 1989. Land-surface temperature measurement from space: physical principles and inverse modeling, *Geosci. Remote Sensing*, 27, 268–278.

Wan, Z. and L. Li, 1997. A physics-based algorithm for retrieving land surface emissivity and temperature from EOS/MODIS data. IEEE Trans Geosci Remote Sensing, 35, 980–996.

Watson, K., 1985. Remote sensing: a geophysical perspective, *Geophysics*, 50, 2595–2610.

Watson, K., 1987. Spectral ratio method for measuring emissivity, *Remote Sensing Environ.*, 42, 113–116.

Watson, K., 1992. Two-temperature method for measuring emissivity, *Remote Sensing Environ.*, 42, 117–121.

Watson, K., 1996. A scan-angle correction for thermal infrared multispectral data using side lapping images, *Geophys. Res. Lett.*, 23, 2421–2424.

Watson, K., F. A., Kruse, and S. Hummer-Miller, 1990. Thermal infrared exploration in the Carlin Trend, northern Nevada, *Geophysics*, 55, 70–79.

Watson, K., L. C. Rowan, T. L. Bowers, C. Anton-Pacheco, P. Gumiel, and S. H. Miller, 1996. Lithologic analysis from multispectral thermal infrared data of the alkalic rock complex at Iron Hill, Colorado, *Geophysics*, 61, 706–721.

Wen-Yao, L., R. T. Field, R. G. Gantt, and V. Klemas, 1987. Measurement of the surface emissivity of turbid water, *Remote Sensing Environ.*, 21, 97–109.

Whitbourn, L. B., R. N. Phillips, G. James, M. T. O'Brien, and M. D. Waterworth, 1990. An airborne multiline CO_2 laser system for remote sensing of minerals, *J. Mod. Opti.*, 37, 1865–1872.

Soil Reflectance

E. Ben-Dor

Tel Aviv University
Tel Aviv, Israel

J. R. Irons

Goddard Space Flight Center
Greenbelt, Maryland

G. F. Epema

Wageningen Agricultural University
Wageningen, The Netherlands

3.1 INTRODUCTION

Soil reflectance data can be acquired in the laboratory, in the field, and from the air. Whereas in the laboratory soil reflectance measurements are done under controlled conditions, in the field, reflectance measurements are encumbered by such problems as variations in viewing angle, illumination changes, and soil roughness. Soil reflectance data acquired from the air involve additional difficulties, such as those resulting from relatively low signal/noise ratio and atmospheric attenuations. The laboratory-based measurements enable an understanding of the chemical and physical principles of soil reflectance and are used widely for that purpose. As the sensitivity of portable field spectrometers develops, field soil spectroscopy will become a basic tool for rapid point-by-point monitoring of the soil environment. Recently, considerable effort has been put into the commercial development, opera-

Remote Sensing for the Earth Sciences: Manual of Remote Sensing, 3 ed., Vol. 3, edited by Andrew N. Rencz.
ISBN: 0471-29405-5 © 1999 John Wiley & Sons, Inc.

tion, and use of air-and spaceborne image spectrometers (see Chapter 11). These advances in technology can provide a near-laboratory-quality spectrum of every pixel in the image and very soon will permit remote sensing of soils with high standards. Understanding soil spectra principles and their limitations is crucial to the application of remote soil sensing procedures. Information about soils from reflectance spectra in the visible/near-infrared (VNIR 0.4 to 1.1 μm) and shortwave-infrared (SWIR 1.1 to 2.5 μm) spectral regions represent almost all the data that passive solar sensors can provide. Although the thermal infrared regions (3 to 5 μm, 8 to 12 μm) also contain diagnostic information about soil materials, the VNIR–SWIR will be the spectral region covered in most detail in this chapter because it deals with soil reflectance and not emittance.

The purpose of this chapter is to provide an overview of the chemicophysical principles of the soil reflectance spectrum in the VNIR–SWIR regions and to cover the basic processes involved in the interaction between soils and electromagnetic radiation. We aim to shed more light on soil reflectance analysis for practical uses. In addition, background information on spectroscopy and radiation interactions with a surface is provided (see also Chapter 1).

3.2   SOIL

3.2.1 Background

Soil is derived from the Latin word *solum*, which means floor. Many definitions for soil exist, but the one that is most suitable for our discussion is that of Thompson (1957): "the upper layer of the earth which may be dug, plowed, specifically, the loose surface material of the earth in which plants grow." Soil is a complex material that is extremely variable in physical and chemical composition. Soils are formed by weathering of exposed rocks and minerals of the Earth's crust and by the decomposition of organic matter deposited by flora and fauna. Soil formation or genesis is strongly dependent on the environmental conditions of both the atmosphere and the lithosphere. Five environmental factors exert the greatest influence on soil genesis: climate, vegetation, living organisms (fauna and flora), topography, and parent materials. The great variability in soils results from the interactions of these factors (Buol et al., 1973). The typical consequence of the soil-forming processes is the development of vertical soil profiles with distinct layers or horizons. Soils are identified, classified, and mapped on the basis of characteristic profiles with distinguishing properties in each horizon. Remote sensing in the reflective portion of the spectrum limits observation to energy scattered by the thin upper surface of soil profiles. This upper surface can be subjected to frequent alteration by tillage, precipitation, erosion, and other surface processes. Further, the soil surface can be masked by vegetation or snow. Still, observations of the soil surface, soil surface variation, and partially obscured soil surfaces can be used to infer soil properties and influences. Knowledge of soil reflectance is also required to understand the effects of soil on observations of vegetation canopies. Consequently, a discussion of soil reflectance follows.

3.2.2 Soil Compositions

Any given soil mixture is made up of all three phases of matter: solid, liquid, and gas. A typical soil may consist of about 50% pore space, with spatially and temporally variant proportions of gas and liquid.

3.2.2.1 SOLID PHASE.

This phase contains organic and inorganic components in a complicated and generic mixture of primary and secondary minerals, organic components, and salts. The solid phase consists of three main particle size fractions—sand (2 to 0.2 mm), silt (0.2 to 0.002 mm), and clay (<0.002 mm)—which together govern two major soil properties: texture and structure. Soil texture is a function of the distribution of these three main components and is generally described in terms of quantities of gravel, sand, silt, and clay. Soil structure (a function of adhesion forces between generally fine particles) is a property that describes the aggregation characteristics of soil. These two properties play a major role in soil behavior and influence some major soil characteristics, such as drainage, fertility, moisture, and erosion. The inorganic portion of the solid phase consists of soil minerals, which are generally categorized as either primary or secondary minerals. Primary minerals are components derived directly from weathering of parent materials that were formed under much higher temperatures and pressures than are found at the Earth's surface. Secondary minerals are formed by geochemical weathering processes of the primary minerals. An extensive description of minerals in the soils environment is given by Dixon and Weed (1989), and readers who wish to expand their knowledge in this area are referred to that text. In general, the dominant primary minerals are quartz, feldspar, orthoclase, and plagioclase. Some layer-silicate minerals, such as mica and chlorite, and ferromagnesium silicates, such as amphibole, peroxide, and olivine, also exist in soils.

The secondary minerals in soils—most of them are often termed clay minerals—are aluminosilicates, such as montmorillonite, illite, vermiculite, sepiolite, kaolinite, and gibbsite. The type of clay minerals is strongly dependent on the weathering stage of the soil and can be a significant indicator of the environmental conditions under which the soil was formed. Other secondary minerals in soils are aluminum and iron oxides and hydroxides, carbonates (calcite and dolomite), sulfates (gypsum), and phosphate (apatite). Most of these minerals are relatively insoluble in water and maintain equilibrium with a water solution. Soluble salts such as halite may also be found in soil, but they are mobile in water. Clay minerals are most likely found in the fine-sized particles of the soil ($<2\mu$m; clay fraction) and are characterized by relatively high specific surface areas (50 to 800 $m^2 \, g^{-1}$). The primary minerals and other nonclay minerals are usually found in both the sand and silt portions and consist of relatively small specific surface areas ($<1 \, m^2 \, g^{-1}$).

In addition to the inorganic components in the solid phase, organic components also exist. Although the organic matter content in mineral soils does not exceed 15% (usually less), it plays a major role in soil chemical and physical behavior (Schnitzer and Khan, 1978). Organic matter is composed of decaying tissues from vegetation and micro and macro faunal bodies. Organic matter in soil can be found in various stages of degradation, from coarse dead to complex fine components called humus

(Stevenson, 1982). The surface horizons of a soil profile typically contain more organic matter than do the subsurface horizons. This fact makes consideration of organic matter an essential component for remote sensing applications, where only the surface layer is observed.

3.2.2.2 LIQUID AND GAS PHASES.

These phases in soils are complementary to the solid phase and occupy about 50% of the soil's total volume. The liquid consists of water and dissolved ions in various amounts. The water molecules either fill the entire pore volume in the soil ("saturated"), occupy only a portion of the pore volume ("wet"), or are adsorbed on the surface areas ("dry"). The composition of the soil's gaseous phase is normally very similar to the composition of the atmosphere, with the exception that the concentration of oxygen and carbon dioxide vary depending on the biochemical activity at the root zone.

3.3 DESCRIBING SOIL PROPERTIES

3.3.1 Background

Pedologists have long used soil color to describe soils, to help classify soils, and to infer soil characteristics. As Baumgardner et al. (1985) stated: "Ever since soil science evolved into an important discipline for study and research, color has been one of the most useful soil variables in characterizing and describing a particular soil." Certain qualitative relationships between color and soil properties are well recognized by pedologists on the basis of their collective observations and on a conceptual understanding of the interaction of visible light with soil material. Even though today's instruments can measure soil reflectance as a function of wavelength, soil color continues to play a major role in modern soil classification and description. In this section we discuss soil color and the reflectance properties of soil in the visible through shortwave-infrared regions of the electromagnetic spectrum.

3.3.2 Soil Color

Soil color results from the brain's perception of the eyes' response to light reflected by soil (Nassau, 1980). The eye responds to electromagnetic energy within the 0.4- to 0.7-µm portion of the wavelength spectrum (visible or light region) in a "sensor-like" sensitivity distribution (Figure 3.1). Orna (1978) stated that color provides the perfect link between an easily observed and described property and an underlying theory. Thus it is not surprising that for years, soil color has been used for qualitative assessment of many soil components, such as organic mater content, iron oxides, and carbonates in both the remote sensing and soil science fields (Escadafal, 1993).

3.3.2.1 SOIL COLOR DETERMINANTS.

Soil color is related to the presence of pigments or chromophores that absorb radiation in different wavelengths and intensities. Organic matter, water, iron oxides,

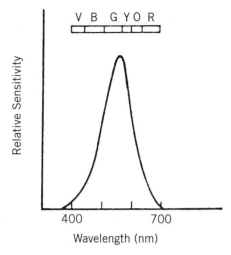

Figure 3.1 Sensitivity of the human eye to reflected photons in the visible spectral region (VIS). V, violet; B, blue; G, green; Y, yellow; O, orange; R, red.

and chemical composition of transition metals in clay minerals are the major components affecting soil color (Leger et al., 1979; Kondratyev and Fedchenko, 1983). In general, the yellow and red colors of soils result from the presence of goethite and hematite, respectively (Karmanova, 1981). Torrent et al. (1983) showed the quantitative relationship between soil color and hematite content. Other iron oxides, such as lepidocrocite (which varies in color between orange and yellow) and ferrihydrate (which is yellow to brown), can also be identified using the Munsell color notation (Schwertmann, 1988). Soil darkness is governed by the presence of humic materials (if shade is not a factor), associated with various clay minerals and humic substances. Black ped facies in soils are often related to a thin coating layer of manganese oxides. The green-blue related colors often encountered in gleyed soils come from "green rust" in Fe^{2+} ions.

Color in soils is also related to mineral compositions in the clay mineral lattice. For instance, green illite generally contains more Fe^{3+} than Fe^{2+} ions, both in the octahedral position. Purple colors in illite are related to the structure of manganese complexes. Most smectites, which contain Fe, have an off-white to green color, but numerous other colors have been observed, including yellow, yellow-green, apple green, blue-green, blue-gray, olivine-green, and brown (Taylor, 1982; Stucki, 1988). In kaolinite, Jepson (1988) described color changes as a result of impurities found in bulk and lattice material (mostly iron and "titanferous" impurities). In biotites, Hall (1941) noted that the relationship between iron, manganese, and titanium is responsible for the color sequence, ranging from red to blue-green.

3.3.3 Color Models

3.3.3.1 MUNSELL COLOR CHART.

The Munsell soil color charts (Munsell Color, 1975) comprise the standard most commonly used to describe soil color visually. The Munsell color system describes colors both in terms of descriptive names and in terms of hue, value, and chroma.

As reviewed by Taylor, *hue* refers to the dominant spectral color in soils (wavelength) and arises from the combination of pigments in minerals (e.g., iron oxides) and organic matter. *Value* refers to the apparent lightness compared to absolute white (pseudoalbedo). This parameter can be influenced by the degree of moisture and particle size distribution as well as by viewing geometry. *Chroma* is a measure of the purity of the hue and is influenced by the spectral distribution of reflected light.

The Munsell soil color chart contains 229 standard color chips arranged on pages of gray background. The color chips are designated by descriptive names and by the Munsell system of color notation. Soil color is described by matching a soil sample visually to the chip peceived to be closest in color. Pendleton and Nickerson (1951) reviewed the appropriate technique for soil color measurement with the Munsell soil color charts and also reviewed the history of soil color description.

The Munsell system designates a color with separate notations for hue, value, and chroma. Although hue, value, and chroma are related to physical quantities, the three parameters are defined on a conceptual basis The numerical scales for Munsell hue, value, and chroma attempt to divide colors into equal steps on the basis of visual perception. The hue scale is divided into 10 families. The families are denoted by five *principal* color names (purple, blue, green, yellow, and red, abbreviated as P, B, G, Y, and R, respectively) and by five *intermediate* color names (purple-blue, blue-green, green-yellow, and yellow-red, abbreviated as PB, BG, GY, YR, and RP, respectively). The hue scale then ranges from 0 to 10 within each family, with 0 and 10 occurring at boundaries between perceptually adjacent hue families (Williamson and Cummins, 1983). The location of a hue within the hue scale is designated by a number followed by a family abbreviation (e.g., 5YR). The value scale ranges from 0 to 10, where black has a value of 0 and white has a value of 10. The chroma scale ranges from 0 out to approximately 20, depending on the hue. A chroma of 0 corresponds to a neutral gray color with no perceptible hue. The hue of a neutral gray color is designated with an N. A color is specified in the Munsell notation by the hue designation followed by the numeric value and chroma designations separated by a virgule (e.g., 5YR 5/3 is the notation for a reddish-brown soil) (Soil Survey Staff, 1975). The Munsell soil color charts contain seven charts, where each chart contains chips within the same hue family plus a chart of low-chroma chips for designating the color of gray.

3.3.3.2 CIE COLOR SYSTEM.

Spectrometers and radiometers permit more physically based characterizations of light beams scattered off soil material in terms of absolute spectral flux or in the relative terms of spectral reflectance. Relating these characterizations to the familiar Munsell notation is not trivial since neither the definitions nor the numeric scales of hue, value, and chroma are physically based. An approach has been developed, however, using the system for color specification developed by the Commission Internationale de l'Eclairage (CIE).

The CIE is an international body of color scientists who recommend methods to establish standards for color measurement (Melville and Atkinson, 1985). In 1931 the CIE defined a three-dimensional coordinate system for the objective measurement and specification of color (CIE, 1931; Judd, 1933). This system has become the most widely recognized method for color specification and instrumental color measurement (Billmeyer and Saltzman, 1981; Williamson and Cummins, 1983; Melville and Atkinson, 1985). The three-dimensional nature of color perception and the development of the CIE system are explained in color science texts such as Billmeyer and Saltzman (1981), Williamson and Cummins (1983), and Wyszecki and Stiles (1982). The system is also discussed briefly in Chapter 6.

The CIE system specifies the color of a reflective (i.e., light scattering) surface as the product of three major factors that affect visual color perception: the spectral distribution of radiant flux in an illuminating beam, the spectral reflectance of the surface, and the spectral sensitivity of the human visual perception (eye and brain). The coordinates of the CIE system are derived in the following manner (ASTM, 1985):

$$X = k \sum_{\lambda=380}^{780} R(\lambda)S(\lambda)\bar{x}(\lambda) \, \Delta\lambda$$

$$Y = k \sum_{\lambda=380}^{780} R(\lambda)S(\lambda)\bar{y}(\lambda) \, \Delta\lambda$$

$$Z = k \sum_{\lambda=380}^{780} R(\lambda)S(\lambda)\bar{z}(\lambda) \, \Delta\lambda \tag{3.1}$$

$$k = \frac{100}{\displaystyle\sum_{\lambda=380}^{780} S(\lambda)\bar{y}(\lambda) \, \Delta\lambda}$$

where k is a normalization factor; X, Y, and Z are the CIE coordinates, referred to as *tristimulus values*; λ refers to wavelength; $R(\lambda)$ represents spectral bidirectional reflectance factors; $S(\lambda)$ represents the relative spectral distribution of flux in the illuminating beam; and $x(\lambda)$, $y(\lambda)$, and $z(\lambda)$ are the color-matching functions of a standard observer defined by the CIE.

The color-matching functions are abstract representations of the spectral sensitivity of an average observer with normal vision. The color-matching functions were defined on the basis of color-matching experiments conducted by perceptual psychologists in the 1920s (Williamson and Cummins, 1983; Wyszecki and Stiles, 1982). Values of the color-matching functions are tabulated by the American Society for Testing and Materials (ASTM, 1985) and by Wyszecki and Stiles (1982). The CIE system eliminated the subjectivity in color specification by establishing a standard observer through the definition of the color-matching functions.

The CIE tristimulus values were related to the Munsell color system by measuring reflectance spectra for all samples in the 1929 *Munsell Book of Color* (Glenn and Killian, 1940; Granville et al., 1943; Kelly et al., 1943). The reflectance spectra were converted to the CIE tristimulus values using the formulas given in equation (3.1). A table was developed from these data relating the second tristimulus value, Y, to the Munsell value. Additionally, a series of graphs called chromaticity diagrams were published showing normalized tristimulus values for the Munsell samples. On the basis of these diagrams, the Munsell designations for the original samples were revised so that whole-number Munsell value and chroma designations would fall along smooth contours within the CIE diagrams (Newhall et al., 1943).

The chromaticity diagram series is published by the American Society for Testing

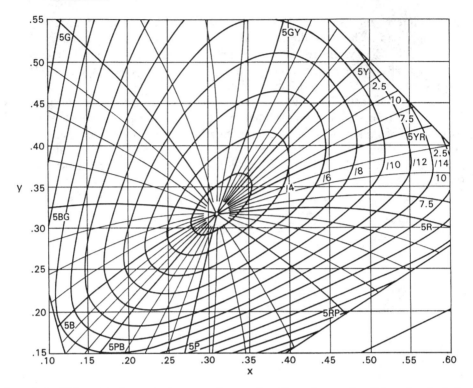

Figure 3.2 CIE chromaticity diagram showing contours of constant Munsell hue and constant Munsell chroma for colors having a Munsell value of 5.

and Materials (ASTM, 1989) and in color science texts (e.g., Wyszecki and Stiles, 1982) for the conversion of CIE chromaticity coordinates to the Munsell notations for hue and chroma. Each diagram corresponds to a specific Munsell value and shows contours of constant Munsell hue and constant Munsell chroma (e.g., Figure 3.2). Tables are also provided that relate the CIE tristimulus value, Y, to the Munsell value. The procedure recommended by the ASTM for converting CIE tristimulus values Y, X, and Z to Munsell hue, value, and chroma requires interpolation between the contours on the diagrams. Computer programs have been written to automate the interpolation (Keegan et al., 1958). A number of investigators have converted soil reflectance spectra to CIE tristimulus values and chromaticity coordinates and then determined the Munsell hue, value, and chroma designations for the soils by interpolation between contours on chromaticity diagrams Shields et al., 1966, 1968; Torrent et al., 1983); (Barron and Montealegre, 1986; Barron and Torrent, 1986; Fernandez and Schulze, 1987; Escadafal et al., 1989).

3.3.3.3 CORRELATION BETWEEN COLOR AND SOIL PROPERTIES.

Various workers have studied the correlation of Munsell's notations with different soil components: McKeague et al. (1971) concluded that no general relationship existed between chroma and dithionite extraction of Fe, or between value and organic matter content. Leger et al. (1979) concluded that color variations in soils are

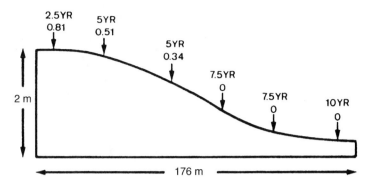

Figure 3.3 Color changes (Munsel parameters) in toposequence (hydrosequence) of Oxisols from Brazil. (After Curi and Franzmeier, 1984.)

the result of changes throughout the entire range of reflected wavelengths rather than changes in specific wavelength ranges. Da Costa (1979) studied various relationships between soil properties and color parameters and found that clay, organic carbon, cation exchange capacity, and water content at different tensions are the most important properties related to value and chroma parameters. Moist and dry values are more correlated with the soil properties (except for organic matter and nitrogen) than is chroma. Silt and sand components are the least important soil properties in determining soil color. Sand is positively correlated with the soil color but only for udic and ustic moisture regimes. Da Costa also concluded that climatic variations in terms of moisture and temperature are expressed by variations in the soil color. Where a soil spectrum can be rearranged into a color space (Berns et al., 1985; Fernandez and Schultz, 1987), the spectrum is highly preferable because it contains unique information that might be overlooked when simply determining the color.

It should be noted that more studies and data are available in the literature about soil color and its relation with soil environment and mineral composition (see, e.g., Westin and Franzee 1976; Taylor, 1982, Melville and Atkinson, 1985) and with satellite information (e.g., Escadafal et al., 1989). To illustrate the effect of an environmental change on soil color we provide Figure 3.3, which represents color changes (hue) along a slope (*toposequence*) and how this relates to various moisture conditions (*hydrosequence*).

3.4 SOIL SPECTRA

3.4.1 Definitions and Limitations

While soil color provides pedologists with a useful concept for recognizing, characterizing, and describing soils, soil color descriptions are limited by the sensitivity of the human eye and the subjectivity of human perception. Modern spectrometers and radiometers allow us to observe more precisely and objectively the intensity of radiation reflected by soils across the wavelength range of natural solar illumination. These instrument allow us to measure, plot, recognize, and analyze soil reflectance spectra. A soil reflectance spectrum is a set of data or a graph that provides the relative intensity of reflected radiation as a function of wavelength. The reflected intensity is expressed relative to the intensity of the illuminating radiation. Soil reflectance values are often determined, from a practical standpoint, by taking a ratio of the energy reflected by a soil surface to the energy reflected by a bright, diffuse reference material. This approach requires the soil and reference surface to be illuminated and observed in exactly the same manner with respect to the positions of the sensor and the sun or other source of illumination (Palmer, 1982, Baumgardner et al., 1985; Jackson et al., 1987).

Soil reflectance data have been acquired in a substantial number of recent remote sensing, field, and laboratory studies (Baumgardner et al., 1985 Koeumov et al., 1992). Most of the studies have focused on the spectral distribution of the scattered radiation, but some data on the directional distribution and polarization state of radiation scattered by soils are also available in the literature. The studies generally demonstrate relationships between spectral reflectance data and certain soil properties that correspond to the well-known relationships with soil color.

To provide a better understanding of soil spectra, in this section we provide background information on the electromagnetic spectrum, radiation interactions with soil, and soil attributes that affect spectral response. This discussion is limited primarily to the region 0.4 through 2.5 µm of the electromagnetic spectrum, which can be separated into three regions: visible (0.4 to 0.7 µm), near-infrared (0.7 to 1.1 µm), and shortwave infrared (SWIR; 1.1 to 2.5 µm).

3.4.2 Background

Our perception of soil is a manifestation of the interaction of solar radiation with soil materials. All of the interactions with soil involve the processes of scattering and absorption. Any discussion of these two processes first requires an appreciation of the nature of electromagnetic radiation.

3.4.2.1 ELECTROMAGNETIC RADIATION.

Electromagnetic radiation is a dynamic form of energy that is capable of propagating through a vacuum and yet becoming apparent only by its interactions with matter (Suits, 1983). Two distinct concepts of electromagnetic radiation have developed over time to describe the observed behavior of radiation and its interactions with matter. The classic concept treats electromagnetic radiation as a continuous transverse wave consisting of oscillating electric and magnetic fields. This concept describes and predicts the propagation and macroscopic optical behavior of radiation but fails to account for certain observed interactions between radiation and matter at atomic and molecular levels. These microscopic interactions are explained by a concept of radiation as a stream of discrete particles, called quanta or photons, which carry precisely determined amounts of energy. Castellan (1983) and Hunt (1980) concisely reviewed the historical development of these dual concepts, and much of the following discussion is drawn from these two reviews.

Maxwell mathematically formulated the wave concept of electromagnetic radiation in 1862. The formulations, now known as Maxwell's equations, consist of a set of partial differential equations that relate the electric and magnetic fields and predict the propagation of their oscillations. The electric and magnetic fields, as characterized by Maxwell's equations, are perpendicular to each other and to the direction of propagation. The equations accurately predict a wide variety of observed phenomena related to the propagation, reflection, refraction, dispersion, diffraction, and interference of electromagnetic radiation (Hunt, 1980).

Maxwell's equations imply that electromagnetic energy is supplied continuously in a wave (Hunt, 1980). This implication proved incompatible with certain experimental observations, notably the blackbody emission experiments of Planck in 1900, which led eventually to the quantum mechanical model of matter. The rules of quantum mechanics state that an atom or molecule can only exist at certain discrete energy states and cannot possess intermediate amounts of energy. Furthermore, an atom or molecule can absorb electromagnetic radiation only if precisely enough energy is provided to promote the atom or molecule from one energy state to a higher state. Similarly, electromagnetic radiation is emitted when an atom or molecule falls from an excited state to a lower-energy state. The amount of energy emitted as radiation exactly equals the energy difference between the two states. The discernment of the emission and absorption of discrete amounts of energy led to the perception of electromagnetic radiation as a stream of discrete particles each containing a specific amount of energy.

The current theory of electromagnetic interactions resolves the dual concepts of electromagnetic radiation. The theory incorporates the quantum mechanical model of matter and is called quantum electrodynamics (Feynman, 1985). According to this current concept, electromagnetic radiation supplies energy in indivisible packets carried by photons. In other words, electromagnetic radiation behaves like particles. The classic wave concept, however, is not abandoned. In the current view, the waves formulated by Maxwell's equations no longer represent a continuous distribution of electromagnetic energy. Instead, the wave expresses the amplitude of the probability of finding a photon at different places and at different times. Quantum electrodynamics does not invalidate the application of Maxwell's equations to macroscopic optical phenomena; the current theory only alters the classical concept of the waves described by the equations.

The wave concept is still used to characterize electromagnetic radiation. Radiation is characterized with respect to the frequency, wavelength, and wavenumber of the probability amplitude waves associated with the photons. The energy carried by a photon is precisely related to its associated wave frequency (i.e., wave cycles per unit time) by the following equation:

$$En = h\nu \tag{3.2}$$

where En is energy, h is Planck's constant (6.626×10^{-34} J·s), and ν is the frequency. Frequency is a fundamental characteristic of electromagnetic radiation; since the energy of a photon is indivisible, the wave frequency is independent of the medium through which the radiation passes (Castellan, 1983). The frequency is related to wavelength and the speed of radiation propagation in a medium:

$$\nu = \frac{c}{\lambda} \tag{3.3}$$

where λ represents wavelength and c is the speed of propagation. The speed of propagation in a vacuum is approximately $3 \times 10^8 \text{m s}^{-1}$. The wavenumber gives the number of wave cycles per unit distance and is the reciprocal of the wavelength. The flux and reflectance of solar radiation are generally expressed or reported as a function of wavelength in literature pertaining to terrestrial remote sensing.

3.4.2.2 RADIATION INTERACTIONS WITH ATOMS AND MOLECULES.

Characteristics of electromagnetic radiation are altered when radiation interacts with matter (e.g., when radiation is scattered from a soil surface). An understanding of the mechanisms responsible for the alterations required a quantum mechanical concept of matter at the atomic and molecular levels. In particular, the previously mentioned concept of discrete energy states is essential to comprehension; electromagnetic radiation can only be emitted or absorbed when an atom or molecule makes a transition between energy states. The energy of an emitted or absorbed photon equals the difference between the energy levels. Furthermore, the energy-level transitions must be accompanied either by a redistribution of the electric charge carried by electrons and nucleic protons or by a reorientation of nuclear or electronic spins before a photon is emitted or absorbed (Hunt, 1980). The absorption or emission of shortwave radiation usually results from energy-level transitions accompanied by charge redistributions involving either the motion of atomic nuclei or the configuration of electrons in atomic and molecular structures.

A. Vibrational Transitions. A portion of the energy possessed by an atomic or molecular system is by virtue of the translational, rotational, and vibrational motion of the atomic nuclei. Nucleic translations and rotations are restricted in most soil materials and thus do not play a major role in soil interactions with solar radiation. Transitions of vibrational motion, however, significantly affect these interactions. The vibrational motions consist of oscillations in the relative positions of bonded atomic nuclei. The oscillations either stretch molecular bond lengths or bend inter-bond angles. Energy-level transitions that involve nuclear vibrations typically result in the absorption or emission of radiation within the infrared portion of the spectrum (Hunt, 1980).

A molecule possesses several modes of vibration, depending on the number and arrangement of atoms in the molecule. A molecule composed of N atoms may possess $3N - 5$ vibrational modes if the atoms are arranged linearly or $3N - 6$ vibrational modes if the bonding is nonlinear (Castellan, 1983).

The laws of quantum mechanics dictate that only discrete energy levels may be associated with each vibrational model of a molecule. The lowest allowable energy level for each mode is referred to as a ground level. Transitions between the energy levels of each result in the emission or absorption of radiation at specific frequencies. The frequencies (or the corresponding wavelengths and wavenumbers) associated with transitions between a ground level and the next-highest energy level are called *fundamental bands*. The absorbed (or emitted) frequencies are called *overtone bands* when a vibrational mode transits from one state to a state more than one energy level above (or below) the original state. *Combination bands* refer to frequencies associated with transitions of more than one vibrational mode. These combined transitions occur when the energy of an absorbed photon is split between more than one

TABLE 3.1 Fundamental, Overtone, and Combination Bands for the Three Vibrational Modes of Water Vapor

Wavelength (nm)	Frequency (s⁻¹)	Wavenumber (cm⁻¹)	Energy State of Each Mode [a]			Remark
			v_1	v_2	v_3	
6270	4.782×10^{13}	1,595	0	1	0	Fundamental
3173	9.450×10^{13}	3,152	0	2	0	First overtone
2738	1.095×10^{14}	3,652	1	0	0	Fundamental
2662	1.126×10^{14}	3,756	0	0	1	Fundamental
1876	1.598×10^{14}	5,331	0	1	1	Combination
1135	2.640×10^{14}	8,807	1	1	1	Combination
942	3.182×10^{14}	10,613	2	0	1	Combination
906	3.307×10^{14}	11,032	0	0	3	Second overtone
823	3.643×10^{14}	12,151	2	1	1	Combination
796	3.767×10^{14}	12,565	0	1	3	Combination
652	4.601×10^{14}	15,348	3	1	1	Combination

Source: After Castellan (1983, Table 25.2).
[a] Ground state corresponds to $v_1 = v_2 = v_3 = 0$.

mode (Castellan, 1983). The fundamental, overtone, and combination bands associated with the vibrational modes of water vapor are listed in Table 3.1.

The vibrational transitions corresponding to the fundamental bands are generally more likely to occur than transitions corresponding to the combination and overtone bands (Castellan, 1983). Absorption features in reflectance and transmittance spectra are therefore usually strongest for the fundamental bands. Infrared spectroscopy is a useful laboratory method for soil analyses because the fundamental bands for most soil minerals occur in the infrared at wavelengths between 2500 and 50,000 nm (White and Roth, 1986). Remote sensing within the solar portion of the spectrum restricts the detection of vibrational transitions in soil materials primarily to the observation of overtone and combination bands (Hunt, 1980).

Combination and overtone bands associated with hydroxyl group (OH) vibrations are, for example, often apparent in soil reflectance spectra. Hydroxyl groups are found on many soil minerals and the exact wavelength location of the associated bands depends upon which hydroxyl-bearing minerals are present in the soil. The one fundamental hydroxyl band due to oxygen–hydrogen stretching is found near 2.8 μm, and the first overtone band due to this stretch is located near 1.4 μm. Absorption at this overtone band is the most common feature in the near-infrared spectra of terrestrial materials (Hunt, 1980). The hydrogen–oxygen stretch can also be coupled with other vibrations in the molecular structure of the soil minerals to create combination band features. Bending at magnesium–hydroxyl (Mg–OH) bonds, coupled with the stretch, results in a combination band near 2.300 μm, and bending at aluminum–hydroxyl bonds coupled with the stretch produces a combination band near 2.200 μm (Hunt, 1980).

B. Electron Transitions. In contrast to the infrared bands associated with vibrational transitions, the bands associated with electron transitions generally occur in the ultraviolet and visible portions of the spectrum. The location of these bands is due to the relatively large gaps between electron energy states. The principles of quantum mechanics dictate that each electron of an atom, ion, or molecule can exist in only certain states corresponding to discrete energy levels.

The state of an electron attached to a particular atom can be specified by four quantum numbers. The first of these is the principal quantum number, n. The principal quantum number designates the electron shell and can have any integer value greater than zero. Shells are divided into orbitals that are designated by the angular momentum quantum number, 1. This second quantum number can have any non-negative integer value less than the value of its associated shell, n. (By historical convention, the orbitals are also designated by a nonalphabetic sequence of letters; s,p,d, and f, which correspond to values of 0, 1, 2, and 3, respectively). Each value of 1 denotes a particular geometric distribution of the electron's charge around the atomic nucleus (Nassau, 1980). Within each orbital designated by 1, the state of an electron can be further specified by the magnetic quantum number, m, and by the spin quantum number, s. The magnetic quantum number can have any positive or negative integer value with a magnitude less than or equal to 1. The spin quantum number may have only two values: plus or minus ½. No two electrons of the same atom can have the same set of quantum numbers, according to the Pauli exclusion principle (Nassau, 1983).

The Pauli exclusion principle limits the number of electrons that can occupy each shell and each orbital within a shell. The first shell ($n = 1$), for example, can hold two electrons in its lone s orbital ($1 = 0$) while the second shell ($n = 2$) can hold a total of eight electrons with two electrons in its s orbital ($1 = 0$) and six electrons in its p oribital ($1 = 1$). The electrons of a filled shell occur in pairs that have the identical first three quantum numbers and different spin quantum numbers, s. Paired electrons in a closed shell exist in stable ground energy states. A relatively large energy quanta from the x-ray or ultraviolet region is usually required to elevate a paired electron to a higher-energy state. Transitions of unpaired electrons, which are usually found in the outermost shell of an atom, generally require less energy and are more typically involved in the absorption and emission of visible radiation by independent atoms. The emission of yellow light by a sodium vapor lamp, for example, is due to transitions of an unpaired electron in the outermost shell of the sodium atom. These unpaired electrons are the valence electrons that participate in chemical bonds (Nassau, 1980).

Chemical bonding creates the polyatomic molecules and crystals that form most soil materials. Bonding also alters the energy states of the valence electrons involved in the bonds. Chemical bonds occur when the valence electrons of an atom are shared in various ways with neighboring atoms. The shared electrons form pairs that often exist in stable ground energy states. As with paired electrons in isolated atoms, the excitation of a paired electron in a chemical bond often requires the high energy of a photon from the ultraviolet region (Nassau, 1980). Many exceptions to this requirement for ultraviolet excitation, however, exist for electrons participating in molecular bonding and crystalline structures.

Numerous mechanisms have been identified to explain the role of electrons in the absorption and emission of radiation by molecules and crystals. Nassau (1983), for example, has divided the mechanisms into 14 categories. Although not all of the electronic excitations require ultraviolet photons, the mechanisms usually require energy quanta greater than the quanta involved in vibrational transitions of bonded atomic nuclei. Consequently, electronic transitions in molecules are always accompanied by vibrational and, if possible, rotational transitions (Hunt, 1980). An in-depth treatment of these mechanisms of electronic transition would require extensive discussions of crystal field and ligand field theory, of molecular orbital theory, and of the group theory of molecular symmetry. Only brief reviews of several mechanisms relevant to the interactions of shortwave radiation with soils are attempted here.

Although the quantum mechanical mechanisms by which radiation interacts with matter are numerous and complex, some of the characteristics of radiation-scattered soils can be attributed to specific mechanisms. In particular, absorption features in soil reflectance spectra are usually a direct expression of specific mechanisms, and interpretations of these features are especially straightforward in the shortwave region of soil mineral reflectance spectra. These mechanisms do not generally involve the three major elemental constituents of soil minerals: silicon, oxygen, and aluminum. The allowed transitions between vibrational and electronic states associated with these constituents do not produce absorption features in the shortwave region (Hunt, 1980). Instead, most of the shortwave absorption features are due either to hydroxyl groups, as discussed previously, or to iron.

Iron is commonly found in soil materials. Iron ions are soluble in certain conditions and are thus widely distributed by the soil solution. Further, iron ions can easily replace aluminum and magnesium ions in octahedral sites of soil mineral crystals and can even replace silica in some tetrahedral sites (Hunt, 1980). Crystal field effects or charge transfers involving iron usually cause the absorption features found in the visible portion of soil reflectance spectra.

C. Radiation Interactions with a Volume of Soil. The process of radiation scattering by soils results from a multitude of quantum mechanical interactions with the enormous number and variety of atoms, molecules, and crystals in a macroscopic volume of soil. In contrast to certain absorption features, most characteristics of the scattered radiation are not attributable to a specific quantum mechanical interaction. The effects of a particular mechanism often become obscure in the composite effect of all the interactions. The difficulty in accounting for the effects of a large number of complex quantum mechanical interactions often leads to the use of non-quantum-mechanical models of electromagnetic radiation. Physicists frequently resort to the classical wave theory or even to geometrical optics to elucidate the effects of a macroscopic volume of matter on radiation.

D. Refractive Indices. The optical properties of a substance are those properties that determine its effect on incident radiation. The fundamental optical properties of a substance are embodied in a wavelength-dependent quantity called the *complex refractive index:*

$$n = n_r - in_i \qquad (3.4)$$

(Note that the symbol n often represents both the principal quantum number and the complex refractive index by convention. This symbol, n, will represent the complex refractive index from this point forward. Also note that the subscript indicating wavelength dependence, λ, has been omitted from the equation above and from the variables of all the following equations for simplification. All of the following equations are for monochromatic radiation.)

The real part of the refractive index, n_r (dimensionless), is most easily explained in relation to refraction. Refraction refers to a change in the path of a beam when it passes from one transparent medium into another. The change is due to a difference in the speed of radiation propagating through the two media. The real part of the complex index of refraction, n_r, is defined for a particular substance as the ratio of the speed of radiation propagation in a vacuum, c_0, to the speed of propagation through the substance, c:

$$n_r = \frac{c_0}{c} \tag{3.5}$$

Snell's law, formulated by the Dutch scientist Wellebrod Snell in Royen in 1621 (Nassau, 1983), relates the indices of refraction of two media, n_1 and n_2, to the change in direction experienced by a beam passing from one nonabsorbing medium into the other:

$$n_1 \sin \theta_i = n_2 \sin \theta_t \tag{3.6}$$

where θ_i is the angle of incidence and θ_t is the angle of refraction.

The imaginary part of the complex index of refraction, n_i (dimensionless), describes the removal of energy from a beam as it is transmitted through a homogeneous volume of material. The removal of energy (or, equivalently, photons) is due to absorption within the volume. According to Beer's law (also known as the Bouguer law or Lambert–Beer law), the amount absorbed by a thin slice of material is directly proportional to the thickness of the slice:

$$-\frac{\partial I}{I} = a \, \partial z \tag{3.7}$$

where I represents specific intensity (flux per unit area per unit solid angle) constant is a a and z is distance (dimension length). If I_0 is the beam intensity at the top of the volume ($z = 0$), the equation above can be solved to yield

$$I_z = I_0 \exp(-az) \tag{3.8}$$

where I_z is the specific intensity carried by the beam at depth z. The imaginary part of the refractive index is related to the constant α by the following equation (Bohren and Huffman, 1983), where λ is wavelength:

$$n_i = \frac{\alpha \lambda}{4\pi} \tag{3.9}$$

Both the imaginary part of the refractive index, n_i, and the constant α are referred to in the literature as the absorption coefficient. To avoid confusion, only the constant a will be called the absorption coefficient in this chapter, and n_i will only be referred to as the imaginary part of the refractive index.

E. Models of Radiation Scattered by Soils. The geometry of both illumination and observation irradiance and radiance plays a major role in deriving the soil spectrum. The bidirectional reflectance distribution function (BRDF) assumes that the radiation source, the target, and the sensor are all points in the measurements space and that the ratio calculated between absolute values of radiance and irradiance is strongly dependent on the geometry of their positions. Theories and models explaining the BRDF phenomena in relation to soil components are widely discussed and covered in the literature (Liang and Townshend, 1966; Hapke, 1981a,b, 1984, 1993; Pinty et al., 1989, Jacquemoud et al., 1992). The following equation describes the basic function of the BRDF value:

$$fr(q,q',f,f') = \frac{dL(q,q',f,f',E)}{dE(q,f)} \qquad (3.10)$$

where L is the reflected irradiance, E the illuminating radiance, and q, q', f', and f are source, sensor zenith angles and source, target azimuth angles, respectively. Whereas the BRDF is better suited for remote sensing applications, in the laboratory, hemispheric and bihemreflectance factors are also used (Baumgardner et al., 1985). To reduce the effects of geometry and to eliminate systematic and nonsystematic measurement interferences, reflectance standards such as MgO, $BaSO_4$ and Halon are often used to correct the reflectance spectrum relatively (Tkachuk and Law, 1978; Young et al., 1980; Weindner and Hsia, 1981). Another factor that affects the soil spectra is the sensor's field of view (FOV) and sun target geometry. If the soil is homogeneous, a small FOV may be sufficient. However, where some variation occurs in the soil, the FOV should be adjusted to cover a representative portion of the soil.

Theories of radiation scattering from semi-infinite and particulate surfaces have been applied to soils in a number of studies. These studies have been reviewed by Smith (1983), by Irons et al. (1989), and by Curran et al. (1990). Most of the investigators formulated equations for the hemispherical reflectance of soils. These formulations are briefly discussed here only as our focus is on the directional distribution of the scattered radiation. More detailed descriptions are provided for models of the directional distribution.

Several studies of soil reflectance (Vincent and Hunt, 1968, Barron and Montealegre, 1986; Barron and Torrent, 1986); have employed the Kubelka–Munk (1931) formulation of the radiative transfer problem. This equation was derived from the solution of two simultaneous differential equations that give the rate of change in the intensity of radiation traveling in opposite directions through a layer of scattering and absorbing particles (Wendlandt and Hecht, 1966):

$$\partial I = -(K + S)I \; \partial x + SJ \; \partial x \qquad (3.11)$$
$$\partial J = (K + S)J \; \partial x - SI \; \partial x$$

where I is the intensity of the radiation traveling downward through the layer (i.e., positive x direction), J the intensity of the radiation traveling upward (i.e., negative x direction), K is twice the fraction of radiation absorbed per unit path length, and S is twice the fraction of radiation scattered backward per unit path length. By setting the appropriate boundary conditions ($I = I_0$ at $x = O$, and $I = O$ and $J = O$ at $x =$

infinity), the equations above can be solved for the hemispherical reflectance of a semi-infinite medium:

$$R_n = \frac{J_0}{I_0} = \frac{1 - \sqrt{K/(K + 2S)}}{1 + \sqrt{K/(K + 2S)}} \tag{3.12}$$

where J_0 and I_0 are the intensities upward and downward, respectively, at $x = 0$.

Emslie and Aronson (1973) extended the Kubelka–Munk approach by formulating six simultaneous differential equations giving the rate of change in intensity in six mutually orthogonal directions through a layer of particles: the upward direction, the downward direction, and four transverse directions parallel to the surface. Assuming a plane-parallel medium where the rate of change was equal in the four traverse directions, they reduced the six equations to two differential equations identical in form to the Kubelka–Munk differential equations but with revised coefficients.

In addition to extending the radiative transfer approach of Kubelka–Munk, Emslie and Aronson (1973) also provided two methods for calculating the scattering and absorption coefficients on the basis of particle sizes, complex refractive indices, and volume fractions. Their method for particles greater or equal in size to the incident wavelength (coarse-particle theory) was based primarily on geometric optics, but classic wave theory was also used to account for additional attenuation by edges and asperity absorption (i.e., absorption by small perturbations or roughness) on coarse particles. Egan and Hilgeman (1978) expanded this coarse particle theory to include attenuation by edge and asperity scattering as well as absorption. The method developed by Emslie and Aronson (1973) for particles much smaller than the incident wavelength (fine-particle theory) combined classic wave theory for coherent scattering with Rayleigh theory for incoherent scattering from a collection of randomly oriented ellipsoidal dipoles. For reflectance from particles of intermediate size, Emslie and Aronson (1973) also included an empirical formula for the weighted average of results from the coarse-particle and fine-particle theories.

Conel (1969) developed another model for particulate media on the basis of the radiative equation. He employed the numeric discrete ordinates method (Liou, 1980) to solve the radiative transfer equation approximately for the flux leaving the surface at an angle of 54.73°. He then derived the following equation for reflectance:

$$R = \frac{u - 1}{u + 1} \tag{3.13}$$

where

$$u^2 = \frac{1 - \omega_0 <\cos \sigma>}{1 - \omega_0} \tag{3.14}$$

ω_0 is the single-scattering albedo, and $<\cos \sigma>$ is the anisotropy factor. Conel (1969) used this model to evaluate the effects of particle size on absorption features in particulate quartz emissivity spectra. The quartz single-scattering albedos and anisotropy factors were determined using Mie scattering theory.

A number of models have been developed that express soil bidirectional reflectance as a function of illumination and viewing direction [i.e., soil bidirectional reflectance distribution function (BRDF) models]. Several different methods have been applied to the development of these models. The methods include the formulation of empirical functions, the calculation of the relative fraction of illuminated and shadowed surface area on rough surfaces, the use of rough surface scattering theory, and the application of radiative transfer theory.

Empirical equations have been applied to soil BRDF modeling. These equations were generally formulated to mimic trends observed in the directional distribution of radiation scattered from particulate or rough surfaces. The empirical Minnaert equation (Minnaert, 1941), for instance, was originally formulated to satisfy the Helmholtz reciprocity principle for lunar observations (the reciprocity principle states that reflectance does not change if the solar illumination direction and the view direction are interchanged). The Minnaert equation was fit to soil reflectance data with good agreement when the data were stratified with respect to view azimuth angle (Irons and Smith, 1990). As another example, Walthall et al. (1985) formulated a three-term equation that expresses bidirectional reflectance as a function of view zenith and azimuth angles for a constant solar illumination direction. Walthall et al. (1985) fit the equation to bidirectional reflectance data for both plant canopies and bare soil surface, with good agreement in each case. Good agreement was also found when the equation was fit to soil reflectance data acquired by Ranson et al. (1991). These empirical equations are useful for describing soil bidirectional reflectance distributions, but equation parameters are not explicitly related to soil properties. The equations cannot be inverted to estimate soil properties directly on the basis of bidirectional reflectance observations, nor can the equations be used to predict reflectance distributions on the basis of soil property measurements in the field.

Other investigators have developed soil BRDF models by representing a soil surface as a collection of opaque geometric objects that cast shadows. This approach required calculation of the fractions of the surface in view and in shadow and determination of the reflectance from the geometric objects. Many of these models owe recognition to the work of Egbert (1977), who formulated equations for horizontal surfaces covered by spheres or cylinders. Otterman and Weiss (1984), for example, formulated a soil BRDF equation by representing soil surfaces as a collection of randomly located vertical facets sitting on a horizontal Lambertian surface. Deering et al. (1990) fit this equation to soil bidirectional reflectance data with good agreement except for observations near the 0° view zenith angle. Both Roujean et al. (1990) and Norman et al. (1985) derived soil BRDF models by considering the shadows cast by rectangular blocks distributed on a plane surface. Norman et al. (1985) found that their equation fit reflectance data satisfactorily for only a limited range of illumination zenith angles.

Several investigators have used spheres as the opaque objects representing soil surface roughness. Curran et al. (1990) described a soil BRDF model originally published in the Soviet literature by Kastrov (1955) that was based on a distribution of uniform spheres. Cierniewski (1987, 1989) formulated an assembly of equations for the calculation of a shadowing coefficient for spheres distributed on a sloping plane. Each of his equations applies to a subset of illumination, observation, and slope directions. Irons et al. (1992) also represented the soil surface by opaque spheres distributed over a horizontal surface, but they then formulated a single equation for the complete soil BRDF. Reflectance distributions predicted by this equation were comparable to multiple-direction observations of freshly tilled soil reflectance (Irons et al., 1992).

Examples are sparse of the application of rough surface scattering theory to the scattering of solar radiation by soils. Becker et al. (1985), however, have applied the two-scale rough surface scattering model of Leader (1979) to the observed distributions of thermal infrared illumination scattered by soil samples. They employed a carbon dioxide laser and an infrared radiator as active thermal infrared illumination sources. They were unable to reproduce observed backscattering peaks with the model and therefore modified the model with a shadowing function. The modification accounted for shadowing by macroscopic objects on the surface and they were able to fit the modified model more closely to their observations. Other applications of rough surface scattering theory to soils can be found in the Soviet literature as reviewed by Curran et al. (1990).

Radiative transfer theory has been more applied extensively to solar radiation scattering by soils. Cooper and Smith (1985), for example, modeled soil reflectance using a Monte Carlo method to follow photons multiply scattered from surfaces having periodic fluctuation in height. Several other models based on approximate solutions to the radiative transfer equation for semi-infinite particulate surfaces have been developed and applied to interpretations of planetary observations (Hapke, 1981, a, b 1984; 1986; Lumme and Bowell, 1981; Helfenstein and Veverka, 1987; Arvidson et al., 1989; Domingue and Hapke, 1989). Pinty et al. (1989) fit the model of Hapke (1981b) to terrestrial bare soil bidirectional reflectance data using a nonlinear least-squares procedure. Ahmad and Deeming (1992) used a similar approach to formulate a solution to the radiative transfer equation for turbid media and also fit their equation to bare soil bidirectional reflectance data. In both cases, these equations fit the data well and reasonable values were retrieved for the equation parameters. These parameters consist primarily of effective or average single-particle scattering parameters such as the single-scattering albedo and the coefficients of scattering phase functions along with parameters related to the macroscopic morphology or roughness of the surface.

Hapke (1981b) modified his model for particulate surface reflectance to account for a sharp reflectance peak in the anti-illumination direction (i.e., at zero phase angle) that is often observed in bidirectional reflectance distributions from random media. Planetary geologists call this peak the *opposition effect*, and it is also known as the *hot spot, retroreflectance peak*, or *heiligenshein*. Hapke (1986) reviewed discussions of the opposition effect in planetary geology literature and attributed the effect to mutual shadowing by particles. In summary, a sensor pointed at a surface from the anti-illumination direction views only illuminated particles or facets; no shadows are in view. The proportion of shadow versus illuminated area increases rapidly as the sensor moves away from the anti-illumination direction.

3.4.3 Soil Spectra

Empirical research on the relationships between soil properties and the characteristics of scattered radiation has focused predominantly on the spectral distribution of the radiation. Numerous investigators have acquired reflectance spectra for soil samples in the laboratory and for soil surfaces in the field (Obukhov and Orlov, 1964; Bowers and Hanks, 1965; Orlov, 1966 Kasumov et al., 1992; Condit, 1970; Mathews et al., 1973; Montgomery and Baumgardner, 1974; Beck et al., 1976; Stoner and Baumgardner, 1981). Studies of soil reflectance spectra were reviewed by Baumgardner et al. (1985), Irons et al. (1989), and Curran et al. (1990). Of these investigators, Stoner and Baumgardner (1981) acquired the most comprehensive database of reflectance spectra for U.S. soils. The database includes spectra for samples representing 239 soil series from the United States and for samples representing an additional seven soil series located within the state of Paraná, Brazil.

Stoner and Baumgardner (1981) observed their samples in the laboratory after

TABLE 3.2 Characteristics of Surface Samples of Five Mineral Soils Presented in Figure 3.4

Soil Attribute	Iron Dominated	Organic Affected	Iron Affected	Minimally Altered	Organic Dominated
			Reflectance Curve Form		
Soil series	(Not given)	Onaway	Talbott	Jal	Drummer
Horizon sampled	AP (0–10 cm)	Ap	Ap	All	Ap
Soil subgroup	Typic Hapolorthox	Alfic Haplorthod	Typic Calciorthid	Typic Calciorthid	Typic Haplaquoll
Sample location	Londrina, Paraná, Brazil	Delta County, Michigan	Rutherford County, Tennessee	Lea County, New Mexico	Champaign Country, Illinois
Climatic zone	Humid hyperthermic	Humid frigid	Humid thermic	Semiarid thermic	Humid mesic
Parent material	Basalt	Glacial drift	Clayey limestone residum	Fine texture alluvium or lacustrine	Loess over glacial drift
Drainage class	Excessively drained	Well drained	Well drained	Well drained	Poorly drained
Textural class	Clay	Fine sandy loam	Silty clay loam	Loamy fine sand	Silty clay loam
Munsell color	Dark red	Dark brown	Strong brown	Brown	Black
Contents (%)					
Organic matter	2.28	3.3	1.84	0.59	5.61
Iron oxide	25.6	0.81	3.68	0.03	0.76
Moisture at 0.1 bar tension	33.1	27.3	28.2	17.0	41.1
Symbol	A	B	C	D	E

Source: After Stoner and Baumgardner (1981).

careful preparations. The samples were first air dried, crushed, sieved (<2 mm), and equilibrated to a uniform moisture tension (10 kPa). The spectral distribution of bidirectional reflectance factors was then determined for each sample with a spectrometer and with reference to a pressed barium sulfate reflectance standard. Reflectance factor measurements were made at 10-nm increments over the 520-to 2320-nm wavelength region. Physical and chemical properties of the samples were also determined.

Stoner and Baumgardner (1981) identified five characteristic curve shapes on the basis of a visual inspection of their soil reflectance spectra. Each shape was associated with certain soil characteristics. The five associated soil classes were: organic-dominated (organic carbon >2%, iron oxide <1% and fine texture), minimally altered affected (organic carbon <2% and iron oxide <1%), iron-affected (organic carbon <%, iron oxide >1% <4%), organic affected (organic carbon >2%, iron oxides <1%), and iron-dominated (iron oxides >4%). Table 3.2 summarizes some of the chemicophysical characteristics of the five soil groups, and Figure 3.4 presents their spectral reflectance curves.

It is obvious that the analysis of Stoner and Baumgardner (1981) yielded only a limited set of information, which prevented a precise classification of the soils from their reflectance spectra. Accordingly, Kimes et al. (1993) developed an artificial intelligence method for classifying Stoner and Baumgardner's data set and were able to provide more information, such as differentiating between high and low organic matter content and fine and coarse texture in the soils. However, Huete and Escadafal (1991) noted that there are often problems in assigning soils to Stoner and Baumgardner's discrete curve classifications. In Figure 3.5 the spectra of six representative soil samples from arid and semiarid areas in the VNIR–SWIR regions are given to show that although the soils were classified according to the U.S. Department of Agriculture as six different soil groups, the spectra are quite similar in the VNIR region and different in the SWIR region (mostly around 2.3 to 2.4 μm). This suggests that important information is still hidden in the soil spectrum and will require de-

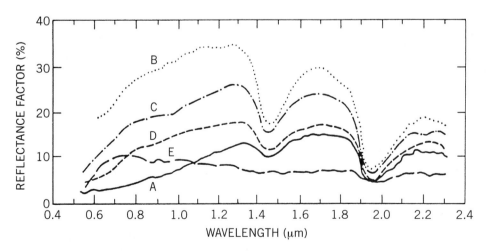

Figure 3.4   Reflectance spectra of five mineral soils that were found to represent more than 400 soil samples. More information about the soils are given in Table 3.2. (After Stoner and Baumgardner, 1981.)

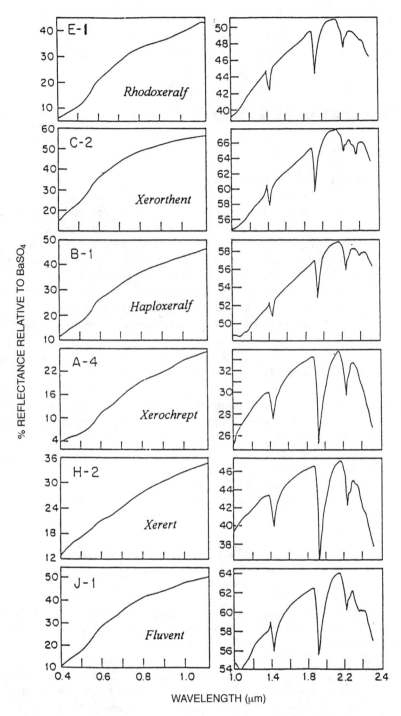

Figure 3.5   Reflectance spectra of six representative soils from arid and semiarid areas in Israel.

velopment of new approaches to analyze the data. The complexity of soil materials does not often permit utilization of simple spectral analysis routines and in many cases does not allow simple use of soil spectral libraries to answer the problem at hand.

3.5 SPECTRAL CHROMOMOPHRES IN SOILS

3.5.1 Background

A chromophore is a parameter or substance (chemical or physical) that significantly affects the shape and nature of a soil spectrum. A given soil sample consists of a variety of chromophores, which vary with environmental conditions. In many cases the spectral signals related to a given chromophore overlap with signals of other chromophore's and thereby hinder the assessment of the affect of a given chromophore. Because of the complexity of the chromophores in soil, it is important to understand the chromophores' physical activity as well as their origin and nature. The spectra of pure minerals are discussed extensively in Chapter 1. In the current chapter, our discussion focuses primarily on factors affecting soil spectra, directly and indirectly, from both chemical and physical chromophores.

3.5.2 Chemical Chromophores

Chemical chromophores are those materials that absorb incident radiation in discreet energy levels. Usually, the absorption process appears on a reflectance spectrum as troughs whose positions are attributed to specific chemical groups in various structural configurations. All features in the VNIR–SWIR spectral regions have a clearly identifiable physical basis. In soils, three major chemical chromophores can be roughly categorized as follows: minerals (mostly clay and iron oxides), organic matter (living and decomposed), and water (solid, liquid, and gas phases).

3.5.2.1 MINERALS.

The spectra of minerals are also covered in Chapter 1. In this section we discuss and highlight the spectral reflectance of common soil minerals that significantly affect soil spectra.

A. Clay Minerals. Clay minerals (also referred to as phyllosilicate minerals) are crys-talline aluminosilicate minerals organized in a layered structure. The crystal structure consists of two basic units: the Si tetrahedron, which is formed by a Si^{3+} ion surrounded by four O^{2-} ions in a tetrahedral configuration, and the Al octahedron, formed by an Al^{4+} ion surrounded by four O^{2-} and two OH^- ions in an octahedral configuration. These structural units are joined together into tetrahedral and octahedral sheets, respectively, by adjacent Si tetrahedral sharing all three basal corners and by Al octahedrons sharing edges. These sheets, in turn, form the clay mineral layer by sharing the optical O of the tetrahedral sheet. Layer silicates are classified into eight groups according to layer type, layer charge, and type of inter-layer cations. The layer type designated 1:1 is organized with one octahedral and one tetrahedral sheet, whereas the 2:1 layer type is organized with two octahedral sheets and one tetrahedral sheet. A one-layer octahedral sheet (1) is also found in highly leached acid soils. The layer silicate charge is a function of isomorphic substitution that occur in both the tetrahedral and octahedral positions during the weathering process. Surface charge density is one of the major factors governing soil behavior, and therefore mineral species and composition are considered to be keys to understanding soil behavior.

B. Origin of Layer Minerals in Soils. All clay minerals are derived from the weathering of primary minerals. Table 3.3 presents some of the clay minerals that exist in soil and their chemical compositions. The occurrence of smectite, vermiculite, illite, or kaolinite is related to the degree of weathering and the chemical nature of the soil environment. Muscovite tends to produce illite, whereas biotite tends to produce vermiculite. Both illite and vermiculite are associated with slightly weathered materials.

Vermiculite requires large amounts of magnesium during clay formation, which would most likely occur in neutral to slightly alkaline soils. Illite occurs to a greater extent than vermiculite in soils of moderate acidity. Illite tends to form smectite as surface potassium ions are removed by the weathering processes and new cation substitution occurs. Smectite minerals (2:1 configuration) are an important component of slightly to moderately weathered soils, which are formed under relatively high pH values and specific Si and Al concentrations in the soil solution. Kaolinite minerals (1:1 configuration) are predominant in highly weathered, leached soils that under stronger weathering and acid conditions turn into gibbsite. Whereas gibbsite is quite rare, smectite and kaolinite are more commonly found in soils. Kaolinite may be formed from a 2:1 mineral during the weathering process and requires an environment where both silica and alumina are accumulated in a ratio favorable to its formation. Illite and vermiculite are associated with youthful materials. Smectite is formed in the middle stage of the weathering process and therefore is most likely found in many soils as the major or secondary mineral. Similarly, the kaolinite component in soils tends to increase with increasing stages of weathering. Soils of warm temperate regions have a high percentage of kaolinite in the clay fraction, whereas cold areas tend to form more illitic-and smectitic-type minerals.

Of all the soil minerals discussed above, smectite is thought to be the most active, because of its high specific surface area and electrochemical reactivity. These characteristics are known to affect many of the soil's properties, as reported by Banin and Amiel (1970) and others.

TABLE 3.3   Representative Clay Minerals in Soil and Their Chemical Compositions

Mineral	*SiO₂*	*Al₂O₃*	*Fe₂O₃*	*TiO₂*	*CaO*	*MgO*	*K₂O*	*Na₂OH*
Kaolinite	45–48	38–40						
Montmorillonite	42–55	0–28	0–30	0–0.5	0–3	0–25	0–0.5	0–3
Illite	50–56	18–31	2–5	0–0.8	0–2	1–4	4–7	0–1
Chlorite	31–33	18–20				35–38		
Gibbsite		65						

Source: After Bear (1964).

TABLE 3.4   Common Nonclay Minerals in Soils and Their Chemical Compositions

Mineral	SiO₂	Al₂O₃	Fe₂O₃	FeO	CaO	MgO	K₂O	Na₂O	P₂O₂
Quartz	100								
Orthoclase	62–66	18–20			0–3		9–15	4–9	
Anorthite	40–45	28–37			10–20			0–2	0–5
Biotite	33–36	13–30	3–17	5–17	0–2	2–20	6–9		
Hornblende	38–58	0–19	0–6	0–22	0–15	2–26	0–2	1–3	
Epidote	35–40	15–35	0–30		20–25				
Apatite					54–55				40–42
Magnetite			69						

Source: After Bear (1964).

C. Nonclay Minerals. Table 3.4 presents some common nonclay minerals commonly found in soils, and their chemical compositions, whereas in Figure 3.6, pure spectra of selected minerals are given. The minerals are divided into five groups: silicates, phosphates, oxides and hydroxides, carbonates, and sulfides and sulfates. The fraction of each mineral in soils is dependent on the environmental conditions and parent materials.

Primary minerals will most likely be found in young soils, where the weathering process is weak. Whereas silicates such as feldspars are rarely found in mature soils, quartz may be also found in some developed soils, depending on the environment, conditions, and parent material. Singer and Ben-Dor (1987) reported that eolian quartz can be found on the surface of Vertisol soils that had developed on basalt. They assumed that quartz which had accumulated on the surface was from fine materials that were transported by the wind. In general, the quartz mineral is spectrally inactive in the VNIR–SWIR region and therefore diminishes other spectral features in the soil mixture. Other nonclay silicate minerals such as feldspars may have some diagnostic absorption features that make the soil spectrum less monotonous.

Oxide group minerals occur in highly weathered areas such as those associated with slopes, highly leached profiles, or in areas of "mature" soils. Phosphate and sulfate minerals can be found in soils as apatite and gypsum, respectively. Although both minerals have unique spectral features, their occurrence in soils may be relatively rare and even nondetectable, whereas other oxides, such as iron, are strongly spectrally active, mostly in the visible region, because of the crystal field and the charge transfer mechanism. The content of free oxides (both iron and aluminum) is low in young soils but increases gradually as soil ages, just as happens with organic matter. Younis et al., (1997) have studied the influence of weathering process on the reflectance spectra of fresh (nonclay) rocks. They concluded that reflectance differences between the fresh and weathered surfaces are highly significant in the VIS–NIR–SWIR spectral region whereas the iron oxides components play an important role in this effect.

3.5.2.2 PHYSICAL MECHANISM OF PHOTON ABSORPTION IN SOIL MINERALS

A. Clay Minerals. Basically, the spectral features of clay minerals in the near-infrared/shortwave-infrared region are associated with overtones and combination

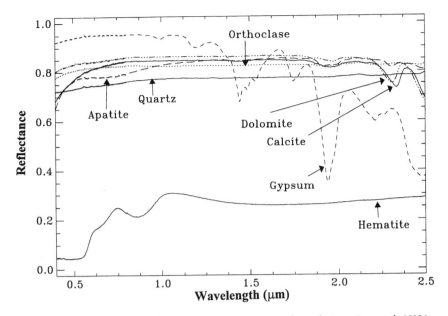

Figure 3.6 Reflectance spectra of representative pure nonclay minerals in soils. (From Grove et al., 1992.)

modes of fundamental vibrations of functional groups. Of all clay minerals, only the hydroxide group is spectrally active in the VNIR–SWIR region (Hunt 1979). The OH group can be found either as a part of the mineral structure (mostly in the octahedral position, which is termed *lattice water*) or as part of a thin water molecule directly and indirectly attached to the mineral surfaces (termed *adsorbed water*). Three major spectral regions are active for clay minerals in general and for smectite minerals in particular: around 1.3 to 1.4 µm, 1.8 to 1.9 µm, and 2.2 to 2.5 µm (Hunt and Salisbury 1970). For Ca–montmorillonite, which represents a common clay mineral in the soil environment, the lattice OH features are found at 1.410 µm (assigned $2v_{OH}$, where v_{OH} symbolizes the stretching vibration around 3.630 cm^{-1}) and at 2.206 µm (assigned $v_{OH} + \delta_{OH}$, where δ_{OH} symbolizes the bending vibration at around 915 cm^{-1}), whereas OH features of free water are found at 1.456 µm (assigned $v_w + 2\delta_w$, where v_w symbolizes stretching vibration at around 3.420 cm^{-1} and δ_w the bending vibration at around 1635 cm^{-1}), 1.910 µm (assigned $v'_w + \delta_w$, where v'_w symbolizes the high-frequency stretching vibration at around 3630 cm^{-1}), and 1.978 µm (assigned for $v_w + \delta_w$). Note that the assignment positions discussed above can change slightly from one smectite to another, depending on chemical composition and surface activity.

The spectra of three smectite end members are given in Figure 3.7 as follows: montmorillonite (dioctahedral, aluminous), nontronite (dioctahedral, iron rich), and hectorite (trioctahedral, manganese) minerals. The OH absorption feature of the $v_{OH} + \delta_{OH}$ in combination mode at around 2.2 µm is slightly but significantly, shifted for each end member. In highly enriched Al smectite (montmorillonite) the Al–OH bond is spectrally active at 2.16 to 2.17 µm. In highly enriched iron smectite (nontronite), the Fe–OH is effective at 2.21 to 2.24 µm, and in highly enriched magnesium smectite (hectorite) the Mg–OH bond is spectrally active at 2.3 µm. Based

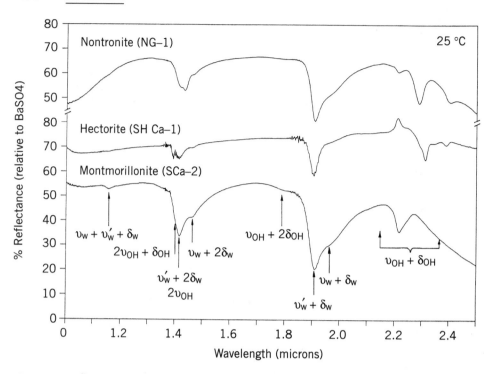

Figure 3.7 Reflectance spectra of three pure smectite end members in the near-infrafed/SWIR region (nontronite = Fe-smectite; hectorite = Mg-smectite; montmorillonite = Al-smectite) at room temperature (25°c). Also given are possible combination and overtone modes for explaining each of the spectral features.

on these wavelengths, Ben-Dor and Banin (1990a) were able to find a significant correlation between the absorbance values derived from the reflectance spectra and the total content of Al_2O_3, MgO, and Fe_2O_3. Except for a significant lattice OH absorption feature at around 2.2 μm in smectite, invaluable information about OH in free water molecules can be measured at around 1.4 and 1.9 μm. Because smectite minerals contributed to the soils relatively high specific surface areas, and those were covered by free and hydrated water molecules, these absorption features can be significant indicators of the water content in soils.

Kaolinite and illite minerals are also spectrally active in the SWIR region, as they both consist of octahedral OH sheets. From Figure 3.8, which presents pure spectra of nonsmectite layer-clay minerals (kaolinite, chlorite, vermiculite, gibbsite and illite), it is observed that different positions and spectral shapes of the lattice OH in the layer minerals affect soil spectra across the SWIR region. These changes are a result of different structural and chemical composition of the minerals. In the case of kaolinite, a 1:1 mineral (one octahedral and one tetrahedral), the fraction of the OH group is higher than in 1:2 minerals (one octahedral and two tetrahedral), and hence the lattice OH signals at around 1.4 and 2.2 μm are relatively strong, whereas the signal at 1.9 μm is very weak (because of relatively low surface areas and adsorbed water molecules). In the case of gibbsite, an octahedral aluminum structure, the 1.4-μm feature is even stronger, but the signal at 2.2 μm is shifted significantly to the infrared region relative to kaolinite. It should be noted that under relatively high

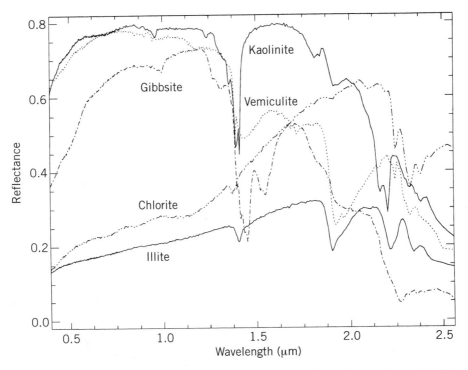

Figure 3.8   Reflectance spectra of representative non-smectite layer clay minerals, in soils. (From Grove et al., 1992.)

signal-to-noise conditions, a second overtone feature of the structural OH ($3v_{OH}$) can be observed at around 0.95 μm in layer OH-bearing minerals (Goetz et al., 1991).

The affinity of water molecules to clay mineral surfaces is correlated to their specific surface area. The specific surface area sequence of the foregoing minerals is smectite > vermiculite > illite > kaolinite > chlorite > gibbsite, which usually provide a similar spectral sequence at the water absorption feature near 1.8 μm (area and intensity). As smectite and kaolinite are clay minerals often found in soils, they can also appear in a mixed-layer formation in which the layers overlap spectrally. Kruse et al. (1991) described a specific case at Paris Basin, France, where intrastratification of smectite/kaolinite (a result of the alkaline weathering process of the flint-bearing chalk) was identified. Figure 3.9 presents the spectra of kaolinite, mixed layer kaolinite/smectite and halloysite (hydrated kaolinite) end members with the two representative spectra from the basin area soils examined by Kruse et al. (1991). The noticeable asymmetrical absorption feature of OH at 2.2 μm was further examined by Kruse to yield a graph that predicts the relative amount of kaolinite in the mixture (Figure 3.10).

B. Carbonates.   Carbonates, especially calcite and dolomite, are found in soils that are formed from carbonic parent materials or in a chemical environment that permits calcite and dolomite precipitation. Carbonates, especially those of fine particle size, play a major role in many of the soil chemical processes most likely to occur at the root zone. A relatively high concentration of fine carbonate particles may cause a fixation of iron ions in the soil and consequently, an inhibition of chlorophyll pro-

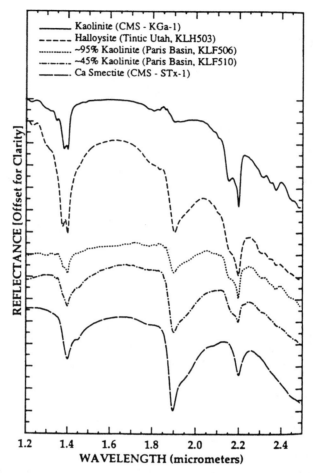

Figure 3.9 Reflectance spectra of kaolinite, mixed-layer kaolinite–smectite, from Paris Basin halloysite, from Tintic Utha and Ca–smectite from Texas. (After Kruse et al., 1991.)

duction. Absence of carbonate in soils, on the other hand, may affect the buffering capacity of the soil and hence negatively affect biochemical and physicochemical processes. The C–O bond, part of the CO_3^- radical in carbonate, is the spectrally active chromophore. Hunt and Salisbury (1971, 1976) pointed out that five major overtones and combination modes are available for describing the C–O bond in the SWIR region. Table 3.5 provides the band positions (calculated and observed) and their spectral assignments. In this table, v_1 accounts for the symmetric C–O stretching mode, v_2 for the out-of-plane bending mode, v_3 for the antisymmetric stretching mode, and v_4 for the in-plane bending mode in the infrared region. Gaffey (1986) has added two additional significant bands: at 2.23 to 2.27 μm (moderate) and 1.75 to 1.80 μm (very weak), whereas Van der Meer (1995) summarized the seven possible calcite and dolomite absorption features with their spectral widths.

It is evident that significant differences occur between the two minerals. This enabled Kruse et al. (1990), Ben-Dor and Kruse (1995), and others to differentiate

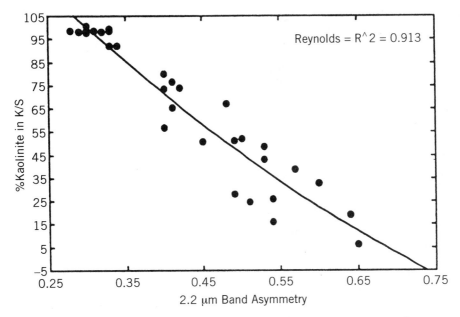

Figure 3.10   Correlation between the asymmetry of the 2.2-μm absorption band and percentage kaolinite from Paris Basin soil samples consisting of interstratification of kaolinite–smectite. (After Kruse et al., 1991.)

between calcite and dolomite formations using airborne spectrometer data with bandwidths of 10 nm. Gaffey and Reed (1987) were able to detect impurities of copper in calcite minerals, as indicated by a broad band between 0.903 and 0.979 μm. However, such impurities are difficult to detect in soils, because overlap with other strong chromophores may occur in this region. Gaffey (1985) showed that impurities of Fe in dolomite shift the carbonate's absorption bands toward longer wavelengths, whereas Mg in calcite shifts the band toward shorter wavelengths. As carbonates in soils are most likely to be impure, it is only reasonable to expect that the carbonates' absorption feature positions will differ from one soil to another.

A correlation between reflectance spectra and the carbonate concentration in soil was demonstrated by Ben-Dor and Banin (1990b). They used a calibration set of soil spectra and their chemical data to find three wavelengths that best predict the calcite content in arid soil samples (1.8, 2.35, and 2.36 μm). They concluded that the strong and sharp absorption absorption features of the C–O bands in the soils examined provide an ideal tool for studying the carbonate content in soils solely from their reflectance spectra. The best performance obtained for quantifying soil carbonate content ranged between 10 and 60%.

C. Organic Matter. Organic matter plays a major role with respect to many chemical and physical processes in the soil environment and has a strong influence on soil reflectance characteristics. Soil organic matter is a mixture of decomposing tissues of plants, animals, and secretion substances. The sequence of organic matter decomposition in soils is strongly determined by the soil's microorganism activity. In the initial stages of the decomposition process only marginal changes occur within the chemistry of the parent organic material. The mature stage refers to the final stage

TABLE 3.5   Position and Width of Carbonate Bands for Calcite and Dolomite

Carbonate Band	Calcite		Dolomite	
	Position (μm)	Width (μm^{-1})	Position (μm)	Width (μm^{-1})
1	2.530–2.541	0.0223–0.0255	2.503–2.518	0.0208–0.0228
2	2.333–2.340	0.0154–0.0168	2.312–2.322	0.0173–0.0201
3	2.254–2.270	0.0121–0.0149	2.234–2.248	0.0099–0.0138
4	2.167–2.179	0.0170–0.0288	2.150–2.170	0.0188–0.0310
5	1.974–1.995	0.0183–0.0330	1.971–1.979	0.0206–0.0341
6	1.871–1.885	0.0190–0.0246	1.853–1.882	0.0188–0.0261
7	1.735–1.885	0.0256–0.0430	1.735–1.740	0.0173–0.0395

Source: Van der Meer (1995), after Gaffey (1986).

of microorganism activity, where new, complex compounds, often called *humus*, are formed. The most important factors affecting the amount of soil organic matter are those involved with soil formation (i.e., topography, climate, time, type of vegetation, and the oxidastate). Organic matter, especially humus, plays an important role in many of the soil properties, such as soil aggregation, soil fertility, soil water retention, ion transformation, and soil color.

Because organic matter has spectral activity throughout the entire VNIR–SWIR region, especially in the visible region, workers have studied organic matter extensively from a remote sensing standpoint (e.g., Kristof et al., 1971). Baumgardner et al. (1970) noted that if the organic matter in soils drops below 2%, it has only a minimal effect on the reflectance property. Montgomery (1976) indicated that organic matter content as high as 9% did not appear to mask the contribution of other soil parameters to soil reflectance. In another study, Schreier (1977) indicated that organic matter content relates to soil reflectance by a curvilinear exponential function. Mathews et al. (1973) found that organic matter correlated with the reflectance values in the range 0.5 to 1.2μm, and Beck et al. (1976) suggested that the region 0.90 to 1.22 μm is suited for mapping organic matter in soils. Krishnan et al., (1980) used a slope parameter at around 0.8μm to predict organic matter content, and Da Costa (1979) found that simulated *Landsat* channels (bands 4, 5, and 6) yield reflectance readings that are significantly correlated with organic carbon content in soils.

The wide spectral range found by different workers to assess organic matter content suggests that organic matter is an important chromophore across the entire spectral region. Figure 3.11 shows a reflectance spectra of coarse organic matter (in the near-infrared/SWIR region) as isolated from an Alfisol and the humus compounds extracted from this organic matter (OM). Numerous absorption features exist that relate to the high number of functional groups in the OM. These can all be explained spectrally by combination and vibration modes of organic functional groups (Elvidge 1990, Chen and Inbar, 1994). Vinogradov (1981) developed an exponential model to predict the humus content in the upper horizon of plowed forest soils by using reflectance parameters between 0.6 and 0.7 m for two extreme end members (humus-free parent material and humus-enriched soil). Schreier (1977) found an exponential function to account for the organic matter content in soil from reflectance spectra.

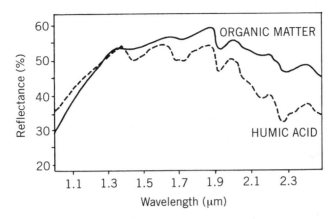

Figure 3.11   Spectral reflectance curves of pure organic matter isolated from Alfisol and its extracted humic acid.

Al-Abbas et al. (1972) used a multispectral scanner with 12 spectral bands covering the range from 0.4 to 2.6 μm from an altitude of 1200 m and showed that a polynomial equation will predict the organic matter content from only five channels. They implemented the equation on a pixel-by-pixel basis to generate an organic content map of a 25-ha field. Dalal and Henry (1986) were able to predict the organic matter and total organic nitrogen content in Australian soils using wavelengths in the SWIR region (1.702 to 2.052μm), combined with chemical parameters derived from the soils. Using similar methodology, Morra et al. (1991) showed that the SWIR region is suitable for identification of organic matter composition between 1.726 and 2.426 μm. Evidence that organic matter assessment from soil reflectance properties is related to soil texture and in more likely due to the presence of clays was given by Al-Abbas (1972) and Leger et al. (1979). Aber et al. (1990) noted that the organic matter, including its decomposition stage, affects the reflectance properties of mineral soil.

Baumgardner et al. (1985) demonstrated that three organic soils with different decomposition levels yielded different spectral patterns (Figure 3.12). A recent study by Ben-Dor et al. (1997), using a controlled decomposition process over more than a year, revealed significant spectral changes across the entire VNIR–SWIR region as the organic matter aged. Figure 3.13 shows a typical spectrum of grape marc (CGM) organic matter during a decomposition process that lasted 392 days. Significant changes can be seen in the slope values across the visible/near-infrared region and within the spectral features across the entire spectrum. Ben-Dor et al. (1997) postulated that some of the analyses traditionally used to assess organic matter content in soils from reflectance spectra may be biased by the age factor. As many soils consist of dry vegetation in one degradation stage or another, assessment of the organic matter using reflectance spectra should consider the vegetations' aging status. Although mineral soil consists of a relatively low content of organic matter (around 0 to 4%), accurate assessment of organic matter content in soils requires high spectral resolution data across the entire VNIR–SWIR region.

D. Water. The various forms of water in soils are all active in the VNIR–SWIR region (on the vibration activity of the OH group) and can be classified into three

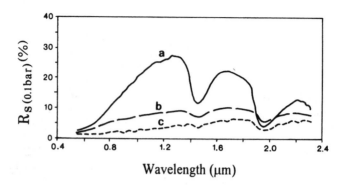

Figure 3.12 Spectral curves of three organic soils exhibiting different levels of decomposition: a, fibric; b, hemic; c, sapric. (After Baumgardner et al., 1985.)

major categories: (1) hydration water where it is incorporated into the lattice of the mineral [e.g., limonite ($Fe_2O_3 \cdot 3H_2O$) and gypsum ($CaSO_4 \cdot 4H_2O$)], (2) hygroscopic water which is adsorbed on soil surface areas as a thin layer, and (3) free water which occupies soil pores. Each of these categories influences the soil spectra differently, providing the capability of identifying the water condition of the soil, and is treated separately below. Three basic fundamentals in the infrared regions exist for water molecules particularly the OH group: ν_{w1} symmetric stretching; δ_w, bending; and ν_{w3}, asymmetric stretching vibrations. Theoretically, in a mixed system of water and minerals, combination modes of these vibrations can yield OH absorption features at around 0.95μm (very weak), 1.2 μm (weak), 1.4μm (strong), and 1.9 μm (very strong) related to $2\nu_{w1} + \nu_{w3}$, $\nu_{w1} + \nu_{w3} + \delta_w$, and $\nu_{w3} + 2\delta_w$, $\nu_{w3} + \delta_w$, respectively.

(1) Hydration water. The hydration water can be seen in minerals such as gypsum as strong OH absorption features at around 1.4 and 1.9 μm (Hunt and Salisbury, 1971; see Figure 3.6).

(2) Hygroscopic water. The hygroscopic (adsorbed) water in soils is adsorbed on the surface areas of clay minerals (especially smectite) and organic matter (especially humus). Early results by Obukhov and Orlov (1964) in the visible region showed that the slope of the spectral curve for soils is not affected by wetting and that the ratio of the reflectance of moist soil to that of dry soil remained practically constant. Sheilds et al. (1968) also pointed out that "moisture has no significant effect on the hue or chroma of several soils." Peterson et al. (1979) observed linear relationships between bidirectional reflectance factors at 0.71 μm of oven-dried soil samples that consisted of water tension between 15 and 0.33 bar. These findings actually suggest that soil albedo is the first factor in the soil spectrum that is altered upon soil wetting (Idso et al., 1975). The primary reason for this is the change in the medium surrounding the particles from air to water, which decreases their relative refractive index (Twomey et al., 1986; Ishida et al., 1991). Based on this idea, Ishida et al. (1991) developed a quantitative theoretical model to estimate the effect of soil moisture on soil reflection. The shape of soil reflectance curves are strongly affected by the presence of water absorption bands at around 1.4 and 1.9 μm, and occasionally, weaker absorption bands at around 0.95 and 1.2 μm. Figure 3.14 represents a Ca–

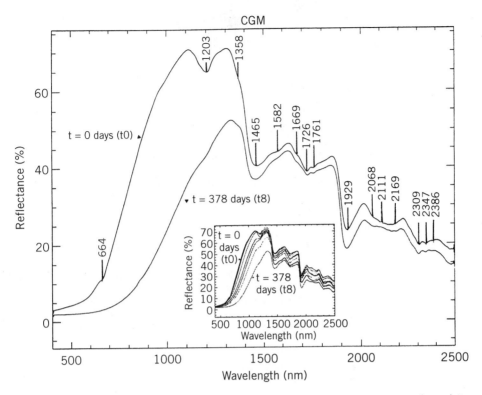

Figure 3.13   Reflectance spectra of two end members representing two extreme composting stages: t0 = 0 days and t8 = 378 days for grape marc material (CGM). Major wavelengths are annotated on the fresh organic matter (t0 = 0) where the inset shows the spectra of all intermediate decomposition stages (after Ben–Dor et al. 1997).

montmorillonite mineral that was exposed to four atmospheric conditions (32, 52, 75, and 100 % relative humidity) for about 20 weeks. It is noticeable that both water absorption feature areas and asymmetry increase with increasing relative humidity. This can be clearly seen in Figures 3.15 and 3.16, where the area and the asymmetry of water OH absorption features are plotted against the relative humidity (curves a and b in both figures). However, notice the presence of a feature at 2.2μm (Figure 3.14), attributed to lattice OH changes (curve c in Figure 3.15), even though free water molecules are assumed not to affect the octahedral OH under natural soil conditions. Because the hygroscopic water in soil is governed by the atmospheric conditions, the significant spectral changes are related to changes in the adsorbed water molecules on the mineral's surfaces. It is interesting to note that a similar observation was by Bowers and Hanks (1965) was based on soils that consisted of different moisture values (ranging from 0.8 to 20.2%). This observation demonstrates that the gas phase (water vapor in this case) in the soil environment plays a major role in the quantitative assessment of both structural and free water OH. Further insight to this problem was provided by Montgomery and Baumgardner (1974) and Montgomery (1976), who pointed out that it was not possible to assess water content in soils quantitatively because of different dry state conditions under which the soils were measured. Using reflectance spectra of several treated smectite

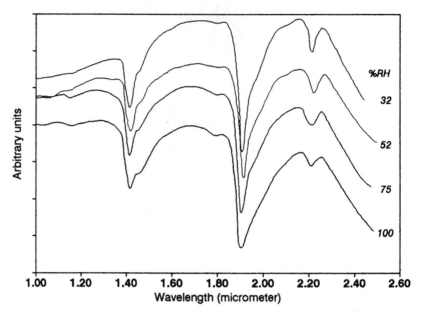

Figure 3.14   Reflectance spectra (in the near-infrared/shortwave infrared region) of Ca–montrmorillonite at various relative humidity (RH) conditions.

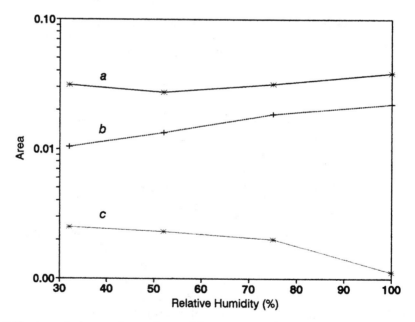

Figure 3.15   Area under the major absorption features of the smectitte spectra from Figure 3.14 versus the relative humidity: a = 1.325 to 1.55 μm, b = 1.825 to 2.15 μm, c = 2.15–2.25 μm.

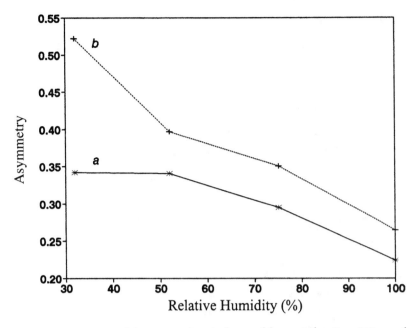

Figure 3.16   Asymmetry parameter of the two water absorption features of the smectite from Figure 3.14 versus the relative humidity: a = 1.325 to 1.55 μm, b = 1.825 to 2.15 μm.

minerals, Cariati et al. (1983) examined shifts of the OH absorption features at 1.4, 1.9, and 2.2μm. They found that vibration properties of the adsorbed water depend strongly on the composition of the smectite structure. In another study, Cariati et al. (1981) pointed out that several kinds of interactions are responsible for the vibration properties of the hygroscopic molecules, where sometimes this may even change with the water content. Because smectite is the most effective clay mineral in the soil environment that affects the reflectance spectrum at the major water absorption features, Cariati's observations may help us to understand the spectral activity of hygroscopic moisture in soils. Further work, however, is still required to implement the results obtained for pure smectite in the complex soil system.

(3) Free pore water.   Free pore water (wet condition) is water that is not in either the hygroscopic phase or filling the entire pore size (saturated condition). The rate of movement of this water into the plant is governed by water tension or water potential gradients in the plant soil system. Water potential is a measure of the water's ability to do work compared to pure free water, which has zero energy. In soils, water potential is less than that of pure free water, due in part to the presence of dissolved salts and the attraction between soil particles and water. Water flow from areas of high potential to lower potential and hence flow from the soil to the root and up the plant occurs along potential gradients. In agricultural systems plant growth occurs with soil water potentials between 15 and 0.3 bar tension (note these are actually negative water potentials); however, water tensions in dessert environments are far greater. Baumgardner et al. (1985) studied the reflectance spectra of a representative soil (Typic Hapludalf by the U.S. Department of Agriculture) over

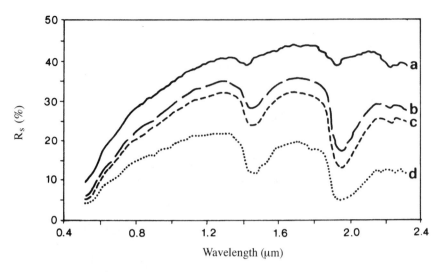

Figure 3.17 Spectra curve of Typic Hapludalf soil at four different moisture tensions: a, oven dry; b, 15 bar; c, 0.3 bar; d, 0.1 bar. (After Baumgardner et al., 1985.)

various water tensions (Figure 3.17). As expected, when water tension decreased (and hence water content increased) the general albedo decreased and the area under 1.4- and 1.9-μm water absorption features at 1.4 and 1.9 μm also decreased. Clark (1981) examined the reflectance of montmorillonite at room temperature for two different water conditions (Figure 3.18) and showed that albedo decreased dramatically from dry to wet material. Other changes related to the water and lattice OH can be observed across the entire spectrum as well. Some of these changes are related directly to the total amount of free and adsorbed water and some, to the increase of the spectral reflectance fraction of the soil (wet) surface. In kaolinite minerals, a similar trend was observed in two moisture conditions; however, the changes around the water OH absorption features were less pronounced than in montmorillonite (Figure 3.19). In montmorillointie, adding water to the sample enhanced the water OH features at 0.94, 1.2, 1.4, and 1.9 μm, because of the relatively high surface area and a corresponding high content of adsorbed water. In kaolinite, the relatively low specific surface area obscured a similar response, and hence only small changes are noticeable. Note that in the montmorillonite, the lattice-OH features at 2.2 μm diminished just as happened with Ca–montmorillonite exposed to different humidity conditions (Figure 3.14), suggesting that the hygroscopic moisture is a major factor affecting the clay minerals' (and soil's) spectra. Clark (1981) also studied a mixture of water in montmorillonite at low temperature that actually simulated a frost situation. In Figure 3.20 it is observed that a noticeable increase in water and lattice-OH features occurred with lower temperatures, emphasizing that the water chromophore in soil is strongly dependent on the soil temperature and water phases. In soils where the entire pore size (or more) is filled with water (in saturated conditions, respectively), it is more likely that the soil reflectance consists of more specular than Lambertian components.

E. Quantitative Assessment of Soil Moisture from Reflectance. Because soil moisture variation changes the soil's albedo values dramatically, this parameter has been

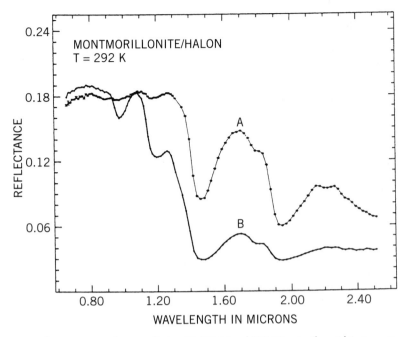

Figure 3.18 Reflectance spectra of montmorillonite with 50% (*A*) and 90% (*B*) water (by weight) at room temperature. (After Clark, 1981.)

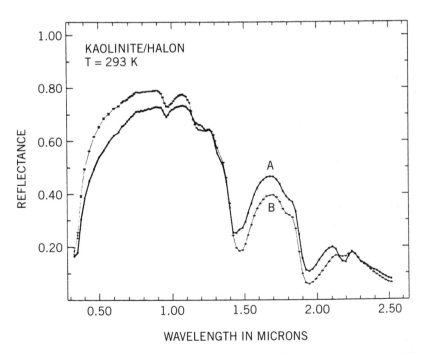

Figure 3.19 Reflectance spectra of kaolinite with 25% (*A*) and 60% (*B*) water (by weight) at room temperature. (After Clark, 1981.)

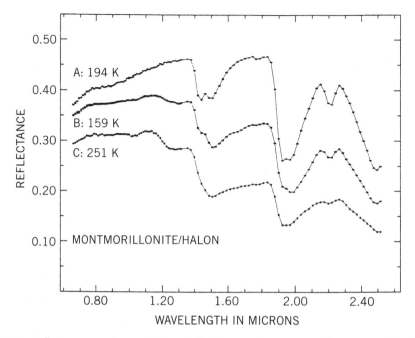

Figure 3.20 Reflectance spectra of montmorillonite mixed with water at low temperatures. The water-ice and bound-water components are 15% and 15% (*A*), 20% and 15% (*B*), and 60% and 15% (*C*) (by weight). (After Clark, 1981.)

widely studied by many workers (Evans, 1948; Brooks, 1952; Condit, 1970; Idso et al., 1975; Twomey et al., 1986). Leger et al. (1979) studied the effect of soil water with other chromophores, such as organic matter and iron oxides, and concluded that the interaction among all three components is more important to understanding the soil spectra than is considering the components individually. Jacquemoud et al. (1992) noted that modeling of soil spectral reflectance as a function of water content is a difficult task as part of the saturation of the 1.9-μm absorption feature. Nevertheless, Downey and Byrne (1986) were able to predict the moisture content of milled peat from reflectance spectra by studying a set of known wet samples at four wavelengths: 2.310, 2.180, 2.100, and 1.680μm. Similarly, Ben-Dor et al. (1991) showed that hygroscopic moisture in arid and semiarid soils can be predicted from two wavelengths in the SWIR region (2.362 and 2.120 μm), and Dalal and Henry (1986) predicted the hygroscopic moisture of Australian soils from three wavelengths: at 1.926, 1.954, and 2.150μm. After examining 160 soils, Condit (1970) noted that although in some cases the soils' spectral shape in the visible region does not change with soil wetting, the near-infrared region is more likely to be changed. Nevertheless, in other soils (mostly sandy soils), the spectral curve shapes have been observed to be significantly changed upon wetting of the soil (Da Costa 1979), which suggests that anomalies may be related to the spectral activity of water in soils.

F. Iron. Iron is the most abundant element on the earth as a whole and the fourth-most-abundant element in the earth's crust. The average Fe concentration in the earth's crust is 5.09 mass, and the average Fe^{3+}/Fe^{2+} ratio is 0.53 (Ronov and Yaroshevsky, 1971). The geochemical behavior of iron in the weathering environment is determined largely by its significantly higher mobility in the divalent than in the trivalent state. Change in the oxidation state, and consequently in mobility, tend to take place at different soil conditions. Major Fe-bearing minerals in the earth's crust are the mafic silicates, Fe-sulfides, carbonates, oxides, and smectite clay minerals. All Fe^{3+} oxides have striking colors, ranging among red, yellow, and brown, due to selective light absorption in the visible range caused by transitions in the electron shell. It is well known that even a small amount of iron oxides can change the soil color significantly. The red, brown, and yellow hue values, all caused by iron, have been used widely in soil classification systems in almost all countries and languages.

A representative soil spectra with various amounts of total Fe_2O_3 is presented in Figure 3.21. The iron's feature assignments in the VNIR region result from the electronic transition of iron cations (3+, 2+), either as the main constituent (as in iron oxides) or as impurities (as in iron smectite). Hunt et al. (1971a) summarized the physical mechanism that allows Fe^{2+} (ferrous) and the Fe^{3+} (ferric) to be spectrally active in the VNIR region as follows: They stated that the ferrous ion typically produces a common band at around 1 µm due to the spin allowed during transition between the E_g and T_{2g} quintet levels into which the D ground state splits into an octachedral crystafield. Other ferrous bands are produced by transitions from the $^5T_{2g}$ to $^3T_{1g}$ at 0.55 µm; to $^1A_{1g}$ at around 0.51 µm; to $^3T_{2g}$ at 0.45 µm, and to $^3T_{1g}$ at around 0.43 µm. For the ferric ion, the major bands produced in the spectrum are the result of the transition from the $^6A_{1g}$ ground state to $4T_{1g}$ at around 0.87 µm, to $^4T_{2g}$ at around 0.7 µm, and to either $^4T_{1g}$ or ^4E_g at around 0.4 µm.

Just as organic matter is an important indicator for soils, iron oxides provide significant evidence that soil is being formed (Schwertmann, 1988). Iron oxide content and species are strongly correlated with the soil weathering process in both the short and long terms. Transformation of iron oxide in soil often occurs during natural soil conditions. Hematite and goethite are common iron oxides in soils, and their relative content in soils is strongly controlled by soil temperature, water, organic matter, and annual precipitation. Hematitic soils are reddish, and goethitic soils are

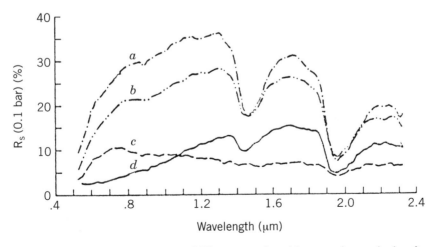

Figure 3.21   Reflectance spectra of soils consisting of different textures but exhibiting iron absorption bands: a, fine sand, 0.20% Fe_2O_3; b, sandy loam; 0.64% Fe_2O_3; c, clay, 25.6% Fe_2O_3; d, silty loam, 0.76% Fe_2O_3. (After Baumgradner et al., 1985.)

yellowish brown. Their reflectance spectra also differ, as can be seen in Figure 3.22. Hematite (α-Fe$_2$O$_3$) has Fe^{3+} ions in octahedral coordination with oxygen. Goethite (α-FeOOH) also has Fe^{3+} in octahedral coordination, but different site distortions, along with oxygen ligand (OH), provide the main absorption features that appear near 0.9 μm. Based on the spectral differences in the visible region of hematite and goethite, Madeira et al., (1997) developed indices to distinguish between these two components in highly leached soils from Brazil. These indices were developed to account for spatial distribution of iron oxides in a large scale using satellite TM data. Based on climatic conditions, Schwertmann drew a tentative line between hematitic and nonhematite soils of the world (Figure 3.23). From a more precise spatial point of view, even local environmental conditions such as organic matter and local weather condition may change the ratio between hematite and goethite, as seen in Figure 3.24.

Leidocrocite (γ-FeOOH), which is associated with goethite but rarely with hematite, is another common unstable iron oxide found in soils. It appears mostly in the subtropical regime and is often found in the upper subsoil position (Schwertmann, 1988). Maghemite (γ-Fe2O3) is also found in soils, mostly in subtropical and tropical regions and has occasionally been identified in soils in humid temperature areas. Ferrihydraite is a highly disordered Fe^{3+} oxide mineral found in soils in cool or temperate moist climate areas, characterized by young iron oxide formations and soil environments relatively rich with other compounds (e.g., organic, silica, etc.).

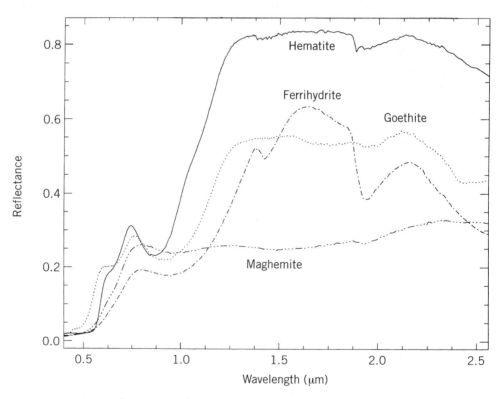

Figure 3.22   Reflectance spectra of representative iron oxide minerals in soils (From Grove et al., 1992.)

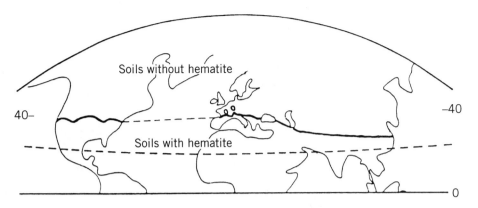

Figure 3.23 A line at approximately 40°N latitude separates soils with and without hematite. (After Schwertmann, 1988.)

Iron associated with the structure of clay minerals is also an active chromophore in both the VNIR and SWIR spectral regions. This can be seen in the nontronite type of mineral already presented in Figure 3.7. Based on the structural OH–Fe features of smectite in the SWIR region, Ben-Dor and Banin (1990a) were able to generate a prediction equation to account for the total iron content in a series of smectite minerals. The wavelengths selected automatically by the method they used were 2.2949, 2.2598, 2.2914, and 1.2661 μm. Stoner (1979) also observed a higher correlation between reflectance in the region 1.55 to 2.32 μm and the iron content in soils, whereas Coyne et al. (1989) found a linear relationship between total iron content in montmorillonite and the absorbance measured in the spectral region 0.6 to 1.1 μm. Ben-Dor and Banin (1995a) used spectra of 91 arid soils and showed that total iron content in soils (both free and structural iron) can be predicted by multiple linear regression analysis and wavelengths at 1.075, 1.025 and 0.425 μm. Obukhov and Orlov (1964) generated a linear relationship between the reflectance values at 0.64 μm and the total percentage of Fe_2O_3 in other soils. Taranik and Kruse (1989) were able to show that a binary encoding technique of the spectral slope values across the VNIR spectra region is capable of differentiating a hematite mineral from a mixture of hematite–goethite–jarosite minerals.

It is important to mention that an indirect influence of the iron on the overall spectral characteristics of soils can often occur. In the case of free iron oxides, it is well known that soil particle size is strongly related to the absolute iron oxide content (Soileau and McCraken, 1967; Stoner and Baumgardner, 1981; Ben-Dor and Singer, 1987). As the iron oxide content increases, the size fraction of soil particles increases as well, because of the cementation effects of the free iron oxides. As a result, problems resulting from various scattering effects are introduced within the soil being examined. Moreover, free iron oxides, mostly in their amorphous state, may coat the soil particles with a film that prevents natural interaction between the soil particle (clay or nonclay minerals) and the sun's photons. Karamanova (1981) found that well-crystallized iron compounds had the strongest affect on the spectral reflectance of soil and that removal of nonsilicate iron (mostly iron oxides) helps to enhance other chromophores in the soil. In this respect, Kosmas et al. (1984) have demonstrated that a second derivative technique in the visible region is a feasible approach

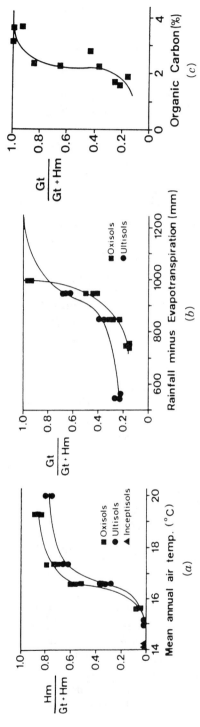

Figure 3.24 Relationship of hematite (Hm) and goethite (Gt) versus several environmental conditions in Brazilian soils: (*a*) mean annual air temperature; (*b*) rainfall minus evapotranspiration; (*c*) organic matter content. (After Schwertmann, 1988.)

154

for differentiating even small features of synthetic goethite from clays and have suggested that such a method may be adopted to assess quantities of iron oxide in mixtures.

It can be concluded from the discussion above that iron in soil is a very strong chromophore and the determination of its content in clay and soils from the reflectance spectra in the entire VNIR–SWIR region is feasible. Based on the complexity of the iron component in the soil environment as well as on the intercorrelation between iron and other soil components, sophisticated methods and relatively high spectral resolution data are absolutely needed to determine iron content using the reflectance spectra.

G. Soil salinity. Soil salinity is one of the major factors affecting biomass production and is the principal cause of soil degradation (Csillag et. al., 1993). Salt-affected areas cover about 7% of the Earth's land surface (Toth et al., 1991) and are located mostly in arid and semiarid regions (Verma et al., 1994). However, salt-affected soils can also be found in subhumid and coastal zone areas associated with hydrogeological structures. Salts in soils are reported to be Na_2CO_3, $NaHCO_3$, and $NaCl$, which are very soluble and mobile components in the soil environment. Typically, saline soils are highly erosive and have poor structure, low fertility, low microbial activity, and other attributes not conducive to plant growth.

The spectral signature of saline soils can be a result of the salt itself, or indirectly, from other chromophores related to the presence of the salt (e.g., organic matter, particle size distribution). Hunt et al. (1971c) reported an almost featureless spectrum of halite (NaCl 433B from Kansas). Although salt is spectrally a featureless property, Hick and Russell (1990) raised a hypothesis that there are certain wavelengths in the VNIR–SWIR region that can provide more accurate information about saline-affected areas. For example, a laboratory spectra of mixtures of SiO_2 and $NaCl + MgCl_2$ showed significant features associated with two of the water absorption bands (at around 1.4 and 1.9 μm, Figure 3.25), due presumably to the high affinity of the salt to water molecules. Hirschfeld (1985) was able to predict the NaCl concentration in a water solution based on four features in the SWIR region at 1.63, 1.70, 2.08, and 2.17 μm. Similarly, Lin and Brown (1992) found that in a water solution, a very good linear correlation could be obtained between the absorption [$\log(1/T)$; T = transmittance] at 1.7809 μm and varying NaCl concentrations (0.1 to 5 M).

Vegetation is an indirect factor that facilitates detection of salt in soils using reflectance measurements (Hardisky et al., 1983; Wiegand et al., 1994). Gausman et al. (1970), for example, pointed out that cotton leaves grown in saline soils had a higher chlorophyll content than that of leaves grown in low-salt soil. Hardisky et al. (1983) used the spectral reflectance of a *Spartina alterniflora* canopy to show a negative correlation between soil salinity and spectral vegetation indices. In the absence of vegetation, the major influence of salt is on the structure of the upper soil surface. In Figure 3.26, saline and nonsaline spectra in the VNIR region are given. As can be seen, the saline soils are relatively higher in albedo than the nonsaline soils. The saline soils had crusted surfaces that tend to be smoother than the generally rough surfaces of the nonsaline soils. Although Gausman et al. (1977) and Rao et al. (1995) reported similar trends in other soils, it should be noted that in relatively high-salt-content soils, the opposite behavior is also reasonable. This is due to the fact that

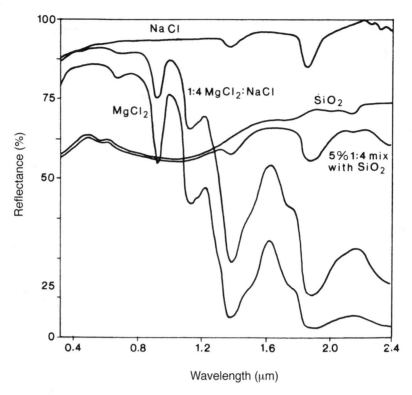

Figure 3.25 Reflectance spectra of sodium and magnesium chlorides with silica mixtures. (After Hick and Russell, 1990.)

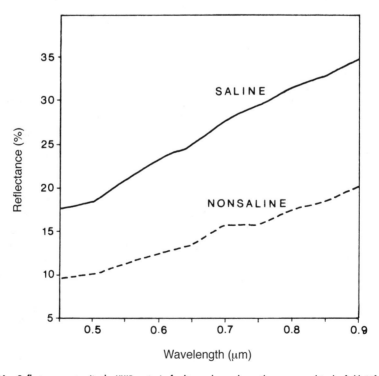

Figure 3.26 Reflectance spectra (in the VNIR region) of saline and nonsaline soils as measured in the field. (After Everitt et al., 1988.)

156

salt is a very hygroscopic material which tends to decrease the soil albedo as water content increases.

Because no direct significant spectral features are found in the VNIR–SWIR region for identifying sodic soil, indirect techniques are thought to be more appropriate for classifying salt-affected areas (Sharma and Bhargava, 1988; Verma et al., 1994). Salt in water is most likely to affect the hydrogen bond in water molecules, causing subtle spectral changes, and based on this, Hirschfeld (1985) suggested that high-spectral-resolution data are required. Support for this idea is given by Szilagyi and Baumgardner (1991), who reported that characterizing the salinity status in soils was feasible with high-resolution laboratory spectra. A relatively high number of spectral channels is required to identify an indirect relationship between salinity and other soil properties that appear to consist of chromophores in the VNIR–SWIR regions. Csillag et al. (1993) analyzed high-resolution spectra from about 90 soils in the United States and Hungary against chemical parameters, including clay and organic matter content, pH, and salt. They state that because salinity is such a complex phenomenon, it cannot be attributed to a single soil property. While studying the capability of commercially available Earth-observing optical sensors, they were able to point out that six broad bands in the visible/near-infrared/SWIR region best discriminate soil salinity (Figure 3.27). These six channels were selected solely on the basis of their overall spectral distribution, which provided complete information about salinity status. Thus it can be concluded that it is necessary to look at the entire spectral region in order to evaluate the salinity level in different environments and unknown soil systems. Mougenot (1993) noted that in addition to an

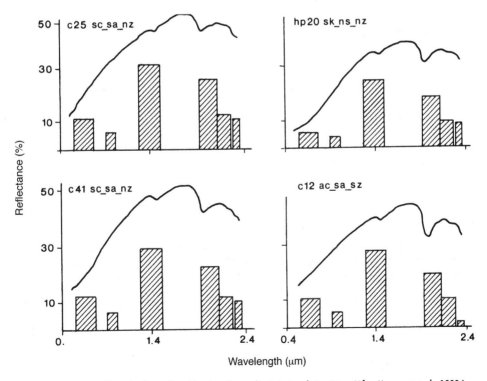

Figure 3.27   Overall weights for six broad bands to better discriminate salinity status. (After Mougenot et al., 1993.)

increase in reflectance with salt content, high salt content may mask ferric ion absorption in the visible region. They concluded that salts are not easily identified in proportions below 10 or 15% Recent study by Metternicht and Zinck (1997) shows that in highly saline soils a spatial distribution of salt and sodium-affected areas can be generated using a transform divergence technique that used six combined TM bands. One more important factor about saline soils is the fact that in modern agriculture, farmers are adding gypsum to sodic soils for soil reclamation (Singh, 1994). The artificial increase of the gypsum content in such soils may alter the soil reflectance spectra significantly (Hunt et al., 1971b) and hence requires attention. In summary, although salt is not a strong and direct chromophore, its interaction with other soil components (water, structure, iron, and organic matter) makes its assessment possible but complicated.

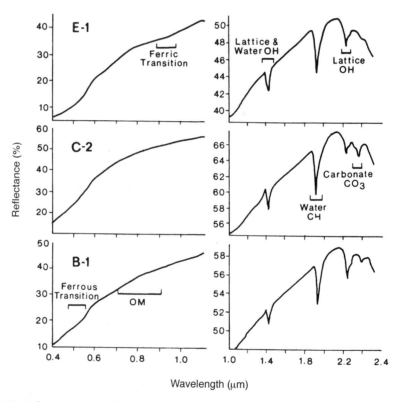

Figure 3.28 Reflectance spectra of three representative arid and semiarid soils with their corresponding chemical values determined in the laboratory. OM, organic matter; EC, electrical conductivity; Fe_2O_3, total iron in soil; Fe(d), iron oxides; HIGF, hygroscopic moisture. Also given are chemically active spectral constituents.

Soil	USDA Name	CaCo₃ (%)	OM (%)	EC (ds/cm)	Fe₂O₃ (%)	Fe(d) (ppm)	HIGF (%)
E-1	Rhodoxeralf	1.83	1.27	0.504	1.3	4302	1.1
C-2	Xerorthent	40.3	1.02	0.55	1.3	1967	2.7
B-1	Haploxeralf	41.9	1.91	0.63	3.8	5232	4.7

3.5.2.3   CHEMICAL CHROMOPHORES: SUMMARY

The chemical chrompohores are related to the vibration of chemical bonds and the charge transfer of ions. Figure 3.28 gives three spectra of representative soils from arid and semiarid areas in the VNIR and SWIR regions, with spectral regions attributed directly to charge transfer and vibration chromophores, respectively. Also provided are the values of selected chromophores properties determined in the laboratory. To summarize the overall chemical chromophore activity in soils, we provide a summary of chromophores associated with soil and geological matter from the literature (Figure 3.29). Also given are the intensities of each chromophore in the

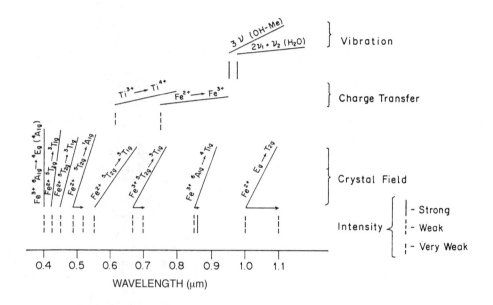

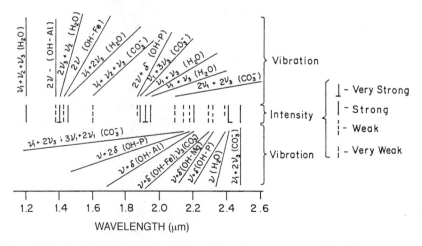

Figure 3.29   Active groups and mechanism in the soil chromphores spectrum. For each possible group the wavelength range and absorption feature intensity is given. The spectrum was generated using information gathered from the literature.

VNIR–SWIR spectral regions as they appear in these studies. The current review demonstrates that high-resolution spectral data can provide additional information, sometimes quantitative, about soil properties strongly correlated to the chromphores (e.g., water, primary and secondary minerals, organic matter, iron oxides, water, and salt).

3.5.3 Physical Chromophores

In addition to chemical chromophores, as discussed above, the reflectance of light from the soil surface is dependent on numerous physical processes. Reflection, or scattering, is clearly described by Fresnel's equation and depends on the angle of incidence radiation and the index of refraction. Generally, physical factors are those parameters that alter a soil spectrum with regard to Fresnel's equation but do not cause changes in the position of the specific chemical absorption. These parameters include particle size, sample geometry, viewing angle, radiation intensity, incident angle, and azimuth angle of the source. Changes in these parameters are most likely to affect the shape of the spectral curve through changes in baseline height and absorption feature intensities. In the laboratory, measurement conditions can be maintained constant; in the field, several of these parameters are unknown and may introduce problems in accurate assessment of the effect of these parameters on soil spectra. Many studies covering a wide range of materials have shown that particle size differences alter the shape of soil spectra (powdered material) (Hunt and Salisbury, 1970; Pieters, 1983; Baumgardner et al., 1985). Specifically, Hunt and Salisbury (1970) quantified effects of about 5% in absolute reflectance due to particle sizes differences and noted that these changes occurred without altering the position of diagnostic spectral features. Under field conditions, aggregate size rather than particle size distributions may be more important in altering soil spectra (Orlov, 1966; Baumgardner et al., 1985). In the field, aggregate size may change over a short time frame due to tillage, soil erosion, eolian accumulation, or physical crust formation (e.g., Jackson et al., 1990). Basically the aggregate size, or more likely, roughness, plays a major role in the shape of field and airborne soil spectra (e.g., Cierniewski 1987, 1989). Escadafal and Huete (1991) showed that five soils with a rough surface exhibited strong anisotropy in reflectance properties. Cierniewski (1987) developed a model to account for soil roughness based on soil reflectance parameters, illumination properties, and viewing geometry for both forward and backward slopes. The model shows that the shadowing coefficient of soil surface decreases with a decrease in soil roughness. For soils on slopes of more than 20° the shadowing coefficient also decreases when the solar altitude increases 0 to 90°. The model indicates that the relationship for soils sloped with a surface roughness lower than 0.5 may be the opposite. Using empirical observations of smooth soil surfaces, Cierniewski showed that the model agreed closely with field observations. A concise summary of the multiple-and single-scattering models of soil particles with respect to the roughness effect is given by Irons et al. (1989).

Viewing angle is also a parameter that significantly affects the reflectance spectra of any object on the Earth's surface (Escadafal and Hute, 1991); however, there are various models to correct for changes in this parameter (Egbert and Ulaby, 1972; Hapke, 1993, Liang and Townshend 1995). In soils the relationship between a shadowing coefficient (simulated), soil roughness factor, solar altitude, and the slope of the soil surface is given in Figure 3.30. It can be clearly seen that as the solar altitude (a) decreases, the shadow parameter (Scm) increases for any rough soil surface factor parameter (RF) and for both forward and backward slopes. It is also evident that as the soil becomes rougher (high RF values), the shadow parameter (Scm) becomes significantly magnified for almost any slope direction. Various authors have attempted to model the geometry of the soil surface to evaluate the effect of different viewing angles. The models considered the soil surface as being composed of spheres (Cierniewski, 1987, 1989), cylinders (Den Dulk, 1989), cubes (Escadafal, 1989), or blocks, ripples, and paraboloids (Mulders et al., 1992). The predictions from these models agreed rather well with empirical observations. The validity of these models depends on their ability to quantify the soil surface accurately in terms of the various shapes, the reflectance differences of the surface constituents, and the reflectance attributed to shadow. Epema (1992) found differences of about 10% in nadir reflectance of moderately rough soils with solar zenith angles of 30° and 60°. Kimes (1983) and Kimes et al. (1985) presented results of the directional distribution of reflectance of bare and vegetated surfaces. For the reflection of bare surfaces, backscattering in the illumination direction dominated. Coulson (1966) found the same results under laboratory conditions for most minerals, but he observed a dominant forward scattering for low-absorbing materials such as gypsum and quartz. Under field conditions, however, even for gypsum and quartz surfaces, backscattering dominates.

A practical solution for evaluating the affects of physical parameters is to evaluate the reflectance of a given target relative to a perfect reflector measured at the same geometry and viewing angle of the target in question. In reality such conditions are impossible to achieve in the field, and complex effects such as particle size effect cannot be removed absolutely by this method. It is postulated that more effort should be expended to account more precisely for physical effects under field conditions (both from a spectroscopy and imaging spectroscopy point of view).

3.5.4 Relationship Between Soil Chromophores and Properties

Soil reflectance spectra are affected directly by chemical and physical chromomophores, as discussed previously. The spectral response is also a product of interaction between these parameters, and a precise understanding of all chemical and physical reactions in soils is necessary. For example, even in a simple mixture of iron oxide, clay, and organic matter the spectral response cannot just be judged by simple linear mixing models of the three end members. There are strong chemical interactions between these components, which in most cases are nonlinear and rather complicated. For example, organic components, mostly humus, affect soil clay minerals in chemical and physical ways. Similarly, free iron oxides may coat soil particles and mask photons that interact with the mineral components or iron oxides (and organic matter as well). Additionally, the coating material may collate fine particles into coarse aggregates that may physically change the soil's spectral behavior. Karmanova (1981) selectively removed the iron oxides from soil samples and concluded that the effect of various iron compounds on the spectral reflectance and color of soils was not proportional to their relative content. Another example of the strong relationship

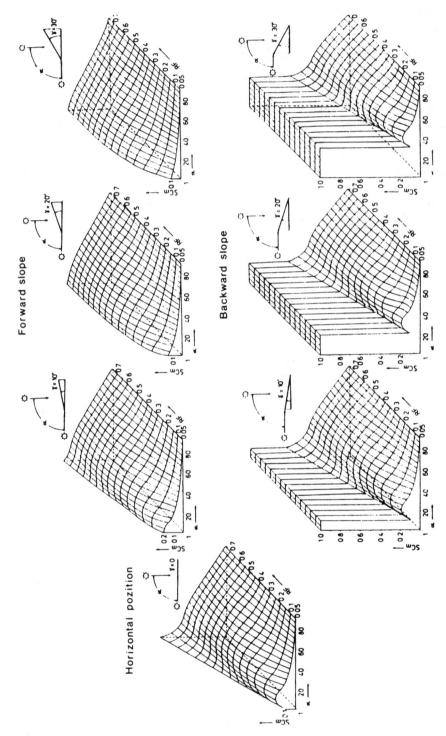

Figure 3.30 Relationship between the simulated shadowing coefficient of a soil surface (SCm) and the rough soil surface factor (BRm), the solar altitude (α), and the slope of soil surface (γ). (After Cierniewski, 1987.)

162

between one chromophore and another is given by Bedidi et al. (1990, 1991), who showed that the normally accepted view of decreasing the soil baseline height with increasing moisture content (visible region) did not hold for lateritic soils (highly leached low-pH soils). They concluded that the spectral behavior of such soils under various moisture conditions is more complex than originally thought. In this respect, Galvao et al. (1995) have shown that spectra from laterite soils (VNIR region) consist of complex spectral features which appear to deviate from other soils. Al-Abbas et al. (1972) found a correlation between clay content and reflectance data in the VNIR–SWIR region and suggested that this was not a direct but an indirect relationship, strongly controlled by the organic matter chromophore. Another anomaly that relates to the interactions between soil chromophores was identified by Gerbermann and Neher (1979). They carefully measured the reflectance properties in the VNIR region of a clay–sand mixture extracted from the upper horizon of a montmorilonic soil and found that adding sand to a clay soil increases the percent of soil reflectance. This observation stands in contrast to what is traditionally expected from adding coarse (sand) to fine (clay) particles in a mixture (soil), which tends to decrease soil reflectance. Similarly, Ben-Dor and Banin (1994, 1995a–c) concluded that intercorrelations between feature and featureless properties play a major role in assessing unexpected information about soil solely from their reflectance spectra in either the VNIR or SWIR regions. Ben-Dor and Banin (1995b) examined arid and semiarid soils from Israel and were able to show that "featureless" soil properties (i.e., properties without direct chromophores such as K_2O, total SiO_2, and Al_2O_3) can be predicted from the reflectance curves due to their strong correlation with "feature" soil properties (i.e., properties with direct chromophores). Csillag et al. (1993) considered the effect of multiple factors affecting the soil spectra in a study of soil salinity. They stated that "salinity is a complex phenomenom and therefore variation in the [soil] reflectance spectra cannot be attributed to a single [chromophoric] soil property." To extract the most information from the soil spectra, they examined chromophoric properties such as organic matter and clay content and ran a principal component analysis to account for the salinity status from the soil reflectance spectra.

These few examples show that soil chromophores do not stand alone in the soil matrix and that spectral anomalies are often found in the soil environment. Studying the soil system by examining all available information about the soil population (spectral and chemical) is the key to understanding soil reflectance spectra and their relationship to soil properties. To some extent this suggests that soil spectra should be judged and examined with caution in order to obtain quantitative information about the soil. This is despite the fact that the chemicophysical mechanism of each the VNIR–SWIR region is well understood.

3.6 FACTORS AFFECTING SOIL REFLECTANCE

3.6.1 Background

Whereas in the laboratory, soil spectra are taken under controlled conditions, in the field, uncontrolled conditions provide an environment that makes soil reflectance interpretation more difficult. Vegetation is one of the major factors masking soil signals and can be termed a biosphere interference. The FAO (1994) states that about 56% of land areas are covered by green vegetation such as forest, pasture, and crops, and the remainder is covered by dry vegetation, snow, and urban and bare soil areas. Within the unvegetated area, only a portion of the soils are characterized by an unaltered surface layer (e.g., as a matter of soil tillage), and hence even partial sensing of the natural soil surface is difficult. This effect can be termed *surface cover interference*. Another important problem in acquiring accurate soil reflectance spectra from air and space is atmospheric interference. Electromagnetic energy interacts with atmospheric gas molecules and aerosol particles that may cause misinterpretation of the "soil spectrum" derived from airborne sensors. These interferences are reviewed briefly in the following section.

3.6.2 Biosphere

3.6.2.1 HIGHER VEGETATION.

Soil is a growing environment for green plants (natural and agricultural) and a sink for decomposing tissues of vegetation and fauna. Because large parts of the world's soils are vegetated (green or dry), the problem of deriving soil spectra from the mixture of soil–vegetation signals is rather complex, especially if dead and live vegetation appear side by side (Aase and Tanaka, 1983). Siegal and Goetz (1977) postulated that "the effect of naturally occurring vegetation on spectral reflectance of Earth materials is a subject that deserves attention." At one extreme are situations where canopy cover is so dense that reflectance from soils is too difficult to interpret. In situations where vegetation cover is only partial, a mixed signal from soil and vegetation occurs, and to some extent the chemical and physical components can be resolved (Murphy and Wadge, 1994). In soil–vegetation mixtures, nonlinear models are typically used to resolve issues of the soil spectra (Goetz, 1992; Ray and Murray, 1996). Otterman et al. (1995) noted that the relaionship between the amount, type, and architecture of vegetation cover and the reflectance properties of the underlying soil is an important issue (e.g., low-albedo soils are those affected most significantly by vegetation). The spectral region 0.68 to 1.3 μm of soils is the region most affected by green vegetation, as a result of the steep reflectance rise caused by vegetation (e.g., Ammer et al., 1991). Dry vegetation does not alter the spectrum in the VNIR region, except for changing the albedo, whereas in the SWIR region spectral features are related to cellulose, lignin, and water content. The low reflectance of green vegetation beyond 1.4 μm indicates that if a soil–vegetation mixture exists, most of the spectral information relates to rock and soil material (Siegal and Goetz, 1977). Two chromophores, water and organic matter, which exist in both plant and soil material, can complicate interpretation of spectra, particularly in the SWIR region. In the green vegetation–soil mixture, liquid water of green and dry vegetation may overlap with the soil water forms. Also, signatures of lignin, cellulose, and protein can significantly affect the soil components in the soil–vegetation mixture. Murphy and Wadge (1994) showed in one case that living vegetation has a significant impact on the SWIR region of soil spectra, whereas decaying vegetation had a greater impact on the 2.2-μm absorption features (see, e.g., reflectance spectra of pure organic matter given in Figure 3.31). Murphy and Wadge (1994) concluded that decaying vegetation tissues had a greater impact on soil spectra than live vegetation, and they suggested greater consideration by workers regarding the effect of dead vegetation on soil spectra.

From a vegetation point of view, Tucker and Miller (1977) postulated that "remotely sensed data of vegetated surfaces could be analyzed more accurately if the contribution of the underlying soils spectra are known." Tueller (1987) and Smith et al. (1990) noted that it is difficult to extract vegetation information when it covers less than 30 to 40% of an area. The normalized differential vegetation index (NDVI) is a parameter commonly utilized to estimate the cover of green vegetation from satellite and airborne data. The index, which is based on the normalized difference between the near-infrared and visible reflectance values, is very sensitive to soil background, atmosphere, and sun-angle conditions. Based on that background, Huete (1988) developed a new index called the soil-adjusted vegetation index (SAVI) which accounts for soil brightness and shadows, and more recently (Liu and Huete, 1995) presented another index, the modified NDVI (MNDVI), which accounts for atmospheric attenuation as well. The SAVI has been shown to minimize soil-related problems significantly in nadir measurements over a variety of plant canopies and densities and in data derived from canopy radiant transfer models (Huete et al., 1991). Figure 3.31 shows the SAVI values versus the background correction factor, L, over different levels of vegetation cover (LAI) for dark and light soils. It is apparent that increasing L values relates to decreasing the SAVI values (when $L=0$, SAVI = NDVI). Huete (1988) noted that the optimal correction factor is achieved at the point where dark and light SAVI values produce the same values. More precise models take into account the vegetation architecture (Otterman et al., 1995) or contain additional correction factors (Rondeaux et al., 1996). Richardson et al. (1975) developed three plant canopy models for extracting plant, soil, and shadow reflectance components of a cropped field. Using such models, Murphy and Wadge (1994) were able to

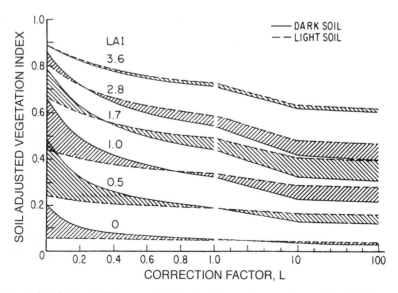

Figure 3.31 Light and dark soil influence on the SAVI values of cotton plant as a function of the shifted origin correction factor. (After Huete et al., 1981.)

separate soil and vegetation spectra by using GER 63-channel imaging spectrometer data (Ben-Dor and Kruse, 1995). Roberts et al. (1993) also incorporated an unmixing procedure to discriminate vegetation litter, and soils using AVIRIS 224-channel imaging spectrometer data (Van, et al. 1993) and were able to account for different soil types using a residual spectrum technique. It can be concluded that soil spectral signatures can be extracted from areas that are partially covered by decaying or live vegetation; however, caution must be taken when assessing the "true" soil reflectance spectra in a vegetation–soil mixture.

3.6.2.2 LOWER VEGETATIONS.

A major vegetation component in arid soil areas, usually ignored by workers, is the biogenic crust. This issue has received more and more attention recently, and its importance to the explanation of anomalies in field soil spectra and satellite data has been shown (Pinker and Karnieli, 1995). The biogenic crust consists primarily of lower, nonvascular plants (microphytic) covering the upper soil surface in a thin layer (Rogers and Langer, 1972; West, 1990). The microphytic community consists of mosses, lichens, algae, fungi, cyanobacteria, and bacteria. Each of these groups has pigments that are spectrally active in the visible region under certain environmental conditions (Figure 3.32) and thus may mask soil features, and more seriously, may be interpreted as the soil signature. O'Neill (1994) showed that spectral features between 2.08 and 2.10 μm of a soil could be attributed to the microphytic crust and speculated that this was due to cellulose. Karnieli and Tsoar (1994) showed that the microphytic crust caused a decrease in the overall albedo in the soils, which led to the false identification of anomalies in arid soils. The spectral response related to the biogenic crust permits linear mixing models, unlike the complex architecture of higher vegetation, which requires nonlinear models to analyze mixed signals. In addition to more basic research and consideration by workers of the biogenic crust

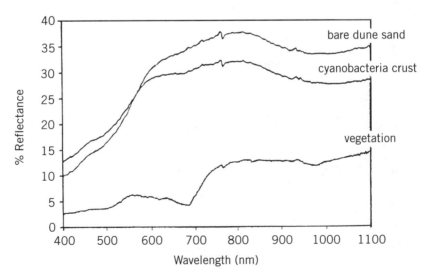

Figure 3.32 Spectral reflectance in the VNIR region of three desert targets: bare soil (sand dune), desert vegetation (sage brush), and biogenic crust (cyanobacteria). (After Karnieli and Tsoar, 1994.)

issues, more quantitative studies are required to account fully for the biogenic crusts effect on the soil spectra.

3.6.3 Lithosphere: Soil Cover and Crust

As Baumgardner stated: "Early remote sensing researchers of soils recognized the fact that soils often formed surface crust that could make a soil appear dry when it was actually wet." Soil crust and cover can be formed by different processes. As outlined above, the biogenic crust is one example of such interference. Eolian material and desert varnish are others. A lithosphere crust that is often found in soil is the *rain crust*. This crust is formed by raindrops (Morin et al., 1981), which cause a segregation of fine particle size at the soil surface. This can increase runoff and lead to soil erosion. The crusting effect is more pronounced in saline soils and well studied with relation to the mineralogical and chemical changes of the soil surface (Shainberg, 1992). The immediate observation after a rainstorm is an enhancement of hue and value of the soil color because of an increase in the fine fraction on the surface. One can assume that the reflectance spectrum of the rain crust would be totally different from that of the original soils, because it contains a greater clay fraction with a different textural component. In the literature, the issue of rain crust as it affects the spectral signature of soils has not received much attention; however, we encourage workers to consider this problem in their studies.

3.6.4 Atmosphere: Gases and Aerosols

The atmosphere's gases and aerosols play a major role in the VNIR–SWIR spectral regions. Across these regions, absorption and scattering of electromagnetic radiation takes place. Water vapor, oxygen, carbon dioxide, methane, ozone, nitrous oxides, and carbon monoxide are the components that are spectrally active across approximately half VNIR–SWIR regions. Some models for retrieving gases and aerosol interferences exist and are widely used by many workers [e.g., LOWTRAN-7 (Kneizys et al., 1988), 5S and 6S codes (Tanre, 1986), and ATRAM (Gao et al., 1993)]. It is beyond the scope of this chapter to discuss these models; however, one should be aware that in many cases the models do not perfectly remove all atmospheric attenuation and may alter the soil spectrum (Boardan and Huntigton 1997). For an example, see Figure 3.33, which provides AVIRIS–ATRAM corrected data and field spectra of the same area (Gao et al., 1993). The greater number of features in the AVIRIS spectrum are presumably artifacts of the correction routine. This problem is most likely to appear in hyperchannel data, where discrete absorption features are more pronounced relative to multichannel data, which practically averages small features into one wide value.

To illustrate the spectral regions under which atmospheric attenuation could affect the soil spectrum, we provide in Figure 3.34 a reflectance spectrum of a playa taken by the AVIRIS sensor over Rogers Dry Lake, California, with correction for solar effects but without removing atmospheric attenuations. The visible region is affected by aerosol scattering (monotonous decay from 0.4 to 0.8 μm) and absorption of ozone (around 0.6 μm), water vapor (0.73 and 0.82 μm) and oxygen (0.76 μm). The

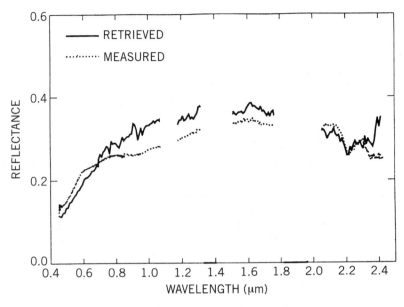

Figure 3.33 Retrieved reflectance spectrum (solid line) from AVIRIS data acquired over an area covered by sericite in the northern Grapevine Mountains, California, and a measured reflectance spectrum (dotted line) in the field using a portable spectrometer. (After Gao et al., 1993.)

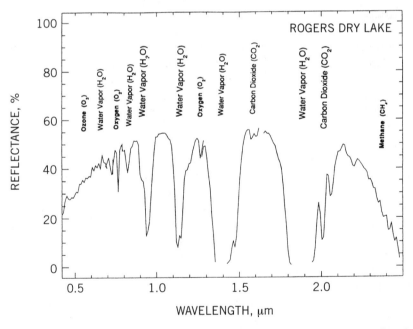

Figure 3.34 AVIRIS spectrum of a playa target at Rogers Dry Lake, California, after removing the solar effect. The major gas absorption's features are annotated to show spectral regions where atmospheric attenuation may overlap with soil features.

near-infrared region is affected by absorption of water vapor (0.94, 1.14, 1.38, and 1.88 μm), carbon dioxides (around 1.56, 2.01 and 2.08 μm), and methane (2.35 μm). As discussed previously, even weak spectral features in the soil spectrum may provide very useful information. Therefore, caution must be taken before applying any quantitative models to soil reflectance spectra derived from air-or spaceborne hyperchannel sensors. Validation of the corrected data is an essential criterion for ensuring that the reflectance spectrum consists of reliable soil information.

3.7   QUANTITATIVE SOIL SPECTROSCOPY

As discussed earlier in this chapter, soil and soil spectra are rather complex phenomena. This prevents a straightforward assessment of reflectance properties by physical theories or models (Liong and Townshend 1996). Discrete absorption bands, caused by chemical activity, allow unique identification of many of the soil chromophores discussed earlier. However, reliance on spectral features to predict mineral abundance in soils is a difficult task (Clark and Roush, 1984; see also Chapter 1). Theories such as those presented by Hapke (1981a,b, 1984, 1986), Pinty and Ramond (1987), Jacquemoud et al. (1992), and Liang and Townshend (1996) have been applied successfully to powders to account quantitatively for multiple scattering and to retrieve the properties of the Earth's surface. All theories used a basic radiative transfer equation to derive scattering of particulate surfaces and include variables for specular and diffuse reflectance radiations, phase dependence, particle scattering, shadowing effects, and viewing and illumination angles. The parameters incorporated in these theories are related directly to physical properties of the surface such as physical structure (i.e., grain size, packing condition, and refraction indices of the soil materials). Nevertheless, Clark and Roush (1984) noted that there is a potential for quantitative conversion of a reflectance spectrum of a multimineral surface to actual mineral abundances using physical models. Because this situation is not simply applied to a soil system, where complex relationships exist between chromophores and because theoretical models do not always agree with reality, (Pinty and Verstraete 1991) empirical quantitative approaches were developed to derive chemicophysical information from the soil spectral data. Basically, the empirical methods are based on the reflectance being equivalent to the transmittance, and on the idea that photons obey Beer's law for a given path length within the surface studied and on absorption coefficients and the concentration of the material. Under laboratory conditions where physical parameters remain constant, no atmospheric attenuation exists, and spectral noise is minimal, a soil spectrum tends to vary with mineralogy.

Under such conditions the empirical relationship between the chemistry and reflectance properties of powders can provide quantitative information about unknown materials solely from their reflectance spectra (Condit, 1972). Manipulation of spectra using derivatives and transformation to log space enables enhancement of weak spectral features and minimizes physical effects (Demetriades-Shah et. al., 1990). A promising quantitative laboratory approach, in the near-infrared/SWIR regions, was developed as a rapid method to analyze moisture in grains (Ben-Gera and Norris, 1968). The method, currently termed near-infrared reflectance analysis (NIRA), is currently widely accepted and used in many disciplines (Stark et al., 1986; Davies and Grant, 1987; Williams and Norris, 1987; Norris, 1988), including remote sensing of vegetation from hyperchannel sensors (Wessman et al., 1988, 1989, Curran et al., 1992, 1997; Gong and Miller, 1992, LaCapra et al., 1996) and other complex mixtures (Honigs, 1984). Basically, the NIRA method assumes that a concentration of a given constituent is proportional to the linear combination of several absorption features. The method is empirical, and although no physical or chemical or assumptions are made, the method has a strong spectroscopy foundation. The NIRA approach has two stages: (1) the calibration stage, where a prediction equation for evaluating a property is developed; and (2) the validation stage, where the preceding stage is validated. The calibration stage uses training samples, which represent the study population in terms of spectral and physicochemical properties. Then a prediction equation is generated based on multiple regression analysis between the soil chemistry data (determined in the laboratory) and selected spectral bands. This calibration equation is further validated in stage 2 against unknown samples and is examined statistically for its prediction performance.

The NIRA concept has been applied successfully to soil, soil minerals, and soil organic materials in the laboratory. Dalal and Henry (1986) showed that organic carbon, nitrogen, and soil moisture can be predicted simultaneously from the reflectance spectra of Australian soils. Morra et al. (1991) applied the NIRA methodology on soil samples of 12 subgroups and established a model to predict the total carbon and nitrogen content solely from the soil reflectance. Recently, Ben-Dor et al. 1997 showed that the nature and shape of the organic matter's reflectance curve varies with time, and that organic matter age can be predicted using the NIRA approach on data in the VNIR–SWIR region. In another study, Ben-Dor and Banin (1990a) applied the NIRA algorithm on SWIR data from several smectite minerals and predicted their total Al_2O_3, Fe_2O_3, MgO, and SiO_2 content. In arid and semiarid soils from Israel, Ben-Dor and Banin (1990b) predicted the $CaCO_3$ content from data in the SWIR region and have shown (E. Ben-Dor, personal communication) the ability to predict quantitatively such soil properties as clay content, specific surface area, cation exchange capacity, hygroscopic moisture, and organic matter content.

Figure 3.35 presents the calibration and validation results of NIRA analysis applied to 91 soils from Israel. Reasonable standard error of calibration (SEC) and performance (SEP) were obtained for the soil properties examined. The NIRA approach has shown that using a relatively low number of spectral channels (25 to 63) is sufficient to predict the following soil properties: hygroscopic moisture (25 channels), organic matter (30 channels), and clay content, cation-exchange capacity, and specific surface area (63 channels). However, prediction of the carbonate content in the soils required 3113 channels. In another study, Ben-Dor and Banin (1995a) showed that the quantitative approach provided by the NIRA routines enables quantitative assessment of SWIR-featureless soil properties, such as total Fe_2O_3, Al_2O_3, SiO_2, free iron oxides, average particle size, and total K_2O. They concluded that the intercorrelation between feature and featureless soil properties is the major mechanism that enables the NIRA to work successfully under so complex a matrix. Examining the NIRA approach for the VNIR spectral region illustrated that this concept is also applicable to monotonous spectral curves with minimal spectral variations (Ben-Dor and Banin, 1994). Recent study by Galvao and Vitorello (1998) revealed similar finding about the relationship between the chemical information of Brazelian soils and their reflectance spectra in the VNIR spectral region. The NIRA approach was able to work in a complex soil system because of the relatively high signal/noise

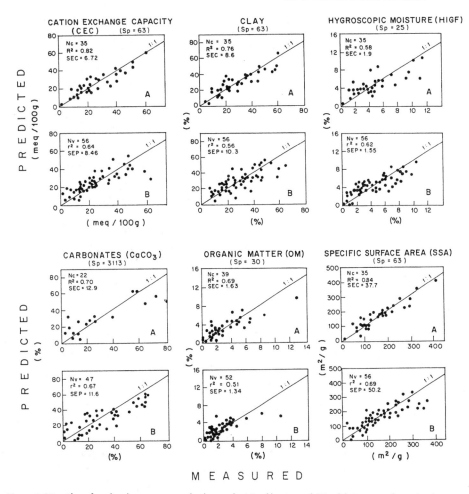

Figure 3.35   Plots of predicted versus measured values at the (*A*) calibration and (*B*) validation stages for each soil property. Nc, number of samples in the calibration stage; Nv, number of samples in the validation stage; SEP, standard error of performance; SEC, standard error of calibration. (After Ben-Dor and Banin, 1995a.)

ratio, high-spectral-resolution capability, constant measurement conditions, and a detailed study of the soil population. The high quality of the spectra used in this study supports the use of sophisticated mathematical manipulations methods such as derivation, conversion to other space, and subtractions. These manipulations can improve the general analysis of results obtained using empirical methods.

The NIRA approach has limitations, as it was based on "older" theories that did not completely represent the scattering process (Chapter 1 of this volume). However, in practice the NIRA model has had several successful applications. Ben-Dor and Banin (1995c) used the NIRA algorithm on laboratory reflectance data (processed to simulate six TM bands in the VNIR–SWIR region). They found that carbonate concentration, loss on ignition, specific surface area, and total SiO_2 in soil could be predicted solely from the TM spectra simulated. These studies show that results based on an empirical approach and no a priori knowledge can be used to analyze soils quantitatively based on reflectance spectra. Other quantitative methods for analyzing soil spectra include those that explain the spectral variability of a given population by vector analysis (Condit, 1972; Price, 1990; Kimes et al., 1993). However, in this approach it is possible that highly correlated channels will be grouped, and this may underestimate subtle but significant spectral features meaningful to the interpretation of soil properties.

It is postulated that although the soil is a complex material and soil spectra may appear featureless, a remarkable amount of information is hidden in the reflected photons, which under controlled conditions can be retrieved. Even though it is difficult to obtain similar findings in field conditions, more studies are strongly urged to create a synergy between physical and empirical models and field and laboratory experiences. Hopefully this will yield better quantitative identification of soil properties from a remote sensing standpoint.

3.8 SOIL REFLECTANCE IN REMOTE SENSING: PAST, PRESENT, AND FUTURE

Many studies have been conducted with the intention of classifying soil and soil properties using optical sensors onboard orbital satellites, such as *Landsat* MSS and TM, SPOT, and NOAA-AVHRR (e.g., Cipra et al., 1980; Mulders, 1987 Frazier and Cheng, 1989; Kierein-Young and Kruse, 1989; Agdu et al. 1990; Moran et al., 1992 White et al., 1997). Qualitative classification approaches have traditionally been used to analyze multichannel data in cases where the spectral information was relatively low. Nevertheless, it has also been possible to obtain usable sets of information about soil type, soil degradation, and soil conditions from "broad" channel sensors by applying sophisticated classification approaches (Price, 1990; Ben-Dor and Banin, 1995c, Galvao and Vitorello, 1998). Over the years, soil spectra have been collected and analyzed in the laboratory both quantitatively and qualitatively by many workers (e.g., Latz et al., 1981; Price, 1995). According to Price (1995), a comprehensive literature survey of these data sets is impractical, as most collections are documented only through internal reports and are not easily obtained. Price (1995) reviewed several spectral data sets; for soils and minerals those of Stoner et al. (1980a,b), Biehl et al. (1984), Satterwhite and Henry (1991), and Grove et al. (1992) are the most appropriate.

Another limitation of soil libraries is that in some cases no chemicophysical information is available with the spectra. Because soil spectra represent rather complicated sets of data, it is critical to have both soil spectra and chemical information in a given data set. In general it can be concluded that if the signal/noise ratio is sufficient and the spectrum consists of a reasonable number of spectral channels, analysis of the spectrum can yield useful information about the chemical characteristics of the soil, even though many of the channels are intercorrelated and the soil matrix is complex. Because of the unknown interactions between soil chromophores, it is difficult to assess the most appropriate wavelengths for explaining the composition of a given soil. The complex interactions between components in soils may cause the theoretical models to be impractical, and hence empirical models may need to be incorporated. It is true that spectral variability can be explained by relatively small and broad spectral bands, but there is no doubt that additional information will provide a better performance. Development of a sophisticated analytical method and a synergy between physical and empirical models may be the keys for retrieving quantitative information about soil properties solely from their airborne reflectance spectra. This option should be the focus for today's workers particularly, as new spectral imaging systems with greater near-laboratory spectral capabilities are emerging and becoming more available.

For soil applications air- and spaceborne imaging spectrometers should consist of a reasonable number of spectral channels across the entire VNIR–SWIR region, which will cover the spectrally active regions of all chromophores with a reasonable bandwidth. Price (1991) believes that a relatively low number of spectral channels (15 to 20) with a bandwidth of 0.04 to 0.10 nm and high signal/noise ratio are those that promise better remote sensing capabilities of soils. Goetz and Herring (1989) preferred more spectral channels (192) but wider bandwidth (about 10 nm) to permit diagnostic evaluation of specific features across the entire VNIR–SWIR region. We believe that for quantitative analysis of soil spectra, the optimal bandwidth and number of channels may be strongly dependent on the soil population and the property examined. There is no doubt, however, that high signal/noise ratio is a crucial factor in quantitative analysis of soil spectra derived from both air and space measurements.

3.9 SUMMARY AND CONCLUSIONS

In summary, it can be concluded that soil spectra carry unique and important information about many of the soil's properties. Because the soil is a very complex system, soil spectra should be judged with caution. Empirical and physical models may yield quantitative information about soils. For each soil population, a separate study should be applied to learn the intercorrelations between all possible chromophores. This is the reason why soil reflectance libraries of the world's soils must be accompanied by additional detailed information about the sampling area (climate, topography, parent material, age, and organic matter) and if possible, detailed chemical and physical data (Stoners et al., 1980a, is a good example). Although workers have shown that soil spectral variation can be gathered into a few broad bands, it is still believed that no substitution is available for the high-spectral-resolution data. This is because even weak spectral features can carry invaluable information about soil properties and conditions. In many cases, only slight spectral differences can be the key for classifying soils based on their spectra. In this respect, if noise is minimal, caution must be used to correct the atmospheric interference appropriately for other environmental conditions, such as relative humidity, water content, slope and aspects, sun angle, shadow, and vegetation coverage, as well as for the intercorrelation between chemicophysical processes in the soil environment.

References

Aase, J. K., and D. L. Tanaka, 1983. Effect of tillage practices on soil and wheat spectral reflectance, *J. Agron.*, 76, 814–818.

Aber, J., C. A. Wessman, D. L. Peterson, J. M. Mellilo, and J. H. Fownes, 1990. Remote sensing of litter and soil organic matter decomposition in forest ecosystems, in *Remote Sensing of Biosphere Functioning*, R. J. Hobbs and H. A. Mooney eds., Springer-Verlag, New York, pp. 87–101.

Agdu, P. A., J. D. Fehrenbacher, and I. Jansen, 1990. Soil property relationships with SPO satellite digital data in east central Illinois, *Soil Sci. Soci. Am. J.*, 54, 807–812.

Ahmad, S. P., and D. W. Deering, 1992. A simple analytical function for bidirectional reflectance, *J. Geophys. Res.*, 97, 18867–18886.

Al-Abbas, H. H., H. H. Swain, and M. F. Baumgardner, 1972. Relating organic matter and clay content to multispectral radiance of soils, *Soils Sci.*, 114, 477–485.

Ammer, U., B. Koch, T. Schneider, and H. Wittmeier, 1991. High resolution spectral measurements of vegetation and soil in field and laboratory, in *Proceedings of the 5th International Colloquium on Physical Measurements and Signatories in Remote Sensing*, Courchevel, France, Vol. 1, pp. 213–218.

Arvidson, R. E., E. A. Guinness, M. A. Dale-Bannister, J. Adams, M. Smith, P. R. Christensen, and R. B. Singer, 1989. Nature and distribution of surficial deposits in Chryse Planitia and vicinity, Mars, *J. Geophys. Res.*, 94, 1573–1587.

ASTM, 1985. *Standard Method for Computing the Colors of Objects by Using the CIE System*, ASTM Stand. E 308–85, American Society for Testing and Materials, Philadelphia, 27. pp.

ASTM, 1989. *Standard Test Method for Specifying Color by the Munsell System*, ASTM stand. D 1535–89, American Society for Testing and Materials, Philadelphia, 27. pp.

Banin, A., and A. Amiel, 1970. A correlation of the chemical physical properties of a group of natural soils of Israel, *Geoderma*, 3, 185–198.

Barron, V., and L. Montealegre, 1986. Iron oxides and color of triassic sediments: application of the Kubelka–Munk theory, *Am. J. Sci.*, 286, 792–802.

Barron, V., and J. Torrent, 1986. Use of Kubelka–Munk theory to study the influence of iron oxides on soil colour, *J. Soil Sci.*, 37, 499–510.

Baumgardner, M. F., S. J. Kristof, C. J. Johannsen, and A. L. Zachary, 1970. Effects of organic matter on multispectral properties of soils, *Proc. Indian Acad. Sci.*, 79, 413–422.

Baumgardner, M. F., L. F. Silva, L. L. Biehl, and E. R. Stoner, 1985. Reflectance properties of soils, *Adv. Agron.*, 38, 1–44.

Bear, F. E., 1964. *Chemicals of the Soil*, 2nd ed., Reinhold, New York, 515 pp.

Beck, R. H., B. F. Robinson, W. H. McFee, and J. B. Peterson, 1976. LARS Inf. Note 081176, Laboratory for Applications of Remote Sensing, Purdue University, West Lafayette, Ind.

Beck, R. H., B. F. Robinson, W. W. McFee and J. B. Peterson. 1976. Spectral characteristics of soils related to the interaction of soil moisture, organic carbon and clay content. LARS Information Note 081176. Laboratory for Applications of Remote Sensing, Purdue University, West Lafayette, IN.

Becker, F., P. Ramanantsizehena, and M.-P. Stoll, 1985. Angular variation of the bidirectional reflectance of bare soils in the thermal infrared band, *Appl. Opt.*, 24, 365–375.

Bedidi, A., B. Cervelle, J. Madeira, and M. Pouget, 1990. Moisture effects on spectral characteristics (visible) of lateritic soils, *Soil Sci.*, 153, 129–141.

Beididi, A., B. Cervelle, and J. Madeira, 1991. Moisture effects on spectral signatures and CIE-color of lateritic soils, in *Proceedings of the 5th International Colloquium on Physical Measurements and Signatures in Remote Sensing*, Courchevel, France, Vol. 1, pp. 209–212.

Ben-Dor, E., and A. Banin, 1990a. Diffuse reflectance spectra of smectite minerals in the near infrared and their relation to chemical composition, *Sci. Geol. Bull.*, 43(2–4), 117–128.

Ben-Dor, E., and A. Banin, 1990b. Near infrared reflectance analysis of carbonate concentration in soils, *Appl. Spectrosc.*, 44(6), 1064–1069.

Ben-Dor, E., and A. Banin, 1994. Visible and near infrared (0.4–1.1mm) analysis of arid and semiarid soils, *Remote Sensing Environ.*, 48, 261–274.

Ben-Dor, E., and A. Banin, 1995a. Near infrared analysis (NIRA) as a rapid method to simultaneously evaluate several soil properties, *Soil Sci. Soc. Am. J.*, 59, 364–372.

Ben-Dor, E., and A. Banin, 1995b. Near infrared analysis (NIRA) as a method to simultaneously evaluate spectral featureless constituents in soils, *Soil Sci.*, 159, 259–269.

Ben-Dor, E., and A. Banin, 1995c. Quantitative analysis of convolved TM spectra of soils in the visible, near infrared and short-wave infrared spectral regions (0.4–2.5μm), *Int. J. Remote Sensing*, 18, 3509–3528.

Ben-Dor, E., and F. A. Kruse, 1995. Surface mineral mapping of Makhtesh Ramon Negev, Israel using GER 63 channel scanner data, *Int. J. Remote Sensing*, 18, 3529–3553.

Ben-Dor, E., and A. Singer, 1987. Optical density of vertisol clays suspensions in relation to sediment volume and dithionite–citrate–bicarbonate extractable iron, *Clays Clay Miner.*, 35; 311–317.

Ben-Dor, E., A. Banin, and A. Singer, 1991. Simultaneous determination of six important soil properties by diffuse reflectance in the near infrared region, in *Proceedings of the 5th International Colloquium on Physical Measurements and Signatures in Remote Sensing*, Courchevel, France, Vol. 1, pp. 159–163.

Ben-Dor, E., Y. Inbar, and Y. Chen, 1997 The reflectance spectra of organic matter in the visible near infrared and short wave infrared region (400–2,500nm) during a controlled decomposition process, *Remote Sensing Environ.* 61:1–15.

Ben-Gera, I., and K. H. Norris, 1968. Determination of moisture content in soybeans by direct spectrophotometry, *Isr. J. Agric. Res.*, 18, 124–132.

Berns, R. S., F. W. Billmeyer, and R. S. Sacher, 1985. Methods for generating spectral reflectance functions leading to color-constant properties, *Color Res. Appl.*, 10, 73–83.

Biehl, L. L., C. S. T. Daughtry, and M. E. Bauer, 1984. *Vegetation and Soils Field Research Data Base: Experiment Summaries*, LARS Tech. Rep. 042832, Laboratory for Application of Remote Sensing, Purdue University, West Lafayette, Ind.

Billmeyer, F. W., Jr., and M. Saltzman, 1981. *Principles of Color Technology*, 2nd ed., Wiley, New York, 181 pp.

Boardman J. W. and F. Huntington, 1997 Mineralogic and geochemical mapping of Virginia City, Nevada using 1995 AVIRIS data *Proceedings of the Twelfth International Conference and Workshops on Applied Geology Remote Sensing*. Denver Colorado I:191–198.

Billmeyer, F. W., Jr., and M. Saltzman, 1981. *Principles of Color Technology*, 2nd ed., Wiley, New York, 181 pp.

Bohren, C. F., and D. R. Huffman, 1983. *Absorption and Scattering of Light by Small Particles*, Wiley, New York, 530 pp.

Bowers, S., and R. J. Hanks, 1965. Reflectance of radiant energy from soils, *Soil Sci.*, 100, 130–138.

Brooks, F. A., 1952. Atmospheric radiation and its reflection from the ground, *J. Meteorol.* 9, 41–52.

Buol, S. W., F. D. Hole, and R. J. McCracken, 1973. *Soil Genesis and Classification*, Iowa State University Press, Ames, Iowa, 360 pp.

Carey, J., 1995. Tilling the soil by satellite, *Bus. Week*, 11, 62–63.

Cariati, F., L. Erre, G. Micera, P. Piu, and C. Gessa, 1981. Water molecules and hydroxyl groups in montmorillonites as studied by near infrared spectroscopy, *Clays Clay Miner.*, 29, 157–159.

Cariati, F., L. Erre, G. Micera, P. Piu, and C. Gessa, 1983. Polarization of water molecules in phyllosylicates in relation to exchange cations as studied by near infrared spectroscopy, *Clays Clay Miner.*, 31, 155–157.

Castellan, G. W. 1983. *Physical Chemistry*, 3rd ed., Addison-Wesley, Reading, Mass., 943 pp.

Chen, Y., and Y. Inbar, 1994. Chemical and spectroscopical analysis of organic matter transformation during composting in relation to compost maturity, in *Science and Engineering of Composting: Design, Environmental, Microbiology and Utilization Aspects*, H. A. J. Hoitink and H. M., Keener, eds., Renaissance Publications, Worthington, Ohio, pp. 551–600.

CIE, 1931. *Proceedings of the 8th Session*, Commission Internationale de l'Éclairage, Bureau Central de la CIE, Paris.

Cierniewski, J., 1987. A model for soil surface roughness influence on the spectral response of bare soils in the visible and near infrared range, *Remote Sensing Environ.*, 23, 98–115.

Cierniewski, J., 1989. The influence of the viewing geometry of bare soil surfaces on their spectral response in the visible and near infrared, *Remote Sensing Environ.*, 27, 135–142.

Cipra, J. E., D. P. Franzmeir, M. E. Bauer, and R. K. Boyd, 1980. Comparison of multispectral measurements from some nonvegetated soils using Landsat digital data and a spectroradiometer, *Soil Sci. Soc. Am. J.*, 44, 80–84.

Clark, R. N., 1981. The reflectance of water–mineral mixtures at low temperatures, *J. Geophys. Res.*, 86, 3074–3086.

Clark, R. N., and T. L. Roush, 1984. Reflectance spectroscopy: quantitative analysis techniques for remote sensing applications, *J. Geophys. Res.*, 89, 6329–6340

Condit, H. R., 1970. The spectral reflectance of American soils, *Photogramm. Eng.*, 36, 955–966.

Condit, H. R., 1972. Application of characteristic vector analysis to the spectral energy distribution of daylight and the spectral reflectance of American soils, *Appl. Opt.*, 11, 74–86.

Conel, J. E., 1969. Infrared emissivities of silicates: experimental results and a cloudy atmosphere model of spectral emission from condensed particulate mediums, *J. Geophys. Res.*, 74; 1614–1634.

Cooper, K., and J. A. Smith, 1985. A Monte Carlo reflection model for soil surfaces with 3D structure, *IEEE Trans. Geosci. Remote Sensing*, 23, 668–673.

Coulson, K. L., 1966. Effects of reflection properties of natural surfaces, *J. Appl. Meteorol.* 10, 1285–1295.

Coyne, L. M., J. L. Bishop, T. Sacttergood, A. Banin, G. Carle, and J. Orenberg, 1989. Near-infrared correlation spectroscopy: quantifying iron and surface water in series of variably cation-exchanged montmorillonite clays, in *Spectroscopic Characterization of Minerals and Their Surfaces*, ACS Symp. Ser. 415, L. M. Coyne, S. W. S. McKeever, and D. F. Blake, eds., American Chemical Society, Washington, D.C., pp. 407–429.

Csillag, F., L. Pasztor, and L. L. Biehl, 1993. Spectral band selection for the characterization of salinity status of soils, *Remote Sensing Environ.*, 43, 231–242.

Curi, N., and D. P. Franzmeier, 1984. Toposequence of Oxisols from the Central Plateau of Brazil, *Soil Sci. Soc. Am. J.*, 48, 341–346.

Curran, P. J., G. M. Foody, K. Ya. Kondratyev, V. V. Kozoderov, and P. P. Fedchenko, 1990. *Remote Sensing of Soils and Vegetation in the USSR*, Taylor & Francis, New York.

Curran, P. J., J. L. Dungam, B. A. Macler, S. E. Plummer, and D. L. Peterson, 1992. Reflectance spectroscopy of fresh whole leaves for the estimation of chemical concentration, *Remote Sensing Environ.*, 39, 153–166.

Curran, P. J., J. H. Kupiec, and G. M. Smith, 1997. Remote sensing the biochemical composition of a slash pipe canopy, *IEEE Trans. Geosci. Remote Sensing*, 35, 415–420.

Da Costa, L. M., 1979. Surface soil color and reflectance as related to physicochemical and mineralogical soil properties, Ph.D. dissertation, University of Missouri, Columbia, Mo., 154 pp.

Dalal, R. C., and R. J. Henry, 1986. Simultaneous determination of moisture, organic carbon and total nitrogen by near infrared reflectance spectroscopy, *Soil Sci. Soc. Am. J.*, 50, 120–123.

Davies, A., and A. Grant, 1987. Review: near infrared analysis of food, *Int. J. Food Sci. Technol.*, 22, 191–207.

Deering, D. W., 1990. Reflectance distribution characteristics of the GRSFE modeling sites, in *Remote Sensing Science for the Nineties: Proceedings of the 10th*

Annual International Geoscience and Remote Sensing Symposium, IGARSS'90, Vol. II, R. Mills, ed., College Park, Md., May 20–24, Institute of Electrical and Electronics Engineers, New York pp. 1357–1360.

Demetriades-Shah, T. H., M. D. Steven, and J. A. Clark, 1990. High resolution derivative spectra in remote sensing, *Remote Sensing Environ.*, 33, 55–64.

Den Dulk, J. A., 1989. The interpretation of remote sensing; a feasibility study, Ph.D. dissertation, Wageningen Agricultural University, Wageningen The Netherlands, 173 pp.

Dixon, J. B., and S. B. Weed, 1989. *Minerals in Soil Environments*, Soil Science Society of America, Madison, Wisc.

Domingue, D., and B. Hapke, 1989. Fitting theoretical photometric functions to asteroid phase curves, *Icarus*, 78, 330–336.

Downey, G., and P. Byrne, 1986. Prediction of moisture and bulk density in milled peat by near infrared reflectance, *J. Food Agric. Sci.*, 37, 231–238.

Egan, W. G., and T. W. Hilgeman, 1978. Spectral reflectance of particulate materials: a Monte Carlo model including asperity scattering, *Appl. Opt.*, 17, 245–252.

Egbert, D. D., 1977. A practical method for correcting bidirectional reflectance variations, in *Proceedings of the Symposium on Machine Processing of Remotely Sensed Data*, West Lafayette, Ind., June 21–23, pp. 178–188.

Egbert, D. D., and T. F. Ulaby, 1972. Effect of angles on reflectivity, *Photogramm. Eng. Remote Sensing*, 348, 556–564.

Elvidge, C. D. 1990. Visible and near infrared reflectance characteristics of dry plant materials, *Int. J. Remote Sensing*, 11, 1775–1795.

Emslie, A. G., and J. R. Aronson, 1973. Spectral reflectance and emittance of particulate materials: 1. Theory, *Appl. Opt.*, 12, 2563–2572.

Epema G. F., 1992. Spectral reflectance in the Tunisian desert, Ph.D. dissertation, Wageningen Agricultural University, Wageningen, The Netherlands, 150 pp.

Escadafal, R., 1989. Charactérisation de la surface des sols arides par observations de terrain et par télédetection. Applications: exemple de la région de Tataouine (Tunisie), Ph.D. dissertation, Université Pierre et Marie Curie, Paris, 317 pp.

Escadafal, R., 1993. Remote sensing of soil color: principles and applications, *Remote Sensing Rev.* 7, 261–279.

Escadafal, R., and A. R., Huete, 1991. Influence of the viewing geometry on the spectral properties (high resolution visible and NIR) of selected soils from Arizona, in *Proceedings of the 5th International Colloquium on Physical Measurements and Signatures in Remote Sensing*, Courchevel, France, Vol. 1, pp. 401–404.

Escadafal, R., M. Girard, and D. Courault, 1988. La couleur des sols: appréciation, mesure et relation avec les propriétiés spectrales, *Agronomy*, 8, 147–154.

Escadafal, R., M. Girard, and D. Courault, 1989. Munsell soil color and soil reflectance in the visible spectral bands of *Landsat* MSS and TM data, *Remote Sensing Environ.* 27, 37–46.

Evans, R. M., 1948. *An Introduction to Color*, Wiley, New York.

Everitt, J. H., D. E. Escobar, A. H. Gerbermann, and M. A. Alaniz, 1988. Detecting saline soils with video imagery, *Photogramm. Eng. Remote Sensing*, 54, 1283–1287.

FAO, 1994. *FAO Year Book*, Vol. 48, FAO Stat. Ser. 125, Food and Agriculture Organization of the Untied Nations, Rome, p. 3.

Fernandez, R. N., and D. G. Schulze, 1987. Calculation of soil color from reflectance spectra, *Soil Sci. Am. J.*, 51, 1277–1282.

Feynman, R. P., 1985. *The Strange Theory of Light and Matter*, Princeton University Press, Princeton, N.J., 158 pp.

Frazier, B. E., and Y. Cheng, 1989. Remote sensing of soils in the eastern Palouse region with *Landsat* thematic mapper, *Remote Sensing Environ.*, 28, 317–325.

Gaffey, S. J., 1985. Reflectance spectroscopy in the visible and near infrared (0.35–2.55mm): applications in carbonate petrology, *Geology*, 13, 270–273.

Gaffey, S. J., 1986. Spectral reflectance of carbonate minerals in the visible and near infrared (0.35–2.55mm): calcite, aragonite and dolomite, *Am. Mineral.*, 71, 151–162.

Gaffey, S. J., and K. L. Reed, 1987. Copper in calcite: detection by visible and near infrared reflectance, *Econ. Geol.*, 82, 195–200

Galvao L. S. and I. Vitorello, 1998 Variability of laboratory measured soil lines of soils from southern Brazil. *Remote Sensing of Environment* 63: 166–181.

Galvao, L. S., I. Vitorello, and W. R. Paradella, 1995. Spectroradiometric discrimination of laterites with principal components analysis and additive modeling, *Remote Sensing Environ.*, 53, 70–75.

Gao, B.-C., K. B. Heidebrecht, and A. F. H. Goetz, 1993. Derivation of scaled surface reflectance from AVIRIS data, *Remote Sensing Environ.*, 44, 165–178.

Gausman, H. W., W. A. Allen, R., Cardenas, and R. L. Bowen, 1970. Color photos, cotton leaves and soil salinity, *Photogramm. Eng. Remote Sensing*, 36, 454–459.

Gausman, H. W., R. W. Leamer, R. J. Noriega, R. R. Rodriguez, and C. L. Wiegand, 1977. Field measured spectrometric reflectances of disked and nondisked soil with and without straw, *Soil Sci. Soc. Am. J.*, 41, 793–796.

Gerbermann, A. H., and D. D. Neher, 1979. Reflectance of varying mixtures of a clay soil and sand, *Photogramm. Eng. Remote Sensing*, 45, 1145–1151.

Glenn, J. J., and J. T. Killian, 1940. Trichromatic analysis of the *Munsell Book of Color*, *J. Opt. Soc. Am.*, 30, 609–616.

Goetz, A. F. A., 1992. Principles of narrow band spectrometry in the visible and IR: instruments and data analysis, in *Imaging Spectroscopy: Fundamentals and Prospective Applications*, F. Toselli and J. Bodechtel, eds., ECSE, EEC, EAEC, Brussels and Luxembourg, pp. 21–32.

Goetz, A. F. H., and M. Herring, 1989. A high resolution imaging spectrometer (HIRIS) for EOS, *IEEE Trans. Geosci. Remote Sensing*, 27, 136–144.

Goetz, F. A. H., P. Hauff, M. Shippert, and A. G. Maecher, 1991. Rapid detection and identifican of OH-bearing minerals in the 0.9–1.0mm region using new portable field spectrometer in *Proceeding of the 8th Thematic Conference on Geologic Remote Sensing*, Denver, Colo., Vol. 1, pp. 1–11.

Gong, P., R. Pu, and J. R. Miller, 1992. Correlation leaf area index of ponderosa pine with hyperspectral CASI data, *Can. J. Remote Sensing*, 18, 275–282.

Granville, W. C., D. Nickerson, and C. E. Foss, 1943. Trichromatic specifications of mean, intermediate and special colors of the Munsell system, *J. Opt. Soc. Am.*, 33, 376–385.

Grove, C. I., S. J. Hook, and E. D. Paylor, 1992. *Laboratory Reflectance Spectra of 160 Minerals, 0.4 to 2.5 Micrometers*, JPL Publ. 92-2, Jet Propulsion Laboratory, California Institute of Technology, Pasadena, Calif.

Hall, A. J., 1941. The relation between color and chemical composition in the bio-tites, *Am. Miner.*, 26, 29–33.

Hapke, B. W. 1981a. Bidirectional reflectance spectroscopy: I. Theory, *J. Geophys. Res.*, 86, 3039–3054.

Hapke, B. W., 1981b. Bidirectional reflectance spectroscopy: 2. Experiments and observation, *J. Geophys. Res.*, 86, 3055–3060.

Hapke, B. W., 1984. Bidirectional reflectance spectroscopy: correction for macroscopic roughens, *Icarus*, 59, 41–59.

Hapke, B. W., 1986. Bidirectional reflectance spectroscopy: 4. The extinction coefficient and the opposition effect, *Icarus*, 67, 264–280.

Hapke, B. W., 1993. *Theory of Reflectance and Emittance Spectroscopy*, Cambridge University Press, New York.

Hardisky, M. A., V. Klemas, and R. M. Smart, 1983. The influence of soil salinity, growth form and leaf moisture on the spectral radiance of *Spartina alterniflora* canopies, *Photogramm. Eng. Remote Sensing*, 49, 77–83.

Helfenstein, P., and J. Veverka, 1987. Photometric properties of lunar terrains derived from Hapke's equation, *Icarus*, 72, 342–357.

Hick, R. T., and W. G. R. Russell, 1990. Some spectral considerations for remote sensing of soil salinity, *Aust. J. Soil Res.*, 28, 417–431.

Hirschfeld, T., 1985. Salinity determination using NIRA, *Appl. Spectrosc.*, 39, 740–741.

Honigs, D. E., G. M. Hieftje, and T. Hirschfeld, 1984. A new method for obtaining individual component spectra from those of complex mixtures, *Appl. Spectrosc.*, 38, 317–322.

Huete, A. R., 1988. Soil adjusted vegetation index (SAVI), *Remote Sensing Environ.*, 25, 47–57.

Huete, A. R., and R. Escadafal, 1991. Assessment of biophysical soil properties through spectral decomposition techniques, *Remote Sensing Environ.*, 35, 149–159.

Huete, A. R., A. Chehbouni, W. Leeuwen, and G. Hua, 1991. Normalization of multidirectional red and NIR reflectance with the SAVI, in *Proceedings of the 5th International Colloquium on Physical Measurements and Signatures in Remote Sensing*, Courchevel, France, Vol. 1, pp. 419–422.

Hunt, G. R., 1980. Spectroscopic properties of rock and minerals in *Handbook of Physical Properties of Rocks*, C. R. Stewart, ed., CRC Press, Boca Raton, Fla., 295 pp.

Hunt, G. R., and J. W. Salisbury, 1970. Visible and near infrared spectra of minerals and rocks: I. Silicate minerals, *Mod. Geol.*, 1, 283–300.

Hunt, G. R., and J. W. Salisbury, 1971. Visible and near infrared spectra of minerals and rocks: carbonates, *Mod. Geol.*, 2, 23–30.

Hunt, G. R., and J. W. Salisbury, 1976. Visible and near infrared spectra of minerals and rocks: XI. Sedimentary rocks, *Mod. Geol.*, 5, 211–217.

Hunt, G. R., J. W. Salisbury, and A. Lenhoff, 1971a. Visible and near-infrared spectra of minerals and rocks: III. Oxides and hydroxides, *Mod. Geol.*, 2, 195–205.

Hunt, G. R., J. W. Salisbury, and C. J. Lenhoff, 1971b. Visible and near-infrared spectra of minerals and rocks: sulfides and sulfates, *Mod. Geol.*, 3, 1–14.

Hunt, G. R., J. W. Salisbury, and C. J. Lenhoff, 1971c. Visible and near-infrared spectra of minerals and rocks: halides, phosphates, arsenates, vandates and borates, *Mod. Geol.*, 3, 121–132.

Hunt, G. R. 1979 Near Infrared (1.3–2.4 µm) spectra of alteration minerals-Potential for use in remote sensing. Geophysics 44:1974–1986.

Idso, S. B., R. D. Jackson, R. J. Reginato, B. A. Kimball, and F. S. Nakama, 1975. The dependence of bare soil albedo on soil water content, *J. Appl. Meteorol.*, 14, 109–113.

Irons, J. R., and J. A. Smith, 1990. Soil surface roughness characterization from light scattering observations, in *Remote Sensing Science for the Nineties: Proceedings of the 10th Annual International Geoscience and Remote Sensing Symposium*, IGARSS'90, Vol. II, R. Mills, ed., College Park, Md., May 20–24, Institute of Electrical and Electronics Engineers, New York, pp. 1007–1010.

Irons, J. R., R. A. Weismiller, and G. W. Petersen, 1989. Soil reflectance, in *Theory and Application of Optical Remote Sensing*, G. Asrar, ed., Wiley Ser. Remote Sensing, Wiley, New York, pp. 66–106.

Irons, J. R., G. S. Campbell, J. M. Norman, D. W. Graham, and W. M. Kovalick, 1992. Prediction and measurement of soil bidirectional reflectance, *IEEE Trans. Geosci. Remote Sensing*, 30, 249–260.

Ishida, T., H. Ando, and M. Fukuhara, 1991. Estimation of complex refractive index of soil particles and its dependence on soil chemical properties, *Remote Sensing Environ.*, 38, 173–182.

Jackson, R. D., S. Moran, P. N. Slater, and S. F. Biggar, 1987. Field calibration of reflectance panels, *Remote Sensing Environ.*, 22, 145–158.

Jackson, R. D., P. M. Teillet, P. N. Slater, G. Fedosjsvs, M. F. Jasinski, J. K. Aase and M. S. Moran, 1990. Bidirectional measurements of surface reflectance for view angle corrections of oblique imagery, *Remote Sensing Environ.*, 32, 189–202.

Jacquemoud, S., F. Baret, and J. F. Hanocq, 1992. Modeling spectral and bidirectional soil reflectance, *Remote Sensing Environ.*, 41, 123–132.

Jepson, W. B., 1988. Structural iron in kaolinites and in associated ancillary minerals, in *Soils and Clay Minerals*, J. W. Stucki, B. A. Goodman, and U. Schwertmann, eds., NATO ASI Ser. D. Reidel, Dordrecht, The Netherlands, pp. 467–536.

Judd, D. B., 1933. The 1931 I.C.I. standard observer and coordinate system for colorimetry, *J. Opt. Soc. Am.*, 23, 359–374.

Karmanova, L. A., 1981. Effect of various iron compounds on the spectral reflectance and color of soils, *Sov. Soil Sci.*, 13, 63–60.

Karnieli, A., and H. Tsoar, 1994. Spectral reflectance of biogenic crust developed on desert dune sand along the Israel–Egypt border, *Int. J. Remote Sensing*, 16, 369–374.

Kastrov, B. G., 1955. On the diurnal change of surface albedo, *Tr. CAO*, 14, 12–22.

Kasumov, O. K., T. A. Nabieva, and O. M. Tereshenkov, 1992. Study of the spectral characteristics of soils of the Azerbaijan SSR, *Sov. Remote Sensing J.* 9, 608–617.

Keegan, H. J., W. C. Rheinboldt, J. C. Schleter, J. P. Menard, and D. B. Judd, 1958. Digital reduction of spectrophotometric data to Munsell renotations, *J. Opt. Soc. Am.* 48, 863.

Kelly, K. L., K. S. Gibson, and D. Nickerson, 1943. Tristimulus specification of the *Munsell Book of Color* from spectrophotometric measurements, *J. Opt. Soc. Am.* 33, 355–376.

Kierein-Young, K. and F. A. Kruse, 1989. Comparison of *Landsat* thematic mapper images and geophysical and environmental reassert imaging spectrometer data for alteration mapping, in *Proceedings of the 7th Thematic Conference on Remote Sensing for Exploration Geology*, Calgary, Alberta, Canada, Vol. 1, pp. 349–359.

Kimes, D. S., 1983. Dynamics of directional reflectance factor distributions for vegetation canopies, *Appl. Opt.* 22, 1364–1372.

Kimes, D. S., W. W. Newcomb, C. J. Tucker, I. S. Zonneveld, J. Van Wijngaarden, J. de Leeuw, and G. F. Epema, 1985. Directional reflectance factor distributions for cover types of northern Africa, *Remote Sensing Environ.* 18, 1–19.

Kimes, D. S., J. R. Irons, E. R. Levine, and N. A. Horning, 1993. Learning class discriminations from a data base of spectral reflectance of soil samples, *Remote Sensing Environ.* 43, 161–169.

Kneizys, F. X., G. P. Abdersen, E. P. Shettle, W. O. Gallery, L. W. Abreu, J. E. A. Selby, J. H. Chetwynd, and S. A. Clough, 1988. *Users Guide to LOWTRAN-7*, AFGL-TR-88-0177, Air Force Geophysics Laboratory, Hanscom AFB, Mass.

Kondratyev, K. Y., and P. P. Fedchenko, 1983. Investigation of humus in soil from their colours, *Sov. Soil Sci.* 15, 108–111.

Kosmas, C. S., N. Curi, R. B. Bryant, and D. P. Franzmeier, 1984. Characterization of iron oxide minerals by second derivative visible spectroscopy, *Soil Sci. Soc. Am. J.*, 48, 401–405.

Krishnan, P., J. D. Alexander, B. J. Butler, and J. W. Hummel, 1980. Reflectance technique for predicting soil organic matter, *Soil Sci. Soc. Am. J.* 44, 1282–1285.

Kristof, S. F., M. F. Baumgardner, and C. J. Johannsen, 1971. *Spectral Mapping of Soil Organic Matter*, J. Pap. 5390, Agricultural Experiment Station, Purdue University, West Lafayette, Ind.

Kruse, F. A., K. Kierein-Young, and J. W. Boardman, 1990. Mineral mapping of Cuprite, Nevada with a 63-channel imaging spectrometer, *Photogramm. Eng. Remote Sensing*, 56, 83–92.

Kruse, F. A., M. Thiry, and P. L. Hauff, 1991. Spectral identification (1.2–2.5mm) and characterization of Paris Basin kaolinite/smectite clays using a field spectrometer in *Proceedings of the 5th International Colloquium on Physical Measurements and Signatures in Remote Sensing*, Courchevel, France, Vol. 1, pp. 181–184.

Kubelka, P., and F. Munk, 1931. Ein Beitrag zur Optik der Farbenstriche, *Z. Tech. Phys.* 12:593.

LaCapra, V. C., J. M. Melack, M. Gastil, and D. Valeriano, 1996. Remote sensing of foliar chemistry of inundated rice with imaging spectrometry, *Remote Sensing Environ.* 55, 50–58.

Latz, K., R. A. Weismiller, and G. E. Van, Scoyoc, 1981. *A Study of the Spectral Reflectance of Selected Eroded Soils of Indiana in Relationship to Their Chemical and Physical Properties*, LARS Tech. Rep. 082181, Laboratory for Application of Remote Sensing, Purdue University, West Lafayette, Ind.

Leader, J. C., 1979. Analysis and prediction of laser scattering from rough-surface materials, *J. Opt. Soc. Am.*, 69, 610–628.

Leger, R. G., G. J. F. Millette, and S. Chomchan, 1979. The effects of organic matter, iron oxides and moisture on the color of two agricultural soils of Quebec, *Can. J. Soil Sci.* 59, 191–202.

Liang, S., and R. G. Townshend, 1996. A modified Hapke model for soil bidirectional reflectance, *Remote Sensing Environ.* 55, 1–10.

Lin, J., and C. W. Brown, 1992. Near-IR spectroscopic determination of NaCl in aqueous solution, *Appl. Spectrosc.*, 46, 1809–1815.

Liou, K. N., 1980. *An Introduction to Atmospheric Radiation*, Academic Press, San Diego, Calif.

Liu, H. Q., and A. Huete, 1995. A feedback based modification of the NDVI to minimize canopy background and atmospheric noise, *IEEE Trans. Geosci. Remote Sensing*, 33, 457–465.

Lumme, K., and E. Bowell, 1981. Radiative transfer in the surfaces of atmosphereless bodies: I. Theory, *Astron. J.*, 86, 1694–1704.

Madeira J., Cervelle A. B. B., Pouget M., and N. Flay 1997 Visible spectrometric indices of hematite (Hm) and goeithite (Gt) content in lateritic soils: the application of a Thematic Mapper (TM image for soil-mapping in Brasilia, Brazil. *International Journal of Remote sensing* 18:2835–2852.

Mathews, H. L., R. L. Cunningham, and G. W. Peterson, 1973. Spectral reflectance of selected Pennsylvania soils, *Proc. Soil Sci. Soc. Am. J.*, 37, 421–424.

McKeague, J. A., J. H. Day, and J. A. Shields, 1971. Evaluation relationships among soil properties by computer analysis, *Can. J. Soil Sci.*, 51, 105–111.

Melville, M. D., and G. Atkinson, 1985. Soil color: its measurement and its destination in models of uniform color space, *J. Soil Sci.*, 36, 495–512.

Metternicht G., and A. Zinck 1997 Spatial discrimination of salt-and sodium-affected soils surfaces. *International Journal of Remote sensing* 18:2571–2586.

Minnaert, M., 1941. The reciprocity principle in lunar photometry, *Astrophys. J.*, 93, 403–410.

Montgomery, O. L., 1976. An investigation of the relationship between spectral reflectance and the chemical, physical and genetic characteristics of soils, Ph.D. thesis, Purdue University, West Lafayette, Ind. (Libr. Congr. No 79-32236).

Montgomery, O. L., and M. F. Baumgardner, 1974. *The Effects of the Physical and Chemical, Properties of Soil and the Spectral Reflectance of Soils*, LARS Inf. Note 1125, Laboratory for Applications of Remote Sensing, Purdue University, West Lafayette, Ind.

Morin, Y., Y. Benyamini, and A. Michaeli, 1981. The dynamics of soil crusting by rainfall impact and the water movement in the soil profile, *J. Hydrol.*, 52, 321–335.

Morra, M. J., M. H. Hall, and L. L. Freeborn, 1991. Carbon and nitrogen analysis of soil fractions using near-infrared reflectance spectroscopy, *Soil Sci. Soc. Am. J.*, 55, 288–291.

Moran, S. M., R. D. Jackson, P. N. Slater, and P. M. Teillet, 1992. Evaluation of simplified procedures for retrieval of land surface reflectance factors from satellite sensor output, *Remote Sensing Environ.*, 41, 169–184

Mougenot, B., 1993. Effect of salts on reflectance and remote sensing of salt affected soils, *Cah. ORSTOM Ser. Pedol.*, 28(1): 45–54.

Mougenot, B., G. F. Epema, and M. Pouget, 1993. Remote sensing of salt-affected soils, *Remote Sensing Rev.*, 7, 241–259.

Mulders, M. A., 1987. *Remote Sensing in Soil Science*, Dev. Soil Sci. 15, Elsevier, Amsterdam, 379 pp.

Mulders, M. A., J. A. Den Dulk, and R. Uijlenhoet, 1992. Description of land surfaces, reflectance measurements and modeling for correlation with remote sensing data, *Sci. Sol*, 30(3), 169–184.

Munsell Color, 1975. *Munsell Soil Color Charts*, MacBeth Division of Kollmorgen Corp., Maryland USA

Murphy, R. J., and G. Wadge, 1994. The effects of vegetation on the ability to map soils using imaging spectrometer data, *Int. J. Remote Sensing*, 15, 63–86.

Nassau, K., 1980. The causes of color, *Sci. Am.*, 243, 106–124.

Nassau, K., 1983. *The Physics and Chemistry of Color*, Wiley, New York, 454 pp.

Newhall, S. M., D. Nickerson, and D. B. Judd, 1943. Final report of the O.S.A. subcommittee on the spacing of the Munsell colors, *J. Opt. Soc. Am.*, 33, 385–418.

Norman, J. M., J. M. Welles, and E. A. Walter, 1985. Contrasts among bidirectional reflectances of leaves, canopies, and soils, *IEEE Trans. Geosci. Remote Sensing*, 23, 659–668.

Norris, K. H., 1988. History, present state, and future prospects for near infrared spectroscopy, in *Analytical Application of Spectroscopy*, C. S. Creaser and A. M. C. Davise, eds., Royal Society of Chemistry, London, pp. 3–8.

Obukhov, A. I., and D. C. Orlov, 1964. Spectral reflectance of the major soil groups and the possibility of using diffuse reflection in soil investigations, *Sov. Soil Sci.*, 2, 174–184.

O'Neill, A. L., 1994. Reflectance spectra of microphytic soil crusts in semi-arid Australia, *Int. J. Remote Sensing*, 15, 675–681.

Orlov, D. C., 1966. Quantitative patterns of light reflectance on soils: I. Influence of particles (aggregate) size on reflectivity, *Sov. Soil Sci.*, 13, 1495–1498.

Orna, M. V., 1978. The chemical origins of color, *J. Chem. Ed.*, 55, 478–484.

Otterman, J., and G. H. Weiss, 1984. Reflection from a field of randomly located vertical protrusions, *Appl. Opt.*, 23, 1931–1936.

Otterman, J., T. Brakke, and A. Marshak, 1995. Scattering by Lambertian-leaves canopy: dependents of leaf-area projections, *Int. J. Remote Sensing*, 16, 1107–1125.

Palmer, J. M. 1982. Field standards of reflectance, *Photogramm. Eng. Remote Sensing*, 48, 1623–1625.

Pendleton, R. L., and D. Nickerson, 1951. Soil colors and special soil color charts, *Soil Sci.*, 71, 35–43.

Peterson, J. B., 1979. *Use Spectral Data to Estimate the Relationship Between Soil Moisture Tension and Their Corresponding Reflectance*, Annu. Rep. OWRT, Purdue University, West Lafayette, Ind., 18 pp.

Pieters, C. M., 1983. Strength of mineral absorption features in the transmitted component of near-infrared reflected light: first results from RELAB, *J. Geophys. Res.*, 88, 9534–9544.

Pinker, R. T., and A. Karnieli, 1995. Characteristic spectral reflectance of semi-arid environment, *Int. J. Soil Sci.*, 16, 1341–1363.

Pinty, B., and D. Ramond, 1987. A simple bidirectional reflectance model for terrestrial surfaces, *J. Geophys. Res.*, 91, 7803–7808

Pinty, N., and M. M. Verstraete, 1991. Extracting information on surface properties from bidirectional reflectance measurements, *J. Geophys. Res.*, 96, 2865–2874.

Pinty, B., M. M. Verstraete, and R. E. Dickson, 1989. A physical model for prediction of bidirectional reflectance over bare soil, *Remote Sensing Environ.*, 27, 273–288.

Price, J. C., 1990. On the information content of soil reflectance spectra, *Remote Sensing Environ.*, 33, 113–121.

Price, J. C., 1991. On the value of high spectral resolution measurements in the visible and near-infrared, in *Proceedings of the 5th International Colloquium on Physical Measurements and Signatures in Remote Sensing*, Courchevel, France, Vol. 1, pp. 131–136.

Price, J. C., 1995. Examples of high resolution visible to near-infrared reflectance spectra and a standardized collection for remote sensing studies, *Int. J. Remote Sensing*, 16, 993–1000.

Ranson, K. J., J. R. Irons, and C. S. T. Daughtry, 1991. Surface albedo from bidirectional reflectance, *Remote Sensing Environ.*, 35, 201–211.

Rao, B. R. M., T. Ravi Sankar, R. S. Dwivedi, S. S. Thammappa, and L. Venkataratnam, 1995. Spectral behavior of salt-affected soils, *Int. J. Remote Sensing*, 16, 2125–2136.

Ray, T. W., and B. C. Murray, 1996. Nonlinear spectral mixing in desert vegetation, *Remote Sensing Environ.*, 55, 59–79.

Richardson, A. J., C. L. Wiegand, H. W. Gausman, J. A. Cullar, and A. H. Gerbermann, 1975. Plant, soil and shadow reflectance components of raw crops, *Photogramm. Eng. Remote Sensing*, 41, 1401–1407.

Roberts, D. A., M. O. Smith, and J. B. Adams, 1993. Green vegetation, nonphotosynthetic vegetation, and soils in AVIRIS data, *Remote Sensing Environ.*, 44, 255–269.

Rogers, R. W., and R. T. Langer, 1972. Soil surface lichens in arid and subarid southeastern Australia: introduction and floristics, *Aust. J. Bot.*, 20, 197–213.

Rondeaux, G., M. Steven, and F. Baret, 1996. Optimization of soil-adjusted vegetation indices, *Remote Sensing Environ.*, 55, 95–107.

Ronov, A. A., and A. A. Yaroshevsky, 1971. Chemical composition of the Earth's crust, in *The Earth's Crust and Upper Mantle*, P. J. Hart, ed., American Geophysical Union, Washington, D.C., pp. 37–57.

Roujean, J. L., M. Leroy, P. Y. Deschamps, and A. Podaire, 1990. A surface bidirectional reflectance model to be used for the correction of directional effects in remote sensing multitemporal data sets, in *Remote Sensing Science for the Nineties: Proceedings of the 10th Annual International Geoscience and Remote Sensing Symposium*, IGARSS'90, Vol. III, College Park, Md. May 20–24, R. Mills, ed., Institute of Electrical and Electronics Engineers, New York, pp. 1785–1789.

Satterwhite, M. B., and J. P. Henley, 1991. *Hyperspectral Signatures (400–2,500nm) of Vegetation, Minerals, Soils, Rocks and Cultural Features: Laboratory and Field Measurements*, ELT-0573, U.S. Army Corps of Engineers, Fort Belvoir, Va.

Schnitzer, M., and S. U. Khan, 1978. *Soil Organic Matter*, Elsevier, Amsterdam.

Schreier H. 1977. Quantitative predictions of chemical soil conditions from multispectral airborne, ground and laboratory measurements. pp. 106–112. IN: 4th Canadian Symposium on Remote Sensing. Quebec City, Quebec, Canada, 613 p.

Schreier, H., 1977. FILL NAME *Proceedings of the 4th Canadian Symposium on Remote Sensing*, Vol. 1, pp. 106–112.

Schwertmann, U., 1988. Occurrence and formation of iron oxides in various pedo-environments, in *Iron in Soils and Clay Minerals*, J. W. Stucki, B. A. Goodman, and U. Schwertmann, eds., NATO ASI Ser., D. Reidel, Dordrecht, The Netherlands, pp. 267–308.

Shainberg, I., 1992. Chemical and mineralogical components of crusting, in *Soil Crusting*, M. E. Sumner and B. A. Stewart, eds., Lewis Publications, Ann Arbor, Mich.

Sharma, R. C., and G. P. Bhargava, 1988. *Landsat* imagery for mapping saline soils and wetlands in north-west India, *Int. J. Remote Sensing*, 9. 39–44.

Shields, J. A., R. J. St. Arnaud, E. A. Paul, and J. S. Clayton, 1966. Measurement of soil color, *Can. J. Soil Sci.*, 4, 83–90.

Shields, J. A., E. A. Paul, R. J. Arnaud, and W. K. Head, 1968. Spectrophotometric measurement of soil color and its relation to moisture and organic matter, *Can. J. Soil Sci.*, 48, 271–280.

Siegal, B. S., and A. F. H. Goetz, 1977. Effect of vegetation on rock and soil type discrimination, *Photogramm. Eng. Remote Sensing*, 43, 191–196.

Singer, A., and E. Ben-Dor, 1987. Origin of red clays layer interbedded with basalt of Golan Heights, *Geoderma*, 39, 283–306.

Singh, A. N., 1994. Monitoring change in the extent of salt-affected soils in northern India, *Int. J. Remote Sensing*, 16, 3173–3182.

Smith, J. A., 1983. Matter–energy interaction in the optical region, p. 61–113. In *Manual of Remote Sensing*, 2nd ed., Vol. I, D. S. Simonett and F. T. Ulaby, eds. American Society of Photogrammetry, Falls Church, Va.

Smith, M. O., S. L. Ustin, J. B. Adams, and A. R. Gillespie, 1990. Vegetation in desert: I. A regional measure of abundances from multispectral images, *Remote Sensing Environ.*, 31, 1–26.

Soileau, J. M., and R. J. McCraken, 1967. Free iron and coloration in certain well-drained Coastal Plain soils in relation to their other properties and classification, *Soil Sci. Soc. Am. Proc.*, 31, 248–255.

Soil Survey Staff, 1975. *Soil Taxonomy: A Basic System of Soil Classification for Making and Interpreting Soil Surveys*, USDA Handb. 436, Soil Conservation Service, U.S. Department of Agriculture, Washington D.C.,

Stark, E., K. Luchter, and M. Margoshes, 1986. Near-infrared analysis (NIRA): a technology for quantitative and qualitative analysis, *Appl. Spectrosc. Rev.*, 24, 335–339.

Stevenson, F. J., 1982. *Humus Chemistry*, Wiley, New York.

Stoner, E. R., 1979. Physicochemical, site and bidirectional reflectance factor characteristics of uniformly-moist soils, Ph.D. thesis, Purdue University, West Lafayette, Ind.

Stoner, E. R. and M. F. Baumgardner, 1981. Characteristic variations in reflectance of surface soils, *Soil. Sci. Soc. Am. J.*, 45, 1161–1165.

Stoner, E. R., M. F. Baumgardner, L. L. Biehl, and B. F. Robinson, 1980a. *Atlas of Soil Reflectance Properties*, Res. Bull. 962, Agricultural Experiment Station, Purdue University, West Lafayette, Ind.

Stoner, E. R., M. F. Baumgardner, R. A. Weismiller, L. L. Biehl and B. F. Robinson,

1980b. Extension of laboratory measured soil spectra to field conditions, *Soil Sci. Am. J.*, 44, 572–574.

Stucki, J. W., 1988, Structural iron in smectite, in *Soils and Clay Minerals*, J. W. Stucki, B. A. Goodman, and U. Schwertmann, eds., NATO ASI Ser., D. Reidel, Dordrecht, The Netherlands, pp. 625–675.

Suits, G. H., 1983. The nature of electromagnetic radiation, in D. S. Simonett and F. T. Ulaby, eds., *Manual of Remote Sensing*, nd ed., Vol. I, American Society of Photogrammetry, Falls Church, Va., pp. 37–60.

Szilagyi, A., and M. F. Baumgardner, 1991. Salinity and spectral reflectance of soils, in *Proceedings of the ASPRS Annual Convention*, Baltimore, pp. 430–438.

Tanre, D., C. Deroo, P. Duhaut, M. Herman, J. J. Morcrette, J. Perbos, and P. Y. Deschamps, 1986. *Simulation of the Satellite Signal in the Solar Spectrum (5S): User Guide*, Laboratory d'Optique Atmospherique, Villenueve D'asc, France.

Taranik, D. L., and F. A. Kruse, 1989. Iron minerals reflectance in geophysical and environmental research imaging spectrometer (GERIS) data, in *Proceedings of the 7th Thematic Conference on Remote Sensing for Exploration Geology*, Calgary, Alberta, Canada, Vol. 1, pp. 445–458.

Taylor, R. M., 1982. Colour in soils and sediments: a review, *Dev. Sedimentol.*, 35, 749–761.

Thompson, L. M., 1957. *Soils and Soil Fertility*, McGraw-Hill, New York.

Tkachuk, R., and D. P. Law, 1978. Near infrared diffuse reflectance standards, *Cereal Chem.*, 55, 981–995.

Torrent, J., U. Schwertmann, H. Fetcher, and F. Alferez, 1983. Quantitative relationships between soil color and hematite content, *Soil Sci.*, 13, 354–358.

Toth, T., F. Csillag, L. L. Biehl, and E. Micheli, 1991. Characterization of semi-vegetated salt-affected soil by means of field remote sensing, *Remote Sensing Environ.*, 37, 167–180.

Tucker, C. J., and L. D. Miller, 1977. Soil spectra contributions to grass canopy spectral reflectance, *Photogramm. Eng. Remote Sensing*, 43, 721–726.

Tueller, P. T., 1987. Remote sensing science application in arid environment, *Remote Sensing Environ.*, 23, 143–154.

Twomey, S. A., C. F. Bohren, and J. L. Mergenthaler, 1986. Reflectance and albedo differences between wet and dry surfaces, *Appl. Opt.*, 25, 431–437.

Van der Meer, F., 1995. Spectral reflectance of carbonate mineral mixture and bidirectional reflectance theory: quantitative analysis techniques for application in remote sensing, *Remote Sensing Rev.*, 13, 67–94.

Vane, G., J. H. Reimer, T. G. Chrien, H. T. Enmark, E. G. Hansen, and W. M. Porter, 1993. Airborne visible/infrared imaging spectrometer (AVIRIS), *Remote Sensing Environ.*, 44, 127–143.

Verma, K. S., R. K. Saeena, A. K. Barthwal, and S. N. Deshmukh, 1994. Remote sensing technique for mapping salt affected soils, *Int. J. Remote Sensing*, 15, 1901–1914.

Vincent, R., and G. Hunt, 1968. Infrared reflectance from mat surfaces, *Appl. Opt.*, 7, 53–59.

Vinogradov, B. V., 1981. Remote sensing of the humus content of soils, *Sov. Soil Sci.*, 11, 114–123.

Walthall, C. L., J. M. Norman, J. M. Welles, G. Campbell, and B. L. Blad, 1985.

Simple equation to approximate the bidirectional reflectance from vegetation canopies and bare soil surfaces, *Appl. Opt.*, 24, 383–387.

Weindner, V. R., and J. J. Hsia, 1981. Reflection properties of pressed polytetrafluoroethylene powder, *J. Opt. Soc. Am.*, 71, 856–862.

Wendlandt, W. W., and H. G. Hecht, 1966. *Reflectance Spectroscopy*, Interscience, New York.

Wessman, C. A., J. D. Aber, D. L. Peterson, and J. M. Meliloo, 1988. Foliar analysis using near infrared spectroscopy, *Can. J. For. Res.*, 18, 6–11.

Wessman, C. A., J. D. Aber, and D. L. Peterson, 1989. An evaluation of imaging spectrometry for estimating forest canopy chemistry, *Int. J. Remote Sensing*, 10, 1293–1316.

West, N. E., 1990. Structure and function of microphytic soil crust in wildland ecosystems of arid to semi-arid regions, *Adv. Ecol. Res.*, 20, 179–223.

Westin, F. C., and C. J. Franzee, 1976. *Landsat* data in a soil survey program, *Soil Sci. Soc. Am. J.*, 40, 81–89.

White K., Walden J., Drake N., Eckardt F. and J., Settle 1997 Mapping the iron oxides content of dune sands, Namib Sand Sea, Namibia, using Landsat Thematic Mapper data. *Remote Sensing of Environment* 62:30–39.

White, J. L., and C. B. Roth, 1986. Infrared spectrometry, in *Methods of Soil Analysis*, Part 1, 2nd ed., A. Klute, ed. *Agronomy*, 9, 291–330.

Wiegand, C. L., J. D. Rhoades, D. E. Escobar, and J. H. Everitt, 1994. Photographic and videographic observations for determining and mapping the response of cotton to soil salinity, *Remote Sensing Environ.*, 49, 212–223.

Williamson, P. C., and K. H. Norris, 1987. Near-infrared technology, in *The Agricultural and Food Industries*, American Association of Cereal Chemists, St. Paul, Minn., 330 pp.

Williamson, S. J., and H. Z. Cummins, 1983. *Light and Color in Nature and Art*, Wiley, New York, 488 pp.

Wyszecki, G., and W. S. Stiles, 1982. *Color Science: Concepts and Methods, Quantitative Data and Formulae*, 2nd ed., Wiley, New York.

Young, E. R, K. C. Clark, R. B. Bennett, and T. L. Houk, 1980. Measurements and parameterization of the bidirectional reflectance feature of $BaSO_4$ paint, *Appl. Opt.*, 19(20), 3500–3505.

Younis M. T., Gilabert M. A., Melia, J., and Bastida, J., 1997 Weathering process effects on spectral reflectance of rocks in assume-arid environment. *International Journal of Remote sensing* 18:3361–3377.

Geobotany: Vegetation Mapping for Earth Sciences

S. L. Ustin

University of California
Davis, California

M. O. Smith

University of Washington, Seattle, Washington

S. Jacquemoud

Université Paris,
Paris, France

M. Verstraete and Y. Govaerts

Space Applications Institute, EC Joint Research Centre
Ispra, Italy

4.1 INTRODUCTION

While the use of remote sensing images for landform mapping is relatively straight-forward, geologic mapping is difficult because most of the Earth's surface, other then extreme desert, alpine, and boreal regions, is covered by vegetation, and direct observation of spectral features associated with mineral deposits and rock types is obstructed. This has led geologists to using the plants themselves as indicators of the

Remote Sensing for the Earth Sciences: Manual of Remote Sensing, 3 ed., Vol. 3, edited by Andrew N. Rencz.
ISBN: 0471-29405-5 © 1999 John Wiley & Sons, Inc.

underlying substrate. The application of remote sensing for mineral mapping has renewed interest in how plant characteristics can provide insight into geologic and soil conditions. From a broad perspective, geobotany may be considered as a study of the integration of the physical and biological processes of interest to earth science. Plants and microbes function at the interface between the soil and atmospheric systems, interacting with all phases of the hydrologic cycle, soil development and erosion, biogeochemical cycling, and geologic weathering. Thus, developing tools for characterizing ecosystems has implications for a wide range of earth science research. Applications for earth science issues ranges from identification of fracture zones, geologic mapping, weathering and erosion, sediment transport, surface and subsurface hydrology, hydrocarbon seepages, soil or water contamination, to many others. The capability for synoptic mapping combined with the nearly ubiquitous cover of the terrestrial surface by plants has created a need for forging better understanding of these relationships.

Plant communities show more spatial variability then can be explained by climates, and relationships between some vegetation and soil types have been known since ancient times. Modern researchers have demonstrated that vegetation is often influenced by mineralization, lithology, and rock fracturing (Cannon, 1960, 1971; Chikishev, 1965; Brooks, 1972), even if these relationships are not always exclusive. Sites with high concentrations of heavy metals, such as mine waste or mill tailings for lead, copper, zinc, nickel, and other metals, may develop unique floras because conditions are toxic to most species (Bradshaw, 1969). Naturally occurring saline or alkaline deposits and serpentine outcrops also develop unique floras (Kruckleberg, 1969). In some cases, leaf concentrations of trace metals may indicate the composition of the underlying soil and may be useful for mineral prospecting (Brooks, 1995a). Plants tolerant of one metal or nutrient are not necessarily tolerant of other excess nutrients; thus indicator species can be developed for certain conditions. Where geologic conditions produce plant characteristics that can be detected spectrally, they can be observed with remote sensing systems. Typically, spectral features rely on vegetation characteristics deemed anomalous for the area under study: either unusually low or high vegetation cover or changes in plant species or community distributions. If enough information is known about plant phonologic cycles or growth patterns, flowering period, early senescence, or other changes in botanical function, these changes can be used to indicate unusual soil or geologic conditions. These relationships are often empirically determined and may be valid only for the specific conditions under which they were developed.

Progress in developing more physical understanding has been slow due to the lack of interdisciplinary understanding between plant mineral nutrition specialists, plant ecologists, geologists, and remote sensing scientists. Nonetheless, if remote sensing is to provide mineral characterization and mapping outside desert areas, there is a critical need to develop collaborations between these disciplines. Ecosystem-based studies, sometimes termed *earth system science*, offer the potential for better understanding the factors controlling plant abundance and distributions through more holistic ecophysiological models connecting plant function with edaphic and climate conditions (Ovington, 1965; Rodin and Bazilevich, 1967; Parton et al., 1987; Running and Coughlan, 1988; McGuire et al., 1992; Prentice et al., 1992; Potter et al., 1993; Perruchoud and Fischlin, 1995).

4.1.1 Chapter Objectives

The primary aim of this chapter is to provide earth scientists with a brief review of plant properties that affect the absorption and scattering of radiation in the visible and infrared spectral region. The focus is on providing sufficient understanding of plant structure and function to understand the spectral characteristics that may be useful in earth science and geobotanical mapping of earth resources. The second aim of the chapter is to introduce several new methods for analyzing remote sensing data that are of particular interest for extracting botanical properties. As we move into the next century, a wide range of new satellite and airborne sensors will become available, and new techniques are needed to analyze these data. These will include hyperspectral sensors with large numbers of contiguous bands and sensors with bands in new spectral regions, including multiband shortwave infrared, thermal infrared, lidar, radar, and others. The case studies illustrate many approaches to geobotany and provide perspective on the range of applications.

4.1.2 Definition of Terms

Geobotany has been defined as the study of plants as related specifically to their geologic environment (Rose et al., 1979) and by Raines and Canney (1980) as a visual survey of vegetation used to define geologic patterns in the landscape. Geobotany is composed of two fields of botanical research: the study of the spatial distribution of plants and plant communities as related to geology and the study of vegetation characteristics (physiology, morphology, anatomy, biochemistry) as related to geology. The first component of geobotany has it roots in studies of holistic plant geography which date back to Darwin (Cain, 1971). The second component is more closely aligned with biogeochemistry. Geochemistry is the study of the chemical composition of the earth and the physical and chemical processes that have produced their distributions. The study of geochemistry has been extended to include biological materials and their interactions in relation to earth chemicals. Although the next two sections describe these fields as separate entities, in the remainder of the chapter we consider geobotany as a more general term that includes spatial distribution of vegetation and biogeochemistry and geology in the broader context of earth sciences.

4.1.3 Geobotany

Some of the earliest modern ecological literature (e.g., Warming, 1892); Schimper, 1898) describe the relationships between plant geography and soils from the perspective of plant function, morphology, and physiology. Warming and Schimper describe plants whose distributions are limited to a particular soil chemistry, such as calcareous, saline, siliceous, serpentine, zinc, and others. Warming emphasized the importance of soil factors in plant distribution and coined the terms used today for describing the association of characteristic plant species with particular edaphic conditions: halophyte, hydrophyte, xerophyte, and so on. While symptoms of deficiency or excess of specific elements have been described by numerous authors, caution must be used in interpreting visual symptoms of plants because different mineral deficiencies or excesses can cause the same symptoms (e.g., chlorosis), some species respond differently to similar exposures, and multiple deficiencies may be present. The books by Walter (1973) and Eyre (1968) provide a good review of the importance of soil chemistry in mineral nutrition and distribution of vegetation.

In North America the term *geobotany* has been used more narrowly to describe a form of mineral prospecting that relies on characteristics of the vegetation to identify the location and extent of ore bodies (Brooks, 1995a). A long list of indicator species have been identified, including those known to "hyperaccumulate" metals (i.e., tolerate and accumulate high concentrations of heavy metals) (Brooks, 1995b). Although the field has developed primarily in this century, Brooks cites several early observations. Roman records show that they were aware that some plant species were good indicators of subsurface water stores. Georgius Agricola (1556) described a pattern of premature plant senescence and stunted growth of plants above ore bodies. Brooks (1995c) reports that Caesalpino (1583) associated *Alyssum bertolonii* with ultramafic soils, and Pope Pius II (1614) suggested that *Ilex aquifolium* might be an indicator of aluminum. Scandinavian miners used indicator plants such as *Lychinis alpina* (pyrite plant) in the seventeenth century to locate ore deposits.

4.1.4 Biogeochemistry

Biogeochemical cycles are the sequences of stages in the transfer and transformation of elements between the lithosphere, hydrosphere, atmosphere, and the biosphere. Carbon, nitrogen, oxygen, sulfur, and water have cycles of global importance through soil, water, biota, and air. In recent years the importance of biologic agents in geochemical cycling and transport has become more widely recognized. Plants, animals, and soil microbes contribute significantly to the rates of geochemical reactions, fates of weathered minerals, and dispersion and transport processes. Elements may be transported vertically to the surface due to reworking by soil organisms or uptake by plant roots. Distribution of elements in the soil may be changed by preferential deposition or accumulation under plant canopies or in the interstices between plants. Organisms mediate the cycling of chemicals that have both atmospheric and geologic phases. Some elements, such as sodium and chloride, are nonvolatile but are transported by the atmosphere over great distances. Others, such as phosphorus, potassium, calcium, and magnesium, do not have an atmospheric phase and are only mildly soluble. These are eventually lost from the lithosphere to the oceans through the mechanism of absorption by plants and eventual leaching during decomposition (Epstein, 1972). Generally, relationships among these ecosystem components are based on understanding the mineral constituents of the soil and the plant's nutrient requirements and tolerances. However, plant–soil relationships are not simple; they are often nonlinear with mineral concentration, tissue age, and are strongly dependent on the water-holding capacity of the soil, pH, and canopy evapotranspiration (Epstein, 1972).

TABLE 4.1   Typical Concentrations of Essential Nutrient Elements in Plant Material Based on Dry Weight of the Tissue

Element	Concentration in Plant (ppm dry weight)	Number Atoms Relative to Molybdenum
Micronutrients		
Molybdenum	0.1	1
Copper	6.0	100
Zinc	20.0	300
Manganese	50.0	1,000
Iron	100.0	2,000
Boron	20.0	2,000
Chlorine	100.0	3,000
Macronutrients	(% dry weight)	
Sulfur	0.1	30,000
Phosphorus	0.2	60,000
Magnesium	0.2	80,000
Calcium	0.5	125,000
Potassium	1.0	250,000
Nitrogen	1.5	1,000,000
Oxygen	45.0	30,000,000
Carbon	45.0	40,000,000
Hydrogen	6.0	60,000,000

Source: Adapted from Epstein (1972).

The concentrations of nutrient elements varies with the plant tissue, age, and growing conditions of the plants. Immobile nutrients such as calcium (i.e., those that become fixed in plant tissue and are not subject to relocation within the plant), continue to accumulate throughout the organ's life and are lost from the plant when the organ senesces and dies. Mobile nutrients (e.g., nitrate and potassium) are those that may be translocated to another organ within the perennial plant. In evergreen plants, mobile nutrients may accumulate until close to the end of the growing season (Mooney and Rundel, 1979; Chapin et al., 1980). Adjusting root uptake kinetics provides a mechanism for regulating nutrient uptake under different soil fertility. Chapin (1988) argues that root kinetics regulate uptake of mobile ions under all soil conditions and immobile ions in fertile soils. For many elements, relative concentrations are determined by stiochiometric requirements defined by plant metabolism. Nutrients can vary in concentration over eight orders of magnitude within a given tissue, as illustrated in Table 4.1. Anomalous tissue concentrations develop when soils have deficiencies or excesses in these elements or high concentrations of other elements. Over extended periods plants respond to nutrient deficiencies, excesses, or imbalances through modifications of growth. Responses can range from reduced or enhanced growth and biomass production, changes in leaf area or root/crown ratios (low nutrient or water status generally leads to increased allocation of biomass to roots relative to shoots), or changes in species distributions (Schlesinger, 1991; Field et al., 1992).

4.2 GEOBOTANICAL REMOTE SENSING

Monitoring and evaluation of terrestrial environments using remote sensing techniques inevitably requires some assessment of the vegetation. Techniques to extract Earth systems information have ranged from visual assessment of simple false color composites to more sophisticated mathematical transformations of the data. Sections 4.2 to 4.6 highlight several analytical methods useful for vegetation analysis and a review of the physical bases for leaf-to-canopy spectral responses.

4.2.1 Historical Perspective

The use of remotely sensed vegetation patterns to map geologic formations began soon after the launch of the first *Landsat* satellites. Cole (1977) used the *Landsat* multispectral scanner (MSS) and aerial photography to map distributions of plant communities and showed that they corresponded to different geologic formations. Lyon (1975a) also used MSS to identify vegetation reflectance anomalies associated with molybdenum mineral deposits. Ballew (1975) correlated anomalous MSS band ratios with surface geochemistry for soils having mercury, gold, silver, lead, copper, and bismuth. Anomalies were due to either rock exposures, reduced vegetation growth, or changes in vegetation type. Lyon (1975b) reviewed the use of early *Landsat* imagery for geobotanical exploration in semiarid regions. Raines et al. (1978) used MSS to identify plant species associated with uranium deposits. Talvitie (1979) discovered the Sokli Carbonatite Massif in Finland using *Landsat* imagery and a Daedalus 11-channel scanner because of better vegetation growth on the nutrient-rich carbonatite.

By 1980, Solomonson et al. (1980) reviewed several geobotanical studies to summarize the relationships between leaf reflectance and metal concentrations. Even at this early stage in remote sensing, their paper showed that the effect of metal accumulation on leaf reflectance varied with the wavelength, metal species (copper, lead, zinc, arsenic, molybdenum, sulfates), and tree species. A recent review by Brooks (1995d) summarizes a large number of early papers on geobotanical exploration using satellite and airborne imagery. Milton (1978) measured field spectra of native species in hydrothermally altered semiarid region. Field spectra of plants and rocks were used to construct a MSS band ratio model to map the plant communities of the East Tintic Mountains, Utah (Milton, 1983). Chang and Collins (1980) and Horler et al. (1980) used laboratory experiments to show that chlorophyll content and reflectance changes when plants are grown in soils with heavy metals. Horler et al. (1981) provide an early review of the use of plant reflectance to detect metal-induced chlorosis. Collins et al. (1980, 1983) were the first to show narrowband spectral changes in airborne data using the 512-band MARK II spectroradiometer. They observed a blue shift due to apparent reduced chlorophyll absorption associated with mineral deposits at the *red edge*, the region between the red and near-infrared wavelengths around 700 nm, where leaf reflectance increases rapidly. Their conclusions were supported by a subsequent experimental study of mineral-stressed plants (Chang and Collins, 1983). Milton et al. (1983) used the Chebyshev polynomial waveform analysis of Collins et al. (1983) on airborne MARK II data over sites of metal anomalies in a hydrothermally altered porphyry gold system in the slate belt of North Carolina and showed that the strongest changes in spectral reflectance occurred over an area of high copper, molybdenum, and tin. They caution, however, that other weakly anomalous areas did not correspond to areas of high metal concentrations. The problem in application of reflectance anomalies to image data is that reflectance depends not only on pigment content but also on plant leaf area index (total leaf area/ground area), percent cover, species composition and canopy architecture, and extrinsic but correlated factors such as topography.

4.2.2 Combined Geographic Information Systems Analysis

Many mines, mineral outcrops or contaminant sources have small point-source locations within a larger geologic and ecologic context. Often, the goal of geobotany is to identify these point sources, due to their contrast with the surrounding vegetation and terrain. Numerous factors create variability in satellite images that are imposed at multiple scales. Topographic patterns, illumination conditions, and atmospheric composition at the time of measurement are factors that vary at the larger scales of variation. Of the factors affecting local-scale vegetation distribution within a regional climate zone, topography is of critical importance. Although the impact of topographic gradients is most apparent at the ecotones or margins of community ranges, its effect on the local microclimate—modifying temperatures and net radiation, precipitation regime, diurnal, and seasonal cycles—is always present. Vegetation type (e.g., deciduous, evergreen, or conifer) and life form (herbaceous, shrub, or trees) are frequently observed to vary at small scales in response to topographic conditions.

Interpretation of remote sensing images can be improved by linking analysis with a geographic information systems (GIS) database. Digital elevation maps (DEMs) can readily be combined with image data sets to assess topographic patterns. For example, Warner et al. (1989) combined AVHRR data with a DEM to assess inter-annual drought conditions in east Africa. Smith et al. (1990b) developed functions between elevation and mean vegetation cover and surface temperature, then compared the *Landsat* thematic mapper (TM)-derived estimates of vegetation density and temperature to find vegetation distribution anomalies related to soil conditions and water availability. Ustin et al. (1996) used the slope, aspect, and elevation data layers in a vegetation classification scheme with spectral mixture fractions. Similarly, other ancillary data in a GIS may be used to improve relationships using newer geostatistical methods (e.g., co-kriging) or to train neural nets. Later in this chapter we introduce wavelet analyses as an image method that can utilize spectral, spatial, and temporal information that are embedded in remotely sensed images.

There are a number of methods to examine images for texture or spatial patterns in vegetation distribution that can be combined with ancillary data to improve estimates. One example of a spatial analysis used for geologic exploration is lineament analysis (Cetin et al., 1993). Anomalous linear features in an image are assumed to represent structural geologic features, particularly joints and faults. These geologic features have been shown to be associated with zones of higher soil permeability and often have different plant species or changes in plant density growing over these fracture zones. However, by themselves lineament interpretations are typically noisy and ambiguous because many lineaments in satellite images are cultural features or random associations. By combining multidate imagery, or using GIS databases, the reliability of lineament analysis can be improved (see Chapter 9 for more discussion).

4.3 SPECTRAL CHARACTERISTICS OF FOLIAGE AND PLANT CANOPIES

4.3.1 Spectra of Leaf Components

Among the input parameters required to run canopy reflectance models, some of them, such as the leaf area index (LAI) and the leaf angle distribution generally have a biophysical meaning; others, such as leaf reflectance or transmittance, cannot be related directly to such characteristics. Although the albedo may vary, the general shape of reflectance curves for green leaves (Figure 4.1) is similar for all species (Gates et al., 1965). This fact requires adopting a different strategy for plants than the direct spectral feature and band shape matching used for geology applications. The spectral shape of plant leaves is controlled by absorption features of specific molecules and the cellular structure. Since all higher plants (e.g., ferns, conifers, and flowering plants) share a common tissue organization and primary metabolism, the differences between species and for plants growing under different conditions are expressed by changes in the proportions of foliar constituents. Thus, interpretations of botanical spectral features requires some understanding of basic plant physiology and function. During the last 50 years, many empirical relationships have been proposed in the literature to explain the leaf optical properties in terms of chlorophyll or water content, leaf internal structure, or surface properties.

Foliar absorptions are caused primarily by photosynthetic pigments in the visible spectrum, water, cellulose, and other carbon-based compounds in the infrared (Fig. 4.1). Cell wall and water–air interfaces account for much of the scattering in the infrared (Gates et al., 1965; Gates, 1970; Gausman, 1977; Gausman et al., 1978). Major pigments include chlorophyll a and b, β-carotene, lutein, and xanthophyll cycle pigments (Lichtenthaler, 1987). It has long been noted that extracted chlorophyll absorption peaks are shifted about 20 nm to shorter wavelengths than observed in leaf reflectance. Seasonal shifts of the long-wavelength edge of the chlorophyll absorption feature (red edge) also have been observed in foliage (Gates et al., 1965; Horler et al., 1983), as have stress-related shifts (e.g., due to heavy metals) (Chang and Collins, 1983, Milton et al., 1983). Despite 20 years of work, defining the relationship between the red edge and chlorophyll concentration remains an area of active research (Curran et al., 1995, Pinar and Curran 1996). This is due partially to difficulties in measuring the suite of pigments and intermediates and conformational changes that occur during handling. In the intact chloroplast, pigments are arrayed in protein complexes associated with thylakoid membranes in the light-harvesting apparatus. Gamon et al. (1990) observed short-term changes in reflectance near the green peak at 530 nm that were due to changes in the distribution of xanthophyll cycle pigments in response to the light environment. This physiologically mediated rapidly reversible conformation change occurs in response to leaf exposure to high light levels and is part of plants' mechanism for light-intensity regulation.

Water dominates leaf absorption in the shortwave infrared (1.0 to 2.5μm) region and major water absorption bands occur at 1.45, 1.94, and 2.7 μm and secondary

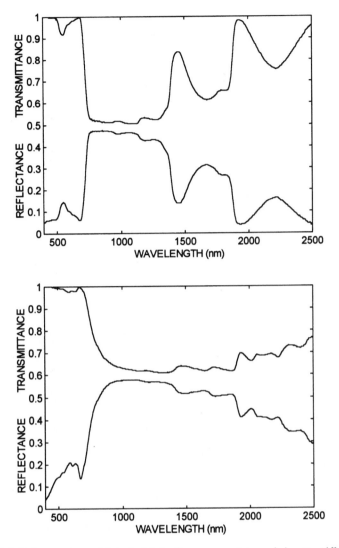

Figure 4.1   Whole leaf spectra showing (*a*) the fresh leaf reflectance, transmission, and absorption (difference) for a corn leaf, and (*b*) the dry leaf reflectance, transmission, and absorption for the corn leaf after drying.

features at 0.96, 1.12, 1.54, 1.67, and 2.20μm (Knipling, 1970; Wooley, 1971, 1973; Wessman, 1990). Protein, cellulose, lignin, and starch affect leaf reflectance (Marten et al., 1985; Weyer, 1985; Card et al. 1988) but have been more difficult to characterize, even in dry leaves (Peterson and Hubbard, 1992; ACCP, 1994; Curran, 1994). Spectra of these organic compounds are composed primarily of C–H, N–H, and O–H bonds that have fundamental molecular absorptions in the region 5 to 8 μm and exhibit overtones and combination band overtones in the shortwave infrared region (Curran 1989; Wessman, 1990). Although absorption spectra of these molecules have been developed (e.g., Card et al., 1988; Wessman, 1990; Barton et al., 1992), they are not identical to the specific absorption coefficients predicted by inversion of the PROSPECT model (Jacquemoud et al., 1996), presumably due to bending and stretching bond vibrations in the intact leaf.

4.3.2 Fluorescence and Thermal Infrared Spectra

Remote sensing studies have looked for ways to directly measure the physiological properties of plants. Remotely sensed fluorescence measurements using active lidar or laser systems provides a mechanism to study plant biochemical constituents and their electronic states over a range of scales from leaves to images. Chappelle and his colleagues performed early remote sensing measurements that showed the potential for detecting plant stresses by measuring laser-induced chlorophyll fluorescence (Chappelle et al., 1984a,b, 1985). Hoge et al. (1983) provided an early study on the feasibility of airborne detection of chlorophyll fluorescence from terrestrial green plants. When photosynthetic efficiency is reduced in chloroplasts and absorbed energy is dissipated via the photosynthetic electron transfer system or the light-harvesting pigment apparatus, increased chlorophyll fluorescence emission occurs. Any increase in fluorescence over the background rate is an indicator of photosynthetic functioning (Valentini et al., 1994). Many environmental conditions have been observed to cause increased fluorescence in plants, including temperature (both excess heating and freezing), water deficit, senescence, air pollution, and soil nutrient deficiencies or excesses. Herbicides that inhibit electron transport also cause increased fluorescence (Lichtenthaler, 1988; Lichtenthaler and Rinderle, 1988). Recent field studies using new near- and far-field laser-induced fluorescence instruments, and techniques have extended earlier studies and continue to show the potential for stress detection (Cecchi et al., 1994; Hoque and Remus, 1994; Kharuk et al., 1994; Krajicek and Vrbova, 1994; Schmuck and Moya, 1994). It is likely that these methods will become more widely used in geobotanical studies to locate point sources of mineral exposures or contamination.

Satellite thermal infrared imagery can be used to map surface temperatures to identify areas susceptible to frosts or to compute a surface energy budget (Jackson, 1982a). These applications are particularly useful for monitoring agricultural production but apply to all vegetated regions. Well-watered plant canopies experience high rates of transpiration under warm sunny sky conditions when leaf temperatures can be several degrees Celsius below the bulk air temperature (Jackson et al., 1977). Jackson (1982b) has shown that soil moisture can be estimated from soil temperatures. Several studies have noted an inverse relationship between the normalized difference vegetation index (NDVI) and surface temperature (Seguin and Itier, 1983; Goward et al., 1985; Kerr et al., 1989; Seguin et al., 1989). These studies have shown that thermal infrared sensing is useful for estimating crop water stress at local and regional scales. Edaphic stresses (e.g., nutrient deficiencies and excesses) can also cause partial stomatal closure, lowering transpiration rates and increasing surface temperatures. Although the use of thermal infrared measurements to locate geobotanical anomolies has had limited application, new multiband, high-spatial-resolution sensors from EOS satellites (e.g., ASTER and MODIS) and airborne sensors indicates that these methods may become more widely used. Lang et al. (1994) have proposed methods for real-time image processing of thermal temperatures. Such techniques are important for natural disaster monitoring. For example, NASA has developed a real-time imaging system for monitoring wildfire temperatures and spread that contributes to more reliable prediction and assessment of hazard (Brass et al., 1996).

4.3.3 Variability at the Canopy Scale

Composition variation in plant communities, the dispersion patterns of plants, and the percentage of canopy cover affect image data at intermediate spatial scales. Canopy structure and composition are defined by the leaf area, number of leaf layers, and leaf angle distributions within the canopy, and the distribution of different canopy components, specifically the stems, green foliage, and litter, and finally, the soil and understory vegetation if the vegetation has multiple layers. While the biochemistry of canopy components affects the scattering and absorption of light, it has little effect on reflectance relative to other sources of landscape scale variation. Both physically based and empirical models, such as those described in Section 4.5, attempt to account for these properties and their interactions in producing at-satellite and top-of-canopy radiances.

The first attempt to represent the bulk transfer of radiation in a plant canopy mathematically probably goes back to the pioneering work of Monsi and Saeki (1953), who described the transmission of light through a cloud of leaves as an extinction process. The solution of this problem is a simple negative exponential of the amount of leaf material present in the canopy, a variable related to the leaf area index. The Kubelka–Munk theory of radiation transfer in layered media was later applied by Allen and Richardson (1968) to the problem of estimating the reflectance and transmittance of plant canopies. This simple approach created an initial paradigm which ultimately led to the development of a suite of instrumental approaches to derive the quantity of leaf material in plant canopies from observations of the downwelling radiation field, including fish-eye photography and other sensors with a wide field of view.

Groundbreaking results were obtained in the 1960s and 1970s by various scholars. It is not possible to do justice to this field and to make an exhaustive review of the many contributions, so we have elected to highlight the work of two outstanding groups. The first concerns D. M. Gates and his collaborators, who contributed largely to the introduction of spectroscopy in plant physiology and fathered the field of biophysical ecology (Gates, 1980). Another notable case is provided by the Estonian school, led by I. V. Ross, who applied the classical methods of radiation transfer developed in astrophysics to the description of the directional reflectance of vegetation canopies (Ross, 1981). Both of these seminal books contain extensive bibliographies and can be used as introductions to these and related fields.

The development of remote sensing techniques during the 1970s and 1980s, and in particular the launch of space instruments dedicated to the observation of the Earth's environments, such as the *Landsat* MSS and, later, the AVHRR series of instruments, stimulated research and promoted new applications. Progress in the monitoring and characterization of plant canopies and ecosystems on the basis of space observations was achieved in various directions, some of which are briefly outlined below. It is important to recognize at the outset, however, that the interpretation of remote sensing data can only be achieved by exploiting the variations of the observed radiometric signals with respect to the independent variables that describe the conditions of observation (Verstraete et al., 1996). The five domains of variation that have proven useful so far in remote sensing correspond to the spatial, temporal, spectral, directional, and polarimetric independent variables of the radiation transfer equation (Gerstl, 1990). Furthermore, the quality and quantity of information derived from these data depend strongly on the degree of sophistication of the tools used in their interpretation. More advanced models, which better take into account the details of the relevant processes, will lead to more appropriate and more accurate information retrieval.

The following general approaches can be distinguished in the utilization of remote sensing for the identification and characterization of plant canopies:

1. The first concerns the spatial analysis of the data. Indeed, right from the start, imaging space sensors provided data on large areas much faster than could be gathered by a field team. Space instruments also provided a spatial resolution sufficient to imagine, and indeed to implement, applications that could take advantage of this new technology. Besides purely cartographic applications, even the primitive (by tomorrow's standards) spectral resolution of early instruments stimulated exercises in pattern recognition and land-surface-type identification. This capability, coupled with the possibility of revisiting the same sites at least periodically, generated a lot of interest in monitoring seasonal and interannual evolution of land surfaces. Today, land cover classification and changes studies remain a priority, as witnessed by the recently adopted IGBP land-use and land-cover change science/research plan (Turner et al., 1995), and remote sensing will continue to provide critical data to address these issues (e.g., Townshend, 1992). The analysis of temporal changes in land surface properties, in particular, has permitted distinguishing different land cover types or the impact of human activities on the landscape (e.g., Lambin and Ehrlich, 1995; Ehrlich and Lambin, 1996).

2. The second is marked by the design and exploitation of simple and largely empirical tools, most notably spectral mixture models and vegetation indices. The former approach assumes that the multispectral observations actually gathered by a remote sensing instrument can be understood as combinations of the spectral signatures of the individual objects present in the scene, called *end members*. Initial studies (e.g., Smith et al., 1990a) assumed that such combinations occurred linearly, but more recent investigations have shown that for canopies, nonlinear terms may be important. Nevertheless, the main outcome of this approach is to provide estimates of the relative importance (often interpreted as coverage) of the elementary surface types in the observed pixels.

   For their part, vegetation indices, which are effectively two-end member linear models, attempt to estimate a given plant or canopy property directly from suitable manipulation of the spectral measurements. Many such indices have been proposed in the literature over the last 20 years. Rather than listing them or reviewing the many applications in which they contributed, we propose to underscore three critical issues related to the proper utilization of these indices. The first concerns the design of appropriate mathematical formulas to address specific questions. Most of the vegetation indices proposed so far were arrived at empirically, but a rational approach to the design of optimal spectral indices has been proposed by Verstraete and Pinty (1996). The second issue relates to the interpretation of these indices. Although most proposed formulas have attempted to take advantage of the high spectral contrast between the red and near-infrared reflectances of vegetation, only recently has there been an effort to quantify the possible physical reasons for the correlations observed (Baret and Guyot, 1991; Pinty et al., 1993; Myneni et al., 1995). The third essential aspect of the proper use of spectral indices relates to their evaluation and therefore to selection of the most relevant formula to address a particular problem. A general solution to this issue has been proposed by Leprieur et al. (1994), who proposed to rank vegetation indices according to a suitably estimated signal/noise ratio. Here again, the development and exploitation of advanced spectral indices will allow the delivery of more reliable information.

3. Work has also been pursued in a third direction: the design and exploitation of models describing explicitly the anisotropy of the reflected radiation field. These bidirectional reflectance distribution function (BRDF) models (see also Chapters 1 and 3) are conceived to describe how the observed reflectance depends on the particular geometry of illumination and observation, given the structural and optical properties of the simulated surface. BRDF models fall into two broad categories: Either they attempt to simulate the observed anisotropy of the target in terms of measurable physical quantities (physically based or causal models), in which case the reliability of the results critically depends on the quality of the model, or they aim at representing the observed effects (empirical or phenomenological models). In either case, the inversion of a BRDF model against a set of directional measurements makes it possible under specific mathematical conditions (e.g., Verstraete et al., 1996) to reconstitute the entire reflectance field and to estimate the directional hemispherical reflectance (albedo) of the surface. The opportunity to evaluate simple physically based BRDF models critically through inversion has been discussed by Pinty and Verstraete (1992).

4.4 VEGETATION REFLECTANCE MODELS

Parallel to the development of new optical remote sensing instruments, radiative transfer modeling methods to extract information about vegetation have arisen during the last decade. Modeling has proven to be a powerful tool (1) to understand the way in which light interacts with plant canopies, and (2) to infer ecological or agronomic characteristics from reflectances measured over vegetated surfaces. The development of computer-based models in the late 1960s has provided better understanding of the interaction of light with plant leaves. This theoretical understanding will lead to development of more reliable remotely sensed information about vegetation health and plant properties relevant to geobotanical exploration. In this section we review recent progress in developing physically based leaf and canopy reflectance models.

4.4.1 Ray Tracing Models

Among various radiative transfer approaches, only ray tracing techniques can account for the complexity of internal leaf structure as it appears in a photomicrograph.

They require a detailed description of individual cells and their unique arrangement inside tissues. The optical constants of leaf materials (cell walls, cytoplasm, pigments, air cavities, etc.) also have to be defined. Using the laws of reflection, refraction, and absorption, it is then possible to simulate the propagation of individual photons incident on the leaf surface. Once a sufficient number of rays have been simulated, statistically valid estimates of the radiation transfer in a leaf may be deduced. The technique has been applied with a number of variants. The first studies were performed at the cell level (Haberlandt, 1914; Gabrys-Mizera, 1976), in particular with epidermal cells the shape of which might influence the path of the incident beams: Convex cells of some plants act as lenses that focus light within the upper region of the palisade parenchyma, which contains many chloroplasts adapted to high light. This phenomenon has been presented primarily as an adaptation to the low-light environment on the tropical forest floor (Bone et al., 1985), but Martin et al. (1989) showed that the epidermal lenses of *Medicago sativa* (alfalfa) could increase absorption of light at low sun angles. Research efforts were also directed toward understanding the transmission path of light through entire leaves: Allen et al. (1973) and later Brakke and Smith (1987) modeled an albino maple leaf by 100 circular arcs and of two media: intercellular space air and cell walls characterized by their index of refraction. The model was used to test the specular and the diffuse nature of reflection at the cell walls. Simulations led to an underestimation of the reflectance and an overestimation of the transmittance in the near-infrared plateau. This was demonstrated shortly afterward by Kumar and Silva (1973), who found that the actual reflectance and transmittance could be better reproduced by adding two more media into the model, cytoplasm and chloroplasts, thereby increasing the internal diffusion. Whatever the approach, the absorption phenomena that characterize leaf optical properties outside the near-infrared plateau had been ignored. Moreover, in all these models, leaves were always described as two-dimensional objects, although the three-dimensional structure of these organs is very important to their physiological function (e.g., for CO_2, H_2O, O_2 diffusion) and to light scattering (Parkhurst, 1986; Vogelmann and Martin, 1993). To address these problems, Govaerts et al. (1996) used Raytran and successfully simulated the optical properties of a virtual three-dimensional dicotyledon leaf. This ray tracing code, designed primarily to calculate photon transport in a plant canopy, requires a three-dimensional representation of leaf internal cellular structure that conforms to the constraints of plant anatomy and physiology.

4.4.2 N-Flux Models

These models, derived from the Kubelka–Munk theory, consider the leaf as a slab of diffusing (scattering coefficient, s) and absorbing (absorption coefficient, k) material. The N-flux equations are a simplification of radiative transfer theory: the solution of these equations yields simple analytical formulas for the diffuse reflectance and transmittance. A two-flux model (Allen and Richardson, 1968) and a four-flux model (Fukshansky et al., 1991; Martinez v. Remisowsky et al., 1992; Richter and Fukshansky, 1996) have been used successfully in the forward mode to calculate the s and k optical parameters of plant leaves. Yamada and Fujimura (1991) later proposed a more sophisticated version in which the leaf was divided into four parallel layers: the upper cuticle, the palisade parenchyma, the spongy mesophyll, and the lower cuticle. The Kubelka–Munk theory is applied with different parameters in each layer, and solutions are coupled with suitable boundary conditions to provide leaf reflectance and transmittance as a function of the scattering and absorption coefficients. But these authors went further, interpreting the absorption coefficient determined in the visible region in terms of chlorophyll content. By inversion, their model became a nondestructive method for the measurement of photosynthetic pigments. The leaf biochemistry has been introduced by Conel et al. (1993), who used a two-flux model to study the influence of water, protein, cellulose, lignin, and starch on leaf middle infrared reflectance. This model was not validated, however. Finally, a very simple model, issued directly from the expression of the reflectance, has been used to estimate the chlorophyll content of wheat leaves (Andrieu et al., 1988).

4.4.3 Plate Models

The first plate model was introduced by Allen et al. (1969), who represented a leaf as an absorbing plate with rough surfaces giving rise to Lambertian diffusion. Parameters used in the model are an index of refraction and an absorption coefficient. This model was successful in reproducing the reflectance spectrum of a compact corn leaf characterized by few air–cell wall interfaces. The same authors rapidly extended the model to noncompact leaves by regarding them as piles of N plates separated by N-1 airspaces (Allen et al., 1970). The solution of such a system, provided in the last century by Stokes (1862), has been extended to N being a real number; this is the generalized plate model. This additional parameter, N, actually describes the leaf internal structure and plays a role similar to that of the scattering coefficient s in the Kubelka–Munk model. The PROSPECT model (Jacquemoud and Baret, 1990) has been designed this way: it requires only three parameters, the structure parameter (N) and the chlorophyll and water contents, to calculate the reflectance and transmittance of any fresh leaf over the whole solar domain. Furthermore, it can be inverted to estimate the leaf biophysical properties. To upgrade PROSPECT, a laboratory experiment associating visible/infrared spectra of plant leaves both with physical measurements and biochemical analyses, was conducted at the Joint Research Centre in Italy, leading to the LOPEX database (Hosgood et al., 1995). Two new parameters, the protein and cellulose + lignin contents, permitted the simulation of dry leaf spectra (Jacquemoud et al., 1996). An exhaustive decomposition of leaf biochemistry was attempted by Fourty et al. (1996), but due to difficulties that each of these authors had in estimating protein content, Baret and Fourty 1997 reduced the biochemistry to the specific leaf area.

4.4.4 Other Models

Tucker and Garatt (1977) proposed an original stochastic model in which the radiation transfer is simulated by a Markov chain. A black maple leaf is partitioned into two independent tissues, a palisade parenchyma and a spongy mesophyll. Four radiation states (solar, reflected, absorbed, and transmitted) are defined, as well as the transition probabilities from one radiation state to another, between the different compartments. These probabilities are set on the basis of the optical properties of the leaf material. Starting with an initial state vector representing the incident radiation, the steady state is computed by iteratively applying the one-step transition matrix, and yields both reflectance and transmittance.

Compared to the canopy level, only a few models use the radiative transfer equation directly at the leaf level. The poor information we have on leaf internal structure and biochemical distribution leads to simplifications that make such an approach less efficient. In Ma et al. (1990), the leaf is described as a slab of water with an irregular surface containing randomly distributed spherical particles. In Ganapol et al. 1998 it is compared to a homogeneous mixture of biochemicals that scatter and absorb light. Each model was able to reproduce a faithful simulation of leaf optical properties.

None of these models are adapted to needle-shaped leaves. The size of individual conifer needles makes the measurement of their optical properties tricky. In practice, only the infinite reflectance of stacked samples can be performed. Dawson et al. (1998) recently designed a model, LIBERTY, which has the capacity of accurately predicting the spectral response of both dried and fresh stacked pine needles.

4.4.5 Summary

The simulation of leaf optical properties has resulted in three categories of models: models based on Monte Carlo ray tracing techniques that require a detailed description of the leaf internal structure; stochastic models using a Markov chain approach, and finally models derived from the radiative transfer equation, where the leaf is considered a slab of diffusing and absorbing material. Each type is able to simulate the reflectance and transmittance of plant leaves accurately and coherently. However, if the goal is to retrieve information about the leaf anatomy or biochemical constituents, only radiative transfer models can be inverted. The latter also have the advantage of being easily coupled with canopy reflectance models, connecting the canopy reflectance directly to the foliar chemistry. Despite advances in understanding the contribution of leaf biophysical characteristics to canopy reflectance, the search for improved leaf optical properties models is not finished, either for remote sensing or physiological purposes.

4.4.6 Models of Canopy Reflectance

Interest in radiation transfer in plant canopies is relatively widespread, but the particular focus depends on the intended application. The biological, ecological, and agronomic communities are concerned about the interaction between solar light and vegetation through the processes of energy absorption, photosynthesis, and productivity. For their part, the meteorological, climatological, and remote sensing communities are more involved with the modeling of scattering processes, the estimation of canopy reflectance, or the albedo of the surface. Since absorption and scattering are two aspects of the same problem, it would be desirable to develop and use common standard models and approaches in both cases. This would facilitate both the exploitation of remote sensing techniques in the first case, and the integration of plant processes in climate models in the second. In the remainder of this section we discuss in somewhat more detail these canopy radiation transfer models, especially the BRDF models, since they constitute the best developed tools to represent the reflectance of the entire canopy.

BRDF models can be classified, depending on the way they represent the transfer of radiation. A first group of models includes all those that ultimately derive from the classical theory of radiation transfer initially developed to describe the propagation of light in stellar atmospheres (Chandrasekhar, 1944, 1960). Models derived from these theories were first extended to take into account the shadowing effect noticeable in relatively dense media and applied to the representation of the reflectance of planetary surfaces (Hapke, 1981; Lumme and Bowell, 1981), and later, of bare soils on Earth (Pinty et al., 1989). Application of these turbid medium concepts to vegetation canopies (e.g., Camillo, 1987; Shultis and Myneni, 1988; Nilson and Kuusk, 1989; Pinty et al., 1990; Knyazikhin and Marshak, 1991), however, required taking into account the finite size of the scatterers, which are much larger than the wavelengths of solar light, and their orientation (Knyazikhin et al., 1992). This led to the development of hot spot models (Marshak, 1989; Verstraete et al., 1990; Jupp and Strahler, 1991; Kuusk, 1991). Another recent development in this direction is the progressive accounting for the effect of the lower boundary condition, the role of the soil reflectance (e.g., Privette et al., 1995; Gobron et al., 1997). These models, however, remain one-dimensional. Although new techniques discussed below provide alternative solutions to this problem, few authors have tried to extend the classical approach of radiation transfer to describe the anisotropy of the reflectance field in terms of the three-dimensional characteristics of the environment, so the pioneering work of Myneni and Asrar (1993) should be underscored. The typical physical parameters used in these models include the single-scattering albedo and phase function (or, equivalently, the reflectance and transmittance of the scatterers), the optical thickness of the medium, and the orientation distribution of the scatterers. Excellent reviews of these approaches have been published (e.g., Goel, 1988; Myneni et al., 1989; Myneni and Ross, 1991; Hapke, 1993; Strahler, 1994).

A second category includes BRDF models designed to take into account the macroscopic properties of vegetation, in particular the size and shape of the trees or assemblages of plants themselves. This line of research dates back at least to the work of Brown and Pandolfo (1969). An important series of contributions was stimulated by Suits (1972), and further work by Suits (1983) and others (e.g., Verhoef, 1984) led to the development of the well-known SAIL model. Early models aimed at estimating the interception of solar light by complex plant arrangements for biological or agronomic purposes (Richardson et al., 1975; Jackson et al., 1979), or to compute the radiation balance of arid regions (Otterman, 1983). Geometric–optic models were developed further by Li and Strahler (1985, 1986, 1992) and Li et al. (1995) to estimate the directional reflectance of these structured surfaces. Geometric models typically describe the reflectance of the surface in terms of the dimension, shape, orientation, and average optical properties of the scattering volumes and surfaces considered (boxes, cylinders, cones, etc.).

A third class comprises a whole range of models exploiting recent advances in computer graphics and visualization techniques. Early approaches involved ray tracing through a series of boxes affected by average optical properties (e.g., Kimes and Kirchner, 1982), while later developments relied on the explicit representation of the position, shape, orientation, and optical properties of all the relevant scatterers in the scene of interest, with such methods as constructive solid geometry or L-systems (Ross and Marshak, 1988; Goel et al., 1991; Govaerts and Verstraete, 1994). The radiation transfer problem itself is then solved through computation of the relative contributions of each object to the reflectance of all others in the case of radiosity methods (Borel et al., 1991; Goel et al., 1991; Gerstl and Borel, 1992), or through the propagation of a large number of individual rays in the case of Monte Carlo ray tracing techniques (e.g., Govaerts and Verstraete, 1996; North, 1996). Both approaches have their own merits; each offers the potential to describe the reflectance of complex and quite realistic scenes, and both can be demanding in terms of input data to set up the scene and in computing resources to generate the solution. The main applications of these models include the establishment of high-quality standards against which other simpler models (Govaerts and Verstraete, 1995) or spectral indices (Goel and Qin, 1994) are evaluated, as well as sensitivity studies to determine the exact role of particular plant canopy properties in the overall reflectance field (Ross and Marshak, 1989). For completeness it should be added that any of the methods mentioned above can be further exploited with artificial intelligence techniques, as proposed by Kimes et al. (1994). Research in this direction is still in its infancy, but significant advantages will most probably be derived from further effort in this area, for instance for the purpose of land cover classification.

The last important class of BRDF models includes empirical functions capable of representing the overall shape of the observed reflectance fields in terms of parameters not associated with any particular physical meaning. Work in this direction also derives from research initially performed in a planetary context, since Minnaert (1941) first proposed a simple trigonometric equation to describe the variations in reflectance of the Moon. Other models and developments in this category include Walthall et al. (1985), Roujean et al. (1992), and Rahman et al. (1993a). The latter was further developed by Engelsen et al. (1996) to support the analysis of MISR data, while the former, together with other models, has been formally expressed as a linear kernel and packaged in a way suitable for the analysis of MODIS data (Wanner et al., 1995). The main advantages of these empirical models are that they may, in principle, be applied to a wide variety of surface types and that they may provide a simple and efficient solution to the estimation of surface albedo. However, they cannot be used to derive a fundamental understanding of the physical processes responsible for the observed anisotropy, nor can they be inverted to retrieve information on the physical state variables which condition the measured signal.

Many of the models surveyed above are likely to be further improved in the near future. The main research issues at this time, however, concern the proper representation of the boundary conditions, and in particular, the accurate accounting of the multiple scattering processes within the canopy as well as between plants, the soil, and the atmosphere. Coupled canopy–atmosphere models have already been proposed (Kriebel, 1978; Tanre et al., 1983; Myneni and Asrar, 1993; Rahman et al., 1993b; Liang and Strahler, 1993, 1995). An alternative approach is to incorporate some of the physically based one-dimensional or empirical models cited earlier as a selectable lower boundary condition in atmospheric radiation transfer codes, as was done by Vermote et al. (1995) with the popular 6S atmospheric radiation transfer code, but much more needs to be done to truly integrate all relevant physical processes into a coherent, comprehensive model.

Further research will be motivated by parallel developments in field instrumentation to characterize the anisotropy of natural surfaces (Deering and Leonoe, 1986; Deering, 1989) and by the upcoming generation of Earth observation satellites to be launched over the next few years. The data generated by these advanced space sensors will certainly contribute to an increase and a diversification of user requirements, as well as stimulate new expectations regarding the exploitation of remote sensing data in a wide range of applications. This unique situation will create both challenges (because a number of instruments may have similar characteristics or comparable performances) and opportunities, such as the synergistic use of multiple sensors.

4.5   DATA ANALYSIS: REMOTE SENSING APPLICATIONS OF GEOBOTANICAL TECHNIQUES

A typical objective of geobotanical remote sensing is the transformation of radiance measurements from imaging sensors over a range of wavelengths into geobotanical parameters [e.g., vegetation type, percent cover, absorbed photosynthetically active radiation (APAR), LAI, biomass]. While the models reviewed above represent a basis for developing methods from first principles, operational techniques remain largely derived from empirical studies. The primary factor that has limited the operational implementation of remote sensing data has been the difficulty in developing models that are spatially and temporally extendible outside the realm of the specific studies in which they were developed. Vegetation indices, classifiers, and simple mixture inversion models, while demonstrated to have local validity in many specific studies over the past decade, have been applied to regional or global analyses with little consideration to their validity at larger scales. Techniques to measure biomass, LAI, and other plant attributes quantitatively are generally limited to the vegetation types in which they were developed and often are not even valid over the full range of a given type.

The development and evolution of geobotanical parameters from remote sensing data typically relies on obtaining positive correlations to local "ground truth" measurements rather than from first principles. The ensemble of established remote sensing techniques includes a number of inherent embedded assumptions that limit the possibility of extending solutions (i.e., where applied outside the region of model calibration). Consequently, the remote sensing literature has numerous results that cannot be used in subsequent studies. One example is the assumption that vegetation end members and/or feature classes can be identified uniquely by multispectral measurements (i.e., that each object class has a unique spectrum), another example is that validation using local *ground-truth measurements* (e.g., measurements from a field experiment or observation of a single image) ensures that measurements are extendible over time or to regional scales. This focus has limited the extendibility of results because the empirical solutions are nonunique (i.e., the analytical framework was not immune to the many factors affecting radiance measurements unrelated to the property of interest) when applied to measurement conditions not identical to those from which they were derived. In other words, although the analytical procedure may have application to other areas, any results, such as coefficients, may not be transferred easily to other areas.

Despite considerable interest in scaling problems, there has been little advance in understanding the scale relationships between laboratory spectra and image pixels outside of radiative transfer models. Increasingly, Earth science research involves large-scale regional-to-global estimates of biophysical processes such as biogeochemical cycling, desertification, erosion, or air pollution. For addressing these problems, we need to apply newer spectral detection approaches. Among these, wavelets and neural networks provide a means for validating remote sensing techniques that extend beyond local ground-truth accuracy assessments. While early sensor technology (e.g., MSS) had only a few broad bands and limited radiometric resolution, these empirical techniques were reasonable approaches to data analysis. Now, however, detector technology has matured, spectral and spatial data can be obtained (at least with airborne sensors) at almost any resolution, data quality has improved, as has our ability to calibrate radiance to surface reflectance using physically based atmospheric radiative transfer algorithms (e.g., MODTRAN or 6S). We must now develop methodologies that can transform the data to geobotanical parameters that filter out nonrelevant factors that cause spectral change. For a transformation to be effective it must satisfy two objectives: (1) be positively correlated to the parameter of interest, and (2) be immune from extrinsic factors causing a response that mimics a given parameter. Verstraete et al. (1996) formalize these arguments and describe 10 approaches for developing the models capable of relating measured radiance to the physical parameter of interest.

4.5.1 Spectral Methods

Many spectral methods employed in vegetation analyses are those described elsewhere (see Chapter 5). Others, such as band ratio vegetation indices, are widely used in remote sensing and a number of reviews are available (Baret and Guyot, 1991; Huete et al., 1994; Verstraete and Pinty, 1996; Verstraete et al., 1996). We highlight several newer spectral and image analysis techniques that are under development and hold promise for addressing the primary needs of geobotanical image analysis. Many computationally simple methods work adequately where analysis goals are focused on visual interpretations of geologic structures [e.g., lineaments, presence or absence of vegetation, or major seasonal conditions (such as wet versus dry season or large interannual changes)]. Where these analyses often fail is in detection of minor landscape changes, where gradients are important, or spectral features associated with the information content are weakly differentiated.

4.5.2 Vegetation Indices

One of the earliest procedures developed for identifying and enhancing the vegetation contribution to remote sensing pixels has been some form of vegetation index (VI), a ratio between the red and the near-infrared spectral regions (Tucker, 1977). Tucker et al. (1986) showed that global composite AVHRR vegetation index images exhibited monthly latitudinal changes in greenness that fluctuated in synchrony with measurements of atmospheric CO_2 concentration, demonstrating a correlation with planetary biospheric activity. However, even the normalized difference vegetation index

(NDVI), probably the most widely used remote sensing index, has been carefully validated at only a few sites (Myneni et al., 1995; Verstratete et al., 1996), including the Kansas FIFE field campaign (Friedl et al., 1995; Sellers et al., 1995), MAC VI (Qi et al., 1995a), and Jasper Ridge (Gamon et al., 1993). Many questions remain uncertain, such as whether vegetation indices are equivalent when extracted at different spatial, temporal, and spectral scales (Gutman, 1991; Qi et al., 1995a). Jackson and Huete (1991) review methods of calculation, including digital counts, satellite radiances, and apparent or surface reflectances. The ratio can be calculated from any sensor having red and near-infrared spectral bands, but each yields different results for the same surface conditions (Huete et al., 1994). In fact, despite many studies, precisely what is measured by VIs is uncertain. Although clearly correlated with chlorophyll absorption by foliage, it has been related to many plant properties, including LAI (Qi et al., 1995a), percent green cover and biomass (Gamon et al., 1993, 1995), productivity (Weigand and Richardson, 1987), APAR (Asrar et al, 1984), and biophysical properties such as photosynthetic capacity (Sellers, 1985). In fact, VIs are sensitive to the life-form (e.g., grasses, shrubs, or trees) and community composition, which affect canopy architecture and plant distribution. Many leaf properties are autocorrelated, accounting for the many plant attributes with which VIs are correlated. Sunview geometry also affects VIs (Epiphanio and Huete, 1995; Qi et al., 1995b). In recent years a number of modified vegetation indices have been developed to minimize some extraneous factors affecting the measurements. The first of these was the NDVI, designed to reduce the influence of variation in illumination conditions. Subsequently, vegetation indices were developed to reduce soil brightness and atmospheric variations (Huete and Jackson, 1987; Kaufman and Tanre, 1992; Pinty and Verstraete, 1992; Huete et al., 1994; Bannari et al., 1996).

4.5.3 Red-Edge Detection

The *red edge* is the reflectance inflection point observed in the spectrum of green plants at the transition between the visible and near-infrared wavelengths. It is the long-wavelength edge of the chlorophyll absorption feature and has been used to describe the variation in leaf and canopy chlorophyll concentration. This boundary point in reflectance typically occurs between 690 and 740 nm in fresh leaves and is determined by the interaction between chlorophyll absorption of red light and the internal scattering process in the leaf (Gates et al., 1965; Horler et al., 1983; Curran et al., 1990). Typically, the wavelength of this inflection is determined by calculating a first or second derivative, but Chappelle et al. (1992) developed a ratio-based analysis. Increased chlorophyll concentration causes a deepening and broadening of the chlorophyll absorption feature, while stresses causing chlorosis usually increase reflectance and narrow the absorption feature, making the inflection point appear to move to longer or shorter wavelengths, respectively (Horler et al., 1983; Curtiss and Ustin, 1989). The relationship has been utilized to detect symptoms of mineral deficiency (Horler et al., 1980; Milton et al., 1991) and air pollution damage (Rock et al., 1988; Ustin et al., 1989; Hoque and Hutzler, 1992).

Using the red edge to quantify environmental stress can be difficult and several authors have noted red-edge changes related to leaf age, developmental stage, and canopy architecture that were partially independent of pigment composition (Gates et al., 1965; Collins, 1978; Shutt et al., 1984; Vanderbilt et al., 1985, Ustin et al., 1989; Rondeaux and Vanderbilt, 1993). These effects could be due to changes in the light-harvesting mechanisms (e.g., altered chloroplast structure or changes in light scattering and refraction due to cell wall thickness or chemistry). The relationship can also be affected by the presence or absence of other pigments, such as the non-photosynthetic leaf pigment anthocyanin (Curran et al., 1991). Railyan and Korobov (1993) examined changes in the red edge of triticale (hybrid of wheat and rye) over the growing season and found that the red-edge wavelength varied with developmental stage. At earlier growth stages, the red edge moved from shorter to longer wavelengths (a shift of about 5 nm was observed), while later they found a shift back to shorter wavelengths. This type of developmental pattern was also observed in pine needles of different ages (Ustin et al., 1989), red spruce and eastern hemlock (Rock et al., 1994), and in loblolly pine and Norway spruce (S. L. Ustin, unpublished data).

Quantitative prediction of pigment composition from the red edge is complicated by interactions among the pigment complexes and the bound proteins in in situ chloroplasts (Curtiss and Ustin, 1989; Railyan and Korobov, 1993). Curtiss and Ustin (1989) found broadening of the chlorophyll absorption band (a red-edge shift to longer wavelengths) in ponderosa pine needles in sites having greater atmospheric ozone exposure. This pattern mimicked increased chlorophyll concentration, except that red reflectance increased under ozone exposures and decreased under higher chlorophyll concentrations. They attributed ozone-induced reflectance changes to an apparent increase in the disorder of the absorbing medium, a pattern consistent with observations that an early sign of ozone injury is granulation of the thylakoid stroma in chloroplasts. These factors, and the correlation between chlorophyll concentration and leaf mass or area at the canopy scale, make red-edge prediction of pigment concentration or state of health of a canopy difficult to interpret. Nonetheless, within a vegetation type, spatial patterns in the red edge coupled with greenness patterns, may be useful to identify areas of potential vegetation stress that could be indicators of soil conditions or contamination.

4.5.4 Multitemporal Vegetation Indices

Despite the tremendous potential of remote sensing to provide repeat observations over time, this technique has not been commonly used. Presumably, the computational and data constraints are less restrictive now than in the recent past, and this area of image analysis will see greater development. Sometimes simple statistical or mathematical approaches are sufficient to identify geobotanical sites because of their defined boundaries and large areal extent or because the site exhibits substantial impact on plant growth. Utilizing multitemporal data is important for many geobotanical applications because the impact may best be seen as a change in the timing of normal phenological events, such as a delay or advance in leaf development or senescence. Bell et al. (1985) observed late leaf flush and premature leaf senescence (Schwaller and Tkach, 1980) in multitemporal MSS images in an area of heavy metal deposits. Seasonal differences in phenological events can be used to identify vegetation types (e.g., DeFries and Townshend, 1994), or separate distributions of deciduous and evergreen woody perennials from annual vegetation (DeFries et al., 1995). Multitemporal NDVI data have been used to estimate crop parameters and separate crops having different phenological patterns (Fischer, 1994). Lambin and Strahler (1994) used change analysis on composited AVHRR NDVI data from the Sahel to identify types of land cover change. Their methodology is under development for the NASA land cover change product for the MODIS sensor. Tucker et al. (1986) related monthly latitudinally averaged NDVI values from AVHRR to measurements of atmospheric CO_2 concentrations that demonstrated the strong coupling between atmospheric and biospheric processes.

4.5.5 Continuum Removal Feature Mapping

One method for identifying and mapping an absorption feature, such as those associated with minerals in imaging spectrometer data, is to remove the convex hull continuum from the spectrum observed (Clark and Roush, 1984). The continuum is the generalized shape of the spectrum in the absence of the particular narrowband absorption feature. This background or continuous spectrum is due to the combination of all the scattering and absorbing characteristics of the material. This approach works well for comparisons to spectral libraries when the continuum is removed by dividing both library and pixel (unknown spectrum) by the continuum reflectance (see Chapters 1 and 5). The technique has been applied to mineral identification (Clark et al., 1990a, b) using the U.S. Geological Survey (USGS) spectral library in the tricorder algorithm (Clark et al., 1993). These analysis tools are available from their Internet site (http://speclab.cr.usgs.gov/). Clark et al. (1995) apply the methodology to the detection of plant stress and species mapping in the San Luis Valley, Colorado, where they were able to detect small spectral shifts of the red edge for crops in various stages of senescence.

The continuum removal method was calculated and regressed against measured leaf water content to estimate of water content of leaf, canopy, and AVIRIS data sets for chaparral communities in the Santa Monica Mountains, California (Ustin et al., 1998). The sensitivity to wildfire in the chaparral varies with canopy water content, among other factors, such as species/community composition and biomass. Measurement of spatial and temporal variation in water content is a potentially significant step for improving fire hazard assessments of wildlands.

The continuum removal technique has also been applied to a study of water content variation at canopy and AVIRIS image scales in salt marshes along the northern shore of the San Francisco Bay estuary (Sanderson et al., 1998; Zhang et al., 1997). The canopy water content of the dominant species varies spatially with interstitial soil salinity. Open water does not express the narrowband water absorption feature of the plant canopy (just the continuum), so the technique is insensitive to water bodies in the marsh. A simple interpolation of field sampling points revealed a spatial pattern that approximated the AVIRIS pixel distribution (Zhang et al., 1997). The accuracy and validity of the water content estimate was investigated in a detailed scaling exercise that compared field spectra and water content samples measured in the salt marsh and AVIRIS derived estimates of water content (Sanderson et al., 1998). This study used a variogram to describe the spatial correlation structure of the canopy water content. Then ordinary kriging estimates were calculated over blocks that corresponded with the coregistered AVIRIS pixels. In this case, co-kriging with more sparsely sampled canopy water content measurements did not improve the model. The relationship was applied to the larger Petaluma River salt marsh, an area approximately 3 km × 6 km. This technique should be considered when there is reason to expect a change in a known absorption feature caused by a geobotanical anomaly or changes in pigment composition or water content of the plant canopy.

4.5.6 Spectral Mixture Analysis and Residuals

Linear spectral mixture analysis (SMA) is useful for vegetation mapping because of the continuous variation in pixel composition at the scale of many satellites. Multiple scattering phenomena in plant canopies creates a nonlinear mixing problem as described in Section 4.3.3. However, a number of studies have examined linear subpixel separation of vegetation and soil spectral components and obtained results in reasonable agreement with supporting field data (Adams and Adams, 1984; Huete, 1986, Smith et al., 1990a,b). A description and review of the methodology are provided in Chapter 5. The procedure assumes that a pixel is a linear combination of the proportional spectral contribution of each of the components. The assumptions of this approach contrast with classification methods that assume no within-class variance. Although end members (the pure spectral types or classes) in SMA may be similar in concept to traditional classification, they are not constructed the same way. That is, end members can be considered as the basic building blocks from which pixels vary in the proportions of common environmental components (soil, vegetation, etc.). Classes defined in clustering algorithms are generally defined by other criteria [e.g., taxonomic category (serpentine, volcanic tuff, corn, soybeans, etc.)]. Variation in proportions of the scene elements, specifically, classes of green foliage, soil, and litter, can reveal important patterns related to soil processes. For example, Smith et al. (1990b) found that the percentage of green foliage could be used to calculate the spatial distribution of evapotranspiration rates for the Owens Valley, California. The valley, about 90 km × 40 km, is predominantly semiarid shrubland on the alluvial fans, with riparian corridors along perennial stream channels and a large halophytic community on the valley floor. Vegetation abundances were generally less then 40% cover, and were to first order, related to soil moisture variation, salinity, soil types, and past disturbance. One feature of SMA is that it minimizes the impact of albedo variation due to topography and maximizes the spectral shape information in the spectral signature. One feature that has been noted is the plant dispersion and biomass contribution to the shadow fraction, which in the absence of topography could provide an additional source of information about the structure and growth dynamics of the vegetation (Smith et al., 1990a, b; Ustin et al., 1993).

Generally, acceptance of the SMA model is dependent on the end-member fractions producing a close fit to the measured pixel reflectance (or radiance). This is estimated by the residuals, often the root-mean-square error, which can be displayed as a gray-scale image, and areas of high error can be visualized in their spatial context. This information has implications for locating and mapping anomalous vegetation growth or small mineral outcrops. If a hyperspectral sensor is used, the specific spectral bands having high residuals can be examined for absorption features associated with particular conditions. Sabol et al. (1992) examined the separability of different materials to better understand conditions when different spectral components can be detected. The primary limitation of simple mixture models is the effect that variability in the spectral end member has on predicted end member fractions.

Roberts et al. (1996) found that for many pixels, reflectance is adequately modeled by combinations of only two or three end members. In fact, solving for more end members in such cases causes instability in the predicted fractions. Roberts et al. 1998 developed a modified SMA procedure that focused on variability within potential end-member classes. In this form of SMA, instead of finding the four or five end members that provide the best overall fit for all pixels in a scene, the approach is to search a spectral library for the two end member combinations that best fit each pixel. All pixels that are unsatisfactorily modeled (i.e., have high residuals) are reanalyzed for three end-member combinations. Few pixels need more than four end members to account for the subpixel mixing. Once the SMA steps are complete, images can be produced to show which end-member spectra were chosen for each pixel. These new images produce maps that approximate the vegetation community distributions for the area.

4.5.7 Foreground/Background Analysis

Foreground background analysis (FBA) is a recent advance in spectral mixing techniques (Harsanyi, 1993; Harsanyi and Chang,·1994; Smith et al., 1994; Pinzon et al., 1998; Ustin et al., 1998), and is equivalent to a single neuron in a neural net. Unique to this technique is the training of the neuron to satisfy the two objectives stated above, where the first objective (correlation to the parameter of interest) is defined by the foreground and the second objective (immunity from extrinsic factors) is determined by the background. The initial development of FBA examples is limited to the spectral dimension and does not consider the spatial context. FBA can be applied to give exactly the same results as residual analysis from SMA; thus SMA is a special case of the FBA.

Pinzon et al. 1998 show how spectral features associated with foliar chemistry, such as water nitrogen and carbon compound content, can be identified in the fresh leaf spectrum. In this analysis, the background is defined as the set of extrinsic factors and the foreground as the objects of interest. For example, in the case of NDVI, the background would be those effects caused by changes in canopy architecture, spectral differences between plants, and the effects of soils of different brightness and elevational differences arising in steep topography. Pinzon et al. 1998 adopted a hierarchical procedure to reduce the nonlinear dependencies of the spectral response. Their procedure relied on the spectral variance to provide an initial separation of the spectra into classes that corresponded with different ranges of the chemical variable. Following this initial classification (which linearized the problem), they trained new vectors for each of the spectral ranges, and these were used to predict chemical concentration of the foliage samples.

4.5.8 Wavelets and Neural Networks: Integrated Approach

Despite questions of consistency and extendibility, parameters derived from remote sensing measurements are used at an increasing rate as inputs to and validation for large-scale biosphere models such as CASA and SiB-II (Potter et al., 1993; Field et al., 1995; Ruimy et al., 1996; Sellers et al., 1996). Wavelet decomposition techniques provide an analytical framework to separate and quantify the effects of scale that have been only speculated about superficially. Parameters computed using wavelet decomposition techniques provide increased immunity to extrinsic factors (e.g., atmosphere, viewing geometry, spectral variability, surface composition) by focusing on specific temporal, spectral, or spatial scales distinct for the parameter of interest. Similarly, when trained properly, neural networks also have increased filter capacity over the past techniques of SMA, classifiers, and indices. The integration of wavelets and neural nets makes possible models that have a much greater capability to separate processes that cause similar effects in spectral measurements.

The continuum removal algorithms of Clarke and Roush (1984) and above are similar to fine spectral scales resulting from wavelet decomposition of images in the spectral domain. The advantage of using wavelets is that we can develop spectral solutions specific to the spatial scales. Unlike minerals that have narrowband absorptions, the spectral scales from vegetation are quite different. The spectra of plants typically have variability at larger spectral scales than most minerals, with much of the variability controlled by the plant phenology and condition. Additionally, at the smaller scales, plant species are more consistent than minerals since plants contain similar metabolic constituents while mineral chemistry is varied.

An overall evolution of simpler image-processing techniques to wavelets and neural networks is necessary to move remote sensing from local observational studies toward an operational science that takes advantage of the nonuniqueness of solutions in remote sensing images. The approach to training neural networks is the key to satisfying application objectives. Constrained energy minimization (CEM) can be viewed as a neural net problem in the simplest form. FBA and CEM techniques are both elementary neural nets comprised of single neurons that differ in the training of a finite impulse response (FIR) filter. In both cases the filter is applied in the spectral domain. Subband coding is an electrical engineering term that refers to filters that utilize information in subbands. For remote sensing applications, these may be spatial, spectral, or temporal scales that have potential for the extraction of geobotanical parameters.

The presence of neural networks and wavelets generalize the analytical framework of many individual techniques developed and used in the past. These new frameworks provide the capability to generate multiple solutions that trade contrast in parameter detection to obtain insensitivity to variable background factors. This trade-off is necessary to create parameters extendible beyond the local study. The training of networks based on remote sensing data is crucial to obtaining accurate regional estimates of parameters. The goal is not to use the most complicated network to develop solutions but to find the simplest network necessary to achieve a specified level of extendibility.

4.6 CASE STUDIES

4.6.1 Case Study: Vegetation Responses to Heavy Metals in Quebec*

Bélanger (1988) provides a synthesis of numerous papers on the geochemical, bio-geochemical, and remote sensing studies undertaken at Thetford Mines, Quebec,

*This case study is from J. R. Bélanger and A. N. Rencz, Geologic Survey of Canada, Ottawa, Ontario, Canada.

Canada. The area is characterized by a very large and highly enriched zone of glacially transported sediments that are enriched in ultrabasic materials (Shilts, 1973, 1975, 1976; Rencz and Shilts, 1980). The ultrabasic rocks (serpentine, periodite, and pyroxenite) have been eroded and dispersed for at least 70 km southeast from the source. Notably enriched values of Ni, Cr, Fe, and Mg in the <2 μm size fraction of till are present (Figure 4.2). The highest concentrations were found within 2 km of the source (>2300 ppm for Ni, 2600 ppm for Cr, and 80 ppm for Co). The patterns of glacial dispersion of the elements is mimicked by a variety of other trace elements that are related to the source of ultrabasic rocks. There is a negative anomaly in the Ca concentrations associated with the absence of carbonate rock debris.

The natural vegetation of the region has been characterized by Marie-Victorin (1964) as deciduous boreal forest. The dominant trees types include black spruce (*Picea mariana*), white spruce (*Picea glauca*), balsam fir (*Abies balsamea*), white birch (*Betula papyrifera*), and trembling aspen (*Populous tremuloides*). The geobotanical studies in this area were designed to investigate the physiology, morphology, and pathology of the dominant plant species. Several species, including coniferous and deciduous trees, were selected for biogeochemical analyses. For each species, foliar

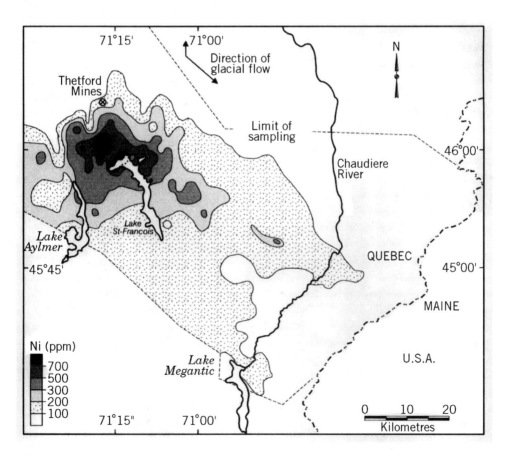

Figure 4.2 Disposal patterns for elements in the clay fraction (<2 μm size) of glacial debris at Thetford Mines, Quebec, Canada. Inset map shows location of study area.

samples were analyzed for trace elements and chlorophyll content. Correlation coefficients were calculated to illustrate relationships between levels of elements in the soil and in plant tissue. There was a relatively high correlation for several elements in several of the tree species, notably for sugar maple and white birch, whereas several plants displayed no obvious correlation: notably, trembling aspen (Table 4.2).

Generally, Bélanger (1988) concluded that correlations between plant samples and soils were relatively weak for Cr and Co but slightly higher for Ni, Mg, and Ca. However, he notes that there was variation between species.

The influence of heavy metals on leaf development was monitored and showed some relationship between chlorophyll and element concentrations in leaf tissue (Rencz and Shilts, 1980), as shown in Table 4.3. Subsequent studies generally found that chlorophyll content was lower in the anomalous areas than the background area and related consistently to low calcium and excessive magnesium. Ca/Mg ratios are in the range 1 to 20 for healthy vegetation (Walker et al., 1955), while anomalously low values around 0.5 were found in birch, aspen, and red maple (Bélanger, 1988).

The utility of remotely sensing this biogeochemical anomaly was investigated in several studies that included ground measurements, airborne measurements, and *Landsat* MSS (Bélanger et al., 1979; Bélanger and Rencz, 1983; Bélanger and Hélie, 1984). The MSS analyses monitored variations in biomass during the growing season using a vegetation index. Additionally, delay in leaf flush (Bell et al., 1985) and premature leaf senescence (Schwaller and Tkach, 1980) were observed. Biomass indices were calculated for four images: June, August, September, and October. The biomass index used a simple infrared/red ratio. Although a straight comparison of the index identifies tonal differences between the dispersal train and background conditions, this approach is insufficient to monitor vegetation types or chlorophyll content because local conditions can vitiate spectral signatures of vegetation. To monitor variations in biomass index and possible anomalous behavior of certain species, scatter diagrams and corresponding maps were plotted to show the relationship between biomass for June versus August, August versus September, and August versus October.

A map of late, early, and normal flush (Figure 4.3) shows a close relationship between geochemical anomalies in the dispersal train and late flush, whereas early and normal flush indicies are located mainly outside the dispersal train. Early flush is characterized by relatively high production of chlorophyll in early June and average

TABLE 4.2 Correlation Coefficients for Co, Ni, and Cr in Soil and Plants Sampled

Species	Co	Ni	Cr	n
Sugar maple	0.81	0.84	0.93	21
White birch	0.50	0.67	0.81	25
Yellow birch	0.27	0.53	0.38	11
Trembling aspen	−0.09	0.10	−0.07	25
Poplar	0.15	0.56	0.55	13
White spruce	0.50	0.26	0.40	24
Black spruce	0.28	0.21	0.52	7
Balsam fir	0.07	0.54	0.16	26

Source: Bélanger (1988).

TABLE 4.3 Correlation Coefficients for Chlorophyll Content and Concentration of Elements in Leaf Tissues

Species	Ni	Cr	Ca	Mg	n
Sugar maple	−0.14	−0.15	0.01	−0.25	17
Aspen	−0.48	—	0.52	−0.19	22
Poplar	−0.78	−0.66	0.74	−0.52	13
Balsam fir	0.30	—	0.45	−0.03	27

Source: Bélanger (1988).

production of chlorophyll in summer. Late flush corresponds to vegetation in which the biomass is low in June but high in August. High biomass index for both June and August correspond to hardwood deciduous species, such as sugar maple and birch, the spectral signatures of which do appear to be affected by stress conditions in this study.

Bélanger (1988) concluded that whereas *Landsat* MSS was effective in describing the geochemical anomaly in the Thetford Mines area, their result was dependent on multidate cloud-free imagery and the fact that the target zone was large and enhanced by glaciation-dispersed mineralized debris extending over a distance of at least 70 km.

4.6.2 Case Study: Combining Digital Elevation Data with Remotely Sensed Imagery for Geobotanical Mapping*

Geobotanical remote sensing remains challenging because the substrate is generally not the primary factor in determining the ecology of a particular location. The regional climate gradient, with local factors such as microclimate and site drainage, and stochastic elements such as disturbance and regeneration, have confounding influences on vegetation patterns. Consequently, ancillary data can be extremely useful as an aid in reducing uncertainty associated with very weak geobotanical associations. DEM data are one of the most used ancillary data sources, because elevation is so important in determining local climate, as are aspect and slope on local microclimate and site drainage, soil development and weathering, and vegetation distribution.

In a study of the East African Rift, Warner et al. (1989) found that NOAA *ETOPO-5* DEM data improved modeling remotely sensed vegetation distribution and the identification of anomalous regions related to tectonic provinces. The vegetation distribution was determined by integrating six NDVI AVHRR multitemporal composite scenes (February, June, and September) from 1984 and 1987 (Figure 4.4, left panel; see color insert). In contrast to the severe drought suffered by most of East Africa in 1984, 1987 was characterized by near-normal precipitation. A second-order general linear model was developed, relating elevation, latitude, and longitude to the NDVI data. All the independent variables had a greater than 99% significance in the model (Figure 4.4, middle panel; color insert).

---

*This case study is from T. A. Warner, Department of Geology and Geography, West Virginia University, Morgantown, West Virginia.

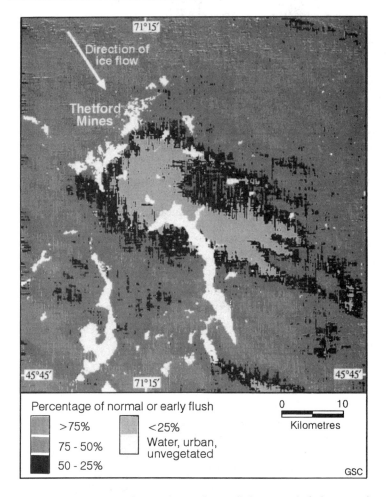

Figure 4.3   Multidate vegetation indicies from *Landsat* MSS showing "flush" times in Thetford Mines, Quebec, Canada.

The model residuals were useful to identify sites where local factors are more important than elevation and the regional climate gradient in determining vegetation growth (Figure 4.4, right panel; color insert). In some places, such as along the coast of the Indian Ocean, this factor can be related to local sources of moisture. However, in many cases the residuals are strongly correlated to geology. Large negative residuals were found in the geologically young Lake Turkana rift basin, as well as on the Pleistocene–Holocene shield complexes and flows, where soil development was limited due to the geologic youth of the area. In these regions the vegetation is anomalously sparse. In contrast, the more clay-rich sedimentary units and the older Miocene–Pliocene basaltic shield complexes support a more dense vegetation than predicted by the model and are therefore associated with large positive residuals. When slope and aspect were included in the model, they did not improve accuracy and were not statistically significant. Yet the observed pattern is one in which steep slopes, particularly those seaward facing, have greater orographic precipitation. Because the slope and aspect are calculated over a matrix of adjacent pixels, this decreased the spatial resolution below the smaller-scale topographic relationships. In addition, aspect cannot be considered in isolation from slope, since it is only on the steeper slopes that aspect is important. These two factors suggest that geobotanical methods that incorporate elevation, slope, and aspect as "logical channels" (Strahler et al., 1978) will not exploit all the information present in the topographic data.

One method to improve topographic data is the TOPOVEG algorithm, developed for analysis of DEM data, *Landsat* TM imagery, and aeromagnetic geophysical data (Warner et al., 1994). The study site for the algorithm was Quetico Provincial Park, Ontario, Canada (Figure 4.5, top panel). Most of this region is underlain by metamorphosed granites and granodiorites, which surround elongate units of metamorphosed mafic lithologies, principally hornblende amphibolite, with occasional serpentinized dunite, gabbro, gabbornorite, and lherzolite. Economic deposits, including gold, silver, copper, nickel, and platinum group metals, are associated with the mafic rocks. Weathering of the aluminum-rich granitic rocks results in soils that tend to be more acid than those derived from the mafic units. Quetico Park is on the southern ecotone of the boreal forest, and is dominated by jack pine (*Pinus banksiana*), black spruce (*Picea mariana*), aspen (*Populus tremuloides*), and paper birch (*Betula papyrifera*). Southern species such as black ash (*Fraxinus nigra*) and American elm (*Ulmus americana*) are generally limited to more fertile sites.

The TM data were converted to nPDF deciduous vegetation index values (Cetin et al., 1993), a measure of the pixel's position on the spectral mixing line between coniferous and deciduous vegetation (Figure 4.5, middle panel). The nPDF (*n*-dimensional probability density function) is a suite of programs centered around a data reduction technique in which the original image bands are combined, based on their Euclidean distance from selected corners of the multispectral data space. An nPDF index was developed, based on the distance of each multispectral pixel from the corner comprising the minimum in bands 4, 5, and 7, and the maximum in the remaining bands. This has the effect of maximizing the difference between the pixels with the geobotanical anomaly compared to the surrounding areas. Thirty-meter elevation data were interpolated from digitized 1: 50,000 topographic maps. These areas are displayed in purple and blue on the color map.

A landscape classification was developed from the digital data, in which slopes greater than 3° were classified into either north-and east-facing, or south-and west-facing groups and slopes less than 3° were classified by drainage characteristics, yielding ecologically relevant site classes. Similarly, the geophysical data were converted from raw measurement units to significant geobotanical classes. The aeromagnetic geophysical data were then classified into anomalously high and background value classes, which were broadly correlated with mafic and granitic rocks, respectively.

The TOPOVEG algorithm is a rule-based classification (based on a higher nPDF deciduous forest index) that predicted an increase in hardwoods on the less acidic south-facing slopes underlain by mafic rocks (Figure 4.5, bottom panel). Other mafic sites and sites underlain by acidic granites favor a more boreal coniferous community (Warner et al., 1991). However, these are generalized patterns, and coniferous and deciduous trees are found at all sites. Thus a direct association between vegetation and substrate results in a noisy association, as shown in the TM maximum likelihood classification (see Figure 4.5). In the TOPOVEG classification, the average nPDF deciduous forest index is calculated for each group of contiguous pixels that forms

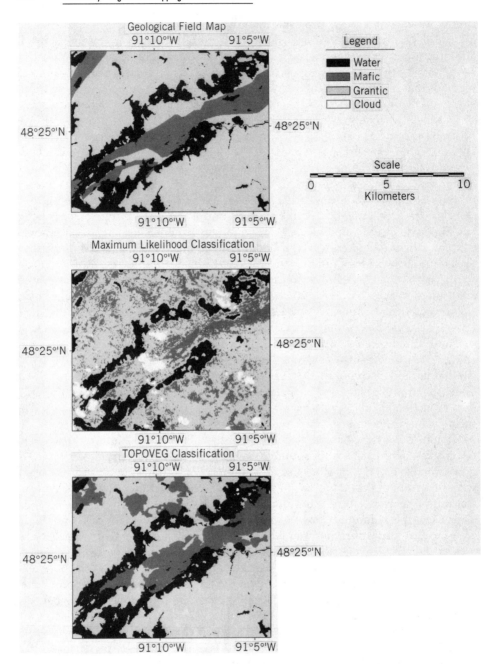

Figure 4.5 (Top) Geological field map of Quetico test site in Quetico Provinicial Park, Ontario, Canada; (middle) maximum likelihood classification of *Landsat* imagery; (bottom) Topoveg classification employing topographic, geobotanical, and aeromagnetic information.

a slope facet. If the south-facing facet is enriched in deciduous species and the site is near a geophysical anomaly, the rock type for the slope facet is classified as mafic. Otherwise, a granitic composition is assumed. The adjacent topographic facets (ridge tops and then north-facing slopes, and finally, wet sites) are assumed to be similar in geology to the adjacent topographic facets. The classification takes place at the landscape scale rather than the pixel scale, yielding greater classification accuracy for the original study site (86%) and for adjacent areas (86 to 79%) compared to the maximum likelihood classification (71 to 64%). More significant, however, is the smoothing of high-frequency noise in the TOPOVEG output, making the classification easier to interpret.

4.6.3 Case Study: Boreal Taiga/Tundra and Fire Scars

Wildfire in boreal forests is the dominant form of environmental disturbance and affects 8 to 12 million hectares per year (Kasischke, forthcoming). In boreal ecosystems, fire frequency and intensity interact with climate to affect the carbon balance of boreal ecosystems. The functioning of these ecosystems is important in the dynamics of the global carbon budget because they store 20 to 25% of the terrestrial carbon stocks and are a major source of methane emissions. The land within the fire scars have increased the permafrost depth due to the extra solar insolation reaching the ground surface, which changes the hydrologic runoff and water yield. Fire scars show accelerated erosion relative to unburned sites. The extent and number of fires is largely influenced by weather patterns, being worse in hot, dry years, and models have predicted 40% increases in fire severity and extent under global warming (Flannigan and Van Wagner, 1991; Wotton and Flannigan, 1993). Typically, fires are caused by lightning, and once started, continue to burn until they are extinguished naturally by the onset of fall–winter weather. The Alaska Fire Service estimates that approximately 18 % of the total Alaskan boreal forest has burned in the past 50 years. There is some evidence, based on the seasonal dynamics of atmospheric CO_2 concentration, that growing seasons are longer and temperatures warmer (Keeling et al., 1996) and that an increase in forest area burned each year has occurred in the past two decades (Stocks et al., 1996). Remote sensing provides the only realistic mechanism to monitor wildfire dynamics, location, and extent of fires. Further, remote sensing data provide a critical input to improve predictions of fire-spread models (Burgan and Rothermel, 1984; Burgan and Shasby, 1984; Clark et al., 1994; Finney, 1995). Optical and radar satellites have been used to map boreal wildfires (French et al., 1996b; Kasischke et al., 1993; Kasischke and French, 1995). Because fire scars are so large, often exceeding 10,000 ha (100 km^2), and the contrast between the burns and unburned landscapes so sharp, only simple methods are needed to detect burns, such as color composites or vegetation index (VI) images. Changes in greenness of the site can be detected at scales from AVHRR (1000 m), to TM (30 m) or *SPOT* (20 m), with differences primarily in the detail of variation within the burn (French et al., 1996b). Fire spread and burn temperatures can be monitored during burns using the thermal channels on the AVHRR or TM satellites. NASA operates a four-channel airborne system, the airborne infrared disaster assessment system (AIRDAS), which has a red SWIR (1.5 to 2.5 μm) and two TIR channels for real-time airborne fire monitoring. Green (1996) reported estimating fire temperatures in the SWIR region using advanced visible/infrared imaging spectrometer (AVIRIS) data due to elevated blackbody emitted radiance. Further, smoke is transmissive in the SWIR wavelengths, making possible ground observation over smoke-covered areas. Radar satellites are important for monitoring boreal systems because cloud cover frequently prevents optical sensors from viewing the ground. Also, rapid regrowth of herbaceous ground cover after fires makes continued detection with optical sensors difficult after the first year or two, when image analysis relies on VI contrasts. French et al. (1996b) found backscatter in C-band systems (e.g., ERS and *Radarsat*) were sensitive to fire scars whereas L-band systems (e.g., JERS) were not. They attributed this to the sensitivity of C-band systems to high soil water content. Harrell et al. (1995) and French et al. (1996b) showed that radar systems were also sensitive to biomass and stand structure in boreal forests. Similar methods would be useful for mapping wildfires in tropical regions; however, in temperate regions such as the western United States, regional fire patterns are difficult to interpret without ambiguity because fire extent is smaller and vegetation cover is heterogeneous over a rugged terrain.

4.6.4 Case Study: Geobotanical Exploration for Hydrocarbon Microseeps in the Eastern Deciduous Forest*

The preliminary analyses of microseeps in Wood and Ritchie Counties, in the dissected plateau of West Virginia, show that by combining methods of geobotanical interpretation and lineament analysis, many random associations can be discounted and a more comprehensive analysis generated. The study site is the historic Volcano Oil Field on the crest of the north/south-trending Burning Springs Anticline, one of the few major structural features (500-m relief) in the region. The crest is almost flat for nearly a mile, and the flanks are characterized by very steep dips, resulting in a box-fold structure at the surface. The oil field was discovered about 1864 and has produced over 2 million barrels of oil. Production was from four sandstone horizons within the Connoquenessing Sandstone, the Pottsville, the Greenbrier, and the Pocono Big Injun Sands. Despite the extensive production, hydrocarbons are still present in parts of the structure. In the 1950s, cores from the Sandhill Well, the first deep well drilled in the center of the Appalachian Basin (West Virginia Geological Survey, 1959), led to the recognition of thrust faulting in the lower and middle Devonian in the subsurface of the anticline.

A geobotanical anomaly associated with the Volcano Oil Field was identified in a late summer/early fall (September) *Landsat* TM image. At the time the image was acquired, trees had not yet begun to senesce. The anomaly is characterized by higher radiance values in bands 4, 5, and 7 and a less prominent reduction in the radiance of bands 1, 2, 3, and 6. The geobotanical anomaly is either weakly developed or absent in scenes acquired in April, July, and October. Maximum contrast between the oaks (*Quercus*), tulip poplars (*Liriodendron tulipifera*), and maples (*Acer*), the dominant species of the area, was obtained in the October image, concurrent with peak fall colors. Unlike the September image, there is little difference between the

*This case study is from T. A. Warner, Department of Geology and Geography, West Virginia University, Morgantown, West Virginia.

area underlain by the Volcano Oil Field and the surrounding region. This suggests that the geobotanical anomaly observed in the September image is a stress-induced feature rather than a difference in community composition. In addition, the stress is apparently only sufficient to cause optical changes immediately prior to senescence.

The October image was used to produce a geobotanical anomaly map based on an *n*-dimensional probality density function (nPDF) transformation (Figure 4.6; see color insert). The nPDF is a suite of programs centered around a data reduction technique in which the original image bands are combined, based on their Euclidean distance from selected corners of the multispectral data space (Cetin et al., 1993). A nPDF index was developed based on the distance of each multispectral pixel from the corner comprising the minimum in bands 4, 5, and 7 and the maximum in the remaining bands. This has the effect of maximizing the difference between the pixels with the geobotanical anomaly, compared to the surrounding areas. Field analysis provided an independent validation. A 4-ft probe extracted soil gas that was analyzed for light hydrocarbons. As expected, there was general agreement between the light hydrocarbon halo over the Volcano Field and the geobotanical anomalies shown in the color map, but there is less agreement in detail.

Lineaments were mapped on the spring image by two independent observers. The lineaments were then categorized based on whether their strike was parallel or near-perpendicular to the strike of the Burning Springs Anticline, the major structural feature controlling the formation of the Volcano Oil Field. The results are displayed as an overlay on the October geobotanical anomaly image (Figure 4.6). A comparison of the lineament analysis with the soil gas data suggests that not all lineaments are of structural significance. Figure 4.7 shows that the soil ethane concentration appears to be highly correlated with distance to cross-strike lineaments, but not with distance to lineaments that are subparallel to the strike. Higher concentrations of ethane are found proximal to the cross-strike lineaments than at greater distances. The combination of the lineament analysis and the geobotanical investigation provides a more comprehensive and less ambiguous interpretation then either do independently.

4.6.5 Case Study: Vegetation Phenology in the Amazon Basin Derived from Wavelet Reconstruction Techniques

The NASA Amazon EOS-IDS team has recently integrated neural net and nonlinear wavelet reconstruction techniques (Smith et al. 1998a, b) to produce monthly cloud-free images of the Amazon Basin from daily AVHRR-global area coverage (GAC) data that are being used as validation for the CASA model. A single finite impulse response (FIR) filter (i.e., neuron) was applied over the five AVHRR band channels to select which day to use for the month (e.g., the most cloud-free day). In contrast to the strategy of selecting the maximum NDVI (Nobre et al., 1991; Shimabukuro et al., 1994; Los et al., 1994; Humberto et al., 1996), the idea was to minimize the local variability in each channel. Selecting the day with the maximum NDVI produced images with greater pixel-to-pixel heterogeneity in the monthly composite images compared to the FIR filter. NDVI at the AVHRR GAC scale (ca. 4 km × 4 km) is highly sensitive to atmospheric variability and the presence of subpixel clouds (Shimabukuro et al., 1994).

A second filter was applied, a one-dimensional median wavelet filter (Donoho

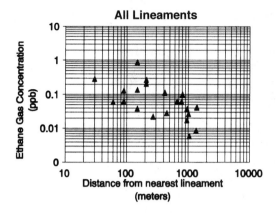

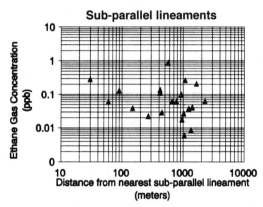

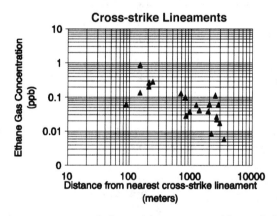

Figure 4.7   Relationship between concentration of ethane and distance to structural features for hydrocarbon microseeps in the Volcano Oil Field, West Virginia.

1993), to each channel in the Amazon FIR filtered data in the temporal direction for each pixel to remove additional residual atmospheric effects. For many months a single pixel will not have a cloud-free day, and this second step of filtering uses temporal patterns to further reduce the effects of variable atmosphere. Even though each pixel was filtered independently over time, the net effect was to further smooth the images in the spatial domain. The median filter was chosen because it is better at removing nonlinear noise caused by clouds that are analogous to Cauchy noise.

Figure 4.8 (see color insert) illustrates a sequence of monthly cloud-free images centered on the Amazon Basin for channels 1, 2, and 3 for the 1983 El Niño year. The 1983 El Niño is different from other years due to the widespread smoke in central and eastern Brazil, as noted by the increased blue on the color composite, which is related to increased reflectance in AVHRR-Ch1. The smoke results from increased burning during the drier periods of El Niño. There were no significant changes on the surface that coincide with the 1983 El Niño. Changes in AVHRR-Ch1 that are independent of AVHRR-Ch2 are often indicative of atmospheric conditions that prevail for the entire month. For comparison, Figure 4.9 shows the maximum NDVI computed over the same period. The NDVI images indicate a significant drop in NDVI that corresponds to the El Niño peak in 1983, especially prominent in the highland terra firme centered around Manous, Brazil. FPAR, used by CASA to determine net primary production (NPP), is computed as a normalized index of NDVI. CASA predicts a reduction of 34% net primary production (NPP), based on the NDVI, in the terra firme areas during the 1983 El Niño. For the 10-year period the

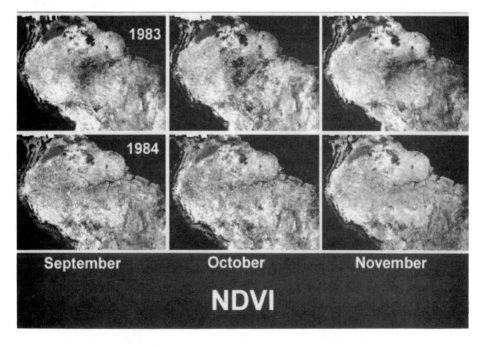

Figure 4.9   Maximum NDVI images for the central Amazon Basin as derived from AVHRR data. NDVI indicates a significant reduction in green vegetation in the highland terra firme coincident with the 1983 El Niño.

calibrated AVHRR-Ch2 provides a more consistent measure of seasonal and spatial variability in the green vegetation because of the reduced effect from atmospheric variation.

In contrast, the color composite of Figure 4.10 (see color insert) reveals an increase in red reflectance (channel 1) that is not accompanied by an increase in NIR (channel 2). Smith et al. (1998a,b) interpreted the reduction of NDVI corresponding to the 1983 El Niño as due to increased smoke in the atmosphere from burning during the drier El Niño periods rather then decreased vegetation abundance as interpreted previously. AVHRR-Ch3 provides an independent measure of subtle phenologic changes in tropical vegetation communities. Normally, the reddish areas designate savannas during the dry period. In June the northern savannas are beginning to green while the savannas south of the equator are senescing. Unusual for 1984 is the reddish color of the upland terra firme forests around Manous, Brazil in the center of the image. This change is evident a year after the 1983 El Niño. To first order, the NPV is expressed in AVHRR-Ch3 for the 10-year AVHRR record. In the savannas, seasonal cycles of Ch3 lag temperature and NDVI by two months, also indicating that Ch3 is a consistent independent spectral measure of the surface.

In contrast to the 1983 year, the 1987 El Niño (Figure 4.11; see color insert) pattern is marked by a seasonal phase shift in the vegetative phenology of the Amazon savannas. The spectral response in both space and time is significantly different between the 1983 and 1987 El Niño periods. These spatial patterns are uniquely different from those resulting from global biosphere models using the standard NDVI inputs (Field et al., 1995). It is considered that the dominant El Niño spectral effects observed by AVHRR are represented by temporal phase shifts in the vegetative phenology of different communities. These phase shifts themselves can be used to further define and map vegetative communities where greenness alone is insufficient to distinguish them. Furthermore, temporal phase shifts are an important geobotanical cue for detecting vegetation stress from other causes (e.g., presence of heavy metals or contaminants).

The mapping of wetlands in the Amazon using SMRR data has been used as a means for validating wetland classifications from AVHRR data (Sippel et al., 1994). The Pantanal wetland, in the Amazon Basin, is easily discernible, and initially one might think that a simple index similar to NDVI could be used to map this wetland. Water beneath the vegetation would reduce the radiance reflected from the canopy, yielding a negative relation between the wetlands presence and albedo in channels 1 and 2. Figure 4.12 illustrates the interaction of spatial scales from the respective coherence between the scales of digital terrain models derived from 37-GHz SMMR data (Sippel et al., 1994) and AVHRR-Ch2. The energy between scales was computed using a Symmlet wavelet decomposition of AVHRR channel 2 and digital terrain images. It was found that a strong negative coherence existed only at the 100-km spatial scale.

Changes in channel 2 at larger and smaller scales have little correspondence with the temporal appearance of wetlands. Spectral variability at the pixel scale in AVHRR-Ch2 has almost no coherence with wetlands. These results indicate that mapping wetlands using AVHRR GAC data in the Amazon basin is a subband filter problem that is clearly scale dependent. Wetlands of 4 km (e.g., the pixel resolution) will not be mapped accurately using these data.

Because of the small extent of many mineral exposures and point sources of con-

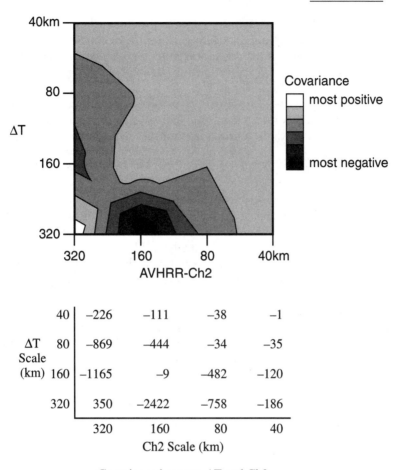

	320	160	80	40
40	−226	−111	−38	−1
80	−869	−444	−34	−35
160	−1165	−9	−482	−120
320	350	−2422	−758	−186

ΔT
Scale
(km)

Ch2 Scale (km)

Covariance between ΔT and Ch2

Figure 4.12 Interaction of spatial scales from the respective coherence between the scales of digital terrain models derived from 37-GHz SMMR data (Sippel et al., 1994) and AVHRR-Ch2 for an area in the Amazon Basin. The energy between scales was computed using a Symmlet wavelet decomposition of AVHRR-Ch2 and digital terrain images. There is a strong negative covariance between the 320-km ΔT scale and the 160-km AVHRR-Ch2. A negative covariance between Ch2 and ΔT is the expected response caused by water inundation and is strong only at the intermediate scales.

tamination, similar scaling problems exist in many remote sensing data sets. Methods such as wavelet decomposition can provide insight into the spatial and spectral scaling of the features of interest, thus avoiding misinterpretation of remote sensing data that is inappropriate for the scale of the process and the image data.

4.6.6 Case Study: Detection of Land-Use Change

A neural network analysis of multidate *Landsat* imagery was used to classify the cutting history in Bluff Creek, California. The time series for Bluff Creek consists of 29 coregistered *Landsat* MSS images and eight *Landsat* TM images that span a 23-

year period (Figure 4.13; see color insert). A foreground background analysis (FBA) framework was used to determine a FIR filter (layer of linear neurons) for each image to (1) minimize variability in the vegetation end member independent of terrain, (2) maximize the contrast between vegetation and cut areas, and (3) minimize the effects of calibration and atmospheric variability. The second layer consisted of a competitive network applied in the time domain to detect the year of cut for each pixel. A drop in the green vegetation fraction in the first layer of neurons from about 1.0 to less than 0.05, indicated that cutting had occurred. With annual images, determination of cutting date for cuts was done with an error of less than 0.4 year and a 0.93 accuracy compared to the Forest Service cutting records (Figure 4.14).

Attempts to retrieve cutting history using conventional remote sensing techniques from single images covering periods of time extending before the *Landsat* satellite program have failed. The cutting history for the Gifford Pinchot National Forest, Washington was supplied by the U.S. Forest Service and we applied spectral mixture analysis (SMA), Foreground background analysis (FBA), and an integrated wavelet/neural net to a 1992 Landsat image. It was found that the simple mixture model applied to single multispectral images had high uncertainties in quantifying seral stage (successional sequence) in this forest. All results indicate a steadily increasing areal coverage with stand age as shown in Figure 4.14 (i.e., area covered by older stands is more than by younger stands). These results differ with Forest Service records (Figure 4.15), which clearly depict that most of the area has been cut in the recent past (i.e., a negative slope between stand age and area covered for each age class). The uncertainty of stand age determined from simple mixture models was highest for the youngest ages (± 20 years, Figure 4.16) and lowest and constant for stands above 60 years of age. End-member fractions did not vary significantly in forest stands older than 60 years of age. Even though younger stands contrast spectrally with older stands, the variability of end-member fractions in old and young stands alike make areal stand age estimates using these methods highly uncertain. This may be due to several factors, (e.g., differences in growth rates between conditions at better and poorer sites, failures of reseeding, or incomplete harvests).

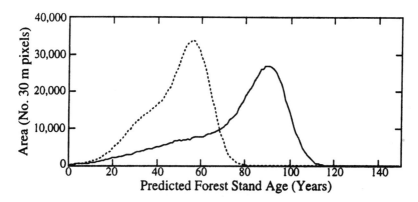

Figure 4.14 Distribution of forest stand age predicted for Gifford Pinchot National Forest using TM data. The solid line is the prediction of stand age for the simple mixture model, and the dotted line is the age distribution predicted using FBA. The areas between Figures 4.10 and 4.1 are different because all pixels near cutting boundaries were omitted in Figure 4.10 in order to remove the effects of shading and shadowing near edges.

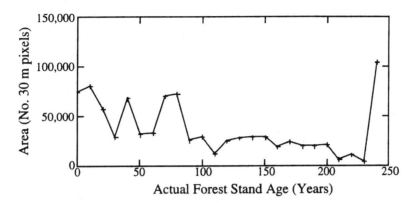

Figure 4.15 Actual forest stand age from Forest Service records. The history was reduced to 10-year steps to smooth out yearly variation due to interannual climate differences and enhance the general trends. Spectral mixture estimates from single images (Figure 4.2) indicate an age structure that is opposite stand-age records. Reduction of Figures 4.10 and 4.11 to two seral stages separated at 12 years produces consistency between Forest Service records and SMA and FBA predictions.

Although topographic effects are important, our results show that spectral variability is the dominant cause of uncertainty in determining stand age using spectral mixture analysis. To deal with this problem, we created linear neurons to minimize the effect of end-member spectral variability (Smith et al., 1994) using FBA. Even though FBA stand-age uncertainty was a factor of 4 lower than SMA age uncertainty, the distribution of apparent stand ages was also inconsistent with Forest Service records. However, neither the SMA and FBA solutions, as determined locally using USFS records, were extendible to full TM scenes covering western Washington and Oregon. This inconsistency in extrapolating to an area larger then the validation area is because the original validation data did not encompass the full range of variation present in northwest conifer forests. In particular, east–west gradients in the NPV fraction (litter, wood, and bark), corresponding to broad soil moisture availability gradients, mimic the spectral variation in older seral stages.

SMA and FBA predictions are consistent with the trend in declining uncertainty in stand-age uncertainty as a function of age (Figure 4.16). The spectral contribution to variance in apparent stand age arises from two main sources: (1) the physical components within the scene, and (2) lighting geometry (e.g., terrain effects). The precision with which stand age can be estimated by either SMA or FBA is highest for young stands and converging to a low standard deviation of <5 years for stands older than about 60 years (Figure 4.16). Although FBA produces more precise estimates for young stands, FBA and SMA yield similar results for mature stands.

The spectral variability unique to stand age is at a finer spatial scale than that arising from terrain and size of clear-cuts. A wavelet decomposition was used to extract two spatial scales (corresponding to 60 m and 120 m) to isolate the spectral variance attributable solely to seral stage. Younger stands have much greater spectral variability than older stands at these spatial scales, due to the compositional variability in the initial stages of regrowth and to less shadowing in gaps between young trees. Utilizing spectral variation from the finest spectral scales to apply SMA and FBA also uncouples stand-age estimates from large-scale (regional) moisture gradi-

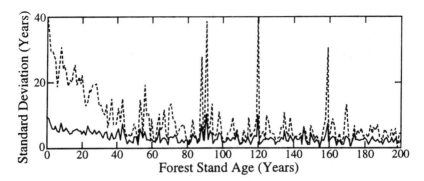

Figure 4.16 Variability in forest stand age estimates for SMA (dashed line) and for FBA (solid line) follow the same trend. The standard deviation decreases for both estimates from young stands until about 60 years. After 50 years the standard deviation is relatively constant. The reduction in stand-age uncertainty between SMA and FBA does not affect the distribution in age classes found in Figure 4.12.

ents and reduces the impact of annuals on the apparent "greenness" during the first few years after cutting (Figure 4.16).

4.7 SUMMARY AND CONCLUSIONS

Environmental resources, contamination and pollution, and impacts of anthropogenic and climate-induced changes are among the concerns of the earth sciences today. Remote sensing, at all scales from the laboratory to space, has significant potential to aid monitoring and detection efforts. This review has attempted to provide prospective on the history of the use of vegetation properties for earth science mapping and illustrate some current approaches to data analysis that may become more widely used as we move from qualitative to quantitative spectral analysis approaches. We have provided a brief review of physical edaphic factors that affect plant species distributions and growth potential. The actual relationships between vegetation characteristics and soil properties are often difficult to resolve with respect to understanding which of many potential interacting factors is significant in a particular locality. Thus both the physical relationships and the image processing steps are subject to errors and misinterpretation. We have provided examples over scales ranging from leaf modeling to regional analysis of satellite images. Resolving the significant spectral and spatial properties over the full range of environmental scales is critical to successful interpretation of remote sensing data for addressing problems in earth sciences. Greater attention to combining spatial and spectral information, topographic data, and other ancillary data sets may reduce ambiguity in image analysis and provide new insight into ecosystem processes and interactions. We illustrate several recent techniques for image processing, from spectral mixture analysis and wavelet modeling, to radiative transfer models of leaves and canopies using a variety of approaches. Although the latter models may be too complex for most earth science applications, these models provide greater confidence for linking empirical models to physical processes. The large number of new satellite sensors, with improved spatial, spectral, and temporal resolution, should open many opportunities for integrated earth science assessments.

References

ACCP, 1994. *NASH Accelerated Canopy Chemistry Programs Final Report, NASA-EOS-IWG*, Accelerated Canopy chemistry Program, October, 19. Durham, N.H.

Adams, J. B., and J. D. Adams, 1984. Geologic mapping using *Landsat* MSS and TM images: removing vegetation by modeling spectral mixtures, in *Proceedings of the 3rd Thematic Conference on Remote Sensing for Exploration Geology*, Vol. 2, Environmental Research Institute of Michigan, Ann Arbor, Mich., pp. 615–622.

Agricola, G., 1556. *De re metallica*, H. C. Hoover and L. C. Hoover, translators, Dover, New York.

Allen, W. A., H. W. Gausman and A. J. Richardson, 1968. Interaction of light with a plant canopy, *J. Opt. Soc. Am.*, 58, 1023–1028.

Allen W. A., H. W. Gausman, A. J., Richardson and J. R. Thomas, 1969. Interaction of isotropic light with a compact leaf, *J. Opt. Soc. Am.*, 59, 1376–1379.

Allen, W. A., H. W. Gausman, and A. J. Richardson, 1970. Mean effective constants of cotton leaves, *J. Op. Soc. Am.*, 60, 542–547.

Allen, W. A., H. W. Gausman, and A. J. Richardson, 1973. Willstätter–Stoll theory of leaf reflectance evaluation by ray tracing, *Appl. Opt.*, 12, 2448–2453.

Andrieu, B., F. Baret, J. Schellberg, and U. Rinderle, 1988. Estimation de spectres de feuilles à partir de mesures dans des bandes spectrales larges, in *Proceedings of the 4th International Colloquium on Spectral Signatures in Remote Sensing*, Aussois France ESA SP-287, pp. 351–356.

Asrar, G., B. M. Fuchs, E. T. Kanemasu, and J. L. Hatfield, 1984. Estimating absorbed photosynthetic radiation and leaf area index from spectral reflectance in wheat, *Agron. J.*, 76, 300–306.

Ballew, G. I., 1975. Correlation of *Landsat-1* multispectral data with surface geochemistry, in *Proceedings of the Tenth International Symposium on Remote Sensing of Environment*, Vol. 2, Environmental Research Institute of Michigan, Ann Arbor, Mich., p. 1045.

Bannari, A., A. R. Huete, D. Morin, and F. Zagolski, 1996. Effects of soil color and brightness on vegetation indexes, *Int. J. Remote Sensing*, 17, 1885–1906.

Baret, F. and T. Fourty. 1997. Estimation of leaf water content and specific weight from reflectance and transmittance measurements. *Agronomie*, 17, (9–10), 455–464.

Baret, F., and G. Guyot, 1991. Potentials and limits of vegetation indices for LAI and APAR assessment, *Remote Sensing Environ.*, 35, 161–173.

Barton, F. E., D. S. Himelsbach, J. H. Duckworth, and M. J. Smith, 1992. 2-Dimensional vibration spectroscopy: correlation of mid-infrared and near-infrared regions, *Appl. Spectrosc.*, 46, 420–429.

Bélanger, J. R., 1988. *Prospecting in Glaciated Terrain: An Approach Based on Geobotany, Biogeochemistry, and Remote Sensing*, GSC Bull. 387, Geological Survey of Canada, Ottawa, Ontario, Canada, 38 pp.

Bélanger, J. R., and R. G. Hélie, 1984. Étude du quaternaire de l'Île du Roi Guillaume à l'aide de la télédétection in *Proceedings of the 8th Canadian Symposium on Remote Sensing, 1983*, Montreal, Quebec, Canada, pp. 685–694.

Bélanger, J. R. and A. N. Rencz, 1983. Prospecting in glaciated terrain: integrating airborne and *Landsat MSS, Adv. Space Res.*, 3, 187–191.

Bélanger, J. R., A. N. Rencz, and W. W. Shilts, 1979. Patterns of glacial dispersal of heavy metals as reflected by satellite imagery, in *Télédétection et Gestion des Ressources, Ste.-Foy*, F. Bonn, ed., Association Québecoise de Télédétection, Quebec City, Quebec, Canada, pp. 59–74.

Bell, R., M. L. Labovitz, and D. P. Sullivan, 1985. Delay in leaf flush associated with heavy metal-enriched soil, *Econ. Geol.*, 80, 1407–1414.

Bone, R. A., D. W. Lee, and J. M. Norman 1985. Epidermal cells functioning as lenses in leaves of tropical rain forest shade plants, *Appl. Opt.*, 24, 1408–1414.

Borel, C. C., S. A. W. Gerstl, and B. J. Powers, 1991. The radiosity method in optical remote sensing of structured 3-D surfaces, *Remote Sensing Environ.*, 36, 13–44.

Bradshaw, A. D., 1969. An ecologist's viewpoint, in *Ecological Aspects of the Mineral Nutrition of Plants*, I. H. Rorison, ed., Blackwell Scientific Publications, Oxford, pp. 415–427.

Brakke, T. W., and J. A. Smith, 1987. A ray tracing model for leaf bidirectional scattering studies, in *Proceedings of the 7th International Geoscience and Remote Sensing Symposium*, IGARSS'87, Ann Arbor, Mich., pp. 643–648.

Brass, J. A., V. G. Ambrosia, and R. Higgins, 1996. AIRDAS development of a unique four channel scanner for disaster assessment and management, in *Proceedings of the 2nd International Airborne Remote Sensing Conference and Exhibition*, San Francisco, June 24–27, Vol. 3, Environmental Research Institute of Michigan, Ann Arbor, Mich., pp. 781–787.

Brooks, R. R., 1972. *Geobotany and Biogeochemistry in Mineral Exploration*, Harper & Row, New York, 290 pp.

Brooks, R. R., 1995a. Specific indicator plants, in *Biological Systems in Mineral Exploration and Processing*, R. R., Brooks, C. E. Dunn, and G. E. M. Hall, eds., Ellis Horwood, New York, pp. 39–70.

Brooks, R. R., 1995b. Plant communities as ore indicators, in *Biological Systems in Mineral Exploration and Processing*, R. R. Brooks, C. E. Dunn, and G. E. M. Hall, eds., Ellis Horwood, New York, pp. 9–38.

Brooks, R. R., 1995c. Introduction to geobotanical prospecting, in *Biological Systems in Mineral Exploration and Processing*, R. R. Brooks, C. E., Dunn, and G. E. M. Hall., eds, Ellis Horwood, New York, pp. 7–8.

Brooks, R. R., 1995d. Remote sensing of vegetation, in *Biological Systems in Mineral Exploration and Processing*, R. R. Brooks, C. E. Dunn, and G. E. M. Hall, eds., Ellis Horwood, New York, pp. 85–116.

Brown, P. S., and J. P. Pandolfo, 1969. An equivalent-obstacle model for the computation of radiation flux in obstructed layers, *Agric. Meteorol.*, 6, 407–421.

Burgan, R. E., and R. C. Rothermel, 1984. *BEHAVE: Fire Behavior Prediction and Fuel Modeling System, Fuel Subsystem*, Gen. Tech. Rep. INT167, Intermountain Forest and Range Experiment Station, USDA Forest Service, Ogden, Utah, 66 pp.

Burgan, R. E., and M. B. Shasby, 1984. Mapping broad-area fire potential from digital fuel, terrain and weather data, *J. For.*, 8, 228–231.

Cain, S. A., 1971. *Foundations of Plant Geography*, Hafner Publishing Company, New York, 556 pp.

Camillo, P., 1987. A canopy reflectance model based on an analytical solution to the multiple scattering equation, *Remote Sensing Environ.*, 23, 453–477.

Cannon, H. L., 1960. Botanical prospecting for ore deposits, *Science*, 132, 591–598.

Cannon, H. L., 1971. The use of plant indicators in ground water surveys, geologic mapping, and mineral prospecting, *Taxon*, 20, 227–256.

Card, D. H., D. L. Peterson, P. A. Matson, and J. D. Aber, 1988. Prediction of leaf chemistry by the use of visible and near infrared reflectance spectroscopy, *Remote Sensing Environ.*, 26, 123–147.

Cecchi, G., P. Mazzinghi, L. Pantani, R. Valentini, D. Tirelli, and P. De Angelis, 1994. Remote sensing of chlorophyll *a* fluorescence of vegetation canopies: 1. Near and far field measurement techniques, *Remote Sensing Environ.*, 47, 18–28.

Cetin, H., T. A. Warner, and D. W. Levandowski, 1993. Data classification, visualization and enhancement using *n*-dimensional probability density functions (nPDF): AVIRIS, TIMS, TM and geophysical applications, *Photogramm. Eng. Remote Sensing*, 59, 1755–1764.

Chandrasekhar, S., 1944. On the radiative equilibrium of a stellar atmosphere: II, *Astrophys. J.*, 100, 76–86.

Chandrasekhar, S., 1960. *Radiative Transfer*, Dover, New York.

Chang, S.-H., and W. Collins, 1980. Toxic effects of heavy metals on plants, in *Proceedings of the 6th William T. Pecora Memorial Symposium*, Sioux Falls, S. Dak. American Society of Photogrammetry and Remote Sensing, Falls Church, Va., pp. 122–234.

Chang, S.-H., and W. Collins, 1983. Confirmation of the airborne biogeophysical mineral exploration technique using laboratory methods, *Econ. Geol.*, 78, 723–736.

Chapin, F. S., III, 1988. Ecological aspects of plant mineral nutrition, *Adv. Miner. Nutri.*, 3, 161–191.

Chapin, F. S., III, D. A. Johnson, and J. D. McKendrick, 1980. Seasonal movement of nutrients in plants of differing growth form in an Alaskan tundra ecosystem: implications for herbivory, *J. Ecol.*, 68, 189–209.

Chappelle, E. W., F. M. Wood, Jr., and J. E. McMurtrey III, and W. W. Newcomb, 1984a. Laser-induced fluorescence of green plants: 1. A technique for the remote detection of plant stress and species differentation, *Appl. Opt.*, 23, 134–138.

Chappelle, E. W., J. E. McMurtrey III, F. M. Wood, Jr., and W. W. Newcomb, 1984b. Laser-induced fluorescence of green plants: 2. LIF caused by nutrient deficiencies in corn, *Appl. Opt.*, 23, 139–142.

Chappelle, E. W., F. M. Wood, Jr., W. W. Newcomb, and J. E. McMurtrey III, 1985. Laser-induced fluorescence of green plants: 3. LIF spectral signatures of five major plant types, *Appl. Opt.*, 24, 74–80.

Chappelle, E. W., M. S. Kim, and J. E. McMurtrey III, 1992. Ratio analysis of reflectance spectra (RARSP): an algorithm for the remote estimation of the concentration of chlorophyll *a*, chlorophyll *b*, and carotenoids in soybean leaves, *Remote Sensing Environ.*, 39, 239–247.

Chikishev, A. G., 1965. *Plant Indicators of Soils, Rocks and Subsurface Waters*, Consultants Bureau, New York, 210 pp.

Clark, K. C., J. A. Brass, and P. J. Riggan, 1994. A cellular automation model of wildfire propagation and extinction, *Photogramm. Eng. Remote Sensing*, 60, 1355–1367.

Clark, R. N., and T. L. Roush, 1984. Reflectance spectroscopy: quantitative analysis techniques for remote sensing applications, *J. Geophys. Res.*, 89, 6329–6340.

Clark, R. N., A. J. Gallagher, and G. A. Swayze, 1990a. Material absorption band depth mapping of imaging spectrometer data using a complete band shape least-squares fit with library reference spectra, in *Proceedings of the 2nd Airborne Visible/Infrared Imaging Spectrometer (AVIRIS) Workshop*, JPL Publ. 90–54, Jet Propulsion Laboratory, California Institute of Technology, Pasadena, Calif., pp. 176–186.

Clark, R. N., T. V. V. King, M. Klejwa, G. Swayze, and N. Vergo, 1990b. High spectral resolution reflectance spectroscopy of minerals, *J. Geophys. Res.*, 95, 12653–12680.

Clark, R. N., G. A., Swayze, A. Gallagher, T. V. V. King, and W. M. Calvin, 1993. *The U.S. Geological Survey, Digital Spectral Library, Version 1: 0.2 to 3.0 µm*, USGS Open File Rep. 93–592, U.S. Geological Survey, Washington, D.C., 1340 pp.

Clark, R. N., T. V. V. King, C. Ager, and G. A. Swayze, 1995. Initial vegetation species and senescence/stress mapping in the San Louis Valley, Colorado using imaging spectrometry data in *Proceedings of the Summitville Forum '95*, H. H. Posey, J. A. Pendleton, and D. Van Zyl, (eds.) CGS Spec. Publ. 38, Colorado Geological Survey, Boulder, Colo., pp. 59–63.

Cole, M. M., 1977. Surface expression of ore bodies, *Trans. Inst. Min. Metall. B*, 86, 169.

Collins, W., 1978. Remote sensing of crop type and maturity, *Photogramm. Eng. Remote Sensing*, 44, 43–55.

Collins, W., S.-H. Chang, and J. F. Kuo, 1980. *Detection of Hidden Mineral Deposits by Airborne Spectral Analysis of Forest Canopies*, NTIS Rep. PB-80-193881, National Technical Information Service, Springfield, Va., 61 pp.

Collins, W., S.-H Chang, G. Raines, F. Canney, and R. Ashley, 1983. Airborne biogeophysical mapping of hidden mineral deposits, *Econ. Geol.*, 78, 737–749.

Conel, J. E., J. van den Bosch, and C. I. Grove, 1993. Application of a two-stream radiative transfer model for leaf lignin and cellulose concentrations from spectral reflectance measurements: Parts 1 and 2, in *Proceedings of the 4th Annual JPL Airborne Geoscience Workshop*, Vol. 1 AVIRIS Workshop, R. O. Green, ed., Oct. 25–29, Washington, D.C., NASA-JPL Publ. 93-26, pp. 39–51.

Curran, P. J., 1989. Remote sensing of foliar chemistry, *Remote Sensing Environ.*, 30, 271–278.

Curran, P. J., 1994. Imaging spectrometry, *Prog. Phys. Geogr.*, 18, 247–266.

Curran, P. J., J. I. Dungan, and H. L. Gholz, 1990. Exploring the relationship between reflectance red edge and chlorophyll concentration in slash pine, *Tree Physiol.*, 7, 33–48.

Curran, P. J., J. I. Dungan, B. A. Maeler, and S. E. Plummer, 1991. The effect of a red leaf pigment on the relationship between red edge and chlorophyll concentration, *Remote Sensing Environ.*, 35, 69–76.

Curran, P. J., W. R. Windham, and H. L. Gholz, 1995. Exploring the relationship between reflectance red edge and chlorophyll concentration in slash pine leaves, *Tree Physiol.*, 15, 203–206.

Curtiss, B., and S. L. Ustin, 1989. Parameters affecting reflectance of coniferous forests in the region of chlorophyll pigment absorption in *Proceedings of the International Geoscience and Remote Sensing Symposium*, (IGARSS '89), Vancouver, British Columbia, Canada, Vol. 4, pp. 2636–2636.

Dawson, T. P., P. J. Curran, and S. E. Plummer, 1998. LIBERTY: modelling the effects of leaf biochemical concentration on reflectance spectra, *Remote Sensing Environ.* 65, 50–60.

Deering, D. W., 1989. Field measurements of bidirectional reflectance, in *Theory and Applications of Optical Remote Sensing*, G. Asrar, ed., Wiley, New York, pp. 14–65.

Deering, D. W., and P. Leonoe, 1986. A sphere-scanning radiometer for rapid directional measurements of sky and ground radiance, *Remote Sensing Environ.*, 19, 1–24.

DeFries, R. S., and J. G. R. Townshend, 1994. NDVI-derived land cover classification at global scales, *Int. J. Remote Sensing*, 15(17), 3567–3586.

DeFries, R., C. Field, A. Fung, C. Justice, S. Los, P. Matson, E. Matthews, H. Mooney, C. Potter, K. Prentice, P. Sellers, J. Townshend, C. Tucker, S. Ustin, and P. Vitousek, 1995. Mapping the land surface for global atmosphere–biosphere models: toward continuous distributions of vegetation's functional properties, *J. Geophys. Res. Atmos.*, 100(20), 867–882.

Donoho, D. L., 1993. Nonlinear wavelet methods for recovery of signals, images, and densities from noisy and incomplete data, in different perspectives on wavelets, I. Daubechies, ed. pp. 173–205. Providence, RI: American Mathematical Society.

Ehrlich, D., and E. F. Lambin, 1996. Broad scale land-cover classification and interannual climatic variability, *Int. J. Remote Sensing*, 17, 845–862.

Engelsen, O., B. Pinty, M. M. Verstraete, and J. V. Martonchik, 1996. *Parametric Bidirectional Reflectance Factor Models: Evaluation, Improvements and Applications*, Tech. Rep. EUR 16426 EN, Joint Research Centre, European Community, Ispra, Italy.

Epiphanio, J. C. N., and A. R. Huete, 1995. Dependence of NDVI and SAVI on sun sensor geometry and effect on FAPAR relationships in alfalfa, *Remote Sensing Environ.*, 51, 351–360.

Epstein, E., 1972. *Mineral Nutrition of Plants: Principles and Perspectives*, Wiley, New York, 412 pp.

Eyre, S. R., 1968. *Vegetation and Soils: A World Picture*, 2nd ed., Edward Arnold, London, 324 pp.

Field, C. B., F. S. Chapin III, P. A. Matson, and H. A. Mooney, 1992. Responses of terrestrial ecosystems to the changing atmosphere: a resource-based approach, *Annu. Rev. Ecol. Systemat.*, 23, 201–235.

Field, C. B., J. T. Randerson, and C. M. Malmstrom, 1995. Global net primary production: combining ecology and remote sensing, *Remote Sensing Environ.*, 51, 74–88.

Finney, M. A., 1995. FARSITE: a fire area simulator for managers, in *The Biswell Symposium: Fire Issues and Solutions in Urban Interface and Wildland Ecosystems*, D. R. Weise, and R. E. Martin, eds., Feb. 15–17, 1994, Walnut Creek, Calif., USFS Gen. Tech. Rep. PSW-GTR-158, pp. 55–56.

Fischer, A., 1994. A model for the seasonal variations of vegetation indices in coarse resolution data and its inversion to extract crop parameters, *Remote Sensing Environ.*, 48, 220–230.

Flannigan, M. D., and C. E. Van Wagner, 1991. Climate change and wildfire in Canada, *Can. J. For. Res.*, 21, 61–72.

Fourty, Th., F. Baret, S. Jacquemoud, G. Schmuck, and J. Verdebout, 1996. Optical properties of dry plant leaves with explicit description of their biochemical composition: direct and inverse problems, *Remote Sensing Environ.*, 56, 104–117.

French, N. F. H., E. S. Kasischke, L. L. Bourgeau-Chavez, P. Harrell, and N. L. Christensen, Sensitivity of ERS-1 SAR to variations in soil water in fire disturbed boreal forest ecosystems, 1996a. *Int. J. Remote Sensing.* 17, (15): 3037–3053.

French, N. F. H., E. S. Kasischke, R. D. Johnson, L. L. Bourgeau-Chavez, A. L. Frick, and S. L. Ustin, 1996b. Using multi-sensor satellite data to monitor carbon flux in Alaskan boreal forests, in *Biomass Burning and Climate Change*, J. L. Levine, ed., Vol. 2, pp. 808–826. MIT Press, Cambridge, Mass.

Friedl, M. A., F. W. Davis, J. Michaelsen, and M. A. Moritz, 1995. Scaling and uncertainty in the relationship between the NDVI and land surface biophysical variables: an analysis using a scene simulation model and data from FIFE, *Remote Sensing Environ.*, 54, 233–246.

Fukshansky, L., N. Fukshansky-Kazarinova, V. Martinez, and A. Remisowsky, 1991. Estimation of optical parameters in a living tissue by solving the inverse problem of the multiflux radiative transfer, *Appl. Opt.*, 30, 3145–3153.

Gabrys-Mizera, H., 1976. Model consideration of the light conditions in noncylindrical plant cells, *Photochem. Photobiol.*, 24, 453–461.

Gamon, J. A., C. B. Field, W. Bilger, A. Bjorkman, A. L. Fredeen, and J. Penuelas, 1990. Remote sensing of the xanthophyll cycle and chlorophyll florescence in sunflower leaves and canopies, *Oecologia*, 85, 1–7.

Gamon, J. A., C. B. Field, D. A. Roberts, S. L. Ustin, and R. Vallentini, 1993. Functional patterns in an annual grassland during an AVIRIS overflight, *Remote Sensing Environ.*, 44, 239–253.

Gamon, J. A., C. B. Field, M. L. Goulden, K. L. Griffin, A. E. Hartley, G. Joel, J. Penuelas, and R. Valentini, 1995. Relationships between NDVI, canopy structure, and photosynthesis in three Californian vegetation types, *Ecol. Appl.*, 5, 28–41.

Ganapol, B., L. Johnson, P. Hammer, C. Hlavka, and D. Peterson, 1998. LEAF-MOD: a new within-leaf radiative transfer model, *Remote Sensing Environ.*, 63(2), 182–193.

Gates, D. M., 1970. Physical and physiological properties of plants, in *Remote Sensing with Special Reference to Agriculture and Forestry*, National Academy of Science, Washington, D.C., pp. 224–252.

Gates, D. M., 1980. *Biophysical Ecology*, Springer-Verlag, New York, 611 pp.

Gates, D. M., H. J. Keegan, J. C. Schleter, and V. R. Weidner, 1965. Spectral properties of plants, *Appl. Opt.*, 4, 11–20.

Gausman, H. W., 1977. Reflectance of leaf components, *Remote Sensing Environ.*, 6, 1–9.

Gausman, H. W., D. E. Escobar, J. H. Everitt, A. J. Richardson, and R. R. Rodriguez, 1978. Spectral properties of plants, *Appl. Opt.*, 4, 11–20.

Gerstl, S. A. W., 1990. Physics concepts of optical and radar reflectance signatures: a summary review, *Int. J. Remote Sensing*, 11, 1109–1117.

Gerstl, S. A. W., and C. C. Borel, 1992. Principles of radiosity method versus radiative transfer for canopy reflectance modeling, *IEEE Trans. Geosci. Remote Sensing*, 30, 271–275.

Gobron, N., B. Pinty, M. M. Verstraete, and Y. Govaerts, 1997. A semi-discrete model for the scattering of light by vegetation, *J. Geophys. Res.*, 102(D8), 9431–9446.

Goel, N., 1988. Models of vegetation canopy reflectance and their use in estimation of biophysical parameters from reflectance data, *Remote Sensing Rev.*, 4, 1–212.

Goel, N. S., W. Qin, 1994. Influences of canopy architecture on relationships between various vegetation indices and LAI and FPAR: A computer simulation. *Remote Sensing Rev.*, 10, 309–347.

Goel, N. S., I. Rozehnal, and R. L. Thompson, 1991. A computer graphics based model for scattering from objects of arbitrary shapes in the optical region, *Remote Sensing Environ.*, 36, 73–104.

Govaerts, Y., and M. M. Verstraete, 1994. Applications of the L-systems to canopy reflectance modelling in a Monte Carlo ray tracing technique: fractals in geosciences and remote sensing, *Proceedings of a Joint JRC/EARSeL Expert Meeting*, Apr. 14–15, European Commission, Ispra, Italy, pp. 211–236.

Govaerts, Y., and M. M. Verstraete, 1995. Evaluation of the capability of BRDF models to retrieve structural information on the observed target as described by a three-dimensional ray tracing code, in *Proceedings of the European Symposium on Satellite Remote Sensing, EUROPTO-SPIE Conference*, Rome, Sept. 26–30, 1994, *Proc. SPIE*, 2314, 9–20.

Govaerts, Y., and M. M. Verstraete, 1996. Modelling the scattering of light in arbitrarily complex media: motivation for a ray tracing approach, in *Final Report of the Ray Tracing Exploratory Research*, Institute for Remote Sensing Applications, Joint Research Centre, European Commission, Ispra, Italy, 26 pp.

Govaerts, Y. M., S. Jacquemoud, M. M. Verstraete, and S. L. Ustin, 1996. Three-dimensional radiation transfer modeling in a dicotyledon leaf, *Appl. Opt.*, 35, 6585–6598.

Goward, S. N., G. D. Cruickshanks, and A. S. Hope, 1985. Observed relation between thermal emission and reflected spectral radiance of a complex vegetation landscape, *Remote Sensing Environ.*, 18, 137–146.

Green, R. O., 1996. Estimation of biomass fire temperature and areal extent from calibrated AVIRIS spectra, in *Proceedings of the 6th Annual Airborne Earth Science Workshop*, March 14–8, JPL Publ. 96–4, vol. 1, Jet Propulsion Laboratory, California Institute of Technology, Pasadena, Calif., pp. 105–114.

Gutman, G. G., 1991. Vegetation indices from AVHRR: an update and future prospects, *Remote Sensing Environ.*, 35, 121–136.

Haberlandt, G., 1914. *Physiological Plant Anatomy*, translated by M. Drummond, Macmillan, London, pp. 615–630.

Hapke, B., 1981. Bidirectional reflectance spectroscopy: I. Theory, *J. Geophys. Res.*, 86, 3039–3054.

Hapke, B., 1993. *Theory of Reflectance and Emittance Spectroscopy*, Cambridge University Press, Cambridge.

Harrell, P., L. L. Bourgeau-Chavez, E. S. Kasischke, N. H. F. French, and N. L. Christensen, 1995. Sensitivity of *ERS-1* and *JERS-1* radar data to biomass and stand structure in Alaskan boreal forests, *Remote Sensing Environ.*, 54, 247–260.

Harsanyi, J. C., 1993. Detection and classification of subpixel spectral signatures in hyperspectral image sequences, Ph.D. dissertation, University of Maryland, College Park, Md., 116 pp.

Harsanyi, J. C., and C. I. Chang, 1994. Hyperspectral image classification and dimensionality reduction: an orthogonal subspace projection approach, *IEEE Trans. Geosci. Remote Sensing*, 32, 779–785.

Hoge, F. E., R. N. Swift, and J. K. Jungel, 1983. Feasibility of airborne detection of laser-induced fluorescence emission from green terrestrial plants, *Appl. Opt.*, 22, 2991–3000.

Hoque, E., and P. J. S. Hutzler, 1992. Spectral blue-shift of the red edge monitors damage class of beech trees, *Remote Sensing Environ.*, 39(1), 81–84.

Hoque, E., and G. Remus, 1994. Native and atrazine-induced fluorescence of chloroplasts from palisade and spongy parenchyma of beech (*Fagus sylvatica* L.) leaves, *Remote Sensing Environ.*, 47, 77–86.

Horler, D. N. H., J. Barber, and A. R. Barringer, 1980. Effects of heavy metals on the absorbance and reflectance spectra of plants, *Int. J. Remote Sensing*, 1, 121–136.

Horler, D. N. H., J. Barber, and A. R. Barringer, 1981. New concepts for the detection of geochemical stress in plants, in *Proceedings of the 8th Annual Conference of the Remote Sensing Society*, Dec. 16–18, 1980, The College of St. Mark and St. John Foundation, Reading, England, pp. 113–123.

Horler, D. N. H., M. Dockray, and J. Barber, 1983. The red edge of plant leaf reflectance, *Int. J. Remote Sensing*, 4, 273–288.

Hosgood, B., S. Jacquemoud, G. Andreoli, J. Verdebout, G. Pedrini, and G. Schmuck, 1995. *Leaf Optical Properties Experiment 93 (LOPEX93)*, Rep. EUR-16095-EN, European Commission, Joint Research Centre, Institute for Remote Sensing Applications, Ispra, Italy.

Huete, A. R., 1986. Separation of soil–plant spectral mixtures by factor analysis, *Remote Sensing Environ.*, 19, 237–251.

Huete, A. R., and R. D. Jackson, 1987. Evaluating evapotranspiration at local and regional scales, *Remote Sensing Environ.*, 23, 213–232.

Huete, A., C. Justice, and H. Liu, 1994. Development of vegetation and soil indices for MODIS-EOS, *Remote Sensing Environ.*, 49, 224–234.

Humberto, R., H. R. Darocha, C. A. Noble, J. P. Bonatti, I. R. Wright, and P. J. Sellers, 1996. A vegetation–atmosphere interaction study for Amazonia deforestation using field data and a single column model, *Q. J. R. Meteorol. Soc.*, 122, 567–594.

Jackson, R. D., 1982a. Canopy temperature and crop water stress, *Adv. Irrig.*, 1, 43–85.

Jackson, R. D., 1982b. Soil moisture inferences from thermal-infrared measurements of vegetation temperatures, *IEEE Trans. Geosci. Remote Sensing*, 20, 282–286.

Jackson, R. D. and A. R. Huete, 1991. Interpreting vegetation indices, *Prev. Vet. Med.*, 11, 285–200.

Jackson, R. D., R. J. Reginato, and S. B. Idso, 1977. Wheat canopy temperature: a practical tool for evaluation of water requirements, *Water Resources Res.*, 13, 651–656.

Jackson, R. D., R. J. Reginato, P. J. Pinter, and S. B. Idso, 1979. Plant canopy information extraction from composite scene reflectance of row crops, *Appl. Opt.*, 18, 3775–3782.

Jacquemoud, S., and F. Baret, 1990. PROSPECT: a model of leaf optical properties spectra, *Remote Sensing Environ.*, 34, 75–91.

Jacquemoud, S., S. L. Ustin, J. Verdebout, G. Schmuck, G. Andreoli, and B. Hosgood, 1996. Estimating leaf biochemistry using the PROSPECT leaf optical properties model, *Remote Sensing Environ.*, 56, 194–202.

Jupp, D. L. B., and A. H. Strahler, 1991. A hot spot model for leaf canopies, *Remote Sensing Environ.*, 38, 193–210.

Kasischke, E. S., forthcoming. Patterns of carbon-based greenhouse gas emissions during biomass burning in an Alaskan boreal forests, *J. Atmos. Chem.*

Kasischke, E. S., and N. F. H. French, 1995. Locating and estimating the areal extent of wildfires in Alaskan boreal forests using multiple season AVHRR NDVI composite data, *Remote Sensing Environ.*, 51, 263–275.

Kasischke, E. S., N. F. H. French, P. Harrell, N. L. Christensen, S. L. Ustin, and D. Barry, 1993. Monitoring of wildfires in boreal forests using large area AVHRR NDVI composite image data, *Remote Sensing Environ.*, 45, 61–71.

Kaufman, Y. J., and D. Tanre, 1992. Atmospherically resistant vegetation index (ARVI) for EOS-MODIS, *IEEE Trans. Geosci. Remote Sensing*, 30, 261–270.

Keeling, C. D., J. F. S. Chin, and T. P. Whorf, 1996. Increased activity of northern vegetation inferred from atmospheric CO_2 measurements, *Nature*, 382, 146–451.

Kerr, Y. H., J. Imbernon, G. Dedieu, O. Hautecoeur, J. P. Lagouarde, and B. Seguin, 1989. NOAA-AVHRR and its use for rainfall and evapotranspiration monitoring, *Int. J. Remote Sensing*, 10, 847–854.

Kharuk, V. I., V. N. Morgun, B. N. Rock, and D. L. Williams, 1994. Chlorophyll fluorescence and delayed fluorescence as potential tools in remote sensing: a reflection of some aspects of problems in comparative analysis, *Remote Sensing Environ.*, 47, 98–105.

Kimes, D. S., and J. A. Kirchner, 1982. Radiative transfer model for heterogeneous 3D scenes, *Appl. Opt.*, 21, 4119–4129.

Kimes, D. S., J. A. Smith, P. A. Harrison, and P. R. Harrison, 1994. Application of techniques to infer vegetation characteristics from directional reflectance(s), in *Proceedings of the 6th ISPRS International Symposium on Physical Measurements and Signatures in Remote Sensing*, Val d'Isere, France, Jan. 17–21, pp. 581–592.

Knipling, E. B., 1970. Physical and physiological basis for the reflectance of visible and near-infrared radiation from vegetation, *Remote Sensing Environ.*, 1, 155–159.

Knyazikhin, Y. V., and A. L. Marshak, 1991. Fundamental equations of radiative transfer in leaf canopies, and iterative methods for their solution, in *Photon–Vegetation Interactions*, R. Myneni and J. Ross, eds., Springer-Verlag, New York, pp. 9–43.

Knyazikhin, Y. V., A. L. Marshak, and R. B. Myneni, 1992. Interaction of photons in a canopy of finite-dimensional leaves, *Remote Sensing Environ.*, 39, 61–74.

Krajicek, V., and M. Vrbova, 1994. Laser-induced fluorescence spectra of plants, *Remote Sensing Environ.*, 47, 51–54.

Kriebel, K. T., 1978. Measured spectral bidirectional reflection properties of four vegetated surfaces, *Appl. Opt.*, 17, 253–259.

Kruckleberg, A. R., 1969. Soil diversity and the distribution of plants with examples from western North America, *Madroño*, 20, 129–154.

Kumar, R., and L. Silva, 1973. Light ray tracing through a leaf cross section, *Appl. Opt.*, 12, 2950–2954.

Kuusk, A., 1991. The hot spot effect in plant canopy reflectance, in *Photon–Vegetation Interactions*, R. Myneni, and J. Ross, eds., Springer-Verlag, New York, pp. 139–159.

Lambin, E. F., and D. Ehrlich, 1995. Combining vegetation indices and surface temperature for land-cover mapping at broad spatial scales, *Int. J. Remote Sensing*, 16, 573–579.

Lambin, E. F., and A. H. Strahler, 1994. Change-vector analysis in multitemporal space: a tool to detect and categorize land-cover change processes using high temporal-resolution satellite data, *Remote Sensing Environ.*, 48, 231–244.

Lang, M. K., G. W. Donohoe, S. H. Zaidi, and S. R. J. Brueck, 1994. Real-time image processing techniques for noncontact temperature measurement, *Opt. Eng.*, 33(10), 3465–3471.

Leprieur, C., M. M. Verstraete, and B. Pinty, 1994. Evaluation of the performance of various vegetation indices to retrieve vegetation cover from AVHRR data, *Remote Sensing Rev.*, 10, 265–284.

Li, X., and A. H. Strahler, 1985. Geometric-optical modeling of a conifer forest canopy, *IEEE Trans. Geosci. Remote Sensing*, 23, 705–721.

Li, X., and A. H. Strahler, 1986. Geometric-optical bidirectional reflectance modeling of a conifer forest canopy, *IEEE Trans. Geosci. Remote Sensing*, 24, 906–919.

Li, X., and A. H. Strahler, 1992. Geometric-optical bidirectional reflectance modeling of the discrete crown vegetation canopy: effect of crown shape and mutual shadowing, *IEEE Trans. Geosci. Remote Sensing*, 30, 276–292.

Li, X., A. H. Strahler, and C. E. Woodcock, 1995. A hybrid geometric optical radiative transfer approach for modeling albedo and directional reflectance of discontinuous canopies, *IEEE Trans. Geosci. Remote Sensing*, 33, 466–480.

Liang, S., and A. H. Strahler, 1993. Calculation of the angular radiance distribution for a coupled atmosphere and canopy, *IEEE Trans. Geosci. Remote Sensing*, 31, 491–502.

Liang, S., and A. H. Strahler, 1995. An analytic radiative transfer model for a coupled atmosphere and leaf canopy, *J. Geophys. Res.*, 100, 5085–5094.

Lichtenthaler, H. K., 1987. Chlorophylls and carotenoids: pigments of photosynthetic biomembranes, *Methods Enzymol.*, 148, 350–382.

Lichtenthaler, H. K., 1988. *Applications of Chlorophyll Fluorescence in Photosynthesis Research, Stress Physiology, Hydrobiology, and Remote Sensing*, Kluwer Academic, Dordrecht, The Netherlands, 366 pp.

Lichtenthaler, H. K., and U. Rinderele, 1988. The role of chlorophyll fluorescence in the detection of stress conditions in plants, *CRC Crit. Rev. Anal. Chem.*, 19, S29–S85.

Los, S. O., C. O. Justice, and C. J. Tucker, 1994. A global 1 degree by 1 degree NDVI data set for climate studies derived from the GIMMS continental NDVI data. *Int. J. Remote Sens.* 15, 3493–3518.

Lumme, K., and E. Bowell, 1981. Radiative transfer in the surfaces of atmosphereless bodies: 1. Theory, *Astron. J.*, 86, 1694–1704.

Lyon, R. J. P., 1975a. Correlation between ground metal analysis, vegetation reflectance and ERTS brightness over a molybdenum skarn deposit, Pine Nut Mountains, western Nevada, *Proceedings of the 10th Remote Sensing of Environment Symposium*, Ann Arbor, Mich., pp. 1031–1044.

Lyon, R. J. P., 1975b. Mineral exploration applications of digitally processed *Landsat* imagery, SRSL Tech. Rep., 72–10, Stanford Remote Sensing Laboratory, Stanford University, Stanford, Calif., pp. 271–292.

Ma, Q., A., Ishimaru, P., Phu, and Y. Kuga, 1990. Transmission, reflection, and depolarization of an optical wave for a single leaf, *IEEE Trans. Geosci. Remote Sensing*, 28, 865–872.

Marie-Victorin, Fr., 1964. *Flore Laurentienne*, Presse Universitaire, Montréal, Québec, Canada, 925 pp.

Marshak, A. L., 1989. The effect of the hot spot on the transport equation in plant canopies, *J. Quant. Spectrosc. Radiat. Transfer*, 42, 615–630.

Marten, G. C., J. S. Shenk, and F. E. Barton II, ed., 1985. *Near Infrared Reflectance Spectroscopy (NIRS): Analysis of Forage Quality*, Agric. Handb. 643, U.S. Department of Agriculture, Washington, D.C., 95 pp.

Martin, G., S. A. Josserand, J. F. Bornman, and J. Volgemann, 1989. Epidermal focusing and the light microenvironment within leaves of *Medicago sativa, Physiol. Plant.*, 76, 485–492.

Martinez v. Remisowsky, A., J. H. McClendon, and L. Fukshansky, 1992. Estimation of the optical parameters and light gradients in leaves: multi-flux versus two-flux treatment, *Photochem. Photobiol.*, 55, 857–865.

McGuire, A. D., J. M. Melilo, L. A. Joyce, D. W. Kicklighter, A. L. Grace, B. Moore III, and C. J. Vorosmarty, 1992. Interactions between carbon and nitrogen dynamics in estimating net primary productivity for potential vegetation in North America, *Global Biogeochem. Cycles*, 6, 101–124.

Milton, N. M., 1978. *Spectral Reflectance Measurements of Plants in the East Tintic Mountains, Utah*, USGS Open File Rep. 78–448, U.S. Geological Survey, Washington, D.C., 121 pp.

Milton, N. M., 1983. Use of reflectance spectra of native plant species for interpreting airborne multispectral scanner data in the East Tintic Mountains, Utah, *Econ. Geol.*, 78, 761–769.

Milton, N. M., W. Collins, S.-H. Chang, and R. G. Schmidt, 1983. Remote detection of metal anomalies on Pilot Mountain, Randolph County, North Carolina, *Econ. Geol.*, 78, 605–617.

Milton, N. M., B. A. Eiswerth, and C. M. Ager, 1991. Effect of phosphorus deficiency on spectral reflectance and morphology of soybean plants, *Remote Sensing Environ.*, 36, 121–127.

Minnaert, M., 1941. The reciprocity principle in lunar photometry, *Astrophys. J.*, 93, 403–410.

Monsi, M., and T. Saeki, 1953. Uber den lichtfaktor in den pflanzengesellschaften und seine bedeutung fur die stoffproduktion, *Jpn. J. Bot.*, 14, 22–52.

Mooney, H. A., and P. W. Rundel, 1979. Nutrient relations of the evergreen shrub, *Adenostoma fasciculatum*, in the California chaparral, *Bot. Gaz.*, 140, 109–113.

Myneni, R. B., and G. Asrar, 1993. Radiative transfer in 3-dimensional atmosphere vegetation media, *J. Quant. Spectrosc. Radiat. Transfer*, 49, 585–598.

Myneni, R., and J. Ross, 1991. *Photon–Vegetation Interactions*, Springer-Verlag, New York, 565 pp.

Myneni, R., J., Ross, and G. Asrar, 1989. A review on the theory of photon transport in leaf canopies, *J. Agric. For. Meteorol.*, 45, 1–153.

Myneni, R. B., F. G. Hall, P. J. Sellers, and A. L. Marshak, 1995. The interpretation of spectral vegetation indices, *IEEE Trans. Geosci. Remote Sensing*, 33, 481–486.

Nilson, T., and A. Kuusk, 1989. A reflectance model for the homogeneous plant canopy and its inversion, *Remote Sensing Environ.*, 27, 157–167.

Nobre, C. A., P. J. Sellers, and J. Shukla, 1991. Amazonian deforestation and regional climate change, *J. Climate*, 4, 957–988.

North, P. R. J., 1996. Three-dimensional forest light interaction model using a Monte Carlo method, *IEEE Trans. Geosci. Remote Sensing*, 34, 946–956.

Otterman, J., 1983. Absorption of insolation by land surfaces with sparse vertical protrusions, *Tellus*, 35B, 309–318.

Ovington, J. D., 1965. Organic production, turnover and mineral cycling in woodlands, *Biol. Rev.*, 40, 295–336.

Parkhurst, D. F., 1986. Internal leaf structure: a three-dimensional perspective, in *On the Economy of Plant Form and Function*, T. J. Givnish, ed., Cambridge University Press, Cambridge, pp. 215–249.

Parton, W. J., D. S., Schimel, C. V. Cole, and D. S. Ojima, 1987. Analysis of factors controlling soil organic matter levels in Great Plain grasslands, *Soil Sci. Soc. Am. J.*, 51, 1173–1179.

Perruchoud, D. O., and A. Fischlin, 1995. The response of the carbon cycle in undisturbed forest ecosystems to climate change: a review of plant–soil models, *J. Biogeogr.*, 22, 759–774.

Peterson, D. L., and G. S. Hubbard, 1992. Scientific issues and potential remote-sensing requirements for plant biochemical content, *J. Imaging Sci. Technol.*, 36, 446–456.

Pinar, A., and P. J. Curran, 1996. Grass chlorophyll and red edge, *Int. J. Remote Sensing*, 17, 351–357.

Pinty, B., and M. M. Verstraete, 1992. On the design and validation of bidirectional reflectance and albedo models, *Remote Sensing Environ.*, 41, 155–167.

Pinty, B., M. M. Verstraete, and R. E. Dickinson, 1989. A physical model for predicting bidirectional reflectances over bare soil, *Remote Sensing Environ.*, 27, 273–288.

Pinty, B., M. M. Verstraete, and R. E. Dickinson, 1990. A physical model of the bidirectional reflectance of vegetation canopies: 2. Inversion and validation, *J. Geophys. Res.*, 95, 11767–11775.

Pinty, B., C. Leprieur, and M. M. Verstraete, 1993. Biophysical canopy properties and classical indices, *Remote Sensing Rev.*, 7, 127–150.

Pinzon, J. E., S. L. Ustin, C. M. Castaneda, and M. O. Smith, 1998. Investigation of leaf biochemistry by hierarchical foreground/background analysis, *IEEE Trans. Geosci. Remote Sensing*, 36, 1–15.

Potter, C. S., J. T. Randerson, C. B. Field, P. A. Matson, P. M. Vitousek, H. A. Mooney, and S. A. Klooster, 1993. Terrestrial ecosystem production: a process model based on global satellite and surface data, *Global Biogeochem. Cycles*, 7, 811–841.

Prentice, I. C., W. Cramer, S. P. Harrison, R. Leemans, R. A. Monserud, and A. M. Solomon, 1992. A global biome model based on plant physiology and dominance, soil properties and climate, *J. Biogeogr.*, 19, 117–134.

Privette, J. L., R. B. Myneni, W. J. Emery, and B. Pinty, 1995. Inversion of a soil bidirectional reflectance model for use with vegetation reflectance models, *J. Geophys. Res.*, 100, 25497–25508.

Qi, J., F. Cabot, M. S. Moran, and G. Dedieu, 1995a. Biophysical parameter estimations using multidirectional spectral measurements, *Remote Sensing Environ.*, 54, 71–83.

Qi, J., M. S. Moran, F. Cabot, and G. Dedieu, 1995b. Normalization of sun/view angle effects using spectral albedo-based vegetation indices, *Remote Sensing Environ.*, 52, 207–217.

Rahman, H., M. M. Verstraete, and B. Pinty, 1993a. Coupled surface–atmosphere reflectance (CSAR) model: 1. Model description and inversion on synthetic data, *J. Geophys. Res.*, 98, 20779–20789.

Rahman, H., B. Pinty, and M. M. Verstraete, 1993b. Coupled surface–atmosphere reflectance (CSAR) model: 2. Semiempirical surface model usable with NOAA Advanced Very High Resolution Radiometer data, *J. Geophys. Res.*, 98, 20791–20801.

Railyan, V. Ya., and R. M. Korobov, 1993. Red edge structure of canopy reflectance spectra of Triticale, *Remote Sensing Environ.*, 46, 173–182.

Raines, G. L., and F. C. Canney, 1980. Vegetation and geology, in *Remote Sensing in Geology*, B. S. Siegal, and A. R. Gillespie, eds., Wiley, New York, pp. 365–380.

Raines, G. L., T. W. Offield, and E. S. Santos, 1978. Remote-sensing and subsurface definition of facies and structure related to uranium deposits, Powder River Basin, Wyoming, *Econ. Geol.*, 73, 1706–1723.

Rencz, A. N., and W. W. Shilts, 1980. Nickel in soils and vegetation of glaciated terrains, in *Nickel in the Environment*, J. O. Nriagu, ed., Wiley, Toronto, Ontario, Canada, pp. 151–185.

Richardson, A. J., E. C. Wiegand, H. Gausman, J. Cuellar, and A. Gerberman, 1975. Plant, soil, and shadow reflectance components of row crops, *Photogramm. Eng.*, 41, 1401–1407.

Richter, T., and L. Fukshansky, 1996. Optics of a bifacial leaf: 1. A novel combined procedure for deriving the optical parameters, *Photochem. Photobiol.*, 63, 507–516.

Roberts, D. A., M. Gardner, R. Church, S. Ustin, G. Scheer, and R. O. Green, 1996. Mapping chaparral in the Santa Monica mountains using multiple endmember spectral mixture models, in *Summaries of the 6th Annual JPL Airborne Earth Science Workshop*, Mar. 4–8, Vol. 1, AVIRIS Workshop, Pasadena, Calif., pp. 197–202.

Roberts, D. A., M. Gardner, R. Church, S. L. Ustin, G. Scheer, and R. O. Green, 1998. Mapping chaparral in the Santa Monica Mountains using multiple endmember spectral mixture models, *Remote Sensing Environ.*, 65, 267–279.

Rock, B. N., T. Hoshizaki, and J. R. Miller, 1988. Comparison of in situ and airborne spectral measurements of the blue shift associated with forest decline, *Remote Sensing Environ.*, 24, 109–127.

Rock, B. N., D. L. Williams, D. M. Moss, G. N. Lauten, and M. Kim, 1994. High-spectral resolution field and laboratory optical reflectance measurements of red spruce and eastern hemlock needles and branches, *Remote Sensing of Environ.*, 47, 176–189.

Rodin, L. E., and N. I. Bazilevich, 1967. *Production and Mineral Cycling in Terrestrial Vegetation*, G. E. Fogg, translator Oliver & Boyd, London.

Rondeaux, G., and V. C. Vanderbilt, 1993. Specularly modified vegetation indices to estimate photosynthetic activity, *Int. J. Remote Sensing*, 14, 1815–1823.

Rose, A. W., H. E. Hawkes, and J. S. Webb, 1979. *Geochemistry in Mineral Exploration*, Academic Press, San Diego, Calif., 657 pp.

Ross, I. V., 1981. *The Radiation Regime and Architecture of Plant Stands*, Junk, Boston. pp. 391.

Ross, J., and A. L. Marshak, 1988. Calculation of the canopy bidirectional reflectance using the Monte-Carlo method, *Remote Sensing Environ.*, 24, 213–225.

Ross, J., and A. L. Marshak, 1989. The influence of leaf orientation and the specular component of leaf reflectance on the canopy bidirectional reflectance, *Remote Sensing Environ.*, 27, 251–260.

Roujean, J.-L., M. Leroy, and P.-Y. Deschamps, 1992. A bidirectional reflectance model of the Earth's surface for the correction of remote sensing data, *J. Geophys. Res.*, 97, 20455–20468.

Ruimy, A., L. Kergoat, C. B. Field, and B. Saugier, 1996. The use of CO_2 flux measurements in models of the global terrestrial carbon budget, *Global Change Biol.*, 2, 287–296.

Running, S. W., and J. C. Coughlan, 1988. A general model of forest ecosystem processes for regional applications: I. Hydrologic balance, canopy gas exchange and primary projection processes, *Ecol. Model.*, 42, 125–154.

Sabol, D. E., J. B. Adams, and M. O. Smith, 1992. Quantitative sub-pixel spectral detection of targets in multispectral images, *J. Geophys. Res.*, 97, 2659–2672.

Sanderson, E. W., M. Zhang, S. L. Ustin, and E. Rejmankova, 1998. Geostatistical scaling of canopy water content in a California salt, *Landscape Ecol.*, 13, 79–92.

Schimper, A. F. W., 1898. *Plant-Geography upon a Physiological Basis*, P. Groom and I. B. Balfour, eds., Oxford University Press, London, 1903, 839 pp.

Schlesinger, W. H., 1991. *Biogeochemistry*, Academic Press, San Diego, Calif., 443 pp.

Schmuck, G., and I. Moya, 1994. Time-resolved chlorophyll fluorescence spectra of intact leaves, *Remote Sensing Environ.*, 47, 72–76.

Schwaller, M. R., and S. J. Tkach, 1980. Premature leaf senescence as an indicator for geobotanical prospecting with remote sensing techniques, in *Proceedings of the 14th International Symposium on Remote Sensing of Environment*, pp. 345–357.

Seguin, B., and B. Itier, 1983. Using midday surface temperature to estimate daily evaporation from satellite thermal IR data, *Int. J. Remote Sensing*, 4, 371–383.

Seguin, B., E. Assad, J. P. Freteaud, J. Imbernon, Y. H. Kerr, and J. P. Lagouarde, 1989. Use of meteorological satellite for water balance monitoring in Sahelian regions, *Int. J. Remote Sensing*, 10, 1001–1017.

Sellers, P. J., 1985. Canopy reflectance, photosynthesis and transpiration, *Int. J. Remote Sensing*, 6, 1335–1372.

Sellers, P. J., M. D. Heiser, F. G. Hall, S. J. Goetz, and D. E. Strebel, S. B. Verma,

R. L. Desjardins, P. M. Schuepp, and J. I. MacPherson, 1995. Effects of spatial variability in topography, vegetation cover and soil moisture on area-averaged surface fluxes: a case study using the FIFE 1989 data, *J. Geophys. Res. Atmos.*, 100, 25607–25629.

Sellers, P. J., D. A. Randall, G. J. Collatz, J. A. Berry, C. B. Field, D. A. Dazlich, C. Zhang, G. D. Collelo, and L. Bounoua, 1996. A revised land surface parameterization (SIB2) for atmospheric GCMS: I. Model formulation, *J. Climate*, 9, 676–705.

Shilts, W. W., 1973. Glacial dispersal of rocks, minerals and trace elements in Wisconsonian till, southeastern Quebec, Canada, *Geol. Soc. Am. Mem.*, 316, 189–219.

Shilts, W. W., 1975. Principles of geochemical exploration for sulfide deposits using shallow samples of glacial drift, *Can. Inst. Min. Metall. Bull.*, 68, 33–80.

Shilts, W. W., 1976. Mineral exploration and till, in *Glacial Till*, R. F. Legget, ed., Spec. Publ. 12, Royal Society of Canada, Ottawa, Ontario, Canada, pp. 205–224.

Shimbukuro, Y. E., B. N. Holben, and C. J. Tucker, 1994. Fraction images derived from NOAA AVHRR data for studying the deforestation in the Brazilian Amazon, *Int. J. Remote Sensing*, 15, 517–520.

Shultis, J. K., and R. B. Myneni, 1988. Radiative transfer in vegetation canopies with anisotropic scattering, *J. Quant. Spectrosc. Radiat. Transfer*, 39, 115–129.

Shutt, J. B., R. R. Rowland, and W. H. Hearlty, 1984. A laboratory investigation of a physical mechanism for the extended infrared absorption ("red shift") in wheat, *Int. J. Remote Sensing*, 5, 92–102.

Sippel, S. J., S. K. Hamilton, J. M. Melack, and B. J. Choudhury, 1994. Determination of inundation area in the Amazon river floodplain using the SMMR 37 GHz polarization difference, *Remote Sensing Environ.*, 48, 70–76.

Smith, M. O., S. L. Ustin, J. B. Adams, and A. R. Gillespie, 1990a. Vegetation in deserts: I. A regional measure of abundance from multispectral images, *Remote Sensing Environ.*, 31, 1–26.

Smith, M. O., S. L. Ustin, J. B. Adams, and A. R. Gillespie, 1990b. Vegetation in deserts: II. Environmental influences on regional abundance, *Remote Sensing Environ.*, 29, 27–52.

Smith, M. O., D. A. Roberts, J. Hill, W. Mehl, B. Hosgood, J. Verdebout, G. Schmuck, C. Koechler, and J. B. Adams, 1994. A new approach to determining spectral abundances of mixtures in multispectral images, *IEEE Trans. Geosci. Remote Sensing, IGARSS '94*, California Institute of Technology, Pasadena, Calif. (CD).

Smith, M., R. Weeks, J. Richey, E. Mayorga and V. Ballester 1998a. Mesoscale changes in the Amazon Basin inferred from a 9-year time series of AVHRR-GAC images. I. Surface Change. Remote Sens. Environ. (submitted).

Smith M, R. Weeks, J. Richey, E. Mayorga, and V. Ballester, 1998b. Mesoscale changes in the Amazon basin inferred from a 9-year time series of AVHRR-GAC images. II. Energy Balance Dynamics. Remote Sens. Environ. (Submitted).

Solomonson, V. V., P. L. Smith, Jr., A. B. Park, W. C. Webb, and T. J. Lynch, 1980. An overview of progress in the design and implementation of *Landsat*-D systems, *IEEE Trans. Geosci. Remote Sensing*, 18, 137–146.

Stocks, B. J., B. S. Lee, and D. L. Martell, 1996. Some potential carbon budget implications of fire management in the boreal forest, in *Forest Management and the*

Global Carbon Cycle, M. J. Apps, and D. T. Price, eds., Springer-Verlag, Berlin, pp. 89–96.

Stokes, G. G., 1862. On the intensity of the light reflected from or transmitted through a pile of plates, *Proc. R. Soc. London*, 11, 545–556.

Strahler, A. H., 1994. Vegetation canopy reflectance modeling: recent developments and remote sensing perspectives, in *Proceedings of the 6th ISPRS International Symposium on Physical Measurements and Signatures in Remote Sensing*, CNES, Jan. 17–21, Val d'Isere, France, pp. 593–600.

Strahler, A. H., T. L. Logan, and N. A. Bryant, 1978. Improving forest cover classification accuracy from *Landsat* by incorporating topographic information, in *Proceedings of the 12th International Symposium on Remote Sensing of Environment*, Vol. 2, Environmental Research Institute of Michigan, Ann Arbor, Mich., pp. 927–942.

Suits, G. H., 1972. The calculation of the directional reflectance of a vegetative canopy, *Remote Sensing Environ.*, 2, 117–125.

Suits, G. H., 1983. The extension of a uniform canopy reflectance model to include row effects, *Remote Sensing Environ.*, 13, 113–129.

Talvitie, J., 1979. Remote sensing and geobotanical prospecting in Finland, *Bull. Geol. Soc. Finland*, 51, 63–73.

Tanre, D., M. Herman, and P. Y. Deschamps, 1983. Influence of the atmosphere on space measurements of directional properties, *Appl. Opt.*, 22, 733–741.

Townshend, J. R. G., 1992. *Improved Global Data for Land Applications*, Tech. Rep., IGBP, Stockholm.

Tucker, C. J., 1977. Red and photographic infrared linear combinations for monitoring vegetation, *Remote Sensing Environ.*, 8, 127–150.

Tucker, C. J., and M. W. Garatt, 1977. Leaf optical properties as a stochastic process, *Appl. Opt.*, 16, 635–642.

Tucker, C. J., I. Y. Fung, D. C. Kealing, and R. H. Gammon, 1986. Relationship between atmospheric CO_2 variations and a satellite derived vegetation index, *Nature*, 319, 195–199.

Turner, B. L., II, D. Skole, S. Sanderson, G. Fischer, L. Fresco, and R. Leemans, 1995. *Land-Use and Land-Cover Change: Science/Research Plan*, Tech. Rep., IGBP, Stockholm.

Ustin, S. L., B. Curtiss, S. N. Martens, and V. C. Vanderbilt, 1989. Early detection of air pollution injury to coniferous forests using remote sensing, in *Transactions: Effects of Air Pollution on Western Forests*, R. K. Olson and A. S. Lefohn, eds., Air and Waste Management Association, Pittsburgh, Pa., pp. 351–378.

Ustin, S. L., M. O. Smith, and J. B. Adams, 1993. Remote sensing of ecological processes: a strategy for developing ecological models using spectral mixture analysis, in *Scaling Physiological Processes: Leaf to Globe*, J. Ehlringer, and C. Field, eds., Academic Press, San Diego, Calif., pp. 339–357.

Ustin, S. L., Q. J. Hart, L. Duan, and G. Scheer, 1996. Vegetation mapping on hardwood rangelands in California, *Int. J. Remote Sensing*, 17, 3015–3036.

Ustin, S. L., D. Roberts, S. Jacquemoud, J. Pinzon, G. Scheer, M. Gardner, C. M. Castaneda, and A. Palacios 1998. Estimating canopy water content of chaparral shrubs using optical methods, *Remote Sensing Environ.*, 65, 280–291.

Valentini, R., G. Cecchi, P. Mazzinghi, Scarascia G. Mugnozza, G. Agati, M. Bazzani, P. De Angelis, F. Fusi, G. Matteucci, and V. Raidmondi, 1994. Remote sensing of chlorophyll *a* fluorescence of vegetation canopies: 2. Physiological significance of fluorescence signal in response to environmental stresses, *Remote Sensing Environ.*, 47, 29–35.

Vanderbilt, V. C., L. Grant, L. L. Beihl, and B. F. Robinson, 1985. Specular, diffuse, and polarized light scattered by two wheat canopies, *Appl. Opt.*, 24, 2408–2418.

Verhoef, W., 1984. Light scattering by leaf layers with application to canopy reflectance modeling: the SAIL model, *Remote Sensing Environ.*, 16, 125–141.

Verma, S. B., P. J. Sellers, C. L. Walthall, F. G. Hall, J. Kim and S. J. Goetz, 1993. Photosynthesis and stomatal conductance related to reflectance on the canopy scale, *Remote Sensing Environ.*, 44, 103–116.

Vermote, E., D. Tanre, J. L. Deuze, M. Herman, and M. M. Morcrette, 1995. *Second Simulation of the Satellite Signal in the Solar Spectrum (6s)*, Tech. Rep., NASA Goddard Space Flight Center, Greenbelt, Md.

Verstraete, M. M., and B. Pinty, 1996. Designing optimal spectral indexes for remote sensing applications, *IEEE Trans. Geosci. Remote Sensing*, 34, 1254–1265.

Verstraete, M. M., B. Pinty, and R. E. Dickinson, 1990. A physical model of the bidirectional reflectance of vegetation canopies: 1. Theory, *J. Geophys. Res.*, 95, 11765–11775.

Verstraete, M. M., B. Pinty, and R. B. Myneni, 1996. Potential and limitations of information extraction on the terrestrial biosphere from satellite remote sensing, *Remote Sensing Environ.*, 58, 201–214.

Vogelmann, T. C., and G. Martin, 1993. The functional significance of palisade tissue: penetration of directional versus diffuse light, *Plant Cell Environ.*, 16, 65–72.

Walter, H., 1973. *Vegetation of the Earth in Relation to Climate and Eco-physiological Conditions*, J. Wieser, translator, Springer-Verlag, New York, 237 pp.

Walker, R. B., H. M. Walker, and P. R. Ashworth, 1955. Calcium–magnesium nutrition with special reference to serpentine soils, *Plant Physiol.*, 30, 214–221.

Walthall, C. L., J. M. Norman, J. M. Welles, G. Campbell, and B. Blad, 1985. Simple equation to approximate the bidirectional reflectance for vegetative canopies and bare soil surfaces, *Appl. Opt.*, 24, 383–387.

Wanner, W., X. Li, and A. H. Strahler, 1995. On the derivation of kernels for kernel-driven models of bidirectional reflectance, *J. Geophys. Res.*, 100, 21077–21089.

Warming, E., 1892. *Oecology of Plants*, P. Groom, and I. B. Balfour, translators, Oxford University Press, London, 1909, 422 pp.

Warner, T. A., C. S. Evans, and J. R. Heirtzler, 1989. Remote sensing of geobotanical trends in East Africa, in *Proceedings of IGARSS '89/12, Canadian Symposium on Remote Sensing: An Economic Tool for the Nineties*, July 10–14, Vancouver, British Columbia, Canada, Vol. 3, pp. 1331–1334.

Warner, T. A., D. J. Campagna, C. S. Evans, D. W. Levandowski, and H. Cetin, 1991. Analyzing remote sensing geobotanical trends in Quetico Provincial Park, Ontario, Canada, using digital elevation data, *Photogramm. Eng. Remote Sensing*, 57, 1179–1183.

Warner, T. A., D. W. Levandowski, R. Bell, and H. Cetin, 1994. Rule-based geobotanical classification of topographic, aeromagnetic and remotely sensed vegetation community data, *Remote Sensing Environ.*, 50, 4151.

Weigand, C. L., and A. J. Richardson, 1987. Spectral components analysis: rationale, and results for three crops, *Int. J. Remote Sensing*, 8, 1011–1032.

Wessman, C. A., 1990. Evaluation of canopy chemistry, in *Remote Sensing of Biospheric Functioning*, R. J. Hobbs, and H. A. Mooney, eds., Springer-Verlag, New York, pp. 135–156.

West Virginia Geological Survey, 1959. *A Symposium on the Sandhill Deep Well, Wood County, West Virginia*, Rep. Invest. 18, West Virginia Geological Society, Morgantown, W.Va., 182 pp.

Weyer, L. G., 1985. Near-infrared spectroscopy of organic substances, *Appl. Spectrosc. Rev.*, 21, 1–43.

Wooley, J. T., 1971. Reflectance and transmittance of light by leaves, *Plant Physiol.*, 47, 656–662.

Wooley, J. T., 1973. Change of leaf dimensions and air volume with change in water content, *Plant Physiol.*, 41, 815–816.

Wooley, B. M., and M. D. Flanningan, 1993. Length of the fire season in a changing climate, *For. Chron.*, 69, 187–192.

Yamada, N., and S. Fujimura, 1991. Nondestructive measurement of chlorophyll pigment content in plant leaves from three-color reflectance and transmittance, *Appl. Opt.*, 30, 3964–3973.

Zhang, M. H., S. L., Ustin, E., Rejmankova, and E. W. Sanderson, 1997. Monitoring Pacific coast salt marshes using remote sensing, *Ecol. Appl.*, 7(3), 1039–1053.

Data Analysis

Spectral Analysis for Earth Science: Investigations Using Remote Sensing Data

John F. Mustard

Brown University
Providence, Rhode Island

Jessica M. Sunshine

Science Applications International Corporation
Chantilly, Virginia

5.1 INTRODUCTION

Remote sensing in the earth sciences has grown remarkably over the last decade. This growth has been spurred by rapid advances in technology that have opened up new avenues of analysis and inquiry. Of these advances, the merging of spectroscopy and imaging has been the most important. Spectroscopy has been used as a quantitative tool in the laboratory for many years and there exists a wealth of understanding and analysis strategies for such data. Although early imaging spectrometer instruments suffered through the usual development problems, these systems are now approaching the spectral resolution and quality of laboratory measurements, blurring the distinction between the two but also bringing some of the more advanced laboratory spectral analysis techniques to bear on complex earth science problems of the field. The most advanced sensors and instruments are currently mounted on aircraft, but there are exciting plans to integrate the best of these into orbiting platforms which will facilitate greater accessibility and wider geographic coverage.

 Broadly speaking, spectral analysis refers to the extraction of quantitative or qual-

Remote Sensing for the Earth Sciences: Manual of Remote Sensing, 3 ed., Vol. 3, edited by Andrew N. Rencz.
ISBN: 0471-29405-5 © 1999 John Wiley & Sons, Inc.

itative information from reflectance spectra based on the albedo- and wavelength-dependent reflectance properties of materials. The goal of this chapter is to highlight some of the major developments in the use of spectral analysis for earth science applications that have evolved over the last decade, since the preceding edition of the *Manual of Remote Sensing*. Many of these developments have been pushed forward through the efforts of NASA and the Jet Propulsion Laboratory in the development of imaging spectroscopy (e.g., Goetz et al., 1985), although other agencies and companies [notably Geophysical Environmental Research (GER) and the Canadian Centre for Remote Sensing] have also contributed substantially in this area. The preceding *Manual of Remote Sensing* contained many well-documented and excellent examples in the use of broadband data analysis techniques (*broadband* refers to sensing systems that have coarse, low spectral resolution, like the *Landsat* thematic mapper). We do not devote much effort to these areas, except to identify the major approaches and the theoretical and spectroscopic bases for their success. The majority of this chapter is devoted to the use of spectroscopic techniques for material identification and quantification.

The foundation for quantitative analysis of remote sensing data through spectral analysis is reflectance spectroscopy. In general, *reflectance* is defined as the ratio of the intensity of the electromagnetic radiation scattered from a surface to the intensity of the radiation incident upon it. When measured as a function of wavelength, reflectance spectra exhibit specific albedo, continuum, and absorption features which are a function of the material properties of the surface measured. The absorption features are related to the chemical composition and mineralogy of the surface, while the continuum and overall albedo are a function of nonselective absorption and scattering as well as broad wavelength selective absorptions. These spectral properties, which cover broad wavelength regions, define the continuum of a spectrum and are partially controlled by both the physical properties of the surface (particle size, roughness, texture, etc.) and the chemical composition.

The physical principles governing reflectance spectroscopy have been elucidated through many theoretical, laboratory, and observational studies over the last three decades. In Chapter 1, Clark provides an excellent review of these principles with examples that illustrate the reflectance properties of materials with different mineralogy, composition, and texture. The goal of spectral analysis is to exploit this understanding and modify it for the analysis of remotely sensed data in order to achieve specific research and/or applied science objectives. The specific objectives may range from mapping of broad geologic units in support of reconnaissance geologic investigation to the detailed analysis of spectral signatures to map compositional variations related to a process. Despite what may appear to be a radically different approaches and methodologies to data analysis, all such investigations exploit the wavelength-dependent properties of materials as outlined by Clark in Chapter 1.

Presently, there is no single, universally accepted methodology for spectral analysis of remote sensing data. This is due in part to the enormous growth in the use of remote sensing for earth science investigations over the last two decades and a parallel growth in the number of techniques for data analysis. Frequently, the specific methodology used in a given investigation is optimized for the local objectives. The most effective techniques, and those that are more readily transported from one field site or data set to another, are those with solid foundations in the basic principles of reflectance spectroscopy. Within the diversity of techniques, we can generalize three basic categories of spectral analysis: definition and mapping of broad-scale units, identification of the presence of specific mineralogic assemblages or lithologic units, and quantification of the amount of material present. The techniques for spectral analysis progress from simple techniques, suitable for discriminating different units but not necessarily providing compositional determinations, to precise deterministic techniques that require high spectral resolution and fidelity as well as a solid foundation in reflectance spectroscopy.

5.1.1 Definition and Mapping of Broad-Scale Units

A basic requirement of many earth science investigations is the existence of maps of surface compositions. The creation of a map involves simplification of small-scale complexity through the identification of similarities and trends and organization of this information into coherent and physically meaningful units. Units are the fundamental means of recording the properties of the surface and conveying this information to others. From the earliest investigations using aerial photography, the definition of broad-scale units was recognized as one the strengths of remote sensing. With the evolution of technology from black-and-white photographs to digital multispectral images acquired from orbit, these capabilities have been widely recognized and utilized. Units are defined on the basis of shared textural and spectral properties as exhibited in the multispectral data. Although some underlying physical basis can be identified, for the most part the specific compositional properties of the units cannot be determined uniquely. Field investigations for verification and ground truthing can be used to assign field-based names to units defined on the basis of remote sensing analyses, which can then be extended throughout the investigation site or across regions. For the most part, the definition and mapping of broad-scale units can be accomplished with broadband sensors through the use of simple spectral analysis techniques (e.g., color composites, band ratios, principal components analysis).

5.1.2 Identification of Specific Minerals, Assemblages, or Lithologic Units

The identification of specific mineral and/or lithologic compositions requires spectral analysis techniques that are based on the principles of reflectance spectroscopy. When the number of spectral channels is small, such as in multispectral systems, determination of broad mineral and surface compositional classes is possible. For example, with broadband sensors it is possible to identify regions likely to be enriched in ferric minerals and to distinguish these from areas enriched in hydrous minerals or vegetated regions. However, due to the lack of spectral resolution, it is not possible to define the unit composition. When the number of spectral channels of an instrument is sufficiently high (often loosely referred to as high spectral resolution) and spectral coverage spans a wavelength range that includes many individual absorption bands such that the spectral properties are well defined (e.g. absorption band shape, strength, etc.), then a determination of mineral type and composition, mineral assemblage, and lithology is possible. Typically, the methods for detailed characterization require ground-truth information in the form of a spectral library. The spectral library must be representative of the material in the field and may include laboratory and/or in situ spectra of materials in the field.

5.1.3 Quantification of the Amount of Material

For many investigations, identification of materials present in a scene and the quantification of their abundance in a spatial context is critical and important information. Mixtures of materials within the field of view of remote sensing instruments is a natural phenomenon, due both to the physical size of the image pixel and because earth processes tend to create mixed surfaces. Weathering of rocks, for example, creates rinds of altered minerals surrounding fresher rock in the interior. Over time, the rind breaks off and is transported away, exposing more fresh rock. Thus rock outcrops may contain mixtures of altered and unaltered minerals at the surface. The transported material may end up in soils, which contain minerals transported from elsewhere combined with those formed in situ and with organic material. As another example, vegetation is tightly coupled to rainfall in semiarid regions. Where there are strong environmental gradients such as in rain shadows or mountain fronts, continuous changes in vegetation abundance occur with no distinct boundaries separating classes or zones. Such situations have been notoriously difficult to characterize and analyze with classification algorithms but are highly amenable to constituent abundance algorithms.

5.2   INFORMATION CONTENT OF REMOTE SENSING DATA

Effective use of spectral analysis for earth science investigations must begin with basic considerations of the overall objectives of a project and the potential information content of the remotely sensed data. These two categories are intimately coupled and careful early planning is an important step in optimizing the value of the results within available resources. It may be tempting to think that because higher spatial and spectral resolution will increase your ability to resolve objects spatially and discriminate their composition, the best approach is to obtain the highest resolution possible over an entire field site. However, data with high spatial and spectral resolution increases the need for human and computational resources for data processing and analysis. For the highest resolutions available, these costs are currently prohibitive if the analysis includes wide area coverage.

Definitions of spatial, spectral, and radiometric resolutions vary considerably because the usage and definitions have evolved from different technical and applications-oriented communities. It is certainly not the intent of this chapter to consider and explain fully all the possible definitions, but it is important to establish the basic meanings for the terms that have been adopted for this discussion. Spatial resolution in earth science applications typically refers to the ability to identify objects and/or determine patterns and spatial arrangements in imaging data. This notion of spatial resolution is a complex concept to define, as it varies with the instantaneous field of view of the detector (IFOV), the Raleigh criterion, spectral contrast in the scene, and so on (e.g., Simonett et al., 1983). However, for simplicity, we adopt the definition based on the geometric properties of the imaging system. The smallest resolution cell in an image is defined by the IFOV and is typically quoted as the pixel size in meters or the angular resolution of the detectors in milliradians. Spectral resolution is determined by the bandwidth of the channels but is also coupled to the spectral coverage of the system. Narrow bandwidths permit a better characterization of the spectral signatures of materials, as long as the spectral coverage *includes* diagnostic absorptions features of the objects of interest. For example, 10-nm bandwidth channels between 0.4 and 1.1 μm would be effective for characterization and discrimination of ferric minerals, but the same wavelength range would be useless for investigations concerned with clay and/or carbonate minerals which have diagnostic absorptions in the wavelength region 2 to 2.5 μm. Radiometric resolution refers to the maximum number of discrete levels available for encoding reflected or emitted radiation. For a typical 8-bit system this corresponds to 256 levels (e.g., *Landsat*), while a 12-bit systems would have a maximum of 4096 levels. Although it is desirable to acquire data with the greatest number of levels, and thus a finer radiometric resolution, there are practical limits related to instrument fidelity, data acquisition and transmission, and storage.

The practical trade-offs between spectral and spatial resolution can be illustrated with a few examples. Consider the objectives for reconnaissance mapping of a large region (e.g., several hundred kilometers on a side or $\geq 10^5$ km^2) for mineral exploration that include the discrimination of rock outcrops from areas of thick soil cover or obscuring vegetation, and a first-order understanding of the distribution of major lithologic units. These objectives may be characterized as defining broad-scale units and mapping their distributions. If remotely sensed data with 1-m spatial resolution and 100 spectral channels were acquired for this area, on the order of 10^{12} bytes of data would need to be processed. Given the basic objectives, this investment is probably not warranted. On the other hand, a more modest data acquisition plan involving a system with 50-m spatial resolution and 10 spectral channels would result in a more manageable data volume of $\geq 10^8$ bytes, and provided that the 10 spectral channels are well placed for discrimination of the basic units, sufficient spectral dimensionality to achieve the basic objectives. As another scenario, consider the need to map alteration assemblages of clay and ferric minerals in a well-exposed outcrop of a small region (e.g., 10 km^2). This would require high spectral resolution data (≥ 10 nm/spectral channels) to distinguish between the different ferric and clay mineralogies. The small size of the region would permit a high spatial resolution as well.

This trade-off between spectral and spatial resolution affects the design and operation of many current and planned remote sensing systems (see Chapter 11) and is well illustrated by the moderate-resolution imaging radiometer (MODIS) (Salomonson and Toll, 1991; King et al., 1992). The MODIS instrument will be used to measure and monitor biological and physical properties, including surface temperature, ocean color, the condition and dynamics of vegetation and land surface covers, and cloud and aerosol properties. This instrument can provide data in up to 36 spectral bands (20 between 0.4 and 3.0 μm and 16 between 3 and 15 μm), and it is designed to obtain global coverage every 1 to 2 days. Although it will have the capability to obtain a maximum spatial resolution of 250 m per pixel at nadir, the large swath width (2300 km), coupled with numerous bands and a finite data rate, precludes the use of this high-resolution mode for the collection of global data sets.

Instead, the instrument can be programmed to maximize spectral and spatial resolution of the data returned for the specific application required. When a greater spectral resolution is required to identify and map compositionally dependent parameters, data would be acquired with a lower spatial resolution. For applications that require higher spatial resolution, the key physical properties will need to be resolved with fewer bands.

5.3   SIMPLE METHODS OF SPECTRAL ANALYSIS

Spectral analysis need not be complicated in order to extract important information from a remotely sensed data set or to achieve a specific set of objectives. All that is really required is that the materials of interest can be discriminated objectively. A number of enhancement and digital processing techniques have been developed with this goal in mind. When applied with a concrete understanding of the physical causes of absorption (see Chapter 1), simple methods are semiquantitative and can be very effective tools. Most of these methods are used in the analysis of broadband sensors much as the *Landsat* thematic mapper (TM), which lack the spectral resolution and coverage required for the more sophisticated and deterministic approaches (described later in this chapter). Nevertheless, the approaches can also be quite effective for systems with higher spectral resolution.

5.3.1 Color Composites

Color composite images are the most basic form of spectral analysis. The selection of specific spectral bands for display in the red, green, and blue image planes of a computer display can be used for rapid first-order analysis of the spectral properties of a scene even in the absence of calibration (whenever possible, 24-bit displays should be used, as significant information may be lost in compressing the full range of color in a three-band color composite to an 8-bit display). When the image data are stretched to maximize the dynamic range displayed, the relative color variations of the surface materials become evident.

As an example, consider the images shown in Figure 5.1 see color insert). These are two color composites of a *Landsat* TM scene over a portion of the Grand Canyon, Arizona. In the top image, the data are displayed in "natural" color, which approximates how the terrain would appear to the human eye [TM band 1 (0.45 to 0.52 µm) in blue, 2 (0.52 to 0.60 µm) in green, and 3 (63 to 0.69 µm) in red]. The earthy tones are certainly familiar colors of typical rocks in this region. In contrast, the bottom image shows the same region but uses two of the infrared channels with *Landsat* TM band 3 (0.63 to 0.69 µm) in blue, 4 (0.76 to 0.90 µm) in green, and 7 (2.08 to 2.35 µm) in red. The very different view actually can be used for rapid assessment of vegetated versus nonvegetated regions as well as certain rock properties. In this color composite, vegetated areas appear green since the reflectance properties of vegetation are a maximum in TM band 4, a minimum in TM band 3, and typically low in TM band 7 (although this depends on the water content of the vegetation). Soils and rocks with abundance ferric oxides exhibit low reflectances in band 3, with moderate to high reflectances in bands 4 and 7, and regions rich in these minerals exhibit the distinctive yellow colors. The rocks in the lower parts of the canyon have higher amounts of amphiboles, which tend to have lower reflectances in the visible to near-infrared, but exhibit increases in reflectance in band 7 and thus appear reddish. There are also some lithologies in these scenes with high reflectances in the visible but declining toward longer wavelengths and appear in colors of blue and cyan. By selecting various combinations of bands, different spectral properties can be emphasized and/or suppressed, which when combined with a basic knowledge of the spectral properties of materials expected in a scene, can provide an excellent first-order assessment of the diversity of units in a region.

An objective method for determining the optimum bands for display has been developed by Crippen (1989). A correlation matrix for the suite of bands in a given scene is calculated. The correlation coefficients for each pairwise channel combination is then used to calculate a ranking index:

$$\text{index} = \sqrt{1 + 2abc - a^2 - b^2 - c^2} \tag{5.1}$$

where a, b, and c are the pairwise correlation coefficients. The maximum index value of 1.0 indicates the three bands that display the maximum spectral variability and an index of 0.0 indicates the minimum. This approach can also be used with band ratios or spectral indices (described below). Although the index provides an objective method to determine the best three bands for maximum spectral diversity in a color composite display, these may not be the best for enhancing the discrimination of materials of interest. For example, the maximum spectral diversity may be controlled by vegetation abundance, while the analysis objective is lithologic discrimination.

5.3.2 Band Ratios

Ratio images, created by dividing one spectral image channel by another pixel by pixel, are widely used in earth science investigations because of the proven ability of ratio images to enhance the discrimination of surface units. When applied with a well-formulated rationale and spectroscopic basis, they can be very effective spectral analysis tools. The principal strengths of ratio images are that the effects of topographic shading are removed, albedo differences are suppressed, and the relative color properties of materials are enhanced. Before calculating a ratio image, the data must be corrected for any system contribution (e.g., dark current) or environmental signal (e.g., atmospheric path radiance) that is additive to the digital numbers (DNs) measured. These additive signals contribute significantly to the calculation of the ratios and will cause erroneous results unless removed.

This is well illustrated in Figure 5.2 (see color insert), which is the same region as that shown in Figure 1 and uses the same spectral bands. The band ratio images and color composite of the ratios shown in Figure 5.2a were created using the raw digital data, with no corrections for scattering. The consequences of this negligent calibration can readily be identified in ratio images. Except for regions dominated by materials with extraordinary spectral contrast (e.g., vegetation), the resultant images are correlated to the albedo of the input data, and topographic shading is still apparent in the ratio image. Note that this effect is enhanced for the ratios using band 1, which has the greatest amount of atmospheric scattering. Compare these with the band ratio images shown in Figure 5.2b, where the data were corrected for atmospheric scattering. Not only are the effects of topographic shading and albedo eliminated from the product, there are dramatic differences in spatial distribution of color units. It is now easier to trace like-color compositional units around the canyon walls, but also highlights the distinct color properties of the topmost rock unit from the vegetated areas of the plateau. Multiplicative corrections would also be required if the absolute value of the ratios was important information, but are not necessary if the objectives are simply to discriminate surface units.

Band ratios have been most used extensively in the analysis of broadband sensors such as *Landsat* multispectral scanner (MSS), TM, and *SPOT*, spurred on by the successes of many of the early studies (Rowan et al., 1974; Goetz et al., 1975; Ashley and Abrams, 1980; Rowan and Kahle, 1982; Powdysocki et al., 1983). Ratios can also be used for the analysis of data from narrowband imaging spectrometer systems, although the continuous spectral coverage permits more sophisticated methods (see below). The most common band ratios exploit two major classes of absorption: visible to near-infrared absorptions due to ferric and ferrous-bearing minerals and infrared absorptions associated with hydrous, hydroxyl, and carbonate minerals. The spectra shown in Figure 5.3a indicate that ferric and ferrous-bearing minerals exhibit diagnostic absorptions in the visible to near-infrared wavelength regions (see Chapter 1). There are distinct differences in spectral properties of the minerals hematite and goethite relative to the nonferric minerals kaolinite and montmorillonite are a consequence of the electronic transition and charge transfer absorptions that occur in the ferric compounds (Sherman and Waite, 1985; Chapter 1, this volume). The same data filtered to the spectral bandpasses for the TM sensor are shown in Figure 5.3b. While the diagnostic features that distinguish kaolinite from montmorillonite are not present, the major differences between the ferric minerals and the hydroxyl-bearing minerals are evident.

The common multispectral ratios for the minerals shown in Figure 5.3 are shown in Table 5.1. TM band 3 is positioned near the first maximum in the spectra of the ferric oxides, and thus the ratio of TM band 3 to band 1 will highlight the strong increase in reflectance from the ultraviolet through the visible. This ratio for materials lacking ferric compounds will be close to unity (e.g., montmorillonite and kaolinite, Table 5.1), while for ferric-bearing compounds the ratio will be much greater than 1 (e.g., hematite and goethite, Table 5.1). In addition, many ferric minerals exhibit an increase in reflectance between 0.9 and 1.7 μm. Thus the ratio of TM band 5 to band 4 will be much greater than 1 for surfaces enriched in these compounds and near or less than unity for other materials (Table 5.1).

With the wavelength coverage of the TM system, it is also possible to highlight the presence of such hydrous minerals as kaolinite, montmorillonite, alunite, and others. These minerals typically show a decrease in reflectance from 1.7 to 2.2 μm due to overtones and combination overtones of water and hydroxyl fundamental absorptions (Chapter 1) (Figure 5.3a). Therefore, the ratio of TM band 5 to band 7 increases with increasing amounts of these minerals and/or the combined strength of these absorption features (Table 5.1). There are many possible band ratio combinations and the specific combinations used should be constructed with reference to the spectral properties of the materials present in the field site. For example, Sultan et al. (1987) used several ratios of TM bands that were based on analysis of the field and laboratory reflectance spectra of lithologies from the Arabian-Nubian Shield in

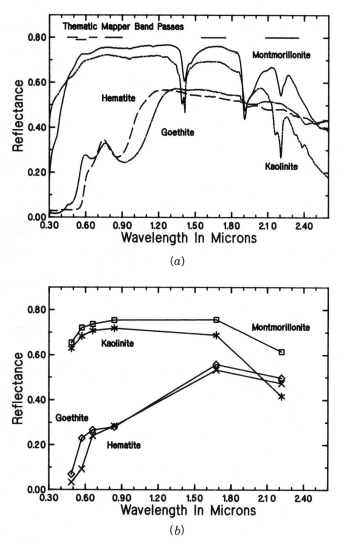

Figure 5.3 Laboratory reflectance spectra of the minerals montmorillonite, kaolinite, hematite, and goethite measured in RELAB (Pieters, 1983; Mustard and Pieters, 1989) at 5-nm spectral resolution (top) at the spectral resolution of the thematic mapper.

Egypt. They showed that the ratio of TM band 5 to band 1 was highly correlated with the volume percent of opaques, 5/7 with hydroxyl-bearing phases, and the product of two ratios, 3/4 and 5/4, with ferrous absorptions in amphiboles. Multiple-band-ratio images can be used to construct color composites that allow the simultaneous analysis of several features. Additional examples of applications using band ratios are presented by Sabine in Chapter 8.

Ratios provide a simple and rapid tool for analyzing spectral properties of surface materials and discriminating different units. When ratios are applied with a good knowledge of the spectral properties of the key materials in the scene, they can pro-

TABLE 5.1 TM Band Ratios for Spectra Shown in Figure 5.3

Mineral	*3/1*	*5/4*	*5/7*	*3/4×5/4*
Montmorillonite	1.1262	0.95793	0.61064	0.60173
Kaolinite	1.1262	1.0033	0.67157	0.65574
Hematite	7.2598	1.8767	0.07343	0.06211
Goethite	3.8719	1.9934	0.14474	0.13756

vide useful first-order information about the surface composition. Ratios do not have a sufficient level of sophistication to make deterministic statements of surface composition, largely because materials may exhibit similar ratio values but have different spectral and compositional causes. However, some spectral features are so strong and characteristic of specific materials (e.g., the red edge in vegetation) that ratios can be used to identify areas with these properties rapidly and therefore can be effective for subsetting data.

5.3.3 Statistical Transforms

Ratio techniques are not effective for discriminating among materials that exhibit more subtle spectral differences. An inherent property of multichannel remote sensing data is that the spectral properties of materials are highly correlated channel to channel and across much of the wavelength region. This is a consequence of the general properties of reflectances of solid surface. Reflectance, which is constrained to lie between 0 and 1.0, changes intensity slowly from channel to channel. Although the different channels are often thought of as independent variables, they are not; there are no negative values, and spectral discontinuities, where reflectances exhibit large changes from one channel to the next, are very rare. Vegetation, which typically transitions from less than 10% reflectance at 0.6 µm to greater than 60% at 0.8 µm, is one exception, although even this change in reflectance occurs over a 200-nm wavelength interval. The discrimination of different lithologies and soils is commonly based on very subtle, but consistent and real, spectral differences. Even band ratios, where the fundamental albedo information has been removed, show high degrees of correlation.

A number of approaches have been developed to maximize the spectral variance of remote sensing data to assist in the discrimination of different surface materials. The most widely used and well developed of these techniques is principal components analysis (PCA) (e.g., Taylor, 1974; Williams, 1983; Loughlin, 1991). In PCA the original data are projected onto a new set of orthogonal coordinates that are defined by the statistical properties of the input data. PCA requires no a priori information and, unlike band ratios or color composites, is based on the entire spectral domain of the data set. Because the new output images are linear, additive combinations of the original input images and depend on the spectral variability within the scene analyzed, the individual images are not readily interpretable in terms of specific absorptions or minerals as are band ratios. However, through the use of ground truth and evaluation of the weightings given to each band in the principal components, it is possible to build more quantitative understanding into this approach (e.g., Smith et al., 1985; Jaumann, 1991; Huete, 1986).

PCA is generally based on the correlation matrix between spectral channels, which is constructed with data from the entire scene, a representative sample of the scene, or a subset of the scene. The correlation matrix is decomposed into its characteristic eigenvalues and eigenvectors using standard matrix decomposition routines, the details of which will not be covered here. Conceptually, this decomposition defines a new set of axes that are linear combinations of the original variables (spectral channels in this case) that maximize the amount of variance that each axis describes and are orthogonal to one another. The components are ordered where the first axis (or principal component) describes the most variance and the last component the least. The amount of the spectral variability contained in each component is given by the eigenvalue, and the relative proportion or contribution of each band to that component is given by the eigenvector. When reflectance or raw radiance data are input, the first principal component is typically a weighted average of all the bands used and is similar in appearance to an albedo image of the scene. When vegetation comprises a significant component in a scene, it is commonly highlighted in the second principal component, which is a consequence of the strong spectral contrast between the visible and near-infrared reflectance spectra of vegetation. The next components typically highlight spectral variability due to rocks and soils.

In broadband systems such as TM, the latter principal components are dominated by random and systemic noise and are usually of little interest for interpretation and analysis. With hyperspectral and imaging spectrometer data, however, the latter components may contain important information. It is usually worthwhile inspecting these results because statistical outliers which comprise only a small proportion of the scene's spectral variance may be contained in these components. Green et al. (1988) have developed a very useful modification of PCA called the minimum noise fraction (MNF) MNF transform. An estimate of the noise covariance matrix is first derived, and then the reflectance data are rotated and scaled to make the noise isotropic with unit variance in all bands. These transformed data are then analyzed using principal components. Because the noise of the data set is prewhitened and distributed equally among the bands, this approach results in a better discrimination of the eigenvalues related to surface-spectral elements from those related to instrument and scene noise. This better orders the large number of components produced with imaging spectrometer data to put spectrally and spatially relevant components first.

PCA may be implemented with a more quantitative spectral basis through a number of different strategies. One method proposed by Crósta and Moore (1989) and Loughlin (1991) uses subsets of the total number of possible spectral channels. The particular bands are chosen to maximize the variance due to a set of absorption features. For TM images, a PCA is performed separately on band combinations 1, 3, 4, 5 and 1, 4, 5, 7. The first set of bands emphasizes spectral variability in the visible regions, which is coupled to ferric and ferrous iron absorptions. The second set emphasizes spectral variability at longer wavelengths and is coupled to hydroxyl absorptions. Each analysis maximizes the variance in the input data for a given suite of absorptions and thus allows a separation of materials based on these different spectral properties. When the full TM band set is used, the spectral variability from these different wavelength regions is combined. PCA has also been applied to data sets with more spectral channels, and in these applications it is possible to interpret the eigenvectors in terms of specific spectral properties and features. For example, Jaumann (1991) performed PCA on a suite of rock and soil laboratory spectra and showed that when plotted as a function of wavelength, the first few eigenvectors, exhibited features that were analogous to mineral absorption bands seen in the spectra of the rocks and soils. He then correlated the eigenvectors to the major element oxide chemistry of the samples using multiple linear regression. These eigenvectors and regression coefficients were then used to map surface composition for a remotely acquired data set. In this example, ground-truth data were used to construct a PCA framework that was then applied to the remote data. Therefore, the remote observations were transformed into a PCA data space defined by laboratory spectra.

Smith et al. (1985) and Pieters et al. (1985) used PCA to model mixing relationships between spectroscopic end members. This approach is predicated on the assumption that the spectral variability contained in the data set is due fundamentally to mixing between a small number of discrete compositional endmembers with unique spectral properties. Other approaches to mixing are discussed in detail below. If the mixing systematics are linear, or can be linearized, the relative distance between end members as measured in the transformed PCA data space is proportional to the abundance of the end members. In a similar approach, Huete (1986) decomposed the spectral variability of laboratory data from a large number of soils into the principal components and showed that the first four eigenvectors, combined linearly, accounted for greater than 98% of the spectral variability of the soil suite. By this approach it is possible to account for much of the variability in diverse soil spectra with a small number (four in this case) of basis eigenvectors. Each of the basis vectors was shown to be correlated to specific soil types, the spectra of which are a function of the composition.

PCA and other statistical transforms are thus effective tools for summarizing the common components of spectral variability of a scene into a small number of variables. The separation of materials in a remote sensing scene by PCA is driven by the spectral variability of the materials and is thus linked to spectral properties. The principal limitation of this approach is that the statistical transforms are scene specific, and therefore the results and methodologies for interpretation are not easily transported to different regions or to scenes acquired during different seasons. However, through the use of laboratory data to define the PCA space (e.g., Smith et al., 1985; Jaumann, 1991; Huete, 1986) or subsetting the data into specific spectral regions (e.g., Crósta and Moore, 1989; Loughlin, 1991), it is possible to bring a stronger measure of spectral understanding to the analysis of data with statistical transforms.

5.4 FEATURE MAPPING AND THE IDENTIFICATION OF MINERALS

The identification and mapping of specific minerals is one of the great strengths of imaging spectroscopy. This capability is important not only for mapping the presence of minerals over wide areas, but also because it can be used to discriminate among minerals that may not be readily identified through inspection in the field. An important distinction of this type of approach from those presented in the preceding discussion is that feature mapping and mineral identification is physically based. It exploits the fact that many minerals exhibit absorption bands that are diagnostic of mineral type and composition (the physical basis for absorption and the relationship to mineral structure and chemistry is presented in Chapter 1). By using as much of the available spectral information as possible, absorption features can be quantitatively characterized and then compared and analyzed in the context of spectral libraries (e.g., JPL, USGS, RELAB), thus linking the remote measurements to a form of ground truth. Because the basic physical interactions in laboratory and remote measurements of spectral properties are the same, this is a valid approach, although there are clearly issues of scale and texture that must be considered. It should be obvious that an explicit requirement for this type of technique is that the remote measurements must be calibrated to reflectance.

There are three basic approaches to mapping and analyzing mineral absorption features. The first simply characterizes the position, strength, and shape of absorption features. The results can then be displayed as absorption feature maps or used as input to analytical systems to compare the results to laboratory spectra characterized in the same way. The second method seeks to compare the complete shape of absorption features from a remote data set to those in a spectral library or database. This allows the direct identification of minerals and combinations of minerals or assemblages. The third approach, which attempts to deconvolve overlapping and superimposed absorptions quantitatively, has applications in the identification of mineral combinations and abundances and for mineral compositions across solid solutions.

5.4.1 Position, Strength, and Shape

If presented with the reflectance spectrum of a mineral, an experienced analyst is able to identify quickly the particular mineral on the basis of the absorption bands present in the spectrum. The key features that are used in this process are the position, shape, and relative strength of all the absorptions present. Most minerals exhibit a number of distinct absorptions that vary in position, shape, and strength, which in combination are a unique indicator of the particular mineral species (e.g., Burns, 1993; Hunt and Salisbury, 1970; Hunt et al., 1973; Chapter 1, this volume). Implicit in this approach is that there are wavelength regions that do not exhibit absorptions and that the absorption features are in essence departures from a smoothly varying and continuous function. This continuous function is commonly referred to as the continuum.

The continuum of spectrum is poorly understood and difficult to define. It may be qualitatively defined as the collective properties of spectral regions exhibiting smoothly varying spectral properties that, taken as a whole, define the upward limit of the general reflectance curve for a material. Like all reflectance data, the shape and properties of a continuum includes both absorption and scattering processes. However, unlike absorption features, the absorption processes are generally nonselective or extremely broad and ill defined (Clark and Roush, 1984). Scattering processes also contribute significantly to the continuum, perhaps best illustrated by the effects of changing particle size on reflectance spectra (e.g., Nash and Conel, 1974; Pieters, 1983; Crown and Pieters, 1987; Chapter 1, this volume). The solid surface of a freshly exposed rock generally exhibits a low reflectance and relatively weak absorption bands. As this rock is broken up into fine particles, the overall reflectance brightens, the absorption bands become deeper, and the continuum becomes accentuated. Further decreases in particle size result in a weakening of the absorptions due to the increase in scattering relative to volume absorption (Hapke, 1993). Except for the transition from a coherent rock to particulate, the continuum generally scales multiplicatively with textural changes and illumination and thus is a relatively fixed or constant feature of a material.

Full and quantitative analysis of absorption features for mineralogic identification currently requires detailed examination and modeling of individual spectra using optical constants and complex scattering codes (Chapter 1). Due to the large size of hyperspectral data sets and the lack of high-quality optical constants for the vast majority of earth materials, this approach is not feasible for terrestrial data sets, and more automated methods that are robust and stable are desirable. A number of methods have been developed over the last decade to automate these procedures (e.g., Green and Craig, 1985; Kruse et al., 1986, 1988; Yamaguchi and Lyon, 1986). The method of Kruse et al. (1988) is well tested and has been integrated into expert systems and analytical packages (e.g., Kruse et al., 1993). This work forms the basis of the following discussion. However, the other techniques are based on the same principles and differ primarily in the methods of implementation.

The basic concepts of this approach are illustrated in Figure 5.4. The initial step in the analysis is the definition of the continuum. This can be achieved automatically by defining the high points in the spectrum through the use of slope and magnitude criteria. A high point is defined simply by the wavelength where the reflectance is higher than the reflectance values at short and longer wavelengths. The additional criterion that the slope is positive on the short-wavelength side and negative on the long-wavelength side of this point over a given wavelength range helps to minimize the selection of spurious points due to noise. It is prudent to use a smoothing function to suppress the effects of noise and calibration error (Kruse et al., 1988). Straight-line segments are drawn between the defined high points and then the actual reflectance and that defined by the straight-line segments are compared for each channel in the data set. The reflectances for the model continuum are the larger of the two compared values. This is considered a model continuum since it is an approximation based on the data and not derived from first principles using optical constants and scattering models. A number of investigators have developed successful approaches to define the continuum, which is actually an important step in many spectral analysis techniques (e.g., Green and Craig, 1985; Clark et al., 1987).

The model continuum approximates the convex hull for the spectrum. The continuum is then removed from the reflectance data by dividing the reflectance by the continuum at each channel (Figure 5.4). The objective of the continuum removal is to isolate the properties of the absorption features from the overall reflectance properties of other components in the signal. The overall reflectance can be well modeled as exponential functions of the absorption coefficient and mean optical path length of each component multiplied together (Clark and Roush, 1984; Hapke, 1993). Therefore, to preserve and properly scale the critical strength, shape, and position parameters, the reflectance must be divided by the continuum. This provides a first-order correction for the effects of illumination that is also a multiplicative factor.

The continuum-removed data are then used to define the wavelength position of the minima, strengths, and asymmetries of discrete absorption bands (Figure 5.4).

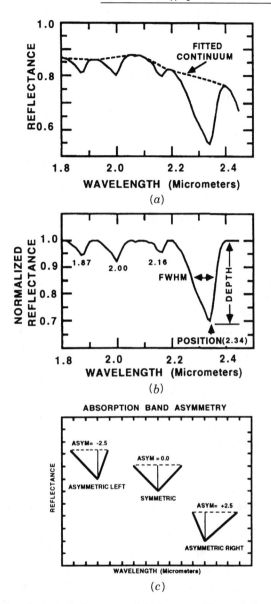

Figure 5.4   Schematic diagram illustrating the basic concepts behind mapping absorption band position, strength, and shape using a laboratory spectrum of calcite as an example. (*a*) definition of the continuum; (*b*) spectral curve after the continuum is removed; the band position is defined as the lowest point in the relative reflectance spectrum, the depth by equation (5.1), and the full width at half maximum (FWHM) as illustrated; (*c*) definition of asymmetry for negative, zero, and positive asymmetry. Reprinted from Remote Sensing of Environment, Vol. 44, F. A. Kruse, A. B. Lekhoff, and J. B. Dietz, Expert system-based mapping in northern Death Valley, California/Nevada using the airborne visible/infrared imaging spectrometer (AVIRIS), pp. 309–335, Copyright 1993, with permission from Elsevier Science.

Band minima are identified in an analogous way to the methods used in determining the high points for the continuum, except it is the low points that are defined and cataloged. If there are more than two band minima identified between a given pair of high points, the absorption band is defined as a multiple band and assigned an order of two or greater, depending on the number of minima encountered. The relative strength of each band is calculated using the formula presented by Clark and Roush (1984):

$$S_1 = \frac{R_c - R_b}{R_c} \quad \text{or} \quad S_2 = 1.0 - \frac{R_c - R_b}{R_c} \tag{5.2}$$

where S is the band strength, R_c the reflectance of the continuum at the wavelength of R_b, and R_b the reflectance at the band minimum. For S_1, band strength is normalized to be between 0.0 and 1.0, where weak bands have a strength near 1.0 and strong bands have values much less than 1.0. S_2, also normalized to be between 0 and 1.0, is essentially the inverse of S_1, and weak bands have strengths close to 0.

Band shape can be parameterized for the purposes of this systematic analysis by two simple parameters: full width at half maximum and asymmetry (Figure 5.4). The width of the feature is the absolute difference in microns between the right and left sides of the absorption, where the reflectances are half the strength of the band. Asymmetry is defined as the base 10 logarithm of the sum of reflectances over the number of channels to the right of the band minimum divided by the sum of reflectances over the number of channels to the left of the band minimum. The use of the $\log_{10}$ preserves the linearity in calculated asymmetry for bands with equal asymmetries to the right and left of the band minimum. The result is that a perfectly symmetric band will have a value of 0, asymmetric to the right will be positive and to the left, negative.

This parameterization allows a rapid reduction in the dimensionality of an imaging spectrometer data set to a suite of key absorption features. The specific parameters for the example shown in Figure 5.4 are presented in Table 5.2. This reduces the spectral data to four discrete absorptions, each parameterized by four variables. Although this is a useful summary of the number and character of absorptions, this information in and of itself does not necessarily provide a link to the surface mineralogy. To make practical use of the results from such an analysis, they must be coupled to ground truth, either through field or laboratory spectra. Kruse et al. (1993) have integrated the band parameterization approach with spectral library information into an expert system for analysis of hyperspectral data. The basic method involves reducing the library data to a table of band parameters and then

TABLE 5.2   Absorption Band Parameters for the Mineral Calcite from Figure 5.4

Band	Wavelength	Depth	FWHM	Asymmetry
1	2.340	0.3001	0.0980	0.3047
2	1.997	0.0788	0.0490	0.5154
3	1.870	0.0603	0.0490	0.6033
4	2.164	0.0452	0.0392	0.1947

using these results to build a hierarchy of facts and rules to catalog the remote data. These decisions cascade from first-order discrimination of vegetated from nonvegetated surfaces, rocks from soils, hydrated minerals from ferric minerals, and so on. Kruse et al. (1993) analyzed the probabilities for 27 different minerals with this approach in a data set for the northern Grapevine Mountains, Nevada, and mapped with reasonable accuracy the distribution of five specific minerals. Although this approach requires high signal/noise ratio (>50:1), it can be an effective starting point in an analysis.

5.4.2 Feature Mapping Using Complete Band Shape

The method described above is rapid and effective for processing the large amounts of data typical of hyperspectral data sets. When coupled to complementary analyses of spectral libraries, it can yield reasonably accurate results for the identification of minerals and their spatial distributions. However, the approach above is sensitive to noise, which limits the applicability. In addition, the method for band parameterization condenses into the four variables the amount of spectral information from a given band. As such, much information concerning the specifics of the band shape are lost from the analysis, which can lead to ambiguity in identifying minerals. Although less of a concern for narrow strong absorptions with simple structure, this becomes a greater problem for absorptions with greater width (e.g., ferric and ferrous absorptions) and complexity (multiple hydroxyl bands with differences in concavity, etc.). Furthermore, the spectral contrast of absorptions in remotely acquired data sets is generally lower than for laboratory or field spectra. This is primarily a consequence of mixing of components at the subpixel scale, and therefore simple comparisons of band strength from remotely acquired spectra to spectral libraries will be complicated.

Clark and others (Clark et al., 1990, Clark and Swayze, 1995) have developed a technique for absorption band mapping that uses the entire suite of spectral information within a given spectral range to define the band. This approach is intimately coupled to data from a spectral library, since the algorithm computes the degree of similarity between the remote data and the library data. A continuum is first removed from both library and remote data using predefined channels on either side of the absorption feature (Figure 5.5a). Averaging of several proximal channels is generally performed to increase the signal/noise ratio for these points. Thus a priori knowledge is a requirement since the analysis must begin with a determination of specific minerals or components that are to be mapped or analyzed in the remote data set.

The continuum is removed from each spectrum around the absorption band by dividing the reflectance by the straight-line continuum, resulting in a new suite of spectra, defined as

$$L_c(\lambda) = \frac{L(\lambda)}{C_l(\lambda)} \quad \text{and} \quad O_c(\lambda) = \frac{O(\lambda)}{C_0(\lambda)} \tag{5.3}$$

where $L(\lambda)$ is the library spectrum as a function of wavelength (λ), C_l is the continuum for the library spectrum, L_c the continuum removed library spectrum, O and O_c the

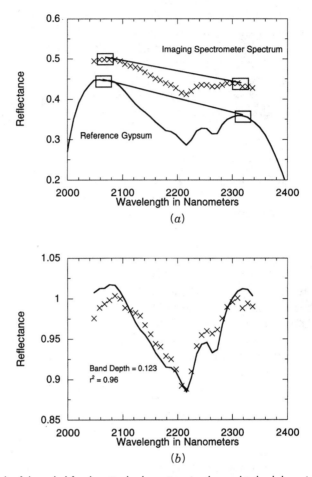

Figure 5.5 Example of the method for absorption band mapping using the complete band shape. (*a*) A spectrum from a remotely acquired data set is shown (imaging spectrometer spectrum) along with a laboratory spectrum of the mineral gypsum (reference gypsum). The laboratory spectrum is convolved to the spectral band passes of the remote system, and then an continuum is removed using points defined on either side of the absorption. The boxes indicate that several channels are used to defined the continuum points. (*b*) The continuum-removed reference spectrum is then fit to the continuum-removed observed spectrum using least squares [equation (5.7)]. From this fit, a band depth can be calculated using equation (5.2) (0.123 for this example), and the quality of the fit is determined using a correlation coefficient ($r^2 = 0.96$). Example applications of this approach for a wide variety of minerals are shown in Figure 5.7.

observed and continuum removed observed spectra, respectively, and C_0 is the continuum for the observed spectrum.

The next step in this approach is to adjust the contrast of the continuum removed library spectrum $[L_c(\lambda)]$ to best fit the continuum removed observed spectrum $[O_c(\lambda)]$ using a simple linear gain and offset adjustment. This is not strictly valid from a theoretical standpoint, as it does not correctly account for band saturation at low albedos or large grain sizes, nor does it properly account for nonlinear affects of intimate mixing on band strength and shape. However, most of the variance due to these effects can be simulated to a first order with this approximation, and the approach is computationally efficient, thus allowing rapid analysis of large volumes of data. A new continuum-removed reference library spectrum, L'_c, is calculated by using an additive constant, k, in the equation

$$L'_c = \frac{L_c + k}{1.0 + k} \qquad (5.4)$$

which can be rewritten into the standard form of the linear equation:

$$L'_c = a + bL_c \qquad (5.5)$$

The constants a and b are defined in terms of k by the equations

$$a = \frac{k}{1.0 + k} \quad \text{and} \quad b = \frac{1.0}{1.0 + k} \qquad (5.6)$$

This can be rearranged to a standard least-squares equation where the values of a and b are determined that best match the spectrum of the continuum-removed reference spectrum, $L_c(\lambda)$, to the continuum-removed observed spectrum, $O_c(\lambda)$:

$$L'_c = a + bL_c \qquad (5.7)$$

Since, by definition, the number of spectral channels will be greater than or equal to three (two to define the continuum, one for the band center), this is an over-determined problem and a solution will always exist. This results in a superposition of the library spectrum by the linear gain and offset into the same reference frame as the observed spectrum (Figure 5.5b). The parameters that can then be derived are the band strength as defined in equation (5.2), and the root-mean-square (RMS) difference between the remote and library spectra. The RMS calculation provides an overall goodness-of-fit measure, while the band depth is proportional to the abundance of the mineral. In an automated analysis of image data, the goodness of fit can be set to a predetermined threshold level, depending on the noise in the observed data set, where an RMS value above this threshold would indicate the presence of the given mineral. These RMS and band depth measures can be combined by multiplying them together to produce a map of the distribution and relative spectral abundance of minerals. Ancillary information such as albedo and continuum properties can also be included in the algorithm. For example, when searching for the mineral magnetite, the algorithm also checks that the albedo is generally low, or if the target material is vegetation, the continuum should have a steep rise from the visible to near-infrared. This additional information permits increased fidelity in discriminating among materials.

An excellent example of this approach is presented by Clark et al. (1995), where they have used this method to map, in detail, the presence of a wide variety of minerals at Cuprite, Nevada (Figure 5.6). Cuprite has been the object of numerous remote sensing studies, beginning with analysis of data from broadband sensors by Ashley and Abrams (1980) and followed by a number of hyperspectral studies (e.g., Goetz and Srivastava, 1985; Hook and Rast, 1990; Kruse et al., 1990; Swayze et al.,

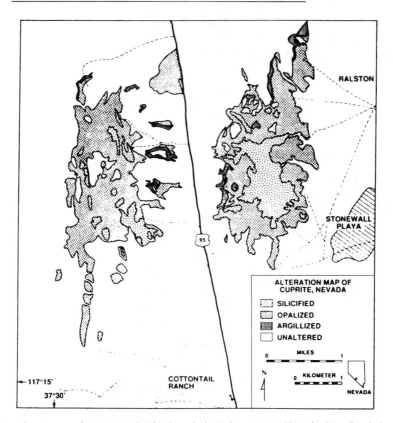

Figure 5.6 Alteration map of Cuprite, Nevada. This shows the basic alteration assemblages for this well-studied geologic site and forms the base map for the results shown in Figure 5.7. A wide variety of alteration minerals have been identified at this site, including quartz, alunite, kaolinite, buddingtonite, and ferric oxides. (After Ashley and Abrams, 1980; Hook and Rast, 1990.)

1992). The well-documented surface mineralogy, excellent lithologic exposures, sparse vegetation cover, and limited soil development have contributed to the selection of Cuprite as a test site for evaluation of sensor and algorithm performance. The geology of the site is characterized by acid sulfate hydrothermal systems in two discrete centers. The eastern center consists of hydrothermal alteration of Tertiary volcanic rocks, while in the western center, Cambrian metasedimentary rocks have been altered. The alteration zones typically follow a concentric pattern from weakly altered argillized rocks on the outside through opalized to silicified rocks in the interior. The extraordinary array of well-exposed lithologies and minerals in this deposit is a consequence of the rapidly changing fluid compositions and temperatures during its formation in what is interpreted to have been a hot spring deposit (Hook and Rast, 1990; Buchanan, 1981).

The image in Figure 5.7 *a* (see the color insert) provides an overview of Cuprite in simulated TM true-color bandpasses. The AVIRIS data have been convolved to approximate the spectral band passes of the TM sensor, and the scene is approximately 17 km long and 10.5 km wide. The mineral map shown in Figure 5.7*b* (color insert) is derived from analysis of AVIRIS data, emphasizing minerals with relatively narrow vibrational absorptions in the spectral region 2 to 2.5 μm (i.e., OH^-, CO_3^-, and SO_4^- bearing minerals; see Chapter 1). The reference mineral spectra are taken from the USGS spectral library, and the map indicates areas in the AVIRIS scene that contain absorptions that match those selected from the spectral library. Regions in black indicate an absence of these minerals. The mineral map shown in Figure 5.7c (color insert) is derived from analysis of the same data but with a focus on electronic absorption features associated with Fe^{2+}- and Fe^{3+}- bearing minerals, such as hematite, goethite, and amphiboles. These absorption features are broader than the vibrational absorptions and are typically found in the spectral region 0.4 to 1.2 μm. These two figures show 41 different mineral categories out of more than 250 specific materials that were mapped in this scene (although not all materials were present). By combining the two mineral maps (b and c) it is possible to construct detailed mineral assemblage maps. This is a tremendous achievement in discriminating among minerals and surface materials with remote sensing and begins to demonstrate the level information available from high-quality imaging spectroscopy data sets.

Consider a little more carefully the results shown in Figure 5.7. In Figure 5.7b, Clark et al. (1995) have been able to resolve subtle shifts in absorption positions related to solid solution series and element substitutions and mapped minerals with these characteristics (e.g. high, medium, and low aluminum muscovite; K-alunites of different temperatures). This level of mineralogic discrimination requires excellent radiometric resolution, precision, and accuracy as well as a thorough spectral library. The discrimination of dickite from kaolinite is a good example of this, as they have very similar absorptions (Chapter 1). In Figure 5.7c the primary focus is on minerals with electronic absorption features associated. Again, solid solution substitutions in Fe^{2+}- and Fe^{3+}- bearing minerals affect the absorption band properties, as does the degree of crystallinity of the minerals (e.g., large, medium, fine, and nanophase hematite). This field site has been analyzed extensively over the years, and the mineral maps presented here have been confirmed through field and laboratory studies (Swayze et al., 1992, Swayze and Clark, 1995). The results shown in Figure 5.7 are a dramatic example of the high level of precision in mineral mapping that is possible with high-quality imaging spectrometer data (in this case, 1995 AVIRIS data; see Chapter 11) and well-characterized library spectra.

5.4.3 Absorption Band Modeling

The approaches to absorption feature mapping described above rely on correlations between the parameterization of features observed in remotely acquired data to those in spectral libraries or ground-truth data sets. These methods are therefore excellent for applications for which spectral libraries representative of the materials in the scene exist. They also include a certain degree of flexibility to handle noisy data and complexities introduced by mixtures or particle size effects. However, when presented with mineralogic compositions that are different from those found in spectral libraries and/or with several minerals in a mixed surface, simple matching of absorption features can prove to be inadequate. Mineral identification is particularly difficult when overlapping absorptions combine, obscuring the minima of one or more of the diagnostic absorption bands. Models that account more completely for the processes of absorption by focusing on individual absorption bands rather then combined absorption features are more easily extended beyond the scope of preexisting libraries.

One absorption band-based approach, developed by Huguenin and Jones (1986), examines the various higher-order derivatives of spectra to identify the location of individual absorptions. This method assumes that each absorption is symmetric around its band center, but unlike other approaches discussed below, does not require that absorptions have a specific shape. Band centers are identified where the second derivative of the spectrum is negative, the fourth derivative is positive, and the fifth derivative is zero. Like any derivative analysis, this method is highly sensitive to noise. Therefore, the Huguenin and Jones approach is critically dependent on its incorporation of an intelligent smoothing algorithm, the details of which remain proprietary to the authors. Nonetheless, the approach is capable of resolving overlapping band centers separated by as little as 0.3 to 1.0 of the full width at half maximum (assuming Gaussian shaped absorptions). Results derived with this approach are very complementary to that of Kruse et al. (1988) and Clark et al. (1990).

Many workers (e.g., Burns, 1970; Smith and Strens, 1976; Farr et al., 1980; Clark, 1981; Singer, 1981) have approximated absorption bands with Gaussian distributions of the form

$$g(x) = s\exp\left[\frac{-(x - \mu)^2}{2\sigma^2}\right] \tag{5.8}$$

where x is wavenumber (energy), s the strength or amplitude of the distribution, μ the center (mean) wavenumber, and σ the width. However, the simple Gaussian approximation has been shown to be an inadequate physical model for absorption bands (Sunshine et al., 1990). This is particularly well illustrated for the absorption bands in the mineral orthopyroxene. The 0.9-μm absorption is due to electronic transitions from a single site in the orthopyroxene structure (Burns, 1970) and thus should be modeled by a single Gaussian, if the absorption is in fact Gaussian in form. However, two Gaussian distributions are required to fit the feature adequately (e.g., Singer, 1981; Roush and Singer, 1986).

The underlying assumption of the Gaussian model is that energy (wavenumber) is the random variable. However, the center of a band, the location of the maximum probability of absorption, is a function of the average crystallographic and molecular properties of the material. Similarly, the width of absorptions is primarily a consequence of perturbations to these average properties. Fundamentally, the energy of absorption is determined by the average bond length of the atoms defining the absorption site. Thus, to the first order the random variable should be variations in bond length (Sunshine et al., 1990).

Since the average bond length is randomly or Gaussian distributed, the key determining factor for the shape of absorption is how bond-length variations are mapped into wavelength or energy. According to the crystal field theory for electronic absorptions (Marfunin, 1979; Burns, 1993), absorption energy (e.g., wavelength), and bond length are related by a power law. Sunshine et al. (1990) empirically determined the optimum mapping of the bond length to energy by fitting an absorption due to a single absorption, the 0.9-μm orthopyroxene absorption. This analysis revealed that energy and bond length are inversely proportional, suggesting that Coulombic potential energy dominates the absorption site. This inverse mapping of bond lengths to energy leads to a modified Gaussian distribution, $m(x)$, of the form

$$m(x) = s\exp\left[\frac{-(x^{-1} - \mu^{-1})^2}{2\sigma^2}\right] \qquad (5.9)$$

The difference between Gaussian and modified Gaussian distribution is reflected in the symmetry of the band models.

Having formulated a physically based mathematical description for isolated absorption bands, Sunshine et al. (1990) were able to model spectra accurately as a series of modified Gaussian distributions superimposed on a continuum. In the modified Gaussian model (MGM), the continuum is modeled as a straight line in energy (cm^{-1}) that sits above the general spectral curve. It should be noted that continua used in most other methods (e.g., Kruse et al., 1988; Clark et al., 1990) use reflectance maxima to define the continuum. This difference is critically important for the case of overlapping absorptions, where the true reflectance maxima are obscured. Under the MGM each absorption is characterized by three model parameters: a band center (μ), a bandwidth (σ), and a band strength (s), while the continuum is characterized by the slope and offset of a straight line in energy. The model calculations are performed in natural log reflectance and energy, and all parameters, including the continuum, are fit simultaneously using a stochastic least-squares method (Tarantola and Valette, 1982).

Examples of MGM solutions to a low-calcium pyroxene, a high-calcium pyroxene, and mixtures of the two minerals are shown in Figure 5.8. Each panel in this figure includes (top to bottom) the residual error between log of the modeled spectrum and log of the actual spectrum (offset 10%), the individual modified Gaussian distributions representing absorption bands, the continuum, and the modeled spectrum superimposed on the actual spectrum. [The nonrandom residual error near 0.9 μm in (a) is symmetric with respect to the modeled absorption. Unlike an asymmetric residual, which generally indicates the presence of an additional absorption band, this symmetric error is a characteristic of minor band saturation, as discussed by Sunshine and Pieters (1993).] Each of the pure minerals has a unique MGM fit to the spectrum, where the strength, center, and width of the absorptions are characteristic of that mineral and controlled by its chemistry and crystal structure. In the mixed spectra it is not immediately apparent that there are two mineral phases, as there is only one band minimum in both the 1-and 2-μm regions. Yet, using the MGM, it is possible to deconvolve these overlapping absorptions successfully. Furthermore, modified Gaussian distributions from the solutions to the mixed spectra match in position and width the modified Gaussians calculated for the pure minerals. The strengths of the absorptions vary in proportion to the abundance of each mineral in the mixture.

Sunshine and Pieters (1993) analyzed absorption band systematics in pyroxene mixtures using the MGM and derived a particle-size-independent relationship between the relative band strengths of the 1-and 2-μm bands and the abundance of these minerals in the mixtures. Sunshine et al. (1993) subsequently applied this relationship to spectra of natural samples that contained variable amounts of pyrox-

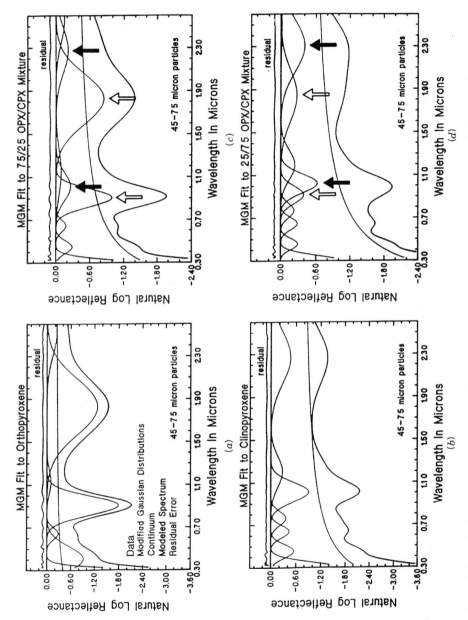

Figure 5.8 Modified Gaussian model (MGM) fits of reflectance spectra: (a) MGM fit to orthopyroxene size separate; (b) MGM fit to clinopyroxene size separate; (c) MGM fit to 75% orthopyroxene–25% clinopyroxene mass fraction mixture; (d) MGM fit to 25% orthopyroxene–75% clinopyroxene mixture. In (c) and (d) open arrows correspond to absorptions from the low-Ca orthopyroxene phase. Solid arrows correspond to absorptions from the high-Ca clinopyroxene phase.

enes and other minerals. They were able to deconvolve successfully the mafic mineral abundances in these more complex natural lithologies. The MGM has also been used to model the variations in absorption band properties in solid solution series (Sunshine, 1994; Mustard, 1992). In these applications it was shown that the modified Gaussian distributions vary systematically in position and strength with mineral composition and that relationships could be derived that related the absorption band properties to composition. The study of Sunshine (1994) examined the changes in electronic transition absorptions across the olivine solid solution series. Mustard (1992) focused on the actinolite solid solution series, where absorptions due to electronic transitions, as well as charge transfers and vibrations, were present. Although the MGM model was not developed specifically to model hydroxyl or charge transfer absorptions, the general principles behind the MGM model apply. Therefore, for minerals that exhibit systematic changes in vibrational absorptions as a function of composition and/or environment of formation, the MGM may be an effective tool for characterizing these absorptions and relating them to composition and environment of formation.

It should be noted that in contrast to other methods described above, the MGM neither relies on a library of spectra that may or may not reproduce all permutations and combinations of absorption bands, nor requires knowledge of end-member spectra, as is needed for both linear and nonlinear mixing described below. Instead, the MGM derives compositional information directly from a measured spectrum. This is particularly useful in applications where the exact mineralogy, major and minor element chemistry, and physical state (particle size, alteration state, etc.) of materials is unknown or may not exist in library spectra.

5.5   FULL SPECTRAL MAPPING

As discussed above, mapping minerals based on the absorption features in their spectra is quite powerful. Although such methods are strongly linked to composition, they are by definition focused only on specific wavelength regions and therefore neglect contributions from the rest of the spectrum. Feature mapping is particularly problematic for many materials (e.g., rocks), whose spectra do not contain well-defined absorptions. Instead of diagnostic features, these spectra are characterized by their continuum shapes and/or very broad absorptions. To address some of these concerns, a number of alternative and complementary approaches have been developed that compare spectra simultaneously over their full wavelength range.

There are two general classes of approaches to mapping based on the entire spectral signature. The first group of methods are spectral similarity searches. Under a spectral similarity search, a scene is examined to determine which pixels are most "similar" to a specific spectrum of interest or reference spectra. As discussed below, there are numerous ways of objectively defining the similarity between two spectra, all of which can be used to produce spectral similarity maps. A second category of mapping methods that uses the full spectral response are spectral detectors or matched filters. Here a spectrum of interest is detected in a scene by highlighting pixels with similar spectral properties simultaneously while repressing all other background spectral signatures. Spectral detection maps can be produced by many approaches, each of which is based on a different characterization of the spectrum of interest and the background.

5.5.1 Spectral Similarity Maps

One of the fundamental uses of imaging spectroscopy is to determine the spatial extent of specific materials based on their spectral response. For example, one application of interest might be to remotely map the location of a particular outcrop. This can be accomplished by first identifying a representative spectrum of the outcrop from either a known location within the scene or from field data and/or a spectral library. (The use of spectra from field instruments or a library does, of course, require that the scene be calibrated to absolute reflectance and that the reference spectrum be resampled to the same wavelengths as the remote data.) The scene can then be searched pixel by pixel for spectra that are, based on some predefined metric, most like the representative outcrop spectrum. A similarity image is thus generated, with high values being most similar to the outcrop spectrum. Typically, a threshold value must then be set to define a tolerance level for what is considered similar enough (within the noise) to be mapped as the outcrop. There are a variety of algorithms that have been used to define the similarity between two spectra. As discussed below, the diversity in approaches stem from the fact that a spectrum can be viewed from many different perspectives in addition to its physical and/or compositional meaning.

One approach to defining similarity is to compare the difference between two spectra at each wavelength mathematically. Although a variety of metrics could be imagined, it is common to use the RMS distance:

$$\text{RMS} = \frac{\left[\sum_{i=1}^{N} (R_i - P_i)^2 \right]^{1/2}}{N} \tag{5.10}$$

where N is the number of bands in the spectrum, R_i the value of the spectrum of interest or reference spectrum at the ith band, and P_i the value of the spectrum of a pixel of the scene at the ith band. The RMS difference can be calculated for each pixel, with low RMS values indicating areas that are most similar to the spectrum of interest. To produce a map, some maximum threshold RMS value would be chosen. This RMS distance (or RMS error) is often used a measurement of goodness of fit (e.g., Clark and Swayze, 1995, as discussed above) and is a good choice for defining overall spectral similarity from a statistical perspective.

An alternative viewpoint is to consider spectra geometrically. The spectral response as a function of wavelength, while compositionally significant, can also be interpreted geometrically as a vector in N-space (where N is the number of bands). Visualizing spectra as vectors leads naturally to at least two metrics for defining similarity. One approach is to use the vector dot product as a similarity measurement. The dot product between two vectors is

$$\mathbf{R}_i \cdot \mathbf{P}_i = \sum_{i=1}^{N} R_i P_i \tag{5.11}$$

The dot product is proportional to the projection of one vector onto the other. Thus pixels with large vector projections onto the spectrum of interest could be considered most similar. After the appropriate thresholding, the dot product could then be used to produce a similarity map.

A complementary vector-based approach is to measure the angular distance between two vectors. Under the so-called *spectral angle mapper* (e.g., Kruse et al., 1993), spectra whose vectors are separated by small angles are considered most similar. By measuring angular distance and ignoring vector length, this metric downplays differences in albedo and as such provides a good complement to the dot product approach.

One other common similarity method for spectral searching is binary encoding. Binary encoding is based on Hammond distances (e.g., Mazer et al., 1988) and comes from the field of digital communications. Under this algorithm, each spectrum or signal is first divided by its average. At each wavelength, if the spectrum is greater than or equal to its average, it is assigned a value of 1. If it is below the spectral average, a value of 0 is assigned. Thus, based on variations in amplitude, a generalized description of the spectrum is encoded into a 1 bit vector (each point or wavelength is either 0 or 1). To preserve additional information on the shape of the spectrum, local spectral slope is also encoded. The local slope is defined at each wavelength, i, as simply

$$R_{i+1} - R_{i-1} \tag{5.12}$$

The second bit is set to 1 if the slope is positive and 0 if the slope is negative. After encoding spectra into this 2-bit scheme, a measurement of similarity known as the *Hamming distance* can be used. The Hamming distance is defined as

$$\sum_{i=1}^{N} B(R)_i \text{ XOR } B(P)_i \tag{5.13}$$

where N is the number of bands in the spectrum, $B(R)_i$ the binary-encoded reference spectrum (or spectrum of interest) at the ith band, $B(P)_i$ the binary-encoded spectrum of a pixel of the scene at the ith band, and XOR the the exclusive OR logical operator. At each wavelength i, the exclusive OR will return a value of 1 if the 2-bit spectra differ and a value of 0 if they are the same. After summing over all bands, the pixels with small Hamming distances (i.e., closest to 0) will be most similar to the reference spectrum. It should be noted that by defining amplitude relative to the mean (to encode the first bit) the binary encoding process is insensitive to albedo differences between spectra. Some variations in the details of the binary encoding method exist. In particular, some implementations use 1-bit encoding based only on the amplitude or only on the local slopes information. The binary encoding of the spectra allows for a very rapid calculation of the XOR logical operation, and as such, this similarity metric is extremely computationally efficient. An excellent application of the binary encoding technique for mapping minerals in the Grapevine Mountains can be found in Kruse et al. (1993).

Each of these similarity measurements in and of themselves is reasonable, straightforward to implement, and can be used to produce a similarity image. However, there are several issues of general concern. First, there is no clear choice as to which method to use. Furthermore, the various methods, each derived from a different perspective, can produce different results. As none of the approaches are strongly compositionally based, there is, in general, no reason to give more weight to any particular method. In some cases, however, a technique may be better suited for a particular scenario. For example, neither binary encoding nor SAM retain albedo information. In situations where albedo is meaningful, the RMS difference may be a better metric. While the choice of methods remains uncertain, it should be noted that obtaining similar results from different methods does provide a greater degree of confidence in the solution. In all cases it is clear that careful evaluation of results from several techniques is warranted. A second concern is that each of the approaches requires that an acceptance threshold be defined in order to produce a map showing the spatial extent of a given material (e.g., an outcrop). While some combination of experience with the method and the particular scene can aid in establishing a reasonable threshold, determining the threshold value remains somewhat arbitrary. As such, care must be taken to examine the sensitivity of the resulting map to changes in the threshold value.

5.5.2 Spectral Detection Maps

The basis of spectral similarity mapping is spectral comparisons and similarity indexes. In contrast, spectral detection mapping centers around the concept of maximizing the signal of the spectrum of interest while suppressing all background spectra, or in the language of signal processing, maximizing the signal/clutter ratio. The key difference between similarity searches and spectral detection algorithms is that the latter accounts directly for the presence of background materials, while the former ignores them. Spectral detection, often referred to as *matched filtering*, can be thought of conceptually as a compromise between processes that maximize the contribution of the signal of interest (e.g., spectral similarity searches) and those that minimize the remaining background. Actual determination of the optimal matched filter depends on assumptions about the state of knowledge of both the spectrum of interest and its background. In some cases approaches require significant knowledge of the spectral properties of both [e.g., foreground–background analysis (FBA); Smith et al., 1994]. Alternatively, some techniques rely on statistical estimates from the scene or generalized spectral characteristics.

One of the more fully developed spectral detection algorithms was recently presented by Harsanyi and others (Harsanyi, 1993; Harsanyi and Chang, 1994). Their technique addresses the specific problem of identifying materials with known spectral responses that are present only at low abundances. Under Harsanyi's low-probability detection (LPD) algorithm, the background spectral signatures are assumed to be unknown and are statistically estimated from the scene. There are many situations where this theoretical framework is applicable, such as the case where a material exists only at a scale below the resolution of the sensor (i.e., has subpixel exposures). This could, for example, describe an outcrop that is either partially covered by vegetation or exposed only in small areas. Any measurement of subpixel exposures would, of course, include a spectral contribution not only from the material of interest (e.g., the outcrop) but also from other scene components (e.g., vegetation) and is thus well suited to the LPD approach.

As with the similarity mapping techniques, the first step of the LPD is to define the spectral signature of interest from within the scene or from a spectral library. (In the case of a library spectrum, it must first be converted to at-sensor radiance or DN). Under the LPD, each pixel in the scene, **P**, is then described as a linear combination of the reference spectrum (or spectrum of interest), **R**, and the unknown background spectral properties, **U**, plus system noise, **n**:

$$\mathbf{P} = \mathbf{R}a_r + \mathbf{U}\mathbf{a}_u + \mathbf{n} \tag{5.14}$$

where a_r is the abundance of the reference spectrum and the vector $\mathbf{a}_u$ is the abundance of unknown background spectra.

An orthogonal subspace projection, **S**, is used to solve equation (5.14), where

$$\mathbf{S} = \mathbf{I} - \mathbf{U}\mathbf{U}^{\#} \tag{5.15}$$

and **I** is the identity matrix and $\mathbf{U}^{\#}$ is the pseudoinverse or $(\mathbf{U}^T\mathbf{U})^{-1}\mathbf{U}^T$. When equation (5.14) is multiplied by this orthogonal subspace projection operator, the contribution of the background, $\mathbf{U}\mathbf{a}_u$, is eliminated and the abundance of the reference spectrum can be determined.

Recall, however, that the background spectral properties, **U**, are unknown. Using the scene itself, it is possible to estimate the background contribution statistically. This is accomplished by calculating the data correlation matrix:

$$\mathbf{C} = \frac{1}{p} \sum_{i=1}^{p} \mathbf{P}_i\mathbf{P}_i^T \tag{5.16}$$

where p is the number of pixels in the scene and $\mathbf{P}_i$ is the spectrum of the ith pixel. The correlation matrix can then be decomposed into a matrix of eigenvectors, **V**, and their associated eigenvalues using a principal component transform, as described in previous sections. As in PCA analysis, the first N eigenvectors can be chosen to be significant. As such, the first N eigenvectors, $\mathbf{V}_N$, can be used as a statistical approximation of the background. $\mathbf{V}_N$ can then be substituted for the unknown background contribution **U** and the orthogonal subspace projection, equation (5.15), can be rewritten as

$$\mathbf{S} \simeq (\mathbf{I} - \mathbf{V}_N\mathbf{V}_N^{\#}) \tag{5.17}$$

Having fully defined the LPD procedure, the definition of *low probability* is clearer. For the LPD approach to work, the spectrum of interest must be a minor scene component in the sense such that it is not included among the primary eigenvectors (the first N) used to estimated the background. As discussed previously, there is no one-to-one correspondence between principal components and spectral properties. Thus the statistical definitions inherently required by LPD may be difficult to comply with, as many materials that are minor components may statistically have properties that are described at least partially by the primary eigenvectors. In addition, care must be taken in evaluating the output of the LPD, in that it may include pixels which have low-probability spectral properties (i.e., are not represented by the principal eigenvectors) but that are spectrally different from the reference spectrum of interest. It is clear, however, that for situations which conform to the theoretical framework of the LPD, the LPD detection scheme is quite powerful.

Several examples of the success of the LPD in discrimination of poorly exposed lithologies are documented by Farrand and Harsanyi (1995). One of the goals of this study was to map the spatial extent of palagonite tuffs found along the rim of Easy Chair Crater (ECC), part of the Lunar Crater Volcanic Field in Nevada, using imaging spectrometer data from a 1992 AVIRIS scene (Figure 5.9a). Spectra of representative lithologies of interest were collected from field samples and measured with the NASA-supported reflectance spectrometer facility at Brown University (RELAB) (Pieters, 1983; Mustard and Pieters, 1989) and are shown in Figure 5.10. The principal distinguishing spectral properties among the materials shown in Figure 5.10 are not specific absorption features but the overall continuum. Spatially, the ECC palagonite tuff is confined to a small region along the south and west rim of Easy Chair Crater (Figure 5.9). Spectrally, the ECC tuff is very similar to other background materials as shown in Figure 5.10. In fact, the overall continuum properties can be reasonably well modeled by a spectral mixture of the rhyolite and orange and red cinders. The spectral similarity of the ECC palagonite tuff to the background materials made discrimination by standard spectral mixture analysis problematic (spectral mixture analysis is described in detail below). Either the regions rich in the ECC palagonite tuff were adequately modeled by rhyolite and red cinder end members, or if an actual spectral end member for the tuff was used, it predicted high tuff abundances in regions where the tuff did not exist (Figure 5.9b).

However, the small spatial extent of the ECC palagonite tuff is well suited to the LPD method. Using the reference spectrum of the ECC palagonite tuff (after conversion to radiance units) as the known spectrum of interest, Farrand and Harsanyi (1995) used the LPD algorithm successfully to map the spatial extent of the ECC palagonite tuff (Figure 5.9c). The LPD isolates the ECC palagonite tuff dramatically while suppressing other scene components. The brightest values correspond directly to outcrops of the palagonite as defined by field work. The lower-intensity pixels correspond to smaller outcrops with low abundances of the tuff at the scale of the pixel.

5.5.3 Summary

The use of the full spectral range and resolution for identification and mapping of surface materials can be used to provide useful results. The methods discussed here range from spectral similarity techniques, which provide broad correlations, to the low-probability detection algorithm, which attempts to highlight areas with spectral properties similar to a target material with narrow spectral limits. It is important to note that all these approaches provide results, but they may not necessarily give the same results, because of the different algorithms used. When applying these methods of spectral analysis, it is useful to employ a combination of methods. Where there is a high degree of correlation between methods, a greater degree of confidence in the

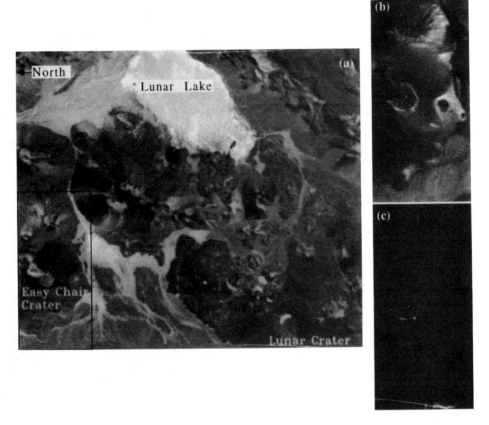

Figure 5.9 (a) Location of the area studied by Farrand and Harsanyi (1995) using the low-probability detection (LPD) technique. This is an image of the Lunar Lake volcanic field as seen by the AVIRIS sensor at a wavelength of 1.62 μm. The area shown in (b) and (c) is focused around the Easy Chair Crater, located in the lower left hand corner of the image. North is to the left side of the image and the image is approximately 12 km wide; (b) fraction image for mixture modeling using ECC palogonite tuff. The distributions are more extensive than known from field observations; (c) low probability detection algorithm results for ECC palogonite tuff against a background of other materials. This more accurately displays the distribution of the ECC palogonite tuff. Reprinted from W. H. Farrand and J. C. Harsanyi, Journal of Geophysical Research, Vol. 100, 1565–1578, 1995, copyright by the American Geophysical Union.

material identification is obviously indicated. As with all the approaches to spectral analysis discussed here, results should be cross-referenced and linked to ground-truth information.

5.6 MIXTURE MODELING

Many of the approaches described above that exploit the spectral properties of materials to identify surface composition make the implicit assumption that there is a one-to-one correspondence between the observations and ground truth (i.e., labo-

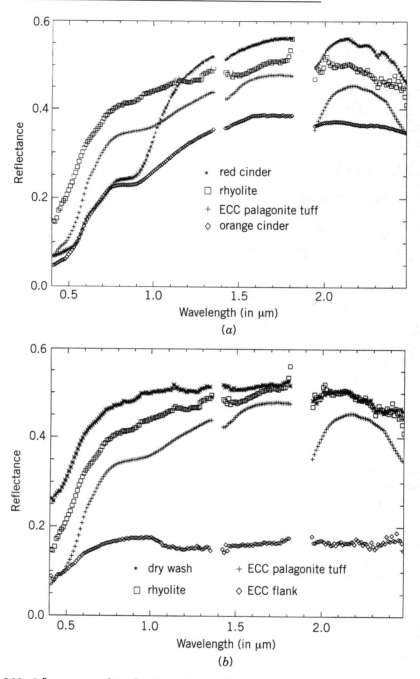

Figure 5.10 Reflectance spectra of Easy Chair Crater palagonite tuff compared to various background materials: (*a*) laboratory reflectance of volcanic materials exposed at the field site of the Lunar Lake volcanic field; (*b*) laboratory spectrum of the target material, ECC palagonite tuff, compared with the AVIRIS spectral properties other common components in the scene.

ratory or field measurements of spectral properties of small samples are representative of surfaces measured remotely). The fact that at the scale of most observing systems, the pixel contains a mixture of materials is generally treated as a second-order effect. As a second-order effect, it is accommodated through adding complexity to the approaches, such as mineral combinations or simply considered a consequence of working with natural scenes. Certainly, the principal goal for mapping in the earth sciences is to simplify the complexity observed on the ground into a set of well-defined units that can then be organized and used to understand and analyze larger-scale associations. However, anyone with experience in field observations will recognize that continuous variations in the proportions of components are the norm when viewed at scales that approximate the typical spatial element of a sensor. These may simply be in the proportions of components that are of little interest to the analyst (e.g., vegetation abundance when mapping geology or variations in illumination), or they may comprise a critical signal that could be the main objective of an investigation (e.g., changing proportions of minerals due to a facies change in a sedimentary sequence).

The concept of the mixed pixel has certainly been recognized for many years (e.g., Horowitz et al., 1975; Jackson, 1983; Huete et al., 1984). However, the idea that mixing at the subpixel scale is a natural consequence of earth processes and an inherent feature of remote sensing data sets that can be quantitatively exploited has only evolved over the last two decades. Among the researchers who have contributed to this effort (e.g., Mustard and Pieters, 1987a; Boardman and Goetz, 1991; Farrand, 1991; Li and Strahler, 1986), Adams and co-workers (Adams and Adams, 1984; Adams et al., 1986, 1993; Gillespie et al., 1990; Smith et al., 1990a,b; Sabol et al., 1992) have conducted some of the more structured analyses and multifaceted approaches to the problem.

The basic premise of mixture modeling is that within a given scene, the surface is dominated by a small number of common materials that have relatively constant spectral properties. If most of the spectral variability within the scene is a result of varying proportions of these common components (end members), it follows that the spectral variability captured by the remote sensing system can be modeled by mixtures of these components. It has been documented that the reflectance spectrum of a mixture is a systematic combination of the reflectance spectra of the components in the mixture (e.g., Nash and Conel, 1974). The systematics are basically linear if the components are arranged in spatially distinct patterns, analogous to the squares on a checkerboard (e.g., Singer and McCord, 1979). In this case the scattering and absorption of electromagnetic radiation is dominated by a single component on the surface, and thus the spectrum of a mixed pixel is a linear combination of the end-member spectra weighted by the areal coverage of each end member in the pixel. If, however, the components of interest are in an intimate association, like sand grains of different composition in a beach deposit, the mixing systematics between these different components are nonlinear. The spectral properties of the different end members become convolved in this case, because the electromagnetic radiation interacts with more than one end member as it is multiply scattered in the surface.

The question of whether linear or nonlinear processes dominate the spectral signatures of mixed pixels is still an unresolved question. It probably depends on a number of factors and conditions of the scene. A discussion of the implications of nonlinear mixing is included near the end of this section. Nevertheless, the linear approach has been demonstrated in numerous applications to be an insightful technique for interpreting the variability in remote sensing data and a powerful means for converting spectral information into data products that can be related to the physical abundance of materials on the surface. With this as background, we discuss spectral mixture analysis in more detail in the following sections, with some examples.

5.6.1 Theoretical Framework

The principal objective of spectral mixture analysis is to define a coherent set of spectral end members that are representative of physical components on the surface and that model the spectral variability inherent in a given scene. The theoretical limit to the number of end members is defined by the dimensionality of the data and constraints of the mixture inversion. This results in a total number of possible end members equal to the number of spectral channels in the data set, plus 1. However, the number of end members that may be practically defined is far fewer, typically ranging from three to seven, depending on the number of channels and the spectral variability of the scene components, even for hyperspectral imaging spectrometer data sets. The reason for this is that the theoretical limit requires that the spectral channels be independent variables and that the spectral end members be linearly independent. Channel-to-channel variance in spectral data sets is highly correlated; thus some of the spectral information is redundant. In addition, materials of different physical composition may exhibit similar spectral properties over a given wavelength range or have spectral properties that can be defined mathematically by linear combinations of other components. Such materials would therefore not be resolved. Strategies to overcome these limitations, such as multiple-end-member models, are discussed by Adams et al. (1993).

A practical way to approach the definition of end members is to apply a field-based framework in which *classes* of materials are considered (e.g., soils, vegetation, lithologies, shade) in the context of the basic technical constraints imposed by the data set to be examined (e.g., number of spectral channels, wavelength range, pixel size). It is also practical to begin with a few well-defined and spectrally distinct components, and build complexity into the model as is warranted by subsequent analyses. This is illustrated conceptually in Figure 5.11. Here we consider a scene sampled at the spectral and spatial resolution of the *Landsat* thematic mapper that consists of three basic components: vegetation, a rock unit represented by gabbro, and shade. The end-member shade is one of the more interesting and novel features of spectral mixture analysis. Natural surfaces are never uniformly illuminated, and variations are caused by topographic slope changes, which result in both shading and shadow that vary with the seasons and the diurnal cycle, as well as shadows at subpixel scales caused by small-scale topography, trees, shrubs, and boulders. The shade end member mixes with the other end members in the scene in proportion to the amount of this variation in illumination and may be considered as a neutral multiplicative scaling factor.

The first stage in a mixture analysis is to define a suite of image end members (selected from the image data). An image end member (IE) is one that is contained within a scene and has a maximum abundance of the physical end member it is most

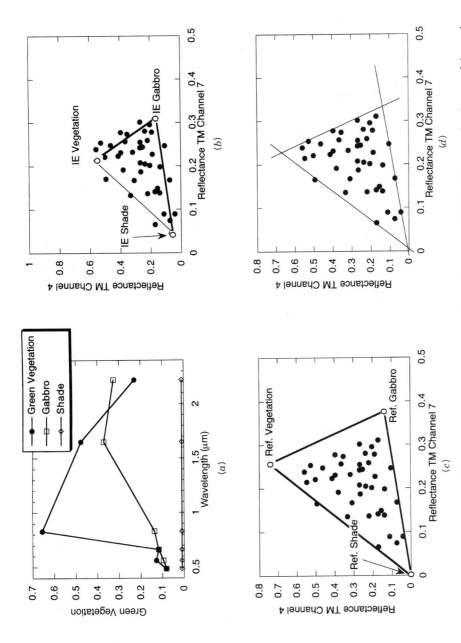

Figure 5.11 Conceptual diagrams illustrating the basic concept and terms of spectral mixture analysis: (*a*) visible–near infrared spectra of three end members—green vegetation, gabbro, and shade—at the spectral resolution of the thematic mapper sensor; (*b*) scatter plot of TM band 7 versus TM band 4 of data representative of a scene containing the variable proportions of the end members; image end members (IE) are represented by the open circles; (*c*) same scatter plot but with the reference end members plotted and the proper mixing space defined; (*d*) automated techniques for the identification of mixing end members can use the convex hull of the data cloud to define the vertices of the intersection where the reference end member should be. (After Adams, 1993.)

closely associated with. This is illustrated in Figure 5.11*b* where the IE are represented by the open circles. They may be selected objectively (e.g., Adams et al., 1993; Boardman, 1993; Tompkins et al., 1997) or based on criteria such as field knowledge or other analysis methods (e.g., ratios, PCA). For a typical scene, it is unlikely that a single pixel or group of pixels can be selected that corresponds to a pure component (the pixel is comprised of 100% of the end member), although some pixels will probably be relatively pure. Conceptually, the image end members are those that best bound the data cloud, as shown in Figure 5.11*b*. Once selected, these end members are used to solve the following equation for each pixel in the image:

$$\mathrm{DN}_b = \sum_{i=1}^{N} F_i \mathrm{DN}_{ib} + E_b \quad \text{and} \quad \sum_{i=1}^{N} F_i = 1.0 \qquad (5.18)$$

where DN_b is the intensity of a given pixel in bandpass or wavelength b, F_i the fractional abundance of end member i, DN_{ib} the intensity of image end member i at wavelength b, N the number of end members, and E_b the error of the fit for bandpass b. The second equation is a constraint that the fractions sum to 1.0. If M is the number of channels in the data sets, this results in M equations in N unknowns and may be solved using standard linear least-squares inversion.

The fractions calculated for each of the data points (solid circles) shown in Figure 5.11*b* correspond to the relative distance from the image end member (IE) points, subject to the constraint that the sum of the fractions equals 1. Because the image end members are themselves mixtures of other materials, there will exist fractions in the results that will be greater than 1.0 or less than 0, although the sum of the fractions will still equal 1. Data points that fall within the lines joining the IEs would have positive fractions, while those that fall outside the lines would have negative fractions of the IE opposite this line. For example, those pixels that plot outside (to the right) of the line joining the image end members gabbro and vegetation will have negative shade fractions. Thus negative fractions or fractions that are greater than 1.0 do not necessarily indicate an error in the method of application.

The validity of a mixture model solution using a particular suite of end members can be tested using three basic criteria: the root-mean-square (RMS) error of the fit of the end-member spectra to each of the pixels in the image, the spatial patterns and coherency of the fraction images, and the absolute values of the fractions. The RMS error is computed using the standard sum of the squares of the difference between the model and the data, normalized to the number of spectral channels. In general, an average RMS error is calculated for the entire scene as well as for each pixel, and this second measure can then be scaled and displayed for analysis (see the example below). If the average error is within the level of system noise, an adequate solution is indicated. The RMS error image can be analyzed for indications of where specific regions of the input data are not well modeled by the end members selected and these areas can be used to refine the selection of end members for subsequent analyses.

The fraction images can be displayed by scaling the fractions to fit the 8-bit dynamic range of typical computer displays [e.g., $100(F_i + 1)$ scales the fractions such that 0% equals a DN of 100 and 100% equals a DN of 200]. The spatial distribution and abundances of the end members represented in the fraction images should conform to a priori expectations or conventional spectral parameterizations for the scene. For example, the spatial patterns and abundance of vegetation should be similar to conventional vegetation indices (e.g., the NDVI), while the shade image should be inversely correlated to the average albedo. Furthermore, the fraction images should have a certain level of coherency. A typical result when too many end members are used than is warranted by the data, or if two end members are used that are not spectrally distinct, is that the fraction images will exhibit little continuity in the fractions, with rapid changes between high and low values.

The absolute values of the fractions can be used to determine if the image end members selected properly bound the spectral variability of the data set. If there are significant regions of one or more fraction images that show either extreme positive or negative values, the end members are not well selected. Although the fractions will still sum to unity (that is a constraint in the original inversion), and the RMS error may be small, the specific image end members should nevertheless be reselected.

The results using image end members provide a first-order perspective on the mixing relationships, and if the end members are relatively pure, the fractional abundances will be similar to the actual abundances. However, it is often desirable to use library or reference end members to calculate the fractions. This provides a better link to spectral libraries and therefore ground truth, but is also essential if the fractional values are intended to be used in multitemporal studies or across scenes that were acquired under different illumination conditions or times of years. If the data are accurately calibrated to reflectance, the image end members can be related to the library reference spectra by

$$\mathrm{DN}_b = \sum_{i=1}^{N} \mathrm{F}_i \mathrm{R}_{ib} + E_b \quad \text{and} \quad \sum_{i=1}^{N} \mathrm{F}_i = 1.0 \qquad (5.19)$$

where R_{ib} is the reflectance of library end member i in bandpass b. This is illustrated in Figure 5.11c, where it can be seen that the spectral variability defined by the reference end members now fully bound the variability of the image data. In this example there would be no negative fractions or fractions greater than 1.0.

The primary result of the spectral mixture analysis is the fraction images which show the distribution and abundance of the end-member components in the scene. They can be used in a variety of ways to analyze the composition of the surface and provide insight into processes. Because the raw radiance data have been converted to data products that make intuitive sense from a field perspective, the fraction images can be used to identify lithologic units and map their distribution. The fraction images can be manipulated to suppress information extraneous to the task and enhance that which is more relevant. For example, the shade image is generally dominated by illumination effects that are not relevant to compositional mapping. The suite of fraction images can be normalized to a shade-free scene, thus allowing the abundance relationships among key components to be better illustrated. This would be accomplished by dividing each of the fraction images by (1.0 − shade). A similar procedure could be employed to remove the fractional contributions of vegetation to the scene, thus emphasizing rock and/or soil relationships.

The correspondence between the spectral abundance of materials determined from spectral mixture analysis and actual physical abundance on the ground depends on a number of factors. Smith et al. (1990b) have shown that green leaf abundance determined from leaf area index is highly correlated to the abundance of the green vegetation end member from spectral mixture analysis of arid regions, and Mustard and Pieters (1987b) have shown that mineral abundances from mixture modeling of hyperspectral data are well correlated with physical abundances in terrains where the minerals exhibit unique and diagnostic absorptions. Sabol et al. (1992) have analyzed the general question of detectability for a variety of different materials, spectral imaging systems, and performance parameters. The detectability increases with the signal/noise ratio of the measurements, number and placement of spectral channels, and the spectral contrast of the materials. For high-spectral-resolution data and well-differentiated end members, the detectability may be very low, approaching a few percent, while for low-spectral-resolution systems with little spectral contrast among the end members, the detectability may be a few tens of percent. In a similar manner, Lawler and Adams (1994) performed an analysis of system and environmental noise for some typical measurement platforms and showed that formal uncertainties in the fractional abundances may be incorporated directly into the mixture inversion equations.

One of the great strengths of hyperspectral systems (see Chapter 11) is that direct material identification is possible (see also the discussion above). Materials that are not common in a scene but that have diagnostic absorptions distinct from the suite of end members used to model the mixture systematics can be detected and mapped in spectral mixture analysis through the use of band residual images. As discussed above, the RMS error provides a summary of the degree to which the mixture model fits the actual data. However, it does not indicate if this is distributed randomly across the wavelength region or concentrated in a few narrow absorptions. Band residuals are calculated as simply the difference between the model spectrum and the actual data, and stored as a separate file. Gillespie et al. (1990) used this approach to identify small, localized occurrences of Fe^{2+}-bearing epidote in metavolcanic outcrops in Owens Valley. This type of analysis could be a component of the approaches discussed above for mapping the presence of specific minerals or materials that are not part of the spectral mixture model.

5.6.2 Example

The basic procedures and results of spectral mixture analysis are best illustrated by way of example (Mustard, 1994). Figure 5.12 (see color insert) is a *Landsat* TM scene of a well-exposed geologic terrain within the Cape Smith Fold and Thrust Belt of northern Quebec. It was constructed in the Early Proterozoic (ca. 1920–1840 my) (Parrish, 1989) during northward underthrusting of Superior Province basement (Hoffman, 1985; St.-Onge and Lucas, 1990), resulting in a large stack of normal sequence thrust faults which were reimbricated by later out-of-sequence thrusts and folded during the latest period (St.-Onge et al., 1988). The region covered by this scene includes the Archean basement (lower right corner) and fluvial to deepwater sediments (rest of the image). Many gabbro and layered peridotite–gabbro sills have intruded these sediments and the strength contrast between the sills and the sediments has resulted in the well-developed folds shown by the resistant sill units. The layered sills can be up to 500 m thick and have a consistent internal stratigraphy. A basal chilled margin of melano gabbro is overlain by pyroxene-rich peridotite that grades to peridotite and then into cyclically interlayered olivine pyroxenite and peridotite. Massive gabbro and pyroxene gabbro cap the sequence. The general chemistry and magmatic stratigraphy are very similar to those of differentiated komatitic flows nearby, and these sills are the target of mineral exploration for platinum group elements (Hynes and Francis, 1982; Barnes et al., 1992; St.-Onge and Lucas, 1990).

The TM color image in Figure 5.12 shows the well-exposed outcrops of gabbro and ultramafic sills in shades of blue, granodiorite in whitish tan, vegetated areas in reds and browns, and snow in white. The sedimentary units are not well exposed, covered largely by till, soil, and tundra vegetation. In this color composite it is possible to distinguish the outcrops readily from the vegetated regions. The results of an image-based mixture analysis are presented in Figure 5.13, with end-member spectra shown in Figure 5.14. It begins with the selection of three image end members, representing green vegetation, well-exposed gabbro from one of the sills, and shade (using an ideal shade that has a uniform reflectance of 0.01 at all wavelengths). This results in an average RMS error of 4.6 DN. The spatial distribution and abundance of these end members, shown by the fraction images (Figure 5.13*a*), is generally consistent with a priori knowledge, with the rock fractions concentrated on the ridge tops and vegetation on the interridge regions, and the shade image (shown in reverse stretch) approximates the illumination effects. Although the fractions are well bounded (i.e., fractions generally lie between 0 and 1), note that the RMS error image contains a significant amount of spatial information. This shows that the average RMS error of 4.6 DN is not evenly distributed among all pixels. Clearly, the regions well modeled by vegetation, gabbro, and shade (e.g., lakes, some ridges) have a low RMS error. The regions not well modeled exhibit distinct, spatially coherent patterns that can be related back to the color composite in Figure 5.12. For example, some of the well-exposed sills (light blue in color composite) are not well modeled in this solution and thus may represent rock types that have different spectral properties than the gabbro. Note the linear regions of extremely high RMS error that will appear in all subsequent results. These are snow banks and illustrate that some components, although present in the scene, can be ignored because they are not relevant to the overall objectives.

The spatial patterns of the RMS error image, coupled with the color composite, can be used to guide the selection of additional end members. A fourth end member representing tundra vegetation (NPVeg) is selected from the region of high RMS error. The mixture analysis for these four end members is shown in the second row of Figure 5.13. Comparison of the fraction images for vegetation and gabbro between the three and four end-member solutions shows that tundra vegetation was partially modeled by both the green vegetation and gabbro, while the shade image is largely unchanged. Note that the vegetation fraction image has more sharply defined distributions in the four-versus three-end-member case, and that the two vegetation fraction images together account for spectral variability due to the nonlithologic components. For this solution the average RMS error is 2.3 DN. Despite the lower average RMS error, there is still a substantial amount of spatial coherency (i.e., pixels of high RMS are grouped) and information in the error image. In this case the regions of higher error (excluding the snow banks) are concentrated in the areas of the rock outcrops, suggesting that an additional lithologic end member may be warranted.

The last row of Figure 5.13 shows the results of the mixture modeling for a five-end-member case. The fifth end member was selected from the image data in the region of highest error that is also associated with the rock outcrops as indicated by the color composite. In the resulting fraction images, note the change in the abundance patterns in the gabbro end member between the four-and five-end-member case. The distributions are much more sharply defined in the five-end-member solution, and many of the largest decreases are correlated with high-to-moderate abundances in the fraction image for the ultramafic end member. The green vegetation

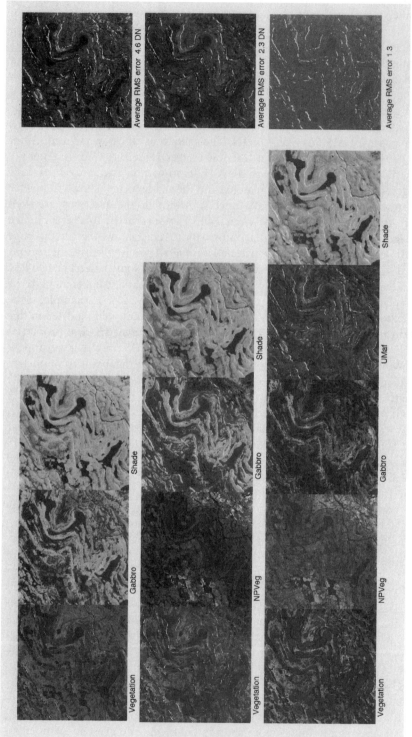

Figure 5.13 Sequential mixing example. This sequence of images illustrates the basic methods of spectral mixture analysis using image end members as applied to the scene shown in Figure 5.12. The top row of images shows the fraction images derived from using three end members (spectra shown in Figure 5.14)—green vegetation, gabbro, and shade—where the shade image is displayed as (1 − shade) to better show how this end-member models primarily illumination effects. The second row of images shows the fraction images derived from using four end members: green vegetation, nonphotosynthetic vegetation (NPVeg), gabbro, and shade. The final row of images shows the results after including an end member for this area (Umaf). The average RMS error for this solution is 1.1 DN.

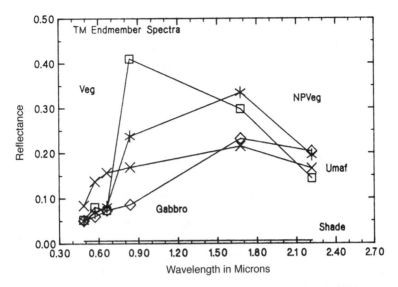

Figure 5.14 End members used in sequential mixing example. These are image end members derived from the scene shown in Figure 5.12. The spectrum of green vegetation (Veg) is shown in the squares, moss and tundra vegetation (with significant amounts of woody and nonphotosynthetic vegetation) by the asterisks (NPVeg), gabbroic rocks (Gabbro) by the diamonds, ultramafic rocks (Umaf) by the crosses, and shade by the solid line along the bottom. Shade is an ideal shade, with a reflectance of 0.01 μm at all wavelengths.

end member is even more sharply defined than the four-and three-end-member solutions. However, the tundra vegetation fraction image exhibits less contrast. Also note that the fraction image for the ultramafic end member shares some characteristics in distributions with the tundra vegetation end member. This is probably due to mimicry, in that some of the regions underlain by sediments with complex vegetation covers have spectral properties that can be modeled adequately by a mixture of tundra vegetation and ultramafic rock. Although in some areas this mixture does occur, it is probably not as widespread as indicated by the fraction images. Because of this it would be prudent to focus the interpretation and analysis of the fraction image for the ultramafic end member only on regions with high abundances having obvious geologic relationships with the gabbroic end member.

5.6.3 Objective Methods of End-Member Selection

In a typical approach to spectral mixture analysis, image end-member selection is achieved through an educated trial-and-error basis. An analyst has some knowledge of the field site or data set and a set of objectives for conducting the analysis. This guides the analyst to select an initial set of image end members to model the spectral image data. On the basis of these first results, the number and physical location of the image end member may be refined until an acceptable solution is achieved, as was demonstrated above (e.g., minimum error, objectives achieved). The analysis may be guided by other data processing techniques that help to understand the dimensionality of the data and inherent spectral variability. However, in many situations a more objective method of determining these essential components is desired. For example, results should be repeatable, and the fraction images should describe realistic physical variables or components in the scene.

Repeatability can be achieved with a straightforward statistical approach such as principal components analysis or through novel applications of convex geometry (e.g., Xu and Greeley, 1992; Boardman, 1993). The latter method has been shown to have utility in determining the number of end members that are possible in a data set and to estimate the spectral properties of those end members. In this approach, illustrated conceptually in Figure 5.11d, the raw radiance or reflectance data are first transformed into ordered principal components using a minimum noise fraction (MNF) algorithm (Green et al., 1988). This determines the number of valid dimensions of the data sets. To this reduced data set, a simplex is fit to the convex hull of the n-dimensional data cloud. The faces of this simplex are regions void in one or more end members and the vertices of the simplex define the spectral properties of the end members. This method is repeatable and has distinct advantages for objective analysis of a data set to assess the general dimensionality and to define end members. The primary disadvantage of this method is that it is fundamentally a statistical approach dependent on the specific spectral variance of the scene and its components. Thus the resulting end members are mathematical constructs and may not be physically realistic.

The simplex method described above may be characterized as an outside-in approach to defining the image end members (i.e., the external surfaces of the data cloud are fit by a simplex). A technique that is the conceptual opposite has been developed by Tompkins et al. (1993, 1997) called modified spectral mixture analysis (MSMA). In this approach the end-member spectra are not prescribed but are treated as unknowns along with the fractional abundances. The model equations are the same as equations (5.10) and (5.11), but because F_i and R_{ib} are both unknowns, they must be solved nonlinearly. The nonlinear inversion employs a damped least-squares technique as presented by Tarantola and Valette (1982). In short, a starting model is provided (a suite of possible end-member spectra, estimated fractional abundances, and the image data to be modeled). Constraints on the solutions are imposed as additional equations (i.e., fractions must sum to 1.0) or as allowable deviations from the starting model (damping of solutions). Both the starting model and the constraints are based on a priori knowledge. Each successive iteration of the equations results in a calculated change in the previous model that will reduce the error of the fit. In essence the end members are "grown" from the inside of the data cloud to best fit the spectral variability, subject to the model constraints.

Both methods for objective end-member determination provide advantages over a more traditional approach. First, the correct number of end members needed to model the data effectively can be determined quantitatively. Second, the end members can be found in a manner that is objective, repeatable, and ultimately can be automated. In addition, because MSMA solutions are governed by the mixing equations and include constraints, it is both physically and statistically valid. In the absence of any a priori knowledge of the end members, this can be extremely useful in calibration. If one or more end members are already known, the overall set of solutions is more tightly constrained, so that hidden or previously unknown end members are defined more accurately. In fact, the identification of poorly represented end members is another significant strength of MSMA.

5.6.4 Linear Versus Nonlinear Models

A fundamental limitation of the linear model is that it is strictly valid only for the situation where the end members are arranged in discrete patches on the surface. This condition is almost never met in nature, and many constituents of interest for earth science investigations exist in soils or at small scales in intimate association with one another. When the end-member materials are intimately mixed on spatial scales smaller than the path length of photons through the medium, light typically interacts with more than one component, and the measured spectrum is a complex convolution of the end-member spectra rather than a simple linear mixture. The nonlinear effects in spectra of particulate mineral mixtures have been recognized for many years (e.g., Nash and Conel, 1974) and is an area of active research for vegetation and canopy studies (e.g., Roberts et al., 1993). A variety of methods have been developed to treat this situation, including the Hapke model for particulate surfaces (Hapke, 1981, 1993) and Camillo (1987), Pinty et al. (1990), and Verstraete et al. (1990) for plant canopies. Further discussion of nonlinear mixing in plant canopies will not be covered here. The photometric model of Hapke (1981, 1993) has been shown to be a powerful and useful model for application to nonlinear spectral mixing. The validity of the model for linearizing the mixture systematics has been demonstrated in laboratory studies of directional-hemispherical reflectance (Johnson et al., 1983, 1992) and bidirectional reflectance (Mustard and Pieters, 1987a, 1989) and shown to be accurate to approximately 5% absolute abundance. The technique has also been applied successfully to imaging spectrometer data for desert soils in Utah (Mustard and Pieters, 1987b).

The effects of nonlinear mixing on reflectance spectra can be quite dramatic, as illustrated in Figure 5.15. This is a two-dimensional plot (reflectance at 0.6 μm plotted against the reflectance at 0.9 μm) of a data cloud generated using a four-end-member mixing scenario. In Figure 5.15a, the 40 mixture points that constitute the cloud were calculated using a linear mixing model and prescribed fractions. Along the planes joining the end members (e.g., line A–C), the mixtures are in 25% increments. Visually, one recognizes this as a linear problem. The mixture points are spread along the lines joining the end members in this reflectance space at intervals proportional to their fractions. Thus the point halfway along a given line (e.g., line A–C) represents a 50:50 mixture of the end members at the vertices of the line. The systematics are very different in Figure 5.15b, where the same four end members and the 40 mixture points are shown, but here the mixture spectra were calculated using the nonlinear mixing model of Mustard and Pieters (1989), adapted from Hapke (1981). The nonlinear effects are clearly indicated by the curvilinear segments joining end members (e.g., A–C, A–D). In addition, the entire data cloud is shifted to the left against the segment C–D and toward the low-albedo end member (D). What drives this shift is the predominance of low-albedo end members in nonlinear mixing situations.

There are some important implications of these differences for spectral mixture analysis. If a linear mixing model is used on data where the systematics are nonlinear, the fractions calculated will be significantly in error. In tests of linear versus nonlinear mixing on laboratory data, the fractions calculated may be in error by as much as 30% absolute (J. F. Mustard, unpublished data). In addition, the linear model can cause considerable ambiguity and false fractions when used on nonlinear mixtures.

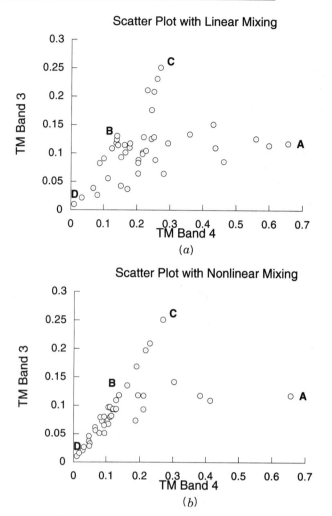

Figure 5.15 Data clouds for modeled mixtures of four end-member spectra (A, B, C, and D): (*a*) mixtures modeled with a linear mixing model; (*b*) mixtures modeled with a nonlinear (Hapke) mixing model. Individual points are reflectance values at 0.6 μm (Var 3) and 0.9 μm (Var 4).

Absorption bands and continua in nonlinear mixtures cannot be fit adequately with a linear model. However, the least-squares approach will minimize fitting errors using any of the end members in the equation. Thus end members not present in a mixture will be calculated to be present simply to minimize the error. In Figure 5.16 the abundances used to prepare the mixtures are shown on a ternary diagram together with calculated abundances using a linear and nonlinear mixture model. It is evident that the nonlinear model accurately predicts the modal abundances of the mixtures from the reflectance spectra. However, the linear model fractions are significantly in error, and for the enstatite–anorthite mixture a component of olivine is predicted which is not in the actual mixtures.

Results for Linear Mixing:

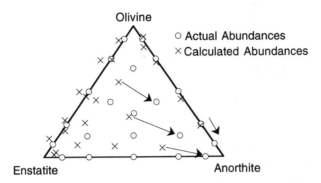

Results for Nonlinear Mixing:

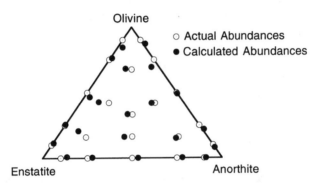

Figure 5.16 Ternary diagrams of mineral abundances. The end-member mineral spectra were used to model the mixture spectra of Figure 5.10. The open circles indicate the known abundances of the prepared mixture. The solid circles represent abundances predicted from a nonlinear mixing model of the mixture spectra. The crosses indicate the predicted abundances derived from a linear mixing model of the same mixture spectra.

Despite the obvious advantages of using a nonlinear approach for intimate mixtures, this has not been widely applied to remotely acquired data of particulate surfaces. There are several reasons for this. It is generally thought that the detailed photometric properties of all end members and surfaces are required to perform the calculations. However, many common particulate materials exhibit quasi-Lambertian behavior when viewed at nadir for incidence angles up to 40°. Therefore, a Lambertian approximation should be adequate for a first-order solution. The calculation of single-scattering albedo from reflectance requires knowledge of the incidence and emergence angles of the end members and remotely acquired spectra. Spacecraft and aircraft pointing information and digital elevation models could be used to generate this information routinely. However, the most important obstacle to the application is that the particle size, composition, and alteration state of the end members are very important controlling parameters of the solutions. Nevertheless, through careful consideration of the nature of the remote data, and as spectral libraries become more well endowed with data, many of the obstacles may be removed.

5.7 SUMMARY AND CONCLUSIONS

Our objectives with this chapter have been to highlight spectral analysis methods and techniques that can be applied to remotely sensed data. We have concentrated primarily on major developments that have occurred since the last *Manual of Remote Sensing*. These developments have been primarily in the application of spectral analysis to hyperspectral imaging data. With the development of imaging spectrometers, it has become possible to utilize quantitative methods that have their origins in the analysis of high-quality and spectral resolution laboratory data. The translation of the laboratory understanding to remote applications has required adjustments in, and/or simplification of, these approaches. This is necessitated in part by the lower quality of the remotely acquired data, the complexity introduced by scaling to the larger spatial scales required for images and image pixels, and to accommodate the inherent variability of natural surfaces. Nevertheless, remote sensing systems continue to improve in quality, and it is not unreasonable to expect extremely high signal/noise ratio and radiometric fidelity from current and future systems in the near term. Thus the degree of distinction between laboratory and image analysis methods is likely to continue to diminish. With this in mind, we have attempted to link the specific approaches and techniques back to the physical fundamentals for reflectance spectroscopy that underlie them. It is our hope, therefore, that although the specific methodologies and techniques will evolve and change, the fundamental theoretical structure will endure.

We began this chapter with a short overview of some broad-scale thematic approaches to the analysis of remotely sensed data (definition and mapping of broad-scale units, identification of the presence of specific mineralogic assemblages or lithologic units, and quantification of the amount of material present). The various spectral analysis techniques presented in this chapter have different strengths and weaknesses relative to the thematic approaches. As was discussed at the beginning, before initiating any analysis strategy using remotely sensed data it is important to define the objectives of the study carefully. This can then help to identify the optimum technique or suite of techniques that will maximize the extraction of information within the available resources. Below are short summaries of the various approaches with respect to the thematic approaches.

1. *Simple methods.*   Color composites, band ratios, and statistical transforms are all categorized as simple methods. The term *simple* is meant to indicate that they do not require high levels of spectroscopic knowledge and can be applied in a relatively straightforward manner. These approaches are very effective for the discrimination of broad-scale spectral units and to identify regional similarities and differences in surface spectral properties. When applied with some a priori knowledge of the surface composition, they can be correlated to specific absorption properties or spectral features in materials (e.g., ferric and hydroxyl absorptions). Simple methods achieve their maximum utility in the analysis of

data from broadband sensors and in areas with high degrees of spectral variability. They can also be effective for subsetting high-spectral-resolution data sets.

2. *Feature mapping/absorption maps.* The analysis and mapping of absorption features is tied closely to the physical processes of absorption and exploits the fact that many materials exhibit unique absorptions that are diagnostic of composition. To be effective the three approaches considered (absorption parameterization, mineral mapping on the basis of absorption features, absorption band modeling) require successively higher levels of knowledge. Absorption parameterization can be automated and used to catalog the presence and properties of all absorptions in a data set, which in and of itself is useful information. The utility of this information is maximized when it can then be used as an input to more detailed analysis with reference to a spectral library, which has been similarly parameterized. The mapping of specific minerals or mineral assemblages on the basis of absorption features requires a good knowledge of the principles of reflectance spectroscopy to properly guide the selection of continuum points and links to the spectral library.

   Absorption band modeling is the most flexible of the approaches under this category but requires an even greater attention to detail. Some of the approaches have been demonstrated to be effective for a wide range of materials and applications (e.g., Clark and Swayze, 1995), while others show promise but have been tested in fewer applications (e.g., Sunshine and Pieters, 1993). As has been demonstrated impressively for Cuprite, Nevada, these approaches can provide a significant enhancement for field mapping, in that specific minerals, compositions, and assemblages can be identified and mapped. These capabilities should become even greater with continued improvement in the quality of hyperspectral data from remote sensing instruments. The primary shortcomings of the methods are that they can be overly focused on absorption features and therefore may miss important features that are not part of a database or modeling experience. Also, these approaches may provide erroneous results if not applied with a concrete knowledge of the field sites and spectroscopy.

3. *Full spectral mapping.* Full spectral mapping may be characterized as spectral comparison techniques whereby the spectrum of a target material is compared against the remote observations and the results indicate where the correlations are the best. In that regard, they can be distinguished from the feature mapping techniques in that the full spectra range and suite of spectral properties are typically employed. The simpler approaches, such as binary encoding, are fast and efficient, although they are not sensitive to the subtleties of the spectral properties. In contrast, the low-probability detection algorithm can be tuned to detect subtle differences between spectra exhibited by the continuum properties. Where such differences are significant, this approach is optimal.

4. *Mixture Modeling.* When this approach is applied with proper knowledge and understanding and in field sites in which mixing is a suitable process to account for the spectral variability, abundance determinations can be estimated reliably. The approach has great utility when the surface is dominated by a few, spectrally distinct components that exhibit continuous changes in relative mixing ratios across the scene. It is also one of the premier approaches for

change detection and monitoring the composition of a surface through time. This is because spectral mixture analysis provides a consistent framework in which to analyze changes in surface abundances that can account for variability in illumination and other scene conditions. The approach is less successful for applications where a large number of discrete materials are expected (e.g., Cuprite, Nevada), due to the fact that wavelengths are highly correlated and to a first-order approximation, albedo and continuum properties dominate the least-squares mixing algorithm.

Future Directions

The developments in remote sensing instrumentation and spectral analysis over the last decade have been outstanding. When the first imaging spectrometers came into operation in the early 1980s, there was tremendous excitement in the remote sensing and spectroscopy communities that these new instruments would greatly improve our ability to identify and map surface compositions. However, these early systems suffered from low signal/noise ratio and other development problems, and many questioned if the incremental improvement in mapping surface compositions remotely was worth the increase in data volume, calibration requirements, and analysis effort. The community of users and instrument developers persisted, however, and as the quality of the data improved, the level of sophistication in data analysis also increased. Now there is little question that hyperspectral remote sensing systems are extremely valuable, and part of that success is due to the development of quantitative spectral analysis techniques.

Hyperspectral imaging data will become more widely used in the next decade as the current and planned instruments become operational, and this increase in data availability will open up many new and interesting applications. A key concept that we have tried to demonstrate in this chapter is that no single spectral analysis method will guarantee success. However, we have emphasized that the specific approach selected must be tied to the objectives of the analysis. If that objective is the identification and discrimination of minerals or lithologies in a complex geologic environment, an algorithm similar to that developed by Clark et al. (1990) would be recommended. If the objectives are surface abundance estimates across a terrain with relatively simple lithologic and surface units, mixture modeling would be a preferred approach. In contrast, the low-probability detection algorithms would provide the greatest opportunities for success in the identification of target materials that are poorly exposed, weakly discriminated spectrally from other components in the scene, and are not common in a given scene.

We expect that with the advent of new systems, data streams, and applications, the greater community of users will develop new and novel approaches to spectral analysis. In general, we expect that the best of these new methods will be soundly based on spectroscopic principles rather than statistical data-driven methods. The advanced algorithms and approaches covered in this chapter are still relatively new and are likely to become more refined over the next few years. The spectroscopic techniques that have emerged from laboratory investigations were developed under carefully controlled conditions but were typically limited in the number of test cases that could be examined. Now as data from remote systems approach the fidelity of laboratory data, the opportunities to assess the utility of these methods in different environments and under a variety of environmental conditions will expand tremendously. We expect that hybrid methods will emerge and that terrain modeling (e.g., explicit incorporation of digital elevation models) and merging of data across wavelength ranges from the visible to the microwave will become more important. There is still much to learn about data exploitation and we look forward to these anticipated developments and the exciting opportunities that will emerge.

References

Adams, J. B., and Adams, J. D. 1984. Geologic mapping using *Landsat* MSS and TM images: removing vegetation by modeling spectral mixtures, in *Proceedings of the 3rd Thematic Conference on Remote Sensing for Exploration Geology, ERIM* 2, pp. 615–622.

Adams, J. B., M. O. Smith, and P. E. Johnson, 1986. Spectral mixture modeling: a new analysis of rock and soil types at the *Viking Lander 1* site, *J. Geophysi. Res.,* 91, 8098–8112.

Adams, J. B., M. O. Smith, and A. R. Gillispie, 1993. Imaging spectroscopy: interpretations based on spectral mixture analysis in *Remote Geochemical Analysis: Elemental and Mineralogical Composition,* C. M. Pieters and P. A. Englert, eds., Cambridge University Press, Cambridge, pp. 145–166.

Ashley, R. P., and M. J. Abrams, 1980. *Alteration Mapping Using Multispectral Images, Cuprite Mining District, Esmeralda County, Nevada,* USGS Open File Rep. 80–367, U.S. Geological Survey, Washington, D.C.

Barnes, S.-J., C. Picard, D. Giovenazzo, and C. Tremblay, 1992. The composition of nickel–copper sulphide deposits and their host rocks from the Cape Smith Fold Belt, northern Quebec, *Aust. J. Earth Sci.,* 39, 335–347.

Boardman, J. W., 1993. Automating spectral unmixing of AVIRIS data using convex geometry concepts, *Summaries of the 4th Annual JPL Airborne Geoscience Workshop,* R. O. Green, ed., JPL Publ. 93–26, vol. 1, Jet Propulsion Laboratory, California Institute of Technology, Pasadena, Calif., pp. 11–14.

Boardman, J. W., and A. F. H. Goetz, 1991. Sedimentary facies analysis using AVIRIS data: a geophysical inverse problem, *Proceeding of the 3rd Airborne Visible/Infrared Imaging Spectrometer (AVIRIS) Workshop,* JPL Publ. 91–28, Jet Propulsion Laboratory, California Institute of Technology, Pasadena, Calif., pp. 4–13.

Buchanan, L. J., 1981. Precious metal deposits associated with volcanic environments in the southwest, in *Relations of Tectonics to Ore Deposits in the Southern Cordillera,* W. R. Dickinson and W. D. Payne, eds., *Ariz. Geol. Dig.,* 24, 237–262.

Burns, R. G., 1970. *Mineralogical Application to Crystal Field Theory,* Cambridge University Press, Cambridge, 224 pp.

Burns, R. G., 1993. *Mineralogical Application to Crystal Field Theory*, Cambridge University Press, Cambridge, 551 pp.

Camillo, P., 1987. A plant canopy reflectance model based on an analytical solution to the multiple scattering equation, *Remote Sensing Environ.*, 23, 453–477.

Clark, R. N., 1981. Water frost and ice: the near-infrared spectral reflectance 0.65–2.5 μm, *J. Geophys. Res.*, 86, 3087–3096.

Clark, R. N., and T. L. Roush, 1984. Reflectance spectroscopy: quantitative analysis techniques for remote sensing applications, *J. Geophys. Res.*, 89, 6329–6340.

Clark, R. N., and G. A. Swayze, 1995. Automated spectral analysis: mapping minerals, amorphous materials, environmental materials, vegetation, water, ice and snow, and other materials: the USGS tricorder algorithm (abstract), *Lunar and Planetary Science XXVI*, pp. 255–256.

Clark, R. N., A. J. Gallagher, and G. A. Swayze, 1990. Material absorption band depth mapping of imaging spectrometer data using complete band shape least-squares bit with library reference spectra, *Proceedings of the 2nd Airborne Visible/Infrared Imaging Spectrometer (AVIRIS) Workshop*, JPL Publ. 90–54, Jet Propulsion Laboratory, California Institute of Technology, Pasadena, Calif., pp. 176–186.

Clark, R. N., et al., 1995. http://speclab.cr.usgs.gov/.

Crippen, R. E., 1989. Selection of *Landsat* TM band and band-ratio combination to maximize lithological information in color composite displays, *Proceedings of the 7th Thematic Conference on Remote Sensing and Exploration Geology*, pp. 917–921.

Crósta, A. P., and J. M. Moore, 1989. Enhancement of *Landsat* thematic mapper imagery for residual soil mapping in SW Minas Gerais State, Brazil: a prospecting case history in greenstone belt terrain, *Proceedings of the 7th Thematic Conference on Remote Sensing and Exploration Geology*, pp. 1171–1187.

Crown, D. A., and C. M. Pieters, 1987. Spectral properties of plagioclase and pyroxene mixtures and the interpretation of lunar soil spectra, *Icarus*, 72, 492–506.

Farr, T. G., B. A. Bates, R. L. Ralph, and J. B. Adams, 1980. Effects of overlapping optical absorption bands of pyroxene and glass on the reflectance spectra of lunar soils, *Proceedings of the 11th Lunar Planetary Science Conference*, pp. 719–729.

Farrand, W. H., 1991, Visible and near-infrared reflectance of tuff rings and tuff cones, Ph.D. thesis, University of Arizona, Tuscon, Ariz., 187 pp.

Farrand, W. H., and J. C. Harsanyi, 1995. Discrimination of poorly exposed lithologies in imaging spectrometer data, *J. Geophys. Res. Planets*, 100, 1565–1578.

Gillespie, A. R., M. O. Smith, J. B. Adams, S. C. Willis, A. F. Fischer III, and D. E. Sabol, 1990. Interpretation of residual images: spectral mixture analysis of AVIRIS images, Owens Valley, California, in *Proceedings of the 2nd Airborne Visible/Infrared Imaging Spectrometer (AVIRIS) Workshop*, R. O. Green, ed., JPL Publi. 90–54, Jet Propulsion Laboratory, California Institute of Technology, Pasadena, Calif., pp. 243–270.

Goetz, A. F. H., and V. Srivastava, 1985. Mineralogical mapping in the Cuprite mining district, Nevada, in *Proceedings of the Airborne Imaging Spectrometer Data Analysis Workshop*, JPL Publ. 85–41, Jet Propulsion Laboratory, California Institute of Technology, Pasadena, Calif., pp. 22–31.

Goetz, A. F. H., F. C. Billingsley, A. R. Gillespie, M. J. Abrams, R. L. Squires, E. N.

Shoemaker, I. Lucchitta, and D. P. Elston, 1975. Application of ERTS images and image processing to regional geologic problems and geologic mapping in northern Arizona; Tech. Rep. 32–1597; Jet Propulsion Laboratory, California Institute of Technology, Pasadena, Calif., 188 pp.

Green, A. A., and M. D. Craig, 1985. Analysis of aircraft spectrometer data with logarithmic residuals, in *Proceedings of the Airborne Imaging Spectrometer Data Analysis Workshop*, JPL Publ. 85–41, Jet Propulsion Laboratory, California Institute of Technology, Pasadena, Calif., pp. 111–119.

Green, A. A., M. Berman, P. Switzer, and M. D. Craig, 1988. A transformation for ordering multispectral data in terms of image quality and implications for noise removal, *IEEE Trans. Geosci. Remote Sensing*, 26, 65–74.

Hapke, B., 1981. Bidirectional reflectance spectroscopy: 1. Theory, *J. Geophys. Res.*, 86, 3039–3054.

Hapke, B., 1993. *Theory of Reflectance and Emittance Spectroscopy*, Cambridge University Press, Cambridge; 455 pp.

Harsanyi, J. C., 1993. Detection and classification of subpixel spectral signatures in hyperspectral image sequences, Ph.D. thesis, University of Maryland Baltimore County, Md. 116 pp.

Harsanyi, J. C., and C. I. Chang, 1994. Detection of low probability subpixel targets in hyperspectral image sequences with unknown backgrounds, *IEEE Trans. Geosci. Remote Sensing*, 32, pp. 779–785.

Hoffman, P. F., 1985. Is the Cape Smith Belt (northern Quebec) a klippe? *Can. J. Earth Sci.*, 22: 1361–1369.

Hook, S. J., and M. Rast, 1990. Mineralogic mapping using airborne visible/infrared imaging spectrometer (AVIRIS) shortwave infrared (SWIR) data acquired over Cuprite, Nevada, in *Proceedings of the 2nd Airborne Visible/Infrared Imaging Spectrometer (AVIRIS) Workshop*, R. O. Green, ed., JPL Publ. 90–54, Jet Propulsion Laboratory, California Institute of Technology, Pasadena, Calif., pp. 199–207.

Horowitz, H. M., J. T. Lewis, and A. P. Pentland, 1975. *Estimating Proportion of Objects from Multispectral Scanner Data, Final Report*, NASA-CR-141862, U.S. Government Printing Office, Washington, D.C., pp. 108

Huete, A. R., 1986. Separation of soil–plant spectral mixtures by factor analysis, *Remote Sensing Environ.*, 19, 237–251.

Huete, A. R., R. D. Jackson, and D. F. Post, 1984. Spectral response of a plant canopy with different soil backgrounds, *Remote Sensing Environ.*, 17, 37–53.

Huguenin, R. L., and J. L. Jones, 1986. Intelligent information extraction from reflectance spectra: absorption band positions, *J. Geophys. Res.*, 91, 9585–9598.

Hunt, G. R., and J. W. Salisbury, 1970. Visible and near-infrared spectra of minerals and rocks: I. Silicate minerals, *Mod. Geol.*, 1, 283–300.

Hunt, G. R., J. W. Salisbury, and C. J. Lenhoff, 1973. Visible and near infrared spectra of minerals and rocks: VI. Additional silicates, *Mod. Geol.*, 4, 85–106.

Hynes, A., and D. M. Francis, 1982. A transect of the early Proterozoic Cape Smith foldbelt, New Québec, *Tectonophysics*, 88, 23–59.

Jackson, R. D., 1983. Spectral indices in *n*-space, *Remote Sensing Environ.*, 17, 37–53.

Jaumann, R., 1991. Spectral-chemical analysis of lunar surface materials, *J. Geophys. Res.*, 96, 22793–22807.

Johnson, P. E., M. O. Smith, S. Taylor-George, and J. B. Adams, 1983. A semiempirical method for analysis of the reflectance spectra of binary mineral mixtures, *J. Geophys. Res.*, 88, 3557–3561.

Johnson, P. E., M. O. Smith, and J. B. Adams, 1992. Simple algorithms for remote determination of mineral abundances and article sizes from reflectance spectra, *J. Geophys. Res.* 97, 2649–2658.

King, M. D., Y. J. Kaufman, W. P. Menzel, and T. Tanré, 1992. Remote sensing of cloud, aerosol, and water vapor properties from MODIS, *IEEE Trans. Geosci. Remote Sensing*, 30, 2–27.

Kruse, F. A., D. H. Knepper, Jr., and R. N. Clark, 1986. Use of digital Munsell color space to assist interpretation of imaging spectrometer data: geologic examples from the northern Grapevine Mountains, California and Nevada, in *Proceedings of the 2nd AIS Data Analysis Workshop*, JPL Publ. 86–35, Jet Propulsion Laboratory, California Institute of Technology, Pasadena, Calif., pp. 132–137.

Kruse, F. A., W. M. Calvin, and O. Seznec, 1988. Automated extraction of absorption features from airborne visible/infrared imaging spectrometer (AVIRIS) and Geophysical Environmental Research imaging spectrometer (GERIS) data, in *Proceedings of the AVIRIS Performance Evaluation Workshop*, JPL Publ. 88–38, Jet Propulsion Laboratory, California Institute of Technology, Pasadena, Calif., pp. 62–75.

Kruse, F. A, K. S. Kierein-Young, and J. W. Boardman, 1990. Mineral mapping at Cuprite, Nevada with a 63 channel imaging spectrometer, *Photogramm. Eng. Remote Sensing*, 56, 83–92.

Kruse, F. A., A. B. Lekhoff, and J. B. Dietz, 1993. Expert system-based mineral mapping in northern Death Valley, California/Nevada, using the airborne visible/infrared imaging spectrometer (AVIRIS), *Remote Sensing Environ.*, 44, 309–335.

Lawler, M. E. and J. B. Adams, 1996. Characterizing image errors using the spectral mixture framework, (*abstract*). Lunar and Planetary Science XXV, Lunar and Planetary Institute, Houston TX, pp. 779–780.

Li, X., and A. H. Strahler, 1986. Geometrical-optical modeling of a conifer forest canopy, *IEEE Trans. Geosci. Remote Sensing*, 23, 705–721.

Loughlin, W. P., 1991. Principal components analysis for alteration mapping, *Photogramm. Eng. Remote Sensing*, 57, 1163–1170.

Marfunin, A. S., 1979. *Physics of Minerals and Inorganic Materials*, Springer-Verlag, New York, 340 pp.

Mazer, A. S., M. Martin, M. Lee, and J. E. Solomon, 1988. Image processing software for imaging spectrometry data analysis, *Remote Sensing Environ.*, 92, 13619–13634.

Mustard, J. F., 1992. Chemical composition of actinolite from reflectance spectra, *Am. Mineral.*, 77, 345–358.

Mustard, J. F., 1994. Lithologic mapping of gabbro and peridotite sills in the Cape Smith fold and thrust belt with thematic mapper and airborne radar data, *Can. J. Remote Sensing*, 20, 222–232.

Mustard, J. F., and C. M. Pieters, 1987a. Quantitative abundance estimates from bidirectional reflectance measurements, *Proceedings of the 17th Lunar Planetary Science Conference, J. Geophys. Res.*, 92, E617–E626.

Mustard, J. F., and C. M. Pieters, 1987b. Abundance and distribution of ultramafic microbreccia in Moses Rock Dike: quantitative application of mapping spectrometer data, *J. Geophys. Res.*, 92, 10376–10390.

Mustard, J. F., and C. M. Pieters, 1989. Photometric phase functions of common geologic minerals and applications to quantitative analysis of mineral mixture reflectance spectra, *J. Geophys. Res.*, 94, 13619–13634.

Nash, D. B., and J. E. Conel, 1974. Spectral reflectance systematics for mixtures of powdered hypersthene, labradorite, and ilmenite, *J. Geophys. Res.*, 79, 1615–1621.

Pieters, C. M., 1983. Strength of mineral absorption features in the transmitted component of near-infrared reflected light: first results from RELAB, *J. Geophys. Res.*, 88, 9534–9544.

Pieters, C. M., J. B. Adams, P. Mouginis-Mark, S. H. Zisk, J. W. Head, T. B. McCord, and M. Smith, 1985. The nature of crater rays: the Copernicus example, *J. Geophys. Res.*, 90(B14), 12393–12413.

Pinty, B., M. Verstraete, and R. Dickinson, 1990. A physical model of the bidirectional reflectance of vegetation canopies: 2. Inversion and validation, *J. Geophys. Res.*, 95, 11767–11775.

Powdysocki, M. H., D. B. Segal, and M. J. Abrams, 1983. Use of multispectral scanner images for assessment of hydrothermal alteration in the Marysvale, Utah mining areas, *Econ. Geol.*, 78, 675–687.

Roberts, D. A., M. O. Smith, and J. B. Adams, 1993. Green vegetation, nonphotosynthetic vegetation, and soils in AVIRIS data, *Remote Sensing Environ.*, 44, 255–269.

Roush, T. L., and R. B. Singer, 1986. Gaussian analysis of temperature effects on the reflectance spectra of mafic minerals in the 1-μm region, *J. Geophys. Res.*, 91, 10301–10308.

Rowan, L. C., and A. B. Kahle, 1982. Evaluation of 0.46–2.36 μm multispectral scanner image of the East Tintic mining district, Utah for mapping hydrothermally altered rocks, *Econ. Geol.*, 77, 441–452.

Rowan, L. C., P. H. Wetlaufer, A. F. H. Goetz, G. C. Billingsley, and J. H. Stewart, 1974. *Discrimination of Rock Type and Detection of Hydrothermally Altered Areas in South-Central Nevada by the Use of Computer-Enhanced ERTS Images*, USGS Prof. Pap. 883, U.S. Geological Survey, Washington, D.C., 35 pp.

Sabol, D. E., J. B. Adams, and M. O. Smith, 1992. Quantitative sub-pixel spectral detection of targets in multispectral images, *J. Geophys. Res.*, 97, 2659–2672.

Salomonson, V. V., and D. L. Toll, 1991. The moderate resolution imaging spectrometer–nadir (MODIS-N) facility instrument, *Adv. Space Res.*, 11, 231–236.

Sherman, D. M., and T. D. Waite, 1985. Electronic spectra of Fe^{3+} oxides and oxide hydroxides in the near-IR to near-UV, *Am. Mineral.*, 70, 1261–1269.

Simonett, D. S, R. G. Reeves, J. E. Estes, S. E. Bertke, and T. T. Sailer, 1983. The development and principles of remote sensing, in *Manual of Remote Sensing*, Vol. 1, 2nd ed., R. N. Colwell, ed., American Society of Photogrammetry, Falls Church, Va., pp. 1–32.

Singer, R. B., 1981. Near-infrared spectral reflectance of mineral mixtures: systematic combinations of pyroxenes, olivine, and iron oxides, *J. Geophys. Res.*, 86, 7967–7982.

Singer, R. B., and T. B. McCord, 1979. Mars: large scale mixing of bright and dark surface materials and implications for analysis of spectral reflectance, in *Proceedings of the 10th Lunar Planetary Science Conference*, pp. 1835–1848.

Smith, G., and R. G. J. Strens, 1976. Intervalence transfer absorption in some silicate, oxide and phosphate minerals, in *Physics and Chemistry of Minerals and Rocks*, R. G. J. Strens, ed., NATO Adv. Study Inst. Petrophys., Wiley, New York, pp. 583–612.

Smith, M. O., P. E. Johnson, and J. B. Adams, 1985. Quantitative determination of mineral types and abundances from reflectance spectra using principal components analysis, *J. Geophys. Res.*, 90(suppl.), C797–C804.

Smith, M. O, S. L. Ustin, J. B. Adams, and A. R. Gillespie, 1990a. Vegetation in deserts: I. A regional measure of abundance from multispectral images, *Remote Sensing Environ.*, 31, 1–26.

Smith, M. O., S. L. Ustin, J. B. Adams, and A. R. Gillespie, 1990b. Vegetation in deserts: II. Environmental influences on regional abundance, *Remote Sensing Environ.*, 31, 27–52.

Smith, M., D. Roberts, J. Hill, W. Mehl, B. Hosgood, J. Verdebout, G. Schmuck, C. Koechler, and J. Adams, 1994. A new approach to quantifying abundances of materials in multispectral images, *Proceedings of the International Geoscience and Remote Sensing Symposium* (IGARSS).

St. Onge, M. R., and S. B. Lucas, 1990. Evolution of the Cape Smith Belt: Early Proterozoic continental underthrusting, ophiolite obduction and thick-skinned folding, in *The Early Proterozoic Trans-Hudson Orogen of North America*, Spec. Pap. 37, J. F. Lewry, and M. R. Stauffer, eds., Geological Association of Canada, Waterloo, Ontario, Canada, pp. 313–351.

St. Onge, M. R., S. B. Lucas, D. J. Scott, N. J. Bégin, H. Helmstaedt, and D. M. Charmicheal, 1988. Thin-skinned imbrication and subsequent thick-skinned folding of rift-fill, transitional-crust, and ophiolite suites in the 1.9 Ga Cape Smith Belt, northern Quebec, in *Current Research*, Part C, Pap. 88-1C, Geological Survey of Canada, Ottawa, Ontario, Canada, pp. 1–18.

Sultan, M., R. E. Arvidson, N. C. Sturchio, and E. A Guinness, 1987. Lithologic mapping in arid regions with *Landsat* thematic mapper data: Metiq dome, Egypt, *Geol. Soc. Am. Bull.*, 99, 748–762.

Sunshine, J. M., 1994. Inferring the composition of mafic lithologies from multi-and hyper-spectral data sets: implications for remote sensing of terrestrial bodies, Ph.D. dissertation, Brown University, Providence, R.I., 191 pp.

Sunshine, J. M., and C. M. Pieters, 1993. Estimating modal abundance from the spectra of natural and laboratory pyroxene mixtures using the modified Gaussian model, *J. Geophys. Res.*, 98, 9075–9087.

Sunshine, J. M., C. M. Pieters, and S. F. Pratt, 1990. Deconvolution of mineral absorption bands: an improved approach, *J. Geophys. Res.*, 95, 6955–6966.

Sunshine, J. M., L. A. McFadden, and C. M. Pieters, 1993. Reflectance spectra of the Elephant Moraine A79001 meteorite: implications for remote sensing of planetary bodies, *Icarus*, 105, 79–91.

Swayze, G., R. N. Clark, F. A. Cruse, S. Sutley, and A. Gallagher, 1992. Ground-truthing AVIRIS mineral mapping at Cuprite, Nevada, *Summaries of the 3rd Annual JPL Airborne Geoscience Workshop*, R. O. Green, ed., JPL Publ. 92-14, Jet Propulsion Laboratory, California Institute of Technology, Pasadena, Calif., pp. 47–49.

Tarantola, A., and B. Valette, 1982. Generalized nonlinear inverse problems solved using the least squares criterion, *Rev. Geophys. Space Phys.*, 20, 219–232.

Taylor, M. M., 1974. Principal components color display of ERTS imagery, in *Pro-

ceedings of the Third Earth Resource Technology Satellite-1 Symposium, NASA Spec. Publ. SP-351, pp. 1877–1897.

Tompkins, S. T., J. F. Mustard, C. M. Pieters, and D. W. Forsyth, 1993. Objective determination of image end-members in spectral mixture analysis (abstract), *Lunar and Planetary Science XXIV*, Lunar and Planetary Institute, Houston, Texas, pp. 1431–1432.

Tompkins, S., J. F. Mustard, C. M. Pieters, and D. W. Forsyth, submitted 1997. A modified approach to spectral mixture analysis for derivation of end-members, *Remote Sensing Environ.*, 59, 472–489.

Verstraete, M., B. Pinty, and R. Dickinson, 1990. A physical model of the bidirectional reflectance of vegetation canopies: 1. Theory, *J. Geophys. Res.*, 95, 11755–11765.

Williams, R. S., 1983. Geologic Applications, in *Manual of Remote Sensing*, 2nd ed., Vol. II, American Society of Photogrammetry, Falls Church, Va., pp. 1667–1954.

Xu, P., and R. Greeley, 1992. Convex set and linear mixing model (abstract), *Lunar and Planetary Science X XIII*, Lunar and Planetary Institute, Houston, Texas, pp. 1545–1546.

Yamaguchi, Y., and R. J. P. Lyon, 1986. Identification of clay minerals by feature coding of near-infrared spectra, in *Proceedings of the International Symposium on Remote Sensing of Environment, First Thematic Conference for Exploration Geology*, ERIM, pp. 627–636.

Integration and Visualization of Geoscience Data

J. R. Harris, D. W. Viljoen and A. N. Rencz

Geological Survey of Canada
Ottawa, Ontario, Canada

A picture is worth 1,000 words!

6.1   INTRODUCTION

The interpretation of geoscience data, including remotely sensed data, can be undertaken on two basic levels, one that includes rigorous quantitative analysis in which statistical procedures are utilized to extract relevant information, whereas the other comprises qualitative analysis in which the traditional art of photo interpretation is employed to both interpret and extract information from imagery. The former relies on decisions based on mathematical/statistical criteria in which computer software/hardware plays an integral role in the data analysis process. The latter also relies on the computer but not so much in the decision-making process but more in a data preparation role in which the data may be combined (merged) and presented in novel ways to highlight or enhance the data.

Figure 6.1 is a summary of the steps involved in qualitative data processing starting with the formulation of a geologic problem through to the final output images that will hopefully provide new knowledge. The geologist is the key element in the entire process as he or she decides on the particular problem to be solved, the data to be collected, data processing strategies, evaluation of the resultant images, and an assessment as to whether the output images assist in understanding the particular geologic problem. Visualization is a component of all these steps. However, for this chapter we focus on the data combination and 2.5-dimensional visualization com-

Remote Sensing for the Earth Sciences: Manual of Remote Sensing, 3 ed., Vol. 3, edited by Andrew N. Rencz.
ISBN: 0471-29405-5 © 1999 John Wiley & Sons, Inc.

ponent of the visualization oriented analysis and processing stage, shown in Figure 6.1 as it applies to geoscience data displayed in a raster (grid) format.

A raster data set is a two-dimensional array of values (i.e., rows and columns) that can be used to depict an image. Each value can represent a measured value (e.g., reflectance) or a color index for a particular indivisible area of the image (pixel). Pixels are most commonly square or rectangular. Irrespective of shape, all pixels in a given raster have the same size, and the pixel values are represented digitally by the same number of bits. Most common image data types are 1-bit (bit maps), 8-bit (256 levels), and 16-or 32-bit integer or floating point. The across-track sample spacing, along-track scan-line spacing, and field of view of many satellite or airborne remote sensors (e.g., *Landsat*, SPOT, radar) are such that there is a regular grid of samples and continuous coverage for an area (e.g., raster data). Other remote sensors (e.g., airborne magnetic sensors) and sampling methods (e.g., geochemical surveys) collect data along nonparallel flight lines or at points in an irregular or regular spatial pattern, respectively. Interpolation methods such as minimum curvature, inverse distance weighting, or kriging are commonly applied to convert these point data into a continuous surface raster image suitable for visualization and analysis. Thematic data, such as geological units on a map, must be digitized, preprocessed, and stored as polygons in an image analysis system (IAS) or geographic information system (GIS). For some operations the polygons must be converted from a structured set of lines and polygon labels to a raster format (data model transformation). File format conversions, data model transformations (point-to-raster interpolation), and radiometric and geometric correction* are all part of data preprocessing, as shown in Figure 6.1. The IAS or GIS commonly provide software tools for radiometric and geometric corrections (georeferencing). With the data in a common projection [e.g., UTM (Universal Transverse Mercator)], datum (e.g., NAD 83—North American Datum, 1983), and stored in a structure directly accessible by the IAS or GIS, it is possible to apply various visualization-oriented analysis and processing procedures.

With respect to the visualization-oriented analysis and processing stage, shown in Figure 6.1, data combination (merging) involves the manipulation and integration of geoscience data to produce images that highlight spatial relationships between different data. The process of merging data has always been undertaken by earth scientists, although primarily on an analog basis by overlaying maps on a light table. With the advent of more advanced computer IAS (image analysis systems) and GIS (geographic information systems) and the present emphasis on the collection and compilation of digital data, data combination has become less cumbersome, less time consuming, and more flexible. For the purpose of this chapter, *data combination* will refer to those techniques that merge data sets for producing images for visual interpretation. This is different from *data integration*, which implies a more rigorous modeling of the data with the result being a thematic map. A detailed discussion of data modeling is beyond the scope of this chapter. Interested readers are encouraged to consult the following references, which provide detailed accounts and case studies of data modeling of geoscience data for mineral exploration, in particular Harris (1989) Bonham-Carter (1994), Rencz et al. (1994a); Harris et al. (1995), and Wright and Bonham-Carter (1996). Enhancement and classification comprise a suite of im-

---

*Standard image processing texts such as Lillesand and Kiefer (1994), Sabins (1997), and Drury (1993) provide further details on georeferencing, radiometric corrections, and other preprocessing procedures.

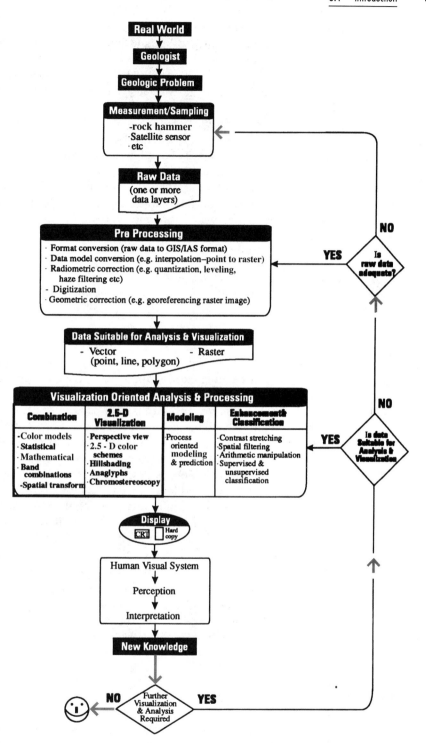

Figure 6.1   Summary of Geoscientific visualization process.

age analysis methods commonly applied to remotely sensed satellite data. These methods are described in standard remote sensing and image analysis texts (e.g., Sabins, 1992; Drury, 1993) and therefore are not discussed in this chapter.

In this chapter we review data combination and visualization processes, particularly as they relate to developments since the early 1980s. The emphasis in this chapter is on 2- and 2.5-dimensional visualization of geoscience data presented by discussions on background theory and a number of case studies. The first part of the chapter provide a brief overview of the human visual system and color models since they are essential to understanding the process of combining data for effective visual communication. A number of data combination techniques and examples of their application are discussed. A brief look at evolving trends including 3- and 4-dimensional and hyperspectral analysis ends the chapter. In Appendix 6A we discuss issues related to display systems, such as monitors and plotters used in visualizing various data discussed in this chapter.

6.2 VISUALIZATION

The process of visualization has been defined in a number of different ways, with some focusing on the computational aspects and others focused on cognitive and/or graphical aspects of visualization (Buttenfield and Mackaness, 1991, p. 432). Buttenfield and Mackaness (1991) provide a definition that embraces all of these aspects: "Visualization is the process of representing information synoptically for the purpose of recognizing, communicating and interpreting pattern and structure. Its domain encompasses the computational, cognitive, and mechanical aspects of generating, organizing, manipulating and comprehending such representations. Representations may be rendered symbolically, graphically, or iconically and are most often differentiated from other forms of expression (textual, verbal, or formulaic) by virtue of their synoptic format and with qualities traditionally described by the term 'Gesalt.' " Thus visualization is a tool for both interpreting image data entered into the computer and for generating images from complex multidimensional data sets (McCormik et al., 1987). Visualization also allows the interpreter to develop mental representations of the data that allow for the identification of patterns and order within the data (MacEachern et al., 1992). Therefore, in the framework of Figure 6.1, visualization is a component of all steps in the analysis of data.

Remote sensing and other forms of raster data can be transformed to produce imagery that can effectively communicate states, processes, and spatial relationships of the real world in ways that were not possible before the computer age. The most effective computer visualization techniques give due consideration to the characteristics of the human visual system.

6.2.1 Models of the Human Visual System

It has been known for many years that the retina of the human eye is composed of color-sensitive receptors that have spectral responses with peaks in the red, green, and blue portions of the spectrum, as shown in Figure 6.2. For reflected light, the

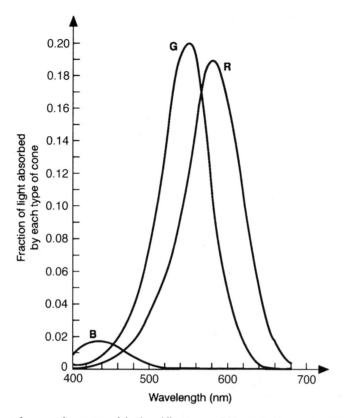

Figure 6.2 Spectral response characteristics of the three different cone cell types in the human retina. (Foley et al., 1990.)

spectral power distribution that impinges on the retina is the result of the interplay between the spectral power distribution of the illuminating source, as shown in Figure 6.3, and the spectral reflectance curves of the observed object, as in Figure 6.4. For direct light sources, such as color computer monitors (CRTs), the color can be decomposed into RGB components that stimulate the different cones in the retina. It seems logical that the human retina creates a "pixelized" image for the visual cortex to interpret where the perceived color at a particular location on the retina is the weighted sum of the nearest cones of each receptor type. This so-called tristimulus model of color vision is simple and attractive to many in the visualization and image processing field since it is analogous to artificial sensors that are used to capture imagery (e.g., camera). The tristimulus model is the basis for the CIE color model, discussed later in this chapter, and is used to define the color gamut of the human eye. This model has been widely accepted for many years even though there are many inconsistencies with respect to visual perception. For example, the color of an object is known to be affected by the color of its surroundings (see Section 6.2.1.2). The tristimulus theory of color perception is valid for color matching experiments where an observer must match a reference color spot with one where the color is controlled by adjusting the contribution of red, blue, and green lights. However, the tristimulus model is not valid for image interpretation and visualization in general.

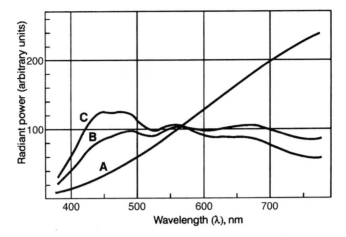

Figure 6.3 Relative spectral power distributions of CIE standard illuminants: A-blackbody radiator at absolute temperature of 2856 K; B-direct sunlight with a correlated color temperature of approximately 4870 K; C-average daylight with a correlated color temperature of approximately 6770 K. (Judd & Wyszecki, 1975.)

6.2.1.1 DISPELLING THE MYTH OF THREE ADDITIVE PRIMARY COLORS.

The Young–Helmholtz theory of color perception (e.g., Sharp and Philips, 1993) postulates that human beings perceive color on the basis of the combined responses of three types of cone receptor on the retina and that any color can be created from mixing the three primary colors; red, green, and blue. However, this may not be the case entirely. RGB primaries can produce a significant subset of the human color gamut but not all. Edwin Land was able to show that full-color images could be

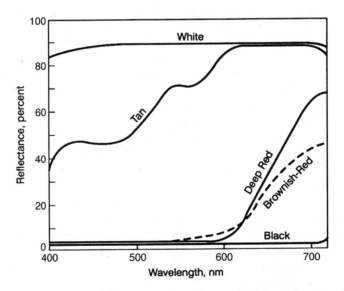

Figure 6.4 Spectral reflectance curves of five imaginary plastics-white, tan, brownish red, deep red, and black. (Judd & Wyszecki, 1975.)

made from a large assortment of wavelength combinations: pairs or trios of wavelength entirely different from the traditional primaries (Friedhoff, 1991).

6.2.1.2   OBJECT/BACKGROUND EFFECTS ON COLOR PERCEPTION.

It is evident that many in the field of visualization believe that the perceived color of an object is the result of the visual cortex evaluating the relative strengths of the signals from the red, green, and blue cone receptors on the retina. However, current theory suggests another option. It seems that the eye is able to see color independently of wavelength. Edwin Land (Friedhoff, 1991) was able to show that in a collage of colored rectangular papers the color of a particular rectangle depended on the arrangement of all the colored papers. The conclusion reached is that the color of a given paper depends on the "context of the light reflected by the larger areas of the scenes," not the wavelength of reflected light. In addition, the brightness of the wavelength components of the illuminating light could be turned up and down without changing the perceived color.

Figure 6.5 and Figure 6.6 are both examples of how brightness and color depend on context. In Figure 6.5, the shading for the entire area is the same, but the perceived brightness of the shading depends on the brightness of the white or black line overlay. This same effect is observed in color images as well. Figure 6.6 (see the color insert) shows how the brightness of the surrounding color affects the perceived brightness of the gray line. A black-and-white photocopy of this figure would reveal that most of the colors shown are quite similar in brightness with the exception of yellow. The color of the gray line as it passes through the yellow is noticeably darker, suggesting that the brightness of the surrounding background affects the perceived color of objects. This reality cannot be explained by the tristimulus theory of color vision, where each location in the visual field is assigned a color based on the weighted sum of responses from the nearest three cone cells. Land's explanation is that the entire scene is implicated in color perception at a particular point and that the eye uses contrast and brightness information in the determination of color.

6.2.1.3   FORM, COLOR, AND BRIGHTNESS.

The human visual system can be subdivided into three systems (form, color, brightness), each with its own distinct function. Perhaps the most important of these is the

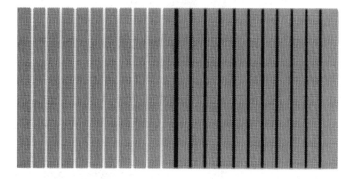

Figure 6.5   Perceived lightness of a gray-tone pattern is affected by whether it is overlaid with white or black lines.

system responsible for brightness contrast and shape perception. Form is responsible for movement, location, and spatial organization. There are a few properties of these systems that make it possible for poorly chosen color schemes to cause conflicts between form (shape) and color systems: (1) the color system is slow and has lower visual acuity, (2) the form system is faster and has better visual acuity but is color blind, and (3) the form and color systems are processed independently.

Figure 6.7 (see the color insert) shows a digital elevation model (DEM) imaged with the same color maps but different elevation range assignments for each color. Color is a "grouping" criterion for the visual system, and this causes different form perceptions in Figure 6.7 for the same mountain. Shading is an important cue for form perception. It is perceived quickly with high spatial resolution, whereas the color system response is slower and has much lower spatial resolution than the form system. The television industry takes advantage of this by transmitting the color content of an image at a lower spatial resolution than the black-and-white component, significantly reducing the bandwidth required to transmit color signals.

The importance of shading for form perception is demonstrated in Figure 6.8 (see the color insert). Forty colors are used in Figure 6.8a, but since this is a continuous surface, similar hues are adjacent and there is very little brightness contrast to help the eye distinguish similar hues. By modulating the intensity with a shaded relief image of the digital elevation model, form perception is improved dramatically, as shown in Figure 6.8b. A fact that may come as a surprise to some is that a shadow can be any hue; it only needs to be darker than the rest of the surface to convey the sense of depth (Livingstone, 1988).

6.2.1.4 COLOR AND DEPTH PERCEPTION.

Depth perception relies on primary and secondary cues (Braunstein, 1976). Primary cues include physiological factors, muscular activity in the eye, and differences between the images on the left and right retina. Secondary cues are pictorial and include relative sizes, hidden objects, perspective, relative clarity of objects in an image, and position relative to the horizon. The secondary cues serve to reinforce perceptions provided by the primary cues. To create images that provide effective three-dimensional illusions, primary cues are most important. Providing primary cues using imagery is achieved by presenting each eye with a slightly different view of the same object or area. A number of methods have been developed to provide these cues, including image stereo pairs, anaglyphs, holography, three-dimensional films, and various devices that modify what we see using reflection, refraction, or diffraction. Some of these methods are discussed later in the chapter.

The cornea and lens of the eye refract light like a prism, where blue light is refracted more than red light. This means that light from blue objects and red objects will fall on slightly different areas of the retina. The structures of the eye communicate to "stabilize" our perception of what is sensed. However, this refraction is evident in a few different ways. One is the clarity with which we can read blue letters compared to red letters, especially at a distance. Another is the tendency of the preconscious and conscious visual system to "decide" which end of the color spectrum is close and which is far. For most viewers, with both eyes open, either Figure 6.9a or b (see the color insert) will initially seem to be a pyramid and the other a hallway. However, you can decide that either the blue or red is closer, and the perception of a pyramid will change to that of a hallway, or vice versa. However, once the decision is made that the blue is farther in (*a*), there is the perception of a hallway and then the initial perception of (*b*) will be that of a pyramid. The implication for creating color 2.5-dimensional images is that colors should be allocated in terms of increasing or decreasing wavelength. If the relationship between the pixel value and wavelength of the color increases so that high values are represented by red, as in Figure 6.9 *a*, using a purple for the highest values, which is common, causes problems with depth perception since the wavelength of the highest values has decreased. In this case it is possible for the conscious visual system to aid with the interpretation of the highest values; however, effective visualization caters to the preconscious system.

6.3   COLOR MODELS

Color models are an artificial framework for describing color. There are many models that have been developed from simple words (red, blue, etc.) to complex formulas that can be used to describe and produce color. Why do we need such models? Simply stated, we need models because terms such as *tan* or *deep red* provide some idea of color but are not sufficient for approximating the spectral power distribution of a light source or reflected light. Color systems (models) are used (1) to define and communicate color more accurately (i.e., quantitative versus qualitative description), and (2) in image analysis, to render different image data simultaneously.

The CIE System (Commission Internationale L'Eclairage) is the globally recognized standard for color specification and is used in applications where quantitative color definition is required. It is the only color system based on color perception and is capable of defining all colors in the gamut of the human visual system. However, since the human visual system is extremely complex, the graphical and mathematical representation of this model is not practical for interactive computer graphics or image processing applications. The RGB (red–green–blue) and CMY (cyan–magenta–yellow) color models are capable of defining a significant subset of the human perceptual color gamut suitable for representing color on computer monitors and color printers, respectively. The IHS (intensity–hue–saturation), HSV (hue–saturation–value), and HLS (hue–lightness–saturation) color models provide a more user-oriented framework for defining and describing color. That is, we commonly describe colors in terms of dominant hue (e.g., green), pale or vibrant (saturation), and lightness or darkness (intensity).

In the realm of image processing, the RGB and IHS models are useful for visualizing relationships between various types of data through color composite imagery that combine different types of geoscience data. In the following section we describe briefly the color models and the mathematical transformations to convert color coordinates between color models. Many examples showing how these color models can be used to combine geoscience data are presented in Section 6.4.

6.3.1 Physicophysical Color Model: The CIE System

On the basis of extensive color-matching experiments, tristimulus color-matching functions were empirically derived that define the proportions of red, green, and blue light required to match stimuli for the full spectrum of visible light as shown in Figure

6.10 (Wright, 1928–1929; Guild, 1931). Each of the three spectral stimuli have negative values for some portions of the visible spectrum. Negative RGB coordinates are conceptually and mathematically awkward to deal with. Conceptually, we must attempt to understand how to add negative light. Using an example from Judd and Wyszecki (1975), suppose that we have a color stimulus (S) that cannot be matched with addition of RGB primaries. However, it is discovered that adding three units of R to S results in a match with $4G$ and $4B$ as follows:

$$S + 3R = 4G + 4B$$

In accordance with principles of color perception proposed by Grassman (1853), this means that a valid mathematical representation of S is

$$S = -3R + 4G + 4B$$

In this way it is possible to define RGB coordinates for colors outside the RGB color gamut and explain the negative RGB coordinates in Figure 6.10.

To facilitate colorimetric calculations, a linear transformation of the RGB color-matching functions was performed to create the XYZ imaginary primaries shown in Figure 6.11. The linear transformation was applied to make x, y, z positive for all wavelengths as well as to facilitate other elements of colorimetric calculation (Judd, 1933). These functions define the CIE 1931 standard colorimetric observer.

The XYZ imaginary primaries are the only tristimulus functions that can define all colors in the color gamut of the human eye. The gamut of a three-stimulus mixture of "real" primaries can be increased by choosing the three stimuli as parts of the spectrum itself, but experiments of this kind have shown that there is no choice of

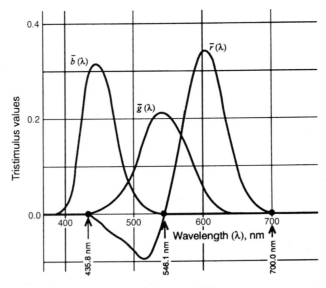

Figure 6.10   Tristimulus valves $r(\lambda)$, $g(\lambda)$, $b(\lambda)$ of spectral stimuli of different wavelengths but constant radiance measured by an average observer with normal color vision. (Judd & Wyszecki, 1975.)

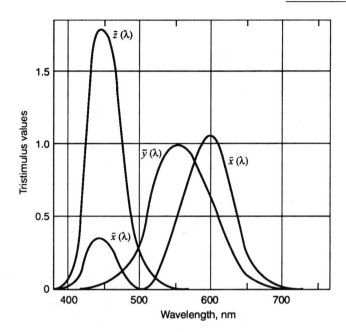

Figure 6.11   Tristimulus valves $\bar{x}(\lambda)$, $\bar{y}(\lambda)$, $\bar{z}(\lambda)$ derived by a linear transformation from the tristimulus valves shown in Figure 6.10. (Judd & Wyszecki, 1975.)

three stimuli that will yield color matches for all other stimuli (Judd and Wyszecki, 1975).

Although the *XYZ* system can define colors, it is not a perceptually uniform color model. That is, the perceptual effect of a unit change in any component is dependent on the starting color in *XYZ* space. Two standardized systems have been developed by the CIE that improve the perceptual nonuniformity of the *XYZ*. Both the L*u*v* and L*a*b* systems are too computationally intensive for interactive computer applications and image display. Interested readers should refer to Judd and Wyszecki (1975) for more information about the CIE standards and the *XYZ* color system.

6.3.2 Hardware-Oriented Color Models: RGB, CMY, and CMYK

The CIE color system is complex and computationally intensive. The RGB color model is simple and easy to implement in computer hardware and software. The RGB color model can be visualized as a three-dimensional Cartesian coordinate system, as shown in Figure 6.12. The diagonal of the cube from black (0,0,0) to white (1,1,1) is achromatic and represents the gray levels of the RGB system. It is important to note that the domain of RGB is dependent on the software or hardware system implementation. For example, a "true-color" computer display usually allocates 8 bits for each RGB component. This means that the domain for *R*, *G*, and *B* ranges from 0 to 255 and the total number of colors that can be represented is 256^3 (16,777,216).

Cyan, magenta, and yellow (CMY) are the subtractive primary colors used in

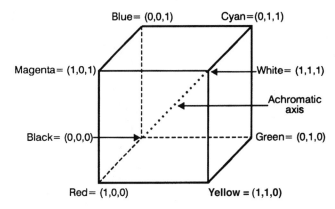

Figure 6.12   RGB (red–green–blue) color model.

printing devices. The color model can be visualized as a three-dimensional Cartesian coordinate system similar to the RGB cube shown in Figure 6.12 except that the RGB and CMY primaries are swapped, respectively, and the achromatic axis is inverted. That is, if CMY are all added in full measure, the result is black. CMY are termed *subtractive primaries* since they selectively remove portions of the spectral distribution of an illuminating source leaving the reflected color. For example, a cyan pigment absorbs light from the red portion of the spectrum, leaving blue and green light. The combination of blue and green light is cyan. Figure 6.13 (see color insert) is a simple RGB ternary triangle that summarizes the relationship between the additive and subtractive primary colors.

Black is often difficult to achieve with CMY, due to properties of the pigments or misalignment of plotter pixels. To compensate for this, plotters often add black separately and the color model becomes CMYK, reflecting this change. The amount of black added to colors is usually the minimum of the CMY primaries (Schneirer, 1993).

Matching colors on a computer monitor to those produced on a plotter depends on the spectral properties of the plotter pigments, the dithering patterns used, and the properties of the monitor. In an ideal world, the color properties of the plotter pigments should be the exact inverse of color monitors. In reality, this is not the case, and "what-you-see-is-what-you-get" (WSIWYG) colors usually require modifying the color monitor's properties to match the colors of the plotter.

6.3.3 User-Oriented Color Models: IHS (HSV, HLS)

A number of color models have been developed to provide a more user-oriented framework for defining color. For example, a color might have RGB coordinates of 200, 200, 120. It is unlikely that this color would be described as an equal mix of red and green with a significant contribution of blue. Viewing this color, a user of an imaging software is more likely to describe this as a pale yellow. This is what is implied by the term *user-oriented color models*. Color in these models is described in terms of hue, saturation, and brightness (intensity), which are defined as follows:

1. *Hue:* the similarity of the color to red, yellow, green, or blue or some combination of any two (e.g., greenish blue)
2. *Saturation:* colorfulness relative to brightness (e.g., pale red)
3. *Brightness (emitted) or lightness (reflected):* perception of dark or light (emitted), black or white (reflected)

A family of related color models have been developed to accommodate this user view of color: intensity–hue–saturation (IHS), hue–saturation–value (HSV), and hue–lightness–saturation (HLS). These models differ in the order of the components, the domain of those components, and the conceptual model that is used. For example, *V* is equivalent to *L* in the HSV and HLS models, respectively. In HLS, the lightness is second, in HSV it is last, so that the order of the coordinates used to define the color are different. The conceptual model of the HSV system is a cone or hexcone, while a double cone or hexcone is used for HLS. With HLS, fully saturated colors ($S = 1$) are possible at $L = 0.5$ in HLS, whereas they are only possible at $V = 1.0$ in HSV. Hue ranges from 0 to 360 in all of these models but the origin ($H = 0$) is usually blue or red. From a users point of view, none of these models provide an advantage over the other.

To be consistent with other material in this chapter related to color model transformations and color model applications, we will use the IHS model and represent it conceptually with a cone, as shown Figure 6.14. Since the domain for *S* is 0 to 1 (or 0 to 100%) for all values of intensity, it could be argued that a cylinder might be a better conceptual model for the IHS color space. However, perceptually the number of colors that can be perceived decreases with intensity and the contributions of hue and saturation become insignificant at an intensity of 0. For these reasons, a cone seems to be a better representation.

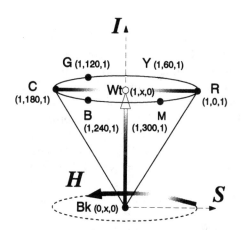

Colors are described by (I,H,S) coordinates.
Achromatic axis for S = 0, H can be any value x.

0 ≤ H ≤ 360
0 ≤ S ≤ 100
0 ≤ I ≤ 100

Figure 6.14 IHS (intensity–hue–saturation) color model.

6.3.4 Color Model Transformations

Color model transformations are used to prepare imagery for display or plotting and for combining various geoscience data, as presented later in the chapter. In this section we present the conceptual and mathematical elements of color model transformations.

6.3.4.1 RGB TO CMY AND CMYK.

The red–green–blue (RGB) to cyan–magenta–yellow (CMY) and CMYK color transform is used for converting additive primaries used for color monitors to subtractive primaries used for color plotters. This is an extremely simple transform involving the following equations:

$$
\begin{array}{ll}
C' = 1 - R & K = \min\,(C',M',Y') \\
M' = 1 - G & C = C' - K \\
Y' = 1 - B & M = M' - K \\
& Y = Y' - K
\end{array}
\tag{6.1}
$$

$$\quad (a) \qquad\qquad\qquad (b)$$

The K refers to the amount of black added to the color. The CMY-to-RGB color transform is the inverse of equation 6.1:

$$
\begin{array}{l}
R = 1 - C \\
G = 1 - M \\
B = 1 - Y
\end{array}
\tag{6.2}
$$

6.3.4.2 RGB TO IHS.

Many different versions of this transform can be found in the literature involving linear transformations (Harrison and Jupp, 1990), color "distances" (Foley, 1981), and double rotation of RGB space (Gillespie et al., 1986). The transform presented here is easiest, conceptually, and most efficient and simplest, mathematically.

Given a color $P(R,G,B)$, the intensity (I) is simply the maximum of (R,G,B). This has been used in the method presented by Foley (1981). In terms of our conceptual model, the color cone shown in Figure 6.14, an intensity of 1 maps a color on the top of the cone irrespective of the number of RGB colors used to create the color. This may be somewhat counterintuitive in that P (0, 1, 1) and Q (0, 0, 1) would have the same intensity even though P has green and blue light sources at full intensity while Q only has blue. In most GIS and IAS, however, this is the framework that is used. It is common to refer to the intensity and saturation components in terms of values between 0 and 100%.

To compute hue (H) and saturation (S), it is necessary to visualize the top of the color cone. From this perspective, the coordinate system used to describe the colors is two-dimensional, as shown in Figure 6.15. Hue will be defined by the angle between OP (angle H) and the X-axis. Saturation will be the distance between the origin (0) and P. We can treat RGB as opposing vectors, at 120° to each other, and

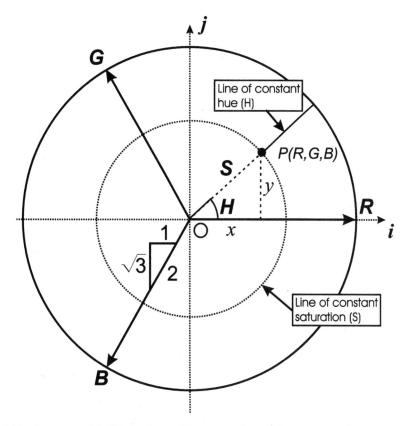

Figure 6.15 Cross section of the IHS cone showing the geometric relationship between intensity, hue, saturation and red, green, and blue coordinates.

the color $P(R,G,B)$ can then be defined in terms of unit vectors $\mathbf{i}$ and $\mathbf{j}$ in the direction of x and y, respectively. The equation that describes P in these terms is

$$P(x, y) = R(\mathbf{i} + \mathbf{j}) + G\left(-\frac{1}{2\mathbf{i}} + \frac{\sqrt{3}}{2\mathbf{j}}\right) + B\left(-\frac{1}{2\mathbf{i}} - \frac{\sqrt{3}}{2\mathbf{j}}\right) \qquad (6.3)$$

where $0 \leqslant R,G,B \leqslant 1$. The $\mathbf{i}$ and $\mathbf{j}$ components can be separated to find x and y of P:

$$x = R - \frac{1}{2G} - \frac{1}{2B} \qquad (6.4)$$

$$y = \frac{\sqrt{3}}{2G} - \frac{\sqrt{3}}{2B} \qquad (6.5)$$

The hue and saturation can now be calculated. The equations for the RGB-to-IHS transformation are

$$I = \max(R, G, B) \tag{6.6}$$

$$H = \tan^{-1}\frac{y}{x} \quad \text{if } H < 0, H = H + 360 \tag{6.7}$$

$$S = \sqrt{x^2 + y^2} \tag{6.8}$$

6.3.4.3 IHS TO RGB.

The equations used in the RGB-to-IHS transformation could be manipulated to convert IHS to RGB. However, the result would be more complex than the following valid and simple solution adapted from Foley (1981).

Figure 6.16 shows the relationship between R, G, B components and hue. From this figure it is apparent that there are six 60° distinct intervals in terms of the contributions of R, G, B to hue. Each of these intervals is characterized by one component at the minimum, one at the maximum, and the other either rising to the maximum or falling to the minimum. The R, G, B components can be in one of the following

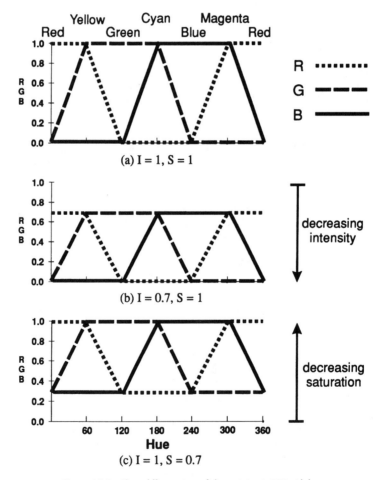

Figure 6.16   Three different views of the variation in RGB with hue.

four "states," and the magnitude of the components can be derived by the associated equations: At the maximum,

$$R, G, \text{ or } B = I \tag{6.9}$$

At the minimum,

$$R, G, \text{ or } B = I(1 - S) \tag{6.10}$$

On the positive slope,

$$R, G, \text{ or } B = I\,[1 - (Sf)] \tag{6.11}$$

On the negative slope,

$$R, G, \text{ or } B = I\,\{1 - [S(1 - f)]\} \tag{6.12}$$

The variable f in equations (6.11) and (6.12) refers to the relative position of the hue into the 60° interval. For example, if $H = 140$, then $f = 1/3$ since the second interval begins at 120°, and therefore

$$\frac{140 - 120}{60} = \frac{1}{3}$$

Referring to Figure 6.16a, Table 6.1 summarizes the appropriate equations to use for each 60° interval. As S decreases, the difference between the minimum and maximum value of R, G, and B decreases. The limit is the achromatic axis, where $S=I$ and H is undefined.

The IHS color model is useful from a user's perspective since it more closely reflects how colors are described. In many situations the IHS model provides advantages over image combination (merging) based on RGB (Buchanan, 1979), as demonstrated by the examples presented later in this chapter.

TABLE 6.1   Equations Used to Calculate RGB Coordinates Given IHS Values

Hue (H)	Equation to Calculate:		
	R	G	B
$0 \leqslant H < 60$	I	$I[1-(Sf)]$	$I(1-S)$
$60 \leqslant H < 120$	$I\{1-[S(1-f)]\}$	I	$I(1-S)$
$120 \leqslant H < 180$	$I(1-S)$	I	$I[1-(Sf)]$
$180 \leqslant H < 180$	$I(1-S)$	$I\{1-[S(1-f)]\}$	I
$240 \leqslant H < 300$	$I[1-(Sf)]$	$I(1-S)$	I
$300 \leqslant H < 360$	I	$I(1-S)$	$I\{1-[S(1-f)]\}$

6.4 DATA COMBINATION

The process of geologic data combination often referred to as data merging or data fusion, must obviously be driven by an overall objective. The nature of the problem will guide the selection of the most appropriate data to combine, and the characteristics of the data will influence the choice of a particular combination technique. The ultimate goal of the combination procedure is to produce images in which the colors can be interpreted in relation to surface features. This implies that the characteristics of the input data are known and that the combination technique preserves the relationship between the input data characteristics and the resulting image tones or colors. This allows for a reasonably accurate assessment of what the image colors represent on the ground. Combination techniques that do not uphold this basic relationship are inherently less useful. The combination or merging process, of course, must make scientific sense in that it is often easy to merge inappropriate data, providing beautiful but uninterpretable images. This problem can be augmented by the particular integration technique chosen (i.e., statistical transforms). A general rule is that the combination process should involve the merging of data that complement each other in a geologically sensible way so that the combined image provides more useful information than all its components.

6.4.1 Combination Techniques

Many techniques are available for combining data. The techniques discussed in this chapter fall under five categories: (1) band combinations (three color composite images), (2) arithmetic combinations, (3) statistical transforms, (4) spatial transforms, and (5) color model transforms.

The first three techniques are standard software-driven tasks available on most IAS or GIS and are not covered here in detail. The fourth technique, based on the wavelet transform (Mallat, 1989), focuses on the spatial properties of the data to be merged. The fifth set of procedures involves associating data sets to components of the RGB and IHS color models, discussed in Section 6.3.

Table 6.2 summarizes different methods for combining data along with references to researchers who have applied these procedures to a variety of geoscience data.

6.4.1.1 BAND COMBINATIONS (COLOR COMPOSITES).

Three channels (bands, images) of raster (gridded) data can be combined using the RGB additive primaries in which each channel is displayed through a different primary color, forming what is traditionally known as a color composite image. RGB color composite images are simple and effective as the mixing of the three primary colors (red, green, blue) can produce a wide range of colors. Any type of digital raster data, including remotely sensed, geophysical, digital terrain, and interpolated point data, may be combined and displayed using this method. For example, two *Landsat* TM channels could be combined with a higher-resolution SPOT or radar image using this technique, although the combined image would lack spectral and spatial contrast (see the next sections).

TABLE 6.2   Summary of Data Combination Techniques

Method	Key References	Data Types Merged
RGB band combinations	Standard method for displaying remotely sensed data in color composite form; references to this technique can be found in any introductory text on remote sensing (e.g., Sabins, 1997; Drury, 1993; Lillesand and Keifer, 1994	Multispectral data (e.g., *Landsat, SPOT,* NOAA)
Arithmetic combinations	Harris and Slaney, 1982; Cliche et al., 1985; Chavez, 1986; Welch and Ehlers, 1987; Harris et al., 1994	Air- and spaceborne radar with *Landsat* data
Statistical transforms	Singh and Harrison, 1985; Shettigara, 1992	Principal component substitution (COS)
Color space transforms (e.g., IHS)	Haydon et al., 1982; Welch and Ehlers, 1987; Harris et al., 1990, 1994; Carper et al., 1990; Chavez et al., 1991; Ehlers, 1991; Rheault et al., 1991; Edwards and Davis, 1994; Rencz et al., 1994b	Air- and spaceborne radar with *Landsat* data; radar with geophysical data; radar with thematic data; *Landsat* data with HCMM and RBV data
Spatial transforms [wavelet transform, high-frequency filter transform (HPF)]	Chavez and Bowell, 1988; Mallat, 1989; Yocky, 1995, 1996; Garguet-Duport et al., 1996	*SPOT* with *Landsat* TM

6.4.1.2   ARITHMETIC COMBINATIONS.

Multiplying, dividing, adding, and subtracting two images is a simple method of combining data and has been used effectively to merge *SPOT* with *Landsat* thematic mapper (TM) data (Chavez and Bowell, 1988) and radar with *Landsat* MSS data (Harris and Slaney, 1982; Cliche et al., 1985). For example, a radar image could be multiplied by three TM channels in turn (e.g., radar × TM1, radar × TM2, and radar × TM3). The arithmetically combined channels could then be displayed using a RGB system, forming a color composite image.

6.4.1.3   STATISTICAL TRANSFORMS.

The principal component transform, which has commonly been applied to *Landsat* MSS data (Gillespie, 1980), is a statistical technique for combining a large number of channels producing a new set of uncorrelated channels (components) that preserves the original variance of the data. Any combination of the components can be displayed through the RGB display system, producing a three-color composite image. This technique and derivations of this technique have been used to merge various types of remotely sensed data (Chavez et al., 1991; Shettigara, 1992). A higher-resolution radar image, for example, could be combined with lower-resolution *Landsat* data by applying a principal component transform to the *Landsat* data producing a series of component images. Any one of these component images can then be replaced by the radar image before an inverse transform is applied. Shettigara (1992) provides further details on a directed principal component technique for merging different data.

Regression techniques have also been used to combine various data (Shettigara, 1992). Again, regression techniques will result in a linearly combined set of new variables that can be displayed using the RGB color model.

6.4.1.4 MULTIRESOLUTION WAVELET DECOMPOSITION.

A fairly new approach for merging data is based on multiresolution decomposition (Levin, 1985; Burt, 1989) whereby a signal (image) is broken down into a series of frequency images using the multiresolution wavelet decomposition (MWD) transform (Mallat, 1989; Daubechies, 1990). Yocky (1995, 1996) and Garguet-Duport et al. (1996) provide details on the wavelet merging method and how it can be applied successfully to merge SPOT with *Landsat* data. They found this technique was superior to other methods for merging remotely sensed data, as it provided minimal distortion of spectral (hue) characteristics.

6.4.1.5 COLOR DISPLAY TRANSFORMS.

Many different color display systems other than the RGB system exist, as discussed earlier. One such system that has seen increased application over the last 10 years is the IHS (intensity–hue–saturation) transform (Pratt, 1978; Haydn et al., 1982; Gillespie et al., 1986; Carper et al., 1990; Harris et al., 1990; Chavez et al., 1991; Ehlers, 1991). This system and its relationship to the RGB system have been discussed in Section 6.3.

6.4.2 Optimizing the Use of IHS Models

The evaluation of specific combination techniques is a subjective procedure given that the end goal is to produce color image maps in which the colors can be interpreted meaningfully. Qualitative factors such as the interpretability and range of colors, effectiveness of combination and the ability to combine two or more channels of data are particularly important factors to consider. The interpretability of colors is a key issue in that the colors on the final image must reflect the original characteristics of the data used to modulate image color or hue, thereby permitting a valid interpretation of the color in relation to the terrain or physical parameter being measured. An interpreter must be able to extract meaningful geologic information based on image hue directly from the image. More information can be displayed in color images as opposed to black-and-white images, as an interpreter can distinguish a broader range of colors than a range of black-and-white shades. As color is the key source for displaying information in combined images, a wide range of colors is obviously desirable. The effectiveness of combination is a subjective measure of how well the spectral and spatial characteristics of the input data have been preserved in the merging process. A number of the combination techniques can distort spectral hues in the final output image. Certain techniques also allow for the combination of more than three channels of data, which may be a useful characteristics in certain circumstances. Harris et al. (1990, 1994) and Rencz et al. (1994b) provide many examples of remotely sensed data combined using the techniques discussed above, as well as an evaluation of these combination techniques with emphasis on the IHS color space algorithm.

Color space transformations, although not without problems, are arguably the most flexible of the techniques for integrating data for the following reasons:

1. Individual control over each color component (intensity, hue, saturation) is possible, thus allowing for greater control over the integration process.
2. Preservation of the spectral and spatial characteristics of the input data can be achieved.
3. Colors reflect the characteristics of the input data and thus can be interpreted meaningfully with respect to terrain features.
4. Up to five channels of data may be combined and still produce an image in which colors can be interpreted meaningfully.

Many IAS and GIS have incorporated these transformations as part of their functionality, thus facilitating the merging of various geoscience data.

6.4.2.1   HUE MANIPULATION.

In cases where only one channel is being used to provide color, arithmetic and IHS combinations produce acceptable results. Figure 6.17c (see the color insert) shows a shaded relief DEM (Figure 6.17b) that has been combined with a surficial geology map (Figure 6.17a) (Dredge and Nixon, 1992) using an arithmetic and IHS transform, respectively. The map area is located in the southern portion of the Melville Peninsula in the Canadian Arctic. The image combined using the arithmetic transform was constructed by multiplying the RGB images comprising the geology map by the shaded DEM. In the IHS transformed image (Figure 6.17c), the shaded DEM has been used to modulate intensity, while the surficial geology map has been used to provide both image hue and saturation. The IHS transformed image provides more vibrant hues, as a result of varying intensity (brightness) values. However, the difference between similar hues have been suppressed somewhat on the IHS transformed image (note the surficial unit displayed in red, orange, and blue and turquoise colors). Table 6.3 shows the RGB and IHS values for each of the map colors. The intensity and saturation values range from 0 to 100, and hue values range from red (0) to purple (360). RGB values range from 0 to 255. When the original intensity is replaced, this can cause a change in brightness of the input hues, causing similar hues to be less separable on the final combined image (i.e., cyan versus turquoise, light green versus dark green, red versus orange; see the values in Table 6.3). The problem with replacing the original intensity with another channel is that the replacement channel is not often equal to the original intensity channel (i.e., the correlation between the two channels can be quite low), which can result in the distortion of hues. This problem can be partially mitigated by using a different saturation value for units displayed with similar hues. The arithmetically combined image, although less dramatic in appearance, preserves the hues more closely, as intensity, hue, and saturation are not separable in RGB space. Additionally, Yocky (1996) and Garguet-Duport et al. (1996) have found that the wavelet transform (MWD) resulted in less hue distortion than that experienced with other methods.

TABLE 6.3   RGB–IHS Values for Figure 6.17

Color	Map Unit	R	G	B	I	H	S
Cyan	Marine delta sediments	156	255	235	80	168	100
Turquoise	Granitic till blanket	57	241	187	58	162	87
Dark green	Carbonate till blanket	142	167	10	35	69	89
Light green	Carbonate till veneer	153	235	15	49	83	88
Red	Paleozoic carbonate	255	6	0	50	1	100
Dark orange	Archean bedrock	255	142	142	78	0	100
Light orange	Felsenmeer	255	210	193	88	5	100
Yellow	Granitic till veneer	255	255	108	71	60	100
Blue	Marine littoral sediments	0	0	230	45	240	100
Purple	Marine offshore veneer	152	79	255	65	265	100

Figure 6.18 (see the color insert) are examples of SPOT and *Landsat* TM imagery of the Rio Grande Valley just south of Alberquerque, New Mexico (Yocky, 1996). Figure 6.18*a* is the original *Landsat* TM image (30-m resolution), Figure 6.18*b* is the panchromatic *SPOT* image (10-m resolution), Figure 6.18*c* is the merged *SPOT* and TM data using the IHS transform, and Figure 6.18*d* shows the data merged using the MWD technique. The colors of the fields by the river are distorted on the IHS merged image (Figure 6.18*c*), due to the difference in reflectivity between the TM and *SPOT* data. Although the IHS merged image incorporates the high spatial details of the *SPOT* data, it also contains many of the intensity characteristics of the *SPOT* data at the expense of the color characteristics of the TM data. The MWD merged image (Figure 6.18*d*) also contains the high spatial details of the *SPOT* data but has retained more of the spectral information from the TM data.

When using three channels to modulate hue, the correlation between the data becomes an important factor. Areas of high correlation in all three channels will be displayed in shades of gray on an RGB color composite image. Given that areas of high correlation will plot along the gray (achromatic axis) in RGB space and will be aligned to the intensity axis during the IHS transformation (Figure 6.12), if the intensity information is simply replaced by another data source (e.g., remotely sensed image), the areas of high correlation (displayed in shades of gray) will be lost in the combined image. This is illustrated in Figure 6.19 (see the color insert), which shows a number of images over the Lake Nipishish district in Labrador, Canada. The lake geochemical composite image (Figure 6.19*b*) is a simple RGB display in which continuous surface maps of copper, zinc, and lead have been displayed through red, green, and blue colors, respectively. These continuous surface maps were interpolated from approximately 400 fairly evenly dispersed sample points using an inverse distanced-weighted (IDW) algorithm. The data were collected and analyzed by Hornbrooke and Friske (1988). Areas relatively high in copper, zinc, and lead are displayed in brighter red, green, and blue colors, respectively, whereas areas relatively high in copper and zinc are shown in yellow (red + green), high in copper and lead in magenta (red + blue), and high in zinc and lead in cyan (green + blue). Areas high in all three elements (e.g., highly correlated) are displayed in white. In the IHS transformed geochemical/*Landsat* TM image (Figure 6.19*c*), intensity has simply been replaced by a TM band 4 image; thus as discussed above, the areas of high correlation (e.g., areas that plot along or close to the gray axis) have not been reproduced on the combined image. To avoid this information loss, the original or calculated intensity can be combined arithmetically, via multiplication or addition, with the data that will be used to replace the intensity. The result of this operation is shown in the IHS transformed image (Figure 6.19*d*), which combines *ERS-1* synthetic aperture radar (SAR) (Figure 6.19*a*) with geochemical data (Figure 6.19*b*). Figure 6.20 is a flow chart summarizing the merging process. All hues, including the areas of high correlation, are reproduced in this image.

Data correlation will affect specific integration techniques in different ways. High correlation between data will generally result in a much narrower color range in an RGB color composite image. This is a problem particularly with *Landsat* TM and MSS data where certain bands are highly correlated.

A number of techniques can be used to decorrelate the data, which has the end effect of increasing the range of colors in the final output image. If the aim is to produce a combined image in which three channels modulate the hue, a decorrelation

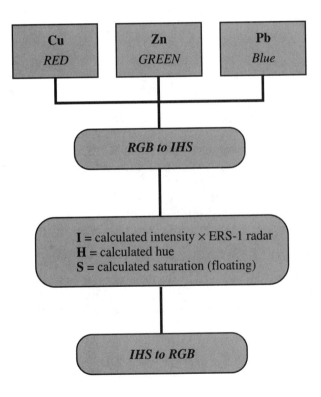

Figure 6.20 Flowchart steps required to produce Figure 6.19*d*.

stretch (Gillespie et al., 1986) can be applied to the data to maximize the color range before input to the IHS transform. This was applied to the gamma ray spectrometer image of the Port Coldwell Alkalic Complex in Canada, shown in Figure 6.21 (see the color insert). The three channels comprising the gamma ray color composite image (uranium, thorium, and potassium displayed in red, green, and blue, respectively) were decorrelated using a decorrelation stretch before being combined with the airborne SAR data. The color range of the decorrelated composite image was much improved over the composite image (not shown) comprising the original channels. This image is discussed further in the following sections.

6.4.2.2 INTENSITY MANIPULATION.

Generally, in the data merging process there is at least one set of achromatic data and a set of chromatic data. As mentioned earlier, shading provides information about form. The objective behind manipulation of intensity, therefore, is to introduce a form (shape) component, which can be of higher spatial resolution than the data used to modulate hue, into a merged image. Typically, a SAR image or derivative product (shaded relief topographic or magnetic image) are best used to modulate intensity. Ideally, to prevent hue distortion the channel used to replace the calculated intensity should be highly correlated with the intensity information. However, this is not often the case and somewhat defeats the purpose of the data combination exercise.

Landsat data in Figure 6.22*d* (see the color insert) has been used to modulate intensity. Although *Landsat* is, on one hand, a logical choice for providing intensity information, as it is characterized by fairly high spatial resolution, the spectral information inherent in the data results in hue distortions. Note the obvious effect that clouds, present on the *Landsat* data (Figure 6.22*a*), have on the transformed image (Figure 6.22*d*). In this case the *Landsat* data could be filtered to emphasize the high-frequency information, and the filtered channel could be used to modulate intensity similar to the HPF (high-pass filter) technique, discussed by Chavez and Bowell (1988) and Chavez et al. (1991). The spatial filter will tend to emphasize the spatial information as opposed to the spectral information, which will reduce the hue distortion problem. However, valuable textural information, such as that found in SAR data, will be de-emphasized. Figure 6.22 is discussed in more detail in the following sections.

Carper et al. (1990) point out that integration methods are generally most effective when the achromatic data are redundant with the color data. In other words, the most effective procedures result from the merging of achromatic data that closely approximates the wavelength range presented by the color data. Edwards and Davis (1994) present a method for combining data in which the achromatic and color data are not redundant, as is often the case when combining color thematic data with a remotely sensed image or a shadowed enhanced digital elevation model (DEM). This method, which they have termed the intensity-adjustment technique, is a modification of the intensity–hue–transformation and intensity replacement technique (Cliche et al., 1985). Edwards and Davis (1994) provide step-by step calculations required for this intensity adjustment method, which basically involves the calculation of a normalized color intensity that is characterized by a piecewise linear relation between the normalized intensities of the input images used to modulate color (hue) and the desired output.

Edwards and Davis (1994) also note that consideration must be given to the frequency distribution of the intensity modulation in the achromatic image. If the intensities of the achromatic image are concentrated at either end of the distribution curve, the intensity substitution method is recommended because it transfers more of the intensity information of the high-resolution achromatic image to the merged color image, a region where the eye is more sensitive. If the intensities of the achromatic image are concentrated at the middle of the distribution curve, the intensity-adjustment method is preferred because it will retain the intensities of the color image (Edwards and Davis, 1994). Thus the intensity-adjustment technique should be used with nonredundant data (i.e., merging shaded DEM with color thematic data) while the intensity substitution method can be used effectively with redundant data. Figure 6.23 (see the color insert) shows a shaded relief image that has been merged with color-coded topographic data (Edwards and Davis, 1994). Figure 6.23*a* shows the merged image using the intensity substitution method (Cliche et al., 1985), and Figure 6.23*b* shows the results using the intensity adjustment technique (Edwards and Davis, 1994). The intensity adjustment method has resulted in a superior color image with a wider range of colors, as the shaded relief image and color-coded topographic data are nonredundant.

6.4.2.3   SATURATION MANIPULATION.

Saturation can be used in a number of different ways, in the data merging process, depending on what is to be emphasized in the final image. A mix of two of the primary colors (red, green, blue) will result in full saturation in the IHS transformed image. The saturation values will plot on the outside surface of the IHS cone (see Figure 6.14). Variable saturation values will result when all three primary colors are mixed in various proportions. In this case saturation values will plot at different radii within the cone. By setting saturation to a constant, a proportionate mix of the intensity and hue data can be achieved. In other words, fully saturated colors reduce the effect of intensity information, while less saturated colors increase the effect of intensity information. Figure 6.24 (see the color insert) comprises a number of IHS transformed magnetic and gravity images of a large portion of northern Ontario, Canada (NTS 1:250,000 sheets 42B, 42A, 41O, 41P) and serves to illustrate the point made above. A shaded relief total field magnetics image has been used to modulate image intensity, while hue information has been supplied by a color-coded gravity image. Three images are presented with constant saturation values, of 100, 50, and 25%, respectively. The image with full saturation emphasizes the gravity information (hue), whereas the less saturated images emphasize the intensity (magnetic) data. Note that as saturation decreases, the colors migrate from pure to more pastel colors, reflecting a migration to an achromatic image. A synthetic saturation file can also be generated in which variable saturation values are used to help distinguish similar hues. A saturation mask can also be generated so that selected areas within the color image are shown in black and white (e.g., saturation of zero). This can be useful for emphasizing anomalies within the image. Saturation can also be enhanced (contrast stretched) to provide a wider color range.

Another form of saturation manipulation is intensity-modulated saturation. This simulates the loss of color saturation that occurs for highly illuminated surfaces. For example, imagine looking at a red can. The color of the can will appear red for those parts of its surface that are moderately or poorly illuminated. The parts of the surface that are facing the source of illumination, however, will appear very pale red or perhaps even white. Simulating this affect can be achieved by dividing a saturation image with an intensity image so pixels with high intensity will reduce the coincident pixel values in the saturation image. This enhances form perception since the human visual system is accustomed to reductions in color saturation in areas of high illumination.

6.4.3 Examples of Combined Images

In the following sections we present examples of different types of raster-formatted geoscience data combined using methods discussed in Section 6.4.1.

6.4.3.1 REMOTELY SENSED DATA.

Figure 6.22 (color insert) presents images of the Antigonish Highlands in Nova Scotia, Canada which combine airborne radar (SAR) and *Landsat* MSS imagery using a simple RGB band combination, arithmetic combination, and IHS color space transformation, respectively. The radar image used in the merging process comprised two separate radar passes, imaged from different look directions. These were arithmetically combined (multiplied and scaled between 0 and 255), producing a radar "product" image. This was a necessary step to achieve maximum enhancement of the terrain, as radar is a directional sensor and terrain enhancement is a function of the orientation of topographic features in relation to the radar look direction (Harris, 1986). Figure 6.22*a* shows an IHS-transformed *Landsat* MSS false color composite image comprising bands 7, 5, and 4 displayed through red, green, and blue colors, respectively. The color range of this image is much broader then the raw color composite as a result of stretching the saturation channel before conversion back to RGB space for display. This decorrelation process is discussed in more detail in Section 6.4.2.1. Figure 6.22*b* was constructed by displaying *Landsat* MSS band 7 through red and two separate radar passes imaged from different look directions through green and blue colors. These channels were normalized (means and standard deviations normalized) before final output to improve the color range. Figure 6.22*c* was produced by multiplying *Landsat* MSS bands 7, 5, and 4 by the SAR product image and scaling the data to 8 bits (255 gray levels) and then displaying the channels through red, green, and blue, respectively. Figure 6.22*d* was constructed by converting *Landsat* MSS bands 7, 5, and 4 from RGB to IHS space and replacing the intensity with an arithmetic combination of the original intensity and the SAR product image. In addition, the saturation channel was contrast stretched to improve the range of colors on the resulting IHS transformed image.

Combining radar and *Landsat* MSS data, from a geological perspective, is logical, as both sensors provide different but complementary terrain information, with the MSS providing information on surface reflectance properties in which vegetation patterns are enhanced while the radar provides information on surface morphology and roughness. These data are useful for mapping structural and lithological patterns (Harris et al., 1993a).

All three techniques preserve the relationship between input data and color; therefore, image colors can be meaningfully related to terrain features. Image hue in

Figure 6.22*b*, which can be interpreted with reference to Figure 6.22*a*, is provided by *Landsat* MSS band 7 (red), radar (green), and the radar (blue), whereas *Landsat* MSS 7, 5, and 4 (displayed in red, green, and blue, respectively) provide the hues in Figure 6.22*c*. In all three images the reddish hues reflect areas of deciduous vegetation growth as a result of strong infrared reflectance in MSS band 7.

Both the arithmetic combination and color display transform provide more effective combination of the higher-resolution radar and lower-resolution MSS data than the simple band combination, as demonstrated by the enhanced relief and higher effective spatial resolution evident in both Figure 6.22 *c* and *d* compared to Figure 6.22*b*. The spectral balance is maintained in these images, as radar inherently provides less spectral information than the MSS data. Radar is therefore best used as a source of intensity information when using the IHS transform to merge data. Figure 6.22*d* (IHS-transformed image), in addition to providing enhanced relief, also provides a slightly wider range of colors and better spectral balance of the input data than the arithmetically combined image (Figure 6.22*c*). This in part results from stretching the saturation component.

Figure 6.22*d* also illustrates the effect of replacing the calculated intensity with another image. In this case the radar image was multiplied by the MSS intensity image. Notice that the hue characteristics of highly correlated areas in all the MSS bands (i.e., clouds) are retained. If the intensity channel was simply replaced by the radar, the clouds would disappear, although the cloud shadows (shown in cyan) would remain.

6.4.3.2 GEOPHYSICAL DATA.

Figure 6.21 (color insert) presents images that are combinations of radar and gamma ray spectrometer data over the Port Coldwell alkalic intrusive complex on the north shore of Lake Superior in Ontario, Canada. The methodology for combining these types of data builds on techniques developed by Broome et al. (1987), Broome (1990), and Harris et al. (1990). The geophysical data [three channel gamma ray spectrometer: equivalent uranium (eU), equivalent thorium (eTH), percent potassium (%K)], and airborne SAR image have been combined using an IHS and arithmetic transform. Figure 6.25 is a flowchart summarizing how Figures 6.21*c* and *d* were produced. The raw gamma ray data were decorrelated using a decorrelation stretch algorithm (Gillespie et al., 1986) available on the image analysis system before being combined with the SAR. The reason for this procedure is discussed in Section 6.4.2.1. Figure 6.21*c* was produced arithmetically by multiplying each gamma ray channel by the SAR image (e.g., SAR × eU, SAR × eTH, SAR × %K), scaling the data, and then displaying the RGB color composite image. The IHS-transformed image, shown in Figure 6.21*d*, was constructed by transforming the three gamma ray channels from RGB to IHS space and replacing the intensity with the SAR data. The calculated intensity was simply replaced by the radar image since the gamma ray data did not display high correlation in all three channels. The calculated saturation was replaced by the total count channel which is a measure of the total radiation emitted from the Earth's surface. The reason for this is to emphasize the granitoids, as they are more radioactive than the volcanic (greenstone) rocks. The granitoids are visible in the bottom left and top right portions of the image. These areas are generally more vibrant as a function of higher saturation, reflecting higher radioactivity.

While both integration techniques offer a fairly wide color range, the IHS trans-

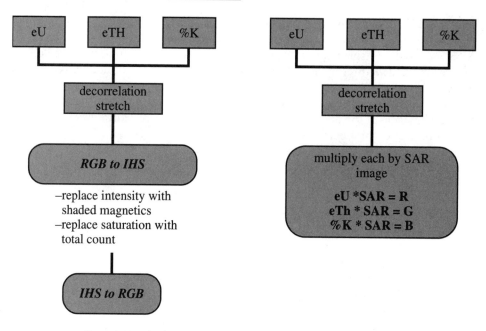

Figure 6.25 Flowchart summarizing required steps to produce Figure 6.21 c and d.

formed image offers a wider color range (due to saturation stretching), and the radar is more predominant than in the arithmetically combined image. This is the result of independent control over hue, intensity, and saturation. However, it should be noted that the colors have been distorted somewhat as a result of the IHS transformation. This was done, in part, by the stretching of the saturation channel to widen the color range of the output image. The other factor, described more completely in Section 6.4.2.1, is that replacement of the original intensity by a substitute channel will also result in some spectral distortion of the image hues.

Color composite combinations of the data (not shown) were less useful, as only three channels (radar plus a maximum of two spectrometer channels) could be combined. The merging was less effective using band combinations in terms of poor spectral contrast and lower effective spatial resolution.

The variation in hues in these images reflects the relative concentration of the radioelements, as eU, eTh, and %K were used to modulate the red, green, and blue components, respectively. Thus areas of higher relative eU emission are bright red, high eTh are green, high %K are blue, relatively high eU and eTh are yellow, and relatively high eU and %K are magenta. These color variations reflect subdivisions in what has been mapped as a homogeneous intrusive body. Some of these radioelement variations correspond directly to mapped lithological units, while others cross-mapped contacts (Graham and Bonham-Carter, 1993). The data assist in delineating zones of very subtle petrological and perhaps compositional zoning that is often difficult to recognize in the field (K. Ford, GSC, personal communication). The radar data enhances topographical and structural features within the intrusive complex as well as the surrounding Archean rocks. The northern and eastern contacts between the complex and the Archean rocks are especially evident appearing as strong topographic lineaments. These lineaments represent faults (K. Ford, GSC, personal communication) that may have exerted some control on the emplacement of each of the different intrusive events apparent on the gamma ray data.

6.4.3.3 THEMATIC DATA.

Figure 6.26 (see the color insert) is an IHS combination of a geology map and airborne magnetic data over volcanic and gneissic terrane in the Mari Lake area in east-central Saskatchewan, Canada. Figure 6.26*a* shows the geological map. Intrusive suites are shown in yellow and pink colors, volcanic rocks in green, and sediments in blue. Figure 6.26*b* shows a shaded relief magnetics image while Figure 6.26*c* is a combination of the two images above using the IHS transform. Image intensity is provided by the shaded magnetics data, while hue is provided by the thematic (lithological) data. This particular combination of data is useful for comparing mapped lithological patterns with magnetic patterns. This is especially relevant from a geologic mapping point of view, as magnetic data are frequently used to map lithology in poorly exposed areas that are difficult to traverse (which characterizes much of this map area). From this perspective, a geologist could use this map to identify areas where the mapped geology does not match the magnetic patterns, indicating that further refinements are required. For example, many of the thinly banded east/west-striking volcanic units correspond with linear magnetic anomalies. However, some of the magnetic anomalies, especially in the bottom half of the image, show no mapped lithological variation. In addition, the magnetic data reveals that some of the linear volcanic units should perhaps be extended beyond their present mapped position. This immediately flags these areas for further field investigation.

6.5   2.5-DIMENSIONAL VISUALIZATION

6.5.1 What Is 2.5-Dimensional?

Images can effectively display two-dimensional features in which X and Y coordinates are used to locate a pixel in two-dimensional space. The pixel value of 3, for example, could be an index to nominal data (e.g., 3 = granite), ordinal data (e.g., 3 = high), interval data (e.g., 3 = 10 to 20), or ratio data (e.g., 3.00 m above sea level). For ordinal, interval, and ratio data, the pixel value can be used as a Z coordinate to visualize the data as a surface. With ordinal and interval data, the pixel value can be used as an index to more than one attribute of the pixel. However, it is not possible to represent more than one Z coordinate simultaneously at any X, Y with a two-dimensional raster data model. 2.5-Dimensional visualization provides more than a two-dimensional plan view of a raster, since a surface can be viewed from many different perspectives, but it cannot model three-dimensional features (e.g., overturned fold) that require simultaneous access to multiple Z coordinates at any X, Y location.

Visualization is discussed in the following section with respect to 2.5-dimensional techniques used to enhance and visualize single raster images or multiraster images that have been combined using methods discussed in Section 6.4.1. A discussion of true three-dimensional modeling and visualization techniques is beyond the scope of this chapter.

6.5.2 Overview of 2.5-Dimensional Visualization Techniques

There are several different ways of visualizing 2.5-dimensional data. These methods can be categorized, and will be discussed, with the following organization:

1. Fixed perspective (plan view)
   a. Single image to both eyes:color, hill shading, color + hill shading (IHS)
   b. Parallax manipulation (single image viewed differently by each eye): chromostereography, anaglyphs
   c. Parallax manipulation (two images viewed individually by each eye): stereogrammetric
2. Variable perspective (oblique view)
   a. Static
   b. Dynamic (fly-through) perspective views (not discussed in this chapter)

Form and color are important aspects of the human visual system, as discussed previously. However, another important aspect for visualization is depth perception. A number of visualization techniques can be applied to raster data-producing images where the appearance of depth is simulated by manipulating grey levels (e.g., tonal differences), hues (e.g., colors), hill shading, and integrated images that utilize both amplitude and slope (hill shading) information. Relief can also be simulated using parallax manipulation techniques that employ different viewing angles obtained by a physical offset in the remote sensing platform (i.e., aerial photography) or by synthetically creating a difference in parallax using a digital elevation model (DEM) or some other potential field image (e.g., anaglyphs). Techniques that simulate a non-nadir viewing position (i.e., perspective view) are also important for certain geologic applications and can be an effective way of visualizing geoscience data. Table 6.4 presents a summary of the more common visualization techniques applied to various types of geoscience data, primarily raster remotely sensed and geophysical data.

6.5.3 Color, Hill Shading, and Integrated Images (IHS)

The use of color for representing 2.5-dimensional data was discussed in some detail in Section 6.2.1.4 and is not reviewed here. Although shaded relief images were used in previous examples, we now look at some of the history and mathematics that create these images.

Hill shading or shadow enhancement has been utilized in cartography as a standard display technique for producing pseudo-2.5-dimensional views of thematic map data for many years (Yoeli, 1965, 1967). Similar techniques have been utilized in geophysics (shaded relief maps) to highlight directional features (see Chapter 13). The methods are based on an illumination model such that light intensities derived

TABLE 6.4   2.5-Dimensional Visualization Techniques

Technique	References	Description
Hill shading (shaded relief maps)	Batson et al., 1975; Broome et al., 1985; McLaren and Kenner, 1989	Commonly applied to topographic and potential field data
Stereogrammetry Stereo pairs	Batson et al., 1976	Introduction of parallax either by different viewing geometries or created artificially; relief can be based on topography or other data set (e.g., magnetic relief)
Analglyphs Chromostereoscopy	Broome et al., 1985; Usery, 1995 Toutin and Rivard, 1995	Depth encoded into image using color; special glasses required to decode the optics to produce depth perception; artificial depth can be encoded by any quantitative data source
Perspective views	Aronoff, 1989; Hearnshaw and Unwin, 1994	Viewing an image obliquely (non-nadir viewing position); viewers position fixed by azimuth (0–360°) and elevation (0–90°); depth provided by topography or other data (e.g., potential field)

from an artificial light source (e.g., "artificial sun") are computed for each pixel of a raster image, often a digital terrain model (Weibel and Heller, 1991). The light intensities (or degree of shading) are produced in a two-step procedure. Initially, surface normals are calculated from elevation data and then an illumination model (commonly a Lambertian model, which assumes diffuse reflection from the illuminated surface) is selected to calculate the amount of backscattered light reflected back. The intensity of the light is proportional to the cosine of the angle between the surface normal and the direction of the light source. Illumination can be calculated rather easily from

$$\cos \gamma = \cos (\text{slope}) \times \cos \alpha + \sin (\text{slope}) \times \sin \alpha \times \cos(\alpha - \text{aspect}) \quad (6.13)$$

where α is the azimuth angle and γ is the incidence angle.

More sophisticated models are available (Foley et al., 1990); however, for visualization these may not be necessary (Weibel and Heller, 1991).

Figure 6.27 is an example of a *Landsat* TM image and an airborne magnetic image over the Mari Lake area of Saskatchewan, Canada, where base metal and gold exploration has been active (see Figure 6.26 for location). The *Landsat* image (Figure 6.27a) was shadow enhanced (Figure 6.26b) using a sun position in the north (0°) at an elevation of 40°, while the airborne magnetic image (Figure 6.27c) was enhanced (Figure 6.27d) from the northeast (315°) using a 30° elevation. The hill shading applied to the magnetic data is more dramatic than the *Landsat* enhancement, as the magnetic data has lower resolution and the nature of magnetic susceptibility

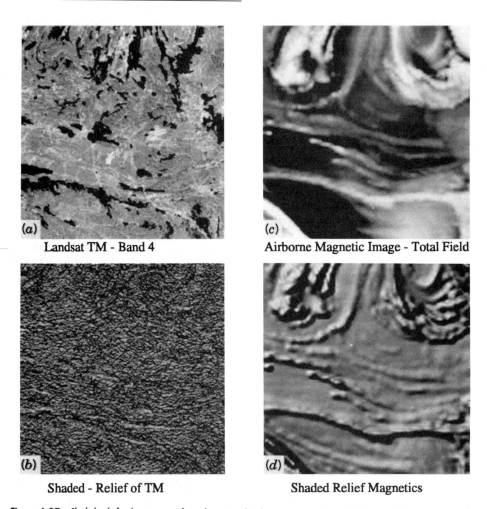

Landsat TM - Band 4   Airborne Magnetic Image - Total Field

Shaded - Relief of TM   Shaded Relief Magnetics

Figure 6.27 Shaded relief enhancements of *Landsat* TM and airborne magnetic data: (*a*) *Landsat* TM band 4 image; (*b*) shaded relief of TM band 4; (*c*) airborne magnetics image; (*d*) shaded relief magnetics image. (Magnetic data courtesy of the Geological Survey of Canada.)

varies continuously and smoothly over space. The reflected solar energy that *Landsat* records is more erratic in a spatial sense and is of a much higher spatial frequency; thus the hill-shading technique tends to emphasize these high-frequency variations, resulting in a less useful enhancement. The hill-shading technique applied to the magnetic data clearly has enhanced the east/west-trending lithologic and tectonic structures that are important for regional exploration in the area, whereas cultural and surficial features are enhanced in the *Landsat* hill-shaded image.

6.5.4 Parallax Manipulation Techniques

Viewing remotely sensed data in stereo greatly aids geologic interpretation, as it is easier to detect and map geologic features (i.e., faults, contacts, dipping beds, etc.).

One has only to view an air photo stereo pair to appreciate the value of relief perception in the interpretation process!

The fact that horizontal separation of the eyes is on average 6 to 7 cm (Drury, 1993), allows objects to be viewed from slightly different positions or angles, thus resulting in parallax, which allows for viewing objects in three dimensions. *Parallax*, which can be defined as the apparent change in position of an object relative to another when it is viewed from different positions (Drury, 1993), forms the basis of stereoscopy. Figure 6.28 defines absolute and relative parallax. Rays extending from the eyes to an object are separated by the angles α_1 and α_2, the size of which is inversely proportional to the distance between the eye and object. Absolute parallax is approximated by the ratio between the eye base (*a*) and the distance to the object and represents the angle at which the eyes must converge to focus the object properly. The two images are mentally fused, resulting in a stereo-model of the object in three dimensions. Relative parallax represents the difference between the two angles of convergence ($\alpha_1/E - \alpha_2/E$ in Figure 6.28). The angles of convergence are different as a result of the varying distance between objects *A* and *B* and each eye.

Parallax has traditionally been achieved by a collection of separate images from different positions along the flight path, as in the case of aerial photography or with different viewing geometry's, as is the case with synthetic aperture radar (SAR) and SPOT data. The *Landsat* series of satellites have sidelap ranging from 10% at the equator to 80% in polar regions and therefore can be viewed as stereo pairs using a standard stereoscope, whereas *SPOT* and *Radarsat* stereo pairs can be generated using slightly different viewing geometries. Parallax can also be artificially introduced into a single image from other geoscience data, such as geophysical or topographic data (Batson et al., 1976), and has been used as a common visualization technique.

The stereoscope has traditionally been used for stereo viewing. However, the anaglyph and polarizing glasses, which allow images to be viewed separately by the left and right eyes, have also been employed. Other more advanced systems include three-dimensional shutter glasses, which are synchronized with alternating left and right images displayed on a CRT, making use of horizontal parallax between the two images to perceive depth. The visual image depth enhancement (VISDEP) (Jones et al., 1984) involves rapidly alternating vertically displaced views of an image on a flat CRT, thus using vertical disparity to generate depth perception. Toutin and Rivard (1995) provide a succinct review of techniques for depth perception.

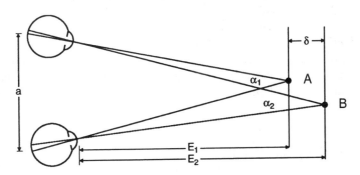

Figure 6.28   Stereoscopic viewing. (Modified from Drury, 1993.)

Stereo pairs can also be viewed using other optical techniques such as the anaglyph, in which each image is projected through a different primary color. Special glasses in which filters match the primary color each image is being projected through ensures that only light from the right image is received by the right eye, and vice versa. Figure 6.29(a) is an anaglyph of LANDSAT TM data over the Mari Lake area in which parallax has been added through the incorporation of a digital elevation model.[2] The stereo complement for each image is calculated by introducing an x-parallax to each pixel to its elevation divided by a stereoscopic factor (viewing elevation) which determines the extent of vertical exaggeration:

$$P_i = \frac{E}{St} \tag{6.14}$$

where P_i is the parallax at each pixel, E the elevation of each pixel derived from the digital elevation data, and St the stereoscopic factor, as discussed above. Batson et al. (1976) provide details of the synthetic stereo process as applied to *Landsat* data. The terrain in this area is fairly flat so that the relief provided in this anaglyph is extremely valuable for identifying subtle structural features. Figure 6-29(b) is also an anaglyph in which parallax has been added based on airborne magnetic response. The "highs" in this image reflect higher magnetite content in the rocks.

A relatively new technique known as chromostereoscopy developed by R. Steenblik and applied by Toutin and Rivard (1994) is also useful for visualizing data. Figure 6.30 (see the color insert) is a magnetic image over the Mari Lake area in Saskatchewan displayed in three dimensions using the ChromaDepth technique, in which depth has been encoded into the image by using color. The image has been enhanced using a shaded relief magnetic image to modulate image intensity. Special optics (viewing glasses) are used to decode the optics to produce depth perception. The artificial depth provide by color variations can be provided by any quantitative data source such as elevation, or a potential field such as gravity or magnetic susceptibility, as has been used in this image. Upon viewing with the appropriate viewing glasses each color is bent (shifted) proportional to its refractive index, generating a differential angular parallax. As depth is assigned to color foreground objects, they will appear red, background objects will be colored blue, and objects in between are positioned according to their position along the electromagnetic spectrum. The image can be created on a single image that has a normal appearance when viewed without glasses; however, when viewed with special glasses the depth can be perceived.

Figure 6.31 (see the color insert) is a combination of an *ERS-1* SAR and airborne magnetics image displayed in three dimensions using the ChromaDepth technique produced by Toutin and Rivard (1994). Initially, an IHS coding system was employed with magnetic data modulating the hue and saturation components and ERS-1 image modulating image intensity. The color scale applied to the magnetic data was manipulated to optimize depth perception by using eight distinct hues, ranging from blue to red.

6.5.5 Perspective Views: Static

One area of research that has had a significant impact on visualization in remote sensing has been the incorporation of elevation data into both the data analysis routines and visualization process. The display of features incorporating visual perspective generally involve integration with topographic data in which elevation defines the Z value. Perspective views require accurate integration of elevation data. Digital topographic data has variously been referred to as digital terrain models (DTMs) or digital elevation models (DEMs) (Weibel and Heller, 1990).

Figure 6.32 (see the color insert) is an example of a perspective view of a *Landsat* TM false color composite image of the Mari Lake region, Saskatchewan, in which topographic (elevation) information in the form of a DEM has supplied the Z value. In certain cases, nontopographic data values have been used to provide the Z information. Examples include a geophysical parameter such as magnetic susceptibility, which results in magnetic relief (shown in Figure 6.31b) as opposed to true topographic relief or in the case of *Landsat* TM data, pseudoelevation data have been generated by treating low magnetic values as low elevation and high magnetic values as higher elevation. In this manner pseudoelevation data can be generated.

6.6   EVOLVING TRENDS

6.6.1   Three- and Four-Dimensional Visualization

Traditional IAS and GIS analysis and visualization of geoscience data has focused on two-or 2.5-dimensional models, but many geologic applications require the collection and analysis of data in three-dimensions. The difference between two-and three-dimensional models have been discussed previously. Recent examples of true three-dimensional GIS applications are reported for oil exploration (Youngmann, 1989), hydrology (Turner, 1989), mining (Bak and Mill, 1989), and geological modeling (Kelk, 1991; Bristow and Raper, 1991). The addition of the third dimension necessitates modification of data models. True three-dimensional rendering systems employ data models that are variations of the traditional two-dimensional raster and vector and include voxels (three-dimensional raster), octrees (three-dimensional quadtree), and topological vector equivalents of two-dimensional vector models such as minimum interpolation algorithms, where values of a three-dimensional object are calculated at the intersection of a gridded block. These three-dimensional surfaces are threaded through the three-dimensional grid and converted to triangulated vectors for display and analysis. Raper (1989) provides a comprehensive review of various three-dimensional models. Turner (1989) identified four major difficulties in the application of true three-dimensional GIS: (1) incomplete and sometimes conflicting data, (2) subsurface environment is characterized by extremely complex spatial relationships, (3) very expensive to collect enough data to resolve these complex relationships and subsurface uncertainties, and (4) the relationships between rock property values and the volume of rock over which they are being averaged (scale effect) are usually unknown.

Presently, true three-dimensional GIS models are not handled by common GIS systems and require separate software and hardware to perform the often complicated three-dimensional analysis and image rendering. The visualization tools provided by such systems include zooming, rotation, panning, and peeling, while analysis tools include three-dimensional grid operators and interpolators as well as hierarchical filters. Turner (1989) and Raper (1989) provide informative reviews of three-dimensional geoscience applications.

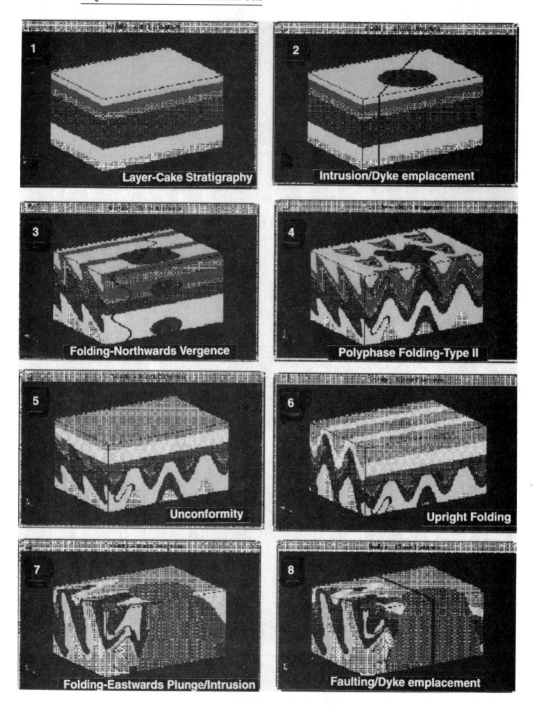

Figure 6.34   Block model of complex structural deformation through time.

Figure 6.33 (see the color insert) (Lewry et al., 1994) is an example of a geologic map over the Precambrian Shield in northern Saskatchewan that has been combined with a three-dimensional seismic trace using IBM Data Explorer (DX) 3.1.0. A vertical section consisting of geology interpreted from seismic reflection data is shown in relation to mapped surface geology. Gravity values are shown both along the surface trace of the transect as well as in three-dimensions above the transect trace.*

The fourth dimension, that of time, is also important in many geologic problems. New tools for visualizing changing geology through time are also in a stage of rapid development. Figure 6.34 is an example of a three-dimensional geologic model produced using a package called NODDY (Jessel, 1993). The series of block diagrams illustrate progressive deformational events through time applied to originally undeformed "layer-cake" stratigraphy. This image set is useful not only for visualizing complex structures but also for incorporating field data into temporal structural models.

6.6.2 Hyperspectral Analysis

Data from airborne sensors such as AVIRIS provide hyperspectral information with relatively small pixel size (see Chapter 11). The data represent a "cube" of information consisting of lines by samples by spectral bands (Kruse et al., 1993). Systems are available for the simultaneous viewing of this information in the spatial and spectral domains (Kruse et al., 1993) as demonstrated in Figure 6.35. The image on the left demonstrates spatial information on one band. The user interactively roams across the image-selecting spectrum at specific locations. The curves on the right reflect the n dimensions of spectral data represented at a point. In this example there were 224 channels of AVIRIS data. The image also illustrates the analytical results of finding the closest spectral match to the spectrum observed in the field. In this example the mineral at A is spectrally analogous to muscovite (based on a spectral matching to library specimens). Similarly, minerals at B and C are identified as kaolinite and alunite, respectively. Further details of the analytical procedures used in this analysis are provided in Chapter 5, and further information on hyperspectral data in Chapter 11. This example illustrates the simultaneous visualization of spatial and spectral data and the end product of analysis to predict the mineralogy at a given point.

6.7 SUMMARY AND CONCLUSIONS

In this chapter we have reviewed various techniques for combining and visualizing geoscience data. The majority of the techniques discussed are included as software routines on many commercial IAS and GIS software and are straightforward to apply. However, the driving force behind these procedures is the particular geologic problem where the most important goal is to produce images that provide useful, complementary information that can be interpreted qualitatively or, in some cases,

---

*Gravity data supplied by M. Thomas (Geological Survey of Canada); three-dimensional visualization by D. W. Desnoyers (Geological Survey of Canada).

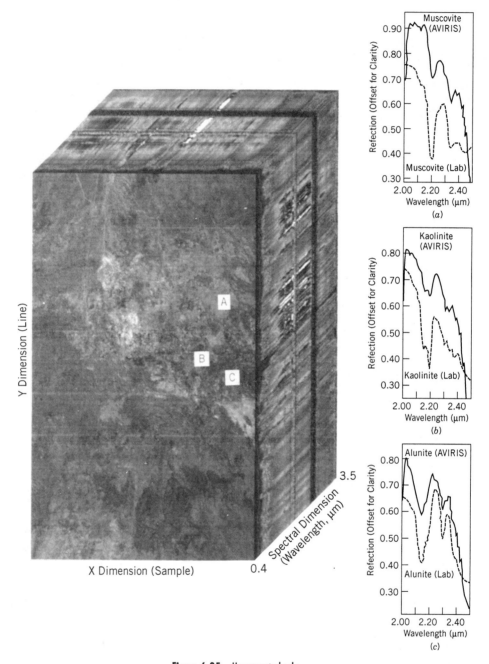

Figure 6.35 Hyperspectral cube.

quantitatively. This contrasts with producing "pretty pictures" only as the colors produced by the techniques discussed in this chapter can be related meaningfully to spectral, chemical, and physical properties of the Earth's surface.

Data visualization, which includes the merging or combining of various geoscience data, is evolving rapidly with the advent of more powerful software and hardware tools. However, as we have attempted to illustrate in this chapter, there are many algorithms and methods for combining and visualizing raster data, each comprising many different parameters. Change in any one of these parameters (e.g., intensity, hue) may cause an appreciable change in the appearance of the resulting imagery. Therefore, the black-box approach, where the user simply feeds data into a computer algorithm and waits for the resulting output, should be used with caution. The geologic problem to be solved and the geoscientist merging the data are the driving forces behind the visualization process, as shown in Figure 6.1. The algorithms used to combine and visualize data are only the tools, which must be used intelligently, to achieve meaningful results.

6A

Display Systems

6A.1. HARDWARE

6A.1.1 Monitors

Image processing includes the monitor and devices for hard-copy output. The video output processor generates a signal driving a display device. The most popular display device is the cathode ray tube (CRT). The quality of the picture is dependent on several characteristics of the device, particularly the scanning techniques and the phosphor (Huang and Chao, 1993).

6A.1.1.1 SCANNING TECHNIQUES.

In remote sensing a raster scanning process is used in which an electron beam scans the CRT in a defined pattern of parallel horizontal lines. The beam current is in an on–off state for monochrome pictures or as a variable current for continuous gray-tone pictures. Scanning can be in a noninterlaced or interlaced fashion. In the interlaced procedure a complete image is generated in two steps. First the even lines are displayed, then the odd lines are displayed. Each step is performed at the refresh rate of the monitor, currently around 1/60 of a second, so that a complete image is presented every $\frac{1}{30}$ of a second. Generally, a 30-Hz interlaced refresh rate is considered adequate; faster rates are required for static image viewing to reduce flicker. The preferred mode for viewing images is noninterlaced, as the entire image is painted at the same time.

6A.1.1.2 PHOSPHOR

Display screens contain a phosphor layer that is less than 10 μ in thickness. Phosphors are typically based on zinc sulfides; however, other materials, notably rare earth elements, are now being used. High-energy electrons excite the phosphor, and visible light referred to as *cathodoluminescence* is emitted. The intensity of illuminated light is proportional to the current of the electron beam. Color on the screen is produced from triads of phosphors: red, green, and blue. Three guns of electron beams independently excite the phosphors and render the desired output color, the final color being a result of the combined emitted luminescence from the phosphors.

The actual pixel size on the screen is determined by the dot size of the electron beam. The size may be defined as the width between points where the beam current drops to a specified fraction ($\approx 37\%$) of the maximum value. The minimum size of the dot is dictated by the current required to produce acceptable phosphor brightness. The beam spot diameter increases linearly with the square root of the beam current (Huang and Chao, 1993). Currently, monitors for personal computers have color display capabilities up to 1600×1400 pixels with 256^3 color possibilities.

6A.1.2 Hard-Copy Devices

There is a range of plotting devices, depending on the desired output product. The most appropriate devices for display of color remotely sensed images are raster devices, including inkjet plotters, laser printers, and thermal wax plotters, which rely on electrostatic principles (Bonham-Carter, 1994). In this process pixels from the image are represented by one or a matrix of plotter pixels. Inkjet plotters produce a raster of dots by injecting small dots of ink onto a paper medium. The inks are the subtractive colors: cyan, magenta, and yellow. Color is then a product of the combination of cyan, yellow, and magenta. Most plotters also include black, as the imperfect superimposition of CMY on one dot usually produces a brown tinge.

In the plotting process a dot of fixed size (typically, around 0.2 mm for most plotters) is either ON or OFF. Therefore, in a CMY system, there are only eight different possibilities, as shown in Table 6A.1. To render a display with a larger color palette, a matrix of dots is used to represent one image pixel in a process referred to as *dithering*. Note that an increase in color possibilities is done at the expense of the size of plotter pixel, a situation unlike video displays, where the intensity of a pixel can be varied; therefore, there is not a trade-off between the number of colors and the size of the plotter pixel.

Typically, dithering matrices are 2×2, 3×3, or 4×4, depending on the number of colors required. In a 2×2 dithering cell there are five possible dot patterns (no dots, one dot, two dots, etc.). As this pattern is available for each of CMY, there are potentially 5^3 different colors possible. Similarly, for a 4×4 matrix there would be 17^3 potential possibilities. This number could be significantly higher; however, ordering of the dot patterns is critical. For example, if only one dot were to be placed in the 4×4 matrix, there are 16 different locations for the dot; however, inappropriate ordering can lead to visually distracting patterns such as herringbone or twill

TABLE 6A.1   Eight Possible Combinations of Cyan, Magenta, and Yellow and Resulting Color

Off	Off	Off	White
Off	Off	On	Yellow
Off	On	Off	Magenta
On	Off	Off	Cyan
On	On	Off	Blue
On	Off	On	Green
Off	On	On	Red
On	On	On	Black

TABLE 6A.2   A 4 × 4 Dithering Cell and Surrounding Cells[a]

16	2	13	3	16	2	13	3
5	11	8	10	5	11	8	10
4	14	1	15	4	14	1	15
9	7	12	6	9	7	12	6
16	2	13	3	16	2	13	3
5	. 11	8	10	5	11	8	10
4	14	1	15	4	14	1	15
9	7	12	6	9	7	12	6

[a] Number refers to the number of dots in a cell and a predefined position for the dot (see Bonham-Carter, 1994).

(Bonham-Carter, 1994). There are set patterns for the placement of a dot as illustrated in Table 6A.2 (Harrison and Jupp, 1990). The order was designed to avoid horizontal, vertical, and diagonal line structures. Note that the sum of four adjacent cells sums to 34 either within one cell or across a cell. This type of structure is based on a "pandiagonal magic square" or "nasik" (see Bonham-Carter, 1994).

Data for plotting are typically defined in RGB space, so these data must be converted to CMY before plotting. The conversion from RGB to CMY must be done effectively, as color perception of substractive coloring is not usually the same as additive coloring. For example, in the additive system subtle changes in intensity are perceived more readily in darker than lighter shades, whereas the situation is reversed for subtractive coloring.

Color filmwriters do not use dithering to render color. In this process a small dot is used to expose photographic film. This is useful to produce high-resolution high-quality image products, but these are expensive.

6A.2   DISPLAY LIMITATIONS: STATIC IMAGES

The display of static images is potentially limited by the spatial capabilities and colors available of the visualization media. There is, of course, the problem that data usually contain more than three channels, and consequently, it is impossible to display all the available data.

The color requirements for an image will depend on the quantization level of data in the image. Data encoded with 8 bits of data requires 256 (2^8) different gray tones or colors for complete representation, whereas 10-bit data require up to 1024 (2^{10}) different gray levels. Typically, satellite information such as *Landsat* data has 8 bits, and airborne information usually has more than 8 bits, with some systems, typically geophysical data, having up to 32 bits of information. Considering that current display systems for personal computers and workstations (as described above) can support 24-bit graphics, there is the possibility of three 8-bit color channels. Consequently, each channel can be represented with any of 256 different levels. Similarly, color plotting devices can support 16^3 (4096) different colors with a 4 × 4 dithering cell. For most practical purposes these capabilities of display media will not limit the representation of colors. The human eye is able to discriminate only about 16 shades of grey, and therefore the inclusion of higher levels of details may not be necessary.

The size requirement for a viewing media depends on pixel size, size of data set, plotter pixel size, and most important, the scale. Several relationships that are important to the consideration of plotting size are

$$dec = \frac{npl}{nd} \tag{6A.1}$$

$$scale = \frac{npl}{nd} \times \frac{plsz}{dsz} \times \frac{1}{1000} \tag{6A.2}$$

$$scale = dec \times \frac{plsz}{dsz} \times \frac{1}{1000} \tag{6A.3}$$

where dec represents decimation, npl is the number of pixels plotted, nd the number of pixels in data, plsz the size of plotter pixel, and dsz the size of data pixel.

In this definition decimation means that data will be dropped out (decimation of <1) or data will be duplicated (>1). Only when decimation is 1 is there a one-to-one relationship between the data and what is seen on the viewing media. Currently, screens for personal computers can display up to 1600×1400 pixels; however, sizes of 1240×1024 or less are more common. At this size, images are normally decimated; however, dynamic zoom and roam capabilities permit full resolution at least over a limited part of the image. For these reasons, high spatial resolution combined with dynamic zoom screen display size is not normally a limitation in visualization of static images.

Decimation factors used in hard-copy plotting are usually related to the selection of an appropriate scale. For example, in plotting *Landsat* TM data the only scale for which there is no decimation is approximately 37,000/1, assuming a 30-m TM pixel combined with a plotter pixel of 0.2021 mm and a 4×4 dithering cell (therefore, an effective plotting pixel of 0.8082 mm). At scales larger than this (<37,000), extra data pixels will need to be inserted. Similarly, for other types of data there is only one scale factor for which no decimation is calculated (assuming that plotter pixel size remains constant).

References

Aronoff, S., 1989, *Geographic Information Systems: A Management Perspective*, WDL Publications, Ottawa, Ontario, Canada, 294 pp.

Bak, P. R. G. and J. B. Mill, 1989. Three dimensional representation in a geoscientific resource management system for the minerals industry, in *Three Dimensional Applications in Geographical Information Systems*, J. Raper, ed., Taylor & Francis, London, pp. 155–181.

Batson, R. M., K. Edwards, and E. M. Eliason, 1975. Computer-generated shaded-relief images, *J. Res. U.S Geol. Surv.* 3(4) pp. 401–408.

Batson, R. M., K. Edwards, and E. M. Eliason, 1976. Synthetic stereo and Landsat pictures, *Photogramm. Eng. Remote Sensing*, 42, 1279–1284.

Bonham-Carter, G. F. 1994. *Geographic Information Systems for Geoscientists: Modeling with GIS*, Pergamon/Elsevier, New York, 398 pp.

Braunstein, R. L., 1976. *Depth Perception Through Motion*, Academic Press, San Diego, Calif.

Bristow, C. S. and J. F. Raper, 1991. Modeling 3D reservoir geometry: a new approach using IVM, in *Advances in Petroleum Geology*, Spec. Publ., Geological Society of London, London.

Broome, J. H., 1990. Generation and interpretation of geophysical images with examples from the Rae Province, northwestern Canada Shield, *Geophysics*, 55(8), pp. 977–997.

Broome, J. H., R. Simard, and D. Teskey, 1985. Presentation of magnetic anomaly map data by stereo projection of magnetic shadowgrams, *Can. J. Earth Sci.* 22(2) pp. 311–314.

Broome, J. H., J. M. Carson, J. A. Grant, and K. L. Ford, 1987. *A Modified Ternary Radioelement Mapping Technique and Its Application to the South Coast of Newfoundland*, GSC Pap. 87–14, Geological Survey of Canada, Ottawa, Ontario, Canada.

Buchanan, M. D., 1979. Effective utilization of colour in multidimensional data representations, *Proc. SPIE*, 199, 9–19.

Burt, P. J., 1989. Multiresolution techniques for image representation, analysis, and "smart" transmission, *Proceedings of the SPIE on Visual Communication and Image Processing IV*, Vol. 1199, pp. 2–15.

350

Buttenfield, B. P., and W. A. Mackaness, 1991. Visualization, in *Geographical Information Systems*, Vol. 1, *Principles*, D. Maguire, M. F. Goodchild, and D. Rhind, eds., Wiley, New York, 649 pp.

Carper, W. J., T. M. Lillesend, and R. W. Kiefer, 1990. The use of the intensity–hue–saturation transformations for merging SPOT panchromatic and multispectral image data, *Photogramm. Eng. Remote Sensing*, 56(4), 459–467.

Chavez, P. S., 1986. Digital merging of *Landsat*-TM and digitized NHAP data for 1:24,000-scale image mapping, *Photogramm. Eng. Remote Sensing*, 52(10), 140–146.

Chavez, P. S., and J. Bowell, 1988. The comparison of the spectral information content of *Landsat* thematic mapper and *SPOT* for three different sites in the Phoenix, Arizona region, *Photogramm. Eng. Remote Sensing*, 54(12), 1699–1708.

Chavez, P. S., S. C. Sides, and J. A. Anderson, 1991. Comparison of three different methods to merge multi-resolution and multi-spectral data: *Landsat* TM and SPOT Panchromatic, *Photogramm. Eng. Remote Sensing*, 57(3), 295–303.

Cliche, G., F. Bonn, and P. Teillet, 1985. Integration of the *SPOT* panchromatic channel into its multi-spectral mode for image sharpness enhancement, *Photogramm. Eng. Remote Sensing*, 51(3), 311–316.

Daubechies, I., 1990. The wavelet transform, time-frequency localization and signal analysis, *IEEE Trans. Inform. Theory*, 36(5), 961–1005.

Dredge, L. A., and F. M. Nixon, 1992. *Surficial Geology, Northern Melville Peninsula, NWT*, Map 1782A, 1: 200,000, Geological Survey of Canada, Ottawa, Ontario, Canada.

Drury, S. A., 1993. *Image Interpretation in Geology*, 2nd ed., Chapman & Hall, London, 243 pp.

Edwards, K., and P. A. Davis, 1994. The use of intensity–hue–saturation transformation for producing colour shaded-relief images, *Photogramm. Eng. Remote Sensing*, 60(11), 1369–1374.

Ehlers, M., 1991. Multi-sensor image fusion techniques in remote sensing, *ISPRS J. Photogramm. Remote Sensing*, 46(3), 19–30.

Foley, J. D., 1981. *Fundamentals of Interactive Computer Graphics*, Addison-Wesley, Reading, Mass.

Friedhoff, R. M., 1991. *Visualization: The Second Computer Revolution*, W. H. Freeman, New York.

Foley, J. D., A. van Dam, S. K. Feiner and J. F. Hughes (1990). Computer Graphics: Principles and Practice, 2nd edition, Addison-Wesley, Reading, Mass.

Garguet-Duport, B., J. Girel, J. Chassey, and G. Pautou, 1996. The use of multi-resolution analysis and wavelets transform for merging *SPOT* panchromatic and multispectral image data, *Photogramm. Eng. Remote Sensing*, 62(9), 1057–1066.

Gillespie, A. R., 1980. Digital techniques of image enhancement In *Remote Sensing in Geology*, B. S. Siegal and A. R. Gillespie, eds., Wiley, New York, pp. 139–226.

Gillespie, A. R., A. B, Kahle, and R. E. Walker, 1986. Colour enhancement of highly correlated images: 1. Decorrelation and HSI contrast stretches, *Remote Sensing Environ.* 20, 209–235.

Graham, D. F., and G. F. Bonham-Carter, 1993. Airborne radiometric data: a tool for reconnaissance geological mapping using a GIS, in *Proceedings of ERIM's 9th Thematic Conference on Remote Sensing*, Pasadena, Calif., pp. 43–54.

Grassman, H., 1853. Zur Theorie der Farbenmischung, *Poggendorffs Ann.*, 89, 69.

Guild, J., 1931. The colourimetric properties of the spectrum, *Philos. Trans. R. Soc. London*, A230, 149.

Harris, J. R., 1986. A comparison of lineaments interpreted from remotely sensed data and airborne magnetics and their relationship to gold deposits in Nova Scotia, *Proceedings of the 5th Thematic Conference on, Remote Sensing for Exploration Geology*, Calgary, Alberta, Canada, pp. 233–249.

Harris, J. R., 1989. Data integration for gold exploration in eastern Nova Scotia using a GIS, *Proceedings of Remote Sensing for Exploration Geology*, Calgary, Alberta, Canada, pp. 233–249.

Harris, J. R., and V. R. Slaney, 1982. A comparison of *Landsat* and *Seasat* imagery for geological mapping in difficult terrain, in *Proceedings of the International Symposium on Remote Sensing of Environment, 2nd Thematic Conference, Remote Sensing for Exploration Geology*, Fort Worth, Texas, pp. 805–814.

Harris, J. R., R. Murray, and T. Hirose, 1990. IHS Transform for the integration of radar imagery and other remotely sensed data, *Photogramm. Eng. Remote Sensing*, 56(12), 1631–1641.

Harris, J., C. Bowie, A. N. Rencz, D. Viljoen, and P. Huppé, 1993a. *Presentation of Geoscientific Map Products: Technical Document*, GSC Open File Rep. 2742, Geological Survey of Canada, Ottawa, Ontario, Canada.

Harris, J., S. Gupta, G. Woodside, and N. Ziemba, 1993b. Integrated use of a GIS and three-dimensional, finite-element model: San Gabriel Basin groundwater flow analysis, in *Environmental Monitoring with GIS*, M. F. Goodchild, B. O. Oarks, and L. T. Steyaert, eds., Oxford University Press, Oxford, pp. 168–172.

Harris, J. R., C. Bowie, A. N. Rencz, and D. Graham, 1994. Computer enhancement techniques for the integration of remotely sensed, geophysical and thematic data for the geosciences, *Can. J. Remote Sensing*, 20(3).

Harris, J. R., L. Wilkinson, J. Broome, and S. Fumerton, 1995. Mineral exploration using GIS-based favourability analysis, Swayze greenstone belt, northern Ontario, in *Proceedings of 1995 Canadian Geomatics*, Ottawa, Ontario Canada (CD-ROM).

Harrison, B. A., and D. L. Jupp, 1990. *Introduction to Image Processing*, Division of Water Resources, CSIRO, Canberra, New South Wales, Australia, 255 pp.

Haydn, R., G. W. Dalka, J. Henkel, and J. E. Bare, 1982. Application of the IHS colour transform to the processing of multi-sensor data and image enhancements in *Proceedings of the International Symposium on Remote Sensing of Arid and Semi-arid Lands*, Jan., pp. 599–616.

Hearnshaw, H. M., and D. J. Unwin, eds., 1994. *Visualization in Geographical Information Systems*, Wiley, Toronto, 243 pp.

Hornbrooke, E. H. W., and P. W. B. Friske, 1988. *Regional lake sediment and water geochemical reconnaissance data, Province of Newfoundland [Labrador]*. GSC Open File Rep. 1636, Geological Survey of Canada, Ottawa, Ontario, Canada.

Huang, H. K. and P. S. Chao, 1993. Architecture and Ergonomics of Imaging Workstations, Chapter 11 *in* The Perception of Visual Information, edited by W. S. Hendee and P. N. T. Wells, Springer-Verlag, New York pp 316–334.

Jessel, M., 1993. *Structural Geophysics using Noddy*, AMIRA P418 Release 1.

Jones, E. R., A. P. McLauring, and L. Cathey, 1984. VISDEP™: visual image depth enhancement by parallax induction, in *Advances in Display Technology IV, Proc. SPIE*, 457, 16–19.

Judd, D. B., 1933. The 1931 I.C.I. standard observer and coordinate system for colourimetry, *J. Opt. Soc. Am.*, 23, 359.

Judd, D. B., and G. Wyszecki, 1975. *Colour in Business, Science, and Industry*, Wiley, New York.

Kelk, B., 1991. 3-D GIS for the geosciences, Computer and Geosciences, 17

Kruse, F. A., A. B. Lefkoff, J. W. Boardman, K. B. Heidebrecht, A. T. Shapiro, P. J. Barloon, and A. F. H. Goetz, 1993. The spectral image processing system (SIPS): interactive visualization and analysis of imaging spectrometer data, *Remote Sensing Environ.*, 44, 145–163.

Levin, M. D., 1985. *Vision in Man and Machine*, McGraw-Hill, New York.

Lewry, J. F., Z. Hajnal, A. Green, S. B. Lucas, D. White, M. R. Stauffer, K. E. Ashton, W. Weber, and R. Clowes, 1994. Structure of a Paleoproterozoic continent–continent collision zone: a Lithoprobe seismic reflection profile across the Trans-Hudson Orogen, Canada, *Tectonophysics*, 232, 143–160.

Lillesand, T. M., and Kiefer R. W., 1994. *Remote Sensing and Image Interpretation*, Wiley, New York, 612 pp.

Livingstone, M. S., 1988. Art, illusion and the visual system, *Sci. Am.*, 258 (1), 78–85

MacEachern, A. M., in collaboration with B. Buttenfield, J. Campbell, D. DiBiase, and M. Monmonier, 1992. Visualization, in *Geography's Inner Worlds: Pervasive Themes in Contemporary American Geography*, R. F. Abler, M. G. Marcus, and J. M. Olsen, eds., Rutgers University Press, New Brunswick, N.J., pp. 99–137.

Mallat, S. G., 1989. A theory for multiresolution signal decomposition: the wavelet representation, *IEEE Trans. Pattern Anal. Mach. Intell.*, 2, 674–693.

McCormik et al., 1987. McCormik, B. H., T. A. Defanti, and M. D. Brown. 1987 Visualization in Scientific computing. SIGGRAPH Computer Graphics Newsletter 21(6)

Mclaren, R. A, and T. J. M. Kennie, 1989. Visualization of digital terrain models: techniques and applications *in* Raper J. F. (ed) Three Dimensional Applications in Geographical Information Systems. Taylor and Francis, London, pp. 79–98

Pratt, W. K., 1978. *Digital Image Processing*, A Wiley-Interscience Publication, Wiley, New York, 749 pp.

Raper, T., ed., 1989. *Three Dimensional Applications in Geographical Information Systems*, Taylor & Francis, London, 189 pp.

Rencz, A. N., J. R. Harris, G. P. Watson, and B. Murphy, 1994a. Data integration for mineral exploration in the Antigonish Highlands, Nova Scotia, *Can. J. Remote Sensing*, 20 (3), 258–267.

Rencz, A. N., J. Harris, J. Glynn, G. Labelle, P. Huppé, and H. Press GSC Open File Rep. 2741 Geological Survey of Canada) 1994b. *Presentation of Geoscientific Map Products*, Ottawa, Ontario, Canada.

Rheault, M., R. Simard, C. Garneau, and V. R. Slaney, 1991. Sar Landsat TM-Geophysical data integration utility of value-added products in geological exploration. *Can. J. Remote Sensing*, 17 (2), 185–190.

Sabins, F., 1997. *Remote Sensing: Principles and Interpretation*, 2nd ed., W. H. Freeman, New York, 449 pp.

Schneier, B., 1993. Colour models: RGB isn't the only game in town, *Dr. Dobb's J.* July.

Sharp, P. F., and R. Philips, 1993. Physiological optics, in *The Perception of Visual Information*, W. S Hendee, and P. N. T. Wells, eds., Springer-Verlag, New York, pp. 1–29.

Shettigara, V. K., 1992. A generalized component substitution technique for spatial enhancement of multispectral images using a higher resolution data set, *Photogramm. Eng. Remote Sensing*, 58 (5), 561–567.

Singh, A., and A. Harrison, 1985. Standardized principal components, *Int. J. Remote Sensing*, 6, 883–896.

Toutin, T., and B. Rivard, 1995. A new tool for depth perception of multi-source data, *Photogramm. Eng. Remote Sensing*, 61(10), 1209–1211.

Turner, A. K., 1989a. The role of three-dimensional information systems in subsurface characterization for hydrogeological applications, in *Three Dimensional Applications in Geographical Information Systems*, J. F. Raper, ed., Taylor & Francis, London, pp. 115–127.

Turner, A. K. (ed), 1989b. Three Dimensional Modeling with Geoscientific Information Systems, NATO ASI Series C: Mathematical and Physical Sciences, Vol. 354, Kluwer Academic Publishers, 443 pp.

Usery, E. L., 1995. Virtual stereo display for three dimensional geographic data, *Photogramm. Eng. Remote Sensing*, 59 (12), 1737–1744.

Walker, J. 1986. The Amateur Scientist: The hyperscope and the pseudoscope aid experiments on three-dimensional vision, Scientific American, Vol. 255, No. 3, pp. 134–138

Weibel, R., and M. Heller, 1991. Digital terrain modelling, in *Geographical Information Systems*, Vol. 1, *Principles*, D. Maguire, M. F. Goodchild, and D. Rhind, Wiley, New York, pp. 269–297.

Welch, R., and W. Ehlers, 1987. Merging multiresolution *SPOT* HRV and *Landsat* TM data, *Photogramm. Eng. Remote Sensing*, 53 (3), 301–303.

Wright, W. D., 1928–1929. A re-determination of the trichromatic coefficients of the spectral colours, *Trans. Opt. Soc. London*, 30, 141.

Wright, D. F., and G. F. Bonham-Carter, 1996. VHMS favourability mapping with GIS-based integration models, Chisel Lake–Anderson Lake area, in *EXTECH 1: A Multidisciplinary Approach to Massive Sulphide Research in the Rusty Lake–Snow Lake Greenstone Belts, Manitoba*, G. F. Bonham-Carter, A. G. Galley, and G. E. M. Hall, eds., GSC Bull. 426, Geological Survey of Canada, Ottawa, Ontario, Canada, pp. 339–376, 387–401.

Yocky, D. A., 1995. Image merging and data fusion using the discrete two-dimensional wavelet transform, *J. Opt. Soc. Am. A.*, 12 (9), 1834–1841.

Yocky, D. A., 1996. Multi-resolution wavelet decomposition image merger of *Landsat* thematic mapper and *SPOT* panchromatic data, *Photogramm. Eng. Remote Sensing*, 62 (9), 1067–1074.

Yoeli P, 1965. Analytical hill shading, Surveying a Mapping 25, pp 573–579

Yoeli P, 1967. Mechanisation in analytical hill shading, Cartographic Journal 4, pp 82–88

Youngman, C., 1989. Spatial data structures for modeling sub-surface features, in *Three Dimensional Applications in Geographical Information Systems*, J. F. Raper, ed., Taylor & Francis, London, pp. 129–136.

Applications

Stratigraphy

Harold R. Lang

Jet Propulsion Laboratory,
California Institute of Technology,
Pasadena, California

7.1   INTRODUCTION

In their classic textbook, Krumbein and Sloss (1963) provided a comprehensive treatise on stratigraphy, with no mention of remote sensing whatsoever. The index of the last edition of the *Manual of Remote Sensing* (Colwell, 1983) contains only one stratigraphic entry (p. 2437), and that was in reference to counting craters on the Moon (p. 2390). In a more recent stratigraphic textbook, Miall (1990, pp. 328–330) observed that "careful thought should be given to relative costs and benefits of focussing on remote sensing during preliminary exploration versus devoting more resources to fieldwork. The two are, of course, not mutually exclusive, but should be complementary." According to Baltuck (1991), "the use of remote sensing to document stratigraphy from space" had been demonstrated by 1991.

These observations suggest that during the 1980s, remote sensing emerged as a workable stratigraphic tool. Accordingly, recent geological remote sensing textbooks (e.g., Prost, 1993; Berger, 1994) contain chapters devoted specifically to the application of remote sensing methods to stratigraphy. More detailed review of the stratigraphic and remote sensing literature shows that remote sensing methods have been used successfully in at least five areas of stratigraphic research (Table 7.1).

The purpose of this chapter is to introduce readers who have expertise in stratigraphy and/or geologic remote sensing to this relatively new approach to stratigraphic analysis. Items 2 to 5 of Table 7.1 are emphasized. The focus is on mapping methods that incorporate passive remote sensing surveys of the Earth's land surface, acquired

Remote Sensing for the Earth Sciences: Manual of Remote Sensing, 3 ed., Vol. 3, edited by Andrew N. Rencz.
ISBN: 0471-29405-5 © 1999 John Wiley & Sons, Inc.

Table 7.1   Areas of Stratigraphic Research Where the Utility of Visible–Infrared Remote Sensing Methods Has Been Demonstrated

1. *Sedimentary processes:* Crowley and Hook, 1996; Gillespie et al., 1984; Harris and Kowalik, 1995; Kahle et al., 1984; Millington et al., 1987; Quarmby et al., 1989; Townsend et al., 1989
2. *Sedimentary petrology:* Ferrari et al., 1996; Gaffey, 1985 and 1986; Galvao and Vitorello, 1995; Hunt and Salisbury, 1975 and 1976; Janza, 1975; Lang et al., 1990; Rowan et al., 1991, 1992 and 1995; Sgavetti et al., 1995; Van der Meer, 1994
3. *Mapping lithostratigraphic units:* Chapman and Sable, 1960; Jansma and Lang, 1996; Jansma et al., 1991; Johnson et al., 1991; Kahle et al., 1984; Lang et al., 1985; Lang and Paylor, 1994; Leith and Alvarez, 1985; Paylor et al., 1989; Rowan et al., 1970; Sabins, 1969; Salmon and Vincent, 1974; Sgavetti, 1992; Sgavetti et al., 1995
4. *Lithologic/time-stratigraphic correlation:* Beratan et al., 1990; Lang et al., 1987 and 1991; Lang and Paylor, 1994; Paylor et al., 1985; Sgavetti, 1992; Sgavetti et al., 1995
5. *Mapping lithofacies:* Boardman, 1991; Chapman and Sable, 1960; Ernst and Paylor, 1996; Ferrari et al., 1996; Kraus, 1992; Lang et al., 1990; Lang and Paylor, 1994; Rowan et al., 1970 and 1992; Sgavetti, 1992; Sgavetti et al., 1995; Stucky and Krishtalka, 1991

from aircraft or satellite, made across visible and thermal infrared wavelengths. Relevant principles are reviewed, and papers that document relevant methodologies are identified, so that readers can pursue topics that are beyond the scope of this review article.

7.2 BASIC CONCEPTS

7.2.1 Purpose of Stratigraphy

Stratigraphy is "that branch of geology which treats the formation, composition, sequence, and correlation of the stratified rocks . . ." (American Geological Institute, 1976, p. 411). Stratigraphy has evolved from emphasis on constructing, classifying and interpreting stratigraphic columns (e.g., Krumbein and Sloss, 1963) into an interdisciplinary field emphasizing analysis and modeling of depositional systems, sedimentary basin formation and evolution, sedimentary tectonics, the burial/thermal history of strata, and paleogeography (e.g., Miall, 1990). The overall goal is to determine earth history.

Mapping (both surface and subsurface) has always been a primary source of basic stratigraphic information. This information includes:

1. *Sequence:* characterizing vertical changes in color, grain size, mineralogy, lithology, paleontology, bedding characteristics, primary sedimentary structures, and other physical and chemical properties of strata; measured at specified localities and documented in stratigraphic columns.
2. *Correlation:* determining correspondence of strata at different localities based on petrology, lithology, paleontology, relative stratigraphic position, and/or age.
3. *Facies:* identifying bodies of strata having common "aspects" (such as color,

grain size, mineralogy, bedding, source terrain, diagenesis, and/or other lithological or paleontological criteria) that suggest deposition in similar environments.

4. *Geometry:* measuring the three-dimensional form of strata and/or facies determined by constructing cross sections, fence diagrams, block diagrams, and/or isopachous maps.

7.2.2 Stratigraphic Value of Remote Sensing

Compared to those emphasizing structural geology, investigations aimed specifically at obtaining basic stratigraphic information using remote sensing data other than aerial photographs are relatively rare and recently reported in the remote sensing and geological literature. For the most part, geologic remote sensing experiments in sedimentary terrain have been constrained by the assumption, tacit or implied, that the stratigraphic utility of remote sensing lies in providing maps that match those produced by field geologists (e.g., Williams, 1983).

Duplication of conventional field mapping may not be an advantageous application of remote sensing surveys for stratigraphic analysis. For example, correlation and facies analysis may be enhanced by remotely mapping unconventional units, identified by their geomorphic expression and spectral characteristics at wavelengths beyond human vision, rather than mapping conventional units which are defined on the ground, without benefit of the synoptic perspective afforded by observations from aircraft or satellite. Furthermore, the digital data from modern remote sensing instruments provide consistent, objective, gridded, geophysical measurements of the landscape. Gridded surface data cannot be obtained practically using any other stratigraphic method. These attributes of remote sensing make the technique uniquely useful for stratigraphic investigations that can benefit from surface mapping.

7.2.3 Definition of Mapping Units

The basic procedure for surface stratigraphic mapping is tracing contacts of stratigraphic units across the landscape and plotting these contacts on a cartographic base. Experience in the field of stratigraphy over the last 200 years has resulted in codification of formal rules for defining and naming these mapping units (North American Commission on Stratigraphic Nomenclature, 1983). According to the code, a lithostratigraphic unit "generally conforms to the Law of Superposition and commonly is stratified and tabular in form . . . and . . . recognized and defined by observable rock characteristics; boundaries may be placed at clearly distinguished contacts or drawn arbitrarily within a zone of gradation" (p. 855). Formations are "the basic [formal] lithostratigraphic units used in describing and interpreting the geology of a region" (p. 858). A formation's most important property is that it is mappable, based on direct field observations of lithological characteristics. According to the code, units that are defined and mapped using indirect measurements, such as geophysical well logs, seismic reflection surveys, or any other "remote sensing" instrument are useful, but are considered informal.

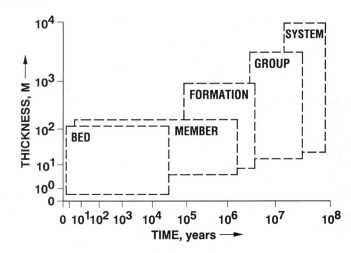

Figure 7.1   Typical ranges of thickness and time represented by formal lithostratigraphic units (based on the author's experience in North American sedimentary basins).

Formations can be subdivided into thinner, formal lithostratigraphic units, called *members* or *beds*, or lumped together in thicker, formal units, called *groups* or *systems*. Although there are no code requirements for the absolute thickness or time represented by any of these formal lithostratigraphic units, typical ranges of values do exist in practice (Figure 7.1).

When examined in the context of the code, the units that have been mapped using visible to infrared remote sensing data are informal lithostratigraphic units, equivalent to beds, members, formations, and groups. For example, Chapman and Sable (1960) used aerial photographs to map shale and sandstone intervals in Cretaceous strata of northern Alaska, at 1:64,000 scale. Their mapping units corresponded to previously unmapped beds and members of the Kukpowruka and Toruk Formations. Sabins (1967) used thermal infrared images to map what he called "stratigraphic-radiometric units" in the Indio Hills of southern California. He showed that these units corresponded to previously unmapped sandstone, siltstone, and conglomerate beds in the Plio-Pleistocene Imperial and Palm Springs Formations. Lang and Baird (1981) and Lang et al. (1985) used the term *spectral stratigraphic unit* for a member of the Eocene Wasatch Formation that they discovered and mapped in Wyoming, at 1:24,000 scale, with multispectral visible/short-wavelength infrared aircraft image data. Sgavetti (1992) proposed a complex photostratigraphic nomenclature and coined the terms *photostratigraphic* and *photofacies* units for the beds and members that she mapped in the south-central Pyrenees of Spain, at 1: 18,000 scale, using the approach described by Chapman and Sable (1960). Because these units are commonly unconformity bound, Sgavetti (1992) suggested that sequence- and seismic-stratigraphic concepts are directly applicable to analysis of photostratigraphic units. These and other examples cited in Table 7.1, items 3 and 4, show that remotely mapped, informal lithostratigraphic units can be incorporated into the formal lithostratigraphic nomenclature and assigned ages using standard methods of correlation (e.g., Lang et al., 1987; Lang and Paylor, 1994; Sgavetti et al., 1995).

7.3 STRATIGRAPHIC ANALYSIS OF VISIBLE–INFRARED IMAGE DATA

Two types of sensors have provided visible–infrared images for stratigraphic mapping. Aerial cameras loaded with black-and-white or color film that is sensitive to visible (0.4 to 0.7 μm) and/or near infrared (0.7 to 1.0 μm) radiation have provided pictures for geological mapping for over 70 years. Optomechanical instruments, with anywhere from 1 to 224 channels covering visible through thermal infrared wavelengths (0.4 to 12.0 μm), have operated on aircraft since the 1960s. Global acquisition of this type of data in digital format began in 1972 with launch of the Landsat series of satellites.

Existing Earth-observing satellites carry optomechanical instruments that provide digital data in fewer than eight channels, with bandpasses of about 0.5 μm. This spectral resolution is generally insufficient to allow identification of individual minerals, as illustrated in Figure 7.2 and discussed in Section 7.3.2. Such identification is only possible today using imaging spectrometer aircraft instruments. These instruments have spectral resolution of 0.1 μm or better and make measurements in as many as 224 contiguous channels.

Another sensor characteristic that is important for stratigraphic applications is spatial resolution. This parameter determines the maximum mapping scale that can be achieved. Figure 7.3 provides guidelines for such a determination. The figure shows an empirical relationship between spatial resolution and maximum potential mapping scale and is based on the author's experience and results reported in the references cited in Table 7.1, items 3 to 5.

Sensors that operate at visible–thermal infrared wavelengths are restricted to measurements of the Earth's surface. Earth materials in only the top few micrometers influence the measurements; except in essentially vegetation-free, dry environments,

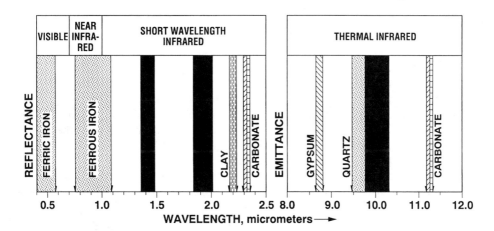

Figure 7.2 Visible to infrared wavelength positions of spectral bands that are characteristic of important sedimentary rock-forming minerals. Wavelengths shown in black near 1.4 and 1.9 μm are not available for remote sensing of the Earth's surface due to atmospheric water absorption of electromagnetic radiation, and near 10.0 μm due to atmospheric ozone absorption. (Compiled from Hunt, 1977; Hunt and Salisbury, 1975, 1976; Lang et al., 1987, 1990.)

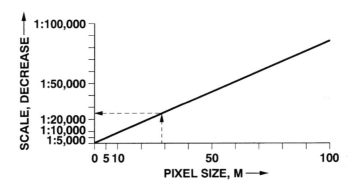

Figure 7.3 Nomograph relating spatial resolution (pixel size) to the maximum image scale that is potentially useful for photostratigraphic interpretation. Use of the nomograph is illustrated by the dashed line: *Landsat* thematic mapper data, with a 28.5-m pixel size, is potentially useful at approximately 1:25,000 or smaller scales.

where thermal infrared measurements can be influenced by bedrock buried by sediments up to 1.5 m thick (Nash, 1985, 1988). Remotely sensed measurements made across visible and thermal infrared wavelengths, therefore, respond directly only to water, ice, vegetation, soil, alluvium, and weathered bedrock (modified by sensor, atmospheric, topographic, and cultural effects) (Figure 7.4). Because the focus of stratigraphy is bedrock, these effects must be understood and accounted for when using remote sensing data for stratigraphic mapping. Thorough discussion of all the items on Figure 7.4 is beyond the scope of this review but is provided in almost any geological remote sensing textbook (e.g., Sabins, 1987; Prost, 1993; Berger, 1994). Other chapters of this volume also contain relevant information (e.g., Chapters 1 through 6 and 11).

With thorough understanding of these factors, the stratigrapher can use two approaches to obtain stratigraphic information from visible through infrared remote sensing data: (1) photogeology and (2) spectroscopy–thermometry.

7.3.1 Photogeology

Photogeological interpretation of aerial photographs has long been a standard complement to field mapping (Ray, 1960; Compton, 1962). The method relies on the basic stratigraphic principles of superposition and cross-cutting relationships (Boulter, 1989) and concepts of geomorphology (Thornbury, 1965) to map bedrock stratigraphy. Knowledge about the weathering and erosional characteristics of strata allows the photointerpreter to infer lithology. For example, in most environments sandstone and limestone tend to be resistant to weathering and therefore hold up steep slopes and form ridges, while shales tend to be less resistant to weathering and therefore hold up low slopes or valleys. Image tone, color, and texture also provide important clues to lithology. Chapman and Sable (1960) provide an excellent example of applying these concepts to the problem of mapping strata and inferring lithology.

Although photogeology was developed originally for analysis of aerial photo-

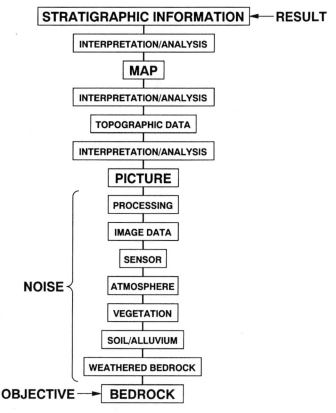

Figure 7.4   Potentially important layers of material, data, and analysis between the objective of remote stratigraphic surveys (bedrock strata) and the desired result (lithostratigraphic information).

graphs, the procedure is equally applicable to interpreting remote sensing images acquired by sensors operating at wavelengths beyond those sampled by film. For example, Sgavetti et al. (1995) integrated results of local photogeological interpretation of aerial photographs and regional photogeological interpretation of *Landsat* thematic mapper images covering a 6000-km^2 area of northern Somalia. Ridge-forming photogeological units, primarily limestones, provided important stratigraphic marker beds in their assessment of the 1.2-km-thick Jurassic–Eocene sequence. Because of political instability in the area, field access was limited. Nevertheless, their remote assessment of the regional stratigraphic framework provided major new insights regarding passive margin development and subsequent rifting in the region.

When topographic information is available, the attitudes of stratigraphic contacts mapped photogeologically and the thicknesses of stratigraphic units can be calculated using standard geologic map analysis methods. Knowledge of the *x, y, z* coordinates of at least three points on both the top and bottom of a stratigraphic unit is required (Figure 7.5). This procedure, referred to as the *three-point problem*, is a standard method for analyzing geologic maps that are plotted on topographic contour base maps (Compton, 1962; Boulter, 1989). It has also been used for analyzing strati-

PLAN (MAP) VIEW

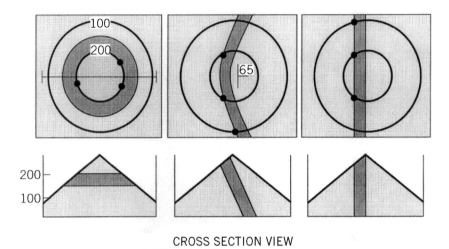

CROSS SECTION VIEW

Figure 7.5 Simplified maps and cross sections, of arbitrary scale, illustrating the dependence of the map trace and outcrop width of a lithostratigraphic unit (the dark band) on the unit's dip and strike. Knowledge of the x–y–z coordinates of at least three points on both the top and bottom of the unit provides sufficient geometric information to determine the unit's dip, strike, and thickness.

graphic contacts that have been transferred from aerial photograph interpretations to topographic contour base maps (Ray, 1960). Stereo image pairs obtained by film cameras or optomechanical sensors also have provided the required topographic information (Ray, 1960; Berger et al., 1992; Prost, 1993). With the advent of digital elevation models (DEMs), these same geometric procedures have been automated for computer analysis of coregistered digital remote sensing images and DEMs (McGuffie et al., 1989; Chorowicz et al., 1991; Morris, 1991).

7.3.2 Spectroscopy–Thermometry

Spectroscopy is used to infer the mineralogy of materials on the Earth's surface using multispectral visible through infrared radiance measurements. This approach is based on (1) knowledge about the interaction of electromagnetic radiation and matter, and (2) laboratory measurements of the spectral properties of rocks and minerals. The laboratory spectroscopic studies have usually been supported by mineralogical determinations using x-ray diffraction (XRD) methods. Field spectroscopic and temperature measurements have also proven to be useful for assessing the in situ spectral properties of rocks (e.g., Lee, 1975; Milton, 1987). Most of the reports cited below and in Table 7.1 incorporate these types of laboratory and field measurements. This information is the basis for the design of sensors and the development of methods for image processing that produce pictures portraying spectral differences due to lithologic variations.

Radiance measurements by visible to infrared sensors record spectral reflectance and emittance variations that are determined by the mineralogy of sedimentary rocks. Diagnostic features are recorded as minima (called *bands*) in reflectance or emittance spectra (Hunt, 1977). At thermal infrared wavelengths, radiance measurements are also influenced by the temperature of rocks.

The wavelengths of some major mineral bands that have been recognized in the spectra of weathered sedimentary rocks are identified in Figure 7.2. It should be clear from the figure that only a few minerals control the spectral properties of weathered sedimentary rocks.

Determinative mineralogy using visible–thermal infrared data should be considered as just another geochemical technique, and as such is applicable to stratigraphic analysis as described by Odin et al. (1982). Over visible through near-infrared wavelengths (Figure 7.2), the primary mineralogic influence on the spectral reflectance of sedimentary rocks is the iron oxidation state of minerals. Common iron-bearing, sedimentary-rock-forming minerals include hematite, goethite, "limonite," and jarosite. Thus remote measurements over this wavelength interval are particularly useful for mapping redbeds. This was demonstrated by Salmon and Vincent (1974), who mapped hematite-bearing redbeds of the Triassic Chugwater Group in Wyoming, successfully using *Landsat* data.

In the short wavelength–infrared region (Figure 7.2), the primary mineralogic influence on the spectra of sedimentary rocks is by clay and carbonate minerals. Spectral reflectance features between 2.15 and 2.25 µm allow distinction of kaolinite, smectite, and illite. Gypsum, anhydrite, and zeolite minerals such as analcime also exhibit spectral bands over these wavelengths. Carbonate bands exist near 2.35 µm due to calcite, and near 2.30 µm due to dolomite. Lang et al. (1987) used this information to determine the stratigraphic distribution of clay species, gypsum, calcite, and dolomite throughout a 1-km-thick Paleozoic–Mesozoic section in Wyoming, based on diagnostic bands recorded by 128-channel imaging data covering 1.2 to 2.4 µm. In another Wyoming study, Boardman (1991) mapped a dolostone member of the Permian Goose Egg Formation, based on the presence of a dolomite band at 2.30 µm in 224-channel imaging data covering 0.4 to 2.4 µm. Van der Meer (1994) used 63-channel imaging data, covering 0.4 to 2.4 µm, to map dolomitization of calcite limestones in southern Spain, based on shift of the carbonate band from 2.35 µm (due to calcite) to 2.30 µm (due to dolomite).

Using laboratory measurements, Rowan et al. (1991, 1995) showed that the total organic carbon (TOC) content and organic matter (OM) maturity of Paleozoic, Cretaceous, and Paleogene shale samples from Nevada and Colorado could be quantified using visible–thermal infrared spectral data. Based on these results, Rowan et al. (1992) mapped OM maturity of the Mississippian Chainman Shale Formation in Nevada, with images created from *Landsat* thematic mapper channels at 0.8 and 1.7 µm. TOC and OM measurements have become standard stratigraphic tools for determining the thermal and burial history of strata (Miall 1990). These results represent another stratigraphic application of near-short wavelength infrared spectral data, in addition to determinative mineralogy.

Over thermal infrared wavelengths (Figure 7.2), sensor radiance measurements combine both spectral (emittance) and temperature information. Spectral emittance bands include those due to gypsum, quartz (and other silicates), and carbonates. A carbonate band shifts from 11.3 µm for calcite to 11.2 µm for dolomite. Using this information, Lang et al. (1987) mapped quartz-rich sandstone, carbonate, and gypsum beds in Paleozoic–Cretaceous strata on the margin of the Wind River basin, Wyoming, with six-channel thermal infrared image data covering 8.1 to 11.7 μm. Data from the same aircraft sensor led to discovery of a dendritic network of quartz-rich channel sandstone beds that were deposited by an Eocene river system near the center of the same basin (Stucky and Krishtalka, 1991). Lang et al. (1991) used the same type of data in another Wyoming basin to map a member of the Cretaceous Cody Shale Formation. Subsequent biostratigraphic studies showed that the new member coincides with a regional planktic foraminiferal biozone and is the local record of a late Coniacian (88 Ma) eustatic sea-level highstand.

In addition to these examples of passive measurement results, active (laser) measurements have also been used for spectroscopic (reflectivity) mapping in the thermal infrared. For example, Kahle et al. (1984) used an airborne laser system to map the composition of modern alluvial fans and Paleozoic dolostone and orthoquartzite beds in Death Valley, California.

Thermal inertia (TI) variations of sedimentary rocks also influence radiances measured by sensors operating in the region 8.0 to 12.0 μm. TI is a measure of a material's resistance to changes in temperature (Kahle, 1980). As shown by Figure 7.6, TI differences distinguish sediments from sedimentary rocks and distinguish different sedimentary lithologies, including shale, limestone, sandstone, and dolostone. Water,

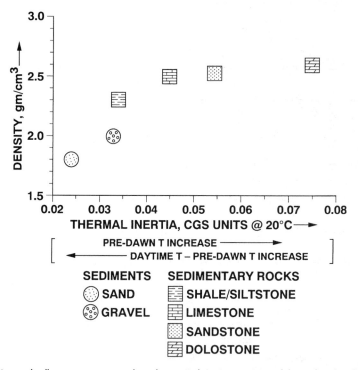

Figure 7.6 Scatter plot illustrating variations in thermal inertia (cal cm^{-2} s$^{-\frac{1}{2}}$ ° C^{-1}) and density for selected sediments and sedimentary rocks with different textures and mineralogies. Also shown are potential trends of (1) nighttime temperatures and (2) daytime minus nighttime temperatures, recorded by a thermal infrared sensor. (Compiled from Janza, 1975; Kahle, 1980; Sabins, 1980.)

with a TI of 0.038 (Janza, 1975; not shown on Figure 7.6), influences the thermal properties of sediments and sedimentary rocks in all but the most dry environments.

Although TI images can be calculated from calibrated thermal radiance images using models that incorporate topography, albedo, and atmospheric data (e.g., Kahle, 1980, 1987); in practice, simple nighttime thermal radiance images or images created from differences in daytime and nighttime thermal radiance measurements have found greatest use in mapping sedimentary rocks (e.g., Sabins, 1967, 1969, 1980; Rowan et al., 1970; Wolfe, 1971).

Sabins (1987) reviewed several relevant examples of images and results from nighttime thermal infrared aircraft surveys covering sedimentary terrains. These examples include mapping shale and sandstone beds in a Tertiary turbidite sequence, in the Temblor Range of central California; mapping individual Tertiary lake beds that are covered by a 1-m-thick veneer of windblown sand and gravel and are therefore invisible on aerial photographs, in the Imperial Valley of southern California; and mapping Precambrian chert and dolostone beds that are covered by a thin veneer of soil and are therefore invisible on aerial photographs, in the western Transvaal, South Africa. Rowan et al. (1970) used similar data to map previously unrecognized limestone and dolostone members of Paleozoic formations in the Arbuckle Mountains, Oklahoma.

7.4   RECIPE FOR STRATIGRAPHIC MAPPING WITH REMOTE SENSING DATA

Lang et al. (1987) proposed a recipe for remote stratigraphic mapping. They successfully used the procedure to map at 1:24,000 scale and analyze a 1-km-thick, Pennsylvanian to Cretaceous, carbonate and siliciclastic sequence exposed on the margin of the Wind River foreland basin in Wyoming. Lang and Paylor (1994) further illustrated procedures involved in the recipe. Essentially the same procedure had been used earlier by Leith and Alvarez (1985) in their 1:250,000-scale assessment of a 10-km-thick, Mesozoic to Neogene sequence exposed in the Tadjik basin of the former Soviet Union. Sgavetti et al. (1995) and Ferrari et al. (1996) adopted the method in their analysis of Jurassic–Eocene strata in Somalia. In simplified form, this recipe involves five steps:

1. Acquisition of appropriate remote sensing data covering the area of interest.
2. Reconnaissance photogeologic interpretation of small-scale images to identify regional stratigraphic markers and to locate relatively undeformed reference localities for detailed stratigraphic interpretation.
3. Detailed photogeologic interpretation of large-scale images covering reference localities to determine resistance to erosion, attitude, sequence, thickness, and spectral characteristics of the photogeologically defined stratigraphic units. This step requires topographic information. Results are used to construct a photogeologic map and stratigraphic columns.
4. Correlation of stratigraphic columns constructed in step 3 with stratigraphic columns defined elsewhere in the image, and with existing conventional surface and subsurface sections. Correlation with conventional sections provides a

means for assigning formal lithostratigraphic names and geologic ages to the informal stratigraphic units that were defined using image data. The photogeologic map and stratigraphic columns from step 3 can then be labeled with formation names obtained from the correlations to create a conventional geologic map.

5. Field work to check results, conduct conventional field studies, and collect field spectra and/or surface samples for laboratory XRD and spectroscopic analyses.

7.5 SUMMARY AND CONCLUSIONS

Review of dozens of reports, published primarily since 1980, shows that remote sensing has emerged as a workable stratigraphic tool. Stratigraphic studies that incorporate image data acquired from aircraft or satellite, over visible to thermal infrared wavelengths (0.4 to 12.0 μm), demonstrate that this new tool is applicable to many areas of stratigraphic research. These research areas include studies of sedimentary processes, petrology, lithostratigraphic units, lithologic/time stratigraphic correlation, and lithofacies.

Characteristics of modern remote sensing image data make them particularly useful for mapping sedimentary rocks. They provide a synoptic perspective and consistent, objective, gridded, geophysical measurements of the landscape at wavelengths beyond those of human vision or film. These data can contribute to any stratigraphic investigation that would benefit from surface mapping. According to the stratigraphic code, units that have been mapped remotely are informal lithostratigraphic units. They can be tied to the formal stratigraphic nomenclature and dated through methods of correlation.

Two approaches are used in concert for remote mapping of sedimentary rocks: (1) photogeology and (2) spectroscopy–thermometry. Photogeology is applicable to interpreting images acquired at wavelengths beyond those sampled by film. When combined with topographic information, photogeology can be used to obtain information remotely about the lithology, sequence, attitude, and thickness of lithostratigraphic units. With coregistered digital topographic and image data, these methods have been automated on desktop personal computers.

The spectroscopic approach is based on results from laboratory and field investigations of the spectral properties of weathered sedimentary rocks. These results show that multispectral remote sensing methods provide a tool for remote determinative mineralogy, provided that data with sufficient spectral resolution (<0.1 μm) are available. Important sedimentary-rock-forming minerals that have been mapped remotely include hematite, smectite, kaolinite, illite, calcite, dolomite, analcime, gypsum, and quartz. The organic carbon content and thermal maturity of shales can also be quantified and mapped remotely.

These capabilities have been applied successfully to strata ranging in age from Precambrian to Holocene that were deposited in depositional environments ranging from continental to deep marine. The major limitation of the spectral approach is that it is applicable primarily in relatively vegetation-free arid regions where bedrock is exposed.

Thermal infrared measurements made at night can also provide bedrock lithologic information in arid regions. The method can be used in areas with as much as 1.5

m of soil cover. Because nighttime temperatures correlate with bedrock density, shale, limestone, sandstone, conglomerate, dolostone, and chert beds have been mapped using thermal data.

Thus remote sensing surveys can provide valuable geometric and compositional information to complement conventional approaches to mapping surface stratigraphy. This method has been used at mapping scales ranging from 1:18,000 to 1:250,000, depending on the spatial resolution of the data and the mapping objectives. Although this new stratigraphic tool has been thoroughly documented, published discussions (e.g., Johnson et al., 1992, versus Monod et al., 1992; and Berger, 1993, versus Wessel, 1993) suggest that it is not yet universally accepted by field mappers. But this slow acceptance is not surprising, based on the history of other stratigraphic techniques such as magnetostratigraphy and chemical stratigraphy.

ACKNOWLEDGMENTS

This chapter includes results of research carried out at the Jet Propulsion Laboratory, California Institute of Technology, under contract with the National Aeronautics and Space Administration (NASA). Work was done as part of the Multispectral Analysis of Sedimentary Basins Project. Reference herein to any specific commercial product, process, or service by trade name, trademark, manufacturer, or otherwise does not constitute or imply endorsement by NASA, the U.S. government or Jet Propulsion Laboratory, California Institute of Technology. The final manuscript benefited from helpful comments from two anonymous referees.

References

American Geological Institute, 1976. *Dictionary of Geological Terms*, rev. ed., Anchor Press/Doubleday, Garden City, N.Y., 472 pp.

Baltuck, M., 1991. Preface, *M. Geol.*, 28(2–3), ii.

Beratan, K. K., R. G. Blom, J. E. Nielson, and R. E. Crippen, 1990. Use of *Landsat* thematic mapper images in regional correlation of syntectonic strata, Colorado River Extensional Corridor, California and Arizona, *J. Geophys. Res.*, 95(B1), 615–624.

Berger, Z., 1993. Geologic stereo mapping of geologic structures with SPOT satellite data: reply, *Am. Assoc. Pet. Geol. Bull.*, 77(4), 662–664.

Berger, Z., 1994. *Satellite Hydrocarbon Exploration Interpretation and Integration Techniques*, Springer-Verlag, New York, 319 pp.

Berger, Z., H. L. Williams, and D. W. Anderson, 1992. Geologic stereo mapping of geologic structures with SPOT satellite data. *Am. Assoc. Pet. Geol. Bull.*, 76(1), 101–120.

Boardman, J. W., 1991. Sedimentary facies analysis using imaging spectrometry, *Proceedings of the 8th ERIM Thematic Conference on Geologic Remote Sensing*, Denver, Colo., pp. 1189–1199.

Boulter, C. A., 1989. *Four Dimensional Analysis of Geological Maps: Techniques of Interpretation*, Wiley, New York, 296 pp.

Chapman, R. M., and E. G. Sable, 1960. *Geology of the Utukok–Corwin Region Northwestern Alaska*, USGS Prof. Pap. 303-C, U.S. Geological Survey, Washington, D.C., pp. 47–167.

Chorowicz, J., J.-Y. Breard, R. Guillande, C.-R. Morasse, D. Prudon, and J.-P Rudant, 1991. Dip and strike measured systematically on digitized three-dimensional geological maps, *Photogramm. Eng. Remote Sensing*, 57(4), 431–436.

Colwell, R. N., ed., 1983. *Manual of Remote Sensing*, Vol. II, 2nd ed., American Society of Photogrammetry, Falls Church, Va., 2440 pp.

Compton, R. S., 1962. *Manual of Field Geology*, Wiley, New York, 378 pp.

Crowley, J. K., and S. J. Hook, 1996. Mapping playa evaporite minerals and associated sediments in Death Valley, California, with multispectral thermal infrared images, *J. Geophys. Res.*, 101(B1), 643–660.

Ernst, W. G., and E. D. Paylor II, 1996. Study of the reed dolomite aided by remotely sensed imagery, Central White–Inyo Range, easternmost California, *Am. Assoc. Petr. Geol. Bull.*, 80(7), 1008–1026.

Ferrari, M. C., M. Sgavetti, and R. Chiari, 1996. Multispectral facies in prevalent carbonate strata of an area of Migiurtinia (northern Somalia): analysis and interpretation, *Int. J. Remote Sensing*, 17(1), 111–130.

Gaffey, S. J., 1985. Reflectance spectroscopy in the visible and near infrared (0.35–2.55 μm): applications in carbonate petrology, *Geology*, 13, 270–273.

Gaffey, S. J., 1986. Spectral reflectance of carbonate minerals in the visible and near infrared (0.35–2.55 microns) calcite, aragonite, and dolomite. *Am. Mineral.*, 71, 151–162.

Galvao, L. S., and Vitorello, I., 1995. Quantitative approach in the spectral reflectance: lithostratigraphy of the Wind River and southern Bighorn Basins, Wyoming, *Int. J. Remote Sensing*, 16(9), 1617–1631.

Gillespie, A. R., A. B. Kahle, and F. D. Palluconi, 1984. Mapping alluvial fans in Death Valley, California, using multichannel thermal infrared images, *Geophys. Res. Lett.*, 11(11), 1153–1156.

Harris, P. M., and W. S. Kowalik, 1995. *Satellite Images of Carbonate Depositional Settings*, AAPG Methods Explor. Ser. 11, American Association of Petroleum Geologists, Tulsa, Okla., 147 pp.

Hunt, G. R., 1977. Spectral signatures of particulate minerals in the visible and near infrared, *Geophysics*, 42, 501–513.

Hunt, G. R., and J. W. Salisbury, 1975. *Mid-infrared Spectral Behavior of Sedimentary Rocks*, Environ. Res. Pap. 520, Air Force Cambridge Research Laboratory, Hanson Air Force Base, Mass., 49 pp.

Hunt, G. R., and J. W. Salisbury, 1976. Visible and near infrared spectra of minerals and rocks: XI. Sedimentary rocks, *Mod. Geol.*, 5, 211–217.

Jansma, P. E. and H. R. Lang, 1996. Applications of spectral stratigraphy to Upper Cretaceous and Tertiary rocks in southern Mexico: Tertiary graben control on volcanism, *Photogramm. Eng. Remote Sensing*. v.62, n.12, 1371–1378.

Jansma, P. E., H. R. Lang, and C. A. Johnson, 1991. Preliminary investigation of the Tertiary Balsas Group, Mesa Los Caballos area, northern Guerrero State, Mexico using *Landsat* thematic mapper data, *Mt. Geol.*, 28(2–3), 137–150.

Janza, F. J., 1975. Interaction mechanisms, in *Manual of Remote Sensing*, Vol. I, R. G. Reeves, ed., American Society of Photogrammetry, Falls Church, Va., pp. 75–179.

Johnson, C. J., H. R. Lang, E. Cabral-Cano, C. G. A. Harrison, and J. A. Barros, 1991. Preliminary assessment of stratigraphy and structure, San Lucas region, Michoacan and Guerrero States, SW Mexico, *Mt. Geol.*, 28(2–3), 121–135.

Johnson, C. J., H. Lang, E. Cabral-Cano, C. Harrison, and J. Barros, 1992. Preliminary assessment of stratigraphy and structure, San Lucas region, Michoacan and Guerrero States, SW Mexico: reply, *Mt. Geol.*, 29(1), 3–4.

Kahle, A. B., 1980. Surface thermal properties, in *Remote Sensing in Geology*, B. S. Siegal, and A. R. Gillespie, ed., Wiley, New York, pp. 257–273.

Kahle, A. B., 1987. Surface emittance, temperature, and thermal inertia derived from thermal infrared multispectral scanner (TIMS) data for Death Valley, California, *Geophysics*, 52(7), 858–874.

Kahle, A. B., M. S. Shumate, and D. B. Nash, 1984. Active airborne infrared laser systems for identification of surface rock and minerals, *Geophys. Res. Lett.*, 11(11), 1149–1152.

Kraus, M. J., 1992. Alluvial response to differential subsidence: sedimentological analysis aided by remote sensing, Willwood Formation (Eocene), Bighorn Basin, Wyoming, USA, *Sedimentology*, 39, 455–470.

Krumbein, W. C., and L. L. Sloss, 1963. *Stratigraphy and Sedimentation*, 2nd ed., W. H. Freeman, New York, 660 pp.

Lang, H. R., and K. W. Baird, 1981. Spectral stratigraphy of the Fort Union/Wasatch Transition, Patrick Draw Geosat petroleum test site, Wyoming, in *Proceedings of the International Geoscience and Remote Sensing Symposium*, Washington, D.C., Vol. 1, pp. 589–594.

Lang, H. R., and E. D. Paylor, 1994. Spectral stratigraphy: remote sensing lithostratigraphic procedures for basin analysis, central Wyoming examples, *J. Nonrenewable Resources*, 3(1), 25–45.

Lang, H. R., W. H. Alderman, and F. F. Sabins, Jr., 1985. Patrick Draw, Wyoming, petroleum test case report, in *The Joint NASA/Geosat Test Case Project Final Report*, Part 2, Vol. II, M. J. Abrams, J. E. Conel, and H. R. Lang, eds., American Association of Petroleum Geologists, Tulsa, Okla., pp. 11-i to 11–112.

Lang, H. R., S. L. Adams, J. E. Conel, B. A. McGuffie, E. D. Paylor, and R. E. Walker, 1987. Multispectral remote sensing as stratigraphic and structural tool, Wind River Basin and Big Horn Basin areas, Wyoming, *Am. Assoc. Pet. Geol. Bull.*, 71(4), 389–402.

Lang, H. R., M. J. Bartholomew, C. I. Grove, and E. D. Paylor, 1990. Spectral reflectance characterization (0.4 to 2.5 and 8.0 to 12.0 micrometers) of Phanerozoic strata, Wind River Basin and southern Bighorn Basin areas, Wyoming, *J. Sediment. Petrol.*, 60(4), 504–524.

Lang, H. R., W. E. Frerichs, A. McGugan, and E. D. Paylor, 1991. Biostratigraphic significance of a new unit, mapped remotely with multispectral thermal infrared data, Late Cretaceous Cody Shale, southern Bighorn Basin, Wyoming, *Mt. Geol.*, 28(2–3), 67–73.

Lee, K., 1975. Ground investigations in support of remote sensing, in *Manual of Remote Sensing*, R. G. Reeves, editor-in-chief), American Society of Photogrammetry, Falls Church, Va., pp. 805–856.

Leith, W., and W. Alvarez, 1985. Structure of the Vakhsh Fold-and-Thrust Belt, Tadjik SSR: geologic mapping on a *Landsat* image base, *Geol. Soc. Am. Bull.*, 96, 875–885.

McGuffie, B. A., L. F. Johnson, R. E. Alley, and H. R. Lang, 1989. IGIS computer-aided photogeologic mapping with image processing, graphics and CAD/CAM capabilities, *Geobyte*, 4(5), 8–14.

Miall, A. D., 1990. *Principles of Sedimentary Basin Analysis*, 2nd ed. Springer-Verlag, New York, 668 pp.

Millington, A. C., A. R. Jones, N. Quarmby, and J. R. G. Townsend, 1987. Remote sensing of sediment transfer processes in playa basins, in *Desert Sediments: Ancient and Modern*, Spec. Publ. 35, L. Frostick, and I. Reid, eds., Geological Society, 369–381.

Milton, E. J., 1987. Principles of field spectroscopy, *Int. J. Remote Sensing*, 8, 1807–1827.

Monod, O., J. Ramirez, M. Faure, H. Sabanero, and M.-F. Campa, 1992. Preliminary assessment of stratigraphy and structure, San Lucas region, Michoacan and Guerrero States, SW Mexico: discussion, *Mt. Geol.*, 29(1), 1–2.

Morris, K., 1991. Using knowledge-base rules to map the three-dimensional nature of geological features, *Photogramm. Eng. Remote Sensing*, 57(9), 1209–1216.

Nash, D. B., 1985. Detection of bedrock topography beneath a thin cover of alluvium using thermal remote sensing, *Photogramm. Eng. Remote Sensing*, 51, 77–88.

Nash, D. B., 1988. Detection of a buried horizon with a high thermal diffusivity using thermal remote sensing, *Photogramm. Eng. Remote Sensing*, 54, 1437–1446.

North American Commission on Stratigraphic Nomenclature, 1983. North American Stratigraphic Code, *Am. Assoc. Pet. Geol. Bull.*, 67(5), 841–875.

Odin, G. S., M. Renard, and C. V. Grazzini, 1982. Geochemical events as a means of correlation, in *Numerical Dating in Stratigraphy*, Parts I and II, G. S. Odin, ed., Wiley, New York, pp. 37–71.

Paylor, E. D., M. J. Abrams, J. E. Conel, A. B. Kahle, and H. R. Lang, 1985. *Performance Evaluation and Geologic Utility of Landsat-4 Thematic Mapper Data*, JPL Publ. 85–66, Jet Propulsion Laboratory, California Institute of Technology, Pasadena, Calif., 68 pp.

Paylor, E. D., H. L. Muncy, H. R. Lang, J. E. Conel, and S. L. Adams, 1989. Testing some models of foreland deformation at the Thermopolis Anticline, southern Bighorn Basin, Wyoming, *Mt. Geol.*, 26(1), 1–22.

Prost, G. L., 1993. *Remote Sensing for Geologists: A Guide to Image Interpretation*, Gordon and Breach, Lausanne, Switzerland, 326 p.

Quarmby, N. A., J. R. G. Townshend, A. C. Millington, K. White, and A. J. Reading, 1989. Monitoring sediment transport systems in a semiarid area using thematic mapper data, *Remote Sensing Environ.*, 28, 305–315.

Ray, R. G., 1960. *Aerial Photographs in Geologic Interpretation and Mapping*. USGS Prof. Pap. 373; U.S. Geological Survey, Washington, D.C., 229 pp.

Rowan, L. C., T. W. Offield, K. Watson, P. J. Cannon, and R. D. Watson, 1970. Thermal infrared investigations, Arbuckle Mountains, Oklahoma, *Geol. Soc. Am. Bull.*, 81(12), 3549–3562.

Rowan, L. C., J. W. Salisbury, M. J. Kingston, N. Vergo, and N. H. Bostick, 1991. Evaluation of visible and near-infrared and thermal-infrared reflectance spectra for studying thermal alteration of Pierre Shale, Wolcott, Colorado, *J. Geophys. Res.*, 96(B11), 18047–18057.

Rowan, L. C., M. J. Pawlewicz, and O. D. Jones, 1992. Mapping thermal maturity in the Chainman Shale, near Eureka, Nevada, with *Landsat* thematic mapper images, *Am. Assoc. Pet. Geol. Bull.*, 76(7), 1008–1023.

Rowan, L. C., F. G. Poole, and M. J. Pawlewicz, 1995. The use of visible and near-infrared reflectance spectra for estimating organic matter thermal maturity, *Am. Assoc. Pet. Geol. Bull.*, 79(10), 1464–1480.

Sabins, F. F., Jr., 1967. Infrared imagery and geologic aspects, *Photogramm. Eng.*, 29, 83–87.

Sabins, F. F., Jr., 1969. Thermal infrared imagery and its application to structural mapping in southern California, *Geol. Soc. Am. Bull.*, 80, 397–404.

Sabins, F. F., Jr., 1980. Interpretation of thermal infrared images, in *Remote Sensing in Geology*, B. S. Siegal and A. R. Gillespie, eds., Wiley, New York, pp. 275–295.

Sabins, F. F., Jr., 1987. *Remote Sensing Principles and Interpretation*, 2nd ed., W. H. Freeman, New York, 449 pp.

Salmon, B. and R. K. Vincent, 1974. Surface compositional mapping in the Wind River Range and Basin, Wyoming by multispectral techniques applied to ERTS-1 data, in *Proceedings of the 9th ERIM International Symposium on Remote Sensing of Environment*, Vol. 3, Ann Arbor, Mich., pp. 2005–2012.

Sgavetti, M., 1992. Criteria for stratigraphic correlation using aerial photographs: examples from the south-central Pyrenees, *Am. Assoc. Pet. Geol. Bull.*, 76(5), 708–730.

Sgavetti, M., M. C. Ferrari, R. Chiari, P. L. Fantozzi, and I. Longhi, 1995. Stratigraphic correlation by integrating photostratigraphy and remote sensing multispectral data: an example from Jurassic–Eocene strata, northern Somalia, *Am. Associ. Petr. Geolo. Bull.* 79(11), 1571–1589.

Stucky, R. K., and L. Krishtalka, 1991. The application of geologic remote sensing to vertebrate biostratigraphy: general results from the Wind River Basin, Wyoming, *M. Geol.*, 28(2–3), 75–82.

Thornbury, W. D., 1965. *Principles of Geomorphology*, Wiley, New York, 618 pp.

Townsend, J. R. G., N. A. Quarmby, A. C. Millington, N. Drake, A. J. Reading, and K. H. White, 1989. Monitoring playa sediment transport systems using thematic mapper data, *Adv. Space Res.*, 1, 177–183.

Van der Meer, F., 1994. Sequential indicator conditional simulation and indicator kriging applied to discrimination of dolomitization in GER 63-channel imaging spectrometer data, *J. Nonrenewable Resources*, 3(2), 146–164.

Wessel, G. R., 1993. Geologic stereo mapping of geologic structures with SPOT satellite data: discussion, *Am. Assoc. Pet. Geol. Bull.*, 77(4), 660–661.

Williams, R. S. (ed.), 1983. Geological applications, in *Manual of Remote Sensing*, 2nd ed., Vol. II, R. N. Colwell, ed., American Society of Photogrammetry, Falls Church, Va., pp. 1667–1953.

Wolfe, E. W., 1971. Thermal IR for geology, *Photogramm. Eng.*, 37, 43–52.

8

Remote Sensing Strategies for Mineral Exploration

Charles Sabine

Geopix
Sparks, Nevada 89436

8.1 INTRODUCTION

8.1.1 Past, Present, and Future

When the second edition of the *Manual of Remote Sensing* (Colwell, 1983) was published, the use of digital imagery from orbital and airborne sensors as a tool for mineral exploration was entering a second decade and was poised at a threshold of new sensor technology and advances in computer hardware and software. Explorationists had achieved spectacular success with the first-generation orbital sensors, the multispectral scanner (MSS), onboard *Landsats 1, 2,* and *3,* in lithologic and alteration mapping (Rowan et al., 1974, 1977; Goetz et al., 1975; Goetz and Rowan, 1981; Segal, 1983), geobotanical applications (Bolivken et al., 1977; Raines et al., 1978; Raines and Canney, 1980), and regional structural analysis (Gold, 1980; Goetz and Rowan, 1981; Rowan and Wetlaufer, 1981). However, the broad bands, limited spectral range, coarse resolution, and monoscopic imagery of MSS greatly constrained geologic interpretation and exploration applications. Airborne multispectral scanners with bands in the shortwave infrared (SWIR) had demonstrated the potential of bands in the 2.2-μm region in detecting hydroxyl-bearing minerals and carbonates associated with hydrothermally altered zones (Abrams et al., 1977, 1983; Rowan and Kahle, 1982; Marsh and McKeon, 1983; Peters; 1983; Podwysocki, et al., 1983). This had led to the addition of a 2.2-μm band (band 7) on the *Landsat 4*

Remote Sensing for the Earth Sciences: Manual of Remote Sensing, 3 ed., Vol. 3, edited by Andrew N. Rencz.
ISBN: 0471-29405-5 © 1999 John Wiley & Sons, Inc.

thematic mapper (TM), which had been launched in 1982. Results from experiments utilizing high-spectral-resolution radiometers aboard aircraft and the space shuttle had demonstrated the potential for the direct identification of certain clay minerals and carbonates from space (Goetz et al., 1982) and spectral anomalies in plant canopies growing in soils over metallic ore deposits (Chang and Collins, 1983; Collins et al., 1983; Labovitz et al., 1983; Milton, 1983; Milton et al., 1983). As a result of these experiments, NASA had begun construction of the first hyperspectral imaging spectrometer, the airborne imaging spectrometer (AIS). Its successor, the airborne visible/infrared imaging spectrometer (AVIRIS) was already in initial planning stages. The successful discrimination of various sedimentary and igneous lithologies, silicified zones, and argillized rocks in multispectral thermal infrared images of the East Tintic Mountains, Utah (Kahle and Rowan, 1980) led to the construction of the thermal infrared multispectral scanner (TIMS) by NASA (Kahle and Goetz, 1983). Acquisition of broadband thermal infrared imagery from orbit by the heat capacity mapping mission had led to the addition of a thermal band on *Landsat 4* (TM band 6). Airborne imaging radar technology had produced the first reliable maps of cloud-shrouded areas of the Amazon region in the RADAMBRAZIL project and the *Seasat* and Shuttle Imaging Radar-A (SIR-A) missions had proven the feasibility of imaging radar from orbit.

In 1983, few if any mining or exploration companies possessed the equipment and expertise to process digital imagery. Most explorationists relied on photographic prints of images, and those with image processing expertise frequently traveled to universities or government laboratories that had the mini- or mainframe computers and software to analyze image data. Personal computers were just coming on the scene and had no practical image processing capability. Computers and image processing software were slow and cumbersome by today's standards; a principal component transformation was often the last task before going home for the night.

The potential that was indicated in 1983 became reality in the 15 years since the second edition was published. Words such as *hyperspectral, interferometry*, and *unmixing* became entrenched in our vocabulary. Thematic mappers aboard *Landsats 4* and *5* fulfilled all expectations, and new earth resource satellites from France, India, and Japan were launched. Orbital imaging radar systems became operational. The success of NASA's airborne imaging spectrometer and its successor, AVIRIS, allowed direct identification and mapping of certain mineral species in remotely sensed data, stimulating commercial development of airborne imaging hyperspectral scanners and compact field spectrometers. The continued success of TIMS led to the inclusion of multispectral thermal infrared bands on some commercial airborne scanners.

The evolution of the personal computer and the advent of Unix-based workstations, together with fast, affordable, and user-friendly software, brought image processing to the desktops of explorationists in large and small companies alike. New spectral analysis software packages have introduced capabilities to perform mineral mapping and spectral unmixing operations on hyperspectral data to produce alteration and lithologic maps at unprecedented levels of refinement. Advancing GIS technology has provided a basis for integrating geochemical and geophysical data with digital imagery to yield new insights into surface and subsurface geology.

The minerals industry has also undergone significant changes since the early 1980s. Exploration, then in recession with forcasts of gloom and doom, is now booming. This change was stimulated in part by discoveries of new deposit types that generated new exploration models and by advances in mining and metallurgical technology that allowed development of previously subeconomic deposits. The most important changes have been geopolitical, however. The end of the cold war and the emergence of third-world countries from military dictatorships have opened enormous regions for exploration. These developments have led explorationists into unfamiliar and uncharted territory with no reliable base maps and little, if any, geologic data. Remote sensing and GIS have played a key role in formulating exploration strategies in these regions.

Some things never change, however. New exploration technologies typically undergo a history of initial euphoria followed by disillusionment and then gradual acceptance as their utility and limitations become understood. In the 1970s MSS was oversold as an exploration tool, and its failure to live up to unrealistic expectations led to rejection of remote sensing as a useful tool by some shortsighted explorationists. During the 1980s, companies willing to invest in innovative strategies gradually developed remote sensing into a useful tool and integrated it into their exploration programs. Other mining companies, traditionally slow to embrace new technology until a competitor shows that it works, soon followed. Exploration geologists, who usually react to "black-box" technology like a cat to rearranged furniture, were also slow to accept remote sensing. Some continue to be hostile toward it, while most have accepted it as a useful source of information.

As this volume is published, geological remote sensing has reached a new threshold as the next generation of orbital systems are prepared for launch. Scanners such as ASTER and MODIS, scheduled for launch in 1998 aboard NASA'S EOS platforms, will provide stereoscopic, high-resolution VNIR, SWIR, and multiband thermal imagery for the entire globe. Specialized data sets from these scanners such as reflectance and emissivity, now used solely for research, will be available as standard products. NASA also intends to launch *Landsat 7*, with its enhanced thematic mapper, in 1997 and is also developing an orbital hyperspectral scanner. The French and Indian governments are also developing the next generations of their *SPOT* and *IRS* satellites. European, Canadian, and Japanese are providing global synthetic aperture radar (SAR) imagery, and NASA is planning a multipolar, interferometric L-band SAR satellite. Innovative scanner designs under development may make it possible for mining and exploration companies to own and operate low-cost airborne hyperspectral systems, revolutionizing airborne remote sensing as the PC revolutionized digital computing.

8.1.2 Scope and Objectives

During the 25 years of remote sensing for exploration, there have been few syntheses of what has been accomplished and what possibilities lie in the future. Colwell's (1983) chapter in the second edition of the *Manual of Remote Sensing*, together with the report of the Joint NASA/Geosat Test Case Project (Abrams et al., 1985) and a special issue of *Economic Geology* (Vol. 78, No. 4, 1983), summarized the state of the art at the end of the first decade. Sabins (1987) and Drury (1993) included sections devoted to resource exploration in their texts, and Legg (1992) published a general introduction to remote sensing and geographic information systems oriented to mineral exploration and mining. Taranik (1988) published a general overview of remote sensing applications for precious metal exploration. Spatz has published comprehensive papers on remote sensing characteristics of porphyry copper and precious metal deposits (Spatz, 1992, 1996; Spatz and Taranik, 1994; Spatz and Wilson, 1994, 1997). For the most part, however, the literature is scattered through various conference proceedings and government reports with a few important papers appearing in regularly published journals.

This chapter does not provide an exhaustive review of the literature, however, but focuses instead on remote sensing strategies that can be employed in an exploration program. The objective is to provide a theoretical and conceptual foundation of remote sensing principals, systems, and techniques that have proven useful and and to explore their application to detecting surface manifestations of mineral deposits, derived from deposit models, in a variety of physical environments. Examples drawn from particularly instructive papers and from the author's own experience will illustrate concepts set forth. Because explorationists are like decathletes among geologists in that they must be proficient in all aspects of earth science, every chapter of this volume is likely to contain useful information that can be applied directly to exploration problems. Throughout this chapter the reader will be referred to other chapters of interest.

A second objective is to write a chapter that will not become obsolete during the next decade. An underlying theme, therefore, is how exploration remote sensing will be done in the twenty-first century. Orbital and airborne systems presently being developed and potential applications of their data products will receive particular attention. Operational uses of image processing techniques now employed for research will also be explored.

8.2 IMAGING SYSTEMS FOR MINERAL EXPLORATION

The decade of the 1990s has seen revolutionary advances in new sensor technology, computers, and image processing software, and that technology has become widely accepted for geologic and mineral resource investigations. Today, digital imagery is used routinely by geoscientists as aerial photography was used two decades ago. Explorationists are also beginning to investigate the utility of imaging spectrometers, multispectral thermal infrared systems and synthetic aperture radar imagery. The next decade will see acquisition of global data at high spatial and spectral resolutions which will be furnished rapidly to geoscientists on CD-ROM or compact high-density data storage devices which can be processed on low-cost, high-efficiency desktop or portable computers. This section highlights some of the imaging systems that have proven useful in mineral exploration as well as technologies in the research and development stage that may become operational for exploration during the next decade. Detailed descriptions of imaging systems appear in Chapters 11, 12, and 13.

8.2.1 Current and Future Satellite Systems

The principal satellite systems being used by explorationists today are the *Landsat* thematic mapper (TM) and multispectral scanner (MSS), the French *SPOT* satellites, the Indian IRS satellites, and the Japanese *FUYO-1* (Table 8.1). *Landsat* satellites have been the workhorses of the minerals industry since the launch of the first MSS aboard *Landsat 1* in July 1972 and will undoubtedly be the satellites of choice in the foreseeable future. The addition of the TM aboard *Landsats 4* and *5* provided improved spatial resolution and additional bands in the visible and shortwave infrared that have proven extremely useful in detecting ferric oxides and hydroxyl-bearing minerals in alteration zones associated with mineral deposits. *Landsat 7*, will carry an enhanced thematic mapper (ETM) that will utilize the same spectral bands as the TM plus a 15-m panchromatic sharpening band.

The French *SPOT* satellite data are not as widely used for exploration as *Landsat* data, except where greater spatial resolution is considered important. The 10-m panchromatic data have been used to sharpen TM imagery, and the multispectral bands have been found by some to be better positioned for ferric oxide detection than TM bands. *SPOT*'s cross-track stereoscopic imaging can be used to make 1:50,000-scale digital terrain models with 20-m elevation intervals. However, a relatively sophisticated knowledge of the imaging geometry is needed to produce digital topographic data from *SPOT* data, and acquisition of multiple data sets is expensive. Future *SPOT* satellites will have a 1600-nm SWIR band and fore-and-aft panchromatic stereoscopic imaging at 5-m spatial resolution with multispectral imaging at 10-m resolution.

TABLE 8.1   Current and Planned Resource Satellites for Mineral Exploration

	MSS	**TM, ETM**[a]	**SPOT**	**FUYO-1**	**ASTER**
Bands	4	7	4	8	14
Swath (km)	185	185	60	75	60
GIFOV (m)					
VNIR	79	30	20	18 × 24	15
SWIR		30		18 × 24	30
TIR		120			90
PAN		15[a]	10		
Bandpasses (μm)					
1	0.50–0.60	0.45–0.52	0.50–0.59	0.52–0.60	0.52–0.60
2	0.60–0.70	0.52–0.60	0.61–0.68	0.63–0.69	0.63–0.69
3	0.70–0.80	0.63–0.69	0.72–0.89	0.76–0.86	0.76–0.86
4	0.80–1.1	0.76–0.90	0.50–0.74	0.76–0.86	1.6–1.7
5		1.55–1.75		1.6–1.71	2.14–2.18
6		10.4–12.5		2.01–2.12	2.18–2.22
7		2.08–2.35		2.13–2.15	2.23–2.28
8		0.50–0.90[a]		2.27–2.40	2.29–2.36
9					2.36–2.43
10					8.12–8.48
11					8.48–8.82
12					8.92–9.28
13					10.2–11.0
14					11.0–11.6
Stereo	None	None	Cross-track	Fore/aft	Fore/aft
SAR				L-band	

[a]Enhanced Thematic Mapper.

The Indian government has launched a series of resource satellites since 1988. The most recent of these, *IRS-1C* launched in 1995, carries a panchromatic sensor with 6-m spatial resolution over a 70-km swath and a multispectral sensor with three VNIR bands at 24-m spatial resolution and one SWIR band with 71-m spatial resolution. *IRS* data are currently available over limited portions of the globe, but global coverage will soon be available as new ground receiving stations come on line and its onboard tape recorder is activated.

With fore-and-aft stereoscopic imaging, 20-m spatial resolution, and high spectral resolution, the Japanese *FUYO-1 (JERS-1)* satellite would probably have been the sensor of choice for most explorationists if the all-important SWIR sensor had not failed. With three bands in the 2.2-μm region covering the spectral range of TM band 7, *FUYO-1* had the potential of providing greater separation of hydrothermal alteration mineral assemblages than is possible with TM. Simulations of *FUYO-1* imagery indicated that the sensor would have been able to distinguish carbonates and propylites with absorption features in the 2.3-μm region from argillic and advanced argillic assemblages with absorption features in the 2.2-μm region (Akiyama et al., 1989a; Freund, 1992).

The advanced spaceborne thermal emission and reflection radiometer (ASTER), scheduled for flight on the *EOS-AM1* platform in 1998, will provide high-resolution stereoscopic VNIR, SWIR, and TIR data globally for the first time. If ASTER performs as specified, it will be extremely important for geological applications. Its SWIR band positions will allow possible separation of carbonates, chlorites, clays and micas, and ammonia-bearing minerals. Although considered an experimental instrument, ASTER data will no doubt be in high demand in the exploration community.

Abrams and Hook (1995) simulated ASTER imagery of Cuprite, Nevada, using AVIRIS and TIMS data resampled to ASTER spatial resolution. Using a three-band decorrelation stretch technique, they were able to distinguish ferric oxides in the VNIR bands, alunite, and kaolinite in the SWIR bands, and opalized and silicified rocks in the TIR bands (Figure 8.1; see the color insert).

CSIRO and an international consortium of mining companies and government geological surveys, working with Autralian government support, is planning the first advanced resource satellite aimed primarily at the needs of the international mining and geological community. Current plans for the proposed *Aries-1* satellite call for 32 contiguous bands in the VNIR (500 to 1000 nm), 32 contiguous bands in the SWIR (2000 to 2500 nm), and three discrete atmospheric correction bands.

The U.S. Department of Energy Sandia Laboratory is developing a multispectral satellite for launch in 1998. This instrument will have 10 bands in the VNIR/SWIR and five bands in the TIR. The Jet Propulsion Laboratory is developing a lightweight multispectral thermal infrared satellite called *Sacajawea* as a follow-on to ASTER. Initial specifications call for five to 10 bands in the range 3 to 5 and 8 to 14 μm with spatial resolution in the range 15 to 30 m.

8.2.2 Airborne Imaging Systems

The principal aircraft systems in use for exploration include sensors manufactured and flown by Geophysical and Environmental Research Corporation, Geoscan, and the Daedalus airborne thematic mapper (Table 8.2). Geoscan MkII is a commercial sensor that was built by Geoscan Pty. Ltd. of Perth, Australia specifically for mineral exploration and resource assessment. It has 46 spectral bands available in the VNIR, SWIR, and TIR, from which 24 can be selected depending on the local geology. The instrument has been disassembled, and no new data acquisitions are planned. Existing nonproprietary Geoscan data are available for many parts of the world, however.

The Geophysical and Environmental Research Corporation (GER) has built and flown a succession of airborne imaging spectrometers that have been used for research and mineral exploration. The earliest version, known as GERIS, had 63 spectral bands in the VNIR and SWIR between 400 and 2500 nm. The original GERIS instrument was succeeded by another 63-channel instrument that included six thermal infrared bands calibrated in flight by onboard blackbody standards. GER is phasing out the 63-channel instrument in favor of their new 79-channel DAIS-7915 sensor, which gathers data in five spectral regions, covering the VNIR, SWIR, and TIR.

The Daedalus airborne thematic mapper (ATM) was built as an airborne simulator for the *Landsat* TM. It scans in 10 channels in the VNIR and SWIR and one broad thermal band.

In 1982, NASA and the Jet Propulsion Laboratory developed the first hyperspectral imaging spectrometer, the airborne imaging spectrometer (AIS). Analysis of data from its first flight over Cuprite, Nevada and subsequent experiments clearly demonstrated that mineral species could not only be discriminated but uniquely identified as well through new spectral matching techniques (Goetz and Srivastava, 1985; Feldman and Taranik, 1988a,b; Hutsinpiller, 1988; Kruse, 1988).

AIS was replaced by the airborne visible/infrared imaging spectrometer (AVIRIS), which advanced hyperspectral imaging to an operational stage of development. AVIRIS has 220 bands with 9.6-nm bandwidths, covering the spectral interval from 410 nm to 2450 nm. AVIRIS continues to collect imagery as a research instrument, and data quality continues to improve. Data acquired since 1995 have signal/noise ratios greater than 400:1 over most of its spectral range, enabling unprecedented mineral mapping capability and subpixel detection of materials at the 5% level (Boardman and Huntington, 1996; Clark and Swayze, 1996; Goetz and Kindel, 1996; Green et al., 1996). Although primarily a research instrument, AVIRIS data can be acquired on a commercial basis, but mission costs are prohibitive for most

TABLE 8.2   Airborne Imaging Systems

	Geoscan MII	GERIS	GER DAIS 7915	ATM	MIVIS
Bands					
VNIR	10	24	32	8	20
SWIR	8	33	40	2	72
TIR	6	6	7	2	10
IFOV (mrad)	2.0	2.2	2.2	2.5	2.0
GIFOV[a] (m)	8	9	9	10	8
Swath[a] (km)	8	6.4	6.4	6	6

[a] At 4,000 m above ground level.

exploration budgets. To reduce costs, several mining companies pooled resources for AVIRIS "group shoots" (Kruse and Huntington, 1996).

The next generation of airborne hyperspectral imagers under development include the Canada Centre for Remote Sensing SWIR full spectrum imager (SFSI) (Hauff et al., 1996; Neville et al., 1996); a wedge imaging spectrometer (WIS) under development by the Santa Barbara Research Center of Hughes Aircraft; the hyperspectral digital imaging collection experimental sensor (HYDICE); and PROBE 1, which is now being flown commercially for mineral exploration by Earth Search Sciences, Inc. (Farrand et. al, 1997). The exploration community can expect compact imaging systems that will fit in light aircraft and produce hyperspectral imagery for real-time field use. Field-portable hyperspectral imagers based on emerging tunable liquid crystal filter technology may soon be available to geologists. Such instruments could image outcrops, pit walls, underground workings, drill core and polished sections as well as aerial scenes and will also measure polarization signatures of reflecting signatures (Denes et al., 1997).

8.2.3 Multispectral Thermal Infrared Systems

Multispectral thermal infrared (TIR) sensors operating in the range 8 to 12 μm are sensitive to spectral emissivity features related to Si–O stretching and bending vibrations in silicate crystal lattices and are therefore useful for mapping compositional variation in silicate rocks (Sabine et al., 1994), as well as silicified zones and jasperoids in mineralized areas.

The rationale for the first multispectral thermal infrared scanner came from the work of Kahle and Rowan (1980) using multiband thermal data from the NASA M2S scanner data over the Tintic mining district of Utah. They showed that the abundance of silicate minerals in sedimentary and igneous rocks could be mapped on the basis of the spectral emissive behavior of silicates and that silicic alteration zones and quartz sandstones were easily detected and mapped. On the basis of that work NASA built the thermal infrared multispectral scanner (TIMS). TIMS collects thermal infrared measurements in six spectral bands: 8.19 to 8.55, 8.58 to 8.96, 9.01 to 9.26, 9.65 to 10.15, 10.34 to 11.14, and 11.29 to 11.56 μm (Palluconi and Meeks, 1985). The instrument is still flown as a research instrument and has seen little commercial application.

Commercially flown GER and Geoscan airborne sensors also have multiband thermal sensing capability. The airborne thematic mapper has a single thermal band. ASTER will be the first satellite system to provide calibrated multispectral thermal imagery on a global basis.

The Aerospace Corporation has developed and flown SEBASS, the first operational thermal infrared imaging spectrometer. SEBASS collects data in 256 spectral bands in the 3.0 to 5.5 μm and 7.8 to 13.5 μm range using a 128 by 128 sensor array with signal-to-noise ratios of 1500 to 2000. The instrument flies aboard a twin Otter aircraft at low altitude, which is useful for district- and deposit-scale mapping.

8.2.4 Synthetic Aperture Radar Systems

Orbital and airborne synthetic aperture radar (SAR) systems are capable of imaging the earth's surface regardless of illumination conditions and weather and are therefore excellent for imaging cloud-shrouded tropical regions. Being sensitive to subtle changes in surface roughness and morphology, SAR data have also proven useful for structural mapping and geomorphic studies. The principal spaceborne SAR systems that have provided data for exploration are *ERS-1* and *ERS-2*, operated by the European Space Agency, the Canadian *Radarsat*, and the SAR imager aboard the Japanese *FUYO-1* (Table 8.3). Designed primarily for oceanographic applications, the vertically polarized antenna and 23° incidence angle of the *ERS* satellites have not proven useful for terrestrial applications. The HH polarization and variable viewing geometry of Radarsat data are providing more useful imagery for geological applications. The *FUYO-1* SAR was designed for terrestrial applications. Its 45° incidence angle is optimal for analyzing terrain morphology without significant distortion, and L-band radar penetrates forest canopies to a greater degree than does C-band radar.

The NASA SIR-C/X-SAR missions of the space shuttle collected multifrequency (X-, C- and L-band), multipolarization (HH, VV, VH, HV) calibrated digital data covering 30% of the earth's land surface between 57° north and south latitude. Multispectral, multipolarization radar data such as this may enable mapping of subtle variations of landscapes, surface cover types, and rocks that is not possible with monochromatic data. Previous shuttle imaging radar missions (SIR-A, SIR-B) also returned imagery that could be useful for exploration. Miranda et al. (1994) carried out a morphostructural analysis of SIR-B imagery covering a portion of the rainforest-covered Guiana Shield in northwestern Brazil, which when integrated with aeromagnetic and limited field data, resulted in a reconnaissance geologic map.

The success of the SIR-C/X-SAR missions have led NASA to propose a low-cost, lightweight, multipolarized L-band system called LightSar, which will offer imaging capabilities in a variety of configurations with spatial resolution as low as 3 m. The Japanese plan to fly their L-band *VSAR* satellite as a follow-on to *FUYO-1*. It will have 20 to 50° incidence angle pointing, 10-m spatial resolution over a 70-km swath and dual polarization (HH or VV).

Engineers at the Brigham Young University Microwave Remote Sensing Laboratory have developed two compact synthetic aperture radar systems that can be operated aboard light aircraft or from a moving truck or rail-mounted vehicle. YSAR is an X-band SAR, and YINSAR is an interferometric SAR capable of generating imagery and digital elevation models with resolutions less than one meter in all three dimensions (Thompson et al., 1998).

Airborne SAR imagery is available for many areas of the world. Large areas of South America, notably Columbia, Venezuela, and Brazil, and the Asian Pacific region were flown in the 1970s and early 1980s by Goodyear Aerospace using an X-band system with 10-m spatial resolution. These data continue to be very useful in

TABLE 8.3   Orbital Synthetic Aperture Radar Systems

	ERS-1, ERS-2	FUYO-1	Radarsat
Wavelength (cm)	5.7, C-band	23, L-band	5.7, C-band
Polarization	VV	HH	HH
Swath (km)	100	75	50–500
Incidence angle (deg)	23	45	20–60
Resolution (m)	28	8–100	18

areas such as the Amazon region which have recently become accessible for exploration. The U.S. Geological Survey provides airborne SAR data covering more than one-third of the United States on CD-ROM, and NASA has obtained imagery of many areas using its AIRSAR system. SAR data covering many parts of the world are also available from commercial vendors such as Intera and the Environmental Research Institute of Michigan (ERIM), which also operate airborne SAR systems on a contract basis.

8.3   INFORMATION CONTENT OF DIGITAL IMAGERY

8.3.1 What Does a Remote Sensing System Measure?

The whole point of using remote sensing for exploration is to extract useful information that can help geologists find, map, and evaluate mineral deposits. Image processing, analysis, and interpretation are steps toward this objective. Processing, whether it takes the form of a simple linear contrast stretch or a more sophisticated principal component transformation, alters data by enhancing certain portions of its information content and suppressing other portions. Analysis and interpretation involves data abstraction in which certain information is ignored in favor of other information. The higher the level of abstraction, the more information is left behind. At each stage of processing, analysis, and interpretation, an analyst must be able to retrace his or her steps back to the data in its original form. It is therefore of paramount importance that an analyst understand the fundamental measurement of a remote sensing instrument.

Remote sensing instruments measure the radiant spectral flux of photons, within discrete wavelength intervals, that are reflected, scattered, and emitted from a finite area of the earth's surface and intervening atmosphere within a discrete solid angle and detected within a certain time interval. The wavelength interval is the bandpass of the instrument; the area of the earth's surface is the ground instantaneous field of view (GIFOV); the time interval corresponds to the dwell time of the instrument; and the solid angle is the instantaneous field of view (IFOV) of the instrument. Multiple bandpasses may be utilized to measure spectral variations of radiance. A series of measurements from many GIFOVs covering a portion of the earth's surface are formatted as two-dimensional arrays of picture elements (pixels) which can be displayed as images. The position of each pixel corresponds to a specific GIFOV in the imaged area and the numerical value assigned to a pixel corresponds to the measured radiant spectral flux for that GIFOV. This value represents a composite formed from components of radiance from all surfaces and surface materials within a GIFOV as well as radiance contributions from the intervening atmosphere. In most data (e.g., *Landsat* TM), the radiance value is scaled to a dimensionless number ranging from 0 to 255 (8-bit data). Some data, such as data that have been converted to physical units of radiance or mathematically transformed, may appear as 16- or 32-bit floating-point numbers.

8.3.2 Image Analysis and Interpretation

Image analysis involves the recognition of spatial and spectral patterns in image data that, through the use of models, can be used to separate components of an image according to topographic and surface cover attributes. Landscape attributes may then be interpreted by a trained geologist to develop geologic information that might pertain to the geomorphology, lithostratigraphy, or structure of the area. This information may be combined with other sources of information (e.g., geologic mapping, geophysical and geochemical surveys) and modeled in a geographic information system to develop thematic maps of favorable terrains and exploration targets.

The recognition of landscape features in an image involves successive steps of data abstraction in which similarities are selected and dissimilarities ignored. At the lowest level of abstraction, each pixel in a single-band image is represented by its tone, a shade of gray that corresponds to the radiance value assigned to that pixel. The arrangement and frequency of change of tones defines textural elements, and tonal variation of a pixel between bands defines a spectral attribute. Groups of neighboring pixels may be identified as textural and spectral units, one or more of which may be modeled as a landscape attribute. Landscape attributes can be broadly classified as elements of topography and surface cover (Table 8.4). Carrying the abstraction to its highest levels, the arrangement, association, distribution, and geologic context of landscape attributes may suggest geologic processes, which in turn may suggest an

TABLE 8.4   Landscape Attributes Associated with Mineral Deposits

Topographic Attributes	Description	Examples
Landforms	Positive or negative landscape elements described in terms of their areal morphology	Volcanic edifices, calderas, lava domes, scarps, playas
Drainages	Negative landscape elements characterized more by a linear pattern than by their area	Structural lineaments, drainage patterns
Surface cover attributes		
Consolidated rocks	Lithified rock outcrops including altered rocks	Host rocks, hydrothermal alteration
Unconsolidated rock material	Unlithified sedimentary material deposited by physical or chemical processes	Sand and gravel, evaporites, placers
Soils and rock weathering products	Materials formed in situ by physical and chemical weathering and by biological processes	Laterites, gossans
Rock coatings	Thin coatings deposited on rock surfaces by chemical and biological processes	Desert varnish, solfataric deposits
Vegetation	Plant material, including algae, lichen, moss, grass, trees, and shrubs	Geobotanical associations, vegetation "stress"
Water	Lakes, ponds, streams, etc.	Crater lakes, placers, evaporites, kimberlites
Culture	Land cover created or modified by human activity	Past and present mining activities
Mixtures	Any combination of surface cover types	

Source: Modified from Taranik (1988).

ore-forming process. The successive steps in the analysis and interpretation of an image can be summarized as follows:

radiance measurement → textural/spectral units → landscape attributes → geologic process → exploration model

8.3.3 Landscape Attributes and Mineral Deposits

Most metallic ore deposit models include common elements that are relevant to remote sensing studies: tectonic setting, host rocks, source rocks, alteration, and structural control. The manifestation of these deposit elements as landscape attributes depends on the type of deposit under consideration, the tectonic history of the region, and the physical environment at the surface. In heavily vegetated humid terrains, geologic features associated with mineral deposits are likely to be expressed as topographic attributes, although certain cover types such as vegetation and soils might be useful indicators of mineralization in certain environments. In tectonically mobile arid and semiarid regions such as northern Chile and Nevada, morphostructural observations can be augmented by spectral characteristics of host rocks and altered zones. In deeply weathered stable cratonic regions (e.g., portions of Australia and northeastern Brazil), the spectral character of residual soils might provide clues to underlying rocks and mineralization.

8.3.3.1 LITHOLOGIC MAPPING.

Lithologic discrimination is reasonably straightforward in sparsely vegetated terrains, and there are numerous examples in the literature in which satellite and airborne imagery have been used for regional lithologic mapping (e.g., Kepper et al., 1986; Podwysocki et al., 1987; Campos-Marquetti and Rockwell, 1989; Fei and Jutz, 1989; Yamaguchi et al., 1989, 1994; Re Kühl, 1992; Trefois et al., 1993). Most approaches entail qualitative distinctions based on overall albedo and spectral characteristics unique to particular rock units. The VNIR bands are useful for mapping color variations in the visible portion of the spectrum, and SWIR bands can be used to map rocks containing carbonates and hydroxyl-bearing minerals.

Hyperspectral sensors can discriminate subtle spectral features of carbonates and hydroxyl-bearing minerals for better lithologic distinction than is possible with broadband multispectral imagery. Kingston (1990), for example, distinguished limestone from dolomite and mapped shale, quartzite, granitic gneiss, and a small skarn deposit in the Mountain Pass district, California, using AVIRIS data. Bowers and Rowan (1996) used a linear spectral unmixing technique to map mineralogic variation in the Ice River Alkaline Complex in British Columbia. Limestone, dolomite, greenstone, evaporites, clay beds, and montmorillonitic or zeolitic tuffs have VNIR/SWIR absorption features that can be distinguished in hyperspectral imagery (Spatz, 1996). Kruse and Boardman (1997) mapped kimberlites in Utah, Wyoming and Colorado from AVIRIS imagery. Kruse (1997) used AVIRIS to map active hot spring deposits at Yellowstone National Park and Steamboat Springs, Nevada. Nusbaum et al. (1997b) mapped geology from AVIRIS imagery of the Tintic Mining District, Utah. Duke (1994) suggests that spectral variation related to Tschermak substitution in muscovite could be used to map metamorphic grade.

Multispectral thermal infrared (TIR) imagery is the most powerful tool for mapping silicate-bearing rocks. A Reststrahlen emissivity minimum, associated with Si–O bond stretching vibration frequency, shifts to shorter wavelengths with decreasing numbers of bridging oxygen ions in silicate lattices. Thus the emissivity minimum shifts from 11 µm to 9 µm in a uniform succession for minerals with isolated tetrahedra, chain, sheet, and framework structures, respectively. Phase equilibria of crystallizing magmas govern mineral associations in igneous rocks so that inosilicates and nesosilicates dominate mafic rocks and tectosilicates comprise felsic rocks. As a result, Reststrahlen features of igneous rock spectra also shift from longer wavelengths for mafic rocks to shorter wavelengths for felsic rocks. Sabine et al. (1994) modeled Gaussian curves to TIMS emissivity spectra of a suite of igneous rocks ranging from leucogranite to anorthosite in a glaciated area in the Sierra Nevada batholith of California. They found that the wavelength of the emissivity minimum determined from the Gaussian curve varied linearly with rock compositional parameters (Figure 8.2), and they used this relationship to create images depicting the quartz and SiO_2 content of the rocks. Watson et al. (1990) discriminated silica-rich rocks from silica-poor rocks in TIMS imagery of the Carlin Mine area, Nevada, and identified previously unknown silicified zones in Tertiary volcanic rocks. They also used the CO_3 reststrahlen feature in TIMS band 6 to distinguish limestone from dolomite. Lahren et al. (1988) used TIMS data to map plutonic and metamorphic rocks in the Northern Sierra accreted terrane in California, and Hook et al. (1994) mapped metamorphic units from TIMS data that were not discernible in color aerial photographs. Rowan et al. (1993) used coregistered TIMS and AVIRIS data to map lithologies in the Iron Hill carbonatite–alkalic igneous rock complex in Colorado, and Crowley and Hook (1996) mapped playa evaporites and associated sediments in Death Valley, California, using TIMS data.

Multispectral thermal infrared bands on commercial GER and Geoscan sensors have not been used extensively for lithologic mapping but have been used to identify silicified zones and jasperoids in mineralized areas. These sensors are not well calibrated, so it is not possible to retrieve accurate emissivity spectra from these data. ASTER, scheduled for launch in 1998, will return calibrated multispectral TIR imagery on a global basis together with high-resolution VNIR/SWIR data.

Lithologic mapping in deeply weathered and vegetated terrains often requires an indirect approach. Using TM imagery, Glikson and Creasey (1995) distinguished peridotite, orthopyroxenite, clinopyroxenite, gabbro, ferrogabbro, anorthosite, mafic granulite, felsic granulite, and granite in the Giles Complex, Australia, on the basis of their weathering products. Macias (1995) also used weathering products to map underlying mafic and ultramafic rocks in Australia with TM data. Bell et al. (1989) found that in boreal forests of northwestern Ontario, there is a higher proportion of coniferous vegetation over metavolcanic rocks and more deciduous trees over metasediments. They also found variations in tree size, spacing, and crown shape to be related to underlying lithology. Harrington (1991) also found that deciduous trees are concentrated over ultramafic bodies as well. Rencz et al. (1996) noted that kimberlite plugs in the Lac de Gras area, Quebec, were preferentially eroded during glaciation, and the resulting depressions were subsequently filled by deep lakes. Reasoning that deep lakes are cooler than shallow lakes, they used the Landsat TM thermal band successfully to identify lakes associated with kimberlites.

In cloud-shrouded tropical rain forests, morphostructural analysis of synthetic

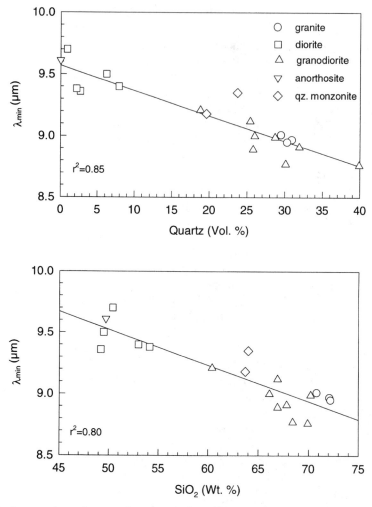

Figure 8.2 Linear correlation of quartz and SiO_2 content of plutonic rocks in the Desolation Wilderness, California, with wavelengths of emissivity minima (λ_{min}) determined from Gaussian curves fitted to TIMS spectra. (Modified from Sabine et al., 1994.)

aperture radar imagery is the only remote sensing tool available for lithologic mapping. Sabins (1987) illustrates this dramatically in a series of SIR-A images of Indonesia in which six kinds of geologic terrain (carbonate, clastic, volcanic, alluvial and coastal melange, and metamorphic) can be mapped according to their distinctive geomorphic style. Sabins also mapped structural lineaments and measured attitudes of bedding from the shapes of hogbacks and cuestas. Miranda et al. (1994) used morphostructural analysis of SIR-B data together with digital aeromagnetic data to develop a reconnaissance geologic map of a portion of the Guiana Shield in northwestern Brazil.

8.3.3.2 HYDROTHERMAL ALTERATION.

Many ore deposits are first detected in the field by the recognition of hydrothermally altered host rocks. Alteration halos may be manifested as spectacular color anomalies such as are produced by interaction of acidic fluids with mafic wall rocks, or they might be very subtle and difficult to detect. Altered rhyolites may be difficult to detect macroscopically, and dolomitization of limestone around Mississippi Valley–type deposits might also be inconspicuous in the field but easily detected in high-resolution SWIR reflectance data. Zonation in alteration halos, manifested by successions of mineral assemblages or spatial gradations in composition and crystallinity of alteration minerals, reflect chemical and temperature gradients in hydrothermal systems and can provide important information relating to the type and position of the ore body as well as the ore-forming process itself.

The identification of hydrothermally altered zones is often one of the first clues to possible mineralization recognized in digital imagery. Some alteration might even be more apparent in imagery than in the field because many alteration minerals have absorption features in their reflectance spectra at wavelengths that are outside the range of human vision. The detection of hydrothermal alteration associated with mineralized systems has been a primary focus of remote sensing geologists since the launch of *Landsat 1* (e.g., Rowan et al., 1974, 1977; Podwysocki et al., 1983). Altered mineralized systems such as those at Cuprite, Goldfield, Yerington, and Virginia City, Nevada, have become standard test sites to evaluate new airborne and satellite sensors, including the NASA airborne imaging spectrometer (Goetz and Srivastava, 1985; Feldman and Taranik, 1988a,b; Hutsinpiller, 1988; Kruse, 1988), AVIRIS (Abrams et al., 1988; Carrere, 1989; Clark et al., 1993; Swayze et al., 1993; Boardman and Huntington, 1997; Nusbaum et al., 1997a), GERIS (Kierein-Young and Kruse, 1989; Rubin, 1989a; Kruse et al., 1990; Spatz and Aymard, 1991), and Geoscan (Davis and Lyon, 1991; Rubin, 1991; Windeler and Lyon, 1991; Windeler, 1993). The majority of alteration studies carried out on a research or operational basis have used *Landsat* TM data because of its two well-positioned bands in the SWIR (Podwysocki et al., 1984; Brickey, 1986; Kepper et al., 1986; Mouat et al., 1986; Collins, 1988; Frei and Jutz, 1989; Haymon et al., 1989; Rockwell, 1989; Cetin and Levandowski, 1991; Loughlin, 1991; Ma et al., 1991; Re Kühl, 1992; Bastianelli et al., 1993; Crósta and Rabelo, 1993; Ferreira and Meneses, 1994; Nash and Wright, 1994; Barniak et al., 1996; Harris et al., 1998).

Hydrothermal alteration is most conveniently described in terms of stable reaction products that occur together in a rock. The mineralogy and distribution of such alteration assemblages associated with intermediate to silicic magmatic intrusions are controlled primarily by chemical gradients in hydrothermal systems, most notably the K^+ and H^+ ion activity, and to a lesser degree by temperature gradients (Guilbert and Park, 1986). Alteration assemblages mapped in numerous deposits associated with active magmatic systems (e.g., copper porphyrys, epithermal gold) strike a remarkably consistent theme. Five distinct alteration types, and some specialized subtypes, are widely recognized (Meyer and Hemley, 1967; Guilbert and Park, 1986). These are typically zonally distributed, from the greatest to the least degree of alteration, as follows:

$$\text{potassic} \rightarrow \text{phyllic} \rightarrow \text{advanced argillic} \rightarrow \text{argillic} \rightarrow \text{propylitic}$$

Potassic and phyllic assemblages normally occur near ore zones, while argillic and propylitic assemblages tend to be in more distal parts of altered zones. This zonation is rarely a neat and exclusive succession of alteration zones, however. Assemblages commonly overlap spatially as a result of changing fluid chemistry and physical conditions through time. Also, alteration assemblages depend largely on the composition of the protolith. Propylitic alteration in rhyolite may include potassic assemblage minerals and little chlorite unless fluids involved seawater to supply the magnesium ion. Chlorites in rhyolitic rocks may also be iron-rich. Table 8.5 summarizes the mineralogy and occurrences of these principal alteration assemblages in mafic to intermediate host rocks, and Figure 8.3 illustrates their environments of formation in terms of temperature and of K^+ and H^+ ion activity.

TABLE 8.5   Hypogene Alteration Assemblages for Porphyry and Epithermal Deposits Associated with Subduction-Related Magmatism

Assemblage	Mineralogy	Occurrence
Potassic	K-feldspar (orthoclase, microcline, adularia), biotite (brown or green), chlorite ± magnetite, hematite, anhydrite, calcite, Fe-bearing carbonates, sericite	Late magmatic alteration within or adjacent to intrusive rock; often associated with stockworks and disseminated mineralization
Phyllic (sericitic)	Sericite (Silica rich, phengite), quartz, pyrite ± topaz, tourmaline, zunyite, K-feldspar, kaolinite, calcite and other carbonates, anhydrite, rutile, apatite, biotite	Overlies potassic zone in most hypogene environments and often associated with sulfide ore
Advanced argillic	Kaolinite group minerals (kaolinite, dickite, halloysite, metahalloysite, nacrite); Pyrophyllite at $T > {\sim}300°C$; Andalusite at $T > {\sim}400°C$ ± diaspore, sericite, quartz, alunite, pyrite, tourmaline, topaz, zunyite, amorphous clays (allophane)	High in hydrothermal systems as upwardly expanding halos above ore shoots; wider in hanging wall; may extend to paleosurface; also occurs as inner zone closest to veins
Argillic	Kaolinite group (kaolinite, dickite, halloysite, metahalloysite), smectite group, illite group, amorphous clays (allophane) ± phengitic sericite, K-feldspar (metastable), biotite (brown, green), chlorite, pyrite, calcite	Replacement of feldspars in aluminosilicate rocks; grades inward to phyllic and outward to propylitic; may overlay a weak potassic core
Propylitic	Chlorite, epidote, zoisite, clinozoisite, carbonates (calcite, ankerite, siderite), albite, pyrite ± sericite, Fe oxides (hematite, jarosite, goethite), smectites, zeolites	Broad, widespread halos distal to ore deposits; may grade into regional low-grade metamorphism of volcanic rocks; epidote increases with depth; usually weak and may be difficult to detect

Source: Meyer and Hemley (1967) and Guilbert and Park (1986).

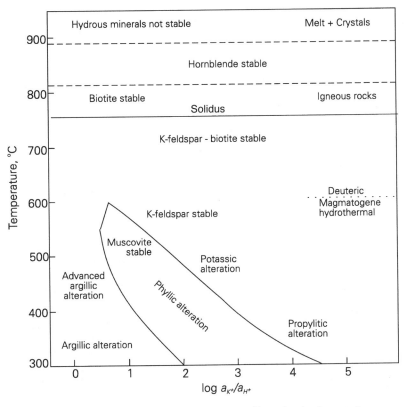

Figure 8.3 Generalized stability fields of some common alteration assemblages plotted as functions of temperature and the ratio of the activities of K+ and H+ at 1 kbar pressure (3 to 4 km depth) and excess SiO₂. (Modified from Burnham and Ohmoto, 1980, and Guilbert and Park, 1986.)

Figure 8.4 illustrates many of the VNIR/SWIR spectral features in the most common and abundant alteration minerals found in hydrothermal systems. In Chapter 1, Clark provides a detailed analysis and review of spectral features of rock-forming and alteration minerals. Most clay minerals, micas, and hydroxyl-bearing minerals such as pyrophyllite and alunite, common to phyllic, argillic, and advanced argillic assemblages, have absorption features associated with Al–OH bonding coupled with fundamental OH stretching modes near 2200 nm. Carbonates and minerals with Mg–OH and Fe–OH bonding, such as epidote, chlorite, biotite, and phlogopite, have absorption features near 2300 nm. These minerals are characteristic of propylites and potassic assemblages. TM band 7 is positioned to detect minerals with absorption features in the range 2200 to 2300 nm, as a group, but its bandpass is too broad to allow distinction of Al–OH, Mg–OH, and carbonate absorption features. Simulations of *FUYO-1* OPS sensor data indicated that the three bands between 2010 and 2400 nm were capable of distinguishing these three spectral groups (Akiyama et al., 1989a; Collins et al., 1992; Freund, 1992). ASTER will have five bands between 2145 and 2430 nm, which may permit further distinction of alteration zones.

The shapes and positions of absorption features in Figure 8.4 are diagnostic of certain minerals and mineral groups, permitting their identification and mapping in hyperspectral data. Subtle changes in the shape, depth, and position of spectral fea-

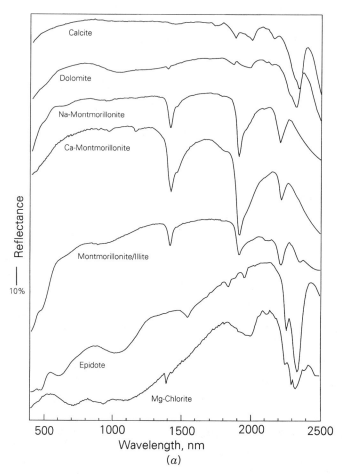

Figure 8.4 Visible and shortwave-infrared reflectance spectra of alteration mineral characteristic of propylitic (*a*), potassic (*b*), phyllic (*c*), argillic (*d*), and advanced argillic (*e*) alteration assemblages. Cation substitution is reflected in wavelength shifts of absorption features in spectra of such minerals as chlorites, micas, montmorillonite, and alunite, and degrees of ordering in crystal structures are reflected in overall shapes of features in illites, alunites, and kaolinite group minerals. (From USGS spectral library.)

tures may also characterize crystallinity (structural ordering of crystal lattices) and cation chemistry. Ordering of the OH⁻ ion in the octahedral layer of kaolinite group minerals (kaolinite, dickite, halloysite) leads to variation in the structure of the doublets near 1400 and 2200 nm. In well-ordered kaolinite spectra (Figure 8.4*b* and *e*), the doublets are sharp and distinct, and in dickite spectra (Figure 8.4*e*) the two parts of the doublets are sharp and nearly equal depth. In spectra of disordered kaolinite and halloysite (Figure 8.4*d*) the doublets are less distinct, and the 2160-nm feature may appear as a weak shoulder.

Illite may occur in any or all zones of a hydrothermal system but varies markedly in its crystallinity depending on temperature and intensity of alteration. Hauff et al. (1991) examined spectra of illites from hydrothermal, metamorphic, and sedimentary environments and found that the 2200-nm feature becomes progressively

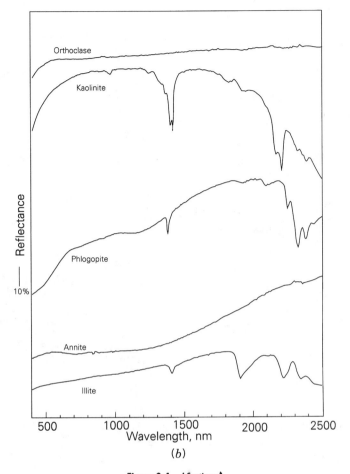

Figure 8.4   (*Continued*)

broader and more asymmetric as illite crystallinity decreases (smectite content increases). Small absorption features between 2300 and 2500 nm in well-ordered illite are absent in disordered illite spectra. These workers used field and laboratory spectra from transects at Preble, Nevada and found well-ordered high-temperature illites within the ore zone and disordered, low-temperature illite with interleaved smectites outside the ore zone. Martinez-Alonzo et al. (1997) also advocate use of illite SWIR spectra to estimate temperature of formation in associated ore bodies. Masinter and Lyon (1991) also found increasing illite crystallinity associated with decalcification and clay recrystallization at the Gold Bar mine, Nevada, using SWIR reflectance spectra. Spectra of well-ordered illite appear in Figure 8.4*b* and *c*, and disordered illite spectra are shown in Figure 8.4*d*.

The position of the muscovite absorption minimum near 2200 nm varies slightly as a function of aluminum content in tetrahedral sites (Duke, 1994). Figure 8.4*c* shows spectra of Fe-rich and Al-rich muscovites as well as the sodium mica paragonite. Fe–Mg substitution in chlorite group minerals leads to shifts in the position of the absorption minimum near 2300 nm and to overall degradation of the spectrum

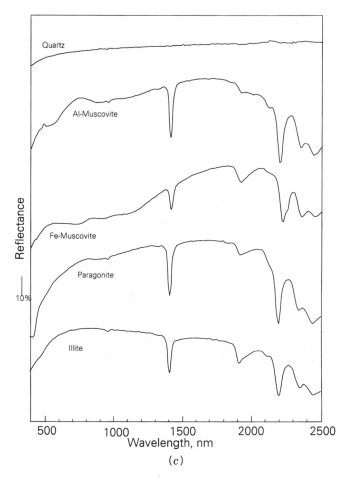

Figure 8.4   (*Continued*)

in the 1400-nm region. Crystallinity and cation chemistry differences can also be traced in the shapes and depths of spectral features in the region of alunite spectra around 1400 and 2200 to 2300 nm. The position of the feature near 2300 nm may be useful in tracking Na–K substitution (Figure 8.4*e*). Shapes and depths of spectral features can, in some cases, be related to temperature of alunite crystallization.

Since the discovery of buddingtonite (ammonium feldspar) at Cuprite, Nevada during analysis of AIS data (Goetz and Srivastava, 1985), ammonium alteration has been of great interest to remote sensing geologists. Buddingtonite was discovered in an active Hg-bearing hot spring deposit at Sulfur Bank in northern California and has also been found in Hg/Au-bearing fossil hot spring deposits at Ivanhoe, Nevada, and Manhattan, California. Other NH_4-bearing species have also been observed in hot spring deposits in Nevada and northern California, including NH_4 micas (Preble and Golconda, Nevada), NH_4 alunite (Ivanhoe, Nevada, and the Geysers, California), and NH_4 jarosite (the Geysers). NH_4-bearing illite occurs in black shales and cherts adjacent to Pb–Zn–Ag exhalative deposits in Alaska (Krohn and Altaner,

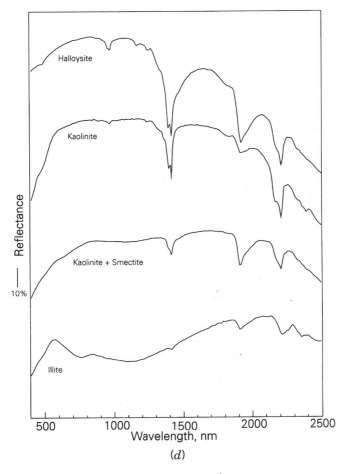

Figure 8.4   (Continued)

1987; Krohn et al., 1988). A buddingtonite locality in the southern Cedar Mountains reported by Krohn (1989) is 10 km long, more than 100 times larger than any other known locality.

Reflectance spectra of ammonium-bearing minerals have diagnostic absorption features based on N–H vibration modes at approximately 1560, 2020, and 2120 nm that are not present in their ammonium-free counterparts (Figure 8.5). Baugh and Kruse (1994) found that the continuum-removed depth of the 2120-nm absorption feature varied linearly with ammonium concentration in buddingtonite from the Cedar Mountains locality and used this relationship to produce quantitative ammonium concentration maps from AVIRIS data. Felzer et al. (1991) also found such a relationship in samples from Cuprite, Nevada but noted that smectite absorption features suppressed the 2120-nm buddingtonite feature and distorted the linear relationship.

8.3.3.3   SUPERGENE LEACHING AND ALTERATION.

In certain environments, oxidizing meteoric waters moving through ore zones and alteration halos above the water table may alter hypogene minerals and produce

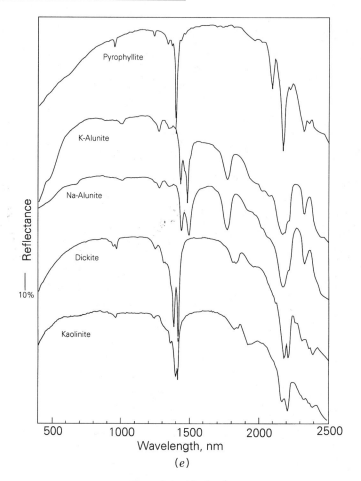

Reflectance

10%

Pyrophyllite

K-Alunite

Na-Alunite

Dickite

Kaolinite

500   1000   1500   2000   2500

Wavelength, nm

(e)

Figure 8.4   (*Continued*)

gossans and leach caps over ore bodies. The mineralogy and chemistry of the entire rock–fluid system determine the mineral assemblages that form in this environment. Pyrite and marcasite, however, are the key minerals in this process because they react readily with water under oxidizing conditions to produce sulfuric acid and ferrous sulfate, which attack other minerals in the system:

$$4FeS_2 + 7H_2O + 14.5O_2 = 2FeO.OH + 6H_2SO_4 + 2FeSO_4$$

Ferric sulfate, a very strong oxidizing agent that contributes significantly to sulfide leaching, may also be produced under strongly oxidizing conditions. Iron may be mobilized as Fe^{2+} but more often is precipitated in situ as goethite, hematite, or jarosite. Copper may be flushed down to the water table and reprecipitated as covellite and/or chalcocite in the zone of secondary enrichment, flushed out of the system entirely, or reprecipitated above the water table as sulfates (brochantite, antlerite), chlorides (atacamite), carbonates (malachite, azurite), or silicates (chryso-

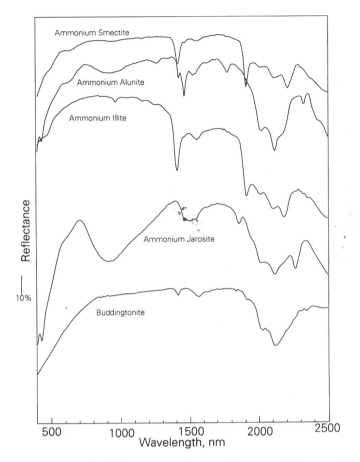

Figure 8.5   Visible and shortwave-infrared reflectance spectra of ammonium-bearing minerals characterized by N–H vibrational absorption features at approximately 1560, 2020, and 2120 nm. (From USGS spectral library.)

cholla). Intense leaching of hypogene alteration minerals (kaolinite, illite, alunite) may produce leach caps of supergene alunite and vuggy chalcedonic silica, which are practically sterile of metal content and difficult to distinguish from advanced argillic alteration assemblages.

The relative abundance of goethite, hematite, and jarosite in limonitic cappings can provide clues to the mineralogy and sulfide content of underlying ore zones. Anderson (1981), for example, found that in cappings associated with porphyry copper deposits, goethite is related to the total sulfide content of the hypogene or supergene sulfide zone, and hematite is related to chalcocite content. Jarosite is an indicator of original pyrite content. Colors of leach caps at visible wavelengths may provide clues to limonite mineralogy. Jarosite imparts pale yellow stains to cappings; goethite, yellow-orange; and hematite, red-orange.

Weathering of leached outcrops can produce thin rinds of vuggy, chalcedonic silica that may be difficult to distinguish from supergene or hypogene silica. Percolation of near-neutral meteoric waters through alteration zones may leave weak overprints of smectite and illite.

Reflectance spectra of some common supergene minerals are shown in Figure 8.6. The VNIR absorption features of the iron-bearing minerals hematite, goethite, and jarosite are associated with electronic transitions of $3d$ electrons in ferric iron and by interactions with surrounding ions. The wavelength of the absorption features is determined principally by lattice parameters such as coordination number, site symmetry, and ligand type (Goetz, 1989). Goethite and jarosite both have ferric iron absorption features near 900 nm, but goethite has two additional features near 660 and 500 nm with a peak near 750 nm. Jarosite is more reflective overall with its peak reflectance near 700 nm. It also has a Fe–OH absorption feature near 2270 nm. The ferric iron absorption feature of hematite is near 860 nm, with its reflectance peak near 730 nm. Alunite can form in supergene as well as hypogene environments. Disordered illite and smectite form by supergene weathering.

Given the importance of supergene processes in mineralized areas, there has been surprisingly little research aimed at characterizing supergene alteration assemblages

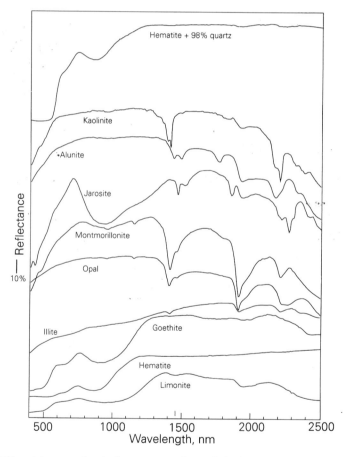

Figure 8.6   Visible and shortwave-infrared reflectance spectra of minerals that form by supergene alteration and weathering. Minerals containing ferric iron (hematite, goethite, limonite, jarosite) have diagnostic absorption features in the range 850 to 1000 nm. Spectra of illite, montmorillonite, alunite, and kaolinite are from low-temperature or diagenetic species. (From USGS spectral library.)

in digital imagery. Most remote sensing work in mineralized areas employs broadband imagery which can map ferric iron minerals as a class but cannot identify mineral species. Fraser (1991), however, described a ratioing technique for discriminating hematite and goethite in TM imagery. Hyperspectral airborne scanners such as AVIRIS have the spectral resolution to distinguish mineral species, and signal/noise ratios are sufficient to allow detection of subtle spectral features. Taranik et al. (1991) used 63-channel GERIS data to map iron oxides in the Cripple Creek district, Colorado. Previous field mapping indicated that the highest concentration of supergene iron minerals was near faults and fractures that were conduits for mineralizing fluids near the center of the district. Jarosite predominated in the most intensely altered areas, giving way to jarosite + goethite in less altered zones. Goethite predominated in the periphery of the district, and hematite was pervasive in unaltered granite and gneiss outside the mineralized area. The spectral resolution of the GERIS data was sufficient to identify three classes based on iron oxide mineralogy: (1) hematite dominant, (2) hematite + goethite, and (3) jarosite + goethite. These classes defined a districtwide zonation in the imagery that corresponded well with the field map. The jarosite + goethite zone occupied the center of the district grading outward to hematite + goethite with the hematite zone lying outside the district.

Agar and Villanueva (1997) used GERIS imagery to distinguish hypogene alteration from more areally extensive supergene alteration in south-central Peru. Their work led to improved target selection and discovery of a new lead-zinc-silver deposit.

8.3.3.4 SKARNS.

Recrystallization and metasomatism of carbonate-bearing rocks around plutons of intermediate composition produce coarse-grained assemblages of Ca–Fe–Mg–Mn silicates called skarns. Mineralized skarns are the principal source of tungsten and also contain large deposits of copper, iron, molybdenum, and zinc and minor deposits of cobalt, gold, silver, lead, bismuth, tin, beryllium, and boron (Einaudi, et al., 1981). They are also sources of industrial minerals such as graphite, asbestos, wollastonite, magnesite, phlogopite, talc, and fluorite. The mineralogy of eight major skarn types are summarized in Table 8.6; many skarn minerals have diagnostic features in their VNIR, SWIR, and TIR spectra. Skarns are typically zoned. Garnet-bearing skarns are proximal to intrusive contacts and are copper-rich; pyroxene-bearing skarns occur distally and are richer in gold, arsenic, bismuth, and tellurium.

Windeler and Lyon (1991) delineated a dolomitized limestone unit in the Ludwig skarn near Yerington, Nevada, by tracking the shift in wavelength of the carbonate absorption feature near 2300 nm using Geoscan MkII imagery. Windeler (1993) also used the Geoscan thermal bands to distinguish two types of garnet–pyroxene alteration in the Ludwig skarn. Goosens (1991) used *Landsat* TM imagery in conjunction with aeromagnetic and airborne radiometric data to map granitic intrusions and associated skarns in Spain.

8.3.3.5 STRUCTURAL MAPPING.

Structural control is an important element in many ore deposit models, such as polymetallic vein systems, Carlin-type deposits, and epithermal hot spring deposits (Cox and Singer, 1986). For this reason, lineament mapping is part of every remote sensing

TABLE 8.6   Comparative Mineralogy of Major Skarn Types

Type	Tectonic Setting	Prograde	Mineralogy Retrograde	Ore
Calcic Fe	Large to small mafic stocks and dikes; oceanic island arc; rifted continental margins	Grandite, salite, ferrosalite, epidote, magnetite	Amphibole, chlorite (ilvaite)	Magnetite (chalcopyrite, cobaltite, pyrrhotite)
Magnesian Fe	Small granitic stocks, dikes, sills; continental margin; synorogenic	Forsterite, calcite, spinel, diopside, magnetite	Amphibole, humite, serpentine, phlogopite	Magnetite (pyrite, chalcopyrite, sphalerite, pyrrhotite
Calcic W	Granitic plutons, batholiths; continental margin; synorogenic to late orogenic	Grandite/spessartine-almandine garnet, hedenbergitic pyroxene, idocrase, wollastonite	Hornblende, biotite, plagioclase, epidote	Scheelite, molybdenite, chalcopyrite, pyrrhotite, pyrite
Calcic Cu	Granitic stocks, dikes; continental margin; synorogenic to late oroogenic	Andraditic garnet, salitic pyroxene, wollastonite	Actinolite, chlorite, smectites	Chalcopyrite, bornite, pyrite, hematite, magnetite
Calcic Zn–Pb	Large stocks and dikes of granodiorite to granite, diorite to syenite; continental margin; synorogenic to late orogenic	Johannsenitic pyroxene, andraditic garnet, bustamite, idocrase	Mn-actinolite, ilvaite, epidote, chlorite	Sphalerite, galena, chalcopyrite, arsenopyrite
Calcic Mo	Stocks of quartz monzonite to granite; continental margin; late orogenic; rare	Hedenbergitic pyroxene, grandite garnet, quartz	Amphibole, chlorite	Molybdenite, scheelite, bismuthinite, pyrite, chalcopyrite
Calcic Sn	Granite stocks and batholiths; continental margin; late orogenic to post-orogenic; rare	Malayite, danburite, datolite, grandite, idocrase	Amphibole, mica, chlorite, tourmaline, fluorite	Cassiterite, arsenopyrite, stannite, pyrrhotite
Magnesian Sn	Granite stocks and batholiths; continental margin; late orogenic to post-orogenic; rare	Spinel, fassaite, forsterite, phlogopite, magnetite, humite, ludwigite, paigeite	Cassiterite, fluoborite, magnetite, micas, fluorite	Cassiterite, arsenopyrite, pyrrhotite, stannite, sphalerite

Source: Modified from Einaudi et al. (1981).

geologist's repertoire. This enthusiasm for lineaments is not shared by all field geologists, however. Typically, the problem with lineament maps is that they are presented in a raw form with little structural analysis or interpretation. Products such
as these quickly find their way to the backs of field geologists' file drawers.

Rowan and Bowers (1995) examined many of the factors that contribute to the
difficulty of integrating lineament maps into mineral exploration and assessment
models. One of the main difficulties is the a priori assumption that lineaments are
topographic and tonal expressions of geologic structures, whereas some may reflect
a more general structural fabric such as joints and foliation, and others may have no
structural origin at all. A second problem relates to scale and resolution of imagery
in relation to the dimensions of ore-controlling structures. Satellite imagery is good
for identifying regional structural zones that may influence the location of mineral
deposits but may fail to resolve ore-controlling structures themselves. As spatial resolutions of satellite imagery continue to approach that of airborne imagery and air
photography, this difficulty is lessening. A third problem is that lineament maps are
indiscriminant with regard to the age and type of structure in relation to the timing
of mineral deposition. Structures that postdate mineralization usually cannot be distinguished from those that control mineralization from imagery alone. The combined
result of these factors is generally poor correlation between mineralized zones and
lineaments.

Rowan and Bowers' (1995) analysis of lineaments in the Reno quadrangle, Nevada and California, provides a good model for modern lineament analysis. They
combined lineaments mapped from *Landsat* TM and airborne synthetic aperture
radar (SAR) imagery into a single data set and imported into a geographic information system (GIS) together with databases on known mines and prospects and
areas mapped as favorable for epithermal precious metal deposits in a previous mineral assessment study. The GIS permits quantitative study of the data sets, including
analysis of lineament orientation, lineament density, and proximity to known deposits. They found that the spatial resolution of the TM and SAR imagery was
sufficient to identify ore-controlling structures but that spatial resolution together
with illumination conditions were also the chief limiting factors controlling the variability of correspondence with known deposits. They also concluded that lineament
density could be a useful parameter for assessing resource favorability.

Bonham-Carter (1985) described a statistical technique for assessing spatial relationships between lineaments and mineral deposits. His method compares a calculated distribution describing the frequency of occurrence of known deposits within
specified distances from lineaments with a similar distribution based on random
points. The technique revealed strong correlations between *Landsat* lineaments with
specific orientations and gold occurrences in the Timmins–Kirkland area, Ontario
(Bonham-Carter, 1985). Rencz and Watson (1988) also used the technique to reveal
a spatial relationship between certain lineament directions and mineral occurrences
in northeastern New Brunswick. Their analysis verified a relationship with northeast-
trending Acadian F1 fold axes determined from independent geologic mapping, and
it also revealed equally important associations with two other lineament directions,
one of which corresponded with late- or post-tectonic extensional faults.

Lineament studies have contributed to a number of exploration and mineral assessment programs and to regional mapping efforts as well. Jiang et al. (1994) found
that arcuate and circular features in *Landsat* TM imagery, together with aeromag-

netic and geochemical data, were useful in targeting porphyry copper deposits in poorly accessible, forested regions of northeastern China. Fernandez-Alonso and Tahon (1991) used images derived from airborne radiometric and aeromagnetic data coregistered to *Landsat* TM imagery to reveal not only surficial structures in western Rwanda but deeper structures as well. Hutsinpiller (1988) mapped lineaments and alteration patterns in the Virginia Range, Nevada from *Landsat* TM imagery and high-altitude air photos and found lineament patterns to be consistent with models of right-lateral and normal faulting in the Walker Lane. Hutsinpiller also found that lineament intersections were twice as dense in altered areas as unaltered areas. Farrand and Seelos (1996) used linear distributions of kaolinite, alunite, and jarosite in mineral maps derived from AVIRIS imagery to map previously unrecognized faults in the Summitville district, Colorado.

Synthetic aperture radar (SAR) imagery obtained at moderate depression angles is very sensitive to subtle topographic features and can be extremely useful for structural mapping. Borengasser and Taranik (1985) identified a previously unrecognized fault in the Candelaria district, Nevada, from SIR-B imaging radar data which assisted in the discovery of additional ore (Spatz, 1996). Miranda et al. (1994) mapped structural lineaments in a heavily forested region of the Guiana shield, Brazil. Premmanee (1989) found close correspondence between lineaments derived from *Landsat* TM and airborne C-band SAR imagery and published joint and foliation data in a portion of the Sudbury structure, Ontario. Singhroy et al. (1993) found the depression angle of *ERS-1* C-band imagery to be too steep to be useful for structural mapping in the Canadian shield.

Broadband thermal imagery can reveal structural information that may not be apparent in the reflected bands. Torres et al. (1989) identified a major east–west lineament in northern Chihuahua, Mexico using the Landsat thermal band (band 6) and also identified areas of high geothermal heat flow. Sabine and Re Kühl (1996) found the thermal band of airborne thematic mapper data useful for identifying lineaments in the northern Carlin Trend, Nevada. The airborne data were acquired in late spring when grasses had dried except in zones where soils retained water along drainage bottom and fractures. These zones were cooler than adjacent areas and thus appeared darker in the thermal band.

8.3.3.6 GEOBOTANICAL REMOTE SENSING.

Vegetation is a nuisance. It obscures outcrops and obstructs exploration efforts. But in many parts of the world that are actively being explored for mineral deposits, vegetation cover is a fact of life. Boreal forests cover vast areas of favorable terrain in Canada, Scandinavia, and Russia, and tropical rain forests conceal hidden mineral wealth in South America, Africa, and Southeast Asia. Remote sensing scientists working in areas such as these have decided not to fight the issue and are using spectral and spatial characteristics of vegetation cover to provide clues to the geological substrate. This section highlights a few applications of geobotany for mineral exploration. A more detailed review of geobotanical remote sensing appears in Chapter 4.

Spectral properties of the leafy portion of vegetation are illustrated in Figure 8.7. Chlorophyll absorbs electromagnetic radiation at 470 and 680 nm. These features, separated by a reflectance peak between 520 and 600 nm, impart the green color to plant leaves. The abrupt reflectance increase between 700 and 800 nm is known as the *red edge*. The broad reflectance peak in the NIR is controlled by leaf cellulose,

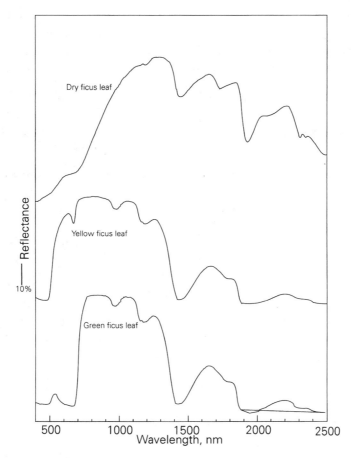

Figure 8.7   Visible and shortwave-infrared reflectance spectra of green, yellow, and dried ficus leaves. (From CSIRO spectral library.)

and the reflectance minima near 1400 and 1900 nm are controlled by leaf water content. Loss of chlorophyll during fall senescence, illustrated by the yellow leaf spectrum in Figure 8.7, results in diminished amplitude of the chlorophyll *a* feature at 680 nm and a shift of the red edge to shorter wavelengths. In the dry leaf spectrum of Figure 8.7, leaf pigment and water features have nearly disappeared, and the spectrum is dominated by cellulose features.

Spectral changes such as those illustrated in Figure 8.7, including the red-edge shift, have also been observed in single-leaf spectra of plants grown in soils doped with heavy metals in greenhouse experiments (Horler et al., 1980, 1983; Milton et al., 1988, 1989, 1991; Mouat et al., 1994). These experiments also produced morphological changes in plants, such as stunting and biomass reduction.

Attempts to detect red-edge shifts in natural plant communities growing in heavy-metal-rich soils over mineralized zoned using field spectrometers and hyperspectral scanners have met with conflicting and often baffling results. Shifts to shorter and to longer wavelengths have been reported by various workers. Singhroy and Kruse (1991) found that the depth and width of the 670-nm chlorophyll absorption feature is a more reliable indicator of concentrated heavy metals than the red edge. The feature is consistently shallower and narrower in both coniferous and deciduous species growing over mineralized areas in Canadian boreal forests.

Remotely sensed measurements of vegetated areas are complicated by canopy effects and mixed spectral signatures. Measurements are composites from thousands or millions of leaves in varying orientations representing more than one species, together with responses from woody plant material, dry plant litter, and soil. Nevertheless, spectral and spatial anomalies have been detected in forested areas over mineralized zones. Rencz and Sangster (1989) detected a halo of anomalous vegetation around the Gaspe copper deposit in Quebec that corresponds with but is larger than mapped alteration zones and aeromagnetic anomalies. The anomalous vegetation had significantly greater reflectance in TM bands 4 and 5 than vegetation in nearby unmineralized areas. Nash and McCoy (1993) found higher overall albedoes in big sage growing over gold placers in the La Sal Mountains, Utah and correlated the anomaly to elevated gold values in leaves. They also noted that gold content in big sage leaves sampled in autumn was higher than in leaves sampled in the spring. Curtis and Maecher (1991) attributed spectral differences in stands of lodgepole pine growing in soils with elevated trace metal content to canopy architectural factors such as branch density, needle loss, and needle length. Payás et al. (1993) found that intensely altered metavolcanic rocks of the Carolina slate belt, North Carolina, were characterized by diminished number of plant species, dominance of chestnut and oak, and differences in tree structure. Intense leaching of underlying rocks produced thin rocky soils deficient in Ca and Mg, important plant nutrients. Similar vegetation anomalies occur over intensely leached volcanic rocks in the Virginia City area, Nevada, shown in Figure 8.8 (see the color insert). In these zones, the natural vegetation cover, consisting of sagebrush, juniper, and various grasses and shrubs, gives way to mostly barren areas with scattered Jeffrey pine.

8.4 IMAGE PROCESSING TOOLKIT

There are three common approaches to image processing. One is the "million monkeys" approach, in which, through trial and error, an image processor tries every conceivable combination and permutation of bands, ratios, principal components, and whatever else he or she can think of in the hope that something interesting will appear on the monitor. A second approach is the "oracle" method, in which images are divined through the magic of computer technology and delivered to the geologist with little evidence of human intervention. Neither of these approaches is likely to produce images that will yield useful information to the project geologist. A more reasoned thematic approach, favored here, tailors the image processing to specific landscape attributes of interest to the field geologist in an intuitively interpretable format. This approach may result in one or more images that illustrate specific geologic aspects of a scene (e.g., lithology, alteration, structure) and which meet the needs of the field geologist.

A bewildering array of image processing techniques is available to an image analyst. Most are highly specialized and rarely applied. The basic toolkit for routine image processing for mineral exploration is quite small. It includes color composite images, usually with some kind of contrast stretch, band ratios, principal component and intensity–hue–saturation transformations, and various spectral and spatial classifiers. Some applications may also require calibration of data to physical units of radiance, reflectance, or emissivity. In the following section we outline the use of some of these tools and cites some examples of their application.

8.4.1 Preprocessing

Image preprocessing includes various radiometric corrections and noise removal algorithms that may be necessary before data can be processed and interpreted. It also includes image registration, or geocoding, and data calibration.

8.4.1.1 GEOCODING.

Geocoding is a good idea. It is essential if imagery is to be merged or integrated with other forms of imagery or with other kinds of data in a geographic information system. Most satellite imagery can be ordered geocoded to a latitude–longitude or Universal Transverse Mercator grid. Data can also be geocoded as a value-added product by a private firm, or it can be done in-house. Registration and geocoding involves selection of well-located control points that can be identified in both the imagery and the map, taking care that the point is a reasonably permanent feature. Road intersections are good choices; a point bar in a stream would be a bad choice. One of several algorithms may then be used to resample the data to conform to the map projection. Care should be taken to ensure that resampling does not signficantly alter the spectral or spatial content of the image.

Geocoding of satellite imagery is fairly straightforward, but time consuming, in most modern image processing software. Because of platform instability, relief displacement, and foreshortening at high scan angles, registration and geocoding of most airborne imagery are nearly impossible. A low root-mean-square error might create the impression that an airborne image is accurately registered, but this value applies only to the control points. Large residual errors may remain distributed throughout the image.

8.4.1.2 CALIBRATION.

Radiance measurements obtained by most satellite and airborne sensors include contributions from the solar irradiance spectrum, atmospheric attenuation and scattering, and viewing geometry effects in addition to surface reflectance. For many purposes these effects can be safely ignored when working with satellite data, but atmospheric contributions can seriously affect results of some image processing operations. Crippen (1988a) demonstrated the importance of adjusting TM data for atmospheric effects in constructing band-ratio images. The most important atmospheric influence on satellite data is path radiance due to atmospheric scattering in the visible bands. Path radiance is greatest in the blue part of the visible spectrum and decreases rapidly in the green and red bands. It is negligible in the infrared. This effect can easily be removed using the dark pixel subtraction method, in which a constant is determined for each visible band using the darkest nonnull pixel and subtracted from all radiance values in the scene. Constants can be determined from bodies of deep, clear, still water, assuming that water is nonreflective at all wavelengths. Any radiance values from bodies of water can be attributed to path radiance. In scenes where no water is present, constants can be determined by one of five other methods: iterative ratioing, radiance-to-reflectance conversion, regression method, covariance matrix method, and regression intersection (Crippen, 1988a). Campos-Marquetti and Rockwell (1989) and Gilabert et al. (1994) describe equations and procedures for reducing TM data to surface reflectance.

Mineral mapping and spectral matching operations on hyperspectral data absolutely require data conversion from raw radiance to surface reflectance. An exception is the Geoscan airborne multispectral scanner, which measures relative radiance, which approximates surface reflectance by adjusting offsets and gains for each channel in flight immediately prior to data acquisition. Rubin (1991) found that for most purposes Geoscan data need no further processing to be treated as surface reflectance measurements. Techniques for converting raw data to surface reflectance include log residuals (Green and Craig, 1985; Rubin, 1989a; Yamaguchi et al., 1989; Marsh, 1993; Yamaguchi et al., 1994), internal average reflectance normalization (Crowley et al., 1988; Kruse, 1988; Pendock et al., 1991; Ben-Dor and Kruse, 1994), empirical line (Roberts et al., 1985; Abrams et al., 1988; Gardiner et al., 1988; Kierein-Young and Kruse, 1989; Lyon and Honey, 1989ab; Kruse et al., 1990; Rockwell, 1991; Baugh and Kruse 1994; Yamaguchi et al., 1994), flat-field correction (Goetz and Srivastava, 1985; Feldman and Taranik, 1988a; Hutsinpiller, 1988; Carrere, 1989), and atmospheric modeling (van den Bosch and Alley, 1990; Green et al., 1993; Rowan et al., 1996).

The choice of which hyperspectral data calibration technique is most applicable depends on the sensor, data quality, and surface conditions in the image area. Ideally, reflectance spectra should be measured from natural or artificial homogeneous ground targets during data acquisition and atmospheric data measured simultaneously from a radiosonde. This is rarely possible. Ground spectra are usually measured at a later date if a field spectrometer is available and suitable surfaces exist within the scene. Rast et al. (1991) compared flat-field, log-residual, and atmospheric modeling using LOWTRAN 7 radiative transfer code constrained with local weather station data to retrieve surface reflectance measurements from AVIRIS data over Cuprite, Nevada. They found that atmospheric modeling was the most effective method, and the flat-field method was the least effective. Farrand et al. (1994) compared the LOWTRAN 7 method with empirical line and spectral mixture analysis techniques to correct AVIRIS data over the Lunar Craters volcanic field, Nevada and found atmospheric modeling to be least effective, with the empirical line and mixture analysis methods both providing good results. The poor performance of the atmospheric modeling approach was attributed to noise and a lack of accurate atmospheric water vapor data. Clark et al. (1995) examined four methods of calibrating AVIRIS data over Cuprite: (1) a public-domain radiative transfer code called ATREM; (2) a MODTRAN-based method developed at the Jet Propulsion Laboratory by Robert Green, (3) an empirical line method using known ground sites, and (4) a combined approach using radiative transfer and ground calibration. The combined approach produced the best results. Boardman and Huntington (1996) achieved excellent results calibrating 1995 AVIRIS data to apparent surface reflectance using a two-step method. The ATREM radiative transfer model was applied in the first step to remove the gross effects of solar irradiance and atmospheric scattering and absorption. The second step empirically estimates and removes cumulative gain errors remaining from the first step based on modeled pixel spectra derived from the AVIRIS data with no a priori knowledge of ground target reflectance.

Multispectral thermal infrared data must also be reduced to physical units of radiance to be useful for quantitative measurements of kinetic temperature or emissivity of the ground surface. TIMS data are calibrated by convolving raw measurements with readings from internal thermal reference sources, calibrated thermal response functions for each band, and an atmospheric radiative transfer model (Sabine et al., 1994). Measurements of spectral emissivity or surface temperature can then be retrieved from the data by solving Planck's law for each band (Kahle et al., 1980; Realmuto, 1990). Emittance spectra may also be extracted from TIMS surface radiance data by the alpha coefficients (Kealy and Gabell, 1990) and thermal log residual (Hook et al., 1990) methods and by the spectral ratio and two-temperature methods of Watson (1992a,b). Surface emissivity and temperature measurements cannot be retrieved accurately from GERIS and Geoscan thermal bands because of the lack of onboard calibration. Lyon and Honey (1991) were able to retrieve what they called *thermal exitance* measurements from Geoscan data by measuring radiance of pans of water that were placed on the ground under the sensor immediately after data acquisition.

8.4.2 Color Composite Images

A color composite image in which separate spectral bands are displayed in the blue, green, and red channels, is the most common form of data presentation. Although the procedure is simple, considerable care must be taken in choosing bands and their display colors in order that the phenomena of interest are well enhanced and displayed in colors that are intuitively reasonable to the geologist (e.g., ferric oxides red, vegetation green, water blue). Various statistical procedures based on band variance have been described, but Crippen (1989) argued that band variance is as much a product of instrument gain and noise as surficial spectral variation. He favored using correlation coefficient matrices to select the least redundant bands and found that TM bands 1, 4, and 7 maximally differentiate lithologic information in most scenes in arid environments, but that a 1–5–7 combination is preferable in scenes where hydroxyl-bearing minerals predominate. The TM 1–4–7 combination based on correlation analysis has been found useful for lithologic discrimination by other authors (e.g., Eriksen and Cowan, 1989; Frei and Jutz, 1989; Torres et al., 1989; Re Kühl, 1992; Trefois et al., 1993). Crósta and Moore (1989) found that TM bands 4, 3, and 1, displayed in red, green, and blue, respectively, worked well in discriminating ferric-oxide-rich soils in a deeply weathered and vegetated terrain in south-central Brazil, and Davidson et al. (1993) used a TM 1–3–5 (R–G–B) combination to discriminate lithologies in the Troodos Massif, Cyprus. Kowalczyk and Logan (1989) favor a TM 5–4–3 (R–G–B) combination for routine photogeologic interpretation. For geobotanical studies, band combinations that display variation in chlorophyl absorption phenomena in vegetation are chosen (e.g., Ager et al., 1989).

8.4.3 Band Ratios

The simple act of dividing numerical values in one band by those in another for each pixel to produce ratio images is the most widely used (and abused) technique for discriminating lithologies and other surface cover types in a scene. The method produces images in which radiance variation that is proportionally constant from band to band (e.g., terrain illumination, albedo) is suppressed and the more interesting radiance variation attributable to spectral reflectance of geological materials is enhanced. Appearances are deceiving; ratioing is more complicated than it appears.

Crippen (1988a) explained the pitfalls of band ratioing and emphasized the importance of adjusting raw TM data for path radiance and sensor calibration offsets. Failure to apply these adjustments before ratioing can cause unexpected results. In a TM 4/1 ratio image of the Eagle Mountain iron mine in California made from unadjusted data, the mine and tailings are not visible and topography is inverted. The same ratio applied to adjusted data produced an image in which the mine and tailings area are anomalously bright and topographic expression is largely removed (Crippen, 1988a). Terrain illumination differences due to topography can produce color variations in unadjusted ratio images for spectrally identical surface materials (Crippen, 1988a; Crippen et al., 1988).

Band selection for ratio images depends on the spectral properties of the surface material of interest and its abundance relative to other surface cover types. Usually, the numerator band is chosen in which the material is highly reflective and a band covering an absorption feature for that material is chosen as the denominator. Correlation analysis of TM data by Crippen (1989) showed that for arid areas, a combination that includes one ratio from the short-wavelength bands (TM 1–4), one from the long-wavelength bands (TM 5, 7), and one ratio with bands from each group produces the best discrimination of geological materials. His analysis and experience of numerous investigators have shown that TM 3/1 best discriminates ferric oxides and TM 5/7 discriminate materials containing hydroxyl-bearing minerals and carbonates with absorption features in the region 2.2 to 2.3 μm (Brickey, 1986; Hutsinpiller, 1988; Kowalczyk and Logan, 1989; Re Kühl, 1992; Bennett, 1993; Sabins and Miller, 1994). TM 4/3 accentuates the red edge in vegetation, and a TM 5/4 ratio helps distinguish rocks and soils from vegetation and may enhance silicified or bleached zones of alteration. The TM 5/4 ratio has also been used by some investigators to highlight ferric oxides in areas where vegetation is a problem (e.g., Davidson et al., 1993; Glikson and Creasey, 1995). Davis et al. (1989) used a color composite of TM ratios 1/4, 3/4, and 4/7 to discriminate phosphorite beds in Saudi Arabia. In *SPOT* and *Landsat* MSS imagery, a 2/1 ratio is used to enhance ferric oxides, and a 3/2 ratio accentuates vegetation.

Field geologists have found ratio images difficult to use and interpret because of their lack of topographic expression and surface albedo information. Crippen et al. (1988) devised one method of restoring topographic and albedo information, called directed band ratioing, which involves adjusting band data so that ratio values increase with increasing bispectral radiance for all surface materials. Resulting images depict topography and surface albedo as intensity, which varies independently of hue. Four-component color composite imaging is another solution. In this procedure, each ratio image is multiplied by a single-band image that depicts topographic and albedo variation (e.g., TM band 5 or PC 1) after first equalizing the means of the four components (Crippen, 1988b; Ford et al., 1990; Knepper, 1993). The method can also be used with principal component color composites, and coregistered images from other sources (e.g., SPOT, SAR) can be used for the topographic element.

The TM 5/7 ratio has become a standard procedure for portraying the distribution of hydroxyl-bearing minerals and carbonates in alteration zones, but this ratio also responds to leaf water content in vegetation, which also appears bright in 5/7 images. Elvidge and Lyon (1985) describe a method of estimating the contribution of vegetation to the 5/7 ratio using a spectral indexing approach. Using atmospherically corrected airborne thematic mapper data from the Virginia City area, Nevada, they extracted pixel values from training sets selected from known vegetated areas and plotted the 1.65 μm/2.22 μm ratio against the 0.83 μm/0.66 μm ratio. These ratios correspond approximately to TM 5/7 and TM 4/3. The training sets plot in a distinct linear trend from which a baseline can be computed by linear regression. Pixels from hydrothermally altered areas with abundant hydroxyl-bearing minerals plot in a separate cluster that lies above the regression line and trends normal to it. Linear equations extracted from these plots can be used to estimate the contribution of vegetation to the 1.65 μm/2.22 μm ratio. Fraser and Green (1987) describe a similar approach which they called *directed principal components*.

Kowalczyk and Logan (1989) describe a clay index algorithm for processing TM data based on Elvidge and Lyon's approach. After applying a path radiance correction, they run a supervised classification using training sets from vegetated areas, areas of outcrop and soil, and areas of water. Cross-plotting and regressing TM 5/7 against TM 4/3 for these training pixels, they compute a *clay index* (CI):

$$CI = (TM\ 5/7) - k_1 - [k_2(TM\ 4/3)] \tag{8.1}$$

where k_1 is the intercept of the vegetation line and k_2 is its slope. The clay index can be computed for each pixel and displayed as a gray scale or density-sliced image.

8.4.4 Principal Components

Most of the variation of radiant spectral flux measured by a sensor depends on slope aspect relative to solar illumination and albedo effects at the surface. Very little of this variation arises from spectral reflectance features of surface materials. Principal component analysis (PCA) is a powerful means of suppressing irradiance effects that dominate all bands so that geologically interesting spectral reflectance features of surface materials can be examined. PCA is also useful in reducing the dimensionality of multispectral data and the high degree of band-to-band correlation that is inherent in such data sets.

A principal component (PC) transformation entails data rotation and translation to a new set of statistically independent orthogonal axes which are the principal components (eigenvalues of the covariance matrix) of the data set. The origin of the new coordinate system is the mean of the original data set. The number of PCs is the same as the number of spectral bands used in the transformation. For each PC, new numerical values are assigned to each pixel according to its new coordinates and may have no relationship to values in the original data. PC 1 is usually dominated by illumination effects arising from differences in slope aspect relative to the sun and provides a good noise-free representation of topography. Spectral reflectance features of surface materials and noise appear in higher-numbered PCs. Because PCA is a statistical procedure, the information in each PC is scene dependent. PCA has been used extensively to extract information related to host rocks, mineralization, and alteration from TM imagery (e.g., Mouat et al., 1986; Amos and Greenbaum, 1987; Eriksen and Cowan, 1989; Frei and Jutz, 1989; Rodriguez and Glass, 1991; Bennettt, 1993) and from hyperspectral imagery (Feldman and Taranik, 1988b; Hutsinpiller, 1988; Ager et al., 1989)

8.4.4.1 DECORRELATION STRETCH.

The decorrelation stretch is a useful application of principal components in situations where the spectral content of images is highly correlated between bands. In such situations, color composite images appear pale and washed out, and conventional stretches do little to improve color saturation. The technique is used routinely to enhance multispectral thermal imagery to suppress surface temperature effects, which dominate the radiant spectral flux measured in all bands, and enhance subtle spectral features arising from emissivity variations (Gillespie et al., 1986; Gillespie, 1992). It has also been widely used to enhance *Landsat* TM imagery (Frei and Jutz, 1989; Bennett, 1993; Davidson et al., 1993) and hyperspectral imagery (e.g., Akiyama, 1989b; Hook et al., 1991; Abrams and Hook, 1995).

In this procedure, three or more bands of data are transformed to principal components and stretched to equalize the variance along three statistically independent axes. Usually, a linear or Gaussian stretch is employed. The stretched data are then transformed back to the approximate original RGB coordinate system for display as a color composite image. Resulting images represent variations in spectral reflectance, or emissivity in thermal data, by hue and overall albedo, or surface temperature, by intensity.

8.4.4.2 FEATURE-ORIENTED PRINCIPAL COMPONENT SELECTION.

A particularly useful and easy technique for enhancing spectrally unique surface cover types within an image was developed by Crósta (Crósta and Moore, 1989) to distinguish ferric-oxide-rich soils derived from mafic and ultramafic rocks and associated sulfide ore bodies from soils derived from granitic migmatites. The feature-oriented principal component selection (FPCS) method allows the identification of PCs that concentrate spectral information associated with specific cover types by examining the eigenvector matrix used to calculate the principal components. Crósta focused his study on a vegetated area of deeply weathered rocks surrounding the O'Toole nickel–copper sulfide and platinum-group-element ore deposit in the Morro do Ferro greenstone belt, Minas Gerais State, Brazil. Crósta (1991) also succeeded in discriminating limonitic gossans in a deeply weathered and vegetated terrain around the Mombuca prospect in southeastern Brazil using FPCS with TM data. The same principals were extended to Geoscan data from Archean greenstone terranes in non-arid regions of central and northeastern Brazil (Prado and Crósta, 1994; Hernandes and Crósta, 1994; Crósta et al., 1996). They produced individual PC images containing spectral information that could be related to the distribution of hematite, goethite, calcite–chlorite, muscovite–sericite–kaolinite, and silica. These mineral component images were then used to map spectral features related to hydrothermal alteration associated with gold mineralization.

Loughlin (1991) modified the FPCS technique by reducing the number of TM bands used to generate PCs, ensuring that certain materials will not be mapped and increasing the likelihood that spectral information associated with materials of interest will be concentrated in a single PC image. Loughlin found that in areas of sparse vegetation, hydroxyl-bearing minerals could be uniquely mapped into either PCs 3 or 4 derived from TM bands 1, 4, 5, and 7. Ferric oxides could be mapped into PC 4 if TM bands 1, 3, 4, and 5 are used. This modification has been used successfully to map alteration from TM imagery in Bolivia (Bastianelli et al., 1993) and Cyprus (Davidson et al., 1993).

8.4.4.3   SELECTIVE PRINCIPAL COMPONENT ANALYSIS.

Selective principal component analysis (Chavez and Kwarteng, 1989), also known as pairwise principal components (Lamb and Pendock, 1989), uses only two spectral bands. Information that is common to both bands (typically, topographic and albedo) is mapped to the first PC, and information that is unique to either band is mapped to the second PC. The second PC taken from TM bands 1 and 3, for example, would be expected to reveal areas of ferric oxide concentrations, and the second PC from bands 5 and 7 would reveal areas of hydroxyl-bearing minerals and vegetation. Resulting images are easier to interpret than PC color composites because only two bands are used. Also, because it is statistically based, the technique compensates for atmospheric path radiance.

8.4.4.4   DIRECTED PRINCIPAL COMPONENTS.

The technique called directed principal components (Fraser and Green, 1987) is similar in concept to the indexing approach of Elvidge and Lyon (1985) and Kowalczyk and Logan (1989) and is useful in separating the contribution of vegetation to the 1.65 µm/2.22 µm (TM 5/7) ratio from the contribution of hydroxyl-bearing minerals. The method is directed at two input band ratio images (e.g., TM 5/7 and TM 4/3). Information that is common to both ratio images is mapped to PC 1, and that which is unique to hydroxyl-bearing minerals is mapped to PC 2. Fraser (1991) adapted the technique for ferric oxide discrimination by using TM 4/1 and TM 3/1 ratio images for input and successfully discriminated areas of hematite, goethite, and vegetation. Ferreira and Meneses (1994) also found directed principal components useful in mapping areas of ferric oxides and hydroxyl-bearing minerals associated with alteration in a deeply weathered savannah woodland terrain in central Brazil.

8.4.5   Comparison of Image Processing Techniques for Alteration Mapping

The *Landsat* TM images in Figure 8.8 (see the color insert) are a result of an exercise aimed at comparing three image processing techniques that are widely used for detecting and mapping hydrothermally altered zones from TM data. The three techniques are band ratios, selective principal components, and feature-oriented principal components. The area depicted is the Comstock mining district in western Nevada. The historic mining town of Virginia City lies in the Virginia Range near the center the image area, and the Carson River and Washoe valleys occupy the southeastern and northwestern corners, respectively. The generalized geologic map of the area in

Figure 8.9 is adapted from Thompson's (1956) geologic map of the Virginia City Quadrangle.

The Comstock district produced over 6 billion grams of silver and 260 million grams of gold in the middle to late nineteenth century from epithermal veins in Tertiary volcanic rocks. Mineralization is confined largely to the Comstock, Silver City, and Occidental fault zones. Intense acid leaching altered andesitic host rocks to propylitic, argillic, advanced argillic, and sericitic assemblages (Hudson, 1986), producing extensive bleached zones in the Geiger Grade and Flowery areas that are readily visible in the TM imagery. Most of the area is vegetated by sagebrush and other shrubs, juniper, and seasonal grasses. Acid-leached zones are mostly devoid of vegetation and support only scattered Jeffrey pine trees.

Figure 8.8*a* is a color composite of TM bands 1, 2, and 3 in blue, green, and red, respectively, and represents a "true color" view as seen by a human observer from space. Tertiary volcanic rocks appear mostly in shades of red, pink, and lavender in this image. Bleached zones and mine tailings appear white, and vegetation appears green. The Mount Davidson diorite stock, which intrudes the volcanic rocks, underlies the dark green to brown area west of Virginia City.

Figure 8.8*b* is a color composite of TM band ratios 3/1 (blue), 5/4 (green), and 5/7 (red). The 3/1 ratio is sensitive to ferric iron oxides, and rocks with an abundance of these minerals appear blue in the image. The 5/7 ratio is sensitive to hydroxyl-bearing minerals that characterize altered zones, but it is also sensitive to vegetation as well. The 5/4 ratio helps to discriminate vegetated areas (5<4) from areas of bare rock and soil (5>4). The bleached zones in the Geiger Grade and Flowery areas appear as bright areas in shades of pink, yellow, and white. Note the circular feature that appears south of the Geiger Grade area and northwest of Virginia City.

A selective principal components image appears in Figure 8.8*c*. This image is a color composite of second principal components taken from the same band pairs (3–1, 5–4, and 5–7) used in the ratio image and displayed in the same color scheme. It reveals somewhat better definition of the altered zones, which appear light pink, and identifies some possible zones south of Silver City, along the northwest side of the Carson River valley and in the northern part of the image. The image also reveals more detail in the Mount Davidson area than the ratio image, greater definition of the circular feature south of Geiger Grade, and a hint of possible zoning around the Flowery area.

The feature-oriented principal component image in Figure 8.8*d* provides the clearest separation of hydroxyl-bearing minerals (red) and ferric oxides (blue) and suppresses the signal from vegetation. Using the covariance eigenvectors for the band combinations recommended by Loughlin (1991) in Table 8.7, it is apparent that ferric oxides will be mapped to PC 4 in the band 1–3–4–5 combination. Hydroxyl-bearing minerals will be mapped to PC 4 in the band 1–4–5–7 combination and will appear bright because the eigenvector for band 5 has a large positive value and the value for band 7 is strongly negative. Principal component 4 from bands 1, 3, 4, and 5 is mapped to the blue channel to highlight ferric oxides in Figure 8.8*d*, and principal component 4 from bands 1, 4, 5, and 7 is mapped to the red channel to highlight hydroxyl-bearing minerals. The sum of the values in the blue and red channels is mapped to the green channel so that areas with both hydroxyl-bearing minerals and ferric oxides appear white. There is some ambiguity between ferric oxides and vegetation that appears dark blue, but this can be resolved by comparing Figure 8.8*c*

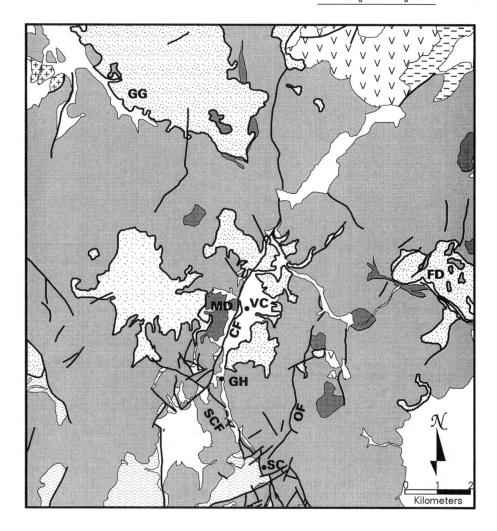

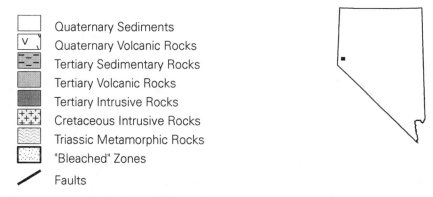

Figure 8.9   Generalized geologic map of the *Landsat* scene in Figure 8.8. CF, Comstock Fault; FD, Flowery District; GG, Geiger Grade; GH, Gold Hill; MD, Mount Davidson; OF, Occidental Fault; SC, Silver City; SCF, Silver City Fault; VC, Virginia City. (After Thompson, 1956.)

TABLE 8.7   Covariance Eigenvectors of Four-Band Data Sets, Virginia City TM Subscene

Band	PC 1	PC 2	PC 3	PC 4
1	0.369	0.631	−0.141	0.668
3	0.422	0.500	−0.160	−0.739
4	0.262	0.072	0.962	−0.010
5	0.786	−0.588	−0.169	0.087
1	0.338	−0.335	0.850	0.225
4	0.241	−0.855	−0.357	−0.289
5	0.783	0.209	−0.352	0.469
7	0.464	0.335	0.161	−0.804

and *d*. The blue halo around the Flowery altered area in part (*d*) appears green in part (*c*) and is therefore more likely due to ferric oxides than to vegetation.

In this area of complex cover types, feature-oriented principal components produced the clearest definition of hydrothermally altered zones, but it still left some ambiguity which had to be resolved with another technique. The lesson is that more than one technique may be needed to resolve all questions in a scene.

8.4.6 Intensity–Hue–Saturation Transformation

Image colors displayed on a monitor or in prints are constructed by addition of their red, green, and blue (RGB) components. Color images can also be described in a polar coordinate system representing intensity (brightness), hue (dominant wavelength), and saturation (purity of color) (IHS). The transformation from RGB space to IHS space is straightforward in most modern image processing systems and has two principal applications in geologic remote sensing: (1) image enhancement and decorrelation, and (2) data integration.

One application is a form of decorrelation stretch in which saturation can be stretched to the full dynamic range independently of hue and intensity and transformed back to RGB space for display. An advantage of this procedure over the principal components decorrelation stretch is that hue does not depend on scene statistics (Gillespie et al., 1986; Sabins, 1987). Kowalczyk and Ehling (1991) applied an adaptive histogram equalization procedure to the intensity channel to compensate for large contrast variations in TM images brought about by illumination effects in vegetated mountainous terrain. Tanaka and Segal (1989) applied an IHS transform to enhance a TM 5/7 and 3/2 ratio composite of the Round Mountain district, Nevada, substituting TM band 5 for the intensity channel to help discriminate vegetation from clay alteration. Yamaguchi et al. (1989) found the IHS transform useful in identifying areas of alunitic alteration in Geoscan data.

The IHS transform is also used to integrate disparate but coregistered data sets into a single image. Legg (1992) combined filtered satellite imagery (intensity), a digitized geological map (hue), and a digital elevation model (saturation) into a single image. He recommends mapping the data set with the most spatial information to intensity and the data set with the greatest dynamic range to hue. Graham and

Bonham-Carter (1993) created an RGB image of airborne gamma ray spectrometer data, which they then transformed to IHS space and substituted radar imagery for intensity and held saturation constant at a maximum value of 255. Carvalho and Araújo (1994) integrated TM imagery with soil geochemistry and geophysical data to characterize a copper–lead–zinc district in Brazil.

8.4.7 Supervised and Unsupervised Classification

Supervised and unsupervised classification of multispectral imagery is a favorite technique of operators who take the "million monkeys" or "oracle" approach to image processing, but it is not favored by project geologists because it so rarely provides useful information. Classification accuracies for lithologic studies are usually poor and cannot compare with performances of experienced image interpreters who can make use of textural and contextual information together with the spectral information and their knowledge of geology. Legg (1992) reported that classification accuracies of multispectral imagery range between 60 and 80%, depending on cover type and sensor. Some new neural net and fuzzy logic classifiers may be able to improve on this accuracy, but current classifiers require too much training data from field measurements to be practical.

Nevertheless, some notable exceptions have shown that classification can produce useful results. Rowan et al. (1987) applied a Bayesian classifier to TM data to map contact metamorphic rocks and residual soil in the Extremadura region of Spain. In this region, Pb–Ag occurrences are found in slates and graywackes in contact aureoles around Hercynian intrusions. Subsequently, Anton-Pacheco (1989) used supervised classification of TM data to map two intensity levels of contact metamorphism in slates as part of a gold exploration program at Albuquerque, Spain.

Spectral classifiers operate on a pixel-by-pixel basis and ignore contextual information that comprises image texture. Various structural and statistical textural classifiers have been tried with varying success (Bishop and Howe, 1989; Wang and He, 1990). A particularly promising approach to textural analysis of radar imagery is based on the semivariogram function (Miranda and MacDonald, 1989; Rubin, 1989). A semivariogram textural classifier developed by MacDonald et al. (1990) has been used to discriminate vegetation types and associated geomorphic surfaces in radar imagery of rain forests in Borneo (Miranda et al., 1992) and Brazil (Miranda and Carr, 1994). The algorithm has also been applied to *Magellan* imagery of Venus (MacDonald and Miranda, 1991).

8.4.8 Imaging Spectroscopy

With improved signal/noise ratios of better than 400:1 over most of the spectral range of 1995 AVIRIS data, imaging spectroscopy has come of age as a useful geological tool. Minerals can be directly identified and mapped by airborne sensors, and in some cases spectral characteristics related to cation chemistry, crystallinity, and temperature of crystallization can be identified. Imaging spectroscopy is a unique concept in remote sensing, and some image processing tools are unique to imaging spectroscopy. Among these are spectral matching and spectral mixture analysis techniques. Additional information on imaging spectroscopy and spectral analysis appears in Chapters 1, 2, and 5.

Spectral matching algorithms compare apparent reflectance spectra from image pixels with reference spectra from a spectral library, field and laboratory spectra, or from reference pixels taken from the imagery. Two widely used algorithms are the spectral angle mapper (Kruse et al., 1993) and tricorder (Clark and Swayze, 1995). Both methods are insensitive to illumination variation due to topography within certain limits. The spectral angle mapper (SAM) treats the pixel and reference spectra as vectors in n-dimensional space, where n is the number of spectral bands used, and computes the arc cosine of the dot product of the two spectra. Small spectral angles represent greater similarity between the reference and pixel spectra. Tricorder, developed at the U.S. Geological Survey Spectroscopy Laboratory, simultaneously analyses for multiple minerals by computing a least-squares fit of all diagnostic spectral features of all materials in the reference file to the pixel spectrum. The reference material with the best overall fit is assigned to the pixel. Figure 8.10 (see the color insert) shows mineral maps of the Cuprite, Nevada, area created by the tricorder from 1995 AVIRIS data at the U.S. Geological Survey Spectroscopy Laboratory. Generalized geologic and alteration maps of the Cuprite area appear in Chapter 5.

Crósta et al. (1998) compared results from SAM and tricorder classifications of 1992 AVIRIS data covering the Bodie and Paramount mining districts in California. They found the tricorder to be more accurate and more sensitive to subtle spectral features than SAM, but also noted that SAM had some distinct advantages in that it was easier to use, faster to compute, and is available in commercial image processing software packages. The computer code can be modified in at least one commercial version of SAM to include some of the features of Tricorder (Bedell R. L, personal communication, 1996).

Spectral matching algorithms classify pixels according to similarity to the single reference material whose reflectance spectrum dominates the signal from a pixel, but surfaces represented by pixels rarely consist of a single material. Pixels are complicated places and contain numerous materials mixed together under varying illumination conditions. Spectral mixture analysis attempts to deconvolve spectra measured from mixed pixels into spectral end members according to a user-defined mixing model. The trick is in choosing the end members.

Spectral end-member identification can be based on field data from locations known to be spectrally "pure," or end members can be generated empirically from the data. One method might entail using spectral matching to identify pixels that most closely match reference spectra. Other empirical end-member identification techniques include the pixel purity index (Boardman et al., 1995) and n-dimensional visualization (Boardman, 1993; Boardman et al., 1995). However end members are identified, the most important aspect of end-member selection is that they be geologically reasonable and pertinent to the geologic problem at hand.

Most spectral mixing analysis algorithms assume a linear mixing model, but this remains an open question. Some evidence suggests that spectral mixtures involving vegetation and some intimately mixed materials do not mix linearly (Clark and Swayze, 1996).

8.5 DEVELOPING A REMOTE SENSING STRATEGY

The remote sensing literature is filled with articles on sensors, data, and image processing techniques that have proven useful in exploration, but it offers little guidance on putting together an integrated exploration program. Most articles are research oriented, aimed toward pushing the performance of a particular sensor to its limits, extracting the maximum amount of information from a data set, or developing and applying new image processing techniques. This research and development bias may lead some remote sensing geologists to believe that a successful exploration strategy is driven by sensor and computer technology rather than by landscape attributes associated with mineral deposits.

Figure 8.11 illustrates this author's approach to exploration remote sensing that is driven not by technology but by attributes of mineral deposit models. At the top of the diagram is the deposit model itself. The choice of model depends on exploration objectives, and multiple models might be considered simultaneously. The deposit model serves as a basis for deducing observable phenomena at the surface (e.g., favorable host rocks, alteration patterns, structural control) that might lend themselves to remote sensing investigation. The list of observable phenomena are then filtered by a set of environmental constraints imposed by nature to produce a new set of landscape attributes (detectable phenomena) that stand a reasonable chance of being detected and exploited in the particular region being explored. The next step is to choose the type of remote sensing data that is most appropriate to detecting these phenomena. This decision is governed by a second filter of data constraints that include factors such as data type, availability, resolution, and scale. Once the data have been selected, the image analyst can employ appropriate image processing techniques to produce thematic images that portray each landscape attribute in an interpretable format. Information from thematic images may then be integrated with information from other sources to produce maps of favorable terrain and exploration targets.

Philosophically, this approach to exploration remote sensing is rooted in classical scientific method, in which deductions are drawn from a working hypothesis and tested by experiment. Here the working hypothesis is the deposit model from which observable and detectable phenomena are deduced and tested by image processing experimentation. The deductive process culminates in an inductive integration stage in which information from various sources is synthesized into a new working hypothesis. The procedure outlined here also implies an integrated team approach to exploration. The remote sensing geologist cannot operate in a vacuum. To be effective, he or she must be able to work interactively with all members of the exploration team, including the exploration manager, project leader, geophysicists, geochemists, and field geologists.

8.5.1 Ore Deposit Models

An ore deposit model is an abstract working hypothesis that seeks to describe the attributes that certain types of ore deposits share in common and to explain processes by which they form. Deposit models provide a conceptual basis for virtually all modern mineral exploration and resource assessment programs. While helping to

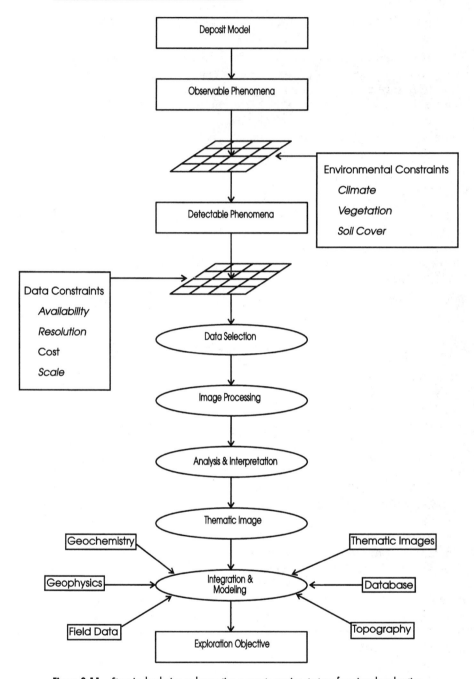

Figure 8.11   Steps in developing and executing a remote sensing strategy for mineral exploration.

guide every aspect of an exploration program, a model may at the same time evolve from its original form as new information is assimilated into it.

8.5.1.1   PORPHYRY ORE MODELS.

Porphyry ore deposits provide good illustrative examples of the importance of deposit models in developing remote sensing strategies. Three types of porphyry deposits are economically important: porphyry copper, porphyry molybdenum, and porphyry gold (Table 8.8). These types are unified by a spatial and genetic association with intrusive stocks and cupolas of felsic porphyry near the transition between plutonic and volcanic environments at the base of stratovolcanoes. Deposits may be intrusion hosted, wallrock hosted, or both. Most types occur in mobile belts near convergent plate margins in continental or island arc settings and tend to be localized on regional faults. Climax-type deposits, however, form in anorogenic extensional settings. Alteration is concentrically zoned around the intrusion, but its type and intensity may vary with deposit type and host rocks.

The lithologic and structural attributes of porphyry deposits that lend themselves to remote sensing studies vary with deposit type and their physical environment. In well-exposed systems, broadband multispectral systems such as *Landsat* TM can

TABLE 8.8   Remote Sensing Characteristics of Porphyry Deposits

Porphyry Type	Characteristics
Porphyry copper	Zoned systems: central potassic zone (K-feldspar, biotite) surrounded by phyllic zone (quartz, sericite, pyrite) and peripheral propylitic zone (chlorite, epidote, albite, calcite); localized on regional strike-slip faults in magmatic belts in convergent plate margin settings; may be copper skarns in wallrock
Calc-alkalic type	Quartz monzonite porphyry host; extensive phyllic zone with abundant pyrite; intense supergene ferric oxide and clay leach zones
Diorite type	Host pluton more mafic; less intense and less extensive phyllic zones and supergene ferric oxides
Porphyry molybdenum	
Granodiorite type	Similar to calc-alkalic porphyry coppers; host plutons more felsic and alteration less extensive and less intense; veinlet and stockwork quartz dominant
Climax type	Silicic anorogenic-type plutons in tensional rift environments; radial and ring-dike structures; weak alteration similar to granodiorite type
Porphyry gold	Similar in petrologic association, alteration and tectonic setting to diorite-type porphyry coppers but smaller; overlain by advanced argillic or acid sulfate zone with abundant alunite, pyrophyllite, pyrite, and supergene ferric oxides; alteration less extensive and less intense than porphyry copper counterparts; potassic zone grades outward to propylitic alteration with weak phyllic alteration between; porphyry copper alteration may be superimposed; may grade upward into high-sulfidation epithermal gold deposits

detect alteration halos and supergene leach caps around porphyry systems as well as regional structures (Spatz and Taranik, 1994; Spatz and Wilson, 1994). Satellite systems with greater spectral resolution in the 2200-nm region (e.g., *FUYO-1* OPS, ASTER) should be capable of differentiating potassic, phyllic, and propylitic alteration in arid settings (Akiyama et al., 1989a; Freund, 1992), while hyperspectral sensors may be able to map distributions of individual minerals and their polymorphs. Multispectral thermal sensors (TIMS, ASTER) should prove useful in mapping zones of silicification and quartz stockworks and may be capable of mapping compositional variation in plutonic hosts (Sabine et al., 1994) and associated skarns (Windeler, 1993).

8.5.2 Environmental Constraints

Environmental parameters determine whether surface phenomena associated with mineral deposits can be detected by remote sensing. The most important of these parameters are climate, soil cover, and vegetation, which combine to determine the amount and quality of exposure in an area. Desert terrains are ideal for remote sensing applications because rocks are well exposed and their spectral and spatial attributes can be measured with relative ease. With little cloud cover and low humidity, a wide selection of image data is usually available, and atmospheric calibration is reasonably straightforward. In humid environments, rocks are obscured by vegetation and soils, and cloud cover is more of a problem. Image analysts must often take an indirect approach to interpretation, relying on geomorphic manifestations of lithologies and mineralized systems. Jiang et al. (1994), for example, found that arcuate and circular features in *Landsat* TM imagery were useful indicators of porphyry copper mineralization in remote forested regions of northeastern China. Cloud cover and humidity in some tropical regions practically preclude all forms of imaging other than radar. Portions of the Amazon region have never been imaged by *Landsat* MSS or TM in more than 25 years of data acquisition,

The tectonic and erosional history of a region is also an important environmental constraint. Stable cratons (e.g., portions of Australia, Brazil, and Canada) may be eroded to plains of low relief, and rocks may be weathered to depths of several meters. Structural and lithologic information might be gleaned from landforms, drainage patterns, and spectral properties of residual soils and weathering products in this type of environment. Australian geoscientists at institutions such as CSIRO's Division of Exploration Geoscience, with industry collaboration, made great progress in developing effective remote sensing tools in deeply weathered terrain (e.g., Fraser et al., 1986, 1987; Abrams and Gabell, 1989; Glikson and Creasey, 1995; Macias, 1995). Tectonically mobile belts are subject to more rapid erosion rates, particularly in humid regions. Near-surface mineralization in epithermal and hot-spring environments is rapidly destroyed in such environments and are preserved only in recently active volcanoes or in buried downfaulted blocks.

Geologic and environmental parameters may combine to produce secondary phenomena associated with mineral deposits that can be detected by remote sensing. Geobotanical phenomena associated with mineralized zones in marginal environments have already been reviewed. Rencz et al. (1996) noted that kimberlite plugs in the District of Mackenzie, Canada, are associated with deep, cold lakes and devised a remote sensing strategy for exploration that employed the *Landsat* TM thermal band. Chris Clark has used *ERS-1* imagery to map ice flow patterns in diamond-bearing glacial till in Quebec as a means of deducing possible source regions (Wagner, 1995).

Geologic and climatic parameters may also combine to produce certain kinds of deposits in certain environments. Nickel laterites and bauxite deposits form by intense leaching in humid tropical environments, and evaporites form in deserts. Chemical weathering of volcanic rocks in humid tropical environments such as in the Philippines produce bouldery clays, and well-sorted gravels suitable for minable placer deposits are rarely present. In the continental margin arc of New Guinea, however, the volcanic cover is limited, and extensive placer deposits formed from lode gold deposits hosted by slates and quartzites (Mitchell and Leach, 1991).

8.5.3 Data Selection

Once detectable phenomena of a particular deposit type have been determined through careful analysis of the interactions between deposit geology and the physical environment, the image analyst must choose the type of data that is appropriate to detecting those phenomena. This decision is constrained by the availability, spatial resolution, spectral resolution, scale, and cost of the data. The first consideration is scale.

Remote sensing objectives often depend on the scale of exploration. Exploration remote sensing at reconnaissance and regional scales takes advantage of synoptic views of large regions provided by satellite imagery. Reconnaissance-scale exploration takes place in early prospecting phases of a project and may involve identification of spectral anomalies as potential exploration targets or regional structural analysis from satellite imagery. As mining companies pursue new exploration opportunities in poorly mapped remote regions of the world, explorationists are finding that satellite imagery may provide the only reliable base map and source of geologic data. *Landsat* TM is the most widely used imagery for reconnaissance work because of its 185-km synoptic view, intermediate spatial resolution, ability to detect hydrothermally altered areas, nearly ubiquitous coverage, and nominal cost. TM imagery has been used extensively for exploration in arid regions throughout North and South America by dozens of mining companies during the last decade and has been credited with putting more than one company on prospective ground for precious and base metals in Argentina, Chile, and Peru. MSS data, also widely available at even lower cost, are still used for reconnaissance-scale studies for structural mapping and detection of ferric oxide concentrations. Some image analysts contend that MSS bands are better positioned than TM bands for ferric oxide detection.

In more difficult regions of dense vegetation and perennial cloud cover, *Landsat* imagery might not be available or practical for exploration at any scale. In terrain such as this, orbital synthetic aperture radar (SAR) imagery is the best alternative for reconnaissance and regional-scale exploration. L-band imagery with moderate incidence angles, such as that provided by *FUYO-1* and *SIR-B*, provides better forest canopy penetration than does shorter-wavelength C-band radar and is excellent for morphostructural analysis of low-relief surfaces (F. P. Miranda, personal communication, 1995). The variable imaging geometry offered by *Radarsat* may allow explorationists to optimize SAR imagery according to their exploration objectives and terrain characteristics. Multifrequency, multipolarization *SIR-C/XSAR* imagery, covering 30% of the earth's land surface between 57° north and south latitudes, may also prove useful for exploration in remote areas. Airborne SAR data are available from the U.S. Geological Survey in 1° × 2° quadrangles for much of the United States.

Regional exploration focuses on confined target areas, such as a mountain range, mineral belt, structural corridor, or volcanic field, in which high potential for undiscovered mineral deposits has been identified. Exploration turns from prospecting to field mapping in conjunction with geophysical and geochemical surveys at this scale, and remote sensing objectives focus on more specific attributes of deposit models. Landsat TM is used widely for regional studies for the same reasons mentioned for reconnaissance studies, but higher spatial resolution data, such as that provided by *SPOT*, becomes more important. Regional data sets of certain kinds of airborne imagery (e.g., GERIS, Geoscan, ATM) are also available for prospective areas in North and South America, Australia, and many other parts of the world. ASTER and ARIES data may find its greatest utility in regional applications. With five spectral bands between 2.1 and 2.4 μm at 30-m spatial resolution, ASTER should be able to distinguish propylitic from argillic and advanced argillic alteration zones in arid regions as well as rocks containing detectable amounts of carbonate- and ammonium-bearing minerals. ASTER's multispectral thermal bands may also aid in mapping silicified zones, silicate-bearing host rocks, and possibly some forms of potassic alteration.

Although satellite data, *Landsat* TM in particular, are used widely to investigate confined target areas at the district or project level, its limited spatial and spectral resolution are often inadequate to provide the kinds of detailed lithologic and structural information that is often required at this scale of exploration. With spatial resolutions of less than 10 m and dozens of spectral bands, airborne multispectral and hyperspectral scanners play their most important role at this scale of investigation. Airborne data are costly, however, compared to satellite data, and project geologists must give careful consideration to the potential benefits of obtaining airborne data in relation to the cost. Some providers maintain archives of nonproprietary data covering certain regions that can be purchased at reasonable cost. To reduce costs, companies might pool resources with other companies to finance a "group shoot" covering several properties in a region. In the next 10 years satellites might provide imagery globally at spatial and spectral resolutions that are now only possible with airborne systems.

The date and time of image acquisition are also important in selecting data. Specific considerations include the influences of sun azimuth and elevation and seasonal variation of ground covers on detectable phenomena. ATM data used in a lineament study of the northern Carlin Trend, Nevada, by Sabine and Re Kühl (1996) were acquired in late spring when grasses had turned brown over most of the area except where soil moisture was retained along drainage bottoms and along some fractures. Few lineaments could be seen in midsummer *SPOT* panchromatic imagery of the same area at approximately the same spatial resolution, presumably because all of the grasses had turned brown and the larger sun elevation angle diminished illumination differences that reveal subtle landforms. Both of these factors combined to reduce the overall contrast in the *SPOT* data to the extent that even major drainages were difficult to discern. The use of 1985 data also permitted analysis of undisturbed terrain that has now been obliterated by mining. Larson et al. (1989) used lineament information from a midwinter, snow-covered TM scene together with aeromagnetic data to define the Ely–Mt. Hamilton–Eureka structural trend in eastern Nevada. Low signal/noise ratios of some hyperspectral and multispectral thermal data preclude data collection during any season but summer, and extensive shadowing may also preclude wintertime imagery in areas of high relief. Flooded rain forests in structurally controlled topographic lows in the Amazon region are easily detected in SAR imagery obtained during the wet season (Miranda et al., 1994).

8.5.4 Integration and Modeling

Modern mineral exploration requires an integrated approach. More often than not, new discoveries are blind ore bodies that lie hidden below the surface under alluvium or unmineralized rock with only the most subtle indication of their presence at the surface. Discoveries are also being made in remote locations where accessibility is limited, if not impossible, but also in well-known districts where mineralization had not been suspected. Analysis and interpretation of digital imagery may play an important role in such discoveries, but rarely is remote sensing credited as the primary means of identifying new deposits. To an ever-increasing degree, *information* from remote sensing studies is analyzed together with information from separate analyses of field, geochemical, geophysical, metallogenic, and other data types in various combinations in order to identify targets for intensive evaluation.

The operative word in the preceding paragraph is *information*. The image analyst who delivers a map with hundreds of spectral "anomalies," which may include iron-stained tuffs and red shales together with hydrothermally altered zones, performs no useful service to the project manager. Lineaments with no structural interpretation associated with them are just lines on a map to the explorationist. The image analyst's job is not complete until he or she, together with other members of the exploration team, has produced fully interpreted thematic images and maps depicting information that can be readily integrated with geochemical, geophysical, field, and other sources of information.

Geographic information systems (GISs) and spatial analysis functionality in some image processing software packages provide powerful tools for analyzing, integrating, and modeling information to produce new insights into possible mineralization in an area. Establishing dynamic links between image processing and GIS software maximizes the capabilities of each and allows explorationists interactively to explore and visualize data, overlay data sets, perform proximity and adjacency analyses, and apply mathematical operators.

Geographic information systems also offer the means to construct empirical predictive models to estimate probabilities of occurrence of undiscovered mineral deposits. Such models are rooted in the deposit model and may be data driven, in which the computer uses training areas, or knowledge driven, in which the computer uses input provided by a team of expert geoscientists. Predictive models may employ relatively simple Boolean logic or simple scoring schemes, or they might use more complex Bayesian or fuzzy membership operators. GIS applications are beyond the scope of this chapter; however, Bonham-Carter (1994) explains modeling of geoscience data with GIS clearly and in considerable detail.

8.5.5 Integrated Exploration at Collahuasi, Chile

TM and *SPOT* imagery played a key role in the discovery of blind copper porphyry systems in the Collahuasi district, northern Chile (Dick et al., 1993). Although exploration had reached an advanced stage in a region of only 280 km², the satellite imagery, in conjunction with IP and aeromagnetic surveys, geologic, structural, and metallogenic data led to a much better understanding of the full extent of the Collahuasi hydrothermal system, which was shown to be roughly circular and 6 km in diameter. A known copper porphyry deposit, which had been the focus of intensive mapping and drilling, was found to be a small part of the larger system. A completely separate zone of hydrothermal alteration was identified at Ujina, 7 km east of the Collahuasi system, and found to be much larger than previously realized. Hydrothermal alteration and structural maps derived from the satellite imagery and field data were used to lay out induced polarization and aeromagnetic surveys, which confirmed additional buried targets beneath alluvium at Collahuasi and beneath ignimbrites at Ujina.

8.5.6 Integrated Alteration and Geochemical Mapping in the Sulphurets–Bruce Jack Lake District, British Columbia, Canada

The work of Rencz et al. (1994) and Harris et al. (1998) provides an excellent example of what can be achieved through an integrated analysis of Landsat TM, geochemical, and gamma ray spectrometer data in glaciated, vegetated terrains at high latitudes. They focused their study on the the Sulphurets–Bruce Jack Lake district, where previous workers had shown that TM mapping could be an effective and useful exploration tool (Ma et al., 1991). Intensive exploration for both precious metal and porphyry copper deposits has led to underground precious metal mines at Bruce Jack Lake and the discovery of the Kerr copper porphyry deposit at Sulphides Lake. Large quartz–pyrite–sericite alteration zones with extensive limonitic surface exposures host porphyry copper–molybdenum and epithermal to mesothermal precious metal systems. Mineralization has been episodic, and various alteration and deposit styles are known to be juxtaposed by faulting.

The study area is mountainous, heavily glaciated, and largely covered by a mixture of coniferous vegetation and alpine grass as well as glacial ice (Figure 8.12*a*; see the color insert). Exposed rock and soil covers approximately 26% of the study area, primarily on the upper flanks and tops of mountains and ridges. These surfaces are mostly bare, with only limited lichen growth, and are not masked by vegetation. The remote sensing strategy, therefore, was to use *Landsat* TM data to identify and map altered rocks and to calibrate and verify these results using lithogeochemical data, gamma ray spectrometer data, and field measurements in a GIS.

A mostly cloud-free TM scene with minimal seasonal snow cover was georeferenced to a topographic map and corrected for atmospheric path radiance using lakes within the study area as reference pixels. Analyses of approximately 1000 surface bedrock samples for 11 major oxides and 34 trace elements were also incorporated into the database, as were interpolated airborne gamma ray spectrometer data. Potassium (%K) data from the gamma ray spectrometer survey were used with lithogeochemical K_2O data to assist in delineating possible zones of potassic alteration and potassium depletion.

Band ratios were used to detect and map hydrothermally altered zones in the TM data in which areas of vegetation, ice, and snow had first been delineated and masked (Figure 12b; color insert). A vegetation mask was created using a biomass ratio (TM band 4/3) to differentiate between vegetated and nonvegetated areas. The ice and snow masks were created from TM band 2 by thresholding to select high-reflectance areas. Figure 8.12b is a two-component color composite image of the areas of outcrop in which the 5/7 ratio is diplayed in red, indicating the presence of hydrous minerals, and the 3/1 ratio is displayed in green, indicating areas of leached and oxidized iron bearing minerals. Yellow areas indicate occurrences of hydrous minerals plus oxidation of iron-bearing sulfides. Ten zones of high 5/7 and 3/1 ratios were then delineated for further analysis. A K-means clustering algorithm was then applied to quantify the ratio imagery more rigorously. Five ratio classes defined various intensities of hydrothermal alteration and began to reveal an east–west zonation in the alteration.

The lithogeochemical data were processed in a GIS and in a separate statistical software package. A point-in-polygon operation intersected the ratio classes with the mapped geological units and lithogeochemical sample points using only the major oxide data. First-order statistics and exploratory box-and-whisker plots for each combination of the intersected categories assisted in the statistical comparison between the TM ratio classes and lithogeochemical data. Alteration indices were calculated from the lithogeochemical data using the database software utilities within the GIS. These indices included:

Sericite index: $K_2O/(Na_2O + K_2O)$
Chlorite index: $(Fe_2O_3 + MgO)/(Fe_2O_3 + MgO + CaO + Na_2O)$
Alkali index: $(Na_2O + CaO)/(Na_2O + K_2O + CaO)$
Iron index: Fe_2O_3/FeO
Hashiguichi index: $Fe_2O_3/(Fe_2O_3 + MgO)$

The sericite index emphasizes areas characterized by enriched potash (K_2O) reflecting sericitic alteration, whereas the chlorite index emphasizes possible areas of more intense chloritization typified by higher ferrous iron and magnesium concentration. The alkali index emphasizes areas of alkali enrichment but can also serve to delineate areas of alkali depletion often associated with base metal and precious metal systems. The iron index, which uses the ratio of ferrous to ferric iron, can potentially emphasize areas of more intense iron oxidation.

A series of continuous surface maps of various oxides, trace elements (Cu, Zn, Pb, Au, As and Sb), and alteration indices were calculated in the GIS to establish spatial patterns in the lithogeochemical data with respect to the ratio images. The trace-element continuous surface maps were combined into RGB color composite images so that variations between trace elements could be clearly displayed.

Analysis of the TM, geochemical, and radiometric data reveal an east–west division in the study area. The red and yellow hues in the eastern portion of the ratio image reflect higher 5/7 values, whereas the western half is characterized by greenish hues reflecting higher 3/1 ratios. These differences correspond generally to volcanic lithologies in the east and pyritized sediments in the west, but they also correlate strongly with two fundamentally different alteration and mineralization systems. The east half of the study area is also characterized by significantly higher SiO_2 and sericite values and lower total iron, K_2O, and alkali index values than the western half. A linear north–south zone of SiO_2 enrichment and alkali depletion appears in the SiO_2 and alkali index continuous surface maps (Figure 8.12c and d; color insert) in the eastern zone. This may represent a zone of intense silicification, quartz veining, and phyllic alteration which may be related to faulting. A linear zone of Au and As enrichment (yellow) and Au, As, and Sb enrichment (white) that appears in the ternary continuous surface map in Figure 8.12e (color insert) corresponds directly with the zone of silicification, sericitic alteration, and alkali depletion. The Cu–Pb–Zn continuous surface map in Figure 8.12f (color insert) is dominated by elevated levels of copper (bright red areas) in the western zone.

This analysis thus reveals evidence that two mineralized systems may be juxtaposed by faulting: a precious metal system in the east and a base metal system in the west. Zones of high TM5/7 values, reflecting more intense phyllic and argillic alteration, tend to dominate the eastern portion of the study area over primarily volcanic rocks, and a linear zone of elevated Au, As, and Sb coincides with a zone of silica enrichment and alkali depletion. Areas of more intense iron oxidation occur in the western half of the area over the pyritic altered metasedimentary rocks and correlate with elevated levels of copper, suggesting a possible base-metal system.

8.6 OPERATIONAL REMOTE SENSING IN THE NEW MILLENNIUM

During the decade of the 1990s, many companies have accepted remote sensing as a fully operational exploration tool, and some have built image processing facilities as sophisticated as some of the leading research laboratories. The impetus to employ remote sensing as an exploration tool has been driven by a number of factors. One such factor has been the growing body of knowledge relating to the capabilities and limitations of remote sensing as an exploration tool. Another factor was the advent of workstation technology and user-friendly commercial image processing software packages. Political and economic changes in developing countries have also created exploration opportunities in remote regions where satellite imagery may provide the only reliable base map and source of geologic data. Perhaps the most important factor, however, was the growing realization during the 1980s that mineral deposits are small-scale manifestations of geologic processes that operate on regional to continental scales. Synoptic views provided by satellite imagery are ideally suited for regional geologic studies involving magmatic arcs, mineral belts and provinces, accreted terranes, and other large-scale geologic features.

A greater diversity of data types in the next decade, together with advancing image processing, GIS, and global positioning system technology, will revolutionize geologic investigations in remote unmapped regions of the world. Explorationists who are just now becoming comfortable with TM and *SPOT* imagery will soon find themselves on new learning curves as new kinds of data become commercially available. The next generation of orbital sensors, due for launch by the end of the century, will provide data to users at spectral and spatial resolutions that have only been possible in airborne sensors. ASTER will measure radiance in five SWIR bands now covered by only one TM band, enabling greater discrimination of hydrothermal alteration assemblages and mineral zoning patterns at regional to project scales. ASTER will also provide global multispectral thermal infrared imagery for the first time, which will offer new capabilities to map silicate-bearing rocks and alteration assemblages characterized by silicate minerals. Spaceborne hyperspectral sensors such as *Aries*, and their successors will provide new capabilities for operational mineral mapping at regional and project scales, while the next generation of airborne instruments will make aerial surveys more affordable for exploration. Innovative designs such as that offered by tunable filters (e.g., Denes et al., 1997) may make it possible to produce rugged and compact imaging spectrometers for direct sale to mining companies for aerial- or ground-based mapping.

Stereoscopic imaging of satellite imagery has enormous potential in mineral exploration for structural and geomorphic mapping and for three-dimensional visualization but has not been applied operationally to any degree. Image processing for stereoscopic viewing requires sophisticated software and specialized equipment. Stereoscopic *SPOT* data have been difficult and costly to obtain and process, and fore-and-aft *FUYO-1* stereo data have also been difficult to obtain for commercial application. The next generation of orbital sensors (e.g., ASTER, *Landsat 7*) will offer a wider selection of stereoscopic data, which may in turn stimulate production of friendly and inexpensive software for stereoscopic imaging. High-spatial-resolution stereoscopic imaging satellites will also enable production of digital elevation models at 1:25,000 scale and 5-m elevation accuracy. This diversity of data types, together with advancing image processing, GIS, and global positioning system technology, will revolutionize geologic investigations in remote unmapped regions of the world.

During the decade of the 1990s, image processing has moved from the desktop to the laptop as notebook-sized personal computers have become faster and more powerful and easy-to-use basic image processing and GIS software linked to GPS technology has become available for use by geologists in the field. New microstorage devices enable easy transport and access to large image data files as well as geophysical, geochemical, and GIS-based databases. Direct satellite uplink and downlink is now possible that will enable transfer of data between file servers in exploration division headquarters and portable computers in remote field camps.

Integrated exploration strategies require integrated image processing, GIS, and statistical modeling functionality. This is now accomplished through dynamic links between image processing and GIS packages, but if software is not chosen with great care to be compatible, dynamic links can lead to forced marriages with offspring of indeterminate parentage. Image processing and GIS functionality have been gradually merging during the past few years, and the next decade may see full integration or fusion of software packages in much the same way that word processing, spreadsheet, and database software is now bundled into integrated packages.

Remote sensing has emerged as a first-order operational exploration tool and is no longer a secondary complement to other geophysical techniques. That position is built on a solid platform of basic and applied research addressing the capabilities and limitations of remote sensing in detecting surface phenomena associated with mineral deposits. As research scientists continue to expand the limits of what is possible in remote sensing, applied scientists in the exploration community will address what is not possible. Research scientists choose test sites such as Cuprite or Goldfield, Nevada, in arid environments where vegetation and soil cover are sparse or nonexistent and success is virtually guaranteed. One does not need remote sensing to find deposits as obvious as these; most have been known for more than a century. New deposits are being found in remote areas where deposits are often obscured by vegetation, soil, alluvium, or unmineralized rock and where surface expressions are subtle and may require great ingenuity to detect. There is still much to learn about how spectral and spatial patterns in vegetation and soils may be used to reveal hidden mineral deposits. We need to know more about spectral properties of alteration mineral assemblages as well as the spectral behavior of individual alteration minerals, and we need to quantify variations in position and shape of spectral features as functions of cation chemistry, temperature, and other physicochemical parameters that relate to environments of ore formation. Questions about the linearity or nonlinearity of spectral mixing need attention, particularly the mixing laws governing intimately mixed materials. As imaging spectrometers become more sensitive, atmospheric radiative transfer models will require refinement, and temporal variations of solar irradiance will become an important issue. Continued development of remote sensing as an exploration tool will depend on the efforts of research scientists who are tuned to the needs of industry and on industry's willingness to support research and development. There is much to learn.

ACKNOWLEDGMENTS

The author is indebted to the many people who offered the benefits of their insight, knowledge, and experience in bringing this chapter to fruition. Andy Rencz performed heroic acts as editor and troubleshooter, and Larry Rowan's thoughtful review greatly improved the manuscipt. Many people provided images and material that were incorporated into this chapter, including Michael Abrams, B. Ballantyne, Richard Bedell, Chris Clark, Roger Clark, Alvaro Crosta, Steve Fraser, J. R. Harris, Gerald Heston, Simon Hook, Tim Minor, Kerry O'Sullivan, David Spatz, and James Taranik. To these people I owe my sincerest thanks. I am also greatly indebted to the Desert Research Institute, which made resources available that enabled completion of this chapter.

References

Abrams, M. J., and A. Gabell, 1989. Geologic mapping and mineral exploration in the Coppin Gap greenstone belt, Australia (abstract), in *Proceedings of the 7th Thematic Conference on Remote Sensing for Exploration Geology*, Vol. I, Environmental Research Institute of Michigan, Ann Arbor, Mich., pp. 361–362.

Abrams, M. J., and S. J. Hook, 1995. Simulated ASTER data for geologic studies; *IEEE Trans. Geosci. Remote Sensing*, 33, 692–699.

Abrams, M. J., R. P. Ashley, L. C. Rowan, A. F. H. Goetz, and A. Kahle, 1977. Mapping of hydrothermal alteration in the Cuprite mining district, Nevada, using aircraft scanner images for the spectral region 0.46 to 2.36 μm; *Geology*, 5, 713–718.

Abrams, M. J., D. Brown, L. Lepley, and R. Sadowski, 1983. Remote sensing for porphyry copper deposits in southern Arizona; *Econ. Geol.*, 78, 591–604.

Abrams, M. J., J. E. Conel, and H. R. Lang, eds., 1985. *The Joint NASA/Geosat Test Case Project: Final Report*; American Association of Petroleum Geologists, Tulsa Okla.

Abrams, M. J., V. Carrere, and A. Gabell, 1988. Mapping hydrothermal alteration using AVIRIS data, in *Proceedings, of the 6th Thematic Conference on Remote Sensing for Exploration Geology*, Environmental Research Institute of Michigan, Ann Arbor, Mich., p. 201.

Agar, R. A. and Villanueva, R., 1997. Satellite, airborne and ground spectral data applied to mineral exploration in Peru, in *Proceedings of the 12th International Conference on Applied Geologic Remote Sensing*, Vol. I, Environmental Research Institute of Michigan, Ann Arbor, Mich., pp. 13–20.

Ager, C. M., N. M. Milton, B. A. Eiswerth, M. S. Power, and S. A. Hauck, 1989. Spectral response of vegetation to metallic elements in northeastern Minnesota, in *Proceedings of the 7th Thematic Conference on Remote Sensing for Exploration Geology*, Vol. I; Environmental Research Institute of Michigan, Ann Arbor, Mich., pp. 173–178.

Akiyama, Y., J. Komai, T. Yokoyama, and K. Okada, 1989a. Preliminary assessment of JERS-1 optical sensor based on the simulated airborne data, in *Proceedings of the 7th Thematic Conference on Remote Sensing for Exploration Geology*, Vol. I; Environmental Research Institute of Michigan, Ann Arbor, Mich., pp. 519–529.

Akiyama, Y., J. Komai, T. Yokoyama, and K. Okada, 1989b. Digital processing and analysis of airborne multispectral data for mapping hydrothermal alteration at Yerington, Nevada, in *Proceedings of the 7th Thematic Conference on Remote Sensing for Exploration Geology*, Vol. II, Environmental Research Institute of Michigan, Ann Arbor, Mich., pp. 969–980.

Amos, B. J., and D. Greenbaum, 1987. Alteration detection using TM imagery: the effects of supergene weathering in an arid environment, in *Proceedings of the 21st Symposium on Remote Sensing of Environment*: Environmental Research Institute of Michigan, Ann Arbor, Mich., p. 795.

Anderson, J. A., 1981. Characteristics of leached capping and techniques of appraisal, in *Advances in Geology of the Porphyry Copper Deposits, Southwestern North America*; S. R. Titley, ed., University of Arizona Press, Tucson, Ariz., pp. 275–296.

Anton-Pacheco, C., 1989. Cartografía digital de rocas en el área de Albuquerque la Codosera, Extremadura, utilizando imagenses *Landsat* thematic mapper; *Proceedings, III Reunión Científica del Grupo de Trabajo en Teledetection*, October 17–19, Madrid, Spain, pp. 157–168.

Barniak, V. J., R. K. Vincent, J. J. Mancuso, and T. J. Ashbaugh, 1996. Comparison of a gold prospect in Churchill County, Nevada, with a known gold deposit in Mineral County, Nevada, from laboratory measurements and *Landsat* TM images, in *Proceedings of the 11th Thematic Conference on Geologic Remote Sensing*, Vol. II; Environmental Research Institute of Michigan, Ann Arbor, Mich., pp. 188–197.

Bastianelli, L., G. D. Bella, and L Tarsi, 1993. Alteration mapping: a case study in mid-south Bolivia, in *Proceedings of the 9th Thematic Conference on Geologic Remote Sensing*, Vol. II; Environmental Research Institute of Michigan, Ann Arbor, Mich., pp. 1133–1144.

Baugh, W. M., and F. A. Kruse, 1994. Quantitative geochemical mapping of ammonium minerals using field and airborne spectrometers, Cedar Mountains, Esmeralda County, Nevada, in *Proceedings of the 10th Thematic Conference on Geologic Remote Sensing*, Vol. II, Environmental Research Institute of Michigan, Ann Arbor, Mich., pp. 304–315.

Bell, R., V. H. Singhroy, C. S. Evans, and S. E. Harrington, 1989. Geologic lithologic mapping in NW Ontario: remote sensing approaches and caveats, in *Proceedings, of the 7th Thematic Conference on Remote Sensing for Exploration Geology*, Vol. II, Environmental Research Institute of Michigan, Ann Arbor, Mich., pp. 819–831.

Ben-Dor, E., and F. A. Kruse, 1994. Mineral mapping of Makhtesh Ramon, Negev, Israel using GER 63 channel scanner data and linear unmixing procedures, in *Proceedings, of the 10th Thematic Conference on Geologic Remote Sensing*, Vol. I, Environmental Research Institute of Michigan, Ann Arbor, Mich., pp. 215–226.

Bennett, S. A., 1993. Use of thematic mapper imagery to identify mineralization in the Santa Teresa district, Sonora, Mexico, *Int. Geol. Rev.*, 35, 1009–1029.

Bishop, M. P., and R. Howe, 1989. Texture algorithm performance analysis for mapping geologic units in the Mojave Desert, in *Image Processing '89*, American Society for Photogrammetry and Remote Sensing, Falls Church, Va., pp. 78–90.

Boardman, J. W., 1993. Automated spectral unmixing of AVIRIS data using convex geometry concepts, in *Summaries of the 4th JPL Airborne Geoscience Workshop*,

JPL Pub. 93–26, Vol. I, Jet Propulsion Laboratory, California Institute of Technology, Pasadena, Calif., pp. 11–14.

Boardman, J. W., and J. F. Huntington, 1996. Mineral mapping with 1995 AVIRIS data, in *6th Aviris Workshop*. See JPL AVIRIS web page, http://makalu.jpl.nasa.gov/docs/workshops/96_docs/toc.htm

Boardman, J. W., and Huntington, J. F., 1997, Mineralogic and geochemical mapping at Virginia City, Nevada using 1995 AVIRIS data in *Proceedings of the 12th International Conference on Applied Geologic Remote Sensing*, Vol. I, Environmental Research Institute of Michigan, Ann Arbor, Mich., pp. 191–198.

Boardman, J. W., F. A. Kruse, and R. O. Green, 1995. Mapping target signatures via partial unmixing of AVIRIS data, in *Summaries of the 5th JPL Airborne Geoscience Workshop*, JPL Publ. 95–1, Vol.I, Jet Propulsion Laboratory, California Institute of Technology. Pasadena, Calif., pp. 23–26.

Bolivken, B., F. Honey, S. R. Levine, R. J. P. Lyon, and A. Prelat, 1977. Detection of naturally heavy-metal poisoned areas by *Landsat 1* digital data, *J. Geochem. Explor.* 8, 457–471.

Bonham-Carter, G. F., 1985. Statistical association of gold occurrences with *Landsat*-derived lineaments, Timmins–Kirkland area, Ontario, *Can. J. Remote Sensing*, 11, 195–210.

Bonham-Carter, G. F., 1994. *Geographic Information Systems for Geoscientists: Modeling with GIS*, Elsevier, New York, 398 p.

Borengasser, M. X., and J. V. Taranik, 1985. Application of shuttle imaging radar-B (SIR-B) data to tectonic analysis of the Candelaria region, Nevada, in *Proceedings of the 4th Thematic Conference on Remote Sensing of the Environment*, Environmental Research Institute of Michigan, Ann Arbor, Mich., pp. 105–112.

Bowers, T. L., and L. C. Rowan, 1996. Remote mineralogic and lithologic mapping of the Ice River Alkaline Complex, British Columbia, Canada, using AVIRIS data, *Photogramm. Eng. Remote Sensing*, 62, 1379–1385.

Brickey, D. W., 1986. The use of thematic mapper imagery for mineral exploration in the sedimentary terrain of the Spring Mountains, Nevada, in *Proceedings of the 5th Thematic Conference on Geologic Remote Sensing*, Environmental Research Institute of Michigan, Ann Arbor, Mich., pp. 607–613.

Burnham, C. W., and Ohmoto, H., 1980. Late stage processes of felsic magmatism, pp. 1–11 *in* S. Ishihara and S. Takenouchi, Eds., *Granitic Magmatism and Related Mineralization*. Soc. Min. Geol. Japan, Special Issue 8, 247 p.

Campos-Marquetti, R., and B. Rockwell, 1989. Quantitative lithologic mapping in spectral ratio feature space: volcanic, sedimentary and metamorphic terrains, in *Proceedings of the 7th Thematic Conference on Remote Sensing for Exploration Geology*, Vol. I, Environmental Research Institute of Michigan, Ann Arbor, Mich., pp. 471–484.

Carrere, V., 1989. Mapping alteration in the Goldfield mining district, Nevada, with the airborne visible/infrared imaging spectrometer (AVIRIS), in *Proceedings of the 7th Thematic Conference on Remote Sensing for Exploration Geology*, Vol. I, Environmental Research Institute of Michigan, Ann Arbor, Mich., pp. 365–378.

Carvalho, O. A. J., and A. H. Araújo, 1994. Characterization of Cu–Pb–Zn orebodies through geophysical, geochemical, geological and TM-*Landsat* data integration in Palmeiro(')polis volcano-sedimentar [*sic*] sequence—Brazil, in *Proceedings* of the 10th Thematic Conference on Geologic Remote Sensing, Vol. I, Environmental Research Institute of Michigan, Ann Arbor, Mich., pp. 104–114.

Cetin, H., and D. W. Levandowski, 1991. An integrated study of remotely sensed and geophysical data for mineral exploration in Lincoln County, Nevada: a new approach to mapping hydrothermal alteration using thematic mapper data, in *Proceedings of the 8th Thematic Conference on Geologic Remote Sensing*, Vol. II, Environmental Research Institute of Michigan, Ann Arbor, Mich., pp. 1271–1278.

Chang, S. H., and W. Collins, 1983. Confirmation of the airborne biogeophysical mineral exploration technique using laboratory methods, *Econ. Geol.*, 78, 723–736.

Chavez, P. S., and A. Y. Kwarteng, 1989. Extracting spectral contrast in *Landsat* thematic mapper image data using selective principal component analysis, *Photogramm. Eng. Remote Sensing*, 55, 339–348.

Clark, R. N., and G. A. Swayze, 1995, Mapping minerals, amorphous materials, environmental materials, vegetation, water, ice and snow, and other materials: the USGS tricorder algorithm, in *Summaries of the 5th Annual JPL Airborne Earth Science Workshop*, JPL Publ. 95–1, Vol. 1, Jet Propulsion Laboratory, California Institute of Technology, Pasadena, Calif., pp. 39–40.

Clark, R. N., and G. A. Swayze, 1996. Evolution in imaging spectroscopy analysis and sensor signal-to-noise: an examination of how far we have come, in *6th AVIRIS Workshop*.

Clark, R. N., G. A. Swayze, and A. Gallagher, 1993. Mapping minerals with imaging spectroscopy, in *Advances Related to United States and International Mineral Resources: Developing Frameworks and Exploration Technologies*; USGS Bull. 2039, U.S. Geological Survey, Washington, D.C., pp. 141–150.

Clark, R. N., G. A. Swayze, K. Heidebrecht, R. O. Green, and A. F. H. Goetz, 1995. Calibration to surface reflectance of terrestrial imaging spectrometry data: comparison of methods, in *Summaries of the 5th JPL Airborne Geoscience Workshop*, JPL Publ. 95–1, Vol. I, Jet Propulsion Laboratory, California Institute of Technology Pasadena, Calif., pp. 41–42.

Collins, A. H., 1988. Implications of patterns of faulting and hydrothermal alteration from *Landsat* TM images and NHAP aerial photos to mineral exploration and tectonics of the Virginia Range, Nevada, in *Proceedings of the 6th Sixth Thematic Conference on Remote Sensing for Exploration Geology*, Vol. I, Environmental Research Institute of Michigan, Ann Arbor, Mich., pp. 185–198.

Collins, A., M. Freund, P. Kowalczyk, J. Berry, D. Cole, J. Pershouse, L. Burgess, S. Marsh, M. Abrams, T. Munday, J. Huntington, I. Ginsberg, W. Malila, and F. Henderson, 1992. *Evaluation of Simulated JERS-1 OPS Data for Geologic Mapping Using Sites in Utah, Wyoming and Arizona*, GEOSAT Committee, Joint GEOSAT/ERSDAC/JGI/JAPEX Rep. 175.

Collins, W., S. H. Chang, G. Raines, F. Canney, and R. Ashley, 1983. Airborne biogeophysical mapping of hidden mineral deposits, *Econ. Geol.*, 78, 737–749.

Colwell, R. N. (ed.), 1983. *Manual of Remote Sensing*, 2nd ed., American Society of Photogrammetry and Remote Sensing, Falls Church, Va., 2440 pp.

Cox, D., and D. A. Singer, eds., 1986, *Mineral Deposit Models*, USGS Bull. 1693, U.S. Geological Survey, Washington, D.C.

Crippen, R. E., 1988a. The dangers of underestimating the importance of data adjustments in band ratioing, *Int. J. Remote Sensing*, 9, 767–776.

Crippen, R. E., 1988b. Image display of four-dimensional spectral data, in *Proceedings of the 6th Thematic Conference on Remote Sensing for Exploration Geology*, Environmental Research Institute of Michigan, Ann Arbor, Mich., pp. 677–678.

Crippen, R. E., 1989. Selection of *Landsat* TM band and band-ratio combinations to maximize lithologic information in color composite displays, in *Proceedings of the 7th Thematic Conference on Remote Sensing for Exploration Geology*, Vol. II, Environmental Research Institute of Michigan, Ann Arbor, Mich., pp. 917–921.

Crippen, R. E., R. G. Blom, and J. R. Heyada, 1988. Directed band ratioing for the retention of perceptually-independent topographic expression in chromaticity-enhanced imagery, *Int. J. Remote Sensing*, 9, 749–765.

Crósta, A., 1991. High resolution geochemistry and satellite data integration: application to mineral exploration in a tropical environment, in *Proceedings of the 8th Thematic Conference on Geologic Remote Sensing*, Vol. II, Environmental Research Institute of Michigan, Ann Arbor, Mich., pp. 1335–1348.

Crósta, A. P., and J. M. Moore, 1989. Enhancement of *Landsat* thematic mapper imagery for residual soil mapping in SW Minas Gerais State, Brazil: a prospective case history in greenstone belt terrain, in *Proceedings of the 7th Thematic Conference on Remote Sensing for Exploration Geology*, Vol. II, Environmental Research Institute of Michigan, Ann Arbor, Mich., pp. 1173–1187.

Crósta, A. P., and A. Rabelo, 1993. Assessing *Landsat*/TM for hydrothermal alteration mapping in central-western Brazil, in *Proceedings of the 9th Thematic Conference on Geologic Remote Sensing*, Vol. II, Environmental Research Institute of Michigan, Ann Arbor, Mich., pp. 1053–1061.

Crósta, A. P., I. D. M. Prado, and M. Obara, 1996. The use of Geoscan AMSS data for gold exploration in the Rio Itapicurú greenstone belt (BA), Brazil, in *Proceedings of the 11th Thematic Conference on Geologic Remote Sensing*, Vol. II, Environmental Research Institute of Michigan, Ann Arbor, Mich., pp. 205–214.

Crósta, A. P., C. Sabine, and J. V. Taranik, 1998. Hydrothermal alteration mapping at Bodie, California using AVIRIS hyperspectral data, *Remote Sensing Environ.*, 65, 309–319.

Crowley, J. K., and S. J. Hook, 1996. Mapping playa evaporite minerals and associated sediments in Death Valley, California, with multispectral thermal images, *J. Geophys. Res.*, 99B, 643–660.

Crowley, J. K., D. W. Brickey, and L. C. Rowan, 1988. Airborne imaging spectrometer data for the Ruby Mountains, Montana: mineral identification by remote sensing in a vegetated metamorphic terrain, in *Proceedings of the 6th Thematic Conference on Remote Sensing for Exploration Geology*, Environmental Research Institute of Michigan, Ann Arbor, Mich., pp. 225–226.

Curtis, B., and A. G. Maecher, 1991. Changes in forest canopy reflectance associated with chronic exposure to high concentrations of soil trace metals, in *Proceedings of the 8th Thematic Conference on Geologic Remote Sensing*, Vol. I, Environmental Research Institute of Michigan, Ann Arbor, Mich., pp. 337–347.

Davidson, D., B. Bruce, and D. Jones, 1993. Operational remote sensing mineral exploration in a semi-arid environment: the Troodos Massif, Cyprus, in *Proceedings of the 9th Thematic Conference on Geologic Remote Sensing*, Vol. II, Environmental Research Institute of Michigan, Ann Arbor, Mich., pp. 845–859.

Davis, W. P., and R. J. P. Lyon, 1991. Production of alteration maps using airborne high-resolution multi-spectral imagery: Virginia City, Nevada, in *Proceedings of the 8th Thematic Conference on Geologic Remote Sensing*, Vol. I, Environmental Research Institute of Michigan, Ann Arbor, Mich., pp. 521–534.

Davis, P. A., K. F. Mullins, G. L. Berlin, A. M. Al-Farasani, and S. M. Dini, 1989. Phosphorite exploration in the Thaniyat and Sanam districts, Kingdom of Saudi Arabia, using *Landsat* thematic mapper data, in *Proceedings of the 7th Thematic Conference on Remote Sensing for Exploration Geology*, Vol. II, Environmental Research Institute of Michigan, Ann Arbor, Mich., pp. 1205–1221.

Denes, L. J., Gottlieb, M., Kaminsky, B., and Huber, D. F., 1997. A spectro-polarimetric imager for scene discrimination: *Proceedings of the 1997 International Symposium on Spectral Sensing Research*, San Diego, California, p. 39–45.

Dick, L. A., G. Ossandon, R. G. Fitch, C. M. Swift, and A. Watts, 1993. Discovery of blind copper mineralization at Collahuasi, Chile, in *Program and Abstracts; Integrated Methods in Exploration and Discovery*, Society of Economic Geologists, Littleton, Colo., pp. AB21–23.

Drury, S. A., 1993. *Image Interpretation in Geology*, 2nd ed., Chapman & Hall, London, 283 pp.

Duke, E. F., 1994. Near infrared spectra of muscovite, Tschermak substitution, and metamorphic reaction progress: implications for remote sensing, *Geology*, 22, 621–624.

Einaudi, M. T., L. D. Meinert, and R. J. Newberry, 1981. Skarn deposits, in *Economic Geology*, 75th anniv. vol., B. J. Skinner, ed., Economic Geology Publishing, El Paso, Texas, pp. 317–391.

Elvidge, C. D., and R. J. P. Lyon, 1985. Estimation of vegetation contribution to the 1.65/2.22 micrometer ratio in the airborne TM imagery of the Virginia Range, Nevada, *Int. J. Remote Sensing*, 6, 137–155.

Eriksen, A. S., and D. R. Cowan, 1989. Further development and application of GIS technology and geostatistics to assist in the exploration for uranium in western North America, in *Proceedings of the 7th Thematic Conference on Remote Sensing for Exploration Geology*, Vol. II, Environmental Research Institute of Michigan, Ann Arbor, Mich., pp. 1285–1307.

Farrand, W. H., and A. Seelos, 1996. Using mineral maps generated from imaging spectrometer data to map faults: an example from Summitville, Colorado, in *Proceedings of the 11th Thematic Conference on Geologic Remote Sensing*, Vol. II, Environmental Research Institute of Michigan, Ann Arbor, Mich., pp. 222–230.

Farrand, W. H., R. B. Singer, and E. Merényi, 1994. Retrieval of apparent surface reflectance from AVIRIS data: a comparison of empirical line, radiative transfer, and spectral mixture methods, *Remote Sensing Environ.*, 47, 311–321.

Farrand, W., Harsani, J., Vance, L., Cox, T., and Jenssen, R., 1997. Application of ESSI Probe 1 data to mineralogic mapping in *Proceedings of the 12th International Conference on Applied Geologic Remote Sensing*, Vol. I, Environmental Research Institute of Michigan, Ann Arbor, Mich., p. 190.

Feldman, S. C., and J. V. Taranik, 1988a. The use of high-resolution remote sensing techniques to define hydrothermal alteration mineralogy, Tybo mining district, Nevada, in *Bulk Minable Precious Metal Deposits of the Western United States*, R. W. Schafer, J. J. Cooper, and P. G. Vikre, eds., Geological Society of Nevada, Reno, Nev., pp. 531–549.

Feldman, S. C., and J. V. Taranik, 1988b. Comparison of techniques for discriminating hydrothermal alteration minerals with airborne imaging spectrometer data, *Remote Sensing Environ.*, 24, 67–83.

Felzer, B., P. Hauff, and A. F. H. Goetz, 1991. Quantitative reflectance spectroscopy using NH_4 absorption bands for buddingtonite and associated minerals at Cuprite, Nevada, in *Proceedings of the 8th Thematic Conference on Geologic Remote Sensing*, Vol. I, Environmental Research Institute of Michigan, Ann Arbor, Mich., pp. 549–562.

Fernandez-Alonzo, M. and A. Tahon, 1991. Litholigic discrimination and structural trends in W-Rwanda (Africa) on images of airborne radiometric and aeromagnetic surveys coregistered to a *Landsat* TM scene, *Photogramm. Eng. Remote Sensing*, 57, 1155–1162.

Ferreira, L. G. J., and P. R. Meneses, 1994. Discrimination of hydrothermally altered zones through visible–near infrared spectrometry and multispectral image processing, in *Proceedings of the 10th Thematic Conference on Geologic Remote Sensing*, Vol. I, Environmental Research Institute of Michigan, Ann Arbor, Mich., pp. 127–137.

Ford, J. P., R. K. Dokka, R. E. Crippen, and R. G. Blom, 1990. Faults in the Mojave Desert, California, as revealed on enhanced *Landsat* images, *Science*, 245, 1000–1003.

Fraser, S. J., 1991. Discrimination and identification of ferric oxides using satellite thematic mapper data: a Newman case study, *Int. J. Remote Sensing*, 12, 635–641.

Fraser, S. J., and A. A. Green, 1987. A software defoliant for geological analysis of band ratios, *Int. J. Remote Sensing*, 8, 525–532.

Fraser, S. J., A. R. Gabell, A. A. Green, and J. F. Huntington, 1986. Targeting epithermal alteration and gossans in weathered and vegetated terrains using aircraft scanners: successful Australian case histories, in *Proceedings of the 5th Thematic Conference on Geologic Remote Sensing*, Environmental Research Institute of Michigan, Ann Arbor, Mich., pp. 63–84.

Fraser, S. J., C. L. Horsefall, A. R. Gabell, J. F. Huntington, and A. A. Green, 1987. Targeting hydrothermal alteration systems in north Queensland using aircraft thematic mapper data, in *Proceedings of the 4th, Australasian Remote Sensing Conference*, September 14–18, Adelaide, Vol. 1, pp. 340–351.

Frei, M., and S. L. Jutz, 1989. Use of thematic mapper data for the detection of gold bearing formations in the eastern desert of Egypt, in *Proceedings of the 7th Thematic Conference on Remote Sensing for Exploration Geology*, Vol. II, Environmental Research Institute of Michigan, Ann Arbor, Mich., pp. 1157–1172.

Freund, M. J., 1992. Geosat test case: Silver Bell, Arizona: comparison of *JERS-1* simulated data with *Landsat* TM, in *Joint GEOSAT/ERSDAC/JGI/JAPEX Simulation Study*, Geosat Committee, Norman, Okla., pp. 65–70.

Gardiner, J. L., R. Birnie, and H. Zantrop, 1988. A field spectrometer and remote sensing study of the Fresnillo mining district, Mexico, in *Proceedings of the 6th Thematic Conference on Remote Sensing for Exploration Geology*, Environmental Research Institute of Michigan, Ann Arbor, Mich., pp. 229–236.

Gilabert, M. A., C. Conese, and F. Maselli, 1994. An atmospheric correction method for the automatic retrieval of surface reflectances from TM images, *Int. J. Remote Sensing*, 15, 2065–2086.

Gillespie, A. R., 1992. Enhancements of multispectral thermal infrared images: decorrelation contrast stretching, *Remote Sensing Environ.*, 42, 147–156.

Gillespie, A. R., A. B. Kahle, and R. E. Walker, 1986. Color enhancement of highly correlated images: I. Decorrelation and HSI contrast stretches, *Remote Sensing Environ.*, 20, 209–235.

Glikson, A. Y., and J. W. Creasey, 1995. Application of *Landsat-5* TM imagery to mapping of the Giles Complex and associated granulites, Tomkinson Ranges, western Musgaves Block, central Australia, *AGSO J. Aust. Geol. Geophy.*, 16, 173–193.

Goetz, A. F. H., 1989. Spectral remote sensing in geology, in *Theory and Applications of Optical Remote Sensing*, G. Asrar, ed., Wiley, New York, pp. 491–526.

Goetz, A. F. H., and B. Kindel, 1996. Understanding unmixed AVIRIS images in Cuprite, NV, using coincident HYDICE data, in *6th AVIRIS Workshop*.

Goetz, A. F. H., and L. C. Rowan, 1981. Geologic remote sensing, *Science*, 211, 781–791.

Goetz, A. F. H., and V. Srivastava, 1985. Mineralogic mapping in the Cuprite mining district, Nevada, in *Proceedings of the Airborne Imaging Spectrometer Data Analysis Workshop*, JPL Publ. 85–41, Jet Propulsion Laboratory, California Institute of Technology, Pasadena, Calif., pp. 22–31.

Goetz, A. F. H., F. C. Billingsley, A. R. Gillespie, M. J. Abrams, R. L. Squires, E. M. Shoemaker, I. Luccchitta, and D. P. Elston, 1975. *Application of ERTS Images and Image Processing to Regional Geologic Problems and Geologic Mapping in Northern Arizona*, JPL Tech. Rep. 32–2597, Jet Propulsion Laboratory, California Institute of Technology, Pasadena, Calif., 188 pp.

Goetz, A. F. H., L. C. Rowan, and M. J. Kingston, 1982. Mineral identification from orbit: initial results from the shuttle multispectral infrared radiometer, *Science*, 218, 1020–1024.

Gold, D. P., 1980. Structural geology, in *Remote Sensing Geology*, B. S. Siegal and A. R. Gillespie, eds., Wiley, New York, pp. 419–483.

Goossens, M. A., 1991. Integration of remote sensing data and ground data as an aid to exploration for granite related mineralization, Salamanca Province, W-Spain, in *Proceedings of the 8th Thematic Conference on Geologic Remote Sensing*, Vol. I, Environmental Research Institute of Michigan, Ann Arbor, Mich., pp. 393–406.

Graham, D. F., and Bonham-Carter, G. F., 1993. Airborne radiometric data—a tool for reconnaissance geologic mapping using a GIS: *Proceedings of the 9th Thematic Conference on Geologic Remote Sensing*, Vol. I, Environmental Research Institute of Michigan, Ann Arbor, Mich., pp. 43–54.

Green, A. A., and M. D. Craig, 1985. Analysis of aircraft spectrometer data with logarithmic residuals, in *Proceedings of the Airborne Imaging Spectrometer Data Analysis Workshop*, JPL Publ. 85–41, Jet Propulsion Laboratory, California Institute of Technology, Pasadena, Calif., pp. 111–119.

Green, R. O., J. E. Conel, and D. A. Roberts, 1993. Estimation of aerosol optical depth, pressure elevation, water vapor and calculation of apparent surface reflectance from radiance measured by the airborne visible/infrared imaging spectrometer (AVIRIS) using a radiative transfer code, in *Imaging Spectrometry in the Terrestrial Environment*, G. Vane, ed., *Proc. SPIE*, 1937, pp. 2–11.

Green, R. O., C. Chovit, and J. Faust, 1996. In-flight calibration and validation of the airborne visible/infrared imaging spectrometer (AVIRIS) in 1995, in *6th AVIRIS Workshop*.

Guilbert, J. M., and C. F. J. Park, 1986. *The Geology of Ore Deposits*; W. H. Freeman, New York, 985 pp.

Harrington, S. E., 1991. Use of *Landsat* TM data in exploration for ultramafic rock bodies in NW Ontario, Canada (abstract), in *Proceedings of the 8th Thematic Conference on Geologic Remote Sensing*, Vol. II, Environmental Research Institute of Michigan, Ann Arbor, Mich., p. 1123.

Harris, J. R., Rencz, A. N., Ballantyne, B., and Sheridon, C., 1998. Mapping of altered rocks using Landsat TM and lithogeochemical data: *Sulphurets-Bruce Jack Lake District, British Columbia, Canada*: Photogrammetric Engineering and Remote Sensing, Vol. 64, pp. 309–322.

Hauff, P., F. Kruse, R. Madrid, S. Fraser, J. Huntington, M. Jones, and S. Watters, 1991. Illite crystallinity: case histories using x-ray diffraction and reflectance spectroscopy to define ore host environments, in *Proceedings of the 8th Thematic Conference on Geologic Remote Sensing*, Vol. I, Environmental Research Institute of Michigan, Ann Arbor, Mich., pp. 447–458.

Hauff, P., P. Kowalczyk, M. Ehling, G. Borstad, G. Edmundo, R. Kern, R. Neville, R. Marois, S. Perry, R. Bedell, C. Sabine, A. Crosta, T. Miura, G. Lipton, V. Sopuck, R. Chapman, M. Tilkov, K. O'Sullivan, M. Hornibrook, D. Coulter, and S. Bennett, 1996. The CCRS SWIR full spectrum imager: mission to Nevada, June, 1995, in *Proceedings of the 11th Thematic Conference on Geologic Remote Sensing*, Vol. I, Environmental Research Institute of Michigan, Ann Arbor, Mich., pp. 38–47.

Haymon, R. M., R. A. Koski, and M. J. Abrams, 1989. Hydrothermal discharge zones beneath massive sulfide deposits mapped in the Oman ophiolite, *Geology*, 17, 531–535.

Hernandes, G. L. S., and A. P. Crosta, 1994. Realce espectral de imagens Geoscan Mk-II para o mapeamento de minerais de alteraco hidrotermal na area do Deposito Aurifero de Riacho dos Machados (MG, Brazil), in *Proceedings, 38 Congresso Brasiliero de Geologia*, Vol. I, pp. 476–477.

Hook, S. J., A. R. Gabell, A. A. Green, P. S. Kealy, and A. B. Kahle, 1990. A comparison of the model emittance, thermal log residual and alpha residual techniques using TIMS data acquired over Cuprite, Nevada, in *Proceedings of the 2nd Thermal Infrared Multispectral Scanner (TIMS) Workshop*, JPL Publ. 90–55, Jet Propulsion Laboratory, California Institute of Technology, Pasadena, Calif., pp. 20–25.

Hook, S. J., C. D. Elvidge, M. Rast, and H. Watanabe, 1991. An evaluation of short-wave-infrared (SWIR) data from the AVIRIS and Geoscan instruments for mineralogical mapping at Cuprite, Nevada, *Geophysics*, 56, 1432–1440.

Hook, S. J., K. E. Karlstrom, C. F. Miller, and K. J. W. McCaffrey, 1994. Mapping the Piute Mountains, California, with thermal infrared multispectral scanner (TIMS) images, *J. Geophys. Res.*, 99B, 15605–15622.

Horler, D. N. H., J. Barber, and A. Barringer, 1980. Effects of heavy metals on the absorbance and reflectance spectra of plants, *Int. J. Remote Sensing*, 1, 121–136.

Horler, D. N. H., M. Dockray, and J. Barber, 1983. The red edge of plant leaf reflectance, *Int. J. Remote Sensing*, 4, 273–288.

Hudson, D. M., 1986. *The Comstock District, Storey County, Nevada*, Field Trip Guideb. 4, Geological Society of Nevada, Reno, Nev. 13 pp.

Hutsinpiller, A., 1988. Discrimination of hydrothermal alteration mineral assemblages at Virginia City, Nevada, using the airborne imaging spectrometer, *Remote Sensing Environ.*, 24, 53–66.

Jiang, D., P. Wang, and F. Meng, 1994. Application of *Landsat* TM data into exploration for porphyry copper deposits in forested area, in *Proceedings of the 10th Thematic Conference on Geologic Remote Sensing*, Vol. II, Environmental Research Institute of Michigan, Ann Arbor, Mich., pp. 611–618.

Kahle, A. B., and L. C. Rowan, 1980. Evaluation of multispectral middle infrared aircraft images for lithologic mapping in the East Tintic Mountains, Utah, *Geology*, 8, 234–239.

Kahle, A. B., D. P. Madura, and J. M. Soha, 1980. Middle infrared multispectral aircraft scanner data: analysis for geologic applications, *Appl. Opt.*, 19, 2279–2290.

Kahle, A. B., and A. F. H. Goetz, 1983. Mineralogic information from a new airborne thermal infrared multispectral, scanner, *Science*, 222, 24–27.

Kealy, P. S., and A. R. Gabell, 1990. Estimation of emissivity and temperature using alpha coefficients, in *Proceedings of the 2nd Thermal Infrared Multispectral Scanner (TIMS) Workshop*, JPL Publ. 90–55, Jet Propulsion Laboratory, California Institute of Technology, Pasadena, Calif., pp. 11–15.

Kepper, J. C., T. P. Lugaski, and J. S. MacDonald, 1986. Discrimination of lithologic units, alteration patterns and major structural blocks in the Tonopah, Nevada, area using thematic mapper data, in *Proceedings of the 5th Thematic Conference on Geologic Remote Sensing*, Environmental Research Institute of Michigan, Ann Arbor, Mich., pp. 97–115.

Kierein-Young, K. S., and F. A. Kruse, 1989. Comparison of *Landsat* thematic mapper images and Geophysical and Environmental Research imaging spectrometer data for alteration mapping, in *Proceedings of the 7th Thematic Conference on Remote Sensing for Exploration Geology*, Vol. I, Environmental Research Institute of Michigan, Ann Arbor, Mich., pp. 349–359.

Kingston, M. J., 1990. Geologic mapping from airborne visible and near-infrared band-depth images, Mountain Pass, California, in *Proceedings of the 10th Annual International Geoscience and Remote Sensing Symposium*, IGARSS'90, Vol. III, pp. 1703–1706.

Knepper, D. H. J., 1993. Satellite image processing for enhanced spectral discrimination and interpretability, in *Advances Related to United States and International Mineral Resources: Developing Frameworks and Exploration Technologies*, USGS Bull. 2039, U.S. Geological Survey, Washington, D.C., pp. 151–153.

Kowalczyk, P., and M. Ehling, 1991. Analysis of TM images using the IHS transform and adaptive equalization, in *Proceedings of the 8th Thematic Conference on Geologic Remote Sensing*, Vol. I, Environmental Research Institute of Michigan, Ann Arbor, Mich., pp. 207–214.

Kowalczyk, P., and K. Logan, 1989. TM processing for routine use in mineral exploration, in *Proceedings of the 7th Thematic Conference on Remote Sensing for Exploration Geology*, Vol. I, Environmental Research Institute of Michigan, Ann Arbor, Mich., pp. 323–329.

Krohn, M. D., 1989. *Preliminary Description of a Mineral-Bound Ammonium Lo-*

cality in the Cedar Mountains, Esmeralda County, Nevada, USGS Open File Rep. 89–637, U.S. Geological Survey, Washington, D.C.

Krohn, M. D., and S. P. Altaner, 1987. Near-infrared detection of ammonium minerals, *Geophysics*, 52, 924–930.

Krohn, M. D., S. P. Altaner, and D. O. Hayba, 1988. Distribution of ammonium minerals at Hg/Au hot spring deposits: initial evidence from near-infrared spectral properties, in *Bulk Minable Precious Metal Deposits of the Western United States*, Geological Society of Nevada, Reno, Nev., pp. 661–679.

Kruse, F. A., 1988. Use of airborne imaging spectrometer data to map minerals associated with hydrothermally altered rocks in the northern Grapevine Mountains, Nevada and California, *Remote Sensing Environ.*, 24, 31–51.

Kruse, F. A., and J. F. Huntington, 1996. The 1995 AVIRIS geology group shoot, in *6th AVIRIS Workshop*. See JPL AVIRIS web page, http://makalu.jpl.nasa.gov/docs/workshops/96_docs/toc.htm

Kruse, F. A., K. S. Kierein-Young, and J. W. Boardman, 1990. Mineral mapping at Cuprite, Nevada with a 63-channel imaging spectrometer, *Photogramm. Eng. Remote Sensing*, 56, 83–92.

Kruse, F. A., A. B. Lefkoff, J. W. Boardman, K. B. Heidebrecht, A. T. Shapiro, P. J. Barloon, and A. F. H. Goetz, 1993. The spectral image processing system (SIPS): interactive visualization and analysis of imaging spectrometer data, *Remote Sensing Environ.*, 44, 145–163.

Kruse, F. A., 1997. Characterization of active hot-springs environments using multispectral and hyperspectral remote sensing *in Proceedings of the 12th International Conference on Applied Geologic Remote Sensing*, Vol. I, Environmental Research Institute of Michigan, Ann Arbor, Mich., pp. 214–221.

Kruse, F. A. and Boardman, J. W., 1997. Characterizing and mapping of kimberlites and related diatremes in Utah, Colorado, and Wyoming, USA, using the Airborne Visible/Infrared Imaging Spectrometer (AVIRIS) *in Proceedings of the 12th International Conference on Applied Geologic Remote Sensing*, Vol. I, Environmental Research Institute of Michigan, Ann Arbor, Mich., pp. 21–28.

Labovitz, M. L., E. J. Masouka, R. Bell, A. W. Siegrist, and R. F. Nelson, 1983. The application of remote sensing to geobotanical exploration for metal sulfides: results from the 1980 field season at Mineral, Virginia, *Econ. Geol.*, 78, 750–760.

Lahren, M. M., R. A. Schweickert, and J. V. Taranik, 1988. Analysis of the northern Sierra accreted terrane, California, with airborne thermal infrared multispectral scanner data, *Geology*, 16, 525–528.

Lamb, A. D., and N. E. Pendock, 1989. Band prediction techniques for the mapping of hydrothermal alteration, in *Proceedings of the 7th Thematic Conference on Remote Sensing for Exploration Geology*, Vol. II, Environmental Research Institute of Michigan, Ann Arbor, Mich., pp. 1317–1329.

Larson, L. T., T. P. Lugaski, and W. Aymard, 1989. Integration of geological, geophysical and thematic mapper remote sensing data in relation to the geologic occurrence of precious and base metal deposits in the Ely–Hamilton–Eureka, Nevada area, in *Proceedings of the 7th Thematic Conference on Remote Sensing for Exploration Geology*, Vol. II, Environmental Research Institute of Michigan, Ann Arbor, Mich., pp. 1309–1316.

Legg, C. A., 1992. *Remote Sensing and Geographic Information Systems: Geologic Mapping, Mineral Exploration and Mining*, Ellis Horwood, Chichester, West Sussex, England, 256 pp.

Loughlin, W. P., 1991. Principal component analysis for alteration mapping, *Photogramm. Eng. Remote Sensing*, 57, 1163–1170.

Lyon, R. J. P., and F. R. Honey, 1989a. Relating ground mineralogy via spectral signatures to 18-channel airborne imagery obtained with the Geoscan MkII advanced scanner: a 1989 case history from the Leonora, western Australia, gold district, in *Proceedings of the 7th Thematic Conference on Remote Sensing for Exploration Geology*, Vol. I, Environmental Research Institute of Michigan, Ann Arbor, Mich., pp. 331–348.

Lyon, R. J. P., and F. R. Honey, 1989b. Spectral signature extraction from airborne imagery using the Geoscan MkII advanced airborne scanner in the Leonora, Western Australia, gold district, in *Proceedings of the International Geophysics and Remote Sensing Symposium*, IGARSS'89), Vancouver, British Columbia, Canada, pp. 2925–2930.

Lyon, R. J. P., and F. R. Honey, 1991. Extraction of thermal exitance spectra (TIR) from Geoscan MkII scanner imagery: spectra derived from airborne (mobile) and ground (stationary) imagery compared with spectra from direct exitance spectrometry, in *Proceedings of the 8th Thematic Conference on Geologic Remote Sensing*, Vol. I, Environmental Research Institute of Michigan, Ann Arbor, Mich., pp. 117–130.

Ma, J., V. R. Slaney, J. R. Harriss, D. F. Graham, S. B. Ballantyne, and D. C. Harris, 1991. Use of *Landsat* TM data for mapping of limonitic and altered rocks in the Sulpherets area, British Columbia, in *Proceedings of the 14th Canadian Symposium on Remote Sensing*, Calgary, Alberta, Canada, pp. 419–422.

MacDonald, J. A., and F. P. Miranda, 1991. Application of the semivariogram textural classifier (STC) for terrane recognition using *Magellan* data: preliminary results, in *Proceedings of the 22nd Lunar and Planetary Science Conference*, Lunar and Planetary Science Institute, Houston, Texas, pp. 839–840.

MacDonald, J. A., F. P. Miranda, and J. R. Carr, 1990. Textural image classification using variograms, *Proc. SPIE*, 1301, 25–39.

Macias, L. F., 1995. Remote sensing of mafic–ultramafic rocks: examples from Australian Precambrian terranes, *AGSO J. Aust. Geol. Geophys.*, 16, 163–171.

Marsh, S. E., and J. B. McKeon, 1983. Integrated analysis of high-resolution field and airborne spectroradiometer data for alteration mapping, *Econ. Geol.*, 78, 618–632.

Marsh, S. H., 1993. Discrimination between sulphate, carbonate and hydrosilicate minerals using *Landsat* thematic mapper data, in *Proceedings of the 9th Thematic Conference on Geologic Remote Sensing*, Vol. II, Environmental Research Institute of Michigan, Ann Arbor, Mich., pp. 677–685.

Martinez-Alonzo, S. E., Atkinson, W. W., Goetz, A. F. H., and Kruse, F. A., 1997. Short wave infrared (SWIR) spectrometry of illite ("sericite") to estimate temperature of formation of hydrothermal mineral deposits *in Proceedings of the 12th International Conference on Applied Geologic Remote Sensing*, Vol. II, Environmental Research Institute of Michigan, Ann Arbor, Mich, pp. 426–429.

Masinter, R. A., and R. J. P. Lyon, 1991. Spectroscopic confirmation of increasing illite ordering with hydrothermal alteration of argillaceous ore in the Gold Bar mine, Eureka Co., Nevada, in *Proceedings of the 8th Thematic Conference on Geologic Remote Sensing*, Vol. I, Environmental Research Institute of Michigan, Ann Arbor, Mich., pp. 563–571.

Meyer, C., and J. J. Hemley, 1967. Wallrock alteration, in *Geochemistry of Hydrothermal Ore Deposits*, H. L. Barnes, ed., Holt, Rinehart and Winston, New York, 670p.

Milton, N. M., 1983. Use of reflectance spectra of native plant species for interpreting airborne multispectral scanner data in the East Tintic Mountains, Utah, *Econ. Geol.*, 78, 761–769.

Milton, N. M., W. Collins, S. H. Chang, and R. G. Schmidt, 1983. Remote detection of metal anomalies on Pilot Mountain, Randolph County, North Carolina, *Econ. Geol.*, 78, 605–617.

Milton, N. M., M. S. Power, and C. M. Ager, 1988. *Spectral Effects of Heavy Metals on Greenhouse-Grown Hosta ventricosa*, USGS Open File Rep. 88-57, U.S. Geological Survey, Washington, D.C., 13 pp.

Milton, N. M., C. M. Ager, B. A. Eiswerth, and M. S. Power, 1989. Arsenic- and selenium-induced changes in spectral reflectance and morphology of soybean plants, *Remote Sensing Environ.*, 30, 263–269.

Milton, N. M., B. A. Eiswerth, and C. M. Ager, 1991. Effect of phosphorus deficiency on spectral reflectance and morphology of soybean plants, *Remote Sensing Environ.*, 36, 121–127.

Miranda, F. P., and J. R. Carr, 1994. Application of the semivariogram textural classifier (STC) for vegetation discrimination using SIR-B data of the Guiana Shield, northwestern Brazil, *Remote Sensing Rev.*, 10, 155–168.

Miranda, F. P., and J. A. MacDonald, 1989. A variogram study of SIR-B data in the Guiana Shield, Brazil, in *Image Processing '89*, American Society for Photogrammetry and Remote Sensing, Falls Church, Va., pp. 66–77.

Miranda, F. P., J. A. MacDonald, and J. R. Carr, 1992. Application of the semivariogram textural classifier (STC) for vegetation discrimination using SIR-B data of Borneo, *Int. J. Remote Sensing*, 13, 2349–2354.

Miranda, F. P., A. E. McCafferty, and J. V. Taranik, 1994. Reconnaissance geologic mapping of a portion of the rain-forest-covered Guiana Shield, northwestern Brazil, using SIR-B and digital aeromagnetic data; *Geophysics*, 59, 733–743.

Mitchell, A. H. G., and T. M. Leach, 1991. *Epithermal Gold in the Philippines: Island Arc Metallogenesis, Geothermal Systems and Geology*; Academic Press, London, 457 p.

Mouat, D. A., J. S. Myers, and N. M. Milton, 1986, An integrated approach to the use of *Landsat* TM data for gold exploration in west-central Nevada, in *Proceedings of the 5th Thematic Conference on Geologic Remote Sensing*, Environmental Research Institute of Michigan, Ann Arbor, Mich., pp. 431–443.

Mouat, D., J. Lancaster, N. Milton, M. Chesley, J. Hatzell, W. Lassetter, and J. Picone, 1994. Effects of arsenic on spectral reflectance of greenhouse-grown soybean and loblolly pine, in *Proceedings of the 10th Thematic Conference on Geologic Remote Sensing*, Vol. I, Environmental Research Institute of Michigan, Ann Arbor, Mich., pp. 49–59.

Nash, G., and R. McCoy, 1993. The use of high resolution spectroscopy in the determination of geobotanical anomalies associated with gold placers, La Sal Mountains area, Utah, in *Proceedings of the 9th Thematic Conference on Geologic Remote Sensing*, Vol. II, Environmental Research Institute of Michigan, Ann Arbor, Mich., pp. 767–777.

Nash, G. D., and P. M. Wright, 1994. The use of simple GIS techniques to create improved hydrothermal alteration maps from TM data, Steamboat Springs and Virginia City quadrangles, Nevada, USA, in *Proceedings of the 10th Thematic Conference on Geologic Remote Sensing*, Vol. II; Environmental Research Institute of Michigan, Ann Arbor, Mich., pp. 132–141.

Neville, R. A., N. Rowlands, R. Marois, and I. Powell, 1996. The CCRS SWIR full spectrum imager (SFSI), in *Proceedings of the 11th Thematic Conference on Geologic Remote Sensing*, Vol. I; Environmental Research Institute of Michigan, Nusbaum et al 1997a 1997b Ann Arbor, Mich., pp. 417–425.

Nusbaum, R., Weiner, M., and Coombs, C., 1997a. A GIS used to correlate hydrothermal alteration maps from AVIRIS data with geology, Goldfield, Nevada, USA in *Proceedings of the 12th International Conference on Applied Geologic Remote Sensing*, Vol. I, Environmental Research Institute of Michigan, Ann Arbor, Mich., pp. 251–257.

Nusbaum, R., Nettles, J., and Stearns, S., 1997b. Imaging spectroscopy used for geologic mapping of the Tintic Mining District, Utah, USA in *Proceedings of the 12th International Conference on Applied Geologic Remote Sensing*, Vol. II, Environmental Research Institute of Michigan, Ann Arbor, Mich., pp. 438–442.

Palluconi, F. D., and G. R. Meeks, 1985. *Thermal Infrared Multispectral Scanner (TIMS): An Investigator's Guide to TIMS Data*; JPL Publ. 85–32; Jet Propulsion Laboratory, California Institute of Technology, Pasadena, Calif., 14 pp.

Payás, A., R. G. Schmidt, and A. Bonet, 1993. The effects of hydrothermally altered bedrock on natural forest vegetation in the Snow Camp–Saxapahaw area, North Carolina, and the resulting expressions in *Landsat* TM imagery, in *Advances Related to United States and International Mineral Resources: Developing Frameworks and Exploration Technologies*; USGS Bull. 2039, U.S. Geological Survey, Washington, D.C., pp. 189–200.

Pendock, N., A. A. De Gasparis, and M. A. Brown, 1991. Neural net identification of mineral spectra: a case study using airborne multispectral imagery over Cuprite, Nevada, in *Proceedings of the 8th Thematic Conference on Geologic Remote Sensing*, Vol. I; Environmental Research Institute of Michigan, Ann Arbor, Mich., pp. 573–586.

Peters, D. C., 1983. Use of airborne multispectral scanner data to map alteration related to roll-front uranium migration; *Econ. Geol.*, 78, 641–653.

Podwysocki, M. H., D. B. Segal, and M. J. Abrams, 1983. Use of multispectral scanner images for assessment of hydrothermal alteration in the Marysville, Utah, mining area; *Econ. Geol.*, 78, 675–687.

Podwysocki, M. H., M. Power, J. Salisbury, and O. Jones, 1984. Evaluation of low-sun illumination *Landsat-4* thematic mapper data for mapping hydrothermally altered rocks in southern Nevada, in *Proceedings of the 3rd Thematic Conference on Remote Sensing for Exploration Geology*; Environmental Research Institute of Michigan, Ann Arbor, Mich., pp. 317–318.

Podwysocki, M. H., W. J. Ehman, and D. W. Brickey, 1987. Application of combined *Landsat* TM and airborne TIMS scanner data to lithologic mapping in Nevada, in *Proceedings of the 21st International Symposium on Remote Sensing of the Environment*; Environmental Research Institute of Michigan, Ann Arbor, Mich., pp. 317–318.

Prado, I. D. M., and A. P. Crósta, 1994. Análise de principais componentes no mapeamento da alteração hidrotermal no Greenstone Belt do Rio Itapicuru (BA,

Brazil) utilizando imagems multiespectrais aeroportadas Geoscan MK-II, in *Proceedings, 38 Congresso Brasiliero de Geologia*, Vol. I, pp. 444–446.

Premmanee, J., 1989. Lineament mapping from *Landsat* TM and airborne C-SAR of part of the Sudbury structure, in *Proceedings of the 7th Thematic Conference on Remote Sensing for Exploration Geology*, Vol. II; Environmental Research Institute of Michigan, Ann Arbor, Mich., pp. 997–1010.

Raines, G. L., and F. C. Canney, 1980. Vegetation and geology, in *Remote Sensing in Geology*; B. S. Siegal, and A. R. Gillespie, eds., Wiley, New York, pp. 365–380.

Raines, G. L., T. W. Offield, and E. S. Santos, 1978. Remote sensing and subsurface definition of facies and structure related to uranium deposits, Powder River Basin, Wyoming; *Econ. Geol.*, 73, 1706–1723.

Rast, M., S. J. Hook, C. D. Elvidge, and R. E. Alley, 1991. An evaluation of techniques for the extraction of mineral absorption features from high spectral resolution remote sensing data; *Photogramm. Eng. Remote Sensing*, 57, 1303–1309.

Realmuto, V. J., 1990. Separating the effects of temperature and emissivity: emissivity spectrum normalization, in *Proceedings of the 2nd Thermal Infrared Multispectral Scanner (TIMS) Workshop*, JPL Publ. 90–55, Jet Propulsion Laboratory, California Institute of Technology, Pasadena, Calif., pp. 31–36.

Re Kühl, G. E., 1992. Landsat Thematic Mapper band ratio discrimination of altered and unaltered volcanic rocks in the Andes of Argentina, Bolivia and Chile; unpublished M. S. thesis, Cornell University, Ithaca, N.Y., 169 pp.

Rencz, A. N., and D. F. Sangster, 1989. Characterizing vegetation response to an alteration halo using *Landsat* thematic mapper imagery and aeromagnetic data in the Murdochville area, Gaspésie, Québec, in *Current Research, Part B*, GSC Pap. 90-1B, Geological Survey of Canada, Ottawa, Ontario, Canada, pp. 91–94.

Rencz, A. N., Bowie, C., and Ward, B., 1996. Application of thermal imagery from Landsat data to identify kimberlites, Lac de Gras area, District of Mackenzie, N.W.T. in *Searching for Diamonds in Canada, LeChaimant, A. N., Richardson, D. G., DiLabio, R. N. W., and Richardson, K. A., Eds.*, Geological Survey of Canada, Open File 3228, pp. 255–257.

Rencz, A. N., J. R. Harris, and S. B. Ballantyne, 1994. *Landsat* TM imagery for alteration identification, in *Current Research 1994-E*, Geological Survey of Canada, Ottawa, Ontario, Canada, pp. 277–282.

Rencz, A. N., and Watson, G. P., 1988. Statistical relationship of mineral occurrences with geological and Landsat-derived lineaments, northeastern New Brunswick: in *Current Research*, Part B. Geological Survey of Canada, Paper 88-1B, p. 245–250.

Roberts, D. A., Y. Yamaguchi, and R. J. P. Lyon, 1985. Calibration of airborne imaging spectrometer data to percent reflectance using field spectral measurements, in *Proceedings of the 19th International Symposium on Remote Sensing of Environment*; Environmental Research Institute of Michigan, Ann Arbor, Mich.

Rockwell, B. W., 1989. Hydrothermal alteration mapping in spectral ratio feature space using TM reflectance data: Aurora mining district, Mineral County, Nevada, in *Proceedings of the 7th Thematic Conference on Remote Sensing for Exploration Geology*, Vol. II; Environmental Research Institute of Michigan, Ann Arbor, Mich., pp. 1189–1203.

Rockwell, B. W., 1991. Evaluation of GERIS airborne spectrometer data analysis for disseminated gold exploration: a case study from the Getchell Trend/Potosi mining district, Nevada, USA, in *Proceedings of the 8th Thematic Conference on Geologic Remote Sensing*, Vol. II, Environmental Research Institute of Michigan, Ann Arbor, Mich., pp. 837–850.

Rodriguez, E. P., and C. E. Glass, 1991. Digital analysis for mineral exploration at the Puerto Libertad area, Sonora, Mexico, in *Proceedings of the 8th Thematic Conference on Geologic Remote Sensing*, Vol. II, Environmental Research Institute of Michigan, Ann Arbor, Mich., pp. 851–862.

Rowan, L. C., and T. L. Bowers, 1995. Analysis of linear features mapped in *Landsat* thematic mapper and side-looking radar images of the Reno, Nevada–California $1° \times 2°$ quadrangle: implications for mineral resource studies; *Photogramm. Eng. Remote Sensing*, 61, 749–759.

Rowan, L. C., and A. B. Kahle, 1982. Evaluation of 0.46-to 2.36-μm multispectral scanner images of the East Tintic mining district, Utah, for mapping hydrothermally altered rocks; *Econ. Geol.*, 77, 441–452.

Rowan, L. C., and P. H., Wetlaufer, 1981. Relation between regional lineament systems and structural zones in Nevada; *Am. Assoc. Pet. Geol. Bull.*, 65, 1414–1432.

Rowan, L. C., P. H. Wetlaufer, A. F. H. Goetz, F. C. Billingsley, and J. H. Stewart, 1974. *Discrimination of Rock Types and Detection of Hydrothermally Altered Areas in South-Central Nevada by Use of Computer-Enhanced ERTS Images*, USGS Prof. Pap. 883, U.S. Geological Survey, Washington, D.C., 35 p.

Rowan, L. C., A. F. H. Goetz, and R. P. Ashley, 1977. Discrimination of hydrothermally and unaltered rocks in visible and near-infrared images; *Geophysics*, 42, 522–535.

Rowan, L. C., C. Anton-Pacheco, D. W. Brickey, M. J. Kingston, A. Payas, N. Vergo, and J. K. Crowley, 1987. Digital classification of contact metamorphic rocks in Extremadura, Spain using *Landsat* thematic mapper data; *Geophysics*, 52, 885–897.

Rowan, L. C., K. Watson, J. K. Crowley, C. Anton-Pacheco, P. Gumiel, M. J. Kingston, S. H. Miller, and T. L. Bowers, 1993. Mapping lithologies in the Iron Hill, Colorado, carbonatite–alkalic igneous rock complex using thermal infrared multispectral scanner and airborne visible–infrared imaging spectrometer data, in *Proceedings of the 9th Thematic Conference on Geologic Remote Sensing*, Vol. I; Environmental Research Institute of Michigan, Ann Arbor, Mich., pp. 195–197.

Rowan, L. C., R. N. Clark, and R. O. Green, 1996. Mapping minerals in the Mountain Pass, California area using the airborne visible/infrared imaging spectrometer (AVIRIS), in *Proceedings of the 11th Thematic Conference on Geologic Remote Sensing*, Vol. I, Environmental Research Institute of Michigan, Ann Arbor, Mich., pp. 175–176.

Rubin, T., 1989a. Correlation of imaging spectrometer and ground data for alteration mapping at Yerington, Nevada, in *Proceedings, of the 7th Thematic Conference on Remote Sensing for Exploration Geology*, Vol. I, Environmental Research Institute of Michigan, Ann Arbor, Mich., pp. 315–322.

Rubin, T., 1989b. Analysis of radar image texture with variograms and other simplified descriptors, in *Image Processing '89*, American Society for Photogrammetry and Remote Sensing, Falls Church, Va., pp. 185–195.

Rubin, T. D., 1991. Spectral alteration mapping with imaging spectrometers, in *Proceedings of the 8th Thematic Conference on Geologic Remote Sensing*, Vol. I, Environmental Research Institute of Michigan, Ann Arbor, Mich., pp. 13–25.

Sabine, C., and G. E. Re Kühl, 1996. Lineament analysis of airborne thematic mapper imagery, northwestern Carlin Trend, Elko and Eureka Counties, Nevada, *in Proceedings of the 11th Thematic Conference on Geologic Remote Sensing*, Vol. I, Environmental Research Institute of Michigan, Ann Arbor, Mich., pp. 404–410.

Sabine, C., V. J., Realmuto, and J. V., Taranik, 1994. Quantitative estimation of granitoid composition from thermal infrared multispectral scanner (TIMS) data, Desolation Wilderness, northern Sierra Nevada, California, *J. Geophys. Res.*, 99 (B3), 4261–4271.

Sabins, F. F., 1987. *Remote Sensing: Principles and Interpretation*, 2nd ed., W. H. Freeman, New York, 449 pp.

Sabins, F. F., and R. M., Miller, 1994. Resource assessment: Salar de Uyuni and vicinity, Bolivia, in *Proceedings of the 10th Thematic Conference on Geologic Remote Sensing*, Vol. I, Environmental Research Institute of Michigan, Ann Arbor, Mich., pp. 92–103.

Segal, D. B., 1983. Use of *Landsat* multispectral scanner data for the definition of limonitic exposures in heavily vegetated areas, *Econ. Geol.*, 78, 711–722.

Singhroy, V. H., and F. A. Kruse, 1991. Detection of metal stress in boreal forest species using the 0.67 μm chlorophyll absorption band, in *Proceedings of the 8th Thematic Conference on Geologic Remote Sensing*, Vol. I, Environmental Research Institute of Michigan, Ann Arbor, Mich., pp. 361–372.

Singhroy, V. H., P. D., Lowman, and C. R., Morasse, 1993. Preliminary analysis of *ERS-1* SAR for structural and surficial mapping in the Sudbury basin, Canada, unpublished manuscript presented at the 9th Thematic Conference.

Spatz, D. M., 1992. Remote sensing applied to porphyry copper exploration, in *Porphyry Copper Model Short Course*, Northwest Mining Association, Spokane, Wash.

Spatz, D. M., 1996. Geologic features and remote sensing characteristics of the precious metal systems of the American Cordillera, in *Ore Deposits of the American Cordillera*, Geological Society of Nevada, Reno Nev., pp. 783–802.

Spatz, D. M., and W., Aymard, 1991. Airborne scanner (GERIS) imagery of the Delamar gold mining district, Lincoln County, Nevada, in *Proceedings, Technical Papers, GIS/LIS Fall Convention*, pp. B127–B133.

Spatz, D. M., and J. V. Taranik, 1994. Exploration for copper–molybdenum–gold porphyry deposits using multispectral and hyperspectral aerospace remote sensing techniques, in *Proceedings of the International Society for Photogrammetry and Remote Sensing*, Commission VII Symposium on Resource and Environmental Monitoring, Vol. 30, Part 7a, Rio de Janeiro, Brazil, National Institute of Space Research, pp. 460–473.

Spatz, D. M., and R. T., Wilson, 1994. Exploration remote sensing for porphyry copper deposits, western America cordillera, in *Proceedings of the 10th Thematic Conference on Geologic Remote Sensing*, Vol. I, Environmental Research Institute of Michigan, Ann Arbor, Mich., pp. 227–240.

Spatz, D. M., and Wilson, R. T., 1997. Remote sensing characteristics of the volcanic-associated massive sulfide systems in *Proceedings of the 12th International Conference on Applied Geologic Remote Sensing*, Vol. I, Environmental Research Institute of Michigan, Ann Arbor, Mich., pp. 1–12.

Swayze, G. A., Clark, R. N., Sutley, S., and Gallagher, A., 1992. Ground-truthing AVIRIS mineral mapping at Cuprite, Nevada: *Summaries of the Third Annual*

JPL Airborne Geosciences Workshop, Vol. 1, AVIRIS Workshop. JPL Publication 92–14, p. 47–49.

Tanaka, S. M., and D. B., Segal, 1989. Integrated remote sensing/vector-based GIS technology for gold exploration, Round Mountain district, Nevada, USA, in *Proceedings of the 7th Thematic Conference on Remote Sensing for Exploration Geology*, Vol. II, Environmental Research Institute of Michigan, Ann Arbor, Mich., pp. 1269–1283.

Taranik, D. L., F. A., Kruse, A. F. H., Goetz, and W. W., Atkinson, 1991. Remote sensing of ferric iron minerals as guides for gold exploration, in *Proceedings of the 8th Thematic Conference on Geologic Remote Sensing*, Vol. I, Environmental Research Institute of Michigan, Ann Arbor, Mich., pp. 197–205.

Taranik, J. V., 1988. Application of aerospace remote sensing technology to exploration for precious metal deposits in the western United States, in *Bulk Minable Precious Metal Deposits of the Western United States*, Geological Society of Nevada, Reno, Nev., pp. 551–576.

Thompson, D. G., Arnold, D. V., Long, D. G., Miner, G. F., Jensen, M. A., Karlinsey, T. W., Robertson, A. E., and Bates, J. S., 1998. YSAR and YINSAR: Compact, low-cost synthetic aperture radars: *Proceedings of the European Conference on Synthetic Aperture Radar*, Friedrichshafen, Germany, pp. 27–30.

Thompson, G. A., 1956. *Geology of the Virginia City Quadrangle, Nevada*, USGS Bull. 1042-C, U.S. Geological Survey Washington, D.C., 75 pp.

Torres, V., R. K. Vincent, and P. J. Etzler, 1989. Integrated mineral exploration in northern Mexico using *Landsat*, aerial photography and ground work, in *Proceedings of the 7th Thematic Conference on Remote Sensing for Exploration Geology*, Vol. II, Environmental Research Institute of Michigan, Ann Arbor, Mich., pp. 1227–1237.

Trefois, P., C., Volon, and M., Zaki, 1993. Targeting cobalt mineralisations by remote sensing in the district of Bou-Azzer El Graara (Anti Atlas, Morocco), in *Proceedings of the 9th Thematic Conference on Geologic Remote Sensing*, Vol. I, Environmental Research Institute of Michigan, Ann Arbor, Mich., pp. 183–194.

Van den Bosch, J. M., and R. E. Alley, 1990. Application of *Lowtran 7* as an atmospheric correction to airborne visible/infrared imaging spectrometer (AVIRIS) data, in *Proceedings of the International Geophysical and Remote Sensing Symposium*, (IGARSS'90), College Park, Md., pp. 175–177.

Wagner, M. J., 1995. Back to the future with *ERS-1, Earth Observ. Mag.*, 4 (1), 25–28.

Wang, L., and D. C., He, 1990. A new statistical approach for texture analysis, *Photogramm. Eng. Remote Sensing*, 56, 61–66.

Watson, K., 1992a. Spectral ratio method for measuring emissivity, *Remote Sensing Environ.*, 42, 113–116.

Watson, K., 1992b. Two-temperature method for measuring emissivity, *Remote Sensing Environ.*, 42, 117–121.

Watson, K., F., Kruse, and S., Hummer-Miller, 1990. Thermal infrared exploration in the Carlin Trend, northern Nevada, *Geophysics*, 55, 70–79.

Windeler, D. S., 1993. Garnet–pyroxene alteration mapping in the Ludwig scarn (Yerington, NV) with high-resolution multispectral data, in *Proceedings of the 9th Thematic Conference on Geologic Remote Sensing*, Vol. I, Environmental Research Institute of Michigan, Ann Arbor, Mich., pp. 139–150.

Windeler, D. S., and R. J. P. Lyon, 1991. Discriminating dolomitization of marble in the Ludwig scarn near Yerington, Nevada, using high-resolution airborne infrared imagery, *Photogramm. Eng. Remote Sensing*, 57, 1171–1178.

Yamaguchi, Y., M. Urai, and F. R. Honey, 1989. Application of Geoscan AMSS MkI data to lithologic mapping in Queensland, Australia, in *Proceedings of the 7th Thematic Conference on Remote Sensing for Exploration Geology*, Vol. I, Environmental Research Institute of Michigan, Ann Arbor, Mich., pp. 395–409.

Yamaguchi, Y., S. Tsuchida, T. Matsunaga, and R. J. P. Lyon, 1994. Evaluation of *JERS-1/OPS* data for lithologic mapping in Yerington, Nevada, in *Proceedings of the 10th Thematic Conference on Geologic Remote Sensing*, Vol. II, Environmental Research Institute of Michigan, Ann Arbor, Mich., pp. 91–96.

Hydrocarbon Exploration

J. L. Berry

Shell Exporation
Production Technology Co.
Houston, Texas

G. L. Prost

Gulf Canada
Denver, Colorado

9.1 INTRODUCTION

This chapter is built around the unifying themes of the spatial content and the spectral content of the imagery. Spatial content includes the morphology of the ground and the objects on it, as well as the texture of the Earth's surface as seen on the imagery. Both of these aspects of remotely sensed imagery are used in almost all applications of remote sensing to hydrocarbon (oil and gas) exploration, but one of them usually dominates in any given application. Morphology has been dominant in studies of geologic structure and in logistical analyses, which are probably the two most common applications of remote sensing in the hydrocarbon industry. It also dominates in the use of imagery to detect and evaluate offshore oil slicks caused by natural seepage: this has been an important application of remote sensing in the last few years. The spectral content of imagery has more varied but less frequent uses: stratigraphic analysis, bathymetric studies, and attempts to define microseepage through lithologic alteration or the detection of stressed vegetation.

The literature on remote sensing applied to hydrocarbon exploration now consists of thousands of articles and several books. No attempt has been made to give an

---

Remote Sensing for the Earth Sciences: Manual of Remote Sensing, 3 ed., Vol. 3, edited by Andrew N. Rencz.
ISBN: 0471-29405-5 © 1999 John Wiley & Sons, Inc.

exhaustive bibliography. The most basic set of references is found in the Proceedings of the 2nd through 11th thematic conferences on geologic remote sensing published by the Environmental Research Institute of Michigan (ERIM) in Ann Arbor. Sabins (1986), Drury (1993), Berger (1994), Prost (1994), and Russ (1995) are also fundamental sources of information on geologic interpretation of imagery in general and exploration applications in particular. The Russ reference is included because it gives a good overview of the image processing techniques most commonly used in exploration applications. For marine applications, especially oil slick detection, the Proceedings of the 1st through 3rd thematic conferences on remote sensing for marine and coastal environments, also published by ERIM, are useful.

The chapter will first provide a background for remote sensing and hydrocarbon exploration prior to this edition. A section on spatial analysis (structure mapping, logistical applications) is followed by a section on spectral analysis (geologic mapping, geochemical applications, bathymetric surveys). Each major application of remote sensing is illustrated with case histories drawn from the literature or the files of individual companies.

9.2 BACKGROUND AND HISTORY

9.2.1 Early Installations

After *Landsat* MSS was launched in the early 1970's several companies and institutes established remote sensing departments. The relatively large number of people involved was due partly to the massive computer facilities then required to process the data. These installations typically cost over $500,000 and were based on DEC, Prime, or similar computers. IDIMS was the most popular software package in oil companies.

9.2.2 The Geosat Committee

The Geosat Committee Inc., led by former president F. W. Henderson III, included oil and mining companies, universities, government agencies, and value-added firms in its membership. It worked to promote the Open Skies policy, without which remote sensing would have been severely restricted. It also worked for the inclusion of band 7 (2.08 to 2.35 µm) on *Landsat* TM, so that lithology could be better mapped. It sponsored educational workshops and cooperative research projects aimed at learning how to use the proposed new data sets by using simulated data acquired from aircraft over well-studied test sites and publishing the results. Studies were performed for Landsat TM (Abrams et al., 1984), *JERS-1* (Collins et al., 1992), MOMS and PEPS. The GOSAP (Gulf Off-shore Satellite Applications) Project (Biegert et al., 1994, 1996), which has brought together a group of about 30 industrial, academic and governmental institutions to carry out satellite-, aircraft-, surface ship-, and submarine-based investigations of the Gulf of Mexico, is under way.

9.2.3 Hardware and Software Costs

In the late 1980s the introduction of powerful Unix-based desktop workstations and of equally powerful image processing software led to an order of magnitude decrease in the cost of establishing a remote sensing group. The reduction in personnel requirements was equally dramatic: More companies established remote sensing groups, typically of one or two people.

Personal computers are now approaching workstations in power, and the price of software for personal computers is much less than that for workstations. Today (1997) for roughly $20,000 to $40,000 one can obtain a computer with enough disk capacity, a plotter, scanner, and software to establish a minimal image processing shop. In the Yemen case history (Section 9.3.2.1A) the computer used was a 486-based PC. Future oil company installations may well be PC-based and cost little more than a quarter of the current generation of workstation-based systems.

The reduced cost of doing remote sensing and the growing use of GISs (geographical information systems) have led to an increasing integration of remotely sensed data and its interpretation into the overall exploration process. In some companies this process has been accelerated by the successful transfer of remote sensing expertise into the operating divisions.

9.2.4 Modern Installations

The typical modern installation consists of a SUN or Silicon Graphics workstation networked into the company's wider net. Various printers (e.g., HP650C) can be accessed. The workstation commonly has dual monitors with 2 to 20 gigabytes of disk space. Few installations have high-resolution plotting devices, since they require photographic laboratories, and most large commercial labs have plotters of this kind.

Remote sensing is commonly part of a technology group or research center, but many companies today have dispersed their staff such that image processing is part of an information technology (IT) group, while interpretation staff is either in operational teams or forms a small core that handles training, data acquisition, and archiving. As a result of these trends, groups that numbered as many as 20 people now consist of one or two people and have a similar or higher output than they did 10 or even five years ago.

9.2.5 Imagery Used in Hydrocarbon Exploration

When the preceding edition of the *Manual of Remote Sensing* was published (Colwell, 1983), the only satellite data available were *Landsat* MSS, and synthetic aperture radar (SAR) was in its infancy as a geologic tool. *Landsat 3*, with the thematic mapper aboard, had been launched just in time for a few images to be included in the book. Since then the situation has changed completely: Imagery is available from instruments on U.S., French, European, Canadian, Indian, and Russian satellites (Table 9.1).

TABLE 9.1   Some Recent Satellite Systems

System	Agency	Resolution	Spectral Bands
SIR-C	NASA/JPL	15–40	X-band (3cm), C-band (6cm), and L-band (23cm) radar
ERS-1 & -2	ESA	30 m	C-band (5.6cm) radar
JERS-1	Japan		
	SAR	18 m	L-band (23.5 cm) radar
	OPS	18.3×24.2 m	0.52-0.60, 0.63-0.69, 0.76-0.86, 1.60-1.71, 2.01-2.12, 2.13-2.25, 2.27-2.40 microns
ALMAZ	Russia	13×30 to 27×30 m	S-band (10 cm) radar
RADARSAT	Canada	100–100 m	C-band (5.6 cm) radar
IRS	India		
LISS-1		72 m	0.45-0.52, 0.52-0.59, 0.62-0.68, 0.77-0.86 microns
LISS-2		36 m	0.45-0.52, 0.52-0.59, 0.62-0.68, 0.77-0.86 microns
LISS-3 1-C		5 m	0.5-0.75microns
		25 m	0.52-0.59, 0.62-0.68, 0.77-0.86, 1.5-1.7 microns
MOS-1	Japan	50 m	0.51-0.59, 0.61-0.69, 0.72-0.80, 0.80-1.10 microns
MOMS-1	Germany	20 m	0.57-0.62, 0.82-0.92 microns

In the near future several companies plan to launch instruments with 1-m resolution and a powerful stereoscopic capability.

9.2.6 Spatial Content of Imagery

The structural geology of large areas of the world, particularly basins and fold belts, were investigated using stereo aerial photography in the 1940s, 1950s, and 1960s. Classical studies include those of Kupsch (1956) in Saskatchewan, Blanchet (1957) in the Alberta basin, DeBlieux (1962) in Louisiana, Doeringsfeld and Ivey (1964), Trollinger (1968) in the deep Anadarko basin, and Penny (1975) in the central Rocky Mountains.

Because Landsat did not possess stereo capability, the early Landsat-derived basin studies (see Section 9.3.1.4) relied heavily on the detection of "lineaments," or linear geomorphic features of questionable origin but presumed related to structures. Simplistic studies of this kind, however, rapidly lost credibility in large oil companies

TABLE 9.2 Planned High Resolution and Multispectral/Hyperspectral Satellites

High Resolution Proposed Launch Date	*Satellite Name*	*Company*	*Spatial Resolution*
1999	Early Bird	Earth Watch	3 m and 15 m
1999	Quick Bird	Earth Watch	0.82 and 3.28 m
1999	Carterra 1	Space Imaging EOSAT	1 and 4 m
1999	Clark	TRW	3 and 15 m
1999	Core Software/IAI	Core Software/IAI	1.5 m
1999	OrbView 1	Orbital	1 m, 2 m
1999	SPOT 4	SPOT Image	10 (Pan) and 20 m(MSS)
1999	Landsat 7	NASA	15 and 30 and 60 m

Multispectral/Hyperspectral Proposed Launch Date	*Satellite Name*	*Company*	*Spectral Channels*
1997	Lewis	TRW	mission failed
1998	EOS AM/1 (ASTER)	MITI (Japan)	14
1999	GEROS	GER	>10
2000	ARIES	MIA/CSIRO (Australia)	64
Airborne	DAIS 211–15	GER	211
Airborne	AVIRIS	NASA/JPL	224
Airborne	PROBE-1	ESS1/Mteg. Spec	128

and were replaced by careful mapping of structures on the non-stereo imagery. With the launch of SPOT in 1984 it became possible to acquire stereo-pairs from space, and a renaissance of classical photogeological techniques began in the oil industry.

Tectonic studies (Section 9.3.1) continue to have value to companies entering or reentering geologic provinces. These studies help geologists understand the depositional history of an area, the timing of structural development, and trap styles. A classic early tectonic study using *Landsat* was done on the Himalayan chain and surrounding areas by Molnar and Tapponnier (1975, 1977).

Early in the history of *Landsat* MSS work, Collins, et al. (1973) noted the existence of "hazy" patches on unenhanced black-and-white imagery. These areas, within which ground features seemed to be defocused, appeared to correlate with existing oil and gas fields. For several years some geologists used these "hazy" textural anomalies to guide exploration, but they fell out of favor partly because the phenomenon was never explained (perhaps they were naturally occurring smog over leaky fields, surface disturbance below the system resolution, etc.), and partly because they were not very successful in leading to discoveries.

Radar works by illuminating the surface with microwaves and is thus an instrument that maps surface (and under specific conditions, near-surface) texture and morphology. From the time of the RADAM study of the Amazon Basin side-looking airborne radar (SLAR) was used routinely to image perennially cloud-covered areas. SLAR was replaced around 1980 by synthetic aperture radar (SAR), which has become the basic tool for reconnaissance mapping in tropical cloudy and arctic (dark half the year) areas. *Seasat* radar data was used by Sabins et al. (1980) to map the trace of the San Andreas Fault in apparently flat and featureless alluvium in the Durmid Hills of southern California. In 1983 Sabins used SIR-A radar to detect folds in heavily vegetated and deeply eroded parts of Irian Jaya, Indonesia. Airborne SAR appears likely to be replaced to a large extent in the near future by data from Radarsat, launched by Canada in 1996.

9.2.7 Spectral Content of Imagery

Tonal or color anomalies have been related to uplifts and/or mineralogical changes associated with microseepage. Studies using black/white photography over the Williston Basin of Saskatchewan (Kupsch, 1956) suggested that there was a change in soil tone related to slight topographic rises over buried structures. These changes were thought to reflect a change from high moisture (darker) off-structure to low moisture (lighter tones) over the rise. Similar examples of tonal anomalies noted on satellite imagery can be found in Saunders (1981) and Morgan et al, (1982), among others.

Diagenetic alteration over leaking hydrocarbons was observed at Cement Field, Oklahoma, by Donovan (1974). He noted that redbed sandstones are bleached over the field and assumed that this was a result of reduction of the iron due to hydrocarbon microseepage over the field followed by mobilization and transport away from the field in groundwater. These changes usually occur in the subsurface and are noticeable only with continued erosion. Mineral alteration and the associated bleaching were also documented in redbeds over the Lisbon and Little Valley Fields, Paradox Basin, Utah. Here the light-colored areas are the result of an absence of pyrite and an abundance of clay minerals in the sandstone.

A number of studies have examined the effect of leaking hydrocarbons on plants. Some trees grown on urban landfills die as a result of leaking methane (Flower et al., 1981). This is due to a decrease in soil oxygen and an increase in soil carbon dioxide in the root zone. The Geosat Committee, Inc., used the Patrick Draw, Wyoming, field as a test site in their *Landsat* TM simulation study to evaluate the possibility that vegetation stress due to hydrocarbon seepage could be detected. Although stunted sagebrush corresponded to a tonal anomaly over the field, it was never established that this was a result of seepage (Richers et al., 1982). A study of the Lost River gas field, Virginia, showed that chestnut oaks were inhibited in areas with high soil methane, whereas field maples did well in this environment. Thus vegetation anomalies at Lost River are a result of a change in species.

An interesting use of the spectral content of imagery involves the interpretation of thermal images of the Imler Road Anticline, Imperial County, California (Sabins, 1969). Airphotos over the area show little or no relief and a more-or-less uniform sand cover over the entire area. The thermal infrared image, however, reveals an anticline outlined by warm and cool bands that correspond to changes in the thermal inertia of the underlying sandstones and shales.

A similar study of the spectral content of radar images by Sabins (1984) showed how radar wavelengths are directly related to the resolution of the corresponding image. In Sabin's example, X-band radar (wavelength 3.0 to 3.2 cm) has much finer resolution and shows much more detail of surface roughness than does L-band radar (wavelength 23.5-25 cm) of the Copper Canyon alluvial fan, Death Valley.

9.2.8 Advances Since the Second Edition of the Manual

In the preceding edition of this book, a great deal of attention was paid to the use of *Landsat* MSS data in mapping linear features, or lineaments. In large part this was due to the lack of stereo capability on *Landsat*; the synoptic view provided for the first time by *Landsat* also played a part, in that it allowed the recognition of regional-scale features. Although this technique is still used, its use in hydrocarbon exploration has declined, due partly to the subjectivity of lineament "picking" and partly to the availability of SPOT data and the spectral capabilities of *Landsat* TM. This review will, therefore, not discuss lineament mapping. The reader is referred to the preceding edition.

There has been a great increase in the sophistication with which tectonic and structural geologic theory, practice, and models are incorporated into the process of interpreting remotely sensed imagery. The case histories from Papua New Guinea and Mexico are good examples of this.

The extensive use of geographical information systems (GISs) has led to a great advance in the practice of remote sensing in oil companies. The second edition of this book gave examples of some techniques of data integration, such as the use of the magnetic field to warp imagery so that when viewed through a stereoscope the ground surface had the relief of the magnetic field. Now, however, many disparate kinds of data are entered routinely into GIS databases along with the imagery, and analyzed with it. Typical oil company applications include mapping oil wells on imagery and then updating their locations in a database, or "burning" property boundaries onto an orthorectified image to use as a base map.

The arrival of powerful and relatively inexpensive office workstations, as well as of moderately priced but powerful software packages, has allowed much more extensive end-user image processing, in particular the merging of different data sets and digital photogrammetry using *SPOT* data. These advances are ably summarized in several recent books on the use of remotely sensed data in hydrocarbon exploration, notably those by Berger (1994) and Prost (1994). These are listed at the end of the chapter.

The use of radar interferometry to make extremely accurate topographic maps from airborne or spaceborne data is on the horizon. Perhaps even more exciting is the possibility of measuring earth movements due to tectonic forces or human-induced subsidence directly using interferometry.

9.3   SPATIAL CONTENT OF IMAGERY

The spatial content of remotely sensed imagery has been of much greater utility in hydrocarbon exploration than its spectral content. The geomorphology of an area contains information about the lithology of the rock units present, about their structural history, their uplift history, and their state of weathering. The details of the geomorphology also often affect the ease of access and the cost of doing work in an area. The surface lithologies and their weathering profiles often govern the quality of seismic data that can be acquired in an area.

9.3.1 Structure Mapping

Considerable structural mapping was done on *Landsat* MSS and TM images. These data, however, lacked stereoscopic coverage except in the restricted areas of sidelap between paths. Since the advent of *SPOT* it has been possible to do much more rigorous work using photogrammetric techniques; however, the cost of data has, inhibited this use to some extent. *Landsat* TM data are often merged with SPOT Panchromatic data to provide very high resolution images which also give a great deal of lithologic information, as in the Yemen case history (Section 9.3.2.1B).

Radar data have been used very successfully to map structure in the cloudy tropics, especially in Papua New Guinea and in the sub-Andean foldbelt. Most of these data have been acquired from aircraft, but the launch of *Radarsat in* 1995 is likely to lead to a decline in the use of aircraft radar.

9.3.1.1 APPLICABILITY OF SURFACE STRUCTURE MAPPING.

In many modern exploration plays, remote sensing techniques have little application because the play is either offshore or involves a deep stratigraphic section that is separated from the surface by one or more unconformities or thrust faults. Above these discontinuities the structure is often unrelated to the structure in the interval of interest. However, in onshore plays in frontier areas, even though there may be little structure expressed at the surface, remote sensing can play an important role. This is especially true when the area is inaccessible [see Section 9.3.1.3B(2)]. In these areas, satellite imagery can be used not only to map structure, but also to delineate basinal areas, to locate outcrops for study, and to provide logistical support (Section 9.3.2).

There are other plays, such as the Papuan foreland, in which seismic data are of poor quality, and remote sensing can contribute useful information despite the lack of good surface exposure [see section 9.3.1.3B(1)]. Finally, remote sensing is useful in fracture permeability plays like the Austin Chalk, although conventional structure may not be apparent at the surface.

Remote sensing techniques have direct applicability in plays in which the structure exposed at the surface is directly related to the structure in the target interval. These are of two kinds: exposed tectonic zones, such as the Gulf of Suez, San Joaquin Valley, or Papua New Guinea Fold Belt, and low-relief basins such as the Gulf of Mexico Coastal Plain, Permian Basin of Texas, or the western Canada Basin.

9.3.1.2 CONVENTIONAL STRUCTURAL GEOLOGY.

This may be the most frequently encountered application of remote sensing techniques in exploration for hydrocarbons, especially in the initial stages of regional exploration. However, it is not restricted to the initial stages, as the Papua New Guinea example makes clear.

Initial regional reconnaissance interpretations may still be performed monoscopically and may use *form lines* to give an idea of the structure. However, this kind of interpretation is prone to errors caused by the interaction of topography with the geologic structure: The sense of throw on a fault may be misinterpreted, or anticlines may appear to be synclines, and vice versa.

Once an investigation becomes more focused on a small area, it becomes essential to measure dips, strikes, and fault throws quantitatively. In the absence of stereoscopic imagery or the ability to measure offset of key beds, this requires extensive field campaigns, which can become expensive. If stereoscopic imagery is available, there are now digital techniques for solving three-point problems and for automatically posting the elevations of contacts, as illustrated in the Yemen case history.

This quantitative information is then used to draw cross sections that can be balanced and palinspastically restored. These models of the geologic structure then serve as a guide to the layout and interpretation of seismic programs and to drilling in difficult areas (such as Papua New Guinea).

A. Satellite Stereo and Structure Mapping. Certain plays, such as that in Yemen (see Section 9.3.2.1B), have benefited greatly from the use of *SPOT* stereopairs to prepare structure contour maps of the surface formations. This procedure, at least initially, was inhibited by the cost of imagery and computer time necessary to prepare the digital terrain model.

9.3.1.3   EXPOSED TECTONIC ZONES.

Remote sensing provides a valuable basis for regional geologic and tectonic syntheses. These consist of the examination of the structural features present in a large area in the context of plate tectonic models in order to guide exploration toward those basins likely to contain prolific hydrocarbon systems. A relatively small number of images provides a true view, free of cartographic errors or interpretational biases, of a complete tectonic province. These regional image maps or mosaics provide an excellent graphical base for the interpretation of geological and geophysical information. In addition, they can stimulate unique geologic insights. This approach is one of the best ways to perceive the regional effects of the movements of tectonic plates.

Tectonic syntheses rely heavily on remotely sensed data and use more regional data than do basin studies. Typically, they use satellite gravity data, regional aeromagnetic data, and paleomagnetic data from the literature. Basin studies, on the other hand, ideally integrate well data, seismic data, and surface mapping, especially measured stratigraphic sections. In tectonic studies the emphasis in interpreting the remotely sensed data is on structure, whereas in basin studies equal attention is given to stratigraphic correlation and interpretation.

A. Province and Basin-Scale Projects. Basin analyses are intended to provide an integrated interpretation of all the data available to the interpreter. They have been, from the launch of *Landsat 1* onward, the bread and butter of many remote sensing consulting firms and oil company remote sensing groups. Since the launch of *Landsat 3* and *SPOT 1* they have declined in relative importance as the role of remote sensing in more detailed investigations has increased.

The goal of remote sensing basin analysis is to gain insights into the petroleum systems in a basin by looking at the basin as an integrated unit and bringing to bear all the knowledge that one can glean, using the remotely sensed data as a means of tying it all together in a coherent framework. These insights are expected to lead to new plays in the basin, and hence to discoveries. To this end, basin studies carried out by remote sensing consultants have traditionally included a thorough literature review and have been accompanied by a long written report. Recently remote sensing firms have teamed up with petroleum industry consultancies that have access to subsurface and seismic data. In-house studies by oil companies are generally performed in close cooperation with the client Region and utilizing all available data. The case history of a mature basin (Section 9.3.1.4A 1) illustrates this.

These studies have concerned themselves with all types of basins, ranging in exploration maturity from rank frontier to well-explored areas. Therefore, the way in which the remotely-sensed data are used varies significantly. There may be almost no other information available for a frontier area (e.g., Turpan Basin, Nishidai and Berry, 1981). At the other extreme, there may be a century's worth of field mapping, drill data, and seismic available. However, in many cases these data occur only around existing fields, with large areas containing sparse data. The remotely sensed images or image mosaics then become a base for compiling critical information from outcrop, well, and seismic data, and the task of the remote sensing interpreter is to tie these observations together in a coherent and geologically meaningful way.

B. Regional Case Histories. A number of case histories are given showing the application of remote sensing interpretations to the mapping of structure for use in regional studies and petroleum exploration.

(1) Structure mapping in thrust belts. This example illustrates the iterative nature of remote sensing work in the exploration industry. As newer and better data become available, an area is reinterpreted in more detail, especially if management is convinced that the first pass yielded valuable information. The most successful demonstration of this is a discovery. This study also illustrates the importance of careful geologic work: quantitative estimation of structural attitudes, the preparation of detailed cross sections, and the acquisition of ground truth, including seismic data in critical areas. Finally, an important point is that the remote sensing interpretation led to a coherent structural model for the entire area, a "synoptic overview," that allowed the reinterpretation of more local work and re-ranking of the remaining prospects.

Chevron has been active in using remote sensing techniques, and one of the areas in which they have depended on remote sensing most heavily has been the Papuan Fold and Thrust Belt of Papua New Guinea (Figure 9.1). The area is unusual in that the combination of inhospitable topography and climate with the widespread occurrence of a karstic limestone at the surface make seismic data acquisition almost impossible and remote sensing virtually essential.

This area is part of the continent–ocean collision zone at the northern margin of the Australian plate. Dense jungle, rugged topography, and extremely well-developed karst on the surface of the 1200-m-thick Darai limestone (Figure 9.2; see color insert) make field work of any kind in this area extremely difficult, dangerous and time-consuming: On the Darai plateau it is only possible to cover 1 km a day on the ground. The karstic nature and high velocity of the Darai also prevent the acquisition of usable seismic data in much of the area. Exploration therefore has to be based on photogeologic and surface geologic mapping and structural modeling.

When Chevron started work in the area the only available geologic maps were reconnaissance maps (at 1:250,000 scale) based on very limited field work and on

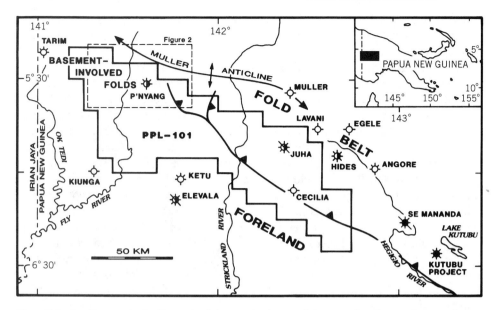

Figure 9.1 Locations map showing P'nyang in relation to regional structural elements and significant exploration wells. Area of Figure 9.2 is shown by dashed line. Solid vertical half circle within well symbol denotes condensate. Areas of multiple wells generalized into one well symbol; well numbers omitted. (From Valenti, 1993.)

photogeology: The planimetric map on which these maps were based was extremely inadequate—major rivers were sometimes plotted 10 to 15 km away from their true positions over considerable reaches, because large areas of ground were obscured by cloud in the original World War II aerial photography: For this reason also the geologic map was rather incomplete.

By 1995 Chevron possessed two airborne radar (SAR) surveys in the area and a new 1:50,000 airphoto survey. The original, proprietary SAR survey acquired in 1985 had north look, normal to the strike of the mountain range, and hence a great deal of the area was in radar shadow. It also had comparatively low resolution by today's standards (Valenti et al., 1996). The later SAR survey, a nonproprietary survey acquired by Intera Technologies of Calgary, Alberta in 1987–1988, had east look, along strike, and thus far less shadow. It was of higher resolution (4 m × 6 m pixels) and was acquired with 55 to 60% sidelap, thus permitting stereoscopic viewing.

A reconnaissance interpretation of the area had been undertaken using the earlier, 1:250,000 SAR coverage (Ellis and Pruett, 1986). The aerial photography was interpreted in detail in 1986–1988, with quantitative estimates of dip and strike, and compiled onto a high-quality 1:100,000-scale topographic base produced by the Royal Australian Survey Corps (Lamerson, 1988). Quantitative estimates of dip cannot be obtained from airborne SAR imagery because the vertical exaggeration varies across the width of the strips (Drury, 1988, pp. 195–196). This aerial photographic work identified surface anticlines, estimated their closures, and ranked the features by prospectivity. On this basis a three-year field campaign (Dekker et al., 1990), consisting of 815 km of helicopter-supported surface traverses, and costing $1.93

million, was carried out (Valenti et al., 1996). This, when combined with the photogeologic work, yielded the first comprehensive geologic map of the Chevron-operated concession area.

Structural analysis based on this work consisted of the construction of serial, true-scale structural cross sections through the anticlinal features (Figure 9.3). Both field and photoestimated dips were used after correction to apparent dip in the line of section. Geometrical analysis was used to determine plunge rate and confirm plunge reversal. Subsurface structure contour maps of the objective Toro sandstone (Figure 9.4) were constructed by hand directly from the cross sections (Valenti et al., 1996).

This work resulted in the discovery of P'nyang, a giant gas field, in August 1990 (Valenti, 1993). The P'nyang field contains up to 3.5 trillion cubic feet (TCF) of gas, and is on a 9 to 13-km-long surface anticline, interpreted as a basement-involved anticline subsidiary to the regional Muller Anticline (Figure 9.1). The existence of a well-developed Toro/pre-Toro reservoir section was confirmed by field study of nearby outcrops in the Muller Range. Similarly, marine source rocks were confirmed (Valenti, 1993). The thick-skinned structural style was confirmed by the penetration of 205-Ma-old granodioritic basement rocks. There is no seismic data over P'nyang or nearby features because the Darai limestone crops out throughout the area (Valenti, 1993).

The discovery of P'nyang led to a second phase of remote sensing interpretation and fieldwork. Detailed manual analysis of the high-resolution second SAR survey data was undertaken to assess its usefulness for lithostratigraphical interpretation and to fill in and refine the previous geologic compilation (the stratigraphic column for the area is given in Figure 9.5). The work was done on film-positive strips on a light table under a mirror stereoscope. Because of the inherent geometric distortions in radar imagery, this method entailed laborious procedures to transfer the data to 1:100,000 topographic base maps, by tracing the drainage, cultural features and prominent radar shadows on the radar imagery and matching them, a small area at a time, with the corresponding features on the topographic maps (Valenti, 1993). A zoom transfer scope would have been faster and more accurate but was not available. Areas of existing surface geology control were used to establish the SAR texture and tone and the surface topographic expression of each lithologic unit. These characteristics were then successfully used to extend mapping beyond the area of control because key lithologies in the succession overlying the Darai limestone were easily recognizable on the radar data.

The results of this exercise were so encouraging that SAR mapping and photogeological intgerpretation were extended to the entire western half of the Papuan Fold and Thrust Belt (PFTB, see Figure 9.1 for area studied), to obtain regional surface geological control, characterize structural styles in different parts of the belt, and constrain regional cross sections and models. An area of 17,000 km2 was mapped and compiled, along with all available published and open-file geologic data, onto ten 1: 100,000 map sheets. Figure 9.6 is a simplified version of part of one of these maps. A 1:250,000-scale tectonic map was then derived from this compilation. This campaign led to recognition of the importance of mass movement processes in the area. Landsliding in the Era beds and Darai limestone (Figure 9.5) in particular, can cause errors in structural interpretation (Valenti, 1993).

A limited seismic program was acquired across the frontal zone of the PFTB in areas where a thin veneer of post-Darai fine-grained clastic sediments allowed the

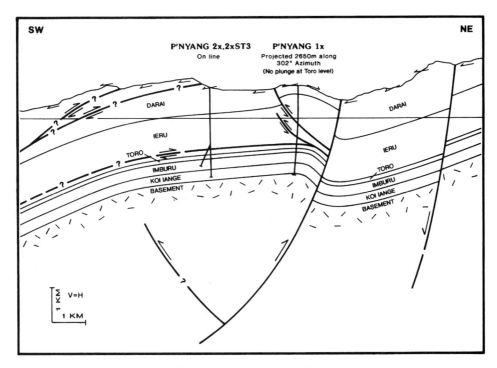

Figure 9.3   Structural cross section, P'nyang Anticline. (From Valenti, 1993.)

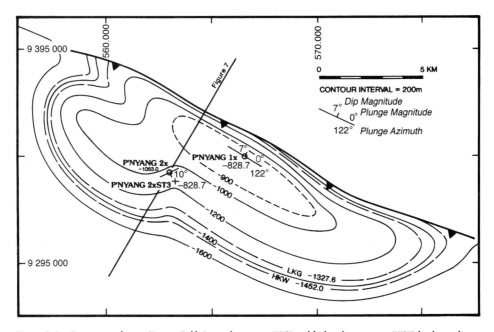

Figure 9.4   Toro structural map, P'nyang Field. Lowest known gas (LKG) and highest known water (HKW) levels are shown. (From Valenti, 1993.)

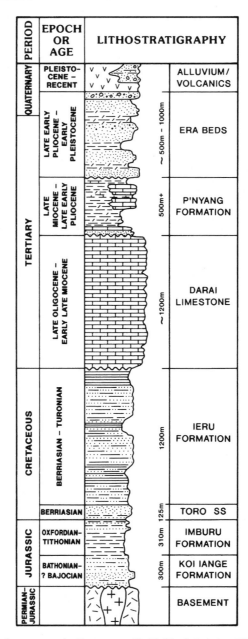

Figure 9.5   Stratigraphic columnar section for P'nyang area. (Modified by M. Little from White et al., 1973, and Rachwal, 1984. (From Valenti, 1993.)

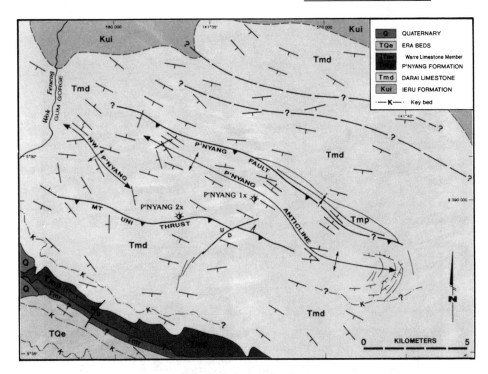

Figure 9.6 Simplified surface geologic map, P'nyang Anticline. Only selected dip/strike symbols are shown; NE–SW linear orientation of symbols indicates locations of field traverses. Small dip/strike symbols denote SAR dips. Dip magnitude numbers omitted for clarity. (From Valenti, 1993.)

acquisition of high-quality data. At Maipe Anticline, near the Irian Jaya border just east of the Tarim well (Figure 9.1), this showed that the basement was involved in thrust faults that extended to the base of Darai. The Darai is passively folded over these thrusts: the surface Maipe Anticline is essentially a basement-involved fault propagation fold (Valenti, 1993).

Five balanced, restorable cross sections were drawn, using the GEOSEC program on a workstation, across the entire width of the PFTB. These honored all the data, including a change in style westward from thin-skinned to basement-involved thick-skinned thrusting, and incorporated the results of depth to detachment calculations. A passive roof duplex structural working model was chosen for the PFTB as a result of this exercise.

Local cross sections and the structure contour maps derived from them were revised to be compatible with the regional model. This particularly affected the interpretation of fault geometry at depth (Figure 9.7). All prospects were reevaluated and reranked. Additional surface geology (Valenti et al., 1996) and, where possible, infill seismic data, were acquired over a number of features. This work forms the foundation for future exploration by Chevron in western Papua New Guinea.

(2) Frontier basins. The Turpan basin in Xinjiang Province lies between the Bogda Shan on the north and the middle Tien Shan on the south (Figure 9.7). The

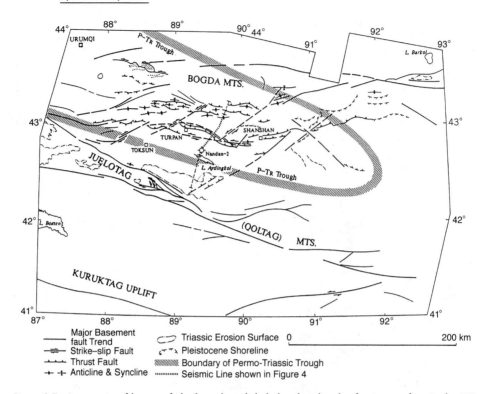

Figure 9.7 Interpretation of basement faults (heavy lines, dashed where buried) and surface structure from *Landsat* MSS, Turpan Basin and surrounding areas. Seismic line (Figure 9.8) and cross section is shown passing through Nandan-2 well. (From Nishidai and Berry, 1991.)

rugged, perennially snow-capped peaks of the Bogda Shan look down on huge co-alesced alluvial fans which lap up against the Huoyan Shan (Flaming Mountain), which runs along the central, east-west axis of the basin. The large playa in the center of the basin, immediately south of the Huoyan Shan, is 154 m below sea level. The southern flanks of the basin slope gently upward to the crystalline core of the Tien Shan, which here has been eroded to a relatively low elevation. The climate is ex-tremely arid, making it an excellent place to use remote sensing methods. When the study began, there was very little literature available in the West. One long seismic line (Figure 9.8) had been published. It was known that three small oil fields had been discovered in the 1950s.

The interpretation was performed on *Landsat* MSS color composites that had been edge enhanced and contrast stretched. *JERS-1* OPS data was available in limited areas. A stereo pair of images was also available. Traditional photogeologic methods were used.

The Turpan basin overlies the northern part of the Tien Shan orogenic belt, which took its final form during the late Carboniferous to early Permian collision of the Tarim plate with the Siberian plate. The Tien Shan in this area consists of high-grade metamorphics with syn-and post-tectonic granite intrusions. It also contains a num-ber of ophiolite belts. The Bogda Shan is composed of low-grade metamorphic

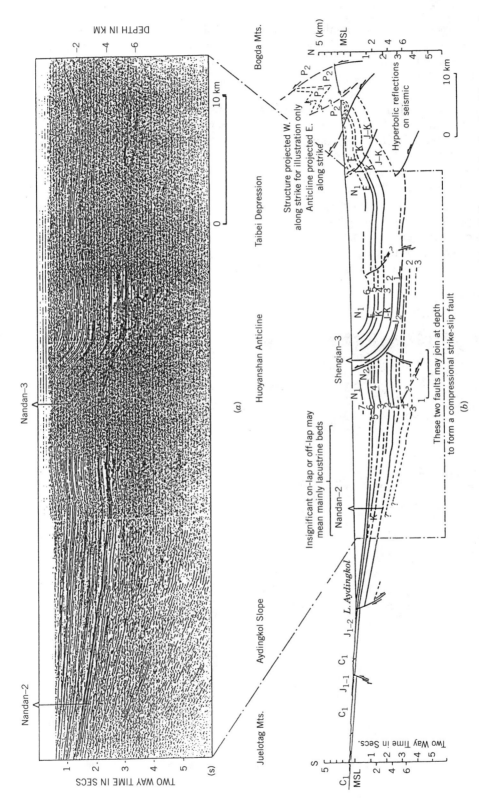

Figure 9.8 (a) North–south seismic line of the central part of the Turpan Basin, (From Lu, 1989.) (b) Hand-migrated line drawing of the seismic line, extended north and south by interpretation of *Landsat* MSS images. (From Nishidai and Berry, 1991.)

oceanic rocks, including basalts, sediments, and melange. The Bogda Shan can be seen on imagery to be thrust northward over the Junggar basin. However, east of Urumqi this thrusting dies out and the range appears to be thrust southward over the Turpan Basin. The Huoyan Shan in the center of the basin is a consequence of this compression, and is a southward vergent anticlinal ridge whose south limb is often overturned and almost invisible on imagery (Figure 9.9). The southern flank of the Turpan Basin appears to be a simple dip slope whose sedimentary cover laps onto the Tien Shan metamorphics (see corner of Figure 9.10).

The sedimentary section in the Turpan Basin is entirely nonmarine and consists of a thin section of Triassic clastic rocks overlain by a thick lacustrine and fluvial Jurassic sequence containing source rocks and coal measures. It is capped with a thin series of Cretaceous clastics. This Mesozoic sequence overlies a horizontal Permo-Triassic erosion surface which, everywhere in Xinjiang, is cut across the late Paleozoic tectonized rocks. The Mesozoic sequence was probably continuous across the mountain ranges now surrounding the Turpan Basin before

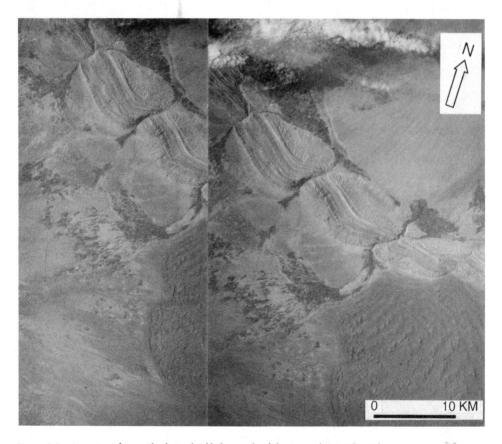

Figure 9.9   Stereo Pair of Space Shuttle Hand-Held Photographs of the Huoyanshan Anticline Belt, Turpan Basin, China.

Figure 9.10   *Landsat* MMS standard false-color composite image of the Yilahu oil field, western part of the Turpan Basin.

their uplift during the Neogene. Within the Turpan Basin the Mesozoic section thickens northward and is buried beneath a northward-thickening section of Neogene lacustrine and fluvial rocks, since this is now a foreland basin to the Bogda Shan.

A key element in the basin interpretation was the use of the seismic section, which was hand-migrated in the area of steep dips on the north flank of the Huoyan Shan. This provided critical insight into the structural style of the basin. A fundamental observation was that the Huoyan Shan's location was determined by a down-to-the-north normal fault which forced the southward-traveling Huoyan Shan thrust to ramp up to the surface. Another section was drawn across the entire basin. Topography from the 1:500,000 Tactical Pilotage Chart (TPC) was carefully profiled and provided critical evidence for the dip in complexly folded rocks on the north side of the basin, which in turn provided critical constraints on the interpretation of structure on this side of the basin.

On the north side of the basin an arcuate area of uplifted alluvial fan extends southeastward from the Bogda Shan toward the Huoyan Shan near the town of Shan Shan. In two places pre-Holocene rocks break the surface of the fans (Figure 9.11). Along the rest of its length this trend is marked by slight incision of the ephemeral drainages, observed on the imagery as an area of darker tone and coarser texture cut by narrow, bright drainages that fan out downstream. On the seismic section there is an apparent splay of the thrust beneath this area, causing a broad, gentle anticline in the strata above it (Figure 9.8). This is interpreted, on the basis of the incised

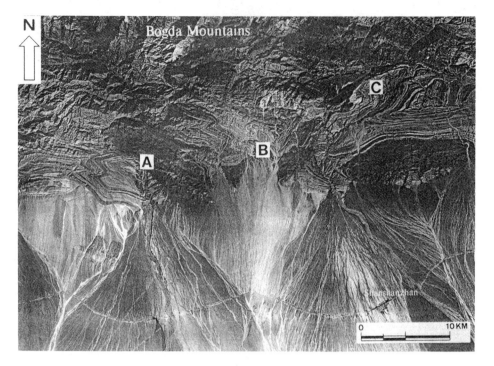

Figure 9.11 Black-and-white print from *Landsat* MSS standard false-color composite image of the Yilahu oil field (at G), western part of the Turpan Basin. (From Nishidai et al., 1993.)

drainages, as a currently active backward-breaking thrust. It is breaking backward because the main thrust is locked by the down-to-the-north basement fault below the Huoyan Shan. This area is believed to have potential structural hydrocarbon traps. It is doubly attractive because this is the deepest area of the basin, and Jurassic source rocks are currently within the oil window due to the enormous load of Neogene fluvial sediments.

Since the interpretation was made, the Chinese have announced the discovery of the Shan Shan complex of fields along this trend. Total reserves to date are about 350 million barrels of oil, and they are currently producing from these fields. By monitoring the area through acquisition of additional imagery we found that the discovery was apparently made as a result of an enormous program of regional seismic. With time, detailed local surveys were added to this dense grid. One of the earlier images shows the discovery well at Shan Shan. Later images show the progressive appearance of many more wells. The reason for this tight drilling grid was apparent from a paper at the 30th International Geologic Conference (Ma and Chen, 1996) discussing the very poor reservoir quality in the area. In summary, careful drawing of cross sections and interpretation of geomorphic anomalies in this frontier basin could have led to the discovery of fields that were later discovered by a regional campaign of seismic acquisition.

9.3.1.4 LOW RELIEF PLAYS.

Considerable use has been made of imagery in low-relief basins, those basins in which the structure has low relief, or in which deep structure has a subtle expression at the surface. An example is provided from the Permian Basin of west Texas.

A. Trans-Pecos and Permian Basin. The goal of this project by Mobil was to understand the tectonostratigraphic evolution of the Trans-Pecos and Permian Basin region of Texas, Mexico, and New Mexico (Markello and Sarg, 1996) (Figure 9.12). The purpose was to understand the controls on the distribution of hydrocarbon accumulations and, if possible, to identify new plays. This is a mature basin and there are, therefore, vast amounts of geological data available. However, the data occur in small areas near oil fields or good surface exposures, whereas large areas have little information. Thus detailed studies that provide an understanding of the overall geologic history are almost entirely local. There are few regional syntheses because large distances, remoteness, difficulty of access to private lands, and lack of surface exposures all hinder documenting regional relationships.

A mosaic of 10 *Landsat* TM images provided a synoptic view of the whole region (Figure 9.13) and a uniform base for compilation of critical structural and stratigraphic relationships that had been documented in the field or subsurface. The interpretation was aimed at delineating faults, folds, and stratigraphic boundaries and extrapolating them between documented data points. It further proved useful as an aid in communicating the results of the project during training classes.

The 10 TM scenes covering the area were acquired in 1982 through 1986. Images were chosen on the basis of their geologic utility and acquisition dates ranged from winter to summer. They were registered individually to USGS 7.5-minute topographic maps, then edge enhanced, and a color composite of bands 7 (R), 4 (G), and 1 (B) was made. The component scenes were color-balanced without sacrificing geologic content and were then mosaicked using a 20-pixel seam crossover algorithm. A decorrelation stretch was then performed on the final file, and each band was written out as a black-and-white negative with a 12.5-μm spot size. These color separations were then composited and printed. The geologic data and interpretations were posted on a series of clear overlays.

The interpreter identified the image expression of documented geologic features of interest at known points and attempted to extrapolate them between these points using the traditional photogeologic keys of tone, texture, and alignment. Previous interpretations had noted many of the same image features but had not positively identified their underlying geologic causes, due to the lack of adequate geologic control. The interpretation was supported by field work in key areas.

Hydrocarbon traps in the Permian Basin are of Permian age. Maturation and migration, however, took place in the Cretaceous. During the Paleozoic era a precursor basin, the Tobosa Basin, occupied the area. It was a passive margin interior sag, with the continental margin lying south of the Marathon Mountains. The basin deepened from Ordovician to Mississippian time, and by the Mississippian an oversteepened carbonate ramp margin rimmed the basin. This is exposed in the Sacramento Mountains near Alamogordo, New Mexico. On the imagery it can be extrapolated to pass northeast toward Roswell, New Mexico, where it is defined at a kink in the Pecos River. To the southwest it is defined on imagery by the southern edge

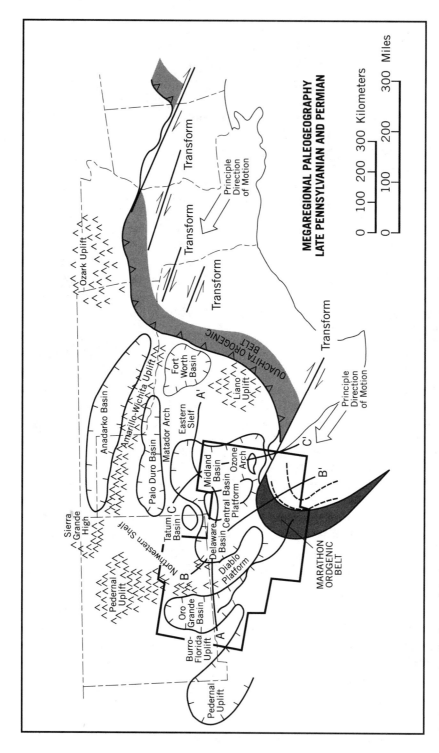

Figure 9.12 Regional tectonic framework for the Trans-Pecos region and Permian Basin of New Mexico and west Texas. The areal extent covered by the composite *Landsat* imagery of Figure 9.12 is outlined. This area covers mountainous terrains of the Trans-Pecos. (From Markello and Sarg, 1996.)

of the White Sands National Monument, at a structural break between the San Andres and Organ Mountains, and at a kink in the Rio Grande (Figure 9.13).

Differentiation of the Tobosa basin occurred during Pennsylvanian orogenesis, and successor basin margins were aligned north-south, almost normal to the Tobosa basin passive margin trends. The successor basins were the Midland, Delaware, Tatum, and Orogrande Basins. The first two are separated by the Central Basin Platform, and the last two by the Diablo Platform. These are both basement horsts. The southern boundary of the area was the Marathon Fold and Thrust Belt. Postorogenic Permian carbonate margins parallel the Pennsylvanian Basin rims and contain most of the reserves of this prolific hydrocarbon province. The Pennsylvanian Basin mar-

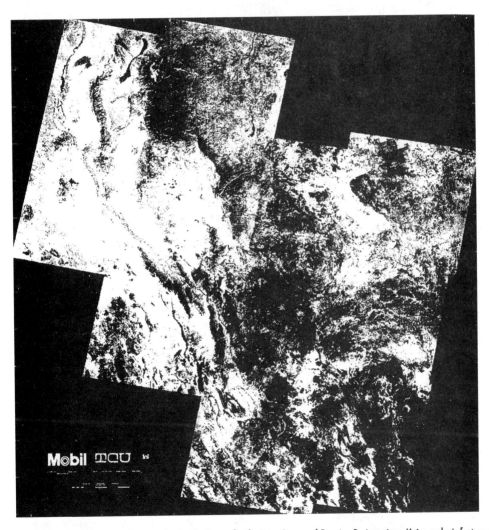

Figure 9.13 Ten-scene, composite *Landsat* image for the Trans-Pecos and Permian Basin regions. Major geologic features are identified, including the sedimentary basins, intervening basement-rooted high blocks, thrust belts/orogens, and rift grabens. (From Markello and Sarg, 1996.)

gins are known from outcrop along the eastern edge of the Diablo Platform and in the Sierra Diablo Mountains. At the northern termination of the Huapache Monocline, which bisects the Delaware Basin and west of which there are no hydrocarbon accumulations in the Delaware Basin, the Pennsylvanian Margin turns northeast, parallel to the Mississippian Ramp Margin. It has been mapped at the surface at another kink in the Pecos River, near Artesia, New Mexico and is known in the subsurface from oil field data east of Artesia. It can similarly be mapped around the Central Basin Platform from subsurface and seismic control. The southern margin of the area was collisional in the Pennsylvanian and can be mapped as outcrop in the Glass and Marathon Mountains (Figure 9.12) and from well control in the Val Verde Basin. The Landsat data were used as a plotting base for all this information and to interpolate between the data points. The same was done for the Permian margins of the basins, which are basinward of the Pennsylvanian margins as they are basinward of the Mississippian ones.

The Jurassic sediments in the Chihuahua Trough, south of the area of investigation are nonmarine to marginal marine clastics and evaporites that were laid down in a rift extending westward from the Gulf of Mexico. No trace was found on the *Landsat* or other surface data of deformation in the Permian Basin area related to this rifting event.

The Late Cretaceous/Early Paleocene Laramide orogeny is manifest in the Trans-Pecos area as the Chihuahuan Fold and Thrust Belt, which is very clearly expressed on the imagery. The eastern limit of deformation is the west flank of the Diablo Platform, approximately along the Rio Grande (Figure 9.12). This basement horst shields the Permian Basin from Laramide deformation. The geometrical relationships responsible for this are visible on the imagery. Finally, Basin and Range extensional faulting of the late Cenozoic can be clearly seen on the imagery.

To sum up, the *Landsat* mosaic provided the fundamental display onto which all of the known surface and subsurface geoscience data, studies, and interpretations were compiled. It provided the continuous surface geologic exposure necessary to interpolate and extrapolate between and away from control data. It illustrates exceptionally well how multiple tectonic events are superimposed on the same geologic terrain. This reinforces the well-known idea that once a basement grain or zone of weakness is established, it will control the location of subsequent tectonic and depositional events. Finally, the *Landsat* display provides an exceptional means of communicating geologic ideas both in the classroom and field.

9.3.15 FRACTURE PERMEABILITY PLAYS.

This work relies on the extension of techniques developed years ago by photogeologists working with low-altitude aerial photographs (i.e., drainage and lineament analysis). However, the best work of this kind has been done using carefully designed models of the local geology to guide interpretation of the surface observations.

Lineament analysis has provided useful information in fracture permeability plays such as the Austin Chalk trend of Texas and the Bakken of the Williston Basin (e.g., Reid, 1988). In these studies, however, one must always be aware of the possibility that the important fracture directions in the reservoir interval may not be the same as those most readily apparent at the surface (Prost, 1988). This may happen because the reservoir was fractured in a stress regime that had ended before deposition of the rocks at the surface, or there may be some factor biasing the interpretation, such as bedding planes in the Cottonwood Creek Field case history below (Section 9.3.1.5.A1).

Fracture studies can also be useful in understanding the nature of a reservoir that may already be producing from fractures. The fractures may be permeability pathways or even, in some cases, permeability barriers. They may compartmentalize the reservoir and lead to the failure of waterflood projects. In the case history below, a particular direction of surface fracturing correlated with well-production statistics, and this information led to savings and efficiencies in planning secondary recovery projects.

A. Conventional Fracture-Enhanced Reservoirs. A fracture study was performed by Amoco over the Cottonwood Creek Field (Figure 9.14) to demonstrate that surface fracture mapping can help plan and evaluate secondary recovery programs where there is a likelihood of fractures controlling the flow of subsurface fluids (Prost, 1996). Fractures can behave as impermeable barriers to flow by preventing the movement of reservoir fluids across the fracture zone and by channeling the fluids along the zone parallel to the trend of the fractures.

The Cottonwood Creek Field in the Bighorn Basin, Wyoming, provides an example of the use of airphotos to map surface fractures related to fracturing in the producing reservoir. The field is a stratigraphic trap in the west-dipping Phosphoria Formation at depths between 1540 and 3100 m. Oil is trapped by the up-dip pinchout of a porous dolomite facies as it grades into an impermeable shale-anhydrite facies. Average porosity in the reservoir is 8 to 10%, and average permeability is 1 millidarcies. Porosities as high as 20% and permeabilities up to 800 millidarcies have been reported (Aud et al., 1991). These variations in porosity and permeability have

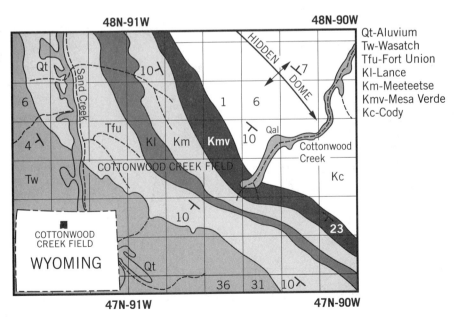

Figure 9.14 Surface geology, Cottonwood Creek Field, Wyoming. (From Prost, 1996.)

been attributed to facies changes, diagenetic pore plugging, vuggy porosity, and fracturing.

Lineaments, assumed to be the surface trace of fractures, were mapped on color-infrared stereo airphotos at a scale of 1:58,000 (National High Altitude Program) over the field and surrounding areas. Lineaments were then digitized and the length-weighted density of fractures was contoured. Maps of the density of all fractures and of fractures within specific trends (Figure 9.15) were compared to production data to determine whether there was any correlation. Production data used includes initial production (IP), porosity-feet, cumulative production to 1986, and ultimate recovery predicted for each well.

Examination of core and production decline curves indicates that some wells do encounter fractured intervals. Many wells also indicate reservoir intervals with high matrix porosity, indicating that this is a dual porosity reservoir.

The map of total fracture density does not correlate well with any of the production data. The primary fracture trend, based on length-weighted statistics, is N30°W–N45°W. Bedding also strikes northwest in this area. Some interpreted fractures may be due to an incorrect interpretation of the strike of bedding as fractures.

Production data were compared to fracture densities within 15 degree azimuth ranges to determine whether there is a correlation to specific trends. The map of northeast fracture density shows zones of intense fracturing that correspond to areas of increased total porosity-feet in the reservoir and high cumulative production (Fig-

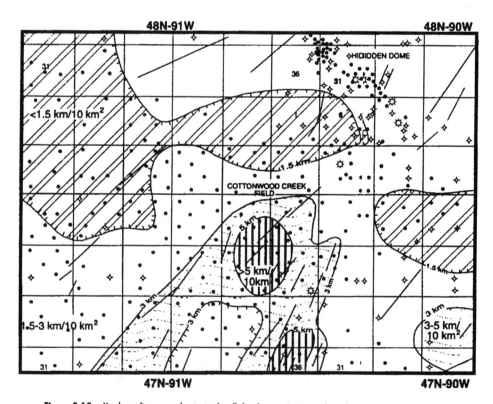

Figure 9.15   Northeast lineament density and well distribution, Cottonwood Creek Field. (From Prost, 1996.)

ures 9.16, and 9.17). Porosity-feet is a measure of reservoir porosity derived from the average interval porosity times the thickness of the interval in feet. It is a good measure of fracture porosity plus matrix porosity. Since fractures generally contain a small percent of a reservoir volume and are drained rapidly, one would expect fracturing to contribute more to the cumulative production measured early in the life of a field, and less to the ultimate recovery predicted for that field. This is the case at Cottonwood Creek Field.

Cottonwood Creek Field contains both matrix and fracture porosity. Whereas the primary (NW) fracture set and total fractures mapped at the surface bear little relation to production, the northeast fracture set appears related to fracture porosity in the producing zones. The fractures were formed during Laramide northeast-directed compression and are maintained open by similarly oriented stresses active at the present time. Mapping of fracture zones allowed reservoir engineers to account for fracturing when planning water floods.

B. Coal Bed Methane. A program by Amoco to evaluate the coal bed methane potential of the Sydney Basin, Australia, used remote sensing to help map fractured areas and optimize their drilling program (Major, 1994). Fractures, known as cleat in coal, are the primary source of porosity and permeability in coal beds. Fracture information is necessary both to locate areas that can deliver gas and to predict directional permeability in order to plan well locations.

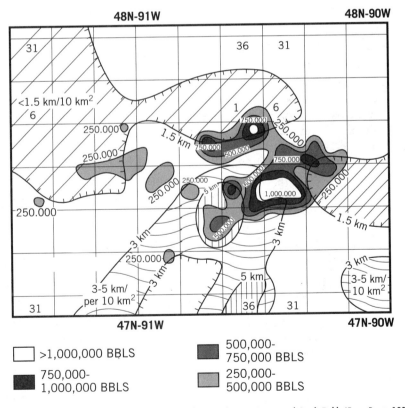

Figure 9.16   Cumulative production and northeast fracture density, Cottonwood Creek Field. (From Prost, 1996.)

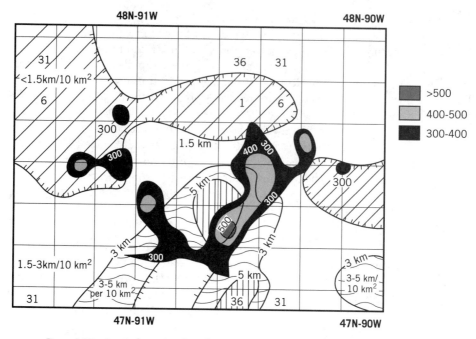

Figure 9.17   Porosity-feet and northeast fracture density, Cottonwood Creek Field. (From Prost, 1996.)

The rocks in the Sydney Basin are gently deformed and nearly flat-lying. Stereo air photos at scales of 1:40,000 and 1:80,000 were used to map the surface trace of fractures (Figure 9.18). These traces, or lineaments, were digitized and length-weighted fracture density was contoured. A field mapping program measured attributes such as orientation, spacing, filling, and vertical continuity of 977 joints at 26 locations in the basin. In most cases a good correlation was found between the orientation of joints in and above the coals and those mapped on imagery.

Air-photo and field measurements were compared to in situ strain measurements by other workers to determine which trends are likely to be open to flow. This suggested that the two dominant trends are capable of forming open joints. On the basis of the remote sensing and maps of coal thickness and quality, a wellsite was chosen at Duncan's Creek near the intersection of north-south and northeast-oriented fractures.

The well was spudded on December 2, 1992 and reached a total depth of 1259 m on January 12, 1993. Evaluation of core indicated that the Upper Permian Illawara Group coal intervals were highly fractured at depths between 785 and 830 m, providing some of the best permeabilities seen in the basin. Pressure fall-off tests also indicated that the objective horizons had good permeabilities. It was concluded that remote fracture mapping, in conjunction with subsurface and field data, can be used to map highly fractured methane-rich coal at depth.

9.3.2 Logistical Applications

There has been an increasing use of remote sensing techniques for logistical planning in the hydrocarbon exploration industry, and it is perhaps the dominant use at pres-

Figure 9.18 Airphoto over the Duncans Creek site, which helped in the mapping of folding and fracturing. The original scale is 1:40,000. The well location is marked just above the photo's center, Duncan's Creek, Cottonwood Creek Field area. (Photo courtesy of Gary Prost. From Major, 1994.)

ent. Projects range from the design of geological field surveys through the design of seismic programs to the planning of pipeline routes. They include bathymetric mapping of coastal waters to assist with shallow seismic surveys and the detection and mapping of old seismic lines and of existing but unrecorded installations.

9.3.2.1 FIELD WORK AND GEOPHYSICAL SURVEYS.

Logistics are perhaps one of the most important applications of remote sensing. Field programs can be optimized (made less costly and time consuming) by the use of accurate and up-to-date base maps, as well as the recognition of roads and trails, obstacles such as deep canyons or badlands, old well sites and seismic lines, and sources of water and aggregate for drilling operations and camp sites, among others. Geophysical surveys, too, can be optimized by recognition of topographic obstacles and surface cover types that require special acquisition and processing.

In mature areas of the United States it is becoming very common for companies to utilize digital ortho maps (DOMs), prepared from specially acquired aerial photography, from the very beginning of a project (Crow, 1996). The DOMs are used in conjunction with global positioning system (GPS) receivers as an up-to-date map base, to correct errors in existing databases, and to verify property boundaries. In developing basins, such as the examples from Yemen (next section), the logistics of planning field operations is even more important.

A. Cost Estimation. In the Yemen Arab Republic Shell, Pecten (Cline, 1996) used *Landsat* TM data to classify the terrain in terms of the cost of acquiring seismic data. This was done on the basis of geomorphology and surface lithology. The proposed concession area was divided into regions of very steep morphology, which would require the use of helicopter transport and areas of smooth topography and soft surficial lithologies, within which the work could be performed more cheaply using truck-mounted Vibroseis, with the aid of bulldozers to smooth the way across gullies (Figure 9.19). The resulting cost-estimate for the proposed seismic program provided information essential in the preparation of a bid for the lease.

A further benefit of this study lay in the fact that old Russian seismic lines, field camps, and well sites could still be identified on the imagery. Since maps of this work were inaccurate and incomplete, the imagery enabled further search of the records in the Yemen government.

B. Logistics and Structure Mapping. Several companies active in Yemen during the period 1991–1993 made extensive use of *SPOT* and *Landsat* TM data to plan their field operations and to interpret structural geology. Typical of these projects was one reported by Clyde Expro PLC, which covers part of the Seiyoun Basin (Jones and Oehlers, 1995; Figure 9.20).

The area is ideally suited to remote sensing techniques: The climate is arid, with 100% exposure of the rocks. Neither adequate topographic nor geologic maps existed, and the relief is sufficiently rugged as to make field operations difficult. This relief consists of an upland plateau that has been dissected to varying degrees by ephemeral streams flowing in narrow, slotlike canyons up to 250 m deep. The lower parts of these canyons are cut into massive limestones of the Umm er Radhuma Formation, whereas the plateau surface generally corresponds with the Rus Formation. Since dips in the area are generally less than 5°, the traces of the formation contacts form very intricate patterns around the sides of the canyons. Several systems of faults are well developed in the surface horizons, and continue into the subsurface where they bound horsts and govern the distribution of traps. Other systems of faults are present only in the subsurface: The surface beds reflect subsurface structure in a subdued, less faulted manner.

Most companies active in this area purchased stereoscopic coverage of SPOT P data with 10-m resolution and *Landsat* TM data (30-m resolution). The data were registered to each other, and the TM was merged with *SPOT* to produce a color image with 10-m ground resolution. The TM band combinations chosen were gen-

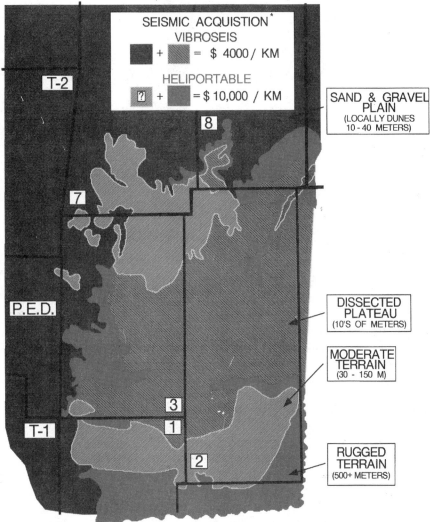

P. D. R. Y. - SHABWA BLOCK
PHYSIOGRAPHIC PROVINCES
(FROM LANDSAT IMAGERY)

* PROVISIONAL

Figure 9.19   Classification of surface types in Shabwa Block, Yemen, in terms of expected cost of seismic acquisition, based lithology and geomorphology.

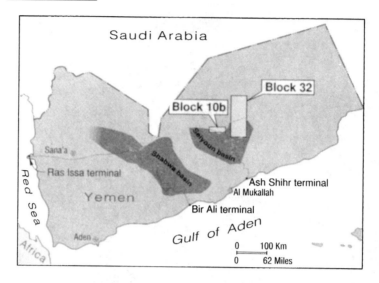

Figure 9.20 Yemen, showing locations of blocks 10b and 32 studied by Clyde Petroleum. (From OGJ, 1995.)

erally 7 (R), 4 (G), and either 1 or 2 (B). Since the rock types present include evaporites, shales, and iron-stained limestones, these band combinations provided excellent lithologic discrimination.

On the merged images one could see nearly every bush in the wadis and every ruin in the highlands. These images were georegistered using ground control points acquired with GPS during a short field campaign. Clyde Expro (Jones and Oehlers, 1995) manually interpreted the imagery for geology using a stereoscope. They were able to map structural features, such as monoclines, with dips as low as 0.2°. Some of these structures were not visible from the ground or air. Features marked as doubtful in the interpretation were flagged and field checked using a helicopter.

Several companies, including British Gas (Fraser, et al., 1996), went further and used the stereo *SPOT* images to produce a digital elevation model (DEM). At the time this work was done, the production of DEMs from *SPOT* data required several hours of computer time for each scene, and was thus an expensive process. In the last two or three years the availability of faster computers has made this process more cost effective. The DEM, once made, was used to generate an orthoimage from one of the *SPOT* scenes. The *Landsat* scene was treated similarly, and the two orthoimages merged. The DEM was also used to generate a topographic contour map.

Structure contour maps could then be drawn on selected outcropping horizons by registering the DEM and the merged image and picking points along the outcrop of the top of a selected horizon. This was done digitally in a GIS (Figure 9.21), and was also done manually by overlaying the two data sets. The first method was faster, but there was considerable noise in the elevation of the formation tops, often due to the steep slopes on which the outcrops occurred. The manual method was slower but lent itself more readily to continuous editing of the picks. Clyde Expro found that the digitally produced structure contour map predicted the elevation of the top Umm er Radhuma to within the ±15-m accuracy.

The images were laminated in clear plastic and bonded onto a heavy linen backing

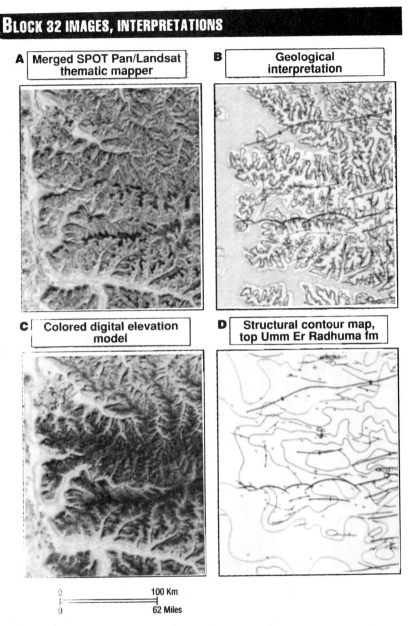

BLOCK 32 IMAGES, INTERPRETATIONS

A | Merged SPOT Pan/Landsat thematic mapper

B | Geological interpretation

C | Colored digital elevation model

D | Structural contour map, top Umm Er Radhuma fm

0   100 Km
0   62 Miles

Figure 9.21   Merged *SPOT/Landsat* TM image, and products derived from it and the *SPOT* stereo pair block 32, Yemen. (From Jones and Oehlers, 1995.)

for use in field operations. This produced easily folded, hard-wearing maps that were simple to manipulate in vehicles and helicopters. These maps were accompanied by access notes describing the tracks, noting every seismic source or receiver omission and each place that a wadi crossing required helicopter assistance, as well as the best means of access to each section of the line for the seismic crews.

The deep, slotlike canyons of Yemen not only make access difficult but also cause degradation of the seismic data due to spurious reflections and cause gaps in the shot and receiver layouts. Seismic data are further degraded by poor geophone coupling in areas of Umm er Radhuma outcrop and by the presence of a thick weathering layer (the water table is below the level of the deepest wadis, perhaps 300 m below the plateau surface). Thus, careful siting of seismic lines to avoid Umm er Radhuma outcrop and minimize canyon crossings was crucially important. In addition, it was necessary to avoid villages and cultivated areas in the wadis. The draft seismic survey plan was therefore overlain on the *SPOT* image and adjusted to satisfy the quality criteria. The lines were then loaded directly into the "mission plan" on a hand-held GPS receiver, which allowed the crews to lay out the lines rapidly and accurately. This technique, according to Jones and Oehlers (1995), increased line layout speed from about 4 km per day to over 8 km per day and a reduction in helicopter time.

The satellite imagery also allowed rapid and accurate determination of the seismic statics correction in this area in which the drilling of upholes is difficult and expensive because of the very deep water table. Geological cross sections were made by overlaying the shotpoint base map on the geological interpretation of the *SPOT* images. Geologic sections and seismic velocities measured in deepwater wells at exploration well sites were then entered into the seismic data processing system, leading to significant improvements in the quality of the processed seismic data without the expense of having a field geologist map out the surface geology in detail along each seismic line. These surface sections also provided useful constraints on the interpretations of the seismic data.

9.3.2.2 PLANNING OF PIPELINE ROUTES, DRILL ROADS, AND OTHER FIELD LOGISTIC.

Remote sensing imagery is uniquely capable of examining large regions and quickly assessing the type of ground cover and land-use patterns in a specific area. When combined with topographic/bathymetric data and cost information, it is a valuable source of information for planning rights-of-way, access into remote areas, tanker routes, and so on.

A. *Routing of the Caspian Oil Pipeline.* Bechtel was asked to do a prototype study of a portion of the proposed Caspian oil pipeline to determine a least-cost route within a predetermined corridor (Feldman et al., 1995). Topographic and geologic maps and remote sensing imagery were used as input, along with costs from previous pipeline projects. Aerial photographs are generally required for pipeline design but were not available. A *Landsat* TM 3–4–5 image was merged with *SPOT* P data using an IHS transformation. Training areas were chosen for specific land-use categories, and a supervised classification was performed to generate a land-use map.

Cost factors were assigned to rock versus soil surfaces, to slope steepness; to stream, wetland, road, and railroad crossings; and to traversing agricultural, urban, and industrial areas. A weighted cost surface, analogous to a topographic surface, was generated for each input parameter. Peaks were areas of relatively high cost, whereas valleys were areas with low costs. These parameter surfaces were then combined to form a cumulative cost surface. The least-cost path could then be traced.

Pipelines are usually designed as straight segments to minimize construction costs.

Through the analysis above, however, it was found that a 42-km straight line path was 14% more costly than the 51-km least-cost path. Most of the difference was attributed to greater costs associated with construction in urban and industrial areas. These results show the advantage of integrating remote sensing images with a GIS analysis in areas where maps and airphotos are restricted or unavailable.

9.4 SPECTRAL CONTENT OF IMAGERY

9.4.1 Stratigraphic Analysis

Hydrocarbon explorationists have not made great use of the spectral content of the data to map stratigraphic variations or structure, except in a few research projects. However, ignoring stratigraphy can severely limit the usefulness of remote sensing projects, as can be seen in the following examples.

9.4.1.1 GEOLOGIC MAPPING.

An example of using air photos, TM imagery, field measurements, and lab spectra to map in arid northern Somalia is provided by Scavetti et al. (1995). The purpose of their work was to generate a preliminary chronostratigraphic framework for mapping and correlating the Jurassic through Eocene section. In so doing they also propose a methodology for this type of stratigraphic remote sensing.

Scavetti defines packages of strata with characteristic stratal patterns, colors, and/ or absorption features, and bounded by surfaces with contrasting erosional profiles. These photohorizons were defined on 1:50,000 air photos, and were tied to control points (lithologic sections) measured in widely-separated areas. The color associations, or multispectral image facies, were defined on *Landsat* TM imagery. Laboratory spectral measurements of fresh and weathered rock surfaces were used to help assign lithologies to the photostratigraphic units.

Photostratigraphic units are generally laterally persistent and are bounded by surfaces that closely approximate time-parallel surfaces that can cross lithostratigraphic units. This indicates 1) that the lithostratigraphy can be time-transgressive, and (2) that some photostratigraphic units are equivalent to sequence stratigraphic units (Figure 9.22). Some environmental determinations are the result of paleontologic analysis of field samples, whereas others are interpreted as, for example, offlap strata in a transgressive system on the basis of the reduced number of stratal packages toward the west within photostratigraphic unit G1 (Figure 9.22).

The photostratigraphic framework, tied to a few field sections, provided a tool for correlating and extending these widely spaced field observations over tens of kilometers. Biostratigraphy tied these units to the existing stratigraphy. The study showed that formations in this area are time-transgressive.

9.4.1.2 MAPPING STRATIGRAPHY AND STRUCTURE.

A joint project by Stanford University and Jet Propulsion Lab determined the utility of color infrared airphotos, NS-001 aircraft imagery, TM, and *SPOT P* data to map geologic units in the White-Inyo Range, California (Ernst and Paylor, 1996). A massive and apparently featureless carbonate, the Reed Dolomite, was selected for investigation using multispectral techniques with the objective of mapping the internal stratigraphy.

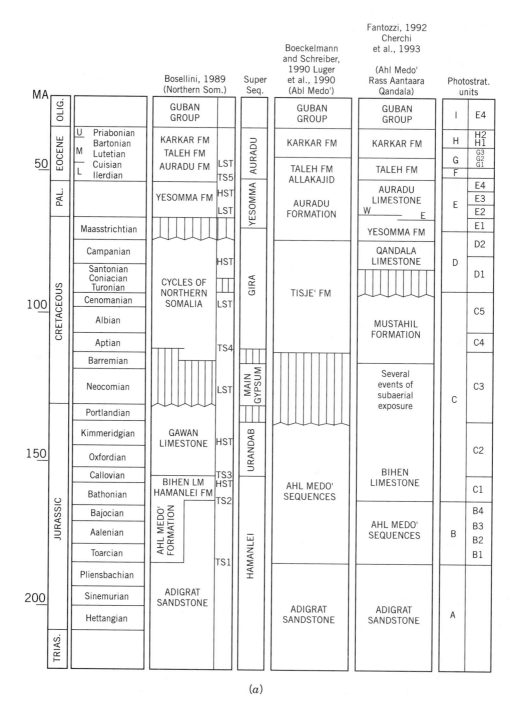

Figure 9.22 (*a*) Stratigraphic framework for northern Somalia; (*b*) correlation scheme across northern part of study. (From Sgavetti et al., 1995.)

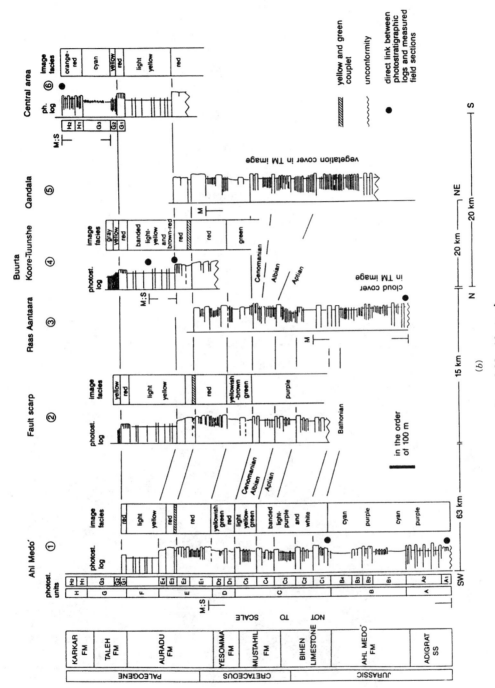

Figure 9.22 (*Continued*)

Wavelength ranges for the TM bands 1-7 are, in order, 0.45-0.52, 0.52-0.60, 0.63-0.69, 0.76-0.90, 1.55-1.75, 2.00-2.36, and 10.4-12.5 (μm. The NS-001 data contained an additional band at 1.00-1.35 (μm (NS-001 band 5). NS-001 data was useful for stratigraphic mapping because of the 10 meter pixel size (compared to a 30 meter TM pixel). The TM scene was processed to a principal components image, and the NS-001 data was processed to a band-ratio image (6/7 red, 6/5 green, 3/1 blue). Although subunits of the dolomite were discriminated on the TM image, they could not be related to discrete spectral features, due to the nature of the principal components algorithm. Spectral differences on the NS-001 image (Figure 9.23) were found to be related primarily to subtle variations in the iron content, but also to carbonate and hydroxyl content and plant cover, both over outcrops and float. The combination of image processing of multispectral data and field mapping permitted the differentiation of subunits in the massive Reed Dolomite. This, in turn, allowed correlation of stratigraphy from one area to another, and permitted the recognition of several previously unmapped structures. Some outcrops remained massive in appearance due to the effects of soil and plant cover. This work demonstrates the capabilities of remote sensing techniques for detecting and stratigraphically discriminating subtle differences in carbonate lithology.

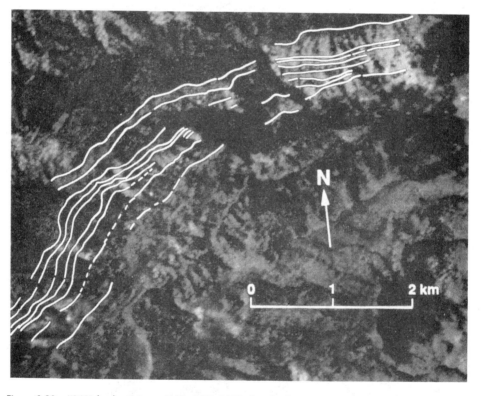

Figure 9.23   NS-001 band ratio image (6/7R, 6/5G, 3/1B) of Reed Dolomite northwest of Birch Creek Pluton. Note the stratigraphic detail. (From Ernst and Paylor, 1996.)

9.4.1.3 MAPPING THERMAL MATURITY IN THE CHAINMAN SHALE, NEVADA.

Rowan et al. (1992) set out to determine whether spectral reflectance, as measured by the *Landsat* TM, could provide information on the maturity of organic matter in the Mississippian Chainman Shale near Eureka, Nevada. The Chainman Shale is considered a regional source rock. Field samples were collected and analyzed using vitrinite reflectance and Rock-Eval pyrolysis. Visible and near-infrared (VNIR) laboratory spectra were obtained from 48 fresh and weathered whole-rock samples. These spectra were used to help guide the analysis of the TM bands.

They found that the shape of a shale reflectance curve changes from concave downward to nearly flat as thermal maturity increases (Figure 9.24). The absolute percent reflectance decreases in all wavelengths between 0.4 and 2.5(μm, with more of a decrease at longer wavelengths.

These workers were able to map differences between mature and supermature areas using a ratio of TM band 4/band 5, after first minimizing the effect of vegetation using the band 4/band 3 ratio. Increasing thermal maturity is characterized by decreasing band 5 reflectance and increasing 4/5 ratio values. Limonitic sandstones, which decrease 4/5 ratio values in organic shales, were screened out by taking only the maximum 4/5 ratio values. Weathering of immature and mature Chainman Shale was found to increase the reflectance in band 5 but had no effect on reflectance of Chainman Shale with vitrinite reflectance values greater than 2.0. Their attempt to correlate TM band reflectance, and ratio combinations, to the percent of total organic carbon (TOC) were unsuccessful. It was felt that this was because the range of TOC values only went to 1.6%. Richer source rock may have more of a spectral response.

This work shows that the VNIR reflectance of a source rock unit can be used to get a feel for the relative maturity of the unit across a region. Such stratigraphic uses of remote sensing will help the energy industry in evaluating the location of the "kitchen" in basins where source rocks outcrop.

9.4.1.4 GEOLOGIC RECONNAISSANCE IN ARCTIC REGIONS.

The northeastern Brooks Range and Arctic National Wildlife Refuge (ANWR) were examined using *ERS-1* synthetic aperture radar (SAR) to evaluate the usefulness of SAR data for geologic mapping in poorly understood arctic regions (Hanks and Guritz, 1997). They generated a low-resolution (90-m) radar mosaic of the entire ANWR, and high-resolution (30 m) individual images. The Porcupine Lake area was used as a test site (Figure 9.25).

The Porcupine Lake area is unvegetated except for tundra on the valley bottoms. The area contains the Continental Divide Thrust Front, which separates regional anticlinoria cored by pre-Mississippian rocks and flanked by Mississippian–Cretaceous Ellesmerian carbonates and clastics on the north from closely spaced thrusting in the Mississippian–Pennsylvanian Lisburne Group carbonates to the south.

The *ERS-1* SAR instrument operates in C-band (6.3-cm wavelength) with a 23° incidence angle. Data were acquired only during summer to minimize variations in backscatter and ground penetration. Images were processed to three stages: (1) radiometric calibration only, (2) terrain correction, and (3) normalized incidence angle images.

Radiometric calibration involved converting SAR brightness values to backscatter

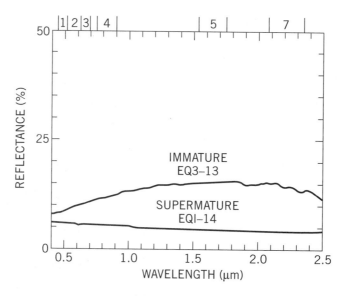

Figure 9.24 Reflectance of mature and immature Chainman Shale. (From Rowan et al., 1992.)

values. The image still, however, has distortion due to layover and foreshortening. These data are difficult to use because one cannot distinguish backscatter effects due to terrain, structure, or lithology.

A terrain correction removes geometric distortions by digitally moving pixels to their correct topographic positions. This is done using a digital elevation model (DEM) of the area. Since the DEM has 90-m resolution, however, the resulting image acquires the lower resolution.

The problem of varying illumination (and thus backscatter) due to slopes at different angles to the radar antenna can be eliminated by making a normalized incidence angle image. One assumes a surface of uniform roughness and uses the DEM to calculate the theoretical backscatter of each pixel based on the incidence angle of the microwaves upon the surface at that point. The resulting artificial SAR image is then subtracted from the terrain-corrected image. Backscatter in the resulting normalized incidence angle image is then due to changes in surface properties, such as roughness and dielectric constant, both of which may be a reflection of the underlying lithology.

Carbonate slopes in this area consist of large, angular rubble and consistently appear as a rough surface on the C-band radar. Shale forms smooth, flat-lying clasts that are often wet or covered with vegetation and thus appear smooth and dark on radar. Conglomerates and sandstones form tabular rubble that forms slopes of intermediate roughness, with some plant cover. Subtle variations, such as surface roughness changes from a shale to a fine-grained sandstone, or between conglomerates and volcanics, cannot be distinguished. This might be overcome by the increase in contrast expected with longer wavelength and greater incidence angles of other systems such as *JERS-1* and *Radarsat*, although optical imagery and ground data are essential to generating a complete geologic map. This work demonstrates the ability of the microwave portion of the spectrum to discriminate between certain

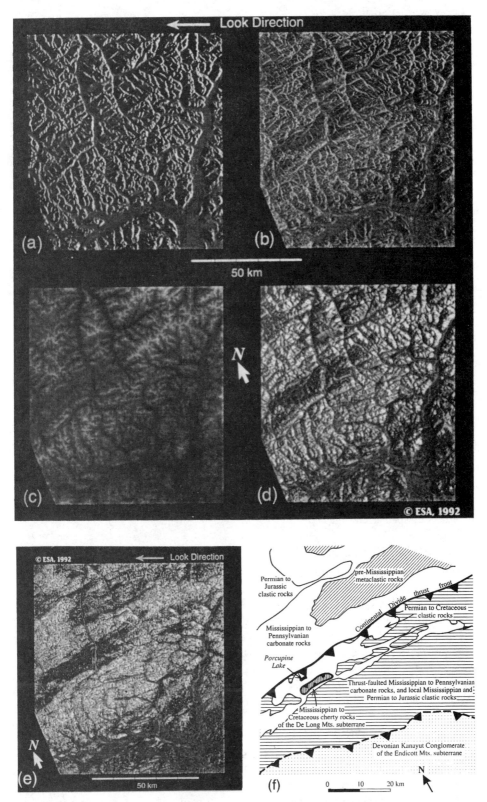

Figure 9.25 Porcupine Lake area: (*a*) Simulated SAR image made from DEM; (*b*) radiometrically-calibrated *ERS*-1 Image; (*c*) DEM; (*d*) terrain-corrected SAR image; (*e*) incidence angle image; (*f*) generalized geologic map. (From Hanks and Guritz, 1997.)

lithologies in the arctic environment. This can assist in mapping both the stratigraphy and structure on local and regional scales.

9.4.2 Geochemical Applications

Many of the world's great oil fields were discovered by drilling alongside evidence of surface seeps. Most of these "macroseeps" onshore were discovered long before the use of modern remote sensing techniques. In addition to obvious phenomena such as burning springs (see, e.g, Nelson and Simmons, 1996, for springs in Indiana), tar pits (California, Trinidad), and mud volcanoes (southern Russia), spectacular surface alteration due to the reducing potential of seeping hydrocarbons is well known from major seep areas as Cement, Oklahoma. Cement was used as a remote sensing test site by Donovan (1974). Although resulting from the same processes, macroseeps are fundamentally different in appearance from microseeps which have been extensively sought using remote sensing techniques.

9.4.2.1 MICROSEEPAGE: THE ALTERATION CHIMNEY MODEL.

Through the years a large number of people have tried to detect hydrocarbon seepage or microseepage directly, using a wide variety of techniques. Recently, many geologists, realizing that the phenomena detected by these techniques are all related to each other, have postulated the existence of an alteration chimney above oil and gas fields due to the vertical microseepage of methane and CO_2 through joints and porosity. The reduction caused by the gas gives rise to a number of phenomena that can be detected by a wide variety of techniques. These include spectral anomalies due to reduction of soil and rock minerals, geomorphic anomalies due to the precipitation of anomalous cement, electrical anomalies due to the alteration of iron minerals to more conducting forms, radioactive anomalies due to the precipitation of uranium minerals from solution, and geobotanical anomalies of several types caused by stress or species changes.

9.4.2.2 SEEPAGE PHENOMENA ONSHORE.

Seepage phenomena onshore have been described in the early literature on remote sensing (see Section 9.2.7). Suffice it to say that onshore seeps manifest themselves as oil-stained rocks, macroseeps (active flows) of oil and/or gas (La Brea, California; Pitch Lake, Trinidad; Maracaibo, Venezuela), mud volcanos (Crimea, Apsheron, Trinidad), altered bedrock and soils (Cement, Oklahoma; Lisbon Valley, Utah), and vegetation stress and changes in species (Lost River, Virginia). These effects, which range from oil on the surface to bleaching of sandstones to deposition of anomalous carbonate cements in the soil, do not require satellite imagery for detection. Still, imagery could be used to help detect hydrocarbon seepage in remote areas.

9.4.2.3 SEEPAGE PHENOMENA OFFSHORE.

The detection of marine oil slicks and the separation of those that are due to seepage from those due to spillage and biological activity relies on the morphology of the slicks as well as their spectral content. For this reason, and because it has become such an important activity in recent years, it is considered separately and at length.

Naturally occurring marine seepage and the resulting slicks have been recognized in the Caspian and Black Seas, the Santa Barbara Channel, and the Gulf of Mexico, among other places (Johnson, 1971). Understanding of the conditions governing visibility of slicks, reinforced by the arrival of radar remote sensing, has made it possible to search for seepage-related slicks worldwide. The existence of such slicks establishes the presence of liquid hydrocarbon charge in a basin before drilling and thus assists in making decisions as to where to direct exploration expenditures.

The variety of manifestations of seepage offshore is perhaps even greater than onshore. Mud volcanoes have been reported on the floor of the Gulf of Mexico and elsewhere, and pockmarks due to gas seepage have been reported in many areas, such as the North Sea, the Barents Sea, and the Persian Gulf (Hovland and Judd, 1988). In some cases, such as the Barents Sea, anomalous concentrations of thermal gas occur within the water column over a wide area in which there are also large pockmarks (craters) on the ocean floor. In other areas, such as the Ragay Gulf in the Philippines, thermal gas is recorded in the water column (Radlinski and Leyk, 1995), but pockmarks have not yet been noted.

Oil and gas seeps on the floor of the Gulf of Mexico and offshore Oregon are associated with distinctive biological communities ("cold-seep communities") consisting predominantly of bacterial (beggiatoa) mats, dense colonies of tubeworms and mussels. These form distinctive hardgrounds (e.g., Flower Gardens in the Gulf of Mexico), and several investigators have recorded the release of large amounts of oil and gas when these are disturbed (MacDonald, Ian, pers. comm.)

In several areas of the world, shallow seismic no-data zones are shown to be associated with areas of seafloor seepage (Hovland and Judd, 1988), and in other cases seismic echoes are associated with bubbles of oil or gas suspended within the water column (Tinkle, et al., 1973; Sweet, 1973). In places where detailed mapping has been done, principally the Gulf of Mexico (MacDonald, et al., 1993; Kornacki, et al., 1994) and the Santa Barbara Channel of California (Mikolaj, et al., 1972; Wilson, et al., 1974), seeps can be demonstrated to occur in linear arrays, and some of these mark the trace of fault zones. In the Santa Barbara Channel and the Gulf of Mexico, the rate of seepage at individual sites has been shown to vary by an order of magnitude over a period of a few months.

A. Seepage at the Sea Surface. Naturally occurring oil slicks on the sea surface have been known in the Caspian Sea and the Gulf of Mexico for centuries. In 1542, after the death of Desoto, the remainder of his expedition sailed through the western Gulf of Mexico and found "a scum cast up by the sea like pitch," and used it to calk their ships (Geyer, 1977). In the early 1900's the Hydrographic Office in New Orleans received reports from several ships of patches of oil, some more than 100 miles long and several miles wide (Geyer, 1977). Early in the *Landsat* program MSS imagery was used with limited success to search for further examples, and with somewhat greater success to map oil spills in Gulf of Mexico (Maul, 1985) and other areas.

After the launch of *Landsat* TM it was realized that the later time of overflight of this satellite, and the greater dynamic range of the TM instrument, allowed the systematic detection and mapping of sea-surface slicks. This process was aided by the award of a NASA EOCAP program to Earth Satellite Corporation to develop the technique and demonstrate its commercial utility. NPA Ltd. In the U.K. were awarded a similar grant by the European Space Agency (ESA) ESA. *SPOT* imagery can be used in the same way, but the cost of data is rather high for the large areas that typically require investigations.

The release of *ERS-1* SAR data allowed the reliable detection of slicks even when cloud cover is considerable, which is often the case in tropical areas of interest and under anticyclonic weather conditions (light winds) that favor the preservation of slicks. In addition, the viewing geometry of *ERS-1* SAR is always optimal for the mapping of slicks. For these reasons a great deal of effort, both by large companies and by several contractors, has been devoted to slick mapping and interpretation since about 1988.

B. Surface Slicks. Oil slicks on the sea surface may readily be seen with the naked eye and are visible because of the absence or suppression of the capillary and very high frequency waves, which changes the amount of light reflected back to the viewer at certain viewing angles. Apparent slicks can be caused by the absence of wind, by the presence of heavy rain, or by the presence on the surface of a layer of surfactant material. This surfactant material may be produced by organisms or may be a component of oil on the surface. Surface oil may be from natural seepage or from spills.

A first requirement for viewing slicks due to surfactants on the sea surface is the presence of sufficient (but moderate) wind to cause ripples to form. This generally occurs at 1.2 ms^{-1}, (2 knots). Further, the occurrence of whitecaps will quickly destroy any slicks that are present, and this generally occurs when the wind reaches about 6 ms^{-1} (12 knots).

Slicks are, however, rarely observable with ease on vertical aerial photographs, whether color or black and white. This is because the image of the sea is dominated by the reflections from individual wave facets, called *highlights*. Highlights are present within slicks as well as on the normal surface. Within slicks they are generally fewer, but larger, and thus in a photograph "the forest is not visible for the trees." An observer can cast an eye about for a better viewing angle (Figure 9.26), and the human eye generally does a better job of integrating observations than does a camera. Cox and Munk (1949) found the solution to this problem during a study of wave slopes for the U.S. Navy: they removed the lens from the camera, thus producing an image which integrated surface reflectance over a wide area, suppressing individual highlights (Figure 9.27). Slicks were readily visible on their oblique photographs (Figure 9.28). Satellite raster data effectively do the same thing, since the reflectance of the sea surface is integrated over the area of each pixel, which typically contains many highlights.

In order to see slicks, however, one still needs sufficient return from the sea surface to be able to detect the contrast between slicked and unslicked surfaces. This depends on both the viewing angle and the dynamic range of the instrument.

At extremely oblique angles of observation there is sufficient reflection of skylight from the slicked and normal surfaces to see this contrast with the naked eye (Figure 9.29). However, satellites generally have a vertical or near-vertical viewing angle, and in this case one has to rely on the scattered reflection of the solar disk (i.e., the glint or glitter pattern) to provide a signal from the sea surface. With 2 to 12-knot winds, wave slopes are generally less than 20°, so this requires a solar elevation of 60° or greater (Figure 9.26) for the detection of slicks. *Landsats 1* and 2 passed overhead too early in the morning (8:30 local time) for this angle to be achieved (except on rare occasions). *Landsats 3* to 5, however, passed over one hour later,

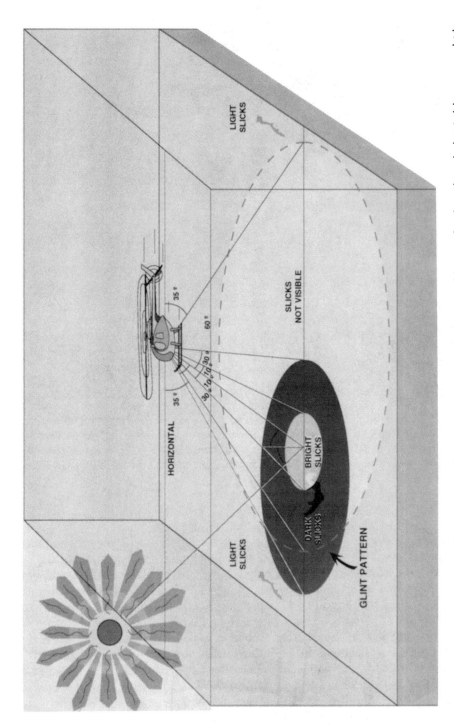

Figure 9.26 Viewing conditions for oil slicks, using airborne platform. Depending on the mean-square slope of the sea surface (hence the wind velocity), slicks appear very bright against the sea surface when viewing within about 10° of the sun's specular point, and dark from 10° to about 30°. They are not visible beyond this. However, when viewing toward and within about 30° below the horizon, slicks appear light against the somewhat darker sea.

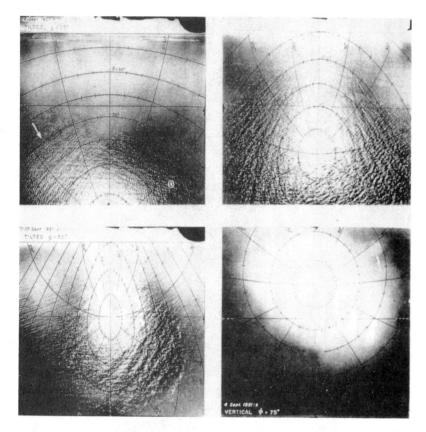

Figure 9.27   (Upper left and right and lower left:) Glitter pattern at solar elevation of $\phi = 75°$, 50°, and 10°, respectively. The superimposed grids consist of lines of constant slope azimuth α (radial) drawn for every 30° and a constant tilt β (quasielliptical) for every 5°. The vessel *Reverie* is encircled in the upper left photograph. The white arrow indicates wind direction. (Lower right:) Photometric photograph. Dashed arrow indicates light circle due to the *Reverie*. (From Cox and Munk, 1949.)

allowing solar elevations of 60° in the summer months over a wide range of latitudes. *SPOT* passes over even later, and is pointable, making it more suitable for the detection of slicks. *Landsat* MSS, with only 6 to 7-bit dynamic range, was not sufficiently sensitive to detect slicks. The 8-bit range of TM and *SPOT* allows the ready detection of slicks: they appear dark against the background of the normal sea surface (generally a digital number below 16 for TM data).

SAR imagery uses the same basic phenomena to detect slicks, but since the wavelength of the radar is comparable with that of the short gravity waves that are suppressed within a slick, the mathematics of scattering phenomena apply. However, it is clear that at angles of incidence greater than 20° there will be few wave facets facing the radar and returns from the sea surface will be relatively low. Therefore, the *ERS* satellites are better suited to the detection of slicks than *JERS-1* SAR, due to a lower angle of incidence (Figure 9.30).

To summarize: to detect slicks of any kind from space requires that wind must be

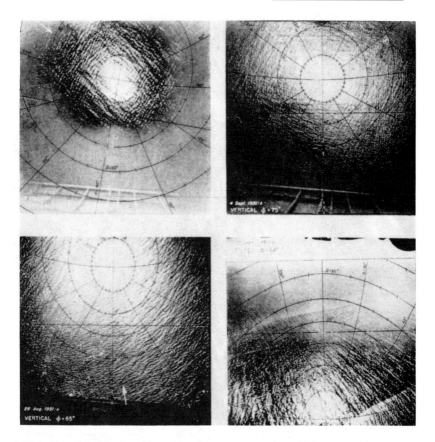

Figure 9.28 (Upper left and right and lower left:) Glitter pattern at wind speeds of 0.7, 3.9, and 14m s⁻¹, respectively. Large rotation of grid in upper left photograph is due to bad yaw, as indicated by off-center position of plane's shadow (within white cross). (Lower right:) Rectangular artificial slick, with near boundary almost through specular point. Brightness of slick sea surface is reduced for large β and somewhat increased for small β. (From Cox and Munk, 1954.)

between 2 and 12 knots, and on visible imagery the angle between the look direction of the satellite and the center of the glint pattern must be less than about 30°.

C. Seepage-Related Slicks and Other Slicks. Slicks related to seepage have a head at a fixed location and a tail extending downwind or downcurrent for a distance that depends on the time that the particular type of oil survives degradation under the prevailing environmental conditions. This survival time, or residence time, of an individual spillet (the slick area due to a single bubble of oil) ranges from 20 minutes to several days. Seepage-related slicks tend to diminish gradually downwind, whereas the widest and thickest part of a slick due to spillage tends to be at the downwind end, since the thickest oil exerts more drag against the wind. The edges of a seepage slick tend to be sharp, except at higher wind speeds in SAR images.

The shape of the slick depends on the wind and the current, integrated over the residence time of the oil in the slick. In practice, only the wind history can be ascertained with any degree of accuracy, since current information is often entirely lacking. In such cases the slick is modeled using the wind history alone, and the currents are

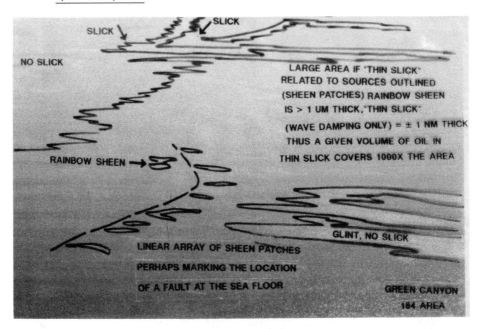

Figure 9.29 Natural oil slicks in the Green Canyon area, Gulf of Mexico, viewed from a helicopter. Geometric viewing relationships are as in Figure 9.26.

deduced. If the current prediction is reasonable (i.e., nearly constant and in agreement with general currents in the area), the slick is judged to be due to seepage.

Slicks due to spillage tend to retain an irregular, equidimensional shape, with some tendency to elongate downwind. The upwind edge often appears to feather out gradually. All of these morphological and textural details of the various types of slicks are the same on VNIR data as they are on radar images. However, the central or downwind area of a spillage slick often contains a region of thicker oil (Figure 9.31) which has a higher reflectivity than the remainder of the slick, and appears slightly more blue than the surrounding seawater on VNIR images.

Slicks due to biological activity in the oceans tend to last longer than seepage-related slicks and are usually concentrated by Langmuir waves and current convergences into complexly folded and interlaced patterns of parallel to subparallel lines. This type of slick almost never occurs as a discrete, limited, elongated body as do seepage-related slicks. In the centers of these narrow, tortuous, and elongated slicks one can often discern, on multispectral imagery, the chlorophyll signature. The spectral signatures of these various types of slicks may be corrected for atmospheric backscatter and instrumental bias and gain. The results, when plotted, allow precise discrimination between thin oil slicks, thick oil slicks, and slicks of biological origin.

Many slicks of medium and heavy grades of oil (e.g., in the Gulf of Mexico) last sufficiently long that they become reworked by oceanic processes such as Langmuir circulation (Figure 9.32; Carnes and Ichiye, 1985; Ichiye, et al., 1985). They eventually become so reworked as to be indistinguishable in morphology from slicks of biologic origin.

Seepage-related slicks should, in theory, be present on every pass of imagery ac-

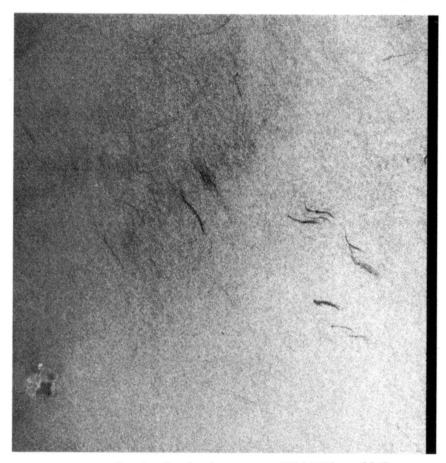

**Garden Banks Area
Gulf of Mexico**

Project GOSAP

10 km

Raw SAR Image
No Processing

ERS-1 SAR Image S-7699
Acquired: 02-Sept-1994 16:49:02
Orbit: 16382
Frame: 1461
Center Lat: N 27.522821
Center Lon: W 93.305634

COPYRIGHT 1994 ESA

F.K. Boyer
Pecten International Company
26 September 1994

Figure 9.30   (*a*) Natural oil slicks in the Garden Banks area, Gulf of Mexico, as seen by *ERS-1* SAR at 9:49 a.m. local time on Sept. 2, 1994; (*b*) Natural oil slicks in the Green Canyon area, Gulf of Mexico, as seen by *ERS-1* SAR at 9:41 a.m. local time September 9, 1993. The bright areas in the north-central part of the scene may be platforms. The vector overlay represents the same slicks as seen on a *Landsat* TM image acquired on July 11, 1993. (Courtesy of ESA and the Geosat Committee, Inc.)

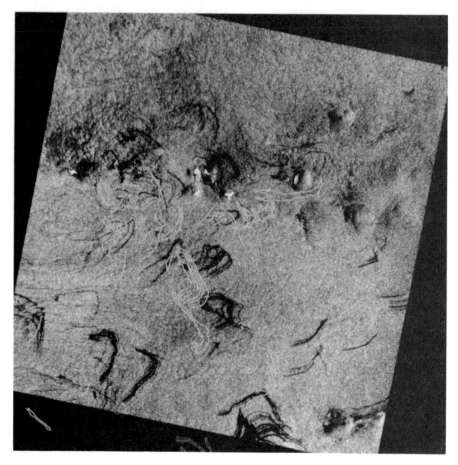

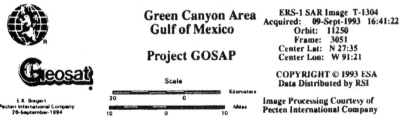

Green Canyon Area
Gulf of Mexico

Project GOSAP

ERS-1 SAR Image T-1304
Acquired: 09-Sept-1993 16:41:22
Orbit: 11250
Frame: 3051
Center Lat: N 27:35
Center Lon: W 91:21

COPYRIGHT © 1993 ESA
Data Distributed by RSI

E K Biegert
Pecten International Company
26-September-1994

Scale

Kilometers
20   0

Miles
10   0   10

Image Processing Courtesy of
Pecten International Company

Figure 9.30(b)   *(Continued)*

quired over the area in which they occur. Slicks of other origins should not. In practice, however, repeatability is not so easy to achieve. Weather reports do not provide wind information precise enough to purchase images based on the optimum speed. Order-of-magnitude variations in seep volume occur with time. Further, the weather in some areas may remain calm for so long that one sees nothing but a tangled mass

Figure 9.31   Close-up of the two patches rainbow sheen shown in Figure 9.28. Oil thickness is from 400 nm to a few microns for rainbow hues to form by interference. (Courtesy of Pecten International Inc.)

of slicks of indeterminate origin. This is especially true of the doldrums belt near the equator.

In heavily traveled shipping lanes bilge pumping may lead to the presence of straight slicks in the same general area on several passes of imagery. Garbage dumping near some cities may have a similar effect. Ship slicks, however, are nearly always distinguishable by their straightness, and garbage is distinguishable by its spectral signature and the irregular variations in the slick along its length.

D. GOSAP project. The Gulf Offshore Satellite Applications Project (GOSAP) is being undertaken by members of the petroleum, marine, and environmental industries under the aegis of the Geosat Committee with support from the European Space Agency (ESA). The goal is to determine how best to use remote sensing technology to address problems faced by exploration and marine engineering organizations. The GOSAP team is evaluating the potential for satellite-based offshore exploration, ocean engineering, and environmental applications using combined satellite and airborne measurements constrained by real-time "sea truth." This pilot project has led to the successful transfer of applications from pure research to commercial viability. Participants estimate that the knowledge of slick detection that they have gained is being used in about 30 hydrocarbon plays each year.

ESA acquired imagery for this project beginning in late 1992. In the summer of 1994 GOSAP achieved the first "top-down" tracking of slicks observed from space,

Figure 9.32 Close-up of oil slick in Green Canyon area, Gulf of Mexico. Specular point of the sun is above the top of the photograph: slick clean water contrast is decreasing in this direction. Drift across the source point of the slick was originally from lower left to upper right but the wind has since changed and is now from left to right and increasing. This has given rise to the generation of Langmuir cells, whose long axes are parallel to the new wind direction. The oil slick is being swept from the diverging limbs of the cells into the convergent zones, and in a few hours this slick will be unrecognizable as a slick due to seepage. (Courtesy of Pecten International Inc.)

actually tracing them to their source at the seafloor. Close cooperation between the participants, ESA, Radarsat International, and Canadian Centre for Remote Sensing (CCRS) allowed the quick release and processing of *ERS-1* radar data. The images were immediately interpreted by Earth Satellite Corporation (a GOSAP member), and a waiting surface vessel was directed to an observed slick. Once on station, the crew collected surface samples and launched a deep-diving submarine, which visited the most active seep sites and pinpointed seepage locations on the seafloor. Here the crew observed and photographed the benthic cold seep communities, acquired geo-chemical samples, made oceanographic measurements, and placed special collecting devices over the active vent in order to obtain measurements of the seepage rate.

In addition to this special real-time effort, GOSAP members collected a large data set of air and spaceborne imagery of Gulf of Mexico slicks. In addition to a large number of *ERS SAR* scenes, this data includes *JERS-1 SAR*, *Landsat* TM, *SPOT*, and Airborne Ocean Color Imager (AOCI) images and space shuttle photographs. Thus it is now possible to begin to determine the variations of seep rate with time and the dependence of seep detection on ambient conditions. In addition, the images contain a great deal of information on other oceanic phenomena, such as using slicks to record current patterns.

Other objectives of the GOSAP consortium involved collecting current data, since the Loop Current has an adverse effect on oil field operations when it enters that part of the gulf in which they are situated. The consortium was also interested in compiling the gravity field of the Gulf of Mexico using radar altimeter data, since gravity is an effective means for understanding deep structures currently being ex-plored for oil. Members were also interested in learning to measure waves and wind, and to use these data as an aid in weather prediction.

9.4.3 Bathymetric Surveys

One of the most difficult environments in which to acquire seismic data is in the near-shore, where water is too shallow for the safe operation of seismic vessels and yet is beyond the reach of onshore acquisition methods. In many areas the problem is compounded by the unavailability of detailed or up-to-date bathymetric charts.

9.4.3.1 WATER DEPTH MAPPING.

Shell overcame the problem of near-shore seismic acquisition on the Mediterranean coast of Egypt west of Alexandria by using *SPOT* imagery to produce a bathymetric map (Figure 9.33). The water in this area was sufficiently clear that on suitably enhanced *SPOT* Multispectral (XS) data the end of the submarine pipeline from the El Alamein terminal to the tanker loading buoy could be seen in 17 m of water. A series of submerged beaches could be mapped along the entire stretch of coast, as well as areas of underwater sand dunes and vegetated hard ground.

A bathymetric map was made by taking advantage of the fact that the depth of light penetration is different for each spectral band. This map allowed the seismic crew to move the acquisition vessel much closer to the coast, without endangering the boat, than they could have otherwise.

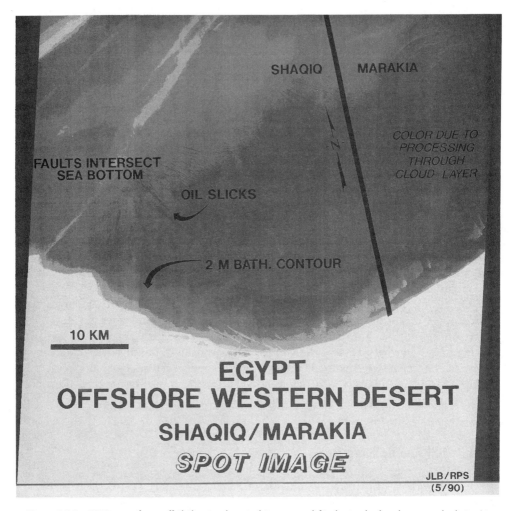

Figure 9.33 *SPOT* image of area off El Alamein oil terminal (on coast at left), density sliced to show water depth. Land is masked white. Red areas in north-west corner are thin clouds. Blue streaks extending northeast from the coastline are submerged coral sand banks. (Courtesy of Pecten International Inc.)

9.5 SUMMARY

Presented here is a review of the uses of remote sensing technology and interpretations in the exploration for hydrocarbons. The emphasis is on advances in remote sensing technology and techniques since the last edition of the Manual. Case histories are taken from the literature showing the latest concepts, techniques, and results of remote sensing projects. Although this can not be a complete review, it is a good survey of the "state-of-the-art" as applied in the energy industry.

The primary improvements in tools are the lower cost and greater power of computers to do the image processing. This has led to a decrease in the size of most energy company remote sensing groups, but has allowed an explosion in the number of small, customer-focused software development, image acquisition, image processing, and image interpretation companies.

Among government-sponsored systems, there has been an increased emphasis on radar satellites. However, sensor systems and their products are increasingly being developed and marketed in the private sector. The trend in optical instrumentation is toward higher spatial (to 1-meter) and spectral resolution (hyperspectral imagery). The chief advancements in software are in the integration of geographic information system (GIS) databases with imagery, allowing the generation of topographic and image base maps corrected to various coordinate systems and with infrastructure easily plotted on the image. These advances also permit the extraction of location information of image features. This leads in turn to the tremendous increase in use of imagery for logistical purposes, that is, for locating wells, planning access, laying out surveys, mapping hazards, and designing of pipeline routes, as well as the making of a more accurate geologic map.

The primary use of remote sensing in the energy industry will probably remain in the evaluation of frontier areas due to difficult access and inhospitable surface conditions for traditional surveys (high topographic relief, jungle or tundra cover, karsting, volcanics, sand, etc.). The applications to survey logistics and infrastructure planning will increase with continued improvements in resolution.

ACKNOWLEDGEMENTS

We would like to acknowledge Shell E & P Technology Company for providing the time and support to work on this chapter. We thank Shell, Chevron, Amoco, Mobil, and Earth Satellite Corporation for providing material included in this chapter.

References

Abrams, M. J., J. E. Conel, and H. L., Lang, 1984. Joint NASA/Geosat Test Case Project: Final Report. American Association of Petroleum Geologists, Tulsa, Okla., in 2 parts and 4 vols.

Aud, W. W., R. B. Sullivan, E. B. Coalson, T. D. Poulson, and P. A. Warembourg 1991. *Acid Refracturing Program Increases Reserves*, SPE Pap. 21821, Cottonwood Creek Unit, Washakie County, Wyo.

Berger, Z., T. H. Lee Williams, and D. W. Anderson, 1992. Geologic stereo mapping of geologic structures with *SPOT* satellite data, Am. Assoc. Petr. Geol. Bull., 76 (1); 101–120.

Biegert, E. K., R. N. Baker, K. E., Schaudt, I. R. MacDonald, B. Tapley, C. K. Shum, J. Amos, J. L. Berry, and A. T. Herring, 1994. Gulf Offshore Satellite Applications Project: GOSAP PP-USA.1; in Proceedings of the First ERS Applications Workshop, Toledo, Spain, June, pp. 417–429.

Biegert, E. K., R. N. Baker, R. V. Sailor, K. E. Schaudt, I. R. MacDonald, B. Tapley, C. K. Shum, J. Amos, J. L. Berry, and J. Hess, 1996. Project GOSAP (PP-USA.1): Gulf Offshore Satellite Applications Project, in Proceedings of the 2nd ERS Applications Workshop, London, Dec. 6–8 pp. 125–128.

Blanchet, P. H., 1957. Development of fracture analysis as an exploration method. *Am. Assoc. Pet. Geol. Bull.*, 41; pp. 1748–1759.

Carnes, M. R., and T. Ichiye. 1985. Numerical modeling of Langmuir circulation and its applications, in Y. Toba, and H. Mitsuyasu, (eds.) The Ocean Surface: Wave Breaking, Turbulent Mixing and Radio Probing. D. Reidel Publ., Dordrecht pp. 471–478.

Chen, Z., D. Ahl, J. Albasini, J. Moody, B. Davis, and M. Oxley, 1996. Evaluation of hyperspectral imagery for detecting microseepage, Proceedings of the 11th Thematic Conference on Geologic Remote Sensing, Las Vegas, Nev. Feb. 27–29, Vol. 1 Environmental Research Institute Ann Arbor Michigan: p. 584.

Collins, A., M. Freund, P. Kowalczyk, J. Berry, D. Cole, D. J. Pershouse, L. Burgess, S. Marsh, Abrams M., T. Munday, J. Huntington, I. W. Ginsberg, W. Malila, and F. W. Henderson, 1992. Evaluation of Simulated JERS-1 OPS Data for Geologic Mapping, Using Sites in Utah, Wyoming, and Arizona: A Report of the Joint GEOSAT/ERSDAC/JGI/JAPEX JERS-1 Petroleum and Mineral Geology Simulation Study, Geosat Committee, Norman, Okla.

Collins, R. J., F. P. McCown, L. P. Stonis, G. Petzel, G. and J. R. Everett, 1973. An evaluation of the suitability of ERTS data for the purpose of petroleum exploration. Third Earth Resource Technology Satellite-1 Symposium, NASA Goddard Space Flight Center, Washington, D.C., 1, pp. 809–22.

Colwell, R. N., ed., 1983. Manual of Remote Sensing, 2nd ed. (2 vols.) American Society of Photogrammetry, Falls Church, Va.

Cronin, B. T., M. K. Ivanov, A. F. Limonov, A. Egorov, G. G. Akhmanov, A. M. Akhmetjanov, E. Kozlova, and shipboard scientific party TTR-5, 1997. New discoveries of mud volcanoes on the Eastern Mediterranean ridge, *J. Geol. Soc. London*, 154; 173–182.

DeBlieux, C., 1962. Photogeology in Louisiana coastal marsh and swamp. Transactions of the 12th Annual Meeting of the *Gulf Coast Association of Geological Societies*, pp. 231–241.

Dekker, L. L., B. M. Clark, H. Evans, G. C. Graf, G. MacFarlane, R. B. Stevens, and R. C., Thomson, 1990. *PPL-101 Field Report of 1987, 1988, 1989 Programs*, unpublished report, Chevron Niugini Pty., Ltd. Papua New Guinea Department of Mining and Petroleum Open-File Rep. F1/R/90–163.

De Oliveira, W. J., and A. P. Crósta, 1996. Detection of hydrocarbon seepage in the São Francisco Basin, Brazil, through *Landsat* TM, soil geochemistry and airborne/field spectrometry data integration, in Proceedings of the 11th Thematic Conference on Geologic Remote Sensing, Las Vegas, Nev., 27–29 Feb., 1996, Vol. I: Environmental Research Inst. Michigan, Ann Arbor: 155–165.

Desjardins, L. 1952. Aerial photos may locate deep-seated salt domes, Oil Gas J., 51(3), 82–84.

Doeringsfeld, W. W. Jr., and J. B. Ivey, 1964. Use of photogeology and geomorphic criteria to locate subsurface structure. Mt. Geol. 1: 183–195.

Donovan, T. J., 1974. Petroleum microseepage at Cement, Oklahoma: evidence and mechanism, Am. Assoc. Pet. Geol. Bull., 58(1): 51–64.

Donovan, T. J., A. R., Barringer, R. S. Foote, and R. D., Watson, 1975. Low-altitude remote sensing experiments at Cement and Davenport Oil Fields, Oklahoma in *Proceedings of the Annual Meeting of the Society of Economic Geologists*, Denver, Colo.

Ellis, J. M., 1986. Geologic Interpretation of Northwestern PPL-18, Papua New Guinea. unpublished report. Chevron Overseas Petroleum Inc., Papua New Guinea Department of Mining and Petroleum Open-File Rep. F1/R/86–126.

Ellis, J. M, 1996. Upgrading remote sensing with GIS technology in support of petroleum operations, in Proceedings of the 11th Thematic Conference on Geologic Remote Sensing, Las Vegas, Nev., Feb. 27–29, 1996, v. I: Environmental Research Inst. Michigan, Ann Arbor: 167–174.

Ellis, J. M., and F. D. Pruett, 1986. Application of synthetic aperture radar (SAR) to southern Papua New Guinea Fold Belt exploration, in Proceedings of the 5th Thematic Conference on Geologic Remote Sensing, Reno, Nev., Environmental Research Inst. Michigan, Ann Arbor: 15–34.

Ernst, W. G., and E. D. Paylor II 1996. Study of the Reed Dolomite aided by remotely sensed imagery, central White-Inyo Range, easternmost California; *Am. Assoc. Petr. Geol. Bull*, 80(7), 1008–1026.

Feldman, S. C., R. E. Pelletier, E. Walser, J. C. Smoot, and D. Ahl, 1995. A prototype for pipeline routing using remotely sensed data and geographic information system analysis, *Remote Sensing Environ*, 53, 123–131.

Flower, F. B., E. F. Gilman, and I. A. Leone, 1981. Landfill gas, what it does to trees and how its injurious effects may be prevented. *J. Arbor*, 7, 43–52.

Fraser, A., P. Huggins, J. Rees, and P., Cleverley, 1996. A satellite remote sensing technique for geological structure horizon mapping, in Proceedings of the 11th Thematic Conference on Geologic Remote Sensing, Las Vegas, Nev., Feb., 27–29 Vol. 2: Environmental Research Institute Mich., Ann Arbor, pp. 665–673.

Geyer, R. A. 1977. Naturally occurring hydrocarbon seeps in the Gulf of Mexico and the Caribbean Sea: Dept. of Oceanography, College of Geosciences, Texas A&M University, 2–3.

Hanks, C. L., and R. M. Guritz 1997. Use of synthetic aperture radar (SAR) for geologic reconnaissance in arctic regions: an example from the Arctic National Wildlife Refuge, Alaska; *Am. Assoc. Pet. Geol. Bull.* 81(1), 121–134.

Horwitz, L., 1972. Vegetation and geochemical prospecting for petroleum, *Am. Assoc. Pet. Geol. Bull.*, 56, 925–940.

Horwitz, L. 1980. Near-surface evidence of hydrocarbon movement from depth. In *Problems of Petroleum Migration*, W. H Roberts, and R. J. Cordell, eds., Am. Assoc. Pet. Geol. Stud. Geol. (10), 241–269.

Hovland, M., and A. G Judd, 1988. Seabed Pockmarks and Seepages. Graham and Trotman, London, p. 293.

Ichiye, T., J. R. McGrath, and M. Howard, 1985. Some dynamic features of Langmuir circulation, in Y. Toba, and Mitsuyasu, H. eds. The Ocean Surface: Wave Breaking, Turbulent Mixing and Radio Probing. D. Reidel Publ., Dordrecht. 479–486.

Johnson, T. C. 1971. Natural oil seepage in or near the marine environment: a literature survey. Coast Guard Office of Research and Development Project No. 714141/002. 30.

Jones, R. F. E., and M. Oehlers, 1995. Successful integration of remote sensing and ground based exploration techniques in an arid environment, Oil and Gas J., 93(10); 47–51.

Kornacki, A. S., J. W. Kendrick and J. L. Berry, 1994. Impact of oil and gas vents and slicks on petroleum exploration in the deepwater Gulf of Mexico, Geo-Ma. Lett. 14; 160–169.

Kupsch, W. O., 1956. Submask geology in Saskatchewan. *Williston Basin Symposium*: pp. 66–75.

Lamerson, P. R., 1988. *Photogeologic Study of PPL-101.* unpublished maps, Chevron Niugini Pty., Ltd. Papua New Guinea Department of Mining and Petroleum Open-file Rep. F1/R/88–52.

Lattman, L. H., and W. W. Olive, 1955. Solution widened joints in Trans-Pecos, Texas, Am. Assoc. Pet. Geol. Bull., 39(10), 2084–2087.

Ma, W., and L. Chen, 1996. Research on characteristics of Jurassic reservoir in the frontier belt of Begede Mountain in Turpan-Hami basin. 30th International Geological Congress Abstracts, Beijing, 2: 849.

MacDonald, I. R., N. L., Jr., Guinasso, J. F. Amos, R. Duckworth, R. Sassen, and J. M. Brooks, 1993. Natural oil slicks in the Gulf of Mexico visible from space. Journal of Geophysical Research 98 (C9): 16, 351–16, 364.

Major, M. J., 1994. Remote view can aid site detection, *AAPG Explorer*, American Association of Petroleum Geologists, Tulsa, Okla., pp. 12–13.

Markello, J. R., and J. F., Sarg, 1996. Phanerozoic Tectono-stratigraphic evolution of the Trans-Pecos and Permian Basin Regions (Mexico, Texas, New Mexico) using landsat imagery, subsurface and outcrop data, in Proceedings of the 11th Thematic Conference on Geologic Remote Sensing, Las Vegas, Nev., Feb., 27–29 Vol. 2 Environmental Research Inst. Michigan, Ann Arbor: 651–664.

Maul, G. A. 1985. Introduction to satellite oceanography. Martinus Nijhoff Publishers, Dordrecht: 386–387.

Mikolaj, P. G., A. A. Allen, and R. S. Schlueter, 1972. Am. Inst. Min. Metall. Pet. Eng. Offshore Technol. Conf., Dallas, paper no. OTC-1549, 1: 1365–1378.

Molnar, P., and P. Tapponnier, 1975. Cenozoic tectonics of Asia: effects of a continental collision, *Science* 189(4201), 419–426.

Molnar, P., and Tapponnier, P. 1977. The collision between India and Eurasia, *Sci. Am.*, 236(4), 30–41.

Morgan, K. M., D. R. Morris-Jones, and D. G. Koger, 1982. Applying Landsat data to oil and gas exploration along the Texas Gulf Coast. *Oil & Gas*, 80 (39), 326–327.

Nelson, J. S., and E. C. Simmons, 1995. Diffusion of methane and ethane through the reservoir cap rock: implications for the timing and duration of catagenesis, Am. Assoc. Petr. Geol. Bull. 79(7); 1064–1074.

Nelson, J. S., and E. C., Simmons, 1996. Diffusion of methane and ethane through the reservoir cap rock: Implications for the timing and duration of catagenesis: reply, Am. Assoc. Petr. Geol. Bull., 80(9); 1486–1488.

Nishidai, T., and J. L. Berry, 1981. *Geological Interpretation and Hydrocarbon Potential of the Turpan Basin (NW China) from Satellite Imagery*, in Proceedings of the 8th Thematic Conference on Geologic Remote Sensing, Denver, CO., April 29–May 2, 1991, vol. 1, 373–389. Environmental Research Institute, Ann Arbor, Michigan.

Penny, F. A., 1975. Surface expression of deep discoveries, central Rockies. *Rocky Mountain Association of Geologists Symposium*, 55–61.

Prost, G. L. 1988. Predicting subsurface joint trends in undeformed strata. Environmental Research Inst. Michigan, Ann Arbor: Proceedings of the Sixth Thematic Conference on Geologic Remote Sensing, Houston, 2: 423–436.

Prost, G. L. (1996). *Airphoto Fracture Analysis of the Cottonwood Creek Field, Bighorn Basin, Wyoming*: SPE Prep. 35289: International Petroleum Conference and Exhibition, Villahermosa, Mexico, pp. 25–32.

Radlinski, A. P., and Z. Leyk, 1995. Formation of light hydrocarbon anomalies in oceanic waters. Geology 23(3); 265–268.

Reid, W. M. 1988. Application of thematic mapper imagery to oil exploration in Austin Chalk, central Gulf Coast Basin, *Texas. Am. Assoc. Pet. Geol. Bull.* 72(2); 239.

Richers, D. M., R. J. Reed, K. C. Horstman, G. D. Michels, R. N. Baker, L. Lundell, and R. W. Marrs, 1982. Landsat and soil-gas geochemical study of Patrick Draw oil field, Sweetwater County, Wyoming. *Am. Assoc. Pet. Geol. Bull.* 66; 903–922.

Rowan, L. C., M. J. Pawlewicz, and O. D. Jones, 1992. Mapping thermal maturity in the Chainman Shale near Eureka, Nevada, with *Landsat* thematic mapper images, *Am. Assoc. Pet. Geol. Bull.* 76(7); 1008–1023.

Sabins, F. F., 1969. Thermal Infrared Imagery and its Application to Structural Mapping in Southern California, Geol. Soc. Amer. Bull. 80, 397–404.

Sabins, F. F. Geologic Interpretation of Space Shuttle Radar Images of Indonesia, Amer, Assoc. Petr. Geol. Bull. 67; 2076–2099.

Sabins, F. F., 1984. Geologic Mapping of Death Valley from Thematic Mapper, *Thermal Infrared, and Radar Images*: Proc. 3rd Thematic Conf., Remote Sensing for Exploration Geol., Colorado Springs, CO., Vol. 1, p. 139–152. Envir. Res. Inst. of Mich., Ann Arbor; Michigan.

Sabins, F. F., R. Blom, and C. Elachi, 1980. Seasat Radar Images of San Andreas Fault, California, Amer. Assoc. Petr. Geol. Bull. 64, 619–628.

Saunders, D. F., 1981. Use of Landsat geomorphic and tonal anomalies in petroleum prospecting. *Unconventional Methods in Exploration Symposium*, Vol. 2, pp. 63–82.

Sgavetti, M., M. C. Ferrari, R. Chiari, P. L. Fantozzi, and I. Longhi, 1995. Stratigraphic correlation by integrating photostratigraphy and remote sensing multispectral data: an example from Jurassic-Eocene Strata, northern Somalia, *Am. Assoc. Pet. Geol. Bull.* 79(11); 1571–1589.

Sweet, W. E. 1973. Am. Inst. Min. Metall. Petrol. Eng. Offshore Technol. Conf., Houston, paper no. OTC-1803, 1: 1667–1672.

Tinkle, A. R., Antoine, J. W., and Kuzela, R. 1973. *Ocean Ind.* 8: 139.

Trollinger, W. V., 1968. Surface evidence of deep structure, Anadarko Basin. *Shale Shaker*, pp. 162–167.

Valenti, G. L., 1993. P'nyang Field: Discovery and Geology of a Gas Giant in the Western Papuan Fold Belt, Western Province, Papua New Guinea. Petroleum Exploration and Development in Papua New Guinea, Proc. Second PNG Petroleum Convention, Port Moresby, 31st May–2nd June 1993. Carman, G. J., and Z., (Eds).

Valenti, G. L., J. C. Phelps, and L. I. Eisenberg, 1996. Geologic remote Sensing for hydrocarbon exploration in Papua New Guinea, in Proceedings of the 11th Thematic Conference on Geologic Remote Sensing, Las Vegas, Nev., Feb., 27–29, Vol. 1: Environmental Research Institute Ann Arbor: Michigan, 97–108.

Wilson, R. D., P. H. Monaghan, A. Osanik, L. C., Price, and M. A. Rogers, 1974. Natural marine oil seepage, *Science* 184 (4139), 857–865.

REFERENCE BOOKS

Berger, Z. 1994. Satellite Hydrocarbon Exploration: *Interpretation and Integration Techniques*, Springer-Verlag, New York: 319 pp.

Drury, S. A., 1993. Image Interpretation in Geology, 2nd ed., Chapman & Hall, London, 283 pp.

Prost, G. L., 1994. Remote Sensing for Geologists: A Guide to Image Interpretation, Gordon and Breach, Lausanne, Switzerland; 326 pp.

Russ, J. C., 1995. The Image Processing Handbook, 2nd ed., CRC Press, Boca Raton, Fla., 674 pp. *This book provides all the practical information necessary to set up a remote sensing group and to carry out the most useful types of image processing for geological work.*

Sabins, F. F., 1986. Remote Sensing: Principles and Interpretation, 2nd ed., W. H. Freeman, New York, 450 pp. *This classic work has a geologic slant.*

Planetary Geology

James F. Bell III

Cornell University,
Ithaca, New York

Bruce A. Campbell

Smithsonian Institution,
Washington, D.C.

Mark S. Robinson

Northwestern University,
Chicago, Illinois

10.1 INTRODUCTION

10.1.1 What Is Planetary Geology?

Planetary geology is the study of surface processes on solid objects in the solar system: planets, satellites, asteroids, comets, and rings. It is a particularly appropriate subject for inclusion in a book on remote sensing, as the vast majority of our current knowledge on the geology of solar system objects has been derived from remote sensing measurements. These measurements have been obtained either using ground-based or Earth-orbital telescopes or robotic space probes equipped with sophisticated cameras or spectrometers. The exceptions are our detailed knowledge of several regions of the Moon from the samples returned by the Apollo and Luna missions and the

Remote Sensing for the Earth Sciences: Manual of Remote Sensing, 3 ed., Vol. 3, edited by Andrew N. Rencz.
ISBN: 0471-29405-5 © 1999 John Wiley & Sons, Inc.

inferences gleaned on the composition and evolution of the early solar system provided by meteorites and cosmic dust samples.

By its very nature, planetary geology is a highly cross-disciplinary subject. To interpret the various types of remote sensing and other data that are available successfully, the planetary geologist must have a good background not only in geology, but also, depending on the specific application, in such diverse subjects as astronomy, mineralogy, geochemistry, geophysics, chemistry, and even biology. In addition, it is becoming extremely important for planetary geologists to be proficient at computer programming and image/spectroscopic processing in order to extract as much information as possible from often limited data sets.

Planetary geology includes a substantial element of exploration and discovery (Figure 10.1). Many times students and researchers in the field will be exposed to completely new terrains or processes that have no clear terrestrial analogs. Examples are abundant in the solar system and include the dominance of impact cratering as a surface modification process on many bodies, large-scale volcanism involving ice or sulfur magmas, small bodies with active atmospheres, and exotic landforms indicative of major planetary climatic changes. In these situations the multidisciplinary nature of the field provides its greatest advantage, because the skilled planetary geologist must be able to apply universal sets of physical laws effectively to new or unusual situations.

In this chapter we provide a brief outline of the field of planetary geology, with particular emphasis on the importance of remote sensing in deriving our current understanding of the solar system. Much of this information has been distilled from excellent introductory textbooks and compilations such as Taylor (1982), Hartmann (1983), Carr et al. (1984), and Greeley (1987) and the exceptional series of planetary reference books published by the University of Arizona Press. References are provided throughout the chapter for readers interested in obtaining more details, and a suggested reading list is included at the end.

10.1.2 Relevance of Planetary Geology

Why is it important to study the geology and surface processes of other objects in the solar system? There are several important reasons. The first involves the quest for knowledge. The hallmark of human nature is to explore the unknown and to push the limits of the available technology in this quest. The exploration of the solar system, including the important task of characterizing the current physical, chemical, and morphologic state of planets, satellites, and so on (hereafter simply referred to as *planetary surfaces*), is an extension of humanity's exploration of the Earth and represents an exciting, educational, and fulfilling endeavor in which everyone can participate.

The second reason involves evolution, not necessarily in the biological sense but in the sense of the physical and chemical evolution of the solar system over time. How did planets form? What was the early solar system like? Why have some objects changed dramatically over time? Planetary geology includes an assessment of the past state of the solar system, for example as preserved in the ancient, scarred surfaces of objects such as the Moon and asteroids, or in the enigmatic polar layered deposits on Mars.

Figure 10.1   Planetary geology involves a substantial amount of exploration and discovery, and most planetary geology research is performed with the use of remote sensing instruments and techniques. Conducting field work is possible, however, as evidenced by the *Apollo* program's successful exploration of the Moon between 1969 and 1972. Here, *Apollo 12* astronaut Alan Bean examines the *Surveyor* spacecraft that had landed at the *Apollo 12* site several years earlier. This was a perfect example of robotic remote sensing observations providing important data used to support follow-up human exploration. The *Apollo 12* lunar module can be seen in the background.

The third, and perhaps most important, reason to study planetary geology is comparative planetology, or the ability to understand processes on or the evolution of one solar system object through comparisons to others. The solar system encompasses a wide diversity of objects, and these objects often allow for very informative "control experiments" to be carried out. For example, in terms of bulk size and density, Earth and Venus are very similar planets, yet the current state of the surfaces of these two bodies is vastly different. Studies of the physical and radiative properties of dust in the Martian atmosphere formed the basis of initial calculations on the effects of impact-or nuclear-generated dust and smoke in the Earth's atmosphere.

Understanding of the processes involved in the formation of the lunar crust is applied routinely to understand the surface composition of Mercury. Many such examples demonstrate the often unexpected utility of comparative planetology toward understanding processes and landforms in the solar system that are often a strong function of environmental conditions (gravity, temperature, atmospheric pressure, etc.).

10.1.3 Specific Measurements

Planetary remote sensing data are acquired using many different techniques, but the specific types of measurements and information to be obtained via these techniques can be grouped into the five general categories described below.

10.1.3.1 GROSS PHYSICAL CHARACTERIZATION.

Physical characterization includes determining the orbital characteristics, mass, shape, size, and density of solar system objects or of specific morphologic features on these bodies. These measurements are obtained from broadband visible, thermal, or radar systems that produce images of objects at various lighting and viewing angles, from radar or laser ranging measurements which determine the topography of planetary surfaces, or from measurements of the Doppler shift of spacecraft radio signals, which are sensitive to the mass, density, and internal density distribution of planetary bodies.

10.1.3.2 MAJOR SURFACE MODIFICATION PROCESSES.

Detailed morphologic study can yield substantial insights into the major surface modification processes that are currently active or that have been active at some time in the past. The chief means used to obtain this information is high-spatial-resolution imaging, perhaps complemented by stereoscopic imaging as well. For most planetary surfaces such imaging has been performed at visible wavelengths at varying resolutions, but for certain situations (e.g., Venus) it is most advantageous or only possible to perform imaging using radar techniques. Additional details are provided in Section 10.3. An understanding of the geologic processes that have acted on a planetary surface is used to determine and define major geologic units. Geologic histories of planetary surfaces are often documented in the form of geologic maps (compiled at various scales dependent on the available data and detail to be illustrated) which portray the three-dimensional surface units that comprise a planetary surface and indicate their relative stratigraphic positions.

10.1.3.3 AGES OF SURFACES.

Ages are determined from remote sensing images either by observation of superposition and other stratigraphic relations between different surface units or by comparing the relative abundances and size distributions of impact craters between units. The former technique allows for qualitative relative age dating only; the latter technique allows for an estimate of absolute ages and age differences for some solar system bodies by scaling of the relative crater density differences to the lunar cratering rate, which is itself tied to an absolute age scale through the detailed study of returned lunar samples (e.g., Wilhelms, 1987). Remote sensing age dating techniques rely on global-scale imaging of planetary surfaces at high enough spatial resolution to allow adequate statistics to be developed on craters over a wide size range. Absolute age correlation for bodies other the Moon is somewhat speculative because the absolute cratering rate for other bodies is simply not known to a high degree of accuracy.

10.1.3.4 SURFACE COMPOSITION AND MINERALOGY.

Determination of the composition of planetary surfaces provides information on their origin and geological/geochemical evolution. This information can be obtained remotely in many ways, including spectroscopic analysis of sunlight reflected from the surface at visible and near-infrared wavelengths, of thermal emission coming from the surface in the infrared, or of high-energy x-rays and gamma rays emanating from the surface and subsurface. Additional details are provided in Section 10.2.4.

10.1.3.5 ATMOSPHERIC CONDITIONS.

Assessment of atmospheric pressure, temperature, circulation, and composition can provide important constraints for use in analyzing planetary geology measurements. Examples include interpretation of sand dune fields on Mars using information on prevalent wind directions and interpretation of surface mineralogic information on Venus in light of the presence of specific chemically reactive species (e.g., H_2SO_4) in the atmosphere. Techniques for collecting these types of observations include imaging, high-resolution spectroscopy, and direct sounding and sampling by planetary landers and probes. A detailed description of methods for collecting planetary atmospheric remote sensing observations is beyond the scope of this chapter. Details can be found, for example, in Hanel et al. (1992).

10.2 BACKGROUND

10.2.1 Physical Data on Planets and Satellites

A summary of physical data on the terrestrial planets and major satellites is presented in Table 10.1. This table also summarizes what is known about the surface and atmospheric composition of these bodies. The inner terrestrial planets are primarily rocky, differentiated bodies composed of silicates, whereas the outer solar system satellites (and Pluto) are primarily icy bodies composed of ices of water, N_2, CO, and other species. This large dichotomy in composition results in substantial differences in the geology of rocky and icy surfaces, and also in substantial differences in the measurement approaches used to study these surfaces remotely.

10.2.2 Pre-spacecraft Telescopic Observations

Prior to the era of detailed spacecraft reconnaissance of the solar system, the primary means of obtaining planetary geology remote sensing data was via ground-based telescopes. Because of the limitations on resolution and image quality achievable by telescopic instrumentation, the scope of telescopic planetary geology was necessarily limited to observations of the Moon and Mars and less frequent observations of

TABLE 10.1 Physical Data on the Terrestrial Planets and Major Planetary Satellites

Planet	Radius (km)	Mass (kg)	Surface Density (g cm^{-3})	Surface Temperature (K)	Surface Gravity (cm s^{-2})	Distance from Sun (AU)	Orbital Period (days)	Rotation Period (days)	Composition Atmosphere	Composition Surface
Mercury	2425	3.29×10^{23}	5.44	100–700	370	0.39	87.97	58.65	—	Silicates (basalt)
Venus	6070	4.87×10^{24}	5.27	740	890	0.72	224.70	244.3[a]	CO_2	Silicates (basalt)
Earth	6378	5.98×10^{24}	5.52	290	982	1.00	365.26	1.0	N_2, O_2	H_2O, silicates (basalt)
Moon	1738	7.17×10^{22}	3.34	100–400	162		27.32	Sync[b]	—	Silicates (basalt)
Mars	3395	6.45×10^{23}	3.95	200–290	374	1.52	686.98	1.03	CO_2	Silicates (basalt?)
Phobos	7	1.26×10^{13}	2.2	≈225	0.3–0.5		0.32	Sync.	—	Carbonaceous rock
Deimos	4	1.80×10^{12}	1.7	≈225	0.3		1.26	Sync.	—	Carbonaceous rock
Jupiter						5.20				
Amalthea	≈86[c]	—	—	≈150	—		0.50	Sync.	—	Carbonaceous rock?
Io	1815	8.94×10^{22}	3.57	≈130	179		1.77	Sync.	Na, SO_2, S	Sulfur, silicates, ices
Europa	1569	4.80×10^{22}	2.97	≈130	132		3.55	Sync.	—	H_2O ice
Ganymede	2631	1.48×10^{23}	1.94	≈150	142		7.15	Sync.	—	H_2O ice, rock
Callisto	2400	1.08×10^{23}	1.86	≈150	122		16.69	Sync.	—	H_2O ice, rock
Saturn						9.54				
Mimas	199	3.67×10^{19}	1.12	≈100	7		0.94	Sync.	—	H_2O ice, rock
Enceladus	249	6.5×10^{19}	1.00	≈100	8		1.37	Sync.	—	H_2O ice, rock

	Radius (km)	Mass (kg)	Density				Period (days)	Rotation	Atmosphere	Composition
Tethys	530	6.12×10^{20}	0.98	≈100	15		1.89	Sync.	—	H_2O ice, rock
Dione	559	1.09×10^{21}	1.49	≈100	22		2.74	Sync.	—	H_2O ice, rock
Rhea	764	2.31×10^{21}	1.24	≈100	28		4.52	Sync.	—	H_2O ice, rock
Titan	2575	1.35×10^{23}	1.88	≈90	135		15.95	Sync.	N_2, CH_4	Ice + organics?
Hyperion	≈130[c]	—	—	≈100	—		27.64	Chaotic	—	H_2O ice, rock
Iapetus	718	1.88×10^{21}	1.03	≈100	15		79.33	Sync.	—	Ice + carbonaceous rock?
Phoebe	≈110[c]	—	—	≈100	—		550.45[a]	0.4	—	Carbonaceous rock?
Uranus						19.18				
Miranda	242	7.1×10^{19}	1.26	≈70	8		1.41	Sync.	—	H_2O ice, rock
Ariel	580	1.44×10^{21}	1.65	≈70	29		2.52	Sync.	—	H_2O ice, rock
Umbriel	595	1.18×10^{21}	1.44	≈70	22		4.14	Sync.	—	H_2O ice, rock
Titania	800	3.43×10^{21}	1.59	≈70	36		8.71	Sync.	—	H_2O ice, rock
Oberon	775	2.87×10^{21}	1.50	≈70	32		13.46	Sync.	—	H_2O ice, rock
Neptune						30.06				
Triton	1353	2.14×10^{22}	2.1	≈45	78		5.88[a]	Sync.	N_2, CH_4	H_2O, N_2, CH_4, CO ices, rock
Nereid	≈170	—	—	≈45	—		360.16	—	—	Ices + rock?
Pluto	≈1150	≈1.2×10^{22}	≈1.9	≈40	53	39.44	90,465	6.39	N_2, CH_4, ?	N_2, CH_4, CO ices
Charon	≈600	≈1.9×10^{21}	1.8–2.2	≈40	35		6.39	—	—	H_2O ice; $CO_2 + CH_4$ ice?

Source: Data compiled from Greeley (1987), Allen (1976), Taylor (1982), Burns and Matthews (1986), Young (1994), Dermott and Thomas (1994).

[a] Retrograde.

[b] Satellite is in synchronous rotation around primary.

[c] Amalthea has an ellipsoidal radius of approximately $127 \times 75 \times 68$ km. Hyperion has an ellipsoidal radius of approximately $164 \times 130 \times 107$ km; Phoebe has an ellipsoidal radius of approximately $115 \times 110 \times 705$ km.

Mercury and large solar system satellites. Despite the limitations, much was learned about these terrestrial bodies through early telescopic drawings and photographs.

The earliest telescopic observations of the Moon allowed for the characterization of the gross morphologic character of the lunar nearside at a spatial resolution as high as 1 km (e.g., Gilbert, 1893; Spurr, 1944; Baldwin, 1949). The obvious differences between the bright, mountainous highlands and the smooth, dark maria were attributed to flooding of depressions by fluid lavas. The nature of the regolith remained uncertain, with some scientists proposing a vast ocean of fine dust that might engulf landing spacecraft. The circular craters were obvious, but a significant degree of controversy raged as to their origin. One proposal suggested that they were formed by volcanic activity, whereas others believed them to be of impact origin. Pioneering work by E. Shoemaker and others in the early 1960s demonstrated the role of cratering on Earth (e.g., Shoemaker, 1963), and the first spacecraft lunar observations supported the interpretation that the Moon's surface processes are dominated by impact events. Color-difference methods and other simple photometric surveys illustrated the range of compositions among lunar basalts and were the precursor of later detailed spectroscopic analyses (e.g., Whitaker, 1966).

Early telescopic observations of Mars concentrated on characterization of surface albedo patterns and on monitoring of surface and atmospheric variability (e.g., Slipher, 1962; Martin et al., 1992). Geologic information was difficult to extract from drawings and photographs having spatial resolutions on the order of hundreds of kilometers, and observations by Schiaparelli, Lowell, Antoniadi, and others demonstrated that the interpretation of features observed on Mars was controversial and often highly subjective (Sheehan, 1988). Nonetheless, it was realized that the Martian surface is divided into at least two gross geologic provinces, bright regions and dark regions, and that the dark regions exhibit much greater evidence for temporal variability. It was also realized that Mars has a dynamic atmosphere that interacts with the surface on diurnal, seasonal, and interannual time scales, as evidenced by observations of local and global dust storms and the waxing and waning of the seasonal polar caps. The composition of the Martian surface was inferred from its color to be dominated by oxidized iron minerals (for a review of the earliest color and spectroscopic measurements, see de Vaucouleurs, 1954).

Spatially resolved telescopic observations of other terrestrial planets and satellites were extremely difficult using pre-space-age technology. Mercurian drawings and photographs compiled just before the turn of the century portrayed enigmatic and controversial surface markings that were never convincingly tied to specific surface geologic processes (see Chapman, 1988). Geologic observations of the surface of Venus at visible wavelengths are precluded by the inability to see below the thick clouds. However, radar echoes from the surface of Venus were first detected in 1961 and were later used to demonstrate that the planet rotates once every 243 days, such that a Venusian "day" lasts longer than its year (Shapiro, 1968).

10.2.3 History of Planetary Exploration Missions

Since the early 1960s several nations have embarked upon an ambitious series of manned and unmanned planetary exploration missions (summarized in Table 10.2). These missions have proceeded along a fairly logical progression of initial flybys (or "hard landings" in the case of the early lunar probes), more detailed orbital reconnaissance, soft landings, and manned exploration (see Nicks, 1985). A summary of the current status of solar system exploration is provided in Table 10.3.

10.2.4 Planetary Remote Sensing Techniques

10.2.4.1 CHOICE OF WAVELENGTH REGIONS FOR PLANETARY APPLICATIONS.

Remote sensing observations for the study of planetary geology span a wide variety of interests and applications. The specific object being observed or goal being addressed drives the choice of which region of the electromagnetic spectrum is appropriate. For example, surface mineralogy can be investigated using spectroscopy from the visible to infrared, because many minerals exhibit diagnostic absorption and emission features at these wavelengths. Geomorphologic information can result from broadband visible orbital imaging in the case of planets with thin to nonexistent atmospheres. For planets with atmospheres, consideration must be made of the opacity of the atmosphere as a function of wavelength; in certain cases, such as for Venus, the only choice for orbital geomorphologic studies is microwave (radar) imaging.

In many cases, the final detailed choice of wavelengths to study may be dictated by instrumental considerations. Charge-coupled device (CCD) cameras, for example, which are now widely available for ground-and space-based study of the planets, are often limited to operation in the region 400 to 1100 nm. Infrared detector technology is continually improving; however, the range of wavelengths available to an instrument is highly dependent on the composition of the detector and on the ability to cool the instrument below some optimal operating temperature.

10.2.4.2 IMAGING SYSTEMS.

Techniques for obtaining planetary images vary according to the specific wavelength region of interest. Early ground-based and spacecraft visible wavelength imaging relied on photographic techniques, often supplemented by the use of broadband color filters. However, photographic film is a nonlinear detector that is extremely difficult to calibrate, so alternatives were sought that allowed for the derivation of more quantitative information from images. A major advance in spacecraft imaging after the film systems of *Lunar Orbiter* was the vidicon, first flown successfully on the *Mariner 4* flyby of Mars in 1965 (Leighton et al., 1967). Vidicon imaging systems work in a two-step process. First, the photoconductive surface is exposed, much like film in a conventional camera, after an electron beam passes over the back of the photoconductor priming it with a negative charge. Photons incident on the photoconductor surface reduce the negative charge in proportion to scene brightness. After this picture forming step, the resultant image on the photoconductor is read out by a scanning electron beam. The areas of the image with a less negative charge will draw more electrons for the beam, and this variation in current is read out as the video signal current. After readout, the image is digitized and sent back to Earth (Leighton et al., 1967; Danielson et al., 1975). Early vidicon images (*Mariners 4, 6, and 7*) were crude and had low photometric accuracy by today's standards. However, they provided a dramatic first look at Mars. Improved vidicon cameras were extremely reliable with a photometric accuracy generally better than 10% (*Mariner 10,*

TABLE 10.2   Summary of Successful Planetary Exploration Missions in the Space Age

Mission Name	Dates	Goals and Results
Luna 2, 9, 13	1959–1968	Lunar hard and soft landings; photography, soil physics
Luna 3	1959	Lunar farside flyby; photography
Mariner 2, 5	1962, 1967	Venus flybys; gravity, radiometry
Ranger 7–9	1964–1965	Lunar hard landings; photography
Mariner 4	1965	Mars flyby; photography
Zond 3, 5–8	1965–1970	Lunar flyby; photography
Luna 10, 12, 14, 19	1966–1971	Lunar orbiter; gravity and magnetic field data, photography
Surveyor I, III, V–VII	1966–1968	Lunar soft landings; photography, soil physics and chemistry
Lunar Orbiter 1–5	1966–1967	Global medium- to high-resolution lunar photography
Venera 4–8	1967–1972	Venus hard and soft landings; photography
Apollo 8, 10	1968–1969	Manned lunar orbiters; photography
Mariner 6, 7	1969	Mars flyby; photography, spectroscopy, atmospheric sounding
Apollo 11, 12, 14–17	1969–1972	Manned lunar exploration and sample return
Luna 16, 20, 24	1970–1976	Unmanned lunar sample return
Lunokhod 1, 2	1970, 1973	Lunar traverse vehicles; covered 20 km and 30 km each
Mariner 9	1971	Mars orbiter; photography, spectroscopy, atmospheric sounding
Pioneer 10	1973	Jupiter flyby; imaging, radiometry
Venera 9, 10	1974, 1975	Venus orbiters and entry probes; atmospheric composition
Pioneer 11	1974, 1979	Jupiter, Saturn flybys; imaging, radiometry, ring science
Mars 5	1974	Mars orbiter; photography, polarimetry
Mariner 10	1974–1975	Venus and Mercury flybys; imaging
Viking 1, 2	1976–1979	Mars orbiters and soft landers; imaging, soil analyses, biological assessments
Pioneer Venus 1, 2	1978	Venus orbiter and entry probes; gravity, atmospheric composition
Venera 11–14	1978–1982	Venus flybys and entry probes; soft landings; photography
Voyager 1	1979, 1980	Jupiter, Saturn system flybys; high-resolution imaging
Voyager 2	1979–1989	Jupiter, Saturn, Uranus, Neptune flybys; high-resolution imaging
Venera 15, 16	1983–1984	Venus orbiters; radar imaging
Vega 1, 2	1985	Venus balloons and atmospheric composition; Halley flybys
Giotto	1986	Halley's Comet flyby; CCD imaging, composition
Phobos 2	1989	Mars orbiter; CCD imaging, imaging spectroscopy
Galileo	1989–present	Venus, Earth, Moon, asteroid flybys, Jupiter orbiter & probe; CCD imaging, spectra
Hubble Space Telescope	1990–present	Earth orbiting telescope; high-resolution synoptic CCD imaging
Magellan	1990—1993	Venus orbiter; global radar mapping and gravity mapping
Clementine	1994	Lunar global multispectral mapping, topography, and gravity

TABLE 10.2 *(Continued)*

Mission Name	Dates	Goals and Results
NEAR	1995–present	C-Type Asteroid flyby, S-Type 1-year orbital mission
Mars Pathfinder	1996–1997	Mars Lander and Rover; CCD imaging, geo-chemistry
Mars Global Surveyor	1996–present	Mars orbiter; CCD imaging, IR spectroscopy, laser altimeter
Cassini	1997–present	Saturn orbiter, Titan radar mapper & entry probe (2004)
Lunar Prospector	1997–present	Lunar orbiter; geochemistry, gravity, topography

Viking, Voyager). The photoconductive surface was sensitive to visible light in a range of about 350 to 650 nm. To obtain multispectral data a filter wheel was employed that allowed for several broadband filters typically placed between about 400 and 600 nm. Color filters led to the mapping of gross color heterogeneities on a scale of 1 to 10 km for Mercury, Mars, and selected outer planet satellites (e.g. Soderblom et al., 1978; Rava and Hapke, 1987).

More recently, the CCD camera has replaced these previous visible wavelength instruments as the detector of choice. CCDs provide excellent dynamic range, linearity, and responsivity from the near-ultraviolet (UV) through the near-infrared (IR) (with some modifications, their sensitivity range can extend from about 200 to 1110 nm). CCDs operate by allowing incident photons to produce a current in silicon-based semiconducting wafers or chips. The chips are divided into thousands or even millions of independent charge-collection regions called pixels ("picture elements"), and these pixels are organized into two-dimensional arrays in order to generate an image. The current in each pixel is a highly linear function of the incident

TABLE 10.3 **Techniques Employed for Planetary Geologic Exploration**

Technique	Mercury	Venus	Moon	Mars	Outer Satellites	Comets	Asteroids
Groundbased	×	×	×	×	×	×	×
Spacecraft							
Flyby	×	×	×	×	×	×	×
Orbiter		×	×	×	[a]	[b]	[c]
Lander		×	×	×		[b]	
Sample return			×	[d]			
Human landing			×				

[a] This includes detailed orbital investigation of the Jovian satellites by the Nasa *Galileo* mission from 1996 to the present, and planned observations of the Saturnian satellites by the NASA/ESA Cassini orbiter and Huygens Titan entry probe during 2004–2008.

[b] Orbital and lander study of the Comet P/Wirtanen are planned for the ESA *Rosetta* mission in 2013–2014.

[c] Orbital study of the asteroid 433 Eros is planned for the NASA *NEAR* mission in 1999.

[d] Current NASA plans call for the first Mars sample return mission to be launched in 2005, with return of samples to Earth by 2008.

photon flux, and most CCDs provide excellent quantum efficiency, the ratio of induced photoelectron current to incident photon flux (e.g., Fowler et al., 1981; Mortara and Fowler, 1981).

High-quality imaging at near-infrared and mid-infrared wavelengths has recently become practical because of advances in infrared-sensitive arrays. Specifically, arrays constructed from indium and antimony (InSb) substrates have had spectacular success in achieving high signal/noise ratio (SNR) and high dynamic range for telescopic and spacecraft imaging applications. Other IR-sensitive substrates, including silicon–arsenic (SiAs), germanium (Ge), and indium–gallium–arsenic (InGaAs), have also been used with good results. A particular advantage of many of these arrays is their ability to operate effectively with only modest cooling requirements. Ge and InGaAs arrays, for example, can operate effectively even at temperatures as high as 250 to 270 K. InSb arrays can produce good data when cooled to liquid nitrogen temperatures (77 K), although they perform even better at liquid helium temperatures (4 K).

Imaging at microwave wavelengths from a few millimeters to several millimeters forms the basis of radar remote sensing systems. Active systems use a powerful transmitter to illuminate the target surface and sensitive receivers to pick up the faint echoes reflected by the surface materials. By carefully measuring the time delay and Doppler shift of returned signals, a radar echo map can be produced. The radar echo is influenced by local slopes, by roughness at the scale of the wavelength, and by the dielectric properties of the surface. By varying the incidence angle, polarization, and wavelength of the illuminating energy, a more complete estimate of the surface properties can be obtained. More details on radar remote sensing can be found in Ostro (1983) and Butrica (1996). Passive microwave systems exploit the fact that all materials radiate energy received from the Sun or internal sources in a characteristic wavelength distribution described by the Planck function. This distribution is strongly dependent on the temperature of the surface and its emissivity (the efficiency of the material in emitting energy to its surroundings). Measurement of the amount of energy radiated from a surface at different wavelengths can thus be used to constrain these and other important surface parameters (e.g., Muhleman, 1972).

10.2.4.3 SPECTROSCOPY.

Spectroscopic remote sensing observations can provide substantially more diagnostic compositional and mineralogic information on planetary surfaces than imaging alone (cf. Chapters 1 and 5). Essentially three types of spectroscopic observations can be obtained: x-ray and gamma ray spectra, reflectance spectra, and thermal emission spectra. Each offers specific advantages for answering certain types of compositional or mineralogic questions.

X-ray and gamma ray spectra provide diagnostic information on the abundances of specific elements in the outermost layers of a planetary surface. These spectroscopic techniques take advantage of the fact that high-energy galactic cosmic ray particles and lower-energy solar x-rays and charged cosmic ray particles penetrate up to several centimeters into the surface and excite x-ray radiation that is characteristic of specific elements. This secondary radiation is then emitted from the surface and can be detected using both scintillator and semiconductor detectors at gamma ray energies and gas-filled proportional counters at x-ray energies. These techniques can provide quantitative information on the abundances of many rock-forming minerals (e.g., Na, Mg, Al, Si, P, S) as well as on the abundance of natural radioactive materials (K, Th, U) and hydrogen. Detailed additional information on x-ray and gamma ray remote sensing techniques can be found in Evans et al. (1993) and Yin et al. (1993).

Reflectance spectroscopy provides diagnostic information on the mineralogy and degree of crystallinity of the uppermost few microns of a planetary surface. This technique involves measuring the spectrum of sunlight reflected from a planetary surface and is thus restricted to the wavelength range where the Sun's flux is highest and where the amount of energy reflected from the object is greater than the amount that is thermally emitted (the typical wavelength range is from 0.3 to 3.5 μm). Many materials, including primary and secondary minerals, exhibit electronic spectral features and vibrational overtone bands at these solar wavelengths. Reflectance spectra reveal absorption features that are characteristic of certain minerals and ices and/or indicate the presence of certain cations. For example, the mineral pyroxene, a common component of basaltic rocks on the Earth, can be detected remotely by the measurement of diagnostic absorption features near 1.0 and 2.0 μm. Variations in the abundances of Fe and Ca in the pyroxene can also be inferred based on subtle shifts in the positions of these bands.

Reflectance spectroscopy is currently the most useful technique for remotely measuring the mineralogy of planetary surfaces, and specific minerals have been identified on the Moon, Mars, and a number of asteroids. Additional detailed information on reflectance spectroscopy theory and techniques can be found in Burns (1993) and Gaffey et al. (1993).

Thermal emission spectroscopy also provides diagnostic information on the mineralogy of planetary surfaces, as well as additional information on surface thermophysical properties such as temperature and thermal inertia. Most of the major rock-forming minerals exhibit their fundamental molecular vibration spectral features at mid-infrared wavelengths, typically from 3.0 to 25.0 μm, where thermal radiation emitted from planetary surfaces at temperatures from 200 to 400 K dominates over reflected sunlight. Unlike reflectance spectra, thermal infrared spectra can exhibit features in both emission and absorption, depending on the nature of the planetary environment. Thermal infrared spectra are more difficult to interpret, but they also allow the potential, through radiative transfer modeling, to infer additional information about a planetary surface such as emissivity, particle size and degree of compaction, and the subsurface temperature profile (e.g., Hapke, 1996). Thermal emission spectra have been used, for example, to obtain remote compositional information on variations in terrestrial basaltic lava flows and to constrain the thermal inertia and rock abundance of the Martian surface (e.g., Christensen, 1986; Kahle et al., 1993). More details on the theory and application of thermal emission spectroscopy can be found in, for example, Salisbury et al. (1991), Hanel et al. (1992), and Salisbury (1993).

10.2.4.4 IMAGING SPECTROSCOPY.

The spectroscopic techniques discussed above have typically been constrained by instrumentation to obtain compositional or mineralogic information only for specific, possibly small places on planetary surfaces. The imaging techniques discussed above have traditionally been used primarily for obtaining morphologic information, at the expense of more detailed compositional or mineralogic data. A recent and important development in remote sensing is the combination of imaging and spectroscopic techniques to allow for the determination of compositional information and the mapping of this information across a planetary surface at high resolution.

Reflectance imaging spectroscopy measurements are currently based on two different techniques. The first technique is to combine an imaging instrument with a modest number (8 to 24) of discrete narrowband filters placed at key wavelengths for the detection of specific minerals and/or ices. The result is a high-spatial-resolution data set at high enough spectral sampling and resolution to allow for the mapping of various spectral units. The trade-off for this method is that not all of the wavelengths can be obtained simultaneously, so spacecraft motion or other variable effects act to introduce uncertainties into the data. The second technique is to combine a spectrometer with a two-dimensional array to allow for the spectra of different spatial locations to be obtained simultaneously. The result is a high-spectral-resolution data set at high-enough spatial resolution to again allow for the mapping of surface units. The trade-off for this method is that only one axis of spatial information can be obtained at a time, so to build an image, the second axis of spatial information must be obtained by scanning the instrument and/or moving the spacecraft. The former approach has been used very successfully by ground-based and Hubble Space Telescope observations of the Moon, Mars, and asteroids, and on the *Galileo, Clementine*, and *Mars Pathfinder* spacecraft missions. The latter technique has also been used successfully by ground-based observers and by instruments on the *Phobos 2, Galileo*, and *NEAR* missions. The choice of technique depends primarily on the importance of obtaining high spectral resolution. For cases where lower resolution is adequate, the discrete filter technique offers excellent image quality; for cases where detailed compositional or mineralogic information is required, spatial image quality must be sacrificed for higher spectral resolution. More details on the theory and implementation of imaging spectroscopy can be found in Adams et al. (1993) and Vane et al. (1993).

10.3 PRIMARY PLANETARY SURFACE MODIFICATION PROCESSES

This section provides a brief overview of the four major surface modification processes that can act on a planetary surface and that can be characterized by planetary geologic remote sensing observations. These processes include the three primary processes most often studied by terrestrial geologists (volcanism, tectonism, and gradation), and a fourth process that often dominates the geology of many planetary surfaces (impact cratering). Most of our knowledge of these processes is based on the study of geologic features on the Earth. As such, an important part of planetary geology is determining how these features may vary due to the action of surface modification processes on other planets.

10.3.1 Impact Cratering

In the context of planetary geology, impact cratering may be the single most important surface modification process at work in the solar system (Figure 10.2), because

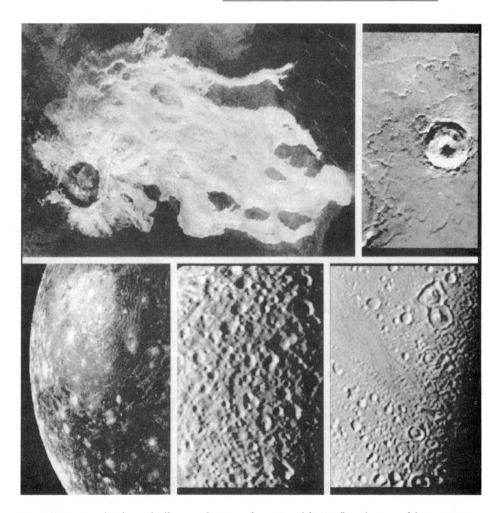

Figure 10.2   Examples of crater landforms on planetary surfaces. (Upper left) *Magellan* radar image of the Venusian impact crater Markham (60 km diameter). The radar-bright flows that extend eastward from the crater rim are probably impact melt material produced during excavation of the crater cavity. (Upper right) *Viking Orbiter* image of the Martian impact crater Yuti (19 km diameter). The morphology of this crater's ejecta blanket suggests the presence of ice or water in the Martian subsurface at the time of impact. (Lower left) *Voyager* image of the Valhalla Basin on the Jovian satellite Callisto. Valhalla is a 2750-km-diameter multiring impact feature on the surface of this ice/rock body. The heavily cratered surface of the Saturnian satellite Mimas (lower middle) can be contrasted in these *Voyager* images to the less densely cratered (younger) and more geologically enigmatic surface of Enceladus (lower right), which is also a Saturnian satellite and which has a similar size and density as Mimas. The geologic explanation of the ridges, smooth plains, and oddly shaped craters on Enceladus, which is only 249 km diameter, is still unknown.

it has operated continuously since the planets originally formed (although at a decreasing rate with time). The detailed study of the role of impacts in the geology of the planets has provided the impetus to understand the cratering process on the Earth as well as the role of cratering in Earth's history.

The impact process involves the transfer of energy from the impactor, known as a *bolide*, to a planetary surface. Bolides can range in size and composition from small meteoroids to larger asteroids and comets to large planetesimals. The amount of energy transferred determines the overall geologic consequences and is a function of the size, velocity, composition, and strength of the bolide and the physical and environmental conditions of the target surface (gravity, composition, atmospheric pressure).

Four physically distinct stages of impact crater growth are recognized to occur (Taylor, 1982): (1) collision and transfer of the bolide's kinetic energy into the surface through a shock wave; (2) rarefaction of the shock wave and decompression of the crustal materials traversed by the compressive shock wave; (3) acceleration of materials disrupted by the rarefaction wave, resulting in excavation of the crater cavity; and (4) subsequent readjustment of the transient crater cavity mostly due to gravitational forces and relaxation of compressed target materials.

Most energy transfer takes place via the compressive shock wave as thermal energy and as kinetic energy through ballistic ejecta and movement of materials within the cavity. Peak pressure may be as high as 5000 kbar (impact velocity of 15 to 25 km/ s^{-1}). Materials under pressures greater than 700 kbar are generally molten (thought to be less than 10% of the excavated volume).

The final crater modification stage is complex and includes fallback of ejecta into the cavity, the emplacement of ejecta blankets preserving an inverted stratigraphy in the region adjacent to the crater, collapse of the walls of the transient crater cavity, rebound of the crater center (uplift), seismic shaking, and the formation of impact melt sheets that drape much of the crater topography (e.g., Figure 10.2). Details on the physics and mechanics of impact processes can be found in Shoemaker (1963), Gault et al. (1968), Hartmann (1977), Roddy et al. (1977), Kieffer and Simonds (1980), and Greeley (1987).

Because impact cratering involves the random collision of objects in the solar system, older surfaces have a higher likelihood of being struck by bolides. The observed density of impact craters per unit surface area thus provides a measure of both the relative age of the surface and the amount of modification experienced by that surface over its geologic history. Considerable effort has been expended trying to determine an absolute age scale based on the technique of counting the number of craters per unit surface area and dividing by an estimate of the flux of incoming bolides per unit time (e.g., Shoemaker, 1966; Shoemaker et al., 1970; Neukum et al., 1975). This age-dating tool has been fairly successful for determining absolute ages of lunar surface units which have been validated by radiometric dating of returned *Apollo* lunar samples. However, impacts are the dominant surface modification process on the Moon, and substantial problems need to be overcome to apply this surface age-dating technique to other solar system objects with more varied geologic histories and cratering rates. For example, not all circular features seen in planetary images may be of primary impact origin (they may be related to volcanic events, karst, secondary cratering events, etc.). Also, studies have shown that surfaces can reach a crater saturation value when the creation of new craters destroys evidence of previously existing craters (Gault, 1970). Such surfaces cannot be calibrated accurately to an absolute age scale. Despite these limitations, impact crater statistics can be used effectively as a means of establishing a relative age sequence among different planetary surfaces and surface units.

10.3.2 Volcanism

Volcanism involves the melting of materials within a planet and the transport or eruption of these materials onto a planetary surface. For most surfaces the melted material is silicate-bearing rock called *magma*. The most common magma on the terrestrial planets is of basaltic composition (e.g., Cas and Wright, 1987). However, volcanism involving molten sulfur or ices of various compositions has been found or postulated to exist on a number of outer solar system objects (Figure 10.3). The heat required to melt planetary materials can be primordial heat retained from the formation of differentiated objects, heat released by radioactive elements in a planetary interior, or heat created by planetary tidal or tectonic processes.

Volcanic landforms exhibit a wide range of morphologies that are a strong function of planetary physical and environmental conditions, physical and chemical properties of the magma, and the style and mechanics of the eruption (Whitford-Stark, 1982). The most common volcanic landforms in planetary geology are basaltic lava flows, shield volcanoes, cinder cones, and lava channels (Figure 10.3). Terrestrial volcanic landforms have been studied in great detail both as a means of obtaining information on the thermal evolution and interior characteristics of the Earth as well as providing a comparative basis for the study of volcanic landforms on other planets (e.g., Greeley and King, 1977; Carr and Greeley, 1980; Basaltic Volcanism Study Project, 1981). Detailed compilations and analyses of planetary volcanic features can be found in Head (1976), Plescia and Saunders (1979), Greeley and Spudis (1981), Whitford-Stark (1982), Wilhelms (1987), and Mouginis-Mark et al. (1992).

10.3.3 Tectonism

Tectonism involves the deformation of the lithosphere of a planet, driven by either internal (volcanic, tidal, radioactive) or external (impact) forces. Tectonic processes manifest themselves as distinctive morphologic features such as faults, fractures, and folds, and the specific characteristics of these features can often be used to infer the local style of deformation that has occurred (Figure 10.4). For example, extensional stretching of the lithosphere often leads to the production of graben and horst topography, whereas compression of the lithosphere often leads to the production of syncline and anticline topography.

The global-scale interpretation of tectonism on other planets has been greatly influenced by the discovery and understanding of plate tectonics on Earth. The Earth's lithosphere is divided into numerous plates of different sizes, each moving at different rates and all floating atop the semimolten upper mantle layer called the asthenosphere (e.g., Cox and Hart, 1986). This results in complex systems of tectonic features that are often difficult to interpret uniquely from regional-or global-scale remote sensing views. In contrast, most other planetary surfaces are thought to be single-plate surfaces with much thicker lithospheres. Global-scale tectonic features on these planetary surfaces are formed by much simpler extensional and compressional forces (such as uplift or cooling), and thus their interpretation via remote sensing techniques is, most often, more straightforward although still complex (e.g., Head and Solomon, 1981).

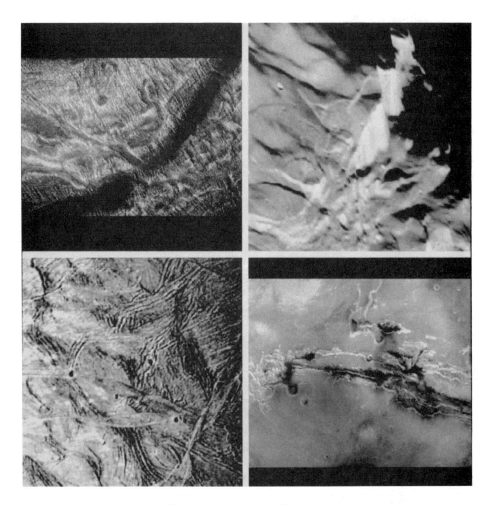

Figure 10.3   Examples of volcanic landforms on planetary surfaces. The diversity of volcanic terrains on Venus rivals even that of the Earth, as evidenced by recent *Magellan* radar images. (Upper left) Flat-topped "pancake" volcanoes, possibly formed by the eruption of extremely viscous lava. Each of these structures is approximately 25 km in diameter with topographic reliefs of up to 750 m. (Upper middle) Large double-summit shield volcano Sapas Mons, located in the Atla Regio area of Venus. This image covers a region 650 km wide. Lava flows originate on the flanks of this volcano and extend for hundreds of kilometers onto the surrounding smooth plains. The summit region of Sapas is radar-bright due to a change in surface chemistry that occurs at high elevations on Venus. Volcanism has also been pervasive on Mars. The summit caldera complex of the shield volcano Olympus Mons, the largest volcanic construct in the solar system, is seen here in a *Viking Orbiter* image (lower right). The caldera is 65 × 80 km across and is composed of a number of coalesced collapse craters with wrinkle ridges on their floors. The ridge and slump morphology of the inner caldera is similar to that seen in the calderas of many terrestrial shield volcanoes. The most volcanically active object in the solar system is Jupiter's innermost moon Io (upper right). In this *Voyager* image, showing many bright and dark flows and vents on Io, the volcano Prometheus is seen erupting along the limb, spewing sulfur and other materials on ballistic trajectories in a plume 50 km high and 300 km across. Volcanism among the icy satellites has also been discovered, as evidenced by, for example, *Voyager* images of the surface of Neptune's large moon Triton (lower left), Europa (Figure 10.17*E*), Enceladus (Figure 10.2), and Miranda (Figure 10.17*D*). This image of Triton shows a region 200 km across having two large smooth-floored craters that have been interpreted as evidence for fluid extrusion of an icy magma of unknown composition.

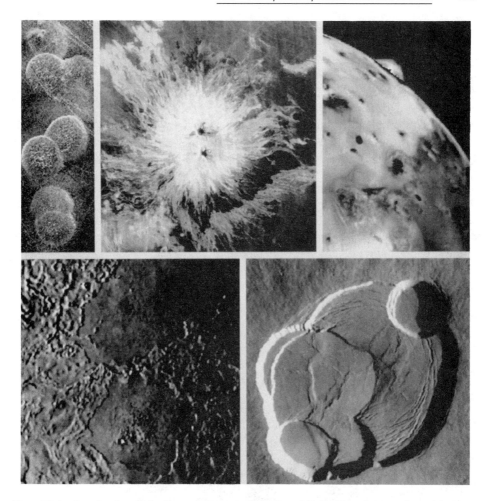

Figure 10.4   Examples of tectonic landforms on planetary surfaces. (Upper left) *Magellan* radar image of part of the interior of Ovda Regio, Venus. This image shows an underlying fabric of NE/SW-trending ridges and valleys, dominated by the large, dark, 20-km-diameter lava-filled valley cutting diagonally across the image. Superposed on these valleys and ridges is a system of NW/SE-trending bright features also of tectonic origin. This typical example testifies to the long history of tectonic deformation experienced by much of the Venusian surface. (Upper right) *Voyager* image of the surface of the Uranian moon Miranda, showing a tectonically modified region 700 m across riddled with ridges, valleys, and scarps. The bright, striated scarp seen extending across the terminator is roughly 5 km high. (Lower left) Portion of a *Voyager* image of complex tectonic terrain on the Jovian moon Ganymede. The region shown is about 50 km across, and shows crisscrossing ridges that may be evidence for either compressional tectonism or for the extrusion of material through a complex system of faults. (Lower right) *Viking Orbiter* mosaic of the Valles Marineris canyon system, thought to have formed from extensional tectonics associated with the huge Tharsis volcanic region to the west. The image shows a region roughly 2000 km across. The maximum width of the canyon system is 200 km, and the maximum floor depth is 7 km.

10.3.4 Gradation

Gradation involves the wearing down of topographic highs and the filling in of topographic lows on a planetary surface. This process of erosion and deposition is driven by gravity and is strongly influenced by environmental factors such as atmospheric pressure and density and surface temperature and composition. The primary means of moving materials by gradation come from the actions of liquid water, ice, wind, and mass wasting (the downslope movement of rock and debris). Each of these processes produces characteristic morphologic features that are often detectable by remote sensing (Figure 10.5). For example, mass wasting results in the formation of landslides, slumps, talus cones, and rock glaciers (e.g., Sharpe, 1968). Water produces channels and other characteristic drainage patterns (e.g., Howard, 1967) and can also lead to the formation of karst topography via dissolution of salts. Ice gradation can produce a variety of glacial and periglacial features, including moraines, cirques, pingos, and polygonal terrain (e.g., Bloom, 1978). The morphologic effects of wind are most often expressed in dunes or yardangs (e.g., Greeley and Iversen, 1985).

The effects of gradation are most pronounced on planets having atmospheres, since an atmosphere is an obvious prerequisite for the formation of aeolian features, and an atmosphere that allows or once allowed liquid water to be stable on the surface is required for the formation of hydrologic and glacial/periglacial features. Aeolian features have been observed on Mars, Venus, and Triton, and gradational morphologies related to both liquid and frozen water have been observed on Mars. Mass wasting does not require the presence of an atmosphere, and features related to mass wasting have been observed on the Moon, Mercury, and Mars. The absence of surface water on Venus precludes most fluvial or other hydrologic features, and mass wasting has thus far been identified primarily in association with steep-sided tectonic troughs and volcanic domes (e.g., Malin, 1992).

10.4 GEOLOGY OF THE TERRESTRIAL PLANETS

This section provides a brief background on the geology of the terrestrial planets: Mercury, Venus, Earth, and Mars as well as the Moon. The information presented here is a starting point containing general background information and a discussion of some previous and current controversies. Much more detailed information on each of these bodies can be found in the references and recommended reading cited at the end of the chapter.

10.4.1 Mercury

Of the terrestrial planets, Mercury is the most poorly understood. Only half of its surface has been imaged by a spacecraft (Figure 10.6) (*Mariner 10*; Danielson et al., 1975), and its close proximity to the Sun makes terrestrial telescopic observations difficult, with only crude consensus emerging on the appearance and origin of albedo features on the planet even after nearly two centuries of observations (see Strom,

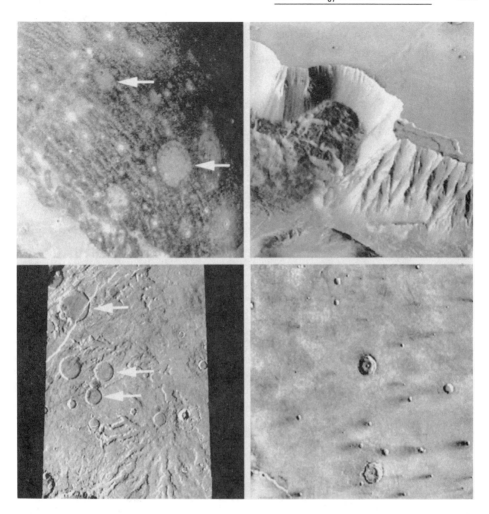

Figure 10.5   Examples of gradational landforms on planetary surfaces. (Upper left) This *Voyager* image shows a dark, densely cratered region on Jupiter's satellite Ganymede. The large bright circular features (arrows), known as "ghost craters" or palimpsests, provide evidence for the viscous relaxation of previously existing topography on the surface of this ice/rock satellite. The larger of the two arrowed craters is 350 km across. (Upper right) Perspective view of a 200-km-wide segment of the Ophir Chasma region of Valles Marineris, Mars, generated from *Viking Orbiter* images and a topographic model. Slumping of the canyon walls has clearly occurred here, moving a large amount of material down the 6-km height of the walls. (Lower left) Degraded rims, lack of interior structure, and the absence of crater ejecta blankets (arrows) provide evidence of gradational processes in this region of the Martian ancient highlands. The middle arrowed crater is 25 km in diameter. (Lower right) Wind streaks on Mars attest to the fact that the atmosphere, although thin, is still capable of redistributing surface materials. Wind streaks like these, from the Syrtis Major region, also help determine the general circulation properties of the Martian atmosphere. The crater in the middle of the image is 25 km in diameter.

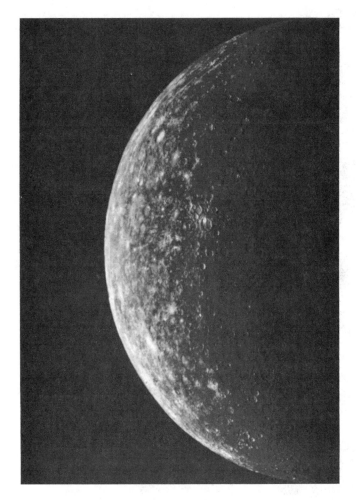

Figure 10.6 *Mariner 10* mosaic of Mercury obtained just prior to the spacecraft's first flyby in 1974. The 35 frames used to construct this mosaic had a spatial resolution varying from 1 to 2 km. These data provided the first view of Mercury, revealing a cratered surface similar to the cratered highlands of the Earth's Moon. The bright crater Kuiper near the center of the image (60 km diameter) is one of the largest young craters on Mercury and defines the Kuiperian stratigraphic unit, analogous to the Copernican unit on the Moon.

1987; Chapman, 1988). Earth-based radar observations were used to resolve large-scale surface features and allowed the determination of its rotation period (Pettengill and Dyce, 1965). Because of its proximity to the Sun, relatively slow rotation period (in two-thirds synchronization with its orbital period), near-zero inclination, and relatively high orbital eccentricity, temperatures on Mercury's surface vary over the largest range in the solar system. For example, equatorial noon temperatures at perihelion may climb as high as 700 K, and equatorial midnight temperatures can drop to 100 K.

A first-order question when studying any planet is the bulk composition of its crust. However, very little is known of the composition or mineralogy of the Mercurian crust. Earth-based visible to near-infrared spectroscopic measurements indicate that Mercury has no gross hemispherical mineralogical differences (e.g., Vilas, 1988). Due to Mercury's high bulk density it is thought to have a large core and a high bulk Fe content, probably twice as much Fe as that of any other planet (cf. Chapman, 1988). However, there is no direct evidence of a high Fe content in the crust (Rava and Hapke, 1987; Vilas, 1988; Cintala, 1992). More recent mid-infrared spectroscopic observations reveal evidence for plagioclase or anorthite minerals on the surface with a lower oxidized iron content than similar minerals on the Moon (Sprague et al., 1994). Perhaps most surprisingly, recent Earth-based radar observations of Mercury have revealed anomalously radar-bright regions near the poles. Analyses have determined that these areas are confined to the interiors of large, permanently shadowed craters, raising the possibility of the presence of subsurface ice in these regions (e.g., Slade et al. 1992, Butler et al., 1993). However, there is skepticism about the possibility of polar ice on Mercury, and alternative hypotheses have been presented to explain the radar bright regions (e.g. Sprague et al., 1995).

The *Mariner 10* flyby images revealed the Mercurian surface to be lunarlike (Figure 10.6). Four primary physiographic units were identified: heavily cratered terrain, intercrater plains, smooth plains, and hilly and lineated terrain (Trask and Guest, 1975; Davies et al., 1978; Spudis and Guest, 1988). The heavily cratered terrain is very similar to the lunar highlands, with densely packed and overlapping craters of all sizes and little evidence for discrete ejecta blankets and secondary crater fields. The origin of the heavily cratered terrain is probably similar to that of the lunar highlands, representing the consequence of heavy bombardment by impacts early in solar system history. The intercrater plains unit covers more than a third of the area imaged by *Mariner 10* and is thus the most extensive physiographic unit yet mapped on Mercury. The unit is characterized by gently rolling plains extensively covered by small impact craters. Its origin is unclear, as it appears to predate the heavily cratered terrain in places but to postdate it elsewhere. Smooth plains resemble the lunar mare in many ways: they have fewer craters than surrounding units, are relatively flat, many have sinuous or arcuate ridges, and they often fill large craters or embay older terrains (cf. Spudis and Guest, 1988). Most of the smooth plains deposits on Mercury are concentrated near the 1300-km-diameter Caloris impact basin. Morphologic and stratigraphic arguments were initially made arguing that these plains are of volcanic origin, like the lunar mare (Murray et al., 1974; Strom et al., 1975) Subsequent analysis proposed that the Mercurian smooth plains are ejecta deposits from a large impact, such as the Caloris basin-forming event (Wilhelms, 1976). Wilhelms deduced that the smooth plains deposits were analogous to the lunar Cayley plains deposits, which were also initially mapped as volcanic deposits—later proven wrong by the return of *Apollo 16* samples. Also, the albedo of the smooth plains is the same as that of the surrounding units, implying that the smooth plains do not have a different composition. The Mercurian hilly and lineated terrain consists of a fragmented assemblage of hills, massifs, and depressions centered on the region antipodal to the Caloris impact basin. This terrain possibly was formed by the focusing of surface and interior seismic energy associated with the Caloris event and the subsequent jostling, lifting, and thrusting of the surface (Schultz and Gault, 1975).

Despite the many similarities, important differences exist between the geology of Mercury and the Moon. For example, impact craters on Mercury have smaller ejecta blankets, brighter ray systems, and deeper, less extensive secondary crater chains.

Most of these differences have been associated with the higher gravity on Mercury relative to the Moon (e.g., Gault et al., 1975; Scott, 1977). Also, morphologic differences between impact craters in the heavily cratered terrain and those in the lunar highlands argue for differences in the physical properties of these surfaces (Cintala et al., 1977). Finally, an extensive system of scarps and ridges on Mercury provides evidence for compressional tectonism on the surface, perhaps resulting from the early cooling and shrinkage of a putative large iron core (e.g., Dzurisin, 1978; Strom et al., 1975).

10.4.2 Venus

With a diameter and bulk density nearly the same as the Earth, Venus is often described as our "sister planet." Remote and *in situ* observations over the past 30 years have shown that this comparison is correct in some ways, but that the two worlds have experienced very different climatic and geologic histories. The surface of Venus is completely hidden beneath a thick blanket of clouds. Earth-based telescopes and visible-wavelength sensors in orbit never view the ground, so our understanding of this planet's topography and surface properties relies almost entirely on remote sensing observations using microwave systems.

Early telescopic observations of Venus revealed the presence of a thick cloudy atmosphere, leading to speculations as to the possibility of life. Spectroscopic analysis demonstrated that the atmosphere was dominated by carbon dioxide, with sulfuric acid clouds forming the opaque shield above the surface. The upper layers of the atmosphere circle the planet every 4 days in a phenomenon known as *superrotation*. These observations made it clear that the atmosphere of Venus differed markedly from that of Earth, but what was the surface like? Because longer-wavelength signals could penetrate the cloud layer and return from the surface, such observations were the logical next step.

The microwave emission from Venus was measured using Earth-based radio telescopes beginning in 1956 (Mayer et al., 1958), leading to the immediate conclusion that the surface of our sister world is far too hot to support life, with a mean temperature of 740 K (above the melting point of lead). This result was confirmed by early spacecraft flybys, including *Mariner 2* in 1962, the first successful interplanetary mission. Further spectroscopic work demonstrated that Venus has little atmospheric water (Janssen et al., 1973), and a series of Soviet descent probes showed that the surface pressure was a crushing 90 times that of Earth (e.g., Moroz, 1983). Winds in the upper atmosphere race along with the 4-day rotation, but near the surface the average speed is only about 1 ms^{-1}. Under this dense atmosphere, there is little change in temperature from day to night. Current models suggest that Venus experienced a "runaway" greenhouse effect early in its history, with the thick CO_2 atmosphere trapping large amounts of outgoing thermal solar radiation. The potential for life was gone, but the study of the geology and properties of Venus using radar developed rapidly.

Earth-based radar maps of the planet improved steadily in resolution from hundreds of kilometers in the 1960s to a current capability of about 1 km. In 1980, the *Pioneer-Venus* (PVO) spacecraft produced the first global topographic database for the planet. This mapping (reviewed by McGill et al., 1983) revealed a surface dominated by vast plains [found by *Venera* landers to probably be composed of basaltic material (Figure 10.7); e.g., Florenskiy et al., 1983; Moroz, 1983], with a globe-girdling belt of highland rises and deformed plateaus in the equatorial region. The highest point on the planet (11 km above the mean elevation) occurs in the Maxwell Montes, which form one flank of a large lava-covered plateau known as Lakshmi Planum. Images collected by the Arecibo and Goldstone systems, and later by the *Venera 15/16* orbital radar mappers, revealed details of the surface which implied large-scale volcanism, tectonic disruption, and modification by impact craters (see the review by Basilevsky and Head, 1988). Perhaps most surprising was the discovery by PVO of regions atop the Venus mountains which have very low microwave emissivities, suggesting a concentration of metallic minerals, surface coatings, or unique soil properties that occur within a narrow range of temperature and pressure (Pettengill et al., 1988).

The *Magellan* mission, which operated in Venus orbit from 1990 to 1993, provided a near-global radar map of the planet at resolutions of about 100 m (Saunders et al., 1992). At the same time, the surface elevation and microwave emission were mapped at somewhat coarser scales. These observations have led to a vigorous and ongoing debate as to the past history and current state of the Venusian lithosphere. The impact crater population can be used to show that the surface is far younger, on average, than that of the Moon or Mars, with a mean age of perhaps 300 to 500 my (e.g., Schaber et al., 1992). A global resurfacing event of this age seems necessary to explain the crater data, but the mechanism by which this occurs and the likelihood of periodic crustal overturn remain contentious. Volcanic eruptions have clearly modified the surface, with large shield volcanoes concentrated along the equatorial highlands and thousands of smaller domes, shields, and cones distributed across the plains (Figure 10.8). Tectonic forces have produced intricate patterns of ridges and fractures on the surface, and a period of widespread early deformation may have led to the development of the elevated tessera plateaus. A major difference from Earth is the lack of plate tectonic processes on Venus. Where midocean ridges and subduction zones form a global system of crustal production and heat loss on Earth, no such features are seen on Venus, leading to significant questions regarding the planet's heat budget (Solomon et al., 1992). The low-emissivity material in the highlands also remains a topic of debate, but is probably some form of surface coating which is cold-trapped at these elevations (e.g., Brackett et al., 1995).

Venus remains an enigmatic neighbor, and the past 30 years of research have only

Figure 10.7   *Venera 13* image of the surface of Venus. The horizon appears tilted at the far left and right edges of the photo, due to the "fisheye lens" used on the panoramic camera. The surface near the lander is composed of relatively flat plates of rock (probably basalt), with dark granular material lying atop and between the plates. The object in the foreground is an instrument cover and is 40 cm across.

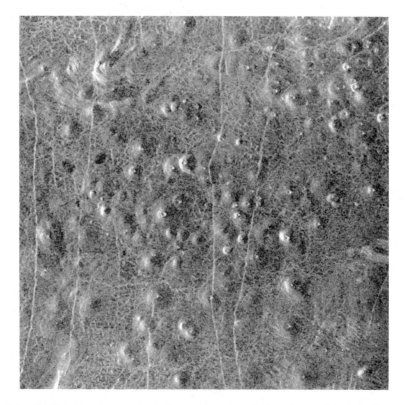

Figure 10.8 *Magellan* radar image of a cluster of volcanic cones in a 450-km-wide part of the Niobe Planitia region of Venus. The individual cones are 2 km in diameter and average 200 m in height. This region of Niobe is also cross-cut by a network of faults and fractures, some younger than the cones.

broadened our knowledge of the differences between the Earth and its sister. Major issues remain to be addressed in terms of atmospheric evolution, the history of the surface and how it is recycled over time, and surface heat flow and mantle processes. Future observations will probably include additional Earth-based radar mapping of the planet, spacecraft atmospheric probes, and perhaps new surface landers or near-surface imaging systems.

10.4.3 Earth

A detailed discussion of terrestrial geology is beyond the scope of this chapter, but it is instructive to include a brief mention of Earth here as part of planetary geology because terrestrial geology provides our primary context for the interpretation of remote sensing data from other planetary surfaces. Remote sensing observations of the Earth can be supplemented with ground truthing and detailed laboratory analyses, providing important information on the types of processes that can be identified using remote sensing and those that cannot.

The earliest and best-known terrestrial remote sensing method is aerial photography, and basic photographic products are now common spacecraft data sets. These images are used for a variety of purposes, including agricultural monitoring, hazards monitoring, geologic studies, urban planning, ocean and coastal watches, and weather prediction. Remote sensing of geologic processes was spurred by the development of camera systems that obtain data in both visible and infrared wavelengths. Because many minerals have characteristic absorption bands in the near-and mid-infrared, many systems such as the *Landsat* thematic mapper (TM) utilize spectral channels that emphasize differences in infrared reflectivity between units. Variations in mineral hydration are often mapped in this fashion, and the same techniques can be used to study vegetation health (by the strength of chlorophyll absorption), ocean color, and urban growth. Thermal infrared remote sensing began with single-channel sensors such as the *Landsat* TM band 6, which provides a measure of radiant flux from a target surface. More advanced instruments utilize multichannel thermal infrared systems to characterize surface mineralogy.

Microwave measurements include both active remote sensing using aircraft and spacecraft radars and passive emission observations at shorter wavelengths. The ability of radar to penetrate clouds makes it ideal for mapping areas where traditional photography is difficult, such as in tropical or mountainous regions, and allows all-season monitoring of surface processes. At shorter wavelengths, radar signals will interact with water in the atmosphere and be reflected back to the receiver, providing the basis for weather observations. Measurement of the Doppler shift of returned echoes allows determination of wind velocities. Orbital radar systems are also used to measure ocean wave spectra (and to infer wind speeds) and to measure surface elevations accurately.

Earth is the largest of the terrestrial planets and is also the most geologically active. The crust is constantly recycled by plate tectonics, wherein new ocean floor is created at midocean ridges and consumed at subduction zones. Convective patterns within the mantle drive the plates above them, leading to a cycle of collisions between the buoyant continental landmasses and producing mountains and other features. Unlike any other terrestrial body, there is abundant liquid water, and the cycling of this water through the atmosphere, crust, and oceans has a major effect on surface weathering and mineralogy. The Earth is our reference point for comparative planetology, and many remote sensing techniques used in planetary exploration are validated through practical terrestrial applications. The development of space technology has led to a fundamental change in studies of terrestrial processes. Where once local events and patterns dominated fields of study, we now can examine the global links between geologic, hydrologic, and atmospheric systems using remote sensing. Optical, infrared, and radar sensors have developed at astonishing rates since the 1950s, and current plans call for satellites capable of monitoring global changes through an array of combined techniques.

10.4.4 Moon

Aside from the Earth, we know more about the Moon than about any other object in the solar system. The Moon has provided a fundamental and critical testing ground for almost all of the currently used planetary geologic remote sensing techniques, from ground-based telescopic observations to robotic orbiters and landers to human exploration and sample return. As well, the proximity and relatively straightforward geology of the airless lunar surface has been important in the theory, development, and testing of the various fundamental tenets upon which the field of planetary geology is based.

A photographic program of telescopic observations of lunar surface geology lasting until the early 1960s provided most of our pre-spacecraft knowledge of the Moon (see Section 10.2.2). Initial spacecraft reconnaissance and photography by the *Luna, Ranger, Zond,* and *Lunar Orbiter* missions substantially expanded our view of lunar geology by providing higher-resolution images (e.g., Figure 10.9) as well as the first images of the far side. Follow-up lander missions in the late 1960s and early 1970s provided detailed high-resolution images of the surface as well as new information on surface physical and chemical properties (see, e.g., Heiken et al., 1991).

The physiography of the Moon can be divided into three primary terrain types: heavily cratered highlands terrains, impact basins and associated structures, and mare regions (e.g., Wilhelms, 1987). The highlands are bright (9 to 12% albedo), rugged, heavily cratered terrains with little or no evidence for discrete ejecta blankets or secondary crater fields. Highlands terrains compose nearly 70% of the near side of the Moon and 98% of the far side (Figure 10.10). While the dominant surface modification process in the highlands is impact cratering, some highlands regions contain relatively smooth, bright plains of uncertain (impact or volcanic) origin. Impact basins are the largest structures on the Moon, consisting of depressions from roughly 300 km to several thousands of kilometers in diameter, usually bordered by one or more multiple-ring mountain belts. The ejecta deposits associated with these basins form the main stratigraphic framework upon which the relative age dating scheme of the lunar surface was first developed (Wilhelms and McCauley, 1970). The lunar geologic time scale has been formally divided into a half dozen periods and epochs bounded primarily by the occurrence of catastrophic impacts, which played a dominant role in shaping the Moon's geology and morphology, but also defined by variations in lunar volcanism and cratering rate (Wilhelms, 1987).

Associated with the basins are extensive areas of surrounding highlands terrain that have been modified by the degradational action of material ejected during basin-forming events. In many cases, discrete basin ejecta deposits or their effects on pre-existing terrain can be traced for several thousand kilometers beyond the basin rim. Mare regions are relatively smooth, dark (5 to 8% albedo), usually circular regions that are almost always confined to the interiors of basins or large craters (Stuart-Alexander and Howard, 1970). Crater counting statistics indicate that these regions are clearly younger than the surrounding highlands and other basin terrains. Their origin as lava flows analogous to terrestrial flood basalts has come primarily from morphologic studies that revealed evidence for volcanic features such as flow fronts, vents, and lava channels. The higher concentration of mare units on the Earth-facing side of the Moon is thought to result from the near side having a thinner crust than the far side (e.g., Bills and Ferrari, 1977), resulting in easier transport of magma through the near-side crust. The thickness of the mare deposits is estimated to be less than 4 km in most place (e.g., Hörz, 1978), so their total volume comprises less than 1% of the lunar crust.

Radar echoes from the Moon were measured in the late 1940s, and many radar astronomy techniques were perfected using lunar returns (e.g., Hagfors, 1964; Hagfors and Evans, 1968). Serious mapping at a range of wavelengths began in con-

Figure 10.9   Oblique view of the 93-km-diameter lunar crater Copernicus obtained by *Lunar Orbiter II* in 1966. This stunning image shows part of the ejecta blanket and bright ray system of Copernicus in the foreground, the relatively smooth floor and 1-km-high central peaks of the crater in the middle, and slumps in the far crater wall, which is 3 to 4 km above the floor. The view is toward the north.

junction with the *Apollo* program, and demonstrated the utility of such data for determining roughness changes at the surface and chemical variations among geologic units (Figure 10.11) (e.g., Zisk et al., 1974; Thompson, 1974).

Earth-based spectroscopic observations, confirmed with returned samples, revealed that the dark mare regions are mostly composed of clinopyroxenes (high calcium) and plagioclase feldspars, with lesser amounts of olivine and ilmenite, while the brighter highlands regions are dominantly anorthite with varying amounts of orthopyroxene (low calcium) and olivine (see reviews in Heiken et al., 1991; Pieters,

Figure 10.10 Global map of Lunar albedo at 750-nm wavelength and 1-km/pixel resolution, from mosaics of images obtained by the *Clementine* UV–visible camera. This image shows the near and far sides of the Moon in Lambert, equal-area projection. The albedo patterns on the near side are the familiar ones seen from Earth and are composed of bright, heavily cratered highlands regions and large, relatively smooth and dark mare regions. The mare cover around 30% of this hemisphere. The mare cover only about 2% of the far side, reflecting a substantial hemispheric asymmetry in crustal thickness.

1993). These minerals possess distinctive spectral signatures in the wavelength range 400 to 2500 nm; thus it has been possible to map their areal distribution for much of the Moon through remote sensing techniques (e.g. Pieters, 1993; Nash et al., 1993).

The most intense phase of lunar exploration occurred during the *Apollo* manned landing program. Twelve astronauts explored the lunar surface at six different sites, using a combination of drilling, sampling, traversing, and other analysis techniques to return 382 kg of rocks and soils to Earth for analysis (Figure 10.1) (Heiken et al., 1991). Some of the most profound results of the Apollo program include the determination of an absolute age-dating chronology for the lunar surface, the identification of the impact (as opposed to volcanic) origin of highlands plains units, and the detailed characterization of lunar regolith mineralogy and mineralogic diversity. Remote sensing instruments in the *Apollo* orbiting command and service module obtained unique photographic, geochemical, topographic, and other data about the Moon (Masursky et al., 1978). The unmanned *Luna* sample return missions between 1970 and 1976 returned 300 g of lunar samples from three other sites, and in situ measurements were conducted by *Surveyor* and *Luna* landers. Samples and data

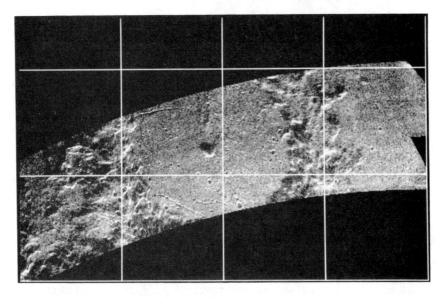

Figure 10.11   Earth-based radar image of the lunar crater Alphonsus (119 km in diameter). The relatively low incidence angle and the polarization state of the echo tend to highlight topographic features. Radar illumination is from the northwest, so we see strong radar returns from mountainsides and crater walls, which trend SW–NE. (Courtesy of S. Zisk, University of Hawaii.)

gathered by these missions have allowed a detailed characterization of the geology and mineralogy of the landing sites and their surrounding terrains. Many of the results obtained from these focused investigations have been used to infer details on the origin of the Moon and its global-scale geology and composition (e.g., Hartmann et al., 1986; Heiken et al., 1991; Wilhelms, 1987).

Spectroscopic observations and detailed analyses of samples from the six *Apollo* missions reveal that the surface of the Moon is a volatile-free, chemically reducing environment subject to extreme ranges of surface temperature and prolonged periods of solar wind and galactic cosmic ray bombardment. Under these conditions, there is a limited range of rocks and minerals that can form and/or exist stably on the lunar surface. The materials on the lunar surface can be classified into two main categories:

1. *Soil/regolith.* Fine-grained rock and mineral fragments are produced by constant impact comminution at all size scales. Soils are divided into mature and immature. *Mature soils*, because of their longer exposure ages, have enhanced abundances of glass and agglutinates (particle aggregates bonded by impact-produced glass) and are typically darker and redder than their immature counterparts. *Immature*, or *fresh soils*, have much lower abundances of agglutinates (higher relative abundances of crystalline mineral phases).

2. *Rocks.* The classical albedo division of the Moon into highlands (bright) and mare (dark) can also be used to describe the two major sets of lunar rock types. *Mare rocks* are basaltic lavas that are primarily composed of pyroxene, plagioclase, olivine and opaque metallic oxides of varying composition, all of which may have widely varying iron, titanium, and aluminum abundances.

Highlands rocks are dominated by plagioclase (anorthosites), pyroxene, and olivine, and have typically been pulverized and vitrified by impacts into unsorted, welded aggregates of material called *breccias*. Other, rarer lunar phases such as quartz and potassium feldspar have been detected in returned samples, but for the most part the mineralogy of the Moon is made up primarily of plagioclase, pyroxene, olivine, and ilmenite. In this regard the Moon is a much less complex geologic object than the Earth or Mars.

The most recent phase of lunar remote sensing has involved the use of more advanced telescopes and instrumentation (such as imaging spectroscopy) for increased spatial and spectral resolution studies of the near side (e.g., Jaumann, 1991; Johnson et al., 1991; Bell and Hawke, 1995; Blewett et al., 1995), as well as multispectral imaging carried out by the *Galileo* (Belton et al., 1992a) and *Clementine* spacecraft (Nozette et al., 1994) (Figure 10.10). These investigations have revealed substantial crustal heterogeneity on the Moon and allow characterization of lunar surface geology and mineralogy at higher spatial resolution than ever before. In particular, analyses of the global high-resolution *Clementine* data to determine mineralogic and compositional abundances, which is just now beginning, are certain to enhance our view of lunar geology fundamentally (e.g., Lucey et al., 1995).

10.4.5 Mars

Mars has fascinated both scientists and the general public because it is the most Earthlike planet in the solar system. Mars has a dynamic atmosphere (although only 1% as dense as Earth's), seasons similar to the Earth's (although roughly twice as long), and a rotation period very close to 1 Earth day. It has towering volcanoes, vast canyons, water-carved channels, frequent and sometimes global-scale dust storms, and polar caps that wax and wane with the seasons. It has been the focus of major ground-based, Hubble Space Telescope (HST), and spacecraft remote sensing and *in situ* observations and missions, and it is an obvious choice for future human exploration missions early in the twenty-first century.

The earliest geologic studies of Mars came from ground-based telescopic observations during periods when Mars and Earth were closest and the spatial resolution approached 100 km under the best terrestrial atmospheric conditions (see the review by Martin et al., 1992, and Section 10.2.2). These studies revealed two major surface units: bright and dark regions, along with detailed characterizations of the growth and decay of the polar caps. The bright and dark regions are organized into a series of albedo patterns that have been fairly constant for at least hundreds of years, except for usually short-time-scale seasonal or interannual variations due to dust storms. Speculations on the composition of the surface centered on oxidized iron (Fe^{3+}), based on the extremely red color of the bright regions, found to be similar to that of many oxidized terrestrial desert rocks and soils (de Vaucouleurs, 1954; Mutch et al., 1976). Spectroscopic observations in the 1960s and 1970s began to quantify the color differences among various regions and indicated that the dark regions contain the mineral pyroxene, linking them to a volcanic origin (e.g., Adams and McCord, 1969; McCord and Adams, 1969; Singer et al., 1979). More recent ground-based and HST telescopic observations in the 1980s and 1990s have revealed even more evidence for mineralogic variability among and between bright and dark areas on Mars (see the review in Bell, 1996).

Spacecraft remote sensing observations beginning in the mid-1960s expanded our view of the geology of Mars and provided evidence that Mars was once much more geologically and climatically active than it is today. Flybys by the *Mariner 4, 6,* and *7* spacecraft provided photographic and/or spectroscopic data for parts of the southern hemisphere, including the south polar cap. These observations revealed a heavily cratered surface much older than had been expected and indicated that the seasonal polar ice deposits are dominantly CO_2 rather than solely H_2O ice (NASA, 1967, 1969). Orbital visible and thermal-infrared imaging and spectroscopic studies by the *Mariner 9* and *Viking* missions revealed a much more geologically complex surface that has been clearly shaped by volcanic, tectonic, impact, and gradational processes (Hartmann and Raper, 1974; Carr et al., 1980). Perhaps most intriguing was the discovery of ancient, dendritic valley network systems that are thought to have been caused by the action of liquid water on the surface (Figure 10.12). The valley networks, along with other morphologic features indicating catastrophic flooding events and changes in erosion rates (Figure 10.13), provide evidence that major climate changes have occurred on Mars (e.g., Pollack et al., 1987; Carr, 1996). Some of the

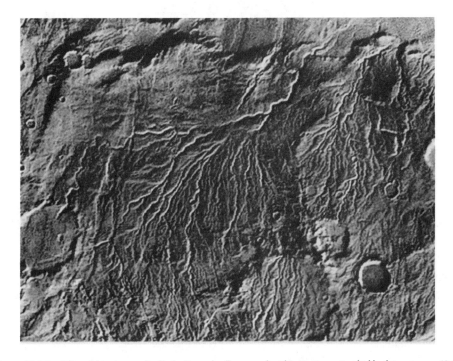

Figure 10.12   *Viking Orbiter* image of a finely dissected valley network within Martian ancient highlands terrain near 42°S, 92°W. The region shown is about 200 km across. Valley networks like this on Mars are similar to terrestrial drainage systems, although less developed and lacking small-scale streams. This has led to the interpretation of these features as being formed by groundwater flow (sapping) rather than runoff (rainfall). However, the majority of Martian valley network features occur in the oldest surface units, suggesting that the Martian climate may have been substantially different (warmer, wetter) in the past than it is today.

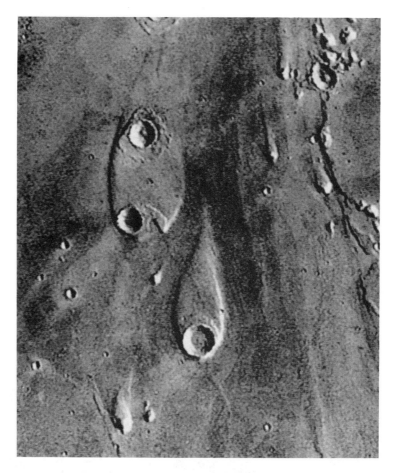

Figure 10.13 Streamlined islands near the mouth of the Ares Vallis outflow channel in Chryse Planitia, Mars. This image shows two teardrop-shaped scarps with heights of 400 m (upper island) and 600 m (lower island) formed by the erosive power of the flood that passed through this region early in Martian history. The craters at the heads of the islands are 8 km (upper) and 10 km (lower) in diameter, and represent the compacted obstacles that caused the water to be diverted. The 10-km-diameter crater at the top of the image still has a well-defined ejecta blanket, indicating that it formed sometime after the flooding event.

evidence for climatic variations is quite controversial, such as claims for the existence of Martian glaciers or oceans, and will require additional high resolution remote sensing and in situ investigations to resolve. The experiments aboard the two *Viking* lander missions (1976 to 1980) were primarily designed to search for evidence of organic materials that could indicate the presence of current or past life on Mars. While the primary data showed that the surface regions examined have only trace levels of organic material (the ppb level or less), high-resolution imaging and a variety of other secondary direct soil sampling experiments provided basic information on the chemical composition, magnetism, and other physical properties of the soil (Figure 10.14) (Hargraves et al., 1977; Toulmin et al., 1977; Clark et al., 1982). More recent spacecraft remote sensing measurements of Mars came from the *Phobos 2*

Figure 10.14 First image from the Martian surface returned by the second *Viking Lander*. The lander footpad is visible in the lower right, and the entire image covers a region about 1 m wide and 2 m long. The rocks visible in the scene are 10 to 20 cm in size and appear vesicular and/or fluted by wind. The *Viking* camera was a line-scanning device, so the first scans obtained for this image (left side) show brightness variations caused by dust settling after landing. Some of the materials kicked up by the landing can be seen inside the lander footpad bowl.

spacecraft in 1989, which obtained thermal-infrared images and near-infrared imaging spectroscopic data for a limited region of the Martian surface, providing new information on the surface thermophysical properties and mineralogy (Bibring et al., 1990; Murchie et al., 1993; Mustard et al., 1993). A new decade-long NASA program of Mars spacecraft exploration was recently initiated with the successful 1997 *Mars Pathfinder* rover and lander mission and the return of spectacular new orbital imagery from the *Mars Global Surveyor* spacecraft. *Mars Pathfinder* provided nearly three months of surface geologic and geochemical measurements of a landing site at the mouth of an ancient Martian outflow channel (Smith et al., 1997; Rieder et al., 1997), and *Mars Global Surveyor* has begun returning images of targeted sites at 5–10 meters per pixel spatial resolution (Malin et al., 1998), as a prelude for full-scale mapping of the planet at resolutions up to 1.5 meters/pixel that is expected to begin in 1999. These missions will be followed by additional orbiters, landers, and rovers to be launched in 1998, 2001, and 2003, with the eventual NASA goal of launching a sample return mission in 2005. Most scientists believe that such detailed robotic exploration and sample return missions are necessary precursors before NASA and/or other national space agencies can mount even more ambitious human exploration missions to Mars, perhaps as early as 2015 to 2020.

The general physiography of Mars is divided into numerous different terrain types, the most extensive being heavily cratered regions, plains, and volcanoes (Mutch et al., 1976; Carr, 1981). The heavily cratered regions occur mostly in the southern hemisphere and are separated from smoother and possibly topographically lower northern plains regions by a global-scale boundary scarp. The origin of this north–south hemispheric dichotomy is unexplained, but gravity measurements and morphologic studies point to a thinner crust in the northern lowlands, possibly related to a giant impact early in Martian history (see Smith and Zuber, 1996) or perhaps a Martian version of plate tectonics (Sleep, 1994). Most of the large impact basins on Mars are found throughout the heavily cratered regions as well (e.g., Schultz et al., 1982), attesting to its great age. Unlike the heavily cratered regions on the Moon and Mercury, however, this terrain on Mars appears to have been substantially eroded by the actions of wind and water and perhaps ice over geologic time. Many of the craters have degraded rims and/or flat, filled floors, and some have branching networks of valleys putatively of fluvial origin (e.g., Carr, 1981).

Volcanism has been an important surface modification process on Mars, as evidenced by the number and variety of volcanic features and constructs (Figure 10.15). The most extensive region of volcanism occurs in the Tharsis plateau, thought to be a vast basaltic construct, and is dominated by four huge shield volcanoes, including Olympus Mons, the largest volcano in the solar system. Other large-scale volcanic landforms occur in Elysium Planitia, in the southern highlands near the Hellas impact basin, and also along the highlands/lowlands boundary scarp. Martian volcanic edifices exhibit a variety of morphologies. The classic Martian shield volcano, Olympus Mons, has a profile grossly similar to that of many terrestrial basaltic shields, although a much larger scale (width 600 km, height 27 km). The nearby edifice, Alba Patera, is over 1000 km across, yet exhibits only a few kilometers of relief. Its flanks exhibit regions that are indicative of pyroclastic deposits and classic examples of effusive lava flows. Additionally, many channels that have been proposed to have been formed by surface runoff are carved in its flanks. These channels may have been induced by eruptive activity (Mouginis-Mark et al., 1992).

Other important physiographic features include channels that debauch into basins

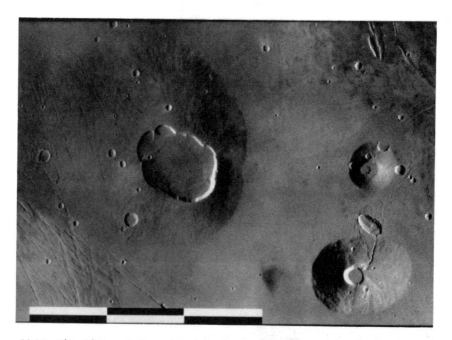

Figure 10.15 *Viking Orbiter* mosaic of some of the "lesser Tharsis" volcanoes Ceraunius Tholus (bottom left), Uranius Tholus (center left), and Uranius Patera (center). These volcanoes are in the Tharsis region of Mars, near 25°N, 95°W. The true areal extent of these three volcanoes is not known because they are embayed by younger lavas from the volcano Ascraeus Mons, 900 km to the southwest. These volcanoes are significantly older than the major Tharsis volcanoes (Olympus, Arsia, Pavonis, Ascraeus Mons), and display a greater variety of morphologies, indicative of both effusive and explosive emplacement (Robinson, 1993). Lighting is from the west (left) for the left side of the mosaic, and from the east for the right side. Each segment of the scale bar is 100 km.

and other lowlands and which were probably formed by vast flooding events; a vast system of tectonically created canyons stretching for more than 4000 km along the equator near the Tharsis volcanic region; fretted, knobby, and "chaotic" terrains that have been extensively modified by degradational processes; polar ice caps composed of both water and CO_2 ices; and polar layered terrains consisting of alternating bands of bright and dark material that may be an indicator of seasonal or longer-term cyclical climatic changes. A more complete discussion of these various units and of the general geology and stratigraphy of Mars is not possible here, but detailed reviews can be found in Mutch et al. (1976) and Carr (1981).

Geologic study of Mars over the past decade has concentrated on the continued refinement and analysis of the *Viking* data, and initial analyses of *Mars Pathfinder* and *Mars Global Surveyor* data. In addition, continued ground-based and HST observations are providing additional constraints on surface mineralogy and surface–atmosphere interactions.

10.5 GEOLOGY OF ASTEROIDS, COMETS, AND METEORITES

There are currently over 8000 asteroids and nearly 600 comets in the solar system with well-determined orbits. In addition, there are a small number of planetary satellites (such as Phobos and Deimos) that are almost certainly captured asteroids. These objects represent only a small fraction of the total number of asteroids and comets, as only the largest, brightest, and nearest objects have been observed. The total number of main-belt asteroids (dominated by small objects between 2 and 4 AU) is likely in the many tens of thousands, and there are perhaps an order of magnitude more objects in the Kuiper Belt beyond the orbits of Neptune and Pluto (Jewitt and Luu, 1995). The total number of comets (originating in the Öort cloud, a huge sphere of comets between approximately 40,000 to 50,000 AU) is estimated to be between 10^{10} to 10^{12} (Kresák, 1983). Only a handful of these objects have been studied in enough detail to determine their geological characteristics (e.g., Figures 10.16 and 10.17A and B). Meteorites, which are samples of asteroids, comets, and even planets, are amenable to much more detailed laboratory characterization, and thus can, by inference, provide substantial information on the geology of their parent bodies.

Most remote sensing observations of asteroids and comets have been conducted using telescopes and multispectral imaging or spectroscopy (e.g., A'Hearn, 1982; Gaffey et al., 1989). These observations have revealed that asteroids, and perhaps even comets, occur in distinct compositional "families," based on their colors, surface mineralogies, and orbital characteristics. Spectroscopic evidence for pyroxene, olivine, spinel, iron–nickel metal, hydrated silicates, and organic compounds has been found on asteroids (Gaffey et al., 1989; Bell et al., 1989). Silicate minerals, organic compounds, H_2O, CO_2, and CO ices, and a variety of molecular and ionic species have been found on comets, which contain more abundant volatile phases than asteroids (see reviews by Spinrad, 1987; A'Hearn, 1988).

A more detailed, though indirect assessment of the surface and interior conditions of asteroids and comets is provided by laboratory studies of the physical, chemical, mineralogic, and isotopic properties of meteorites (Kerridge and Matthews, 1988). Like asteroids, meteorites have been classified based on their chemical and petrologic

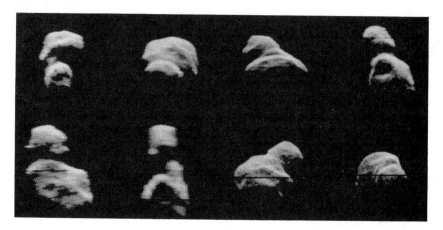

Figure 10.16   Delay-Doppler radar images of the Earth-crossing asteroid 4179 Toutatis. This image sequence shows the rotation of Toutatis over 8 days (start, upper left; end, lower right), and reveals the asteroid to be in an extremely slow, non-principal-axis rotation state. Toutatis was found to be several kilometers long and heavily cratered, with a bifurcated or possibly even binary shape. (From Ostro et al., 1995.)

characteristics (e.g., Sears and Dodd, 1988). These analyses reveal that meteorite parent bodies are diverse and many have undergone a variety of physical and chemical alteration processes since their formation. By inference, then, we should expect to find asteroids and perhaps comets that are differentiated, or have been thermally altered, or have been volcanically active, or are aggregates of several bodies formed via impact processes. A small group of meteorites of lunar and Martian origin have also been identified, and these objects provide additional (although not completely

Figure 10.17   Examples of the range of sizes, shapes, and surface characteristics among small bodies and outer solar system satellites. (*A*) Most asteroids and some planetary satellites are small, rocky bodies having irregular shapes and heavily cratered, ancient surfaces. The four such objects shown here are the Martian satellites Phobos (upper left) and Deimos (lower left), and the main belt asteroids 951 Gaspra (center) and 243 Ida (right). For scale, Ida is roughly 60 km long and all four objects are shown at the correct relative scale. Phobos and Deimos images from the *Viking Orbiter;* Gaspra and Ida images from the *Galileo* spacecraft. (*B*) The nucleus of Comet Halley, as seen by the *Giotto* spacecraft during its 1986 flyby. The nucleus is approximately 16 km long, is irregular in shape, and is very dark (albedo = 0.03) except for isolated regions of intense outburst activity. (*C*) The volcanically active surface of Jupiter's moon Io, as imaged by the *Voyager* spacecraft in 1979. This image shows the volcano Ra Patera (center right), with dark flows up to 300 km long. The lava flows may be composed primarily of silicates, or sulfur, or a combination of both. (*D*) The icy, tortured surface of the Uranian moon Miranda, seen here by *Voyager 2* in 1986. The processes that formed the juxtaposed, sharply bounded chevron-shaped region at the lower left as well as the darker region of grooved terrain at the upper right are unknown, but may be related to ice volcanism and/or intense impact disruption of the satellite. The image is roughly 150 km across. (*E*) The young, icy surface of the Jovian moon Europa is covered by a network of crisscrossing dark bands of unknown origin but probably related to tectonic stresses and/or ice volcanism. Europa has very little tectonic relief, fueling speculation on the existence of a liquid water layer below the solid, cracked crust. This image was taken by the *Voyager* spacecraft in 1979. (*F*) Portion of the south polar region of Neptune's moon Triton, imaged by *Voyager* in 1989. Dark streaks across the south polar cap may be the result of recent geyserlike eruptions of gas, dust, and ice venting from beneath the cap into the satellite's near-vacuum atmosphere and frigid environment (surface temperature 38 K). The diameter of Triton is 2700 km, and this scene has a resolution of 3 km/pixel.

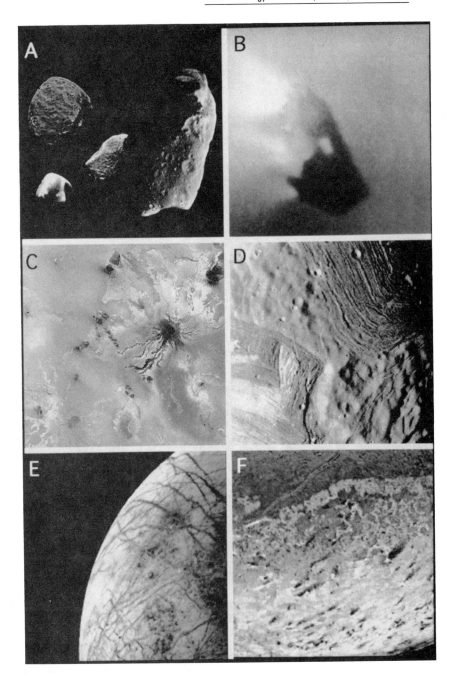

understood) ground-truth data on these planetary surfaces (Eugster, 1989; McSween, 1994).

Spacecraft missions have provided the first close-up images of asteroids and comets, including the *Viking Orbiter* investigations of the Martian moons, the remote sensing observations during the *Giotto* and *Vega 1/2* flybys of comet Halley, and the *Galileo* and Near Earth Asteroid Rendezvous (NEAR) Mission asteroid flybys. Images of Phobos and Deimos, which may be asteroids captured by Mars, reveal irregularly shaped worlds with surface geologies dominated by impact cratering (Figure 10.17A) (Thomas et al., 1992). However, evidence for other geologic processes, such as mass wasting and possibly tectonism, is apparent even on these small bodies. High-resolution images of the nucleus of comet Halley by the *Giotto* and *Vega* missions (Figure 10.17B) revealed a complex and irregular surface that is quite dark except for small localized regions of intense outgassing (Keller et al., 1986; Sagdeev et al., 1986). The surfaces of the main-belt asteroids 951 Gaspra, 243 Ida, and 253 Mathilde were revealed by the *Galileo* and NEAR spacecraft to also be complex and irregular (Figure 10.17A) (Belton et al., 1992b, 1994, Veverka et al., 1997). Impact cratering is the dominant surface modification process on these small objects, and evidence for local-to hemispheric-scale compositional heterogeneity has been found using impact craters both as a surface age dating tool and as a mineralogic probe of the asteroids' interiors.

10.6 GEOLOGY OF THE OUTER SOLAR SYSTEM

An exhaustive review of the geology of the planets and satellites in the outer solar system is beyond the scope of this chapter. Instead, in this section we provide a brief description of the type of geologic information that exists for these objects as well as their possible geologic histories as inferred from the types of morphologic features that have been seen.

The geology of the solid-surface planets and satellites in the outer solar system differs from that of the inner planets. This is because the surfaces and much of the interiors of most of these objects are composed primarily of ice instead of rock. However, at the extremely low temperatures of the outer solar system, ice acts effectively like a rock, so there are some generic similarities in surface geomorphology in response to the same types of tectonic and mass-wasting surface modification processes that act on the terrestrial planets (e.g., Figure 10.17E). There are also a few exceptional places in the outer solar system (Io, Triton, Titan) where volcanic, aeolian, and possibly other more exotic surface modification processes are active as well.

Impact cratering into icy surfaces yields morphologic features that are similar in many ways to impacts into rocky surfaces, but which also exhibit important differences related to variations in gravity, material strength, and composition (cf. Greeley, 1987). Many small icy satellites show evidence of tectonism, manifested as graben, faults, or enigmatic mélangelike regions (e.g., Squyres and Croft, 1986). Jupiter's moon Io (Figure 10.17C) and perhaps Neptune's moon Triton (Figure 10.17F) are the only volcanically active outer solar system objects known to date (Nash et al., 1986; Kirk et al., 1995), although possible evidence for volcanism with water as the "magma" exists from imaging of some of the other icy satellites (Figure 10.17D and

E) (e.g., Allison and Clifford, 1987; Crawford and Stevenson, 1988; Jankowski and Squyres, 1988).

Saturn's moon Titan possesses a thick atmosphere, and there is much speculation that aeolian and/or liquid hydrocarbon surface modification processes and possibly chemical weathering processes are active (e.g., Morrison et al., 1986). However, Titan's thick clouds have prevented detailed study of its surface, and only recently have images been obtained from the HST revealing that Titan does appear to possess a solid surface with evidence for distinct albedo features (Smith et al., 1994). Some of Titan's mysteries are sure to be unlocked from radar mapping and *in situ* imaging of Titan to be performed by the NASA/ESA *Cassini* orbiter and *Huygens* Titan entry probe in 2004.

Pluto possesses a thin atmosphere, but evidence for aeolian activity or other geologic processes on Pluto and its moon Charon are currently only speculative, although ground-based and HST observations do indicate that these bodies appear to have surface markings (e.g., Buie et al., 1997). Even less is known about the geology of the large planetesimals recently discovered beyond the orbits of Neptune and Pluto (Jewitt and Luu, 1995). These bodies inhabit the inner detectable edge of a disk (of perhaps 10^5 primitive objects) known as the Kuiper Belt. Many of these Kuiper Belt objects are large (> 200 km diameter), so there is good reason to believe that their surfaces preserve evidence of impacts or possibly other tectonic or volcanic processes that were active during the early history of the solar system.

Finally, it is worth mentioning that the study of planetary rings, which are known to exist around all the gas giant outer planets, falls within the realm of planetary geology. Rings are probably the result of the impact of previously existing satellites or captured planetesimals, and their shapes and physical characteristics reflect a complex interplay of gravity, tides, and atmospheric drag (e.g., Greenberg and Brahic, 1984; Sicardy and Brahic, 1990). The composition of planetary rings ranges from water ice in Saturn's rings to possibly organic-bearing silicates in the rings of Uranus, and provides clues to the origin and evolution of the ring systems around each giant planet.

10.7   FUTURE PLANETARY GEOLOGIC EXPLORATION

As demonstrated in Table 10.3, most of the initial geologic reconnaissance of the solar system has been carried out either by ground-based or spacecraft flyby missions. Important specific gaps in this basic first exploration phase remain, including as-yet unvisited Pluto and Charon, one-half of Mercury not imaged by *Mariner 10*, and different types of asteroids and comets than observed by the *Galileo, Giotto*, NEAR, and *Vega* spacecraft. Plans are currently being formulated for a "Pluto fast flyby" mission sometime in the early twenty-first century, and various asteroids and comets are being targeted for serendipitous flyby studies as part of several upcoming NASA and ESA spacecraft missions as well as the New Millenium spacecraft technology demonstration program.

More detailed geologic exploration of planetary surfaces requires orbital and landed robotic spacecraft, and this phase has already begun in earnest for Venus, the Moon, and Mars. The *Galileo* orbiter is currently completing its four-year "tour" of the Jovian system, and preliminary analses of the imaging and spectroscopic data from the Galilean satellites have already revealed fundamental new information on the geologic and geophysical properties of these bodies (McKinnon, 1997). Similar data for the Saturnian satellites will be obtained by the *Cassini* orbiter mission early in the next decade. A small lunar orbital spacecraft called *Lunar Prospector* is currently obtaining a global geochemical, gravitational, and topographic map of the Moon in order to provide benchmark datasets against which to test various theories of lunar origin and evolution (Binder et al., 1998). The *Near Earth Asteroid Rendezvous* mission will spend a year in orbit around the S-type asteroid 433 Eros in 1999, obtaining imaging, spectroscopic, geochemical, magnetic, gravity, and topographic data in unprecedented detail (Cheng et al., 1997; Veverka et al., 1997). Early in the next decade the NASA *Comet Nucleus Tour* mission will perform close flybys of three diverse Earth-approaching comets, and the NASA *Stardust* mission will return a sample of cometary material to Earth for detailed laboratory study. The ESA *Rosetta* mission will rendezvous with the periodic comet Wirtanen in 2011 and spend several years in orbit studying how the cometary surface evolves as it approaches closer to the Sun. Other small, focused, low-cost orbital missions to Mercury, asteroids, comets, and the outer solar system are also being studied. Clearly this is an exciting time for planetary geologists!

Perhaps the most ambitious plans, though, have been reserved for Mars. A veritable fleet of robotic spacecraft will be launched to study the planet over the next decade. The first two, NASA's *Mars Pathfinder* lander and rover and the *Mars Global Surveyor* orbiter, have already returned a wealth of new geologic, geochemical, and geophysical information about the red planet. Following these missions, NASA expects to launch two spacecraft each in 1998, 2001, and 2003 in order to obtain higher-resolution global geochemical, mineralogic, and imaging data for the Martian surface as well as to explore several other landing sites in great detail using landers and rovers. These efforts are being planned to lead to the first Mars sample return missions around 2005, and eventually to human exploration around 2015.

ACKNOWLEDGMENTS

We thank David Crown for providing a thorough review of an earlier draft of this chapter, and we thank Peter Thomas for assistance with Table 10.1. We are also grateful to Rick Kline and Diogo Bustani for assistance with the figures for this chapter. This review was supported in part by grants from the NASA Planetary Geology and Planetary Astronomy programs.

References

Adams, J. B., and T. B. McCord, 1969. Mars: interpretation of spectral reflectivity of light and dark regions, *J. Geophys. Res.*, 74, 4851–4856.

Adams, J. B., M. O. Smith, and A. R. Gillespie, 1993. Imaging spectroscopy: interpretation based on spectral mixture analysis, in *Remote Geochemical Analysis: Elemental and Mineralogical Composition*, C. M. Pieters and P. A. J. Englert, eds., Cambridge University Press, Cambridge, pp. 145–166.

A'Hearn, M. F., 1982. Spectrophotometry of comets at optical wavelengths, in *Comets*, L. L. Wilkening, ed., University of Arizona Press, Tucson, Ariz., pp. 433–460.

A'Hearn, M. F., 1988. Observations of comet nuclei, *Annu. Rev. Earth Planet. Sci.*, 16, 273–293.

Allen, C. W., 1976. *Astrophysical Quantities*, Athlone Press, London, 310 pp.

Allison, M. L., and S. M. Clifford, 1987. Ice-covered water volcanism on Ganymede, *J. Geophys. Res.*, 92, 7865–7876.

Baldwin, R. P., 1949. *The Face of the Moon*, University of Chicago Press, Chicago, 273 pp.

Basaltic Volcanism Study Project, 1981. *Basaltic Volcanism on the Terrestrial Planets*, Pergamon Press, Tarrytown, N.Y., 1286 pp.

Basilevsky, A. T., and J. W. Head III, 1988. The geology of Venus, *Annu. Rev. Earth Planet. Sci.*, 16, 295–317.

Bell, J. F., III, 1996. Iron, sulfate, carbonate, and hydrated minerals on Mars, in *Mineral Spectroscopy: A Tribute to Roger G. Burns*, Geochemical Society Special Publication 5, M. D. Dyar, C. McCammon, and M. W. Schaefer, eds. pp. 359–380.

Bell, J. F., III, and B. R. Hawke, 1995. Compositional variability of the Serenitatis/Tranquillitatis region of the Moon from telescopic multispectral imaging and spectroscopy, *Icarus*, 118, 51–68.

Bell, J. F., D. R. Davis, W. K. Hartmann, and M. J. Gaffey, 1989. Asteroids: the big picture, in *Asteroids II*, R. P. Binzel, T. Gehrels, and M. S. Matthews, eds., University of Arizona Press, Tucson, Ariz., pp. 921–948.

Belton, M. J. S., J. W. Head III, C. M. Pieters, R. Greeley, A. S. McEwen, G. Neukum, K. P. Klaasen, C. D. Anger, M. H. Carr, C. R. Chapman, M. E. Davies, F. P. Fanale, P. J. Gierasch, R. Greenberg, A. P. Ingersoll, T. Johnson, B. Paczkowski,

C. B. Pilcher, and J. Veverka, 1992a. Lunar impact basins and crustal heterogeneity: new western limb and far side data from Galileo, *Science*, 255, 570–576.

Belton, M. J. S., J. Veverka, P. Thomas, P. Helfenstein, D. Simonelli, C. Chapman, M. E. Davies, R. Greeley, R. Greenberg, and J. Head, 1992b. *Galileo* encounter with 951 Gaspra: first pictures of an asteroid, *Science*, 257, 1647–1652.

Belton, M. J. S., C. R. Chapman, J. Veverka, K. P. Klaasen, A. Harch, R. Greeley, R. Greenberg, J. W. Head, A. McEwen, D. Morrison, P. Thomas, M. E. Davies, M. H. Carr, G. Neukum, F. P. Fanale, D. R. Davis, C. Anger, P. Gierasch, A. P. Ingersoll, and C. B. Pilcher, 1994. First images of 243 Ida, *Science*, 265, 1543–1547.

Bibring, J.-P., M. Combes, Y. Langevin, C. Cara, P. Drossart, T. Encrenaz, S. Erard, O. Forni, B. Gondet, L. V. Ksanfomality, E. Lellouch, P. Masson, V. I. Moroz, F. Rocard, J. Rosenqvist, C. Sotin, and A. Soufflot, 1990. ISM observations of Mars and Phobos: first results, *Proceedings of the 20th Lunar and Planetary Science Conference*, pp. 461–471.

Bills, B. G., and A. J. Ferrari, 1977. A lunar density model consistent with topographic, gravitational, librational, and seismic data, *J. Geophys. Res.*, 82, 1306–1314.

Binder, A. B., W. C. Feldman, G. S. Hubbard, A. S. Konopliv, R. P. Lin, M. H. Acuna, and L. L. Hood, 1998. Lunar Prospector Searches for polar ice, a metallic core, gas release events, and the Moon's origin, EOS, Trans. A.G.U., 79, 897.

Blewett, D. T., B. R. Hawke, P. G. Lucey, G. J. Taylor, R. Jaumann, and P. D. Spudis, 1995. Remote sensing and geologic studies of the Schiller–Schickard region of the Moon, *J. Geophys. Res.*, 100, 16959–16978.

Bloom, A. L., 1978. *Geomorphology: A Systematic Analysis of Late Cenozoic Landforms*, Prentice Hall, Upper Saddle River, N.J., 510 pp.

Brackett, R. A., B. Fegley, and R. E. Arvidson, 1995. Volatile transport on Venus and implications for surface geochemistry and geology, *J. Geophys. Res.*, 100, 1553–1563.

Buie, M. W., and R. P. Binzel, 1997. Surface appearance of Pluto and Charon, in *Pluto and Charon*, D. Tholen, A. Stern, and M. S. Matthews, eds., University of Arizona Press, Tucson, Ariz., pp. 269–294.

Burns, R. G., 1993. Origin of electronic spectra of minerals in the visible–near infrared region, in *Remote Geochemical Analysis: Elemental and Mineralogical Composition*, C. M. Pieters and P. A. J. Englert, eds., Cambridge University Press, Cambridge, pp. 3–29.

Burns, J. A., and M. S. Matthews, 1986. *Satellites*, University of Arizona Press, Tucson, Ariz., 1021 pp.

Butler, B. J., D. O. Muhleman, and M. A. Slade, 1993. Mercury: full-disk radar images and the detection and stability of ice at the north pole, *J. Geophys. Res.*, 98, 15003–15023.

Butrica, A. J., 1996. *To See the Unseen: A History of Planetary Radar Astronomy*, NASA Spec. Publi. 4218, U.S. Government Printing Office, Washington, D.C., 301 pp.

Carr, M. H., 1981. *The Surface of Mars*, Yale University Press, New Haven, Conn., 232 pp.

Carr, M. H., 1996. *Water on Mars*, Oxford University Press, New York, 229 pp.

Carr, M. H., and R. Greeley, 1980. *Volcanic Features of Hawaii: A Basis for Com-*

parison with Mars, NASA Spec. Publ. 403, U.S. Government Printing Office, Washington, D.C., 211 pp.

Carr, M. H., W. A. Baum, K. R. Blasius, G. A. Briggs, J. A. Cutts, T. C. Duxbury, R. Greeley, J. Guest, H. Masursky, B. A. Smith, L. A. Soderblom, J. Veverka, and J. B. Wellman, 1980. *Viking Orbiter Views of Mars*, NASA Spec. Publ. 441, U.S. Government Printing Office, Washington, D.C., 182 pp.

Carr, M. H., R. S. Saunders, R. G. Strom, and D. E. Wilhelms, 1984. *The Geology of the Terrestrial Planets*, NASA Spec. Publ. 469, U.S. Government Printing Office, Washington, D.C., 317 pp.

Carr, M. H., M. J. S. Belton, K. Bender, H. Breneman, R. Greeley, J. W. Head, K. P. Klaasen, A. S. McEwen, J. M. Moore, S. Murchie, R. T. Pappalardo, J. Plutchak, R. Sullivan, G. Thornhill, and J. Veverka, 1995. The *Galileo* imaging team plan for observing the satellites of Jupiter, *J. Geophys. Res.*, 100, 18935–18956.

Cas, R. A. F., and J. V. Wright, 1987. *Volcanic Successions Modern and Ancient*, Allen & Unwin, London, 528 pp.

Chapman, C. R., 1988. Mercury: introduction to an end-member planet, in *Mercury*, F. Vilas, C. R. Chapman, and M. S. Matthews, eds., University of Arizona Press, Tucson, Ariz., pp. 1–23.

Cheng, A. F., A. G. Santo, K. J. Heeres, J. A. Landshof, R. W. Farquhar, R. E. Gold, and S. C. Lee, 1997. *Near Earth Asteroid Rendezvous:* mission overview, *J. Geophys. Res.* 102, 23, 695–23, 708.

Christensen, P. R., 1986. The spatial distribution of rocks on Mars, *Icarus*, 68, 217–238.

Cintala, M. J., 1992. Impact-induced thermal effects in the lunar and mercurian regoliths, *J. Geophys. Res.*, 97, 947–973.

Cintala, M. J., C. A. Wood, and J. W. Head, 1977. The effects of target characteristics on fresh crater morphology: preliminary results for the Moon and Mercury, *Proceedings of the 8th Lunar, and Planetary Science Conference*, pp. 3409–3425.

Clark, B. C., A. K. Baird, R. J. Weldon, D. M. Tsusaki, L. Schnabel, and M. P. Candelaria, 1982. Chemical composition of Martian fines, *J. Geophys. Res.*, 87, 10059–10067.

Cox, A. and R. B. Hart, 1986. *Plate Tectonics: How IT Works*, Blackwell Scientific, Boston, 392 pp.

Crawford, G. D., and D. J. Stevenson, 1988. Gas-driven water volcanism in the resurfacing of Europa, *Icarus*, 73, 66–79.

Danielson, G. E., K. P. Klaasen, and J. L. Anderson, 1975. Acquisition and description of *Mariner 10* television science data at Mercury, *J. Geophys. Res.*, 80, 2357–2393.

Davies, M. E., S. E. Dwornik, D. E. Gault, and R. G. Strom, 1978. *Atlas of Mercury*, NASA Spec. Publ. 423, U.S. Government Printing Office, Washington, D.C., 128 pp.

Dermott, S. F., and P. C. Thomas, 1994. The determination of the mass and mean density of Enceladus from its observed shape, *Icarus*, 109, 241–257.

de Vaucouleurs, G., 1954. *Physics of the Planet Mars*, Faber and Faber, London, 365 pp.

Dzurisin, D., 1978. The tectonic and volcanic history of Mercury as inferred from studies of scarps, ridges, troughs, and other lineaments, *J. Geophys. Res.*, 83, 4883–4906.

Eugster, O., 1989. History of meteorites from the Moon collected in Antarctica, *Science*, 245, 1197–1202.

Evans, L. G., R. C. Reedy, and J. I. Trombka, 1993. Introduction to planetary remote sensing gamma ray spectroscopy, in *Remote Geochemical Analysis: Elemental and Mineralogic Composition*, C. M. Pieters and P. A. J. Englert, eds., Cambridge University Press, Cambribge, pp. 167–198.

Florenskiy, K. P., A. T. Bazilevskiy, G. A. Burba, O. V. Nikolayeva, A. A. Pronin, A. S. Selivanov, M. K. Narayeva, A. S. Panfilov, and V. P. Chemodanov, 1983. Panorama of *Venera 9* and *10* landing sites, in *Venus*, D. M. Hunten, L. Colin, T. M. Donahue, and V. I. Moroz, eds., University of Arizona Press, Tucson, Ariz., pp. 137–153.

Fowler, A., P. Waddell, and L. Mortara, 1981. Evaluation of the RCA 512 × 320 charge-coupled device CCD imagers for astronomical use, in *Solid State Imagers for Astronomy*, SPIE Vol. 290, pp. 34–44.

Gaffey, M. J., J. F., Bell, and D. P. Cruikshank, 1989. Reflectance spectroscopy and asteroid surface mineralogy, in *Asteroids II*, R. P. Binzel, T. Gehrels, and M. S. Matthews, eds., University of Arizona Press, Tucson; Ariz., pp. 98–127.

Gaffey, S. J., L. A. McFadden, and D. B. Nash, 1993. Ultraviolet, visible, and near-infrared reflectance spectroscopy: laboratory spectra of geologic materials, in *Remote Geochemical Analysis: Elemental and Mineralogical Composition*, C. M. Pieters and P. A. J. Englert, eds., Cambridge University Press, Cambridge, pp. 43–71.

Gault, D. E., 1970. Saturation and equilibrium conditions for impact cratering on the lunar surface: criteria and implications, *Radio Sci.* 5, 273–291.

Gault, D. E., W. L. Quaide, and V. R. Oberbeck, 1968. Impact cratering mechanics and structures, in *Shock Metamorphism of Natural Materials*, B. M. French and N. M. Short, eds., Mono Books, Baltimore, pp. 87–99.

Gault, D. E., J. E. Guest, J. B. Murray, D. Dzurisin, and M. C. Malin, 1975. Some comparisons of impact craters on Mercury and the Moon, *J. Geophys. Res.*, 80, 2444–2460.

Gilbert, G. K. 1893. The Moon's face: a study of the origin and its features, *Philos. Soc. Wash. Bull.*, 12, 241–292.

Greeley, R., 1987. *Planetary Landscapes*, Allen and Unwin, Inc., Winchester, MA, 275 pp.

Greeley, R., and J. D. Iversen, 1985. *Wind as a Geological Process*, Cambridge University Press, Cambridge, 333 pp.

Greeley, R., and J. S. King, 1977. *Volcanism of the Eastern Snake River Plain, Idaho: A Comparative Planetary Geology Handbook*, NASA CR-15462, U.S. Government Printing Office, Washington, D.C., 1308 pp.

Greeley, R., and P. Spudis, 1981. Volcanism on Mars, *Rev. Geophys.*, 19, 13–41.

Greenberg, R., and A. Brahic, 1984. *Planetary Rings*, University of Arizona Press, Tucson, Ariz., 801 pp.

Hagfors, T., 1964. Backscattering from an undulating surface with applications to radar returns from the Moon, *J. Geophys. Res.*, 69, 3779–3784.

Hagfors, T., and J. V. Evans, 1968. Radar studies of the Moon, in *Radar Astronomy*, McGraw-Hill, pp. 219–270.

Hanel, R. A., B. J. Conrath, D. E. Jennings, and R. E. Samuelson, 1992. *Exploration of the Solar System by Infrared Remote Sensing*, Cambridge University Press, Cambridge, 458 pp.

Hapke, B., 1996. A model of radiative and conductive energy transfer in planetary regoliths, *J. Geophys. Res.*, 101, 16817–16831.

Hargraves, R. B., D. W. Collinson, R. E. Arvidson, and C. R. Spitzer, 1977. The *Viking* magnetic properties experiment: primary mission results, *J. Geophys. Res.*, 82, 4547–4558.

Hartmann, W. K., 1977. Cratering in the solar system, *Sci. Am.* 236, 84–99.

Hartmann, W. K., 1983. *Moons and Planets*, Wadsworth, Belmont, Calif., 509 pp.

Hartmann, W. K., and O. Raper, 1974. *The New Mars: The Discoveries of Mariner 9*, NASA Spec. Publ. 337, U.S. Government Printing Office, Washington, D.C., 179 pp.

Hartmann, W. K., R. J. Phillips, and G. J. Taylor, 1986. *Origin of the Moon*, Lunar and Planetary Institute, Houston, Texas, 781 pp.

Head, J. W., 1976. Lunar volcanism in space and time, *Rev. Geophys. Space Phys.*, 14, 265–300.

Head, J. W., and S. C. Solomon, 1981. Tectonic evolution of the terrestrial planets, *Science*, 213, 62–76.

Heiken, G. H., D. T. Vaniman, and B. M. French, 1991. *Lunar Sourcebook: A User's Guide to the Moon*, Cambridge University Press, Cambridge, 736 pp.

Hörz, F., 1978. How thick are lunar mare basalts? *Proceedings of the 9th Lunar and Planetary Science Conference*, pp. 3311–3331.

Howard, A. D., 1967. Drainage analysis in geological interpretation: a summation, *Am. Assoc. Pet. Geol. Bull.*, 51, 2246–2259.

Jankowski, D. G., and S. W. Squyres, 1988. Solid-state ice volcanism on the satellites of Uranus, *Science*, 241, 1322–1325.

Janssen, M. A., R. E. Hills, D. D. Thornton, and W. J. Welch, 1973. Venus: new microwave measurements show no atmospheric water vapor, *Science*, 179, 994–997.

Jaumann, R., 1991. Spectral-chemical analysis of lunar surface materials. *J. Geophys. Res.* 96, 22793–22807.

Jewitt, D. C., and J. X. Luu, 1995. The solar system beyond Neptune, *Astron. J.*, 109, 1867–1876.

Johnson, J. R., S. M. Larson, and R. B. Singer, 1991. Remote sensing of potential lunar resources: I. Near-side compositional properties. *J. Geophys. Res.*, 96, 18861–18882.

Kahle, A. B., F. D. Palluconi, and P. R. Christensen, 1993. Thermal emission spectroscopy: application to the Earth and Mars, in *Remote Geochemical Analysis: Elemental and Mineralogical Composition*, C. M. Pieters and P. A. J. Englert, eds., Cambridge University Press, Cambridge, pp. 99–120.

Keller, H. U., C. Arpigny, C. Barbieri, R. M. Bonnet, and S. Cazes, 1986. First Halley multicolor imaging results from *Giotto, Nature*, 321, 320–326.

Kerridge, J. F., and M. S. Matthews, ed., 1988. *Meteorites and the Early Solar System*, University Arizona Press, Tucson, Ariz., 1269 pp.

Kieffer, S. W., and C. H. Simonds, 1980. The role of volatiles and lithology in the impact cratering process, *Rev. Geophys. Space Phys.*, 18, 143–181.

Kirk, R. L., L. A. Soderblom, R. H. Brown, S. W. Kieffer, and J. S. Kargel, 1995.

Triton's plumes: discovery, characteristics, and models, in *Neptune and Triton*, D. P. Cruikshank, ed., University of Arizona Press, Tucson, Ariz., pp. 949–990.

Kresák, L., 1983. Comet discoveries, statistics, and observational selection, in *Comets*, L. L. Wilkening, ed., University of Arizona Press, Tucson, Ariz., pp. 56–84.

Leighton, R. B., B. C. Murray, R. P. Sharp, J. D. Allen, R. K. Sloan, 1967. *Mariner Mars, 1964 Project Report: Television Experiment Part 1. Investigator's Report, NASA*, JPL Tech. Rep. 32–884, Jet Propulsion Laboratory, California Institute of Technology, Pasadena, Calif., 178 pp.

Lucey, P. G., G. J. Taylor, and E. Malaret, 1995. Abundance and distribution of iron on the Moon, *Science*, 268, 1150–1153.

Malin, M. C., 1992. Mass movements on Venus: preliminary results from *Magellan* cycle 1 observations, *J. Geophys. Res.*, 97, 16337–16352.

Malin, M. C., M. H. Carr, G. E. Danielson, M. E. Davies, W. K. Hartmann, A. P. Ingersoll, P. B. James, H. Masursky, A. S. McEwen, L. A. Soderblom, P. Thomas, J. Veverka, M. A. Caplinger, M. A. Ravine, T. A. Soulanille, and J. L. Warren, 1997. Early views of the Martian surface from the Mars Orbiter Camera of Mars Global Surveyor, *Science*, 279, 1681–1685.

Martin, L. J., P. B. James, A. Dollfus, K. Iwasaki, and J. D. Beish, 1992. Telescopic observations: visual, photographic, polarimetric, in *Mars*, H. H. Kieffer, B. M. Jakosky, and M. S. Matthews, eds., University of Arizona Press, Tucson, Ariz., pp. 34–70.

Masursky, H., G. W. Colton, and F. El-Baz, 1978. *Apollo over the Moon: A View from Orbit*, NASA Spec. Publ. 362, U.S. Government Printing Office, Washington, D.C., 255 pp.

Mayer, C. H., T. P. McCullough, and R. M. Sloanaker, 1958. Observations of Venus at 3.15-cm wavelength, *Astrophys. J.*, 127, 1–10.

McCord, T. B., and Adams, J. B., 1969. Spectral reflectivity of Mars, *Science*, 163, 1058–1060.

McGill, G. E., J. L. Warner, M. C. Malin, R. E. Arvidson, E. Eliason, S. Nozette, and R. D. Reasenberg, 1983. Topography, surface properties, and tectonic evolution, in *Venus*, D. M. Hunten, L. Colin, T. M. Donahue, and V. I. Moroz, eds., University of Arizona Press, Tucson, Ariz., pp. 69–130.

McKinnon, W. B. 19—. Galileo at Jupiter-meetings with remarkable moons, Nature, 390, 23–26.

McSween, H. Y., Jr., 1994. What we have learned about Mars from SNC meteorites, *Meteoritics*, 29, 757–779.

Moroz, V. I., 1983. Summary of preliminary results of the *Venera 13* and *14* missions, in *Venus*, D. M. Hunten, L. Colin, T. M. Donahue, and V. I. Moroz, eds., pp. 45–68, University of Arizona Press, Tucson, Ariz., pp. 45–68.

Morrison, D. T. Owen, and L. A. Soderblom, 1986. The satellites of Saturn, in *Satellites*, J. A. Burns and M. S. Matthews, ed., University of Arizona Press, Tucson, Ariz., pp. 764–801.

Mortara, L., and A. Fowler, 1981. Evaluation of charge-coupled device CCD performance for astronomical use, in *Solid State Imagers for Astronomy*, SPIE Vol. 290, pp. 28–33.

Mouginis-Mark, P. J., L. Wilson, and M. T. Zuber, 1992. The physical volcanology of Mars, in *Mars*, H. H. Kieffer, B. M. Jakosky, and M. S. Matthews, eds., University of Arizona Press, Tucson, Ariz. pp. 424–452.

Muhleman, D. O., 1972. Microwave emission from the Moon, in *Thermal Characteristics of the Moon, Vol. 28*, J. Lucas, ed., MIT Press, Cambridge, Mass., pp. 51–81.

Murchie, S., J. Mustard, J. Bishop, J. Head, C. Pieters, and S. Erard, 1993. Spatial variations in the spectral properties of bright regions on Mars, *Icarus*, 105, 454–468.

Murray, B. C., M. J. S. Belton, G. E. Danielson, J. E. Davies, D. E. Gault, B. Hapke, B. O'Leary, R. G. Strom, V. Suomi, and N. Trask, 1974. Mercury's surface: preliminary description and interpretation from Mariner 10 pictures, *Science*, 185, 169.

Mustard, J. F., S. Erard, J.-P. Bibring, J. W. Head, S. Hurtrez, Y. Langevin, C. M. Pieters, and C. J. Sotin, 1993. The surface of Syrtis Major: composition of the volcanic substrate and mixing with altered dust and soil, *J. Geophys. Res.*, 98, 3387–3400.

Mutch, T. A., R. E. Arvidson, J. W. Head III, K. L. Jones, and R. S. Saunders, 1976. *The Geology of Mars*, Princeton University Press, Princeton, N.J., 400 pp.

NASA, 1967. *Mariner-Mars, 1964 Final Project Report*, NASA Spec. Publ. 139, U.S. Government Printing Office, Washington, D.C., 346 pp.

NASA, 1969. *Mariner-Mars, 1969 Preliminary Report*, NASA Spec. Publ. 225, U.S. Government Printing Office, Washington, D.C., 145 pp.

Nash, D. B., M. H. Carr, J. Gradie, D. M. Hunten, and C. F. Yoder, 1986. Io, in *Satellites*, J. A. Burns and M. S. Matthews, ed., University of Arizona Press, Tucson, Ariz., pp. 629–688.

Nash, D. B., J. W. Salisbury, J. E. Conel, P. G. Lucey, and P. R. Christensen, 1993. Evaluation of infrared emission spectroscopy for mapping the Moon's surface composition from lunar orbit, *J. Geophys. Res.*, 98, 23535–23553.

Neukum, G., B. König, H. Fechtig, and D. Storzer, 1975. A study of lunar impact crater size distributions, *Moon*, 12, 201–229.

Nicks, O. W., 1985. *Far Travelers: The Exploring Machines*, NASA Spec. Publ. 480, U.S. Government Printing Office, Washington, D.C., 255 pp.

Nozette, S., and 33 others, 1994. The *Clementine* mission to the Moon: scientific overview, *Science*, 266, 1835–1839.

Ostro, S. J., 1983. Planetary radar astronomy, *Rev. Geophys. Space Phys.*, 21, 186–196.

Ostro, S. J., R. S. Hudson, R. F. Jurgens, K. D. Rosema, R. Winkler, D. Howard, R. Rose, M. A. Slade, D. K. Yeomans, J. D. Giorgini, D. B. Campbell, P. Perillat, J. F. Chandler, and I. I. Shapiro, 1995. Radar images of asteroid 4179 Toutatis, *Science*, 270, 80–84.

Pettengill, G. H., and R. B. Dyce, 1965. A radar determination of the rotation of the planet Mercury, *Nature*, 206, 1240.

Pettengill, G. H., P. G. Ford, and B. D. Chapman, 1988. Venus: surface electromagnetic properties, *J. Geophys. Res.*, 93, 14881–14892.

Pieters, C. M., 1993. Compositional diversity and stratigraphy of the lunar crust derived from reflectance spectroscopy, in *Remote Geochemical Analysis: Elemental and Mineralogical Composition*, C. M. Pieters and P. A. J. Englert, eds., Cambridge University Press, Cambridge, pp. 341–365.

Plescia, J. B., and R. S. Saunders, 1979. The chronology of Martian volcanoes, *Proceedings of the 10th Lunar and Planetary Science Conference*, pp. 2841–2859.

Pollack, J. B., J. F. Kasting, S. M. Richardson, and K. Poliakoff, 1987. The case for a wet, warm climate on Mars, *Icarus*, 71, 203–224.

Rava, B., and B. Hapke, 1987. An analysis of the *Mariner 10* color ratio map of Mercury, *Icarus*, 71, 397–429.

Rieder R., T. Economou, H. Wänke, A. Turkevich, J. Crisp, J. Brückner, G. Dreibus, H. Y. McSween, Jr., 1997. The chemical composition of Martian soil and rocks returned by the mobile alpha proton X-ray spectrometer: Preliminary results from the X-ray mode, *Science* 278, 1771–1774, 1997.

Robinson, M. S., 1993. Some aspects of Lunar and Martian volcanism as examined with spectral, topographic, and morphologic data derived from spacecraft images, Ph.D. dissertation, University of Hawaii, Honolulu, Hawaii, 244 pp.

Roddy, D. J., R. O. Pepin, and R. B. Merrill, 1977. *Impact and Explosion Cratering: Planetary and Terrestrial Implications*, Pergamon Press, Tarrytown, N.Y., 1315 pp.

Sagdeev, R. Z., G. A. Avanesov, F. Szabo, L. Szabo, and P. Cruvellier, 1986. Television observations of comet Halley from *Vega* spacecraft, *Nature*, 321, 262–266.

Salisbury, J. W., 1993. Mid-infrared spectroscopy: laboratory data, in *Remote Geochemical Analysis: Elemental and Mineralogical Composition*, C. M. Pieters and P. A. J. Englert, eds., Cambridge University Press, Cambridge, pp. 79–98.

Salisbury, J. W., L. S. Walter, N. Vergo, and D. M. D'Aria, 1991. *Infrared 2.1–25 μm Spectra of Minerals*, Johns Hopkins University Press, Baltimore, 267 pp.

Saunders, R. S., and 26 others, 1992. *Magellan* mission summary, *J. Geophys. Res.*, 97, 13067–13090.

Schaber, G. G., R. G. Strom, H. J. Moore, L. A. Soderblom, R. L. Kirk, D. J. Chadwick, D. D. Dawson, L. R. Gaddis, J. M. Boyce, and J. Russell, 1992. Geology and distribution of impact craters on Venus: What are they telling us? *J. Geophys. Res.*, 97, 13257–13302.

Schultz, P. H., and D. E. Gault, 1975. Seismic effects from major basin formation on the Moon and Mercury, *Moon*, 12, 159–177.

Schultz, P. H., R. A. Schultz, and J. Rogers, 1982. The structure and evolution of ancient impact basins on Mars, *J. Geophys. Res.*, 87, 9803–9820.

Scott, D. H., 1977. Moon–Mercury: relative preservation states of secondary craters, *Phys. Earth Planet. Int.*, 15, 173–178.

Sears, D. W. G., and R. T. Dodd, 1988. Overview and classification of the meteorites, in *Meteorites and the Early Solar System*, J. F. Kerridge and M. S. Matthews, eds., University of Arizona Press, Tucson, Ariz., pp. 3–34.

Shapiro, I. I., 1968. Spin and orbital motions of the planets, in *Radar Astronomy*, McGraw-Hill, New York, pp. 143–183.

Sharpe, C. F. S., 1968. *Landslides and Related Phenomena*, Cooper Square, New York.

Sheehan, W., 1988. *Planets and Perception*, University of Arizona Press, Tucson, Ariz., 324 pp.

Shoemaker, E. M., 1963. Impact mechanics at Meteor Crater, Arizona, in *The Solar System: The Moon, Meteorites, and Comets*, B. M. Middlehurst and G. P. Kuiper, eds., University of Chicago Press, Chicago, pp. 301–336.

Shoemaker, E. M., 1966. Preliminary analysis of the fine structure of the lunar surface in Mare Cognitum, in *The Nature of the Lunar Surface*, W. N. Hess, D. H. Menzel, and J. A. O'Keefe, eds., Johns Hopkins University Press, Baltimore, pp. 23–77.

Shoemaker, E. M., R. M. Batson, A. L. Bean, C. Conrad, Jr., D. H. Dahlem, E. N. Goddard, M. H. Hait, K. B. Larson, G. G. Schaber, D. L. Schleicher, R. L. Sutton, G. A. Swann, and A. C. Watters, 1970. *Preliminary Geologic Investigation of the Apollo 12 Landing Site, Part A*, NASA Spec. Publ. 235, U.S. Government Printing Office, Washington, D.C., pp. 113–156.

Sicardy, B., and A. Brahic, 1990. The new rings: contributions of recent ground-based and space observations to our knowledge of planetary rings, *Adv. Space Res.*, 10, 211–219.

Singer, R. B., T. B. McCord, R. N. Clark, J. B. Adams, and R. L. Huguenin, 1979. Mars surface composition from reflectance spectroscopy: a summary, *J. Geophys. Res.*, 84, 8415–8426.

Slade, M. A., B. J. Butler, and D. O. Muhleman, 1992. Mercury radar imaging: evidence for polar ice, *Science*, 258, 635–640.

Sleep, N. H., 1994. Martian plate tectonics, *J. Geophys. Res.*, 99, 5639–5655.

Slipher, E. C., 1962. *Mars: The Photographic Story*, Sky Publishing Corporation, Cambridge Mass., 168 pp.

Smith, D. E., and M. T. Zuber, 1996. The shape of Mars and the topographic signature of the hemispheric dichotomy, *Science*, 271, 184–188.

Smith, P. H., M. T. Lemmon, J. J. Caldwell, M. D. Allison, and L. A. Sromovsky, 1994. HST imaging of *Titan*, 1994, *Bull. Am. Astron. Soc.*, 26, 1180.

Smith, P. H., J. F. Bell III, N. T. Bridges, D. T. Britt, L. Gaddis, R. Greeley, H. U. Keller, K. E. Herkenhoff, R. Jaumann, J. R. Johnson, R. L. Kirk, M. Lemmon, J. N. Maki, M. C. Malin, S. L. Murchie, J. Oberst, T. J. Parker, R. J. Reid, L. A. Soderblom, C. Stoker, R. Sullivan, N. Thomas, M. G. Tomasko, and E. Wegryn, 1997. First results from the Pathfinder camera, *Science*, 278, 1758–1765,

Smythe, W. D., R. Lopes-Gautier, A. Ocampo, J. Hui, M. Segura, L. A. Soderblom, D. L. Matson, H. H. Kieffer, T. B. McCord, F. P. Fanale, W. M. Calvin, J. Sunshine, E. Barbinis, R. W. Carlson, and P. R. Weissman, 1995. Galilean satellite observation plans for the near-infrared mapping spectrometer experiment on the *Galileo* spacecraft, *J. Geophys. Res.*, 100, 18957–18972.

Soderblom, L. A., K. Edwards, E. M. Eliason, E. M. Sanchez, and M. P. Charette, 1978. Global color variations on the Martian surface, *Icarus*, 34, 446–464.

Solomon, S. C., S. E. Smrekar, D. L. Bindschadler, R. E. Grimm, W. M. Kaula, G. E. McGill, R. J. Phillips, R. S. Saunders, G. Schubert, S. W. Squyres, and E. R. Stofan, 1992. Venus tectonics: an overview of *Magellan* observations, *J. Geophys. Res.*, 97, 13199–13256.

Spinrad, H., 1987. Comets and their composition, *Annu. Rev. Astron. Astrophys.*, 25, 231–269.

Sprague, A. L., R. W. H. Kozlowski, F. C. Witteborn, D. P. Cruikshank, and D. H. Wooden, 1994. Mercury: evidence for anorthosite and basalt from mid-infrared 7.3–13.5 μm spectroscopy, *Icarus*, 109, 156–167.

Sprague, A. L, D. M. Hunten, and K. Lodders, 1995. Sulfur at Mercury, elemental at the poles and sulfides in the regolith, *Icarus*, 118, 211–215.

Spudis, P. D., and J. E. Guest, 1988. Stratigraphy and geologic history of Mercury, in *Mercury*, F. Vilas, C. R. Chapman, and M. S. Matthews, eds., University of Arizona Press, Tucson, Ariz., pp. 118–164.

Spurr, J. E., 1944. *Geology Applied to Selenology: I. The Imbrium Plain Region of the Moon*, Science Press, Lancaster, Pa., 112 pp.

Squyres, S. W., and S. K. Croft, 1986. The tectonics of icy satellites, in *Satellites*,

J. A. Burns and M. S. Matthews, eds. University of Arizona Press, Tucson, Ariz., pp. 293–341.

Strom, R. G., 1987. *Mercury: The Elusive Planet*, Smithsonian Institution Press, Washington, D.C., 207 pp.

Strom, R. G., N. J. Trask, and J. E. Guest, 1975. Tectonism and volcanism on Mercury, *J. Geophys. Res.*, 80, 2478–2507.

Stuart-Alexander, D. E., and K. A. Howard, 1970. Lunar maria and circular basins: a review, *Icarus*, 12, 440–456.

Taylor, S. R., 1982. *Planetary Science: A Lunar Perspective*, Lunar and Planetary Institute, Houston, Texas, 481 pp.

Thomas, P., J. Veverka, J. Bell, J. Lunine, and D. Cruikshank, 1992. Satellites of Mars: geologic history in *Mars*, H. H. Kieffer, B. M. Jakosky, C. W. Snyder, and M. S. Matthews, eds., University of Arizona Press, Tucson, Ariz., pp. 1257–1282.

Thompson, T. W., 1974. Atlas of lunar radar maps at 70-cm wavelength, *Moon*, 10, 51–85.

Toulmin, P., III, A. K. Baird, B. C. Clark, K. Keil, H. J. Rose, Jr., R. P. Christian, P. H. Evans, and W. C. Kelliher, 1977. Geochemical and mineralogical interpretation of the *Viking* inorganic chemical results, *J. Geophys. Res.*, 82, 4625–4634.

Trask, N. J., and J. E. Guest, 1975. Preliminary geologic terrain map of Mercury, *J. Geophys. Res.*, 80, 2461–2477.

Vane, G., J. E. Duval, and J. B. Wellman, 1993. Imaging spectroscopy of the Earth and other solar system bodies, in *Remote Geochemical Analysis: Elemental and Mineralogical Composition*, C. M. Pieters and P. A. J. Englert, eds., Cambridge University Press, Cambridge, eds., pp. 121–143.

Veverka, J., P. Thomas, A. Harch, B. Clark, J. F. Bell III, B. Carcich, J. Joseph, C. Chapman, W. Merline, M. Robinson, M. Malin, L. A. McFadden, S. Murchie, R. Farquhar, N. Izenberg, and A. Cheng, 1997. NEAR's flyby of Mathilde: Images of a C-type asteroid, *Science*, 278, 2109–2114,

Vilas, F., 1988. Surface composition of Mercury from reflectance spectrophotometry, in *Mercury*, F. Vilas, C. R. Chapman, and M. S. Matthews, eds., University of Arizona Press, Tucson, Ariz., pp. 59–76

Whitaker, E. A., 1966. The surface of the Moon, in *The Nature of the Lunar Surface: Proceedings of the 1965 IAU Symposium*, W. N. Hess, D. H. Menzel, and J. A. O'Keefe, eds., Johns Hopkins University Press, Baltimore, pp. 79–98.

Whitford-Stark, J. L., 1982. Factors influencing the morphology of volcanic landforms: an Earth–Moon comparison, *Earth Sci. Rev.*, 18, 109–168.

Wilhelms, D. E., 1976. Mercurian volcanism questioned, *Icarus*, 28, 551–558.

Wilhelms, D. E., 1987. *The Geologic History of the Moon*, USGS Prof. Pap. 1348, U.S. Geological Survey, Washington, D.C., 302 pp.

Wilhelms, D. E., and J. F. McCauley, 1970. *Geologic Map of the Nearside of the Moon*, USGS Map I-703, U.S. Geological Survey, Washington, D.C.

Yin, L. I., J. I. Trombka, I. Adler, and M. Bielefeld, 1993. X-ray remote sensing techniques for geochemical analysis of planetary surfaces, in *Remote Geochemical Analysis: Elemental and Mineralogic Composition*, C. M. Pieters and P. A. J. Englert, eds., Cambridge University Press, Cambridge, pp. 199–212.

Young, L. A., 1994. Bulk properties and atmospheric structure of Pluto and Charon, Ph.D. thesis, Massachusetts Institute of Technology, Cambridge, Mass., 124 pp.

Zisk, S. H., G. H. Pettengill, and G. W. Catuna, 1974. High-resolution radar maps of the lunar surface at 3.8-cm wavelength, *Moon*, 10, 17–50.

Suggested Reading

General

Planetary Science: A Lunar Perspective, S. R. Taylor, Lunar and Planetary Institute, Houston, Texas, 481 pp., 1982.

The Geology of the Terrestrial Planets, M. H. Carr, R. S. Saunders, R. G. Strom, and D. E. Wilhelms, NASA Spec. Publ. 469, U.S. Government Printing Office, Washington, D.C., 317 pp., 1984.

Planetary Landscapes, R. Greeley, Allen & Unwin, Winchester, Mass., 275 pp., 1987.

Exploration of the Solar System by Infrared Remote Sensing, R. A. Hanel, B. J. Conrath, D. E. Jennings, and R. E. Samuelson, Cambridge University Press, Cambridge 458 pp., 1992.

Remote Geochemical Analysis: Elemental and Mineralogical Composition, C. M. Pieters and P. A. J. Englert, ed., Cambridge University Press, Cambridge, 594 pp., 1993.

Mercury

Atlas of Mercury, M. E. Davies, S. E. Dwornik, D. E. Gault, and R. G. Strom, NASA Spec. Publ. 423, U.S. Government Printing Office, Washington, D.C., 128 pp., 1978.

Mercury: The Elusive Planet, R. G. Strom, Smithsonian Institution Press, Washington, D.C., 207 pp., 1987.

Mercury, F. Vilas, C. R. Chapman, and M. S. Matthews, eds., University of Arizona Press, Tucson, Ariz., 794 pp., 1988.

Venus

Venus, D. M. Hunten, L. Colin, T. M. Donahue, and V. I. Moroz, eds., University of Arizona Press, Tucson, Ariz., 1151 pp., 1983.

Magellan at Venus, papers published in two special issues of *J. Geophys. Res.*, Aug. and Oct. 1992.

Venus II, S. W. Bougher, D. M. Hunten, and R. J. Phillips, eds., University of Arizona Press, Tucson, Ariz., 1362 pp., 1997.

Earth

Earth, F. Press and R. Siever, W. H. Freeman, New York, 656 pp., 1986.
Geomorphology from Space, N. M. Short and R. W. Blair, Jr., eds., NASA Spec. Publ. 486, U.S. Government Printing Office, Washington, D.C., 717 pp., 1986.
Theory of the Earth, D. L. Anderson, Blackwell Scientific, Boston, 366 pp., 1989.

Moon

Lunar Remote Sensing and Measurements, H. J. Moore, J. M. Boyce, G. G. Schaber, and D. H. Scott, USGS Prof. Pap. 1046-B, U.S. Geological Survey, Washington, D.C., pp. B1–B78, 1980.
Origin of the Moon, W. K. Hartmann, R. J. Phillips, and G. J. Taylor, Lunar and Planetary Institute, Houston, Texas, 781 pp., 1986.
The Geologic History of the Moon, D. E. Wilhelms, USGS Prof. Pap. 1348, U.S. Geological Survey, Washington, D.C., 302 pp., 1987.
Lunar Sourcebook: A User's Guide to the Moon, G. H. Heiken, D. T. Vaniman, and B. M. French, Cambridge University Press, Cambridge, 736 pp., 1991.

Mars

The New Mars: The Discoveries of Mariner 9, W. K. Hartmann and O. Raper, NASA Spec. Publ. 337, U.S. Government Printing Office, Washington, D.C., 179 pp., 1974.
The Geology of Mars, T. A. Mutch, R. E. Arvidson, J. W. Head III, K. L. Jones, and R. S. Saunders, Princeton University Press, Princeton, N.J., 400 pp., 1976.
The Surface of Mars, M. H. Carr, Yale University Press, New Haven, Conn., 232 pp., 1981.
On Mars: Exploration of the Red Planet from 1958–1978, E. C. Ezell and L. N. Ezell, NASA Spec. Publication 4212, U.S. Government Printing Office, Washington, D.C., 535 pp., 1984.
Mars, H. H. Kieffer, B. M. Jakosky, and M. S. Matthews, eds., University of Arizona Press, Tucson, Ariz., 1498 pp., 1992.
Mars Pathfinder, papers published in a special issue of *Science,* 5 December, 1997.
Mars Global Surveyor, published in a special issue of *Science,* 13 March, 1998.

Asteroids, Comets, Meteorites

Asteroids, T. Gehrels, ed., University of Arizona Press, Tucson, Ariz., 1181 pp., 1979.
Comets, L. L. Wilkening, ed., University of Arizona Press, Tucson, Ariz., 766 pp., 1983.

Meteorites and the Early Solar System, J. F. Kerridge and M. S. Matthews, eds., University of Arizona Press, Tucson, Ariz., 1269 pp., 1988.

Asteroids II, R. P. Binzel, T. Gehrels, and M. S. Matthews, eds., University of Arizona Press, Tucson, Ariz., 1258 pp., 1989.

Icy Satellites and the Outer Solar System

Jupiter, T. Gehrels, ed., University of Arizona Press, Tucson, Ariz., 1254 pp., 1976.

Planetary Satellites, J. A. Burns, ed., University of Arizona Press, Tucson, Ariz., 598 pp., 1977.

Satellites of Jupiter, D. Morrison, ed., University of Arizona Press, Tucson, Ariz., 972 pp., 1982.

Saturn, T. Gehrels and M. S. Matthews, eds., University of Arizona Press, Tucson, Ariz., 968 pp., 1984.

Satellites, J. A. Burns and M. S. Matthews, editors, University of Arizona Press, Tucson, Ariz., 1021 pp., 1986.

Uranus, J. T. Bergstralh, E. D. Miner, and M. S. Matthews, eds., University of Arizona Press, Tucson, Ariz., 1076 pp., 1991.

Neptune and Triton, D. P. Cruikshank, ed., University of Arizona Press, Tucson, Ariz., 1249 pp., 1995.

Pluto and Charon, D. Tholen, A. Stern, and M. S. Matthews, eds., University of Arizona Press, Tucson, Ariz., 728 pp., 1997.

Future Exploration

An Integrated Strategy for the Planetary Sciences: 1995–2010, J. A. Burns, chairman, Committee on Planetary and Lunar Exploration, Space Studies Board, National Research Council, National Academy Press, Washington, D.C., 199 pp., 1994.

Review of NASA's Planned Mars Program, J. A. Burns, chairman, Committee on Planetary and Lunar Exploration, Space Studies Board, National Research Council, National Academy Press, Washington, D.C., 29 pp., 1996.

Visible–Infrared Sensors and Case Studies

F. A. Kruse

Analytical Imaging and Geophysics
Boulder, Colorado

11.1 INTRODUCTION

This chapter of the earth sciences volume of the *Manual of Remote Sensing* is intended as a survey of selected available visible–infrared sensors and their applicability to earth sciences. Because of the author's background, the focus is necessarily on geological applications; however, the principles presented are equally applicable to other areas of terrestrial remote sensing.

The chapter is organized into two complementary sections. The first outlines the basic characteristics of selected sensors ranging from operational satellite systems, through both operational and experimental aircraft systems to planned satellite systems. The information in this section is a summary of the latest data available for the selected instruments at the time, described as of June 1998. Information is compiled from published material, commercial brochures, the *Manual of Remote Sensing*, 3rd Edition "Earth Observing Platforms and Sensors" CD-ROM (Morain and Budge, 1996), Lillesand and Kiefer (1987), Pease (1990), and Kramer (1994). In the second section we provide several case studies that compare and contrast the capabilities of sensors, with particular emphasis on the impact of spatial and spectral resolution on their utility for mapping of Earth surface features and properties.

11.2 SENSOR CHARACTERISTICS

In this section we describe the characteristics of selected remote sensing systems that are generally useful for remote sensing of the land surface. No attempt has been

Remote Sensing for the Earth Sciences: Manual of Remote Sensing, 3 ed., Vol. 3, edited by Andrew N. Rencz.
ISBN: 0471-29405-5 © 1999 John Wiley & Sons, Inc.

made to document all available sensors; only a representative sampling has been included (see Kramer, 1994; Morain and Budge, 1996; ASPRS, 1996 for more comprehensive sensor information). References and contacts are provided for obtaining additional information about the sensor systems described.

11.2.1 Satellite Systems

The launch of *Landsat* in 1972 was a landmark in the use of remote sensing technology for earth observation. Researchers and operational users alike quickly recognized the advantages of imaging from satellite platforms. Remote sensing instruments in earth orbit provide platform stability; a synoptic view; and repetitive, worldwide, multitemporal coverage. The following instrument descriptions outline the basic characteristics of selected satellite sensors and their applicability to geologic remote sensing. Examples are given for some of these sensors in the case histories in Section 11.3.

11.2.1.1 AVHRR.

The advanced very high resolution radiometer (AVHRR), launched in 1978 as *TIROS-N* (later, *NOAA-6),* provides regional-scale coverage (2800 to 4000 km swath width) at 1- and 4-km spatial resolutions (Kidwell, 1991; ASPRS, 1996). AVHRR uses silicon (Si), indium-antimonide (InSb), and mercury-cadmium-telluride (HgCdTe) detectors to provide four or five spectral bands (depending on sensor version) covering the ranges 0.55 to 0.68μm, 0.725 to 1.1μm, 3.55 to 3.93μm, 10.3 to 11.3μm, and 11.5 to 12.5μm (*NOAA-7* through *NOAA 14*). Although its principal use is currently for measuring cloud cover and vegetation indices, it is also valuable for providing basic surface geologic information over truly regional scales. Table 11.1 summarizes the AVHRR satellites:

Contact: Customer Services
U.S. Geological Survey
EROS Data Center
Sioux Falls, SD 57198

TABLE 11.1   *AVHRR* Satellite Information

Number	Dates
TIROS-N	10/19/78–01/30/80
NOAA-6	06/27/79–11/16/86
NOAA-7	08/24/81–06/07/86
NOAA-8	05/03/83–10/31/85
NOAA-9	02/25/85–Present
NOAA-10	11/17/86–Present
NOAA-11	11/08/88–09/13/94
NOAA-12	05/14/91–Present
NOAA-14	12/30/94–Present

Phone: (605) 594-6151
Fax: (605) 594-6589
E-mail (Internet): custserv@edcmail.cr.usgs.gov
URL:http://edcwww.cr.usgs.gov/eros-home.html
Additional Detailed Information: http://edcwww.cr.usgs.gov/glis/hyper/guide/
avhrr

11.2.1.2 IRS.

India's *IRS-1A* and *IRS-1B*, launched in 1988 and 1991, provide spectral and spatial coverage similar to the *Landsat* MSS system (ASPRS, 1996). The linear imaging self scanning sensor (LISS) provides spectral coverage in four bands with 72.5-m spatial resolution (*IRS-1A*/LISS I) and 36.25-m resolution (*IRS-1B*/LISS II). *IRS-1C*, the first of the second generation of IRS series satellites, was launched on December 28, 1995. *IRS 1-C* includes a high-resolution 5.8-m panchromatic band with a 70-km swath and stereo capabilities resulting from ±26° across-track steering. The LISS III multispectral sensor includes bands equivalent to *Landsat* TM bands 2, 3, 4, and 5 with a visible and near-infrared (VNIR)ground resolution of 23.5-m, shortwave infrared (SWIR) resolution of 70-m, and an approximate 140-km swath. *IRS-1C* also includes the two-band (0.62 to 0.68 μm, and 0.77 to 0.86μm) Wide Image Field Sensor (WiFS) with 188-m spatial resolution and approximate 800-km-wide swath. IRS-ID, identical to the IRS-1C sensor, was launched on 29 September 1997. The IRS series of satellites provides basic mapping capabilities similar to those of *Landsat* MSS and TM, with significant spatial enhancement provided by the *IRS-1C* 5.8-m panchromatic band.

Contact: Space Imaging/EOSAT
           12076 Grant Street
           Thornton, Colorado 80241
           Phone: (303) 254-2000
           Toll Free (U.S.): (800) 425-2997
           Fax: (303) 254-2215
           E-mail: info@spaceimaging.com
           Customer Service: (301)552-0537 or (800) 232-9037
           URL: http://www.spaceimage.com

11.2.1.3 *JERS-1* OPS.

The Japanese Earth resources satellite optical sensor [*JERS-1* (OPS)], launched in February 1992, provides eight spectral bands covering the VNIR and SWIR regions of the spectrum (ASPRS, 1996). Bands 1 to 4 are similar to *Landsat* TM bands 1 to 3 covering the ranges 0.52 to 0.60 μm (band 1), 0.63 to 0.69 μm (band 2), and 0.76 to 0.86μm (bands 3 and 4). Bands 3 and 4 provide stereoscopic capabilities. *JERS-1* (OPS) bands 5 to 7 cover critical regions in the SWIR, 1.60 to 1.71 μm (band 5), 2.01 to 2.12 μm (band 6), 2.13 to 2.25μm (band 7), and 2.27 to 2.40μm (band 8) (ASPRS, 1996). *JERS-1*(OPS) provides approximately 18-m × 24-m spatial resolution over a swath width of approximately 75 km. The SWIR bands are of particular interest to geologists because they cover key regions critical to mineralogical discrimination (Yamaguchi, 1987).

Contact: Remote Sensing Technology Center of Japan
             Uni Roppongi Building
             1-9-9, Roppongi,
             Minato-ku, Tokyo, 106 Japan
             URL:http://hdsn.eoc.nasda.go.jp/guide/guide/satellite/satdata/jers_
                e.html
             URL:http://hdsn.eoc.nasda.go.jp/guide/guide/satellite; shsendatalops_
                e.html

Additional Contact: National Space Development Agency of Japan (NASDA)
                          World Trade Center Bldg., 2-4-1,
                          Hamamatsu-cho, Minato-ku, Tokyo 105-8060
                          Telex: J28424 (AAB:NASDA J28424)
                          Phone: 81-3-3438-6000 Fax: 81-3-5402-6512
                          URL:http://yyy.tksc.nasda.go.jp/Home/This/This-e/jers_
                            e.html

11.2.1.4 *LANDSAT MSS.*

Landsat was originally launched with two sensor systems, the three-channel Return Beam Vidicon (RBV) and the four-channel multispectral scanner (MSS) (Lillesand and Kiefer, 1987). The RBV bands were designated as bands 1, 2, and 3, while the MSS bands were designated as 4, 5, 6, and 7. While the RBV was part of the payload on Landsats 1 to 3, the MSS quickly became the primary data source for most users because it was the first system capable of producing multispectral digital data on a global basis. The MSS system has been flown on board all five *Landsat* missions (MSS bands were redesignated as 1 to 4 on *Landsats 4* and *5*). An oscillating scan mirror covering 11.56 degrees scans six contiguous image lines simultaneously every 33-ms. Four arrays of six detectors each acquire spectral images in the ranges 0.5 to 0.6 µm (band 1), 0.6 to 0.7 µm (band 2), 0.7 to 0.87 µm (band 3), and 0.8 to 1.1µm (band 4). The MSS scans six contiguous lines simultaneously from west to east, with the motion of the spacecraft building the along-track image. The instantaneous field of view (IFOV) of the scanner is approximately 79 m on a side. The swath is 185 km wide and images are typically provided in "framed" format of 185 km × 185 km with 10% endlap between successive scenes. The dynamic range of the data is 6 bits. MSS data are generally useful for mapping earth surface features. Typically the data are used to produce color composites that discriminate spectral differences between surface cover types. The use of band 7 (band 4 on *Landsats 4* and *5*) with bands 4 and 5 (bands 1 and 2 on *Landsats 4* and *5*) provides an excellent means of discriminating vegetation from other materials. Landsat MSS's utility for geologic mapping is well documented, ranging from early efforts at alteration mapping (Rowan et al., 1974) to a variety of geologic uses (Goetz and Rowan, 1981; Legg, 1991; Prost, 1994). While long the workhorse of the earth science community because of its global coverage and continuous acquisition since 1972, the MSS sensors' 79-m spatial resolution and the selection of spectral bands are not sufficient for many geologic application's requirements. MSS is most useful for environmental studies that map the general distribution of surface vegetation and geologic characteristics such as iron oxides over long time periods. The MSS does not provide the spectral coverage or resolution required for many geologic applications.

Contact: Space Imaging/EOSAT
            12076 Grant Street
            Thornton, Colorado 80241
            Phone: (303) 254-2000
            Toll Free (U.S.): (800) 425-2997
            Fax: (303) 254-2215
            E-mail: info@spaceimaging.com
            Customer Service: (301) 552-0537 or (800) 232-9037
            URL: http://www.spaceimage.com
Contact: Earth Resources Observation Systems (EROS) Data Center
            Customer Services
            U.S. Geological Survey
            Sioux Falls, SD 57198
            Phone: (605) 594-6151
            Fax: (605) 594-6589
            E.mail: *CUSTERV@EDCMAIL.CR.USGS.GOV*
            URL:http://edcwww.cr.usgs.gov/webglis

11.2.1.5 *LANDSAT* TM.

The Thematic Mapper (TM) sensor was included on *Landsat* beginning with *Landsat* 4 launched in 1982. The *Landsat* MSS bands were also included, but renumbered from 4 to 7 to 1 to 4. The Thematic Mapper includes seven spectral bands covering the region from the visible to the thermal infrared (Table 11.2) (ASPRS, 1996). The IFOV is 28.5 m for bands 1 to 5 and 7, and 120 m for band 6. The total field of view of the sensor is 15.4° (approximately 185 km × 185 km at 705 km altitude). The TM sensor has sixteen detectors per band (four for the thermal infrared). Bands 1 to 4 utilize silicon detectors, while bands 5 to 7 use passively cooled indium antimonide (InSb). Band 6 utilizes mercury-cadmium-telluride (HgCdTe) detectors. The TM sensor uses bidirectional scanning (both west to east, and east-to west) to minimize scan mirror oscillation and increase detector dwell time (Lillesand and Kiefer, 1987). The dynamic range of the TM data is 8 bits, providing increased sensitivity over MSS data. *Landsat* 7, scheduled for launch in 1999 will include an additional 15-m-resolution panchromatic band covering the rnge 0.50 to 0.90 µm. *Landsat* TM presents an order-of-magnitude improvement over MSS for Earth-surface mapping

TABLE 11.2   Landsat Thematic Mapper Bands

Band	Wavelengths (µm)
1	0.45–0.52
2	0.52–0.60
3	0.63–0.69
4	0.76–0.90
5	1.55–1.75
6	10.40–12.50
7	2.08–2.35

because of the increased number of spectral bands and improved spatial resolution. Color composites are still commonly used for analysis of TM data. Color infrared composites using bands 4, 3, and 2 (RGB) and true color composites (not available using *Landsat* MSS) utilizing bands 3, 2, and 1 (RGB) have proven extremely useful. Color-ratio-composites have been used successfully for spectral mapping of a variety of surface materials (Rowan et al., 1974; Goetz et al., 1983; Kruse, 1984a; Paylor et al., 1985; Lang et al., 1987, Prost, 1994). These commonly utilize the 5/7 ratio to discriminate areas of clays, carbonates, and vegetation, and the 3/1 ratio to map areas of iron oxides (Sabins, 1997). The third ratio is often the 3/4 or 4/5, which also helps discriminate vegetation. The relatively high spatial and spectral resolution of the TM data also provide improved opportunities for detailed digital analysis. A wide variety of digital techniques, including contrast enhancements, spectral ratioing, principal components, and both unsupervised and supervised classification, have been used to produce image-maps of surface materials (Sabins, 1997).

Contact: Space Imaging/EOSAT
12076 Grant Street
Thornton, Colorado 80241
Phone: (303) 254-2000
Toll Free (U.S.): (800) 425-2997
Fax: (303) 254-2215
E-mail: info@spaceimaging.com
Customer Service: (301) 552-0537 or (800) 232-9037
URL:http://www.spaceimage.com
Contact: Earth Resources Observation Systems (EROS) Data Center
Customer Services
U.S. Geological Survey
Sioux Falls, SD 57198
Phone: (605) 594-6151
Fax: (605) 594-6589
E-mail: *CUSTSERV@EDCMAIL.CR.USGS.GOV*
URL: *http://edcwww.cr.usgs.gov/webglis*

11.2.1.6 SPOT.

The System Pour l'Observation de la Terre (*SPOT*) designates a series of high-spatial-resolution imaging satellites designed by the Centre National d'Etudes Spatiales (CNES), France. *SPOT 1* was launched in February 1986, *SPOT 2* in January 1990, *SPOT 3* in December 1993, and *SPOT-4* in March 1998 (see http://www.spotimage.fr/). *SPOT* utilizes a linear array sensor and push-broom scanning techniques along with pointable optics (ASPRS, 1996). Its repeat period is 26 days; however, off-nadir pointing allows shorter revisit periods as well as stereoscopic imaging. SPOT's payload is a high-resolution visible (HRV) imaging system operating in either panchromatic (PAN) or multispectral (XS) mode. The PAN sensor on SPOT 1 to 3 operates at 10-m spatial resolution from 0.51 to 0.73μm. The multispectral mode provides 20-m spatial resolution in three spectral bands covering the ranges 0.50 to 0.59 μm, 0.61 to 0.68μm, and 0.79 to 0.89μm. *SPOT-4* makes slight adjustments to the spectral configuration, including changing the 10-m panchromatic band coverage to 0.61 to 0.68µm, and adding an additional band operating in the short-wave infrared portion of the spectrum (SWIR) from 1.5 to 1.75µm. *SPOT-4* also offers a "Vegetation Instrument" operating in the four spectral bands with a resolution of 1-km, with a swath width of 2,250 km. The HRV instrument's field of view is 4.13°, providing a swath width of 60 km for nadir pointing. Two identical HRV sensors provide the capability to image a total swath of 117 km in either PAN or XS mode. Data dynamic range is 8 bits. The high spatial resolution of the SPOT system provides capabilities not available with either the Landsat MSS or TM systems. Photogeologic mapping using either single PAN images or stereoscopic PAN coverage allows mapping at 1:24,000-scale. The multispectral mode images are useful for vegetation mapping and general lithologic mapping, but do not really provide the spectral resolution or spectral coverage required for detailed geologic work. *SPOT/Landsat* merges provide one means of overcoming *SPOT*'s spectral shortcoming and *Landsat* TM's relatively poor spectral resolution.

Contact: SPOT Image Corporation
1897 Preston White Drive
Reston, VA 22091-4368
1897 Preston White Drive
Phone: (703) 715 3100
Fax: (703) 648 1813
URL:http://www.spot.com
Contact: SPOT Image
5, rue des Satellites
BP 4359
F 31030 Toulouse cedex 4
France
Phone: +33 (0)5 62 19 40 40
Fax: +33 (0)5 62 19 40 11
URL: http://www.spotimage.fr/welcoma.htm

11.2.2 Aircraft Sensors

Aircraft remote sensing provides both advantages and disadvantages over satellite, spaceborne remote sensing. The platforms are less stable and the spatial coverage is smaller, however, greater spatial resolutions can be obtained, and the user can maintain closer control over when and how the data are collected. In this section we provide an overview of selected aircraft sensors, with particular emphasis on hyperspectral systems. No attempt has been made to document all available sensors; only a representative sampling has been included (see ASPRS, 1996, for more comprehensive sensor information). References and contacts are provided for obtaining additional information about the sensor systems described.

11.2.2.1 AVIRIS.

The Airborne Visible/Infrared Imaging Spectrometer (AVIRIS) is a 224-channel imaging spectrometer built by Jet Propulsion Laboratory (JPL) and first flown in engineering tests during 1987. The system became fully operational in 1989. AVIRIS covers the range 0.4 to 2.5 μm in 224 approximately 10-nm-wide contiguous bands (Porter and Enmark, 1987; Vane et al., 1993). The sensor is a whiskbroom system utilizing scanning foreoptics to acquire cross-track data. The IFOV is 1 milliradian. Four off-axis double-pass Schmidt spectrometers receive incoming illumination from the foreoptics using optical fibers. Four linear arrays, one for each spectrometer, provide high sensitivity in the 0.4 to 0.7 μm, 0.7 to 1.2 μm, 1.2 to 1.8 μm, and 1.8 to 2.5 μm regions respectively. AVIRIS is flown as a research instrument on the NASA ER-2 aircraft at an altitude of approximately 20 km, resulting in approximately 20-m pixels and a 10.5-km swath width. AVIRIS is easily the best-calibrated aircraft imaging system flown today. Routine calibration consists of spectral and radiometric calibration utilizing both laboratory and field measurements (Vane et al., 1987, 1993; Chrien et al.; Green et al., 1996). Spectral calibration consists of calibration of a laboratory monochromator using standard line emission sources and then recording the response of AVIRIS detectors to specific narrow spectral bandwidths of light from the monochromator. Radiometric calibration consists of using a calibrated laboratory spectroradiometer and a 100-cm-diameter integrating sphere to generate a calibration file to convert AVIRIS data numbers to units of radiance. These calibrations are accomplished before and after each flight season. Additionally, JPL conducts in-flight calibration experiments before, during, and after each flight season to verify the laboratory measurements and monitor inflight instrument performance. AVIRIS is particularly well suited to use for mineralogic and lithologic mapping. Its spectral resolution of 10 nm makes direct mineral identification possible (unique to imaging spectrometers), while its 20-m spatial resolution allows mapping of igneous, metamorphic, and sedimentary features at scales suitable for 1:24,000-scale maps (Kruse et al., 1993a; Vane and Goetz, 1993).

Contact: Jet Propulsion Laboratory
4800 Oak Grove Drive
Pasadena, CA 91109
E-mail (archival data): avorders@makalu.jpl.nasa.gov
URL: http://makalu.jpl.nasa.gov

11.2.2.2 CASI.

The Compact Airborne Spectrographic Imager (CASI) is a pushbroom charge coupled device (CCD) imager operating over a 545-nm spectral range between 0.4 and 1.0 μm in up to 288 programmable spectral channels (see CASI marketing literature and *http://www.itres.com*). CASI uses a 578 × 288 CCD Si detector array. Quantization is 12 bits. The IFOV is 1.25 mrad. The instrument configuration is fully programmable to operate in any of three modes. In spatial mode, the full cross-track resolution of 512 pixels is obtained for up to 19 nonoverlapping spectral bands with programmable center wavelengths and bandwidths. In spectral mode, a 288-point spectrum is measured at 1.9-nm intervals for each of a limited number of points across the swath. A programmable monochromatic image at the full spatial resolution (scene recovery channel) is also acquired. Alternatively, a maximum of 101 adjacent pixels can be configured to record 288 spectral bands. Full frame mode provides for digitization of the entire 512 × 288 pixel array. Because of the data quantities involved, however, this mode is used primarily for laboratory and ground-based studies and the collection of data for instrument calibration. The CASI sensor has been operated since 1988 on light aircraft and helicopter platforms, with spatial resolution ranging from 0.5 to 10 m, depending on flight altitude. CASI's use for geology is limited because of its spectral coverage, which does not include the SWIR wavelengths. The majority of geologic material's unique spectral characteristics occur outside this range. CASI has found widespread use in vegetation mapping and environmental studies, as it is well suited to mapping subtle difference in vegetation reflectance spectra.

Contact: ITRES Research Ltd.
Suite 155, East Atrium
2635–37 Avenue N.E.
Calgary, Alberta
CANADA T1Y 5Z6
Phone: (403) 250-9944
Fax: (403) 250-9916
URL: http://www.itres.com/

11.2.2.3 GEOSCAN MKII AIRBORNE MULTSPECTRAL SCANNER.

The GEOSCAN MkII sensor, flown on a light aircraft, was a commercial airborne system that acquired up to 24 spectral channels selected from 46 available bands. Though currently not flying, significant archival data are available. GEOSCAN covered the range from 0.45 to 12.0 μm using grating dispersive optics and three sets of linear array detectors (Lyon and Honey, 1989). A typical data acquisition for geology resulted in ten bands in the visible/near infrared (VNIR, 0.52 to 0.96 μm), eight bands in the shortwave infrared (SWIR, 2.04 to 2.35 μm), and six bands in the thermal infrared (TIR, 8.64 to 11.28 μm) regions (Lyon and Honey, 1990). The instantaneous field of view was 3.0 mrad with a field of view of 45° from nadir. Radiometeric calibration was accomplished using separate calibration sources for each of the spectral regions. The spatial resolution varied depending on flight altitude, but typically ranged from 3- to 5-m. The sensor was stabilized for pitch and yaw and the data corrected for roll. Resampling was applied to the data to produce a constant pixel size and the data were quantized to 8 bits, with gain and offset applied to adjust for different brightness terrains. Data were recorded directly on optical media and displayed in real time in the aircraft. GEOSCAN's high spatial resolution makes it suitable for detailed geologic mapping (Hook et al., 1991). The relatively low number of spectral bands and, low spectral resolution limit mineralogic mapping capabilities to a few groups of minerals in the absence of ground information. Strategic placement of the SWIR bands, however, does provide more mineralogic information than would intuitively be expected based on the spectral resolution limitations.

Contact: (Mark I and II archival data)
Australian Geological and Remote Sensing
Services (A.G.A.R.S.S. Pty Ltd)
32 Wheelwright Road

Lesmurdie Western Australia 6076
AUSTRALIA
Phone: 61 8 9291 7929
Fax: 61 8 9291 8566
URL: http://www.agarss.com.au/MAIN.HTM

11.2.2.4 GER 63 CHANNEL SCANNER (GERIS).

Geophysical and Environmental Research Inc. (GER) has been developing and operating high-spectral-resolution sensors since the late 1970s. The GER 63-channel airborne scanner (also known as GERIS) has been operational since approximately 1987. This instrument is a Kennedy-type scanner consisting of three grating spectrometers with three individual linear detector arrays (Kruse et al., 1990). The IFOV is selectable at 5.0 and 3.3 mrad, covering a scan angle of 90°. The visible/near infrared (VNIR) spectrometer covers the range 0.4 to 1.0 μm with up to 27 bands, 25 nm wide. The first short wave infrared (SWIR 1) spectrometer covers the range 1.5 to 2.0 μm with one band. The second SWIR spectrometer covers the 2.0 to 2.5 μm region with 29 bands with 17.5-nm widths. One additional band is used to record gyroscopic information. The current version includes an additional spectrometer covering the thermal infrared (TIR) region from 8 to 12.5 μm. A total of 63 bands are selectable out of the available 72 bands. The GER64 swath width and spatial resolution are variable, depending on flight altitude; typical resolutions average about 5 to 15 m. Radiometric calibration includes band/channel position and bandwidth using calibrated light sources, signal-to-noise determinations, and radiometric tests to determine transfer coefficients to radiance. The data are processed for roll utilizing the gyroscope channel with accuracies to 1 pixel. Several related sensors built by GER are also currently operational. The Digital Airborne Imaging Spectrometer series (DAIS-2815, 3715, 7915) provides similar capabilities but with different channels and bandwidths. Current instrument specifications are available from GER. The GER63 data are generally suitable for mineralogic and lithologic mapping, and spatial resolution can be tailored to requirements. Despite good placement of spectral bands, the relatively low number of bands and the spectral resolution of these bands limits mineral identification to a few groups of minerals (Kruse et al., 1990).

Contact: Geophysical and Environmental Research Corp.
           1 Bennett Common
           Milbrook, NY 12545
           Phone: (914) 667-6100
           Fax: (914)-667-6106
           E-mail: info@ger.com

11.2.2.5 HYDICE.

The Hyperspectral Digital Imagery Collection Experiment (HYDICE), developed by Hughes Danbury Optical Systems and operational since 1994, was designed specifically to determine the utility of hyperspectral technology for intelligence, military, and civil applications (Basedow et al., 1995). Hydice is a pushbroom imaging spectroradiometer covering the 0.4 to 2.5 μm range in 210 nominally 10-nm-wide channels utilizing a 320 × 210 element InSb focal plane. Spatial resolution ranges from

1 to 4 meters depending on aircraft altitude. Typical ground swath is 308 pixels covering on the order of 1 km, based on the design flight altitude of six km. Calibration information provided includes linear correction coefficients, measured laboratory response of the calibration unit, center wavelength position for each band, and a bad detector element list. Goetz and Kindel (1996) and Resmini et al. (1996) have demonstated that HYDICE can map mineralogy successfully at high spatial resolution. The trade-off for this capability, however, is the relatively small spatial coverage.

Contact: HYMSMO Program Office
           Spectral Information Technology Applications Center (SITAC)
           11781 Lee Jackson Memorial Highway, Suite 400
           Fairfax, VA 22033
           Phone: (703) 591-8546

11.2.2.6   MIVIS.

The Multispectral Infrared and Visible Imaging Spectrometer (MIVIS) built by Daedalus for CNR, Italy and flown since 1993 is a modular system consisting of four spectrometers that provide 102 spectral channels (Bianchi et al., 1996). MIVIS covers the ranges of 0.43 to 0.83 µm in 20 bands (20-nm-resolution), 1.15 to 1.55 µm in eight spectral bands (50-nm resolution), 1.985 to 2.479 µm in 64 spectral bands (8-nm resolution), and 8.21 to 12.70 µm in 10 bands (400-to 500-nm resolution). Quantization is 12 bits. MIVIS is flown on CNR's CASA C200/212. The IFOV is 2.0 mrads with a digitized FOV of 71° covering 755 pixels per line. Ground resolution varies with flight altitude. The system includes an integrated GPS receiver, and roll, pitch, and heading sensors. Two blackbody sources are used as thermal reference sources. MIVIS has been flown extensively in Europe; however, few geologic data have been collected. The layout and spectral resolution of the instrument indicate that it should be very useful for geological applications.

Contact: Consiglio Nazionale delle Richerche (CNR)
           Project L.A.R.A; Via Monte D'oro 11
           00040 Pomezia (RM), Italy
           Phone: 39 6 9100 312/313/314
           Fax: 39 6 9160 1614
           URL: http://ntserver.iia.mlib.cnr.it/index.htm

11.2.2.7   SFSI.

The Short Wavelength Infrared (SWIR) Full Spectrum Imager (SFSI) developed by the Canadian Center for Remote Sensing (CCRS) and flown since 1994 covers the range 1.2 to 2.4 µm range in 120 bands with a nominal full-width-half-max (FWHM) of approximately 10 nm. The instrument utilizes a 488×512 PtSi detector array, refractive optics, and a transmission grating with one order separating filter (Rowlands and Neville, 1994). In practice, 480 lines $\times$ 496 columns are used. Adjacent lines are summed together on-chip, resulting in 240 spectral bands (lines) $\times$ 496 cross-track pixels. Quantization is 12 bits; however, adjacent bands are again summed, giving 120 bands in the final output data. The cross-track pixel field of view is 0.33 mrad, resulting in a total swath of 9.4°. Ground resolutions are typically better than 10 m. Significant post processing is required to calibrate the data to reflectance. SFSI has been flown in Canada by CCRS and by a consortium of mining companies in the southwestern United States during 1995 (Hauff et al., 1996).

Contact: Borstad Associates Ltd.
                114–9865 West Saanich Road,
                Sidney, British Columbia, CANADA V8L 5Y8
                Phone: (250) 656-5633
                Fax: (250) 656-3646
                E-mail: gary@borstad.com
                URL:http://www.borstad.com/homepage.html

11.2.2.8 TIMS.

The Thermal Infrared Multispectral Scanner (TIMS), designed by the Jet Propulsion Laboratory and operational since 1982, collects thermal infrared data in six channels between 8.2 and 12.6 μm utilizing a six-element HgCdTe detector array (Kahle and Goetz, 1983; Palluconi and Meeks, 1985). Approximate spectral coverage is band 1 (8.2 to 8.6 μm), band 2 (8.6 to 9.0μm), band 3 (9.0 to 9.4 μm), band 4 (9.4 to 10.2 μm), bands 5 (10.2 to 11.2 μm), and band 6 (11.2 to 12.2 μm). The sensor's IFOV is 2.5 mrad, providing spatial resolutions of 8 to 20 m at elevations above terrain of approximately 3000 to 8000 m. A typical data set consists of 638 pixels per line, covering an approximately 5-to 12-km swath at the foregoing spatial resolutions. TIMS is well suited to geologic mapping because of its coverage of the thermal infrared portion of the spectrum, where primary rock-forming minerals have their fundamental absorption features. The presence of fundamental absorption features in this region theoretically makes possible direct detection and mapping of silicate mineralogy. The relatively broad spectral resolution of TIMS, however, reduces this capability to the mapping and detection of silica, regardless of origin and mineralogy (Kahle and Goetz, 1983; Kruse and Kierein Young, 1990; Watson et al., 1990; Rowan et al., 1992). TIMS has been demonstrated as a valuable tool for geological mapping. The ASTER instrument (Abrams and Hook, 1995), scheduled for launch on *EOS Platform AM-1* in 1998, will provide TIMS-like capabilities from orbit.

Contact (archival data):   EDC DAAC User Services
                          EROS Data Center
                          Sioux Falls, SD 57198
                          Phone: (605) 594-6116
                          Fax: (605) 594-6589
                          E-mail: edc@eos.nasa.gov
                          URL: http://edcwww.cr.usgs.gov/landdaac/landdaac.html
Contact (new acquisitions):   NASA Dryden Flight Research Center
                          Edwards Airforce Base, CA 93523
                          Phone: (805) 258-3311
                          URL: http://www.dfrc.nasa.gov/Projects/airsci/general

11.2.3 Planned Systems: U.S. Government.

In this section we provide a brief overview of two US government-planned remote sensing systems. The sensors described here serve to illustrate some of the research activities under way that may lead to future commercial systems with similar capabilities.

11.2.3.1 ASTER.

The Advanced Spaceborne Thermal Emission and Reflectance Radiometer (ASTER) is a planned instrument scheduled to fly on NASA's Earth Observing System *(EOS) AM-1* platform in 1999 (Abrams and Hook, 1995). ASTER is a cooperative effort between NASA and Japan's Ministry of International Trade and Industry, with the collaboration of scientific and industry organizations in both countries. The ASTER instrument consists of three separate instrument subsystems. Each subsystem operates in a different spectral region, has its own telescope(s), and is built by a different Japanese company. The three subsystems combined will collect data in 14 spectral bands in the visible and near-infrared (VNIR), short-wavelength infrared (SWIR), and thermal infrared spectral regions (TIR). The VNIR subsystem operates using a Si detector in three spectral bands, with a resolution of 15 m. A backward-looking telescope provides a second view of the target area in band 3 for stereo observations. The individual spectral bands are separated through a combination of dichroic elements and interference filters. The SWIR subsystem uses PtSi–Si detectors in six spectral bands at 30-m resolution. The TIR subsystem uses HgCdTe for five bands in the thermal infrared region at 90-m spatial resolution. Quantization is 8 bits for bands 1 to 9 and 12 bits for bands 10 to 14. ASTER is particularly well suited to geologic mapping. The combination of VNIR/SWIR/TIR capabilities will provide the ability to produce detailed maps of mineralogy and alteration (Abrams and Hook, 1995). Table 11.3 lists the band characteristics for ASTER.

Contact: Jet Propulsion Laboratory
          4800 Oak Grove Drive, MS 183-501
          Pasadena, CA 91109
          URL: http://asterweb.jpl.nasa.gov/asterhome
Contact: JAROS
          Towa-Hatchobori Building
          2-30-1 Hatchobori Chuo-ku
          Tokyo, 104 Japan

11.2.3.2 MODIS.

The Moderate Resolution Imaging Spectrometer (MODIS) is an imaging spectrometer designed primarily for the measurement of biological and physical processes on regional scales (NASA, 1986; Asrar and Greenstone, 1995). MODIS will have 36 bands covering the visible through thermal infrared ranges at from 250-m to 1-km spatial resolution. Bands 1 and 2 will have 250-m spatial resolution with band centers at approximately 0.645 and 0.858 μm respectively. Bands 3 through 7 will provide 500-m spatial resolution covering the approximate region 0.46 to 2.16 μm in five strategically placed spectral bands. Bands 8 to 16 cover the approximate range 0.4

TABLE 11.3   ASTER Instrument Characteristics

Band	Spectral Range (μm)
	VNIR
1	0.52–0.60
2	0.63–0.69
3	0.76–0.86
	SWIR
4	1.60–1.70
5	2.145–2.185
6	2.185–2.225
7	2.235–2.285
8	2.295–2.360
9	2.360–2.430
	TIR
10	8.125–8.475
11	8.475–8.825
12	8.925–9.275
13	10.25–10.95
14	10.95–11.65

to 0.88 μm in nine spectral bands designed primarily for ocean color mapping, providing 1-km spatial resolution. Bands 17 to 19, again with 1-km resolution, are placed between 0.89 to 0.965 μm for mapping of clouds and atmospheric properties. Bands 20 to 36 cover the thermal infrared range 3 to 5 μm and 11 to 14 μm in 17 bands at 1-km resolution. The MODIS swath width will be approximately 2330 km, thus providing regional overview coverage.

Contact: Goddard Space Flight Center
Greenbelt Road
Greenbelt, MD 20771
Phone: (301) 286-2000
Fax: (301) 286-8142
URL: http://ltpwww.gsfc.nasa.gov/MODIS/MODIS.html

11.2.4 Planned Systems: Commercial

Within the next few years, at least three U.S. organizations will launch high-spatial resolution remote sensing systems. Many other domestic and international systems are currently under design. Companies designing these satellites expect that the high-resolution sensors will supplement or replace orthophotography as a data source for geographic information systems (GISs) because of reduced costs, timeliness, frequent revisit, global availability, regional coverage, and the digital nature of the data. In this section we provide a brief overview of selected planned commercial remote sensing systems.

11.2.4.1 EARTHWATCH.

One of several U.S. commercial satellites, EarthWatch's *EarlyBird* satellite was launched in December 1997, but lost contact with earth receiving stations several days after launch. *EarlyBird* was designed to provide high-spatial-resolution imagery with 3-m panchromatic (0.45 to 0.80 μm) and 15-m multispectral [0.50 to 0.59 μm (green), 0.61 to 0.68 μm (red), and 0.79 to 0.89 μm (near-IR) coverage] (ASPRS, 1996). The panchromatic sensor was to have used a 4 mega-pixel staring focal plane array to cover a frame size of approximately 36 km^2 (6 km × 6 km) per exposure, while the multispectral sensor would have used three 4 mega-pixel staring focal plane arrays to cover an area of approximately 900 km^2 (30 km × 30 km) per exposure in three spectral bands. A second mission, the *QuickBird* scheduled for launch in late 1999 will carry a 0.82-m panchromatic band [0.45 to 0.90 μm) and a 3.28-m multispectral sensor with four bands (0.45 to 0.52 μm (blue), 0.52 to 0.60 μm (green), 0.63 to 0.69 μm (red), and 0.76 to 0.90 μm (near-IR)]. Nominal image size will be 22 km × 22 km. Quantization will be 11 bits. Two *QuickBird* satellites are planned. EarthWatch plans to create a digital global imagery archive and distribution network for data distribution.

Contact: EarthWatch Incorporated
1900 Pike Road
Longmont, CO 80501
Phone: (800) 496-1225
Fax: (303) 702-5562
E-mail: info@digitalglobe.com
URL: http://www.digitalglobe.com/ewhome.html

11.2.4.2 ORBVIEW.

Orbital Imaging Corporation (ORBIMAGE) plans to provide a series of low-cost high-spatial-resolution satellites in support of mapping and surveying, natural resources exploration, governmental functions, and news gathering. ORBIMAGE has already launched *OrbView-1*, a weather information satellite, and *OrbView-2*, carrying the SeaWiFs sensor with six visible channels and two NIR channels. Two additional Orbview sensors are planned. *Orbview-3* will consist of a 1-m resolution panchromatic system (0.45 to 0.90 μm) as well as a 4-m multispectral (0.45 to 0.52 μm (blue), 0.52 to 0.62 μm (green), 0.63 to 0.69 μm (red), and 0.76 to 0.90 μm (near-IR)) system, both with a standard swath width of 8 km. *Orbview-4* will have the same 1-m and 4-pancromatic and multispectral configuration as *Orbview-3*; however, it will also carry a 280-channel hyperspectral imaging instrument. This instrument will combine 8-m spatial accuracy with high spectral accuracy over a 5 km swath. Several imaging modes are available to obtain larger, contiguous swaths.

Contact: ORBIMAGE
21700 Atlantic Boulevard
Dulles, VA 20166
Phone: (703) 406-5800
Fax: (703) 406-5552

E-mail: info@orbimage.com
URL:http://www.orbimage.com/

11.2.4.3 SPACE IMAGING/EOSAT.

Space Imaging/EOSAT plans to provide geometrically corrected and digital ortho-photo products combining the properties of imagery with the geometric accuracy of large-scale maps. The *IKONOS 1* satellite will carry a sensor with a 1-m panchromatic sensor (0.45 to 0.90 μm) and a 4-m multispectral sensor with coverage similar to *Landsat*'s spectral ranges [0.45 to 0.52 μm (blue), 0.52 to 0.60 μm (green), 0.63 to 0.69 μm (red), and 0.76 to 0.90 μm (near-IR)] (ASPRS, 1996). Repeat coverage will be approximately every eleven days. Both fore-aft and cross-track stereoscopic capabilities are provided by ±45° tilt capabilities. Individual scenes will be approximately 11 km × 11 km, with swaths possible up to 100 km in length. Quantization will be 11 bits. IKONOS 1 is scheduled for 1998 launch.

Contact: Space Imaging/EOSAT
           12076 Grant Street
           Thornton, Colorado 80241
           Phone: (303) 254-2000
           Toll Free (U.S.): (800) 425-2997
           Fax: (303) 254-2215
           E-mail: info@spaceimaging.com
           Customer Service: (301) 552-0537 or (800) 232-9037
           URL: *http://www.spaceimage.com*

11.2.4.4 ARIES.

The Australian Resource Information and Environmental Satellite (ARIES) is designed as a commercially sustainable resource information satellite using the latest hyperspectral sensing technology. The *ARIES-1* satellite will deliver detailed geological and mineral information to the international mining industry, anywhere in the world and on demand. Because of its hyperspectral approach, *ARIES-1* will also be capable of providing improved environmental and agricultural resource information. *ARIES-1* will have on the order of 60+ spectral bands, with approximately 32 contiguous bands in the visible and near-infrared (0.4 to 1.1 μm) with approximately 20-nm spectral resolution, and 32 contiguous bands in the shortwave infrared (SWIR) (2.0 to 2.5 μm) with approximately 16-nm spectral resolution. Additionally, several bands will be provided for the 0.94 and 1.14 μm regions to assist with atmospheric correction. Optional contiguous coverage from 1.05 to 2.0 μm with 32-nm spectral resolution is also being considered. Spatial resolution is planned to be 30-m and a 10-m-resolution panchromatic sharpening band is also planned. Several simulation studies have been performed, and examples are available (see URL below). Launch is planned for 1999 to be fully operational in early 2000.

Contact: The ARIES Project Office
           PO Box 17, Mitchell, ACT
           Australia, 2911
           Phone: +61 2 62422613;

Fax: +61 2 62416750;
E-mail: smyers@aries-sat.com.au
URL:http://www.cossa.csiro.au/ARIES

11.3 CASE STUDIES

In this section we provide several case studies that illustrate the capabilities of selected sensors and analysis strategies. The first example compares the effect of both spatial and spectral resolution on visual interpretation of satellite and airborne sensors (*Landsat* MSS, *Landsat* TM, *SPOT*, and AVIRIS). The second case history compares spectra from *Landsat* TM, GEOSCAN, GER63, and AVIRIS data to spectral library measurements and contrasts image expression of spectral characteristics in a simple spectral matching algorithm. The final case history is an end-to-end example of state-of-the-art analysis of hyperspectral imagery using AVIRIS data.

11.3.1 Case Study: *Landsat* MSS and TM/*SPOT*/AVIRIS, Northern Grapevine Mountains, Nevada

A coregistered *Landsat* MSS, *Landsat* TM, *SPOT* Panchromatic, and Airborne Visible/Infrared Imaging Spectrometer (AVIRIS) data set of a portion of the northern Grapevine Mountains, Nevada, provides the basis for comparison of sensor characteristics and capabilities, including spatial and spectral resolution (Kruse, 1988; Kruse and Dietz, 1991). Figure 11.1 (see the color insert) shows *Landsat* color composites, a *SPOT* panchromatic image, and an AVIRIS spectral classification. The area covered by the images is approximately 7 km × 9 km. The *Landsat* MSS false-color composite of bands 7, 5, and 4, the equivalent of a color infrared photograph (Figure 11.1a), illustrates the 79-m MSS spatial resolution, which limits effective mapping using MSS to scales smaller than about 1:250,000. Compare the MSS image to the *Landsat* TM image (Figure 11.1b) and the *SPOT* panchromatic image (Figure 11.1c) with 30-and 10-m spatial resolution, respectively. The *Landsat* TM bands 4, 3, and 2 color composite, again comparable to a color infrared photograph, allows effective mapping at scales as large as 1:24,000. Key landmarks, topographic features, and lithologic information can be seen. The apparent gain in going from TM to *SPOT* 10-m resolution is less dramatic than the MSS-to-TM changes. Most of the features visible in the *SPOT* data (Figure 11.1c) are also visible in the TM data (Figure 11.1b). Note the sharpening of spatial details associated with the improved 10-m resolution *SPOT* data, but the loss of lithologic information caused by *SPOT*'s limited spectral coverage and inability to make color composites (note that *SPOT* XS does allow making color composites at 20-m resolution).

Figures 11.1d, 11.1e, and 11.1f illustrate the effects of both spatial and spectral resolution on lithologic mapping capabilities. Band ratioing is a well-established method for extracting spectral information from multispectral data sets (Rowan et al., 1974). Figure 11.1d shows a color-ratio-composite (CRC) image of *Landsat* MSS ratios 4/5, 6/7, and 5/6 for the northern Grapevine Mountains, Nevada site. Note first how topography is suppressed and how this makes the relatively low spatial resolution nature of the data more apparent. The green areas in the image represent occurrences of iron oxide minerals (Rowan et al., 1974, Kruse, 1984a). Little can be said, however, about other important rock-forming and weathering minerals. Because of its additional spectral bands, the *Landsat* TM color-ratio-composite image provides significant amounts of additional mineralogic information. The 5/7, 3/1, 3/4 CRC was designed to emphasize occurrences of iron oxides, clays/carbonates, and vegetation, respectively (Kruse, 1984b). The effect of improved spatial and spectral resolution is readily apparent (Figure 11.1e). The 3/1 ratio emphasizes the distribution of iron oxides by indicating locations where the characteristic UV-Visible Fe^{+3} absorption feature occurs. The 5/7 ratio emphasizes occurrences of both clays and carbonates based on characteristic absorption features near 2.2 and 2.3 µm, respectively, as well as some vegetation, which has absorption features at wavelengths greater than 2.0 µm caused by molecular water. The 3/4 ratio suppresses the effect of vegetation, by de-emphasizing areas that have the characteristic infrared vegetation peak near 0.80 µm. When combined in a CRC image, these band combinations allow interpretation of areas that appear in various shades of green as iron oxides, red areas on the image typically have either clay or carbonate (although they may also have vegetation), and yellow areas on the image are combinations of iron oxides and clays/carbonates. Note in Figure 11.1e how this combination greatly enhances the ability to map the distribution of these minerals over the capabilities provided by the MSS. In addition to the effects of spectral resolution and coverage, also note how the improved (30-m) TM spatial resolution allows better mapping of smaller occurrences of specific materials. For example, the areas of iron oxide are much more spatially coherent and easier to observe. Additionally, structurally controlled alteration areas only a few pixels wide (yellow bands in the center of Figure 11.11e marked by white arrow) are easily resolved by the TM data.

Figures 11.1e and 11.1f illustrate the difference between *discrimination* of mineralogy and *identification* of mineralogy made possible by high spectral resolution. The TM data shown in Figure 11.1e can only be used to determine broad groups of minerals, because the six visible-infrared bands undersample the available reflectance spectrum (Figure 11.2). A true absorption-feature-based spectral classification is not possible and the CRC image and related multispectral digital classification methods are the only option. These approaches use spectral slopes between widely spaced spectral bands rather than actual spectral features to discriminate areas of similar mineralogy, but do not allow identification of individual materials. Imaging spectrometer or Hyperspectral data, on the other hand, collect continuous spectra in narrow contiguous bands (Goetz et al., 1985), and thus allow direct comparison to laboratory reflectance spectra and the use of individual spectral features for identification and classification. Figure 11.1f shows the results of a spectral angle mapper (SAM) classification of AVIRIS data of the northern Grapevine Mountains, Nevada site. SAM determines the similarity of the AVIRIS reflectance spectrum at every pixel to a library of reference materials (J. W. Boardman, unpublished data; Kruse et al., 1993b, also explanation in Section 11.3.3.10). Compare the yellow regions on the TM CRC to the red, yellow, and green areas on the AVIRIS image. Where *Landsat* TM could only be used to map the occurrence of iron oxides with clays and/or carbonates (yellow areas on Figure 11.1e), the AVIRIS data allows the analyst to assign specific mineral names and can be used to tell the difference not only between clays and carbonates, but to discriminate mineral variation within groups, for ex-

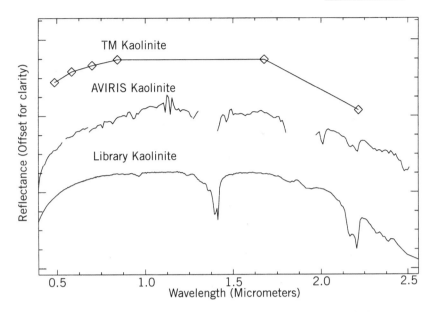

Figure 11.2   Comparison of a library, AVIRIS, and TM spectrum for the mineral kaolinite.

ample, between the two carbonates calcite (red on Figure 11.1f) and dolomite (yellow on Figure 11.1f).

11.3.2 Case Study: *Landsat* TM, AVIRIS, GEOSCAN, GER63, Cuprite, Nevada

11.3.2.1 BACKGROUND.

This example is provided to illustrate the effects of spatial and spectral resolution on information extraction from multispectral/hyperspectral data. Several images of the Cuprite, Nevada, area acquired with a variety of spectral and spatial resolutions serve as the basis for discussions on the effect of these parameters on mineralogic mapping using remote sensing techniques. These images have not been georeferenced, but image subsets covering approximately the same spatial areas are shown. Cuprite has been used extensively as a test site for remote sensing instrument validation (Abrams et al., 1978; Kahle and Goetz, 1983; Kruse et al., 1990; Hook et al., 1991; Swayze, 1997). A generalized alteration map (Figure 11.3) is provided for comparison with the images. Examples from *Landsat* TM, GEOSCAN MkII, GER63, and AVIRIS illustrate both spatial and spectral aspects. All of these data sets have been corrected to reflectance and processed by extracting selected spectral endmembers using averages for regions of interest. The data were then classified using the spectral angle mapper (SAM) algorithm (J. W. Boardman, unpublished data; CSES, 1992; Kruse et al., 1993b). SAM is a simple classification technique that determines similarity to reference spectra. Only three of the numerous materials present at the Cuprite site were used for the purposes of this comparison. Average kaolinite, alunite, and buddingtonite image spectra were selected from known occurrences at Cuprite. The SAM processing was applied to illustrate the similarities and differences between data ac-

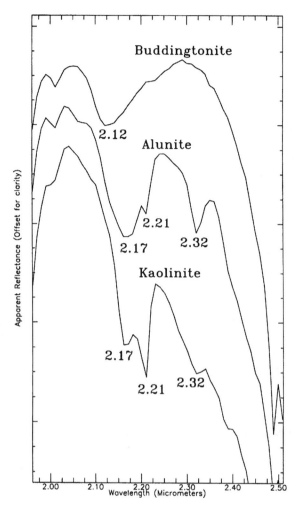

Figure 11.3 Generalized alteration map of the Cuprite, Nevada, site. (From Kruse et al., 1990, After Abrams et al., 1978.)

quired by various sensors; there are numerous other processing strategies that might produce similar or superior results for each specific sensor. Laboratory spectra from the US Geological Survey spectral library (Clark et al., 1990) of the three selected minerals are provided for comparison to the image spectra (Figure 11.4). The following is a synopsis of selected instrument characteristics and a discussion of the images and spectra obtained with each sensor. Refer to Section 11.1 for additional instrument specifics.

11.3.2.2 LANDSAT TM.

The Cuprite TM data were acquired on October 4 1984 and are in the public domain. Figure 11.5 (upper left) shows TM band 3 (0.66 μm) for spatial reference (30-m

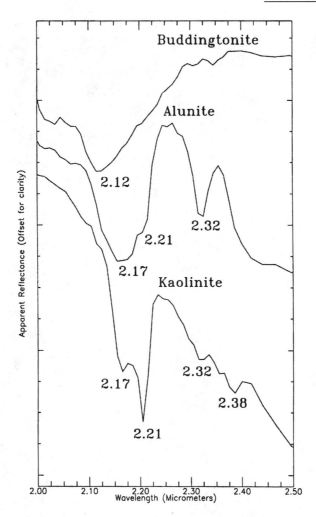

Figure 11.4 Laboratory spectra of kaolinite, alunite, and buddingtonite. Spectra are from the USGS Spectral Library. (Clark et al., 1990.)

resolution). The areas A, B, and C indicate the approximate locations of the reference materials buddingtonite, alunite, and kaolinite, respectively. Figure 11.6 is a plot of the region of interest (ROI) average spectra for these three materials. The small squares indicate the TM band 7 (2.21 μm) center point. The lines indicate the slope from TM band 5 (1.65 μm). Note the similarity of all the "spectra" and how it is not possible to discriminate between the three endmembers shown in Figure 11.4. Because TM only had one band in the critical region at 2.0 to 2.5 μm, the full six-band TM spectrum, which includes visible/near-IR bands, was used in the SAM classification, even though the classifications for the other sensors compared here only used the SWIR region. Figure 11.5 (upper right, lower left, lower right) shows the results of the SAM classification. Compare these images to the alteration map

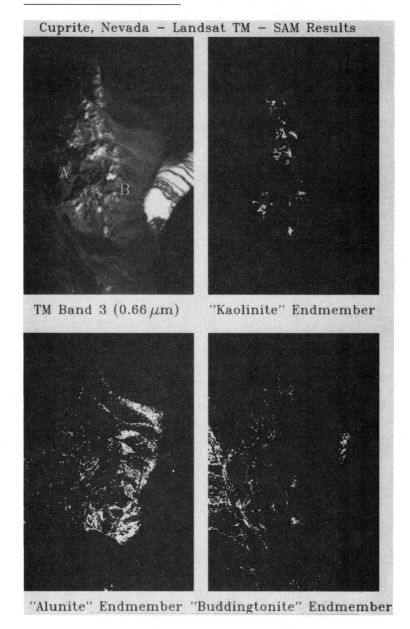

Figure 11.5 Cuprite, Nevada TM images. (Upper Left) TM Band 3 (0.66 μm). The remaining images are the results of SAM classification using image endmembers: (upper right) "kaolinite", (lower left) "alunite", (lower right) "buddingtonite". The full TM spectral range was used for these classifications, so these distributions are not limited to areas with the IR signatures of kaolinite, alunite, and buddingtonite, but to areas that have similar full-range TM spectra.

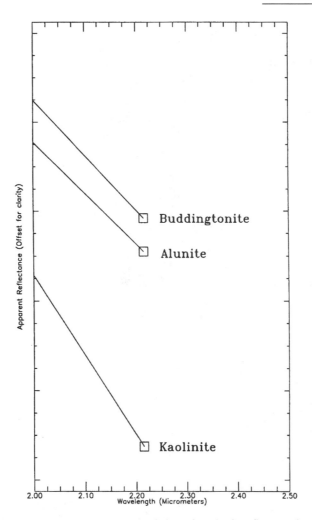

Figure 11.6 Cuprite, Nevada, TM spectra. The squares mark the single TM band 7 reflectance values. The lines connecting the left edge of the plot to the squares indicate the spectral slope from TM band 5 to TM band 7.

shown in Figure 11.3. Note that the "Kaolinite" endmember seems to outline the circular alteration pattern observed for the site, but that the other two endmember images seem totally unrelated to the known alteration pattern.

11.3.2.3 GEOSCAN MKII AIRBORNE MULTISPECTRAL SCANNER.

The GEOSCAN data were acquired in June 1989. Figure 11.7 (upper left) shows GEOSCAN band 3 (0.645 µm) for spatial reference (6-m resolution). The areas A, B, and C indicate the approximate locations of the reference materials buddingtonite, alunite, and kaolinite. Figure 11.8 is a plot of the ROI average spectra for these three materials. Compare these to Figure 11.4 and note that the three minerals appear quite different in the GEOSCAN data, even with the relatively widely spaced spectral

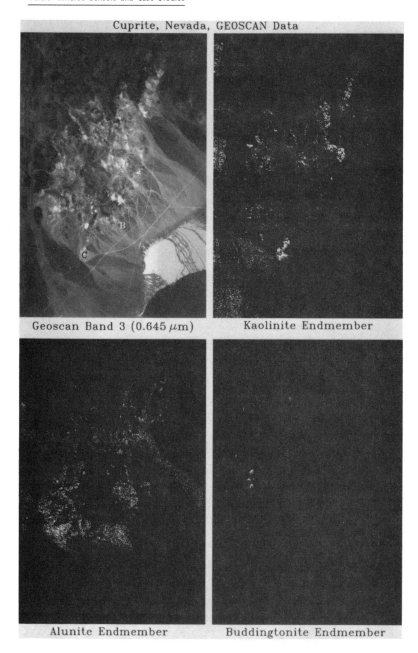

Figure 11.7   Cuprite, Nevada, GEOSCAN images. (upper left) GEOSCAN Band 3 (0.645 μm). The remaining images are the results of SAM classification using image endmembers: (upper right) "Kaolinite", (lower left) "Alunite", (lower right) "Buddingtonite".

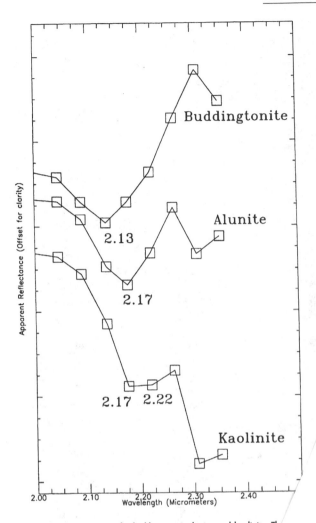

Figure 11.8   Cuprite, Nevada, GEOSCAN spectra for buddingtonite, alunite, and kaolinite. Th⁄ of the GEOSCAN image bands.

bands. Figure 11.7 (upper right, lower left, lower right) sʰ classification. Compare these images to the alteration m the TM classifications shown in Figure 11.5. Note the ⁄ and Alunite distributions mapped using the GEOSC responds to the known alteration pattern observeᵈ discrete, spatially-coherent areas on the buddingtᵒ bution generally corresponds to known occurr⁄ (Goetz et al., 1985, Kruse et al., 1990).

11.3.2.4   GER 63 CHANNEL SCANNER.

The GER63 data described here were acquirᵉ results were previously published in Kruˢ

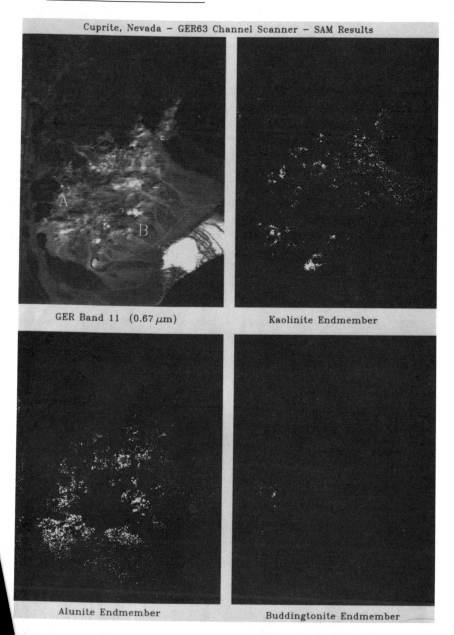

11.9   Cuprite, Nevada, GER63 images. (upper left) GER63 band 11 (0.67 μm). The remaining images are the results classification using image endmembers: (upper right) "Kaolinite", (lower left) "Alunite", (lower right) "Buddingtonite".

GER63 band 11 (0.67 μm) for spatial reference (12-to 22-m spatial resolu-he areas A, B, and C indicate the approximate locations of the reference buddingtonite, alunite, and kaolinite. Figure 11.10 is a plot of the ROI pectra for these three materials. Note that the GER63 adequately discrim-alunite and buddingtonite but does not fully resolve the kaolinite doublet

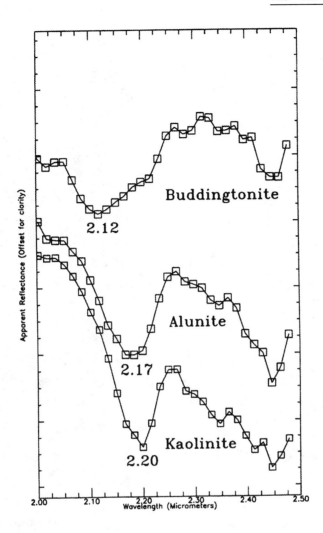

Figure 11.10 Cuprite, Nevada, GER 63 spectra for buddingtonite, alunite, and kaolinite. The squares mark the band centers of the 30 GER63 image bands used in the analysis.

near 2.2 μm shown in the laboratory reflectance data (Figure 11.4). Figure 11.9 (upper right, lower left, lower right) shows the results of the SAM classification. Compare these images to the alteration map shown in Figure 11.3, the TM classifications shown in Figure 11.5, and the GEOSCAN classifications shown in Figure 11.7. Again note the circular distribution of the kaolinite and alunite alteration endmembers, which generally match the known alteration pattern, and the discrete distribution of the buddingtonite endmember.

11.3.2.5 AVIRIS.

The AVIRIS data shown here were acquired during July 1995 as part of an AVIRIS "group shoot" (Kruse and Huntington, 1996). Figure 11.11 (upper left) shows

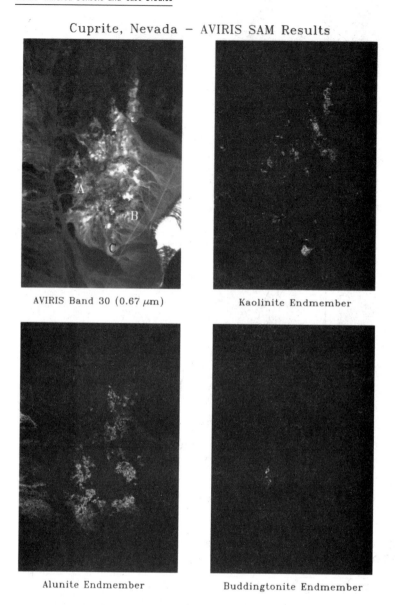

Cuprite, Nevada – AVIRIS SAM Results

Figure 11.11 Cuprite, Nevada, AVIRIS images. (upper left) AVIRIS band 30 (0.67 μm). The remaining images are the results of SAM classification using image endmembers: (upper right) "Kaolinite", (lower left) "Alunite", (lower right) "Buddingtonite".

AVIRIS band 30 (0.67 μm) for spatial reference (20-m spatial resolution). The areas A, B, and C indicate the approximate locations of the reference materials budding-tonite, alunite, and kaolinite. Figure 11.12 is a plot of the ROI average spectra for these three materials. Compare these to the laboratory spectra in Figure 11.4 and note the high quality and nearly identical signatures. Figure 11.11 (upper right, lower left, lower right) shows the results of the AVIRIS SAM classification. Compare these images to the alteration map shown in Figure 11.3, the TM classifications shown in

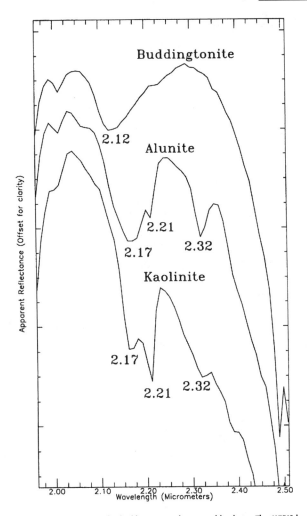

Figure 11.12   Cuprite, Nevada, AVIRIS spectra for buddingtonite, alunite, and kaolinite. The AVIRIS band-centers are too close together (10 nm) to mark on the plot.

Figure 11.5, the GEOSCAN classifications shown in Figure 11.7, and the GER63 classifications shown in Figure 11.9. Note that the AVIRIS data resolve the three alteration minerals into three distinct, spatially coherent zones and that the extracted spectra closely match the laboratory spectra (Figure 11.4) allowing positive identification.

11.3.2.6   DISCUSSION.

There are two aspects emphasized by comparison of these data. The first obvious difference is in the spatial resolution of the various sensors and aspects of the geology that can be observed. The GEOSCAN data have the highest spatial resolution at about 6 m, followed by the GER63 data with approximately 12-m cross-track and

22-m down-track resolution, AVIRIS data with about 20-m resolution, and the TM data with 28.5-m resolution. The GEOSCAN data allow observation of the fine details of the geology and prospecting activities, including individual stratigraphic layers, prospect pits, and minor access roads. The GER63 data have lower spatial resolution and significant geometric (scanning-induced) distortion, and some lithologic information and the distribution of prospect pits cannot be seen. The AVIRIS data appear quite blocky and again, the prospect information is obscured as well as individual stratigraphic layers which are typically thinner than 20 m. The TM data are extremely blocky, and altered areas are lumped together into grossly observable light versus dark areas. Individual beds and prospect pits are not observed, and none of the minor access roads are visible.

Spectrally, it is actually surprising how much information can be obtained from the lower-spectral-resolution instruments in the 2.0 to 2.5 μm region, in particular, the GEOSCAN data. The spectral resolution ranges from greater than 50 nm for TM through approximately 40 nm for GEOSCAN, to 17 nm for GER63, to 10 nm for AVIRIS. While the TM data can not be used to identify individual minerals, the altered areas do have the expected characteristic low band 7 signature. The single 2.21 μm TM band cannot be used for SAM classification or other multispectral classification techniques, and a density slice of the single band is unlikely to produce usable mineralogic information. All six TM bands (excluding the thermal band) were used for the SAM classification, including information from the visible/near-infrared, so it is not surprising that the TM SAM images do not match the known alteration mineralogy very well. The GEOSCAN spectra show distinct minima for the three minerals buddingtonite, alunite, and kaolinite, yet they do not fully resolve the band shapes, particularly for kaolinite. The GER63 data do a better job of showing the mineral spectra band positions and shapes, but the kaolinite doublet is still not resolved. AVIRIS fully resolves the major mineral absorption features, including subtle features in both alunite and kaolinite. Comparison of the GEOSCAN, GER63, and AVIRIS SAM results show that if one already knows what mineral one wishes to map, similar results can be achieved using any of these three sensors. If one wants to identify minerals without a priori knowledge, however, higher spectral resolution is required. Of the sensors shown here, only AVIRIS allows unambiguous identification of all three test minerals by comparison to the laboratory spectra in Figure 11.4. This is a direct result of the high spectral resolution, a convincing argument for the use of hyperspectral sensors similar to AVIRIS for detailed mineralogic mapping. While AVIRIS is currently the only fully operational imaging spectrometer, many other sensors are under development, and high spectral resolution data will soon be widely available.

11.3.3 Case Study: AVIRIS Data Analysis, Goldfield, Nevada

11.3.3.1 INTRODUCTION.

This case study describes AVIRIS processing and results for Goldfield, Nevada, for Airborne Visible/Infrared Imaging Spectrometer (AVIRIS) data collected during an AVIRIS group shoot conducted during summer 1995. At the time of this flight, AVIRIS was the only available instrument with sufficient spectral resolution for direct mineral mapping. AVIRIS is the standard by which future hyperspectral systems will be judged, and the processing described here is one model for processing of other hyperspectral data types. High-quality AVIRIS data were collected during this mission for sites in Arizona, Nevada, California, Utah, Wyoming, and Colorado, USA. The Goldfield site described here consists of two approximately 10 km × 12 km AVIRIS scenes (Figure 11.13).

The Goldfield area was chosen as the shared group site for the 1995 AVIRIS Group Shoot because considerable previous remote sensing and ground information exist for the area. The Goldfield mining district is a volcanic center thought to be a resurgent caldera (Ashley, 1974, 1979; Sabins, 1997). At least two periods of volcanism occurred and the hydrothermal alteration present in the district was caused

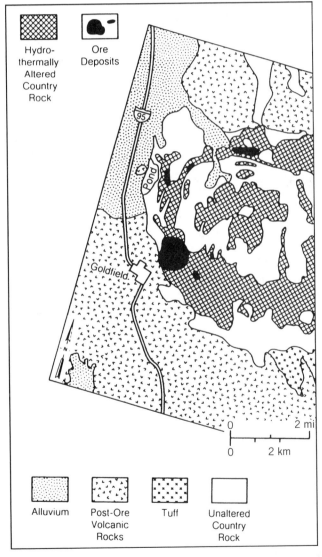

Figure 11.13   Goldfield, Nevada, Location Map and Generalized Geology/Alteration Map (after Sabins, 1997.)

by convective circulation of hydrothermal solutions along a zone of ring fractures and their linear extensions. Rocks exposed at the surface include air-fall and ash-flow tuffs, flows, and intrusive bodies. Hydrothermal alteration is extensive (Ashley, 1974, 1979; Rowan et al., 1974; Sabins 1997). The district exhibits a zoned alteration pattern. The rocks in the area have extensive exposures of alteration minerals including alunite, kaolinite, microcrystalline silica, illite, and montrmorillonite.

The Airborne Visible/Infrared Imaging Spectrometer (AVIRIS) is an imaging spectrometer that simultaneously acquires 224 spectral images at 20-m spatial resolution and 10-nm spectral resolution. Each "scene" covers an approximately 10 km × 12 km area. Although imaging spectrometers collect images, in addition they collect a complete spectrum for each picture element (pixel) of the image, thus allowing detailed mapping based on the spectroscopic characteristics of minerals (Goetz et al., 1985, Boardman et al., 1995). Despite major revisions to the instrument prior to the 1995 flight season, AVIRIS operated without problems during the group shoot flights. Whereas some minor irregularities were observed with the data, the order of magnitude of these is near the noise level in the data (R. O. Green, personal communication). Problems encountered with the data included (1) dropped bits in individual pixels, resulting in anomalous spectra, and (2) multiplexer lag in the instrument, resulting in spatial shifting between bands (bands 13 and 35 were discarded because of a shift of 1 pixel or more). According to Rob Green at Jet Propulsion Laboratory (JPL), "the multiplexer lag artifact is similar to the detector-readout-delay that has been in all AVIRIS data before 1995. The artifact expresses itself very close to the level of the AVIRIS noise." What this means is that the size of the instantaneous field of view (IFOV) of AVIRIS varies slightly as a function of the intensity of the signal. According to Green, the AVIRIS calibration without any correction is still at the 96% level. Based on preview processing, the AVIRIS data were judged to be satisfactory for the purposes of flight sponsors. In general, the data appeared to be of high quality in both the spatial and spectral domains.

A standardized processing and analysis approach has been developed based on nearly 10 years of processing experience with AVIRIS data. While many other approaches are possible; this one is favored by the author and associates. Table 11.4 outlines this approach. In the following sections we describe this approach in more detail.

11.3.3.2 DOWNLOAD QUICKLOOK DATA.

JPL processing to AVIRIS quicklooks was completed within two weeks. Quicklook data were obtained via anonymous FTP from JPL online at: "makalu.jpl.nasa.gov" in the directory "/pub/95qlook." The quicklook data were downloaded for the Goldfield site, spatial coverage was reviewed, and hardcopy output produced. Standard processed, full AVIRIS scenes for each site were ordered from JPL following the preview. Calibrated radiance data (Green et al., 1996) arrived from JPL within two months of the flights. The Goldfield data were processed using the steps discussed below.

11.3.3.3 PREVIEW DATA AND ASSESS QUALITY.

Calibrated radiance data were both spatially and spectrally previewed. The spatial coverage of the site was good and the geometry and coverage of the data were ex-

TABLE 11.4   Standardized AVIRIS Processing Methodology

1. Download quicklook data.
2. Review spatial coverage.
3. Perform preliminary assessment of spatial data quality.
4. Define areas for further processing.
5. Order AVIRIS radiance data for selected scenes.
6. Download radiance data from tape.
7. Perform data quality assessment.
   a. Spatial browsing
   b. Spectral browsing
   c. SNR calculations
8. Correct to apparent reflectance.
   ATREM
   Empirical line correction if ground information available.
   "Effort" correction if no ground information available.
   Spectral browsing
9. Report data problems to JPL.
10. Perform MNF Transform (spectral compression).
11. Calculate Pixel Purity Index Image (limited iterations for identification of bad pixels).
12. Perform masking of bad pixels.
13. Calculate Pixel Purity Index Image (maximum iterations for endmember determination).
14. Conduct N-Dimensional Visualization (Endmember Definition).
15. Compare of endmembers to spectral library for identification.
16. Run Spectral Angle Mapper (SAM).
17. Perform Spectral unmixing and/or Matched Filtering or other advanced mapping methods.
18. Add annotation and create output, and report.

cellent. Spatial browsing of different bands indicated that bands 13 and 35 were spatially offset from the rest of the bands (by one pixel horizontally). Another apparent anomaly was that some pixels exhibited what appeared to be dropped bits in one or more spectral bands. This was only for about 10 to 30 pixels per 614×1024 scene, well less than 1% of the data. These pixels were not used in subsequent processing.

11.3.3.4   ATREM CORRECTION.

The Goldfield AVIRIS data were corrected to apparent reflectance using the ATREM software available from the Center for the Study of Earth from Space (CSES) at the University of Colorado, Boulder. This software can be obtained via anonymous FTP from "cses.colorado.edu" in the directory "pub/atrem. Get the readme file for download instructions. ATREM is an atmospheric model–based correction routine and requires input of data parameters such as the acquisition date and time, the latitude and longitude of the scene, and the average elevation, along with atmospheric model parameters (CSES, 1992). ATREM version 1.31 was used for the Goldfield atmospheric correction. The output of the ATREM procedure is apparent reflectance data and a water vapor image for each scene. Typically, the water vapor image mimics topographic expression. Higher water vapor concentrations occur in the valleys, and lower water vapor concentrations occur over the higher elevations. In the Goldfield case, however, it appears as if there may have been some modulation by clouds. No clouds were visible in the images, however, and based upon review of spectra in areas apparently modulated by clouds, the ATREM apparent reflectance correction appears to have adequately removed water vapor contributions from the spectra.

11.3.3.5 ADJUSTMENTS TO ATREM CORRECTED DATA.

Residual atmospheric features and minor systematic noise in the apparent reflectance corrected data indicate that the quality of the AVIRIS data is better than the models we are currently using to calibrate the data. Empirical models can be used effectively to improve the apparent reflectance data. One means of doing this is a new method called the Empirical Flat Field Optimized Reflectance Transformation (EFFORT) which uses the characteristics of the data themselves without external spectral information (Boardman and Huntington, 1996). This method automatically finds the "flat" spectra in an AVIRIS scene, calculates the least-squares fit between these spectra and a low-order polynomial derived from the spectra, and uses a gain factor to remove the systematic noise. Small systematic noise and atmospheric adsorptions present in every spectrum are effectively removed, resulting in smooth spectra without channel averaging. These spectra are comparable to high-signal-to-noise laboratory spectra.

11.3.3.6 EVALUATION OF APPARENT REFLECTANCE DATA.

Spectral browsing through the apparent reflectance corrected Goldfield data set was used to get an idea of the success of the correction as well as to identify specific minerals. This procedure indicated that the spectral quality of the data was excellent and the apparent reflectance correction adequate, however, several bands were dropped from further analysis. As previously noted, bands at 0.5001 μm (band 13) and 0.6824 μm (band 35) had spatial misregistrations. Additionally, the overlap regions between spectrometers in bands 32 to 34 (0.67 μm) and in bands 97 to 98 (1.26 μm), as well as the spectral regions between approximately 1.3 to 1.4 μm and 1.84 and 1.96 μm, corresponding to the major atmospheric water bands, were masked out during subsequent processing. Individual spectra that contained absorption features attributable to kaolinite, alunite, buddingtonite, muscovite, and calcite could be recognized in the apparent reflectance corrected data.

11.3.3.7 MNF TRANSFORMATION.

The next step of the processing was to perform a "Minimum Noise Fraction" (MNF) Transform to reduce the number of spectral dimensions to be analyzed. The MNF transformation is used to determine the inherent dimensionality of the data, to segregate noise in the data, and to reduce the computational requirements for subsequent processing (Green et al., 1988; Boardman and Kruse, 1994). The MNF transformation is similar to Principal Components (PCs), but orders the data according to decreasing signal-to-noise-ratio (SNR) rather than decreasing variance as in PCs (Green et al., 1988). The MNF transformation can be used to partition the data space into two parts: one associated with large eigenvalues and coherent eigenimages, and a second with near-unity eigenvalues and noise-dominated images. By using only the coherent portions in subsequent processing, the noise is separated from the data, thus improving spectral processing results. For the Goldfield data, the eigenvalue plots fall sharply for the first 10 eigenvalues and then flatten out for the rest of the data. Examination of the eigenimages shows that while the first 10 images contain most of the information, images 11 through 20 still contain coherent spatial detail. The higher numbered MNF bands contain progressively lower signal-to-noise.

11.3.3.8 PIXEL PURITY INDEX (PPI).

Based on the MNF results above, the lower-order MNF bands were discarded and the first 20 MNF bands were selected for further processing. These were used in the "Pixel Purity Index" (PPI), processing designed to locate the most spectrally extreme (unique or different or "pure") pixels (Boardman et al., 1995). The most spectrally pure pixels typically correspond to mixing endmembers (unique materials). The PPI is computed by repeatedly projecting N-dimensional scatterplots back to two dimensions. The extreme pixels in each projection are recorded and the total number of times each pixel is marked as extreme is noted. A PPI image is created in which the digital number of each pixel corresponds to the number of times that pixel was recorded as extreme. A histogram of these images shows the distribution of "hits" by the PPI. A threshold was interactively selected using the histogram and used to select only the purest pixels in order to keep the number of pixels to be analyzed to a minimum. These pixels were used as input to an interactive visualization procedure for separation of specific endmembers.

11.3.3.9 N-DIMENSIONAL VISUALIZATION.

Spectra can be thought of as points in an N-dimensional scatterplot, where N is the number of bands (Boardman, 1993; Boardman et al., 1995). The coordinates of the points in N-space consist of N values that are simply the spectral reflectance values in each band for a given pixel. The distribution of these points in n-space can be used to estimate the number of spectral endmembers and their pure spectral signatures. This geometric model provides an intuitive means to understand the spectral characteristics of materials. In two dimensions, if only two endmembers mix, the mixed pixels will fall in a line in the histogram. The pure endmembers will fall at the two ends of the mixing line. If three endmembers mix, the mixed pixels will fall inside a triangle, four inside a tetrahedron, and so on. Mixtures of endmembers "fill in" between the endmembers. All mixed spectra are "interior" to the pure endmembers, inside the simplex formed by the endmember vertices, because all the abundances are positive and sum to unity. This convex set of mixed pixels can be used to determine how many endmembers are present and to estimate their spectra. The Goldfield AVIRIS data set was analyzed using these geometric techniques. The thresholded pixels from the MNF images above were loaded into an N-dimensional scatterplot and rotated in real time on the computer screen until points or extremities on the scatterplot were exposed. These projections were "painted" using region-of-interest (ROI) definition procedures and then rotated again in three or more dimensions (three or more bands) to determine if their signatures were unique in the AVIRIS MNF data. Once a set of unique pixels were defined, then each separate projection on the scatterplot (corresponding to a pure endmember) was exported to a ROI in the image. Mean spectra were then extracted for each ROI to act as endmembers for spectral unmixing. Using the SWIR data only from 2.0 to 2.4 µm, and the procedure described above, several endmembers were defined for the Goldfield AVIRIS data. These include the minerals calcite, kaolinite, illite/muscovite, an additional unknown clay, alunite, and opaline silica (Figure 11.14). Another mineral, with an "unknown 2.2-μm absorption feature," was also located. Based on the spatial distribution of spectra matching this endmember and known information about the sites, this endmember was identified as representing opaline silica. Unfortunately, similar spectra also occur on alluvial fans away from the altered areas, probably because of weathering and/or spectral mixing. Finally, both "light" and "dark" relatively aspectral endmembers were defined. These endmembers or a subset of these endmembers were used for subsequent classification and other processing.

11.3.3.10 SPECTRAL ANGLE MAPPER (SAM) CLASSIFICATION.

The Spectral Angle Mapper (SAM) is an automated method for comparing image spectra to individual reference spectra (J. W. Boardman, unpublished data; Kruse et al., 1993b). The algorithm determines the similarity between two spectra by calculating the angle in N-dimensions (the spectral angle) between them, treating them as vectors in a space with dimensionality equal to the number of bands. Because this method uses only the vector "direction" of the spectra and not their vector "length", the method is insensitive to illumination. The result of the SAM classification (not shown) is an image showing the best SAM match at each pixel. Additionally, rule images are calculated that show the actual angular distance (in radians) between each spectrum in the image and each reference or endmember spectrum. Darker pixels in the rule images represent smaller spectral angles and thus spectra that are more similar to the endmember spectra. For the purposes of display, the dark pixels are inverted, so that the best matches appear bright. These images present a good first cut of the mineralogy at the sites.

11.3.3.11 SPECTRAL UNMIXING.

While the SAM algorithm does provide a means of identifying and spatially mapping minerals, it only picks the best match to a spectrum. Natural surfaces are rarely composed of a single uniform material, thus it is necessary to use mixture modeling to determine what materials cause a particular spectral "signature" in imaging spectrometer data. Spectral mixing is a consequence of the mixing of materials having different spectral properties within a single image pixel. If the scale of the mixing is large (macroscopic), the mixing occurs in a linear fashion. A simple additive linear model can be used to estimate the abundances of the materials measured by the imaging spectrometer (Boardman, 1991). Each mixed spectrum is a linear combination of the "pure" spectra, each weighted by their fractional abundance within the pixel, a simple averaging. While some intimate non-linear mixing does occur for natural surfaces (eg: soil development resulting in mixing at the scale of mineral grains), the linear model is a good first order approximation in most cases.

To determine the abundances, we must first determine what materials are mixing together to give us the spectral signature measured by the instrument. Selection of endmembers is the most difficult part of linear spectral unmixing. The N-dimensional visualizer approach described above provides one method for using the data themselves to determine endmembers. The ideal spectral library used for unmixing consists of endmembers that, when linearly combined, can form all other observed spectra. This can be presented as a simple mathematical model in which the observed spectrum (a vector) is the result of a multiplication of the mixing library of pure end-

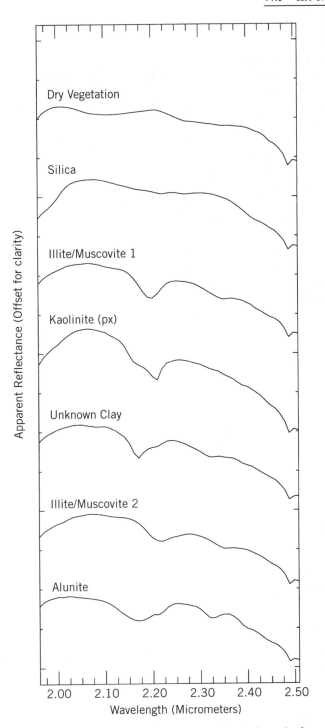

Figure 11.14   Goldfield, Nevada, AVIRIS Endmember spectra. Spectra are offset for clarity. Identifications are based on the USGS Spectral Library. (From Clark et al., 1993.)

member spectra (a matrix) by the endmember abundances (a vector). An inverse of the original spectral library matrix is formed by multiplying together the transposes of the orthogonal matrices and the reciprocal values of the diagonal matrix (Board-man, 1989). A simple vector-matrix multiplication between the inverse library matrix and an observed mixed spectrum gives an estimate of the abundance of the library endmembers for the unknown spectrum.

Linear Spectral Unmixing was used as the final step in producing mineral maps for the Goldfield AVIRIS data. The endmember library defined using the N-dimensional visualization procedure (Figure 11.14) was used in the unmixing process and abundance estimates were made for each mineral. These results can be presented in two ways. First, a set of gray-scale images stretched from 0 to 50% (black to white) provides a means of estimating relative mineral abundances. Selected grayscale abundance image results for a few of the minerals occurring at Goldfield are shown in Figure 11.15. Second, color composite images can be used to highlight specific minerals and mineral assemblages. Pure colors in these images represent areas where the mineralogy is relatively pure. Mixed colors indicate spectral mixing, with the resultant colors indicating how much mixing is taking place and the relative contri-butions of each endmember. For example, in a color composite of selected unmixing results for Goldfield (Figure 11.16; see the color insert), the minerals kaolinite, alu-nite, and muscovite when assigned to red, green, and blue in the color output result in distinctive image colors. These areas were extracted and overlain on a gray-scale image for improved location purposes. Areas that are pure red in this image corre-spond to areas where kaolinite is the spectrally dominant (most abundant) mineral. Areas that are green are dominated by alunite. Areas that are blue contain primarily muscovite. The yellow pixels are an example of mixed pixels, where the contribution of red from kaolinite and of green from alunite results in the mixed yellow color. The color image described above provides an example of how color information can be used to highlight selected minerals. To produce useful mineralogical maps for specific applications, however, these color images, the individual mineral abundance images, and reflectance spectra must be used together to determine the locations and distribution of minerals characterizing specific geologic processes important to those applications.

11.3.3.12 DISCUSSION.

This case study summarizes a standardized processing and analysis approach and results for AVIRIS data of the Goldfield, Nevada site flown during the 1995 AVIRIS geology group shoot. This demonstrates the utility of imaging spectrometers such as AVIRIS to produce detailed, high-quality mineral maps without supporting ground measurements. The group shoot effort demonstrated that a cooperative industry/ NASA effort can provide an efficient means for organizations to share some of the costs and apparent risks of using a new technology while getting data specific to their needs.

11.4 CONCLUSIONS

This chapter has shown the current state of technology with respect to selected air-borne and spaceborne sensors. There is a wide variety of data available including

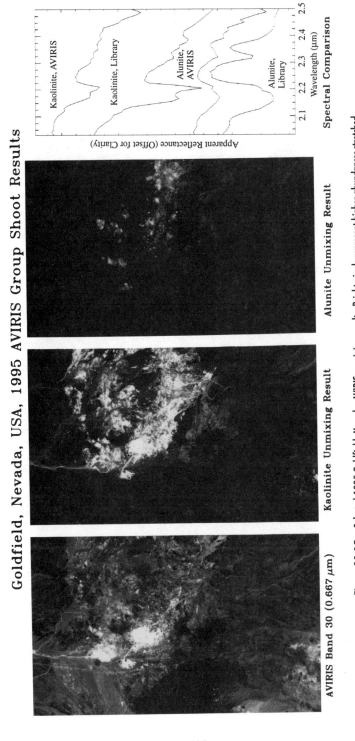

Figure 11.15 Selected 1995 Goldfield, Nevada, AVIRIS unmixing results. Bright pixels represent higher abundances stretched from 0 to 50% (black to white). Representative endmember spectra for kaolinite and alunite are compared to USGS spectral library spectra.

numerous sensors with varying spatial, spectral, and radiometric resolutions. Aircraft sensors described in this chapter act as both operational systems used in a variety of disciplines, and as prototypes for satellite systems. Numerous new systems are planned for the near future. Two trends appear to be developing: (1) sensors with improved spatial resolution (EarthWatch, OrbView, Space Imaging), and (2) sensors with improved spectral resolution (hyperspectral sensors/imaging spectrometers such as *Orbview-4 and ARIES-1*). These technologies are not mutually exclusive. The future probably holds hybrid sensors that use high spatial resolution bands to sharpen data from hyperspectral sensors, as well as hyperspectral sensors with better spatial resolution. These systems will provide near-laboratory-quality spectral information and improved spatial detail.

References

Abram, M., and S. J. Hook, 1995. Simulated ASTER data for geologic studies, *IEEE Trans. Geos. and Remote Sensing*, 33 (3), 692–699.

Abrams, M. J., R. P. Ashley, L. C. Rowan, A. F. H. Goetz, and A. B. Kahle, 1978. Mapping of hydrothermal alteration in the Cuprite Mining District, Nevada using aircraft scanner images for the spectral region 0.46–2.36 µm, *Geology*, 5., 173–178.

Ashley, R. P., 1974. *Goldfield Mining District*: Nevada Bureau of Mines and Geology, Reno, NV, NBMG Rep. 19, pp. 49–66.

Ashley, R. P., 1979. *Relation Between Volcanism and Ore Deposition at Goldfield, Nevada*, Nevada Bureau of Mines and Geology, Reno, NV, NBMG Rep. 33, pp. 77–86.

ASPRS, 1996. Data notebook, in *Proceedings: Land Information in the Next Decade, September, 25–28, 1995*, Vienna, VA, American Society of Photogrammetry and Remote Sensing, Bethesda, Md.

Asrar, G., and R. Greenstone, 1995. *MTPE EOS Reference Handbook*, EOS Project Science Office, NASA Goddard Space Flight Center, Greenbelt, Md., 277 pp.

Basedow, R. W., D. C. Armer., and M. E. Anderson, 1995. HYDICE system: implementation and performance, Proc., SPIE, 2480, 258–267.

Bianchi, R., R. M. Cavalli, L. Fiumi, C. M. Marino, and S. Pignatti, 1996. *CNR LARA Project: Evaluation of two years of airborne imaging spectrometry*, in Proceedings of the 2nd International Airborne Remote Sensing Conference and Exhibition, Environmental Research Institute of Michigan, Ann Arbor, I-534–I-543.

Boardman, J. W., 1989. Inversion of imaging spectrometry data using singular value decomposition: *in Proceedings, IGARSS '89, 12th Canadian Symposium on Remote Sensing*, Vol. 4, pp. 2069–2072.

Boardman, J. W., 1991. Sedimentary facies analysis using imaging spectrometry: A geophysical inverse problem, Unpublished Ph.D. thesis, University of Colorado, Boulder, Co, 212 pp.

Boardman, J. W., 1993. Automated spectral unmixing of AVIRIS data using convex geometry concepts, in *Summaries of the 4th JPL Airborne Geoscience Workshop*, JPL Pub. 93–26, Vol. 1, Jet Propulsion Laboratory, California Institute of Technology, Pasadena, Calif., pp. 11–14.

Boardman, J. W., and J. H. Huntington, 1996. Mineral mapping with 1995 AVIRIS data: in *Summaries of the 6th Annual JPL Airborne Earth Science Workshop*, JPL Pub. 96–4, Vol. 1. AVIRIS Workshop, Jet Propulsion Laboratory, California Institute of Technology, Pasadena, Calif., pp. 9–11.

Boardman, J. W., and F. A. Kruse, 1994. Automated spectral analysis: a geologic example using AVIRIS data, north Grapevine Mountains, Nevada: in *Proceedings of the 10th Thematic Conference on Geologic Remote Sensing*, Environmental Research Institute of Michigan, Ann Arbor, Mich, pp. I-407–I-418.

Boardman, J. W., F. A. Kruse, and R. O. Green, 1995. Mapping target signatures via partial unmixing of AVIRIS data, in *Summaries of the 5th JPL Airborne Earth Science Workshop*, JPL Publication 95–1, Vol. 1., Jet Propulsion Laboratory, California Institute of Technology, Pasadena, Calif., pp. 23–26.

Chrien, T. G., R. O. Green, and M. L. Eastwood, 1990. Accuracy of the spectral and radiometric laboratory of the Airborne Visible/Infrared Imaging Spectrometer, *in Proceedings The International Society for Optical Engineering (SPIE)*, Vol. 1298, pp. 37–49.

Clark, R. N., T. V. V. King, M. Klejwa, and G. A. Swayze, 1990. High spectral resolution spectroscopy of minerals, *J. Geophys. Res.*, 95 (B8), 12653–12680.

Clark, R. N., G. A. Swayze, A. Gallagher, T. V. V. King, and W. M. Calvin, 1993. The U.S. Geological Survey Digital Spectral Library, Version 1:0.2 to 3.0 μm, U.S. Geological Survey, USGS Open File Rep. 93–592, Washington, D.C., 1340 pp.

CSES, 1992. *Atmosphere REMoval Program (ATREM) User's Guide, Version 1.1*, Center for the Study of Earth from Space, Boulder, Colo., 24 pp.

Goetz, A. F. H., and B. Kindel, 1996. Understanding unmixed AVIRIS images in Cuprite, NV using coincident HYDICE data, in *Summaries of the 6th Annual JPL Airborne Earth Science Workshop*, JPL Pub. 96–4, Vol. 1. AVIRIS Workshop, Jet Propulsion Laboratory, California Institute of Technology, Pasadena, Calif., pp. 97–103.

Goetz, A. F. H., and L. C. Rowan, 1981. Geologic remote sensing, *Science*, 211, 781–791.

Goetz, A. F. H., B. N. Rock, and L. C. Rowan, 1983. Remote sensing for exploration: an overview, *Econ. Geol.*, 78 (4), 573–590.

Goetz, A. F. H., G. Vane, J. E. Solomon, and B. N. Rock, 1985. Imaging spectrometry for earth remote sensing, *Science*, 228, 1147–1153.

Green, A. A., M. Berman, B. Switzer, and M. D. Craig, 1988. A transformation for ordering multispectral data in terms of image quality with implications for noise removal, *IEEE Trans. on Geosci. Remote Sensing*, 26(1), 65–74.

Green, R. O., J. E. Conel, J. Margolis, C. Chovit, and J. Faust, 1996. In-flight calibration and validation of the Airborne Visible/Infrared Imaging Spectrometer (AVIRIS), in *Summaries of the 6th Annual JPL Airborne Earth Science Workshop*, JPL Pub. 96–4, Vol. 1. AVIRIS Workshop, Jet Propulsion Laboratory, California Institute of Technology, Pasadena, Calif., pp. 115–126.

Hauff, P. L., P. Kowalczyk, M. Ehling, G. Borstad, G. Edmundo, R. Kern, R. Neville, R. Marois, S. Perry, R. Bedell, C. Sabine, A. Crosta, T. Miura, G. Lipton, V. Sopuck, R. Chapman, C. Tilkov, K. O'Sullivan, M. Hornibrook, D. Coulter, S. Bennett, 1996. The CCRS SWIR Full Spectrum Imager: Mission to Nevada, June, 1995: in *Proceedings of the 11th Thematic Conference on Geologic Remote Sensing, Vol. 1, Environmental Research Institute of Michigan (ERIM)*, Ann Arbor, Mich., pp. I–38 to I–47.

Hook, S. J., C. D. Elvidge, M. Rast, and H. Watanabe, 1991. An evaluation of short-wave-infrared (SWIR) data from the AVIRIS and GEOSCAN instruments for mineralogic mapping at Cuprite, Nevada, *Geophysics*, 56(9), 1432–1440.

Kahle, A. B., and A. F. H. Goetz, 1983. Mineralogic information from a new airborne thermal infrared multispectral scanner: *Science* 222(4619), 24–27.

Kidwell, K. B., 1991, NOAA Polar Orbiter Data Users Guide: NOAA NESDIS National Climate Data Center.

Kramer, H. J., 1994. *Observation of the Earth and its Environment: Survey of Missions and Sensors*, 2nd Ed., Springer-Verlag, Berlin, 580 pp.

Kruse, F. A., 1984a. Munsell color analysis of Landsat color-ratio-composite images of limonitic areas in southwest New Mexico: in *Proceedings of the 3rd International Symposium on Remote Sensing of Environment, Thematic Conference on Remote Sensing for Exploration Geology, 3rd*, Environmental Research Institute of Michigan, Ann Arbor, Mich., pp. 761–773.

Kruse, F. A., 1984b. Evaluation of color-composite-images from Thematic Mapper Simulator data for hydrothermal alteration mapping, Lordsburg mining district, Hidalgo Co., New Mexico (abstract), *Geol. Soc. Am. Abstr. Prog.*, 16(4)..

Kruse, F. A., 1988. Use of Airborne Imaging Spectrometer data to map minerals associated with hydrothermally altered rocks in the northern Grapevine Mountains, Nevada and California: *Remote Sensing Environ.*, 24(1), 31–51.

Kruse, F. A., and J. B. Dietz, 1991. Integration of optical and microwave images for geologic mapping and resource exploration: in *Proceedings of the 8th International Symposium on Remote Sensing of Environment, Thematic Conference on Remote Sensing for Exploration Geology*, Denver, Colorado, Environmental Research Institute of Michigan, Ann Arbor, Mich., pp. 535–548.

Kruse, F. A., and J. H. Huntington, 1996. The 1995 geology AVIRIS group shoot: in *Summaries of the 6th Annual JPL Airborne Earth Science Workshop*, JPL Pub. 96-4, Vol. 1. AVIRIS Workshop, Jet Propulsion Laboratory, California Institute of Technology, Pasadena, Calif., pp. 155–166.

Kruse, F. A., and K. S. Kierein-Young, 1990. Mapping lithology and alteration in the northern Death Valley region, California and Nevada, using the Thermal Infrared Multispectral Scanner (TIMS): in *Proceedings of the 2nd Thermal Infrared Multispectral Scanner (TIMS) workshop, June 6 1990*, JPL Pub. 90-55, Jet Propulsion Laboratory, California Institute of Technology, Pasadena, Calif., pp. 75–81.

Kruse, F. A., K. S. Kierein-Young, and J. W. Boardman, 1990. Mineral mapping at Cuprite, Nevada with a 63 channel imaging spectrometer: *Photogramm. Eng. Remote Sensing*, 56(1), 83–92.

Kruse, F. A., A. B. Lefkoff, and J. B. Dietz, 1993a. Expert system-based mineral mapping in northern Death Valley, California/Nevada using the Airborne Visible/Infrared Imaging Spectrometer (AVIRIS), *Remote Sensing Environ.*, special issue on AVIRIS, May-June, 44,. 309–336.

Kruse, F. A., A. B. Lefkoff, J. B. Boardman, K. B. Heidebrecht, A. T. Shapiro, P. J. Barloon, and A. F. H. Goetz, 1993b. The Spectral Image Processing System (SIPS)-Interactive visualization and analysis of imaging spectrometer data: *Remote Sensing of Environ.*, special issue on AVIRIS, May-June, 44, 145–163.

Lang, H. R., S. L. Adams, J. E. Conel, B. A. McGuffie, E. D. Paylor, and R. E. Walker, 1987. Multispectral remote sensing as stratigraphic tool, Wind River Basin and Big Horn Basin areas, Wyoming: *Am. Assoc. Petr. Geol. Bull.*, 71(4), 389–402.

Legg, C. A., 1991, A review of Landsat MSS image acquisition over the United Kingdom, 1976–1988 and the implications for operational remote sensing: *International Journal of Remote Sensing*, v. 12, no. 1, p. 93–106.

Lillesand, T. M., and R. W. Kiefer, 1987. *Remote Sensing and Image Interpretation*, 2nd ed, Wiley, New York, 721 pp.

Lyon, R. J. P., and F. R. Honey, 1989. Spectral signature extraction from airborne imagery using the Geoscan MkII advanced airborne scanner in the Leonora, Western Australia gold district: in IGARSS'89, *Proceedings of the 12th Canadian Symposium on Remote Sensing*, Vol. 5, pp. 2925–2930.

Lyon, R. J. P., and F. R. Honey, 1990. Thermal infrared imagery from the Geoscan Mark II scanner of the Ludwig Skarn, Yerington, NV, in *Proceedings of the 2nd Thermal Infrared Multispectral Scanner (TIMS) Workshop, JPL Pub. 90–55*, Jet Propulsion Laboratory, California Institute of Technology, Pasadena, Calif., pp. 145–153.

NASA, 1986. *MODIS, Moderate Resolution Imaging Spectrometer, Instrum. Panel Rep.*, IIb, National Aeronautics and Space Administration, Washington, D.C.

Morain, S. A., and A. M. Budge, eds., 1996. *Manual of Remote Sensing* 3rd Ed., Earth observing platforms and sensors, American Society for Photogrammetry and Remote Sensing, Falls Church, Va., CD-ROM Version 1.0.

Palluconi, F. D., and G. R. Meeks, 1985. *Thermal Infrared Multispectral Scanner (TIMS): An investigator's guide to TIMS data*, Jet Propulsion Laboratory Publ. 85–32, Jet Propulsion Laboratory, California Institute of Technology, Pasadena, Calif.

Paylor, E. D., M. J. Abrams, J. E. Conel, A. B. Kahle, and H. R. Lang, 1985. *Performance Evaluation and Geologic Utility of Landsat-4 Thematic Mapper Data*, JPL Publ. 85–66, Jet Propulsion Laboratory, Pasadena, Calif., 68 pp.

Pease, C. B., 1990. Satellite Imaging Instruments: Principles, Technologies, and Operational Systems: Ellis Horwood, New York, 336 pp.

Porter, W. M., and H. E. Enmark, 1987. System overview of the Airborne Visible/Infrared Imaging Spectrometer (AVIRIS), in *Proc SPIE*,, 834, 22–31.

Prost, G. L., 1994. *Remote Sensing for Geologists*, Gordon and, Lausanne, Switzerland, 326 pp.

Resmini, R. G., M. E. Kappus, W. S. Aldrich, J. C. Harsanyi, and M. Anderson, 1996. Use of Hyperspectral Digital Imagery Collection Experiment (HYDICE) sensor data for quantitative mineral mapping at Cuprite, Nevada, in *Proceedings of the 11th Thematic Conference on Geologic Remote Sensing, Environmental Research Institute of Michigan (ERIM)*, Ann Arbor, Mi., pp. I–48 to I–65.

Rowan, L. C., P. H. Wetlaufer, A. F. H. Goetz, F. C. Billingsley, and J. H. Stewart, 1974. Discrimination of rock types and detection of hydrothermally altered areas in south-central Nevada by the use of computer-enhanced ERTS images, USGS Professional Paper 883,, U.S. Geological Survey, Washington, D.C., 35 pp.

Rowan, L. C., K. Watson, and S. H. Miller, 1992. Preliminary analysis of Thermal-Infrared Multispectral Scanner data of the Iron Hill, Colorado, carbonatite-alkalic rock complex, *in Summaries of the 3rd Annual JPL Airborne Geoscience Workshop, June 1–5, JPL Pub. 92–14*, Vol. 2, Jet Propulsion Laboratory, California Institute of Technology, Pasadena, Calif. pp. 28–30.

Rowlands, N. A., and R. A. Neville, 1994. A SWIR imaging spectrometer for remote sensing, in *Proceedings of the SPIE Infrared Technology XX Conference*, July 19–24, SPIE Proc. 2269.

Sabins, F. F., 1997. *Remote Sensing Principles and Interpretation*, Third Edition, W. H. Freeman and Company, New York, 494 pp.

Swayze, G. A., 1997. The hydrothermal and structural history of the Cuprite Mining District, Southwestern Nevada: An integrated geological and geophysical approach: Unpublished Ph. D. thesis, University of Colorado, Boulder, Co, 341 pp.

Vane G., and A. F. H. Goetz, 1993. Terrestrial imaging spectrometry: current status, future trends, *Remote Sensing Environ.*, 44, 117–126.

Vane G., T. G. Chrien, E. A. Miller, and J. H. Reimer, 1987. Spectral and radiometric calibration of the Airborne Visible/Infrared imaging Spectrometer (AVIRIS), in *Proc. SPIE*, 834,. 91–106.

Vane, G., R. O. Green, T. G. Chrien, H. T. Enmark, E. G. Hansen, and W. M. Porter, 1993. The Airborne Visible/Infrared Imaging Spectrometer (AVIRIS), *Remote Sensing Environ.*, 44, 127–143.

Watson, K., F. A. Kruse, and S. Hummer-Miller, 1990. Thermal infrared exploration in the Carlin Trend, northern Nevada: *Geophysics*, 55(1), 70–79.

Yamaguchi, Y., 1987, Possible techniques for lithologic discrimination using the short-wavelength-infrared bands for the Japanese ERS-1, *Remote Sensing Environ.*, 23, 117–129.

Radar: Sensors and Case Studies

Jeffrey J. Plaut

Jet Propulsion Laboratory
California Institute of Technology
Pasadena, California

Benoit Rivard

University of Alberta
Edmonton, Alberta, Canada

Marc A. D'Iorio

Canada Centre for Remote Sensing
Ottawa, Ontario, Canada

12.1 INTRODUCTION

Radar remote sensing has emerged as an important tool in a wide variety of geological applications. As an active sensor, radar does not require solar illumination, and the cloud-penetrating capabilities of radio waves at wavelengths greater than about 1 cm allow imaging in virtually any weather conditions. Compared with optical and infrared phenomena, radar backscatter is sensitive to a vastly different set of properties of the Earth's surface. These include centimeter- to meter-scale surface roughness, topographic slope, and the dielectric properties of the upper layers of the surface materials. Technological advances such as the synthetic aperture, multipolarization and multifrequency radars, and interferometry have given geologists rich and varied data sets to attack problems in many disciplines. In this chapter we review the current

Remote Sensing for the Earth Sciences: Manual of Remote Sensing, 3 ed., Vol. 3, edited by Andrew N. Rencz.
ISBN: 0471-29405-5 © 1999 John Wiley & Sons, Inc.

state of the art in airborne radar remote sensing, a history of spaceborne systems, and case studies that employ a variety of techniques of radar data analysis to geologic problems in a range of environmental conditions.

12.2 SENSORS

12.2.1 Aircraft Systems

Spaceborne synthetic aperture radars (SARs) are particularly useful for studies of extensive areas because they provide coverage with little variation in incidence angle across a wide swath. Aircraft SARs, however, provide higher-resolution data which is used for algorithm development, for testing concepts for future spaceborne systems, and for obtaining data with a variety of imaging geometries and temporal coverage patterns. Many research aircraft SARs are presently available, two of which are described in this section. The first is operated by the Canada Centre for Remote Sensing and is flown on a Convair 580. The second is the AIRSAR system, operated by the NASA/Jet Propulsion Laboratory, and mounted on a DC-8 aircraft.

Commercial systems have included STAR 1 and 2 developed jointly by Intermap Technologies and the Environmental Research Institute of Michigan (ERIM). This partnership is also presently operating IFSAR, an X-band (3-cm) interferometric system.

12.2.1.1 CCRS SAR.

The airborne synthetic aperture radar developed and operated by the Canada Centre for Remote Sensing was commissioned in 1986 to operate in C-band (5.8 cm; Livingstone et al., 1987) and later upgraded with an additional X-band capability in 1988 (Livingstone et al., 1988). The combined C- and X-band system allowed multipolarization (HH and HV or VV and VH) images to be acquired simultaneously at X- and C-bands over a wide range of geometries. The C/X-SAR has been used to support initial *RADARSAT* marketing through projects like GlobeSAR, with airborne imagery providing the basis for simulations of satellite SAR swath configurations, and as a facility for the optimization of future radar satellites (*RADARSAT II* and *III*).

The radar is normally operated in one of three geometries, or modes, known as nadir, narrow swath, and wide swath. The aircraft's operating altitude is variable but about 6.5 km (21,000 ft) is optimum. At this altitude, the high-resolution (6 m × 6 m) real-time imagery can be generated over incidence angles from 0 to 74° (as nadir mode), or 45 to 76° (as narrow swath mode). Lower-resolution (20 m × 10 m) imagery can be obtained over a wide swath of about 63 km, with incidence angles varying from 45 to 85°, to maximize the area of coverage.

Airborne interferometric SAR research began at CCRS in 1989 with a study of the relative merits of stereo SAR versus interferometric SAR for the derivation of both digital terrain models and geocoded SAR imagery. An additional antenna was mounted on the Convair 580 and an experimental "across-track" interferometer was flown for the first time in July 1991. This mode, however, is only one form of SAR interferometry, in which the phase difference between pairs of SAR images is used to derive geophysical information (see Section 12.4).

When pairs of airborne SAR images are combined coherently from separate passes over the same terrain, interference fringes can be observed if nearly the same imaging geometry is maintained between the two passes. The technique, known as *repeat-track interferometry*, was demonstrated for the CCRS airborne SAR in 1990, by flying the C/X-SAR over the same test site, with repeat tracks as close as possible. Using both C and X data from some of the closest combinations of passes, the height of a building was estimated, the movement of a radar reflector between some passes was measured to within an accuracy on the order of 1 mm, and a loss of coherence was observed over a forest canopy because of wind-induced motion in the trees.

In addition to repeat-track interferometry, across-and along-track modes can be employed, using additional receive antennas on one flight line. Across-track interferometry can be used to derive terrain elevation. It is achieved by measuring the phase difference in the data from a remote target, when detected simultaneously by two separate receive antennas displaced in the across-track plane. From a knowledge of the phase difference and the positions of the two receive antennas, which define the baseline, an accurate estimate of the three-dimensional position of image pixels can be obtained. Digital terrain elevation models (DEMs) can then be created.

A second C-band, horizontally polarized receive antenna was mounted on the right-hand side of the Convair 580 in the summer of 1991. Since then, deployments have tested the across-track method with encouraging results for height measurements. Height noise is in the range of 1 to 5 m root mean square with height bias errors dependent on the use of control points. The increasing accuracy of aircraft positioning with differential GPS will allow radar mapping with only a few control points. With improvements in the measurement of aircraft attitude, especially roll, it may be possible to operate without benefit of control points. Evaluation and demonstration of this facility is continuing.

In late 1993 a new double-antenna structure was installed on the CV 580, which allows research into along-track interferometry. By comparing the phase of images from the two receive antennas, which are displaced in the along-track direction, it is possible to measure the velocity of targets moving toward or away from the radar, during only one pass. Potential applications include determination of sea-ice drift, ocean currents, and some ocean wave parameters.

As of 1997 the system has the interferometric modes described above and a C-band polarimetric mode (Livingstone et al., 1995). Further information on the sensor and research on its applications can be accessed on the World Wide Web site of the Canada Centre for Remote Sensing at http://www.ccrs.nrcan.gc.ca.

12.2.1.2 AIRSAR.

The NASA/Jet Propulsion Laboratory's AIRSAR (Airborne Imaging Radar Synthetic Aperture Radar) system (Zebker et al., 1992; Lou et al., 1996) provides an operational testbed for developing advanced techniques of multiparameter SAR observations (Figure 12.1). The resolution of the aircraft system is 10 m (20 MHz bandwidth) or 5 m (40 MHz) in slant range and 1 m in azimuth. Swath widths are 10 to 20 km with an incidence angle range of about 20 to 60°. The system is equipped with full quadpolarization P-, L- and C-band SARs (68, 24, and 5.6 cm, respectively), allowing simultaneous acquisition of multifrequency HH, VV, HV, and VH amplitude and phase data for each resolution cell. The system has several interferometry modes implemented. For along-track interferometry, used primarily to measure surface ve-

Figure 12.1   NASA/JPL's AIRSAR system mounted on a DC-8 aircraft. The primary C-, L-and P-band antennas are mounted on the body of the aircraft, aft of the wing. Along-track interferometry C-and L-band antennas are on the forward body. The inset shows recent upgrade with both C-and L-band secondary antennas for cross-track interferometry for topographic mapping.

locites such as ocean currents, both C- and L-band receive-only antennas are mounted in the front portion of the aircraft. For cross-track interferometry, known as the TOPSAR (topographic SAR) mode, C- and L-band receive-only antennas are mounted on the body of the aircraft, parallel to the primary antennas. The TOPSAR interferometry mode has allowed production of digital elevation models (DEMs) with pixel spacing of 10 m and height accuracies in the range 1 to 5 m. Accurate geolocation of the image products requires precise knowledge of the aircraft position and the interferometric baseline orientation. This is achieved by a combination of systems, including a Honeywell inertial navigation system (INS), a Motorola Eagle four-channel Global Positioning System (GPS) receiver, and an experimental differential GPS system using Turbo Rogue receivers deployed in tandem on the aircraft and on the ground. The airborne SAR data are recorded digitally onboard the aircraft on three high-density digital recorders (HDDRs). All 12 channels (where a channel is single frequency and single polarization) may be recorded on a single recorder with a tape capacity of 15 min; the data rate is 10 Mbps. Data may be processed onboard in a quicklook frame mode or using the real-time processor, which possesses a full swath, single channel with two looks and resolution of 25 to 30 m.

12.2.2 Spaceborne Systems

In 1978, *Seasat* was launched and became the first orbital SAR. Although *Seasat* was designed largely for oceanic observations, the success of this system prompted a series of Shuttle Imaging Radar initiatives (SIR-A, SIR-B) during the 1980s aimed at developing SAR systems for land observations. In the early 1990s, the European Space Agency, Russia, and Japan launched single frequency orbital SARs, which include the remote sensing satellites *ERS-1* and *2*, the Russian *Almaz-1*, and the Japanese Earth Resources Satellite (*JERS-1*). The latest development in orbital SAR has been the launch of the Canadian *RADARSAT* system, also operating at a single frequency. The latest shuttle imaging radar (SIR-C/X-SAR) mission included a dual-frequency antenna in polarimetric mode, which may represent a technology soon to be found in orbital SARs. In this section we describe historically the radars and the mission goals for these SAR systems as well as future planned systems.

12.2.2.1 SEASAT.

Seasat was the first Earth-orbiting satellite designed for remote sensing of the Earth's oceans and had onboard the first spaceborne synthetic aperture radar. *Seasat* was managed by the Jet Propulsion Laboratory and was launched on June 28, 1978 into a nearly circular 800-km orbit with an inclination of 108° (Born et al., 1979; Jordan, 1980; Weissman, 1980). Five complementary experiments, designed to return the maximum information from ocean surfaces, were onboard: (1) a radar altimeter to measure spacecraft height above the ocean surface; (2) a microwave scatterometer to measure wind speed and direction; (3) a scanning multichannel microwave radiometer to measure sea surface temperature; (4) a visible and infrared radiometer to identify cloud, land, and water features; and (5), synthetic aperture radar with fixed look angle to monitor the global surface wave field and polar sea ice conditions.

The SAR sensor operated at L-band frequency (1.275 GHz) and HH polarization (horizontally transmitted and received) (Table 12.1). A look angle of 20° (23° incidence angle) was selected because of the rapid decrease in backscatter at higher incidence angles over the ocean. The swath width was 100 km and the resolution 25 m × 25 m (four looks). *Seasat* imagery clearly demonstrated the sensitivity of SAR to surface roughness and slope. Applications of the data include the study of ocean wave propagation, surface manifestation of internal waves and mesoscale eddies, polar ice motion, land–water boundaries, land-use patterns, mapping geological structures, and soil moisture variations (Born et al., 1979; Gonzalez et al., 1979, Ford et al., 1980; Bernstein, 1982; Kirwan et al., 1983). A massive short circuit in the satellite electrical system ended the *Seasat* SAR mission on October 10, 1978. The brief mission resulted in only partial global coverage. During *Seasat* operations, 14 Earth orbits were completed each day and approximately 42 hours of data were collected.

12.2.2.2 SIR-A, SIR-B.

The success of *Seasat* SAR prompted the first flight of a SAR sensor on the space shuttle. The shuttle was used as a platform to test progressively more complex imaging radars. The Shuttle Imaging Radar A (SIR-A) was launched aboard the space shuttle *Columbia* on November 12, 1981 (Table 12.1). The main goal of the SIR-A

TABLE 12.1 American Spaceborne SAR Missions

	Seasat SAR	SIR-A	SIR-B	SIR-C/X-SAR
General characteristics				
Launch date	June 28, 1978	Nov. 12, 1981	Oct. 5, 1985	Apr. 9, 1994; Sept. 30, 1994
Mission end	Oct. 10, 1978	Nov. 14, 1981	Oct. 13, 1984	Apr. 20, 1994; Oct. 11, 1994
Altitude (km)	800	259	224[a]	204–225
Inclination (deg)	108	38	57	57
Antenna dimensions (m)	10.7 × 2.16	9.4 × 2.16	10.7 × 2.16	L: 12.0 × 2.9 C: 12.0 × 0.7 X: 12.0 × 0.4
Instrument characteristics				
Power (kW)	1.0	1.0	1.2	1.2–7.0
Bandwidth (MHz)	19	6	12	10, 20, 40
Noise equivalent sigma-0 (dB)	−24	−32	−28	L: −40 C: −35 X: −22
Data rate (Mbps)	110		30	L, C: 90 X: 45
Imaging characteristics				
Band (cm, GHz)	L (23.5, 1.275)	L (23.5, 1.275)	L (23.5, 1.275)	X (3.1, 9.6) C (5.8, 5.3) L (23.5, 1.27)
Polarization	HH	HH	HH	quad (L, C), VV (X)
Look (incidence) angle (deg)	20 (23)	47 (50)	15–60 (15–64)	17–63
Range resolution (m)	25	40	58–16	10–26
Azimuth resolution (m)	25	40	25	30
Looks	4	6	4	Variable
Swath width (km)	100	50	10–60	15–90

[a] At 360 km for first 20 orbits, 257 km for 29 following orbits, 224 km for the remainder of the mission.

mission was to expand our understanding of radar signatures of geologic surfaces; a secondary goal was to assess the use of the shuttle as a scientific platform for Earth observations. SIR-A had a design similar to *Seasat*, but the antenna look angle was fixed at 47° to optimize geologic mapping in high-relief terrain and for increased sensitivity to surface roughness. Despite the larger look angle, SIR-A required no more power than *Seasat* because of the lower altitude of the shuttle and because of the higher returns expected for geological targets relative to the ocean's surface. The data were collected and processed optically; therefore, it was not possible to calibrate the SIR-A data.

The SIR-A mission yielded data for geologic studies in high relief superior to that of *Seasat*, which showed extensive layover (Cimino and Elachi, 1982; Settle and Taranik, 1982; Ford et al., 1982). SIR-A data led to the discovery of buried and previously unknown drainage channels in the Saharan desert (McCauley et al., 1982, 1986; Schaber et al., 1986) demonstrating the ability of L-band radar to penetrate several meters of loose sands in hyperarid environments.

The Shuttle Imaging Radar B (SIR-B) (Table 12.1) was the next step in the evolution of NASA's radar remote sensing research program. The radar imagery collected at the fixed look angle of *Seasat* (20°) and SIR-A (47°) provided an insight on the effect of incidence angle on radar backscatter. SIR-B was the first spaceborne SAR with a mechanically tiltable antenna, therefore providing the flexibility of imaging the surface at any look angle between 15 and 60°. The resolution of SIR-B data was improved over that of SIR-A by a factor of 2 (at the 50° incidence angle).

SIR-B was launched on October 5, 1984 aboard the space shuttle *Challenger* into a nominally circular orbit. The average altitude for the first 20 orbits was 360 km; for the next 29 orbits was 257 km; and for the remainder of the mission 224 km. With the addition of a digital data handling system, calibration was made possible and a calibrator was added to supply a known reference signal. Calibration was critical for experiments aimed at determining the variation of backscatter with incidence angle for various terrain types. Approximately 15 multiangle (two or more angles) data sets were collected. Forty-three principal scientists participated in the mission, conducting scientific investigations in geology, renewable resources, oceanography, and calibration techniques. The objectives of their experiments are summarized in (SIR-B Science Team, 1984) and key results of the mission are described in (Elachi, 1986; Elachi et al., 1986; Ford et al., 1986; Cimino et al., 1987; Ford, 1988; Holt, 1988).

12.2.2.3 *ALMAZ-1*.

In March 1991, the Soviet Union launched *Almaz-1*. The year marked the beginning of the permanent presence of nonmilitary SARs in space. *Almaz-1* is an S-band (10 cm) HH polarization SAR (Table 12.2) with a spatial resolution of 15 to 30 m dependent on the incidence angle. The varying look angle was achieved by adjusting the attitude of the satellite. The look-angle range of 20 to 70° is divided into a standard range of 32 to 50°, and two experimental ranges of 20 to 32° and 50 to 70° (the quality of data from the experimental ranges could not be guaranteed). The satellite provided worldwide coverage and could revisit a given location every 1 to 4 days, depending on latitude. Standard scenes vary in size but are approximately 40 km × 40 km and are available in a digital or photographic format. *Almaz-1* ceased operations in 1992.

TABLE 12.2 Non-U.S. Spaceborne SAR Missions

	Almaz-1	ERS-1	JERS-1	RADARSAT
General characteristics				
Launch date	Mar. 31, 1991	July 17, 1991	Feb. 11, 1992	Nov. 4, 1995
Altitude (km)	300	785	570	793–821
Inclination (deg)	73	98.5	98	98.6
Antenna dimensions (m)	1.5 × 1.5	10 × 1	11.9 × 2.4	15 × 1.5
Instrument charateristics				
Power (kW)		1.2	1.3	5
Bandwidth (MHz)		15.5	15	30, 17.3, 11.6
Noise equivalent sigma-0 (dB)		−18	−20.5	−23
Data rate (Mbps)		105	60	85 (recorded)–105 (R/T)
Imaging characteristics				
Band (cm, GHz)	S (10, 3.0)	C (5.7, 5.25)	L (23.5, 1.275)	C (5.7, 5.25)
Polarization	HH	VV	HH	HH
Look (incidence) angle (deg)	30–60	20 (23)	35 (38)	10–60
Range resolution (m)	15–25	20	18	10–100
Azimuth resolution (m)	15	30	18	10–100
Looks	4	4	3	1–8
Swath width (km)	20–45	100	75	45–510

12.2.2.4 *ERS-1, ERS-2.*

The second of the long-duration spaceborne SARs was launched on July 17, 1991 by the European Space Agency. *ERS-1* (Earth Remote Sensing Satellite), the first European satellite to carry a radar imager, was launched into an 800-km altitude and 98.5° inclination orbit. *ERS-1* carries multiple instruments including a scatterometer, which operates globally and continuously measures wind speed and direction and ocean wave parameters, and a radiometer that acquires surface temperature measurements from oceans. The C-band SAR has VV polarization and an incidence angle of 23° (Table 12.2). The swath width is 100 km with 30-m resolution at four looks.

The nature of the satellite's orbit and its complement of sensors enables a global mission providing worldwide geographical and repetitive coverage, primarily oriented toward ocean and ice monitoring, but the SAR imaging capability can provide useful data over land and coastal zones. *ERS-1* is aimed at demonstrating that spaceborne and ground-based technology can satisfy some operational requirements for data products needed within a few hours of the observations being made. Rapid data products can allow significant contributions to meteorology, sea state forecasting, and monitoring of sea ice distribution, all being important for shipping and offshore activities.

In April 1995, ESA launched *ERS-2*, a carbon copy of *ERS-1* with one important difference: *ERS-2*'s payload includes a new instrument designed to measure stratospheric and tropospheric ozone, an important step for environmental studies. During part of the mission *ERS-1* and *ERS-2* were flown in tandem to provide 24-hour repeat data for interferometric studies.

12.2.2.5 JERS-1.

The first Japanese Earth resources satellite (*JERS-1*) was launched on February 11, 1992 (Table 12.2). The imaging radar is flown in conjunction with an optical sensor (OPS). The SAR is distinct from that of *ERS-1* and *Almaz-1*, being an L-band (24-cm wavelength) radar with HH polarization and imaging at a fixed incidence angle of 35°. The angle was selected to optimize land studies, specifically geology and mineral exploration. The spacecraft has a tape recorder, so that data can be collected for any part of the world except poleward of 81.5° latitude. The orbit is solar-synchronous at an altitude of 568 km, and the recurrent period is of 44 days. Among the many applications of *JERS-1* data were studies of earthquake-related ground deformation (e.g., Murakami et al., 1995).

12.2.2.6 SIR-C/X-SAR.

The Shuttle Imaging Radar-C and X-Band Synthetic Aperture Radar (SIR-C/X-SAR) (Figure 12.2, Table 12.1) was flown twice aboard the space shuttle *Endeavour* in 1994 (Evans et al., 1993; Jordan et al., 1995; Stofan et al., 1995). The project is a cooperative experiment between the National Aeronautics and Space Administration (NASA), the German Space Agency (DARA), and the Italian Space Agency (ASI).

SIR-C/X-SAR has provided increased capability over single-wavelength systems by acquiring SAR data simultaneously at L-band (23.5 cm), C-band (5.8 cm), and X-band (3.1 cm). In addition, the SIR-C portion of the radar instrument (L- and C-band) was the first to acquire multipolarization SAR data from a spaceborne plat-

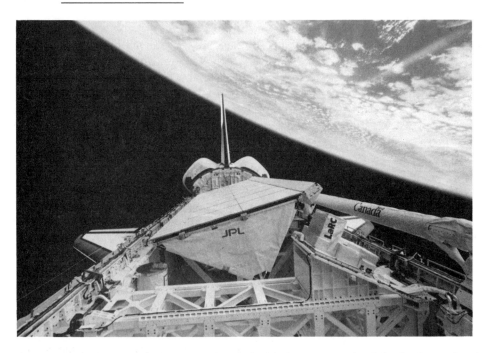

Figure 12.2   SIR-C/X-SAR (Spaceborne Imaging Radar-C / X-band Synthetic Aperture Radar) deployed in the cargo bay of the Space Shuttle *Endeavour* in 1994. The SIR-C antenna platform, above the letters "JPL" contains both C- and L-band panels. The X-band antenna is mounted on a mechanically steerable frame to the right of the SIR-C antennas.

form. Because radar backscatter is most strongly influenced by roughness elements comparable in size to the radar wavelength, the multifrequency capability provides information about the Earth's surface over a wide range of scales not discernible with other systems

The SIR-C beams were electronically steered in the range direction $\pm 23°$ from the nominal 40° off-nadir position without physically moving the large radar antenna. This feature enabled the acquisition of data over a wide range of incidence angles. A mechanical tilt mechanism was used to point the X-SAR antenna to angles between 17 and 63°, in alignment with the L-band and C-band beams. The shuttle attitude was continuously adjusted in the yaw direction to compensate for the Earth's rotation and maintain a zero Doppler frequency at the beam center.

The missions were designed to enable areas to be imaged at multiple incidence angles, important parameters for studying land and ocean processes. Field campaigns provided critical surface measurements to be used in development of algorithms needed to produce geophysical products for assessing global change issues. Several central themes have emerged from the SIR-C/X-SAR program. Research topics include the global carbon cycle, the hydrologic cycle, paleoclimate and geologic processes, ocean circulation and air–sea interactions, and advanced technology. The timing of the two missions (April and October 1994) allowed study of substantial seasonal changes over many of the targets, including characteristics such as leaf density, crop state, soil moisture, river and lake stages, and snow and ice cover.

Each SIR-C/X-SAR mission lasted slightly less than 11 days. Total data collection was approximately 140 hours, or 100 million km^2 of ground coverage. During the October 1994 mission, the shuttle was placed in a 24-hour near-repeat orbit, which allowed acquisition of multiple-pass interferometry data for studies of topography and surface change.

Data processing was conducted at facilities in the United States, Germany, and Italy. Low-resolution 50-m ("survey") processing was performed on the entire data set, and the resulting images were published on CD-ROM and made available on the Internet. Full-resolution (12.5-m) polarimetric processing of the data set is performed by the EROS Data Center, Sioux Falls, South Dakota.

12.2.2.7 *RADARSAT.*

RADARSAT was developed by Canada and launched on November 4, 1995 on a Delta II rocket (Figure 12.3). It was placed in a sun-synchronous orbit at an altitude of 798 km, at an inclination of 98.6°. The sun-synchronous orbit means that the satellite overpasses are always at the same local mean time, which is important to many users. Because of a dawn–dusk orbit *RADARSAT's* solar arrays are in sunlight almost continuously, enabling the satellite to rely primarily on solar rather than battery power. The general characteristics of *RADARSAT* are summarized in Table 12.2. *RADARSAT* offers a variety of beam selections (Luscombe et al., 1993) (Table 12.3) and the unique ability to shape and steer its beam from an incidence angle of less than 20° to more than 50°, to vary its swath from 35 to 500 km using resolutions ranging from 10 to 100 m.

RADARSAT was intended as the first operational radar satellite system capable

Figure 12.3 Canadian radar satellite *Radarsat,* shown in its operational configuration.

TABLE 12.3 *RADARSAT* Modes

Mode	Resolution Range[a] × Azimuth (m)	Looks	Swath Width (km)	Incidence Angle
ScanSAR wide	100 × 100	4–8	510	20–49
ScanSAR narrow	50 × 50	2–4	305	20–40
Wide 1	48–30 × 28	4	165	20–31
Wide 2	32–25 × 28	4	150	31–39
Standard	25 × 28	4	100	20–49
Extended low	63–28 × 28	4	170	10–23
Extended high	22–19 × 28	4	75	50–60
Fine resolution	11–9 × 9	1	45	37–48

[a]Nominal; range resolution varies with range.

of timely delivery of large amounts of data. Over a planned lifespan of five years, the satellite is expected to provide useful information to researchers and operational users working in fields such as agriculture, cartography, hydrology, geology, forestry, oceanography, ice studies, and coastal monitoring (Raney et al., 1991). *RADARSAT* was aimed specifically to provide the first routine surveillance of the entire Arctic region and to cover most of Canada every 72 hours, depending on the swath selected. The information is of use to shipping companies in North America, Europe, and Asia, and to government agencies with ice reconnaissance and mapping mandates. The entire Earth can be covered every 24 days using the standard 100-km beam mode. Data can be downlinked in real time, or stored on one of the two tape recorders until the spacecraft is within range of a receiving station. Processed data are made available to online users within a few hours after the satellite passes over an area.

12.2.3 Future Sensors

Each of the missions launched in the 1990s is unique in its choice of frequency, polarization, resolution, swath width, or orbital parameters. Together they offer a comprehensive data set across the radar portion of the spectrum. Planned radars are pursuing flexibility in swath width, resolution, polarization, frequency, interferometry, and viewing geometries. Future sensors include ASAR on the European satellite for environmental observations Envisat-1, the Japanese VSAR, Russian Almaz series, and the shuttle radar Topography mission (SRTM) (Table 12.4).

12.2.3.1 ASAR:ENVISAT'S ADVANCED SYNTHETIC APERTURE RADAR.

The *ENVISAT-1* mission, which has a planned launch in 1998 (the same year as NASA's *EOS AM-1* platform), shows the commitment of ESA to environmental observations. Compared to *ERS-1* and *2*, *ENVISAT-1* will incorporate a more advanced imaging radar (Karnevi et al., 1993) and better instruments for monitoring the atmosphere and ocean characteristics. The C-band (5.6-cm wavelength) radar represents an evolutionary step from *ERS-2* and will have a variety of swath widths,

TABLE 12.4   Planned SAR Sensors

	ASAR	VSAR	ALMAZ	SRTM
General characteristics				
Launch date	1998	2002	1998	2000
Altitude (km)	800	700	400	233
Inclination (deg)	98	98.1	73	57
Imaging characteristics				
Band (cm, GHz)	C (5.7, 5.25)	L (23.5, 1.275)	3.49 cm	X: (3.1, 9.6)
			9.6 cm	C: (5.8, 5.3)
			70 cm	
Polarization	HH, VV		HH, VV, HV	HH, VV
Look (incidence) angle (deg)	15–45	18–48	25–60	30–55
Swath width (km)	5–40	70–250	20–170	225

including a wide swath mode that has a 100-m pixel size and 405-km swath. The radar will have dual polarization (VV like *ERS-1* and *ERS-2*, and a new HH polarization capability). Also new for a European radar, there will be a variable incidence angle, from 15° to 45°. *ENVISAT*'s orbit has been selected at a 98° inclination at 800-km altitude, which provides 35-day exact repeat coverage (i.e., identical to *ERS-1* and *ERS-2*). The time of equatorial crossing for *ENVISAT* is 10:00 A.M. on the descending pass, which is comparable to the 10:30 A.M. equatorial crossing for the *EOS AM* platforms.

12.2.3.2   VSAR.

The *ALOS* mission, mounted by NASDA of Japan, will include the VSAR sensor operating at L band (23 cm). The mission is planned for launch in 2002, providing continuity to Japan's program of SAR observation, which began with *JERS-1* SAR. VSAR will deliver imagery with highest spatial resolution of 10 m in range and 5 m in azimuth at one look. Because it will operate in two modes, the swath width will either be 70 km (high-resolution mode) or 250 km (low-resolution mode).

12.2.3.3   ALMAZ.

The *Almaz 1B* mission is a Russian initiative planned for launch by 2000. Three SAR instruments are part of the mission, each of which will operate at a different wavelength and offer imagery with different resolutions and swath widths. The sensor operating at 3.5 cm will offer imagery 20 to 35 km in width at a resolution of 5 to 7 m. At the wavelength of 9.6 cm, three modes will be available: detailed (resolution 5 to 7 m, swath 30-55 km), intermediate (resolution 15 m, swath 60 to 70 km), and survey (resolution 15 to 40 m, swath 120 to 170 km). At the longest wavelength of 70 cm, the resolution of the imagery will range between 20 and 40 m for a swath width range of 120 to 170 km.

12.2.3.4   SRTM.

The success of the shuttle-based interferometry experiments during the 1994 flights of SIR-C/X-SAR laid the groundwork for an ambitious global topographic mapping project know as the Shuttle Radar Topography Mission (SRTM). Scheduled for

1999, this mission will combine the existing SIR-C/X-SAR system with a second set of receive-only radar antennas (C- and X-band) to create a fixed-baseline single-pass interferometer. The additional antennas will be mounted on a 60-m rigid "boom" that was derived from space station structural elements (Figure 12.4). The C-band SAR will be operated in a wide swath scanSAR mode, in which four adjacent sub-swaths are illuminated during a single pass of the shuttle. The resulting 225-km-wide swath will allow contiguous mapping of all of the Earth's land surface between 60° north and south latitudes (80% of the total land surface) during a single 11-day space shuttle mission. Data will be used to produce digital elevation models (DEMs) with a horizontal spacing of 30 m and absolute and relative vertical accuracies of 16 and 10 m, respectively. The X-band system will be in a narrow swath, noncontiguous mode, with higher-resolution data nested within the C-band swaths. During the processing phase, swaths of elevation data will be mosaicked and/or averaged to produce digital topographic maps in 2° × 2° latitude–longitude quadrangles. By combining data from ascending and descending orbital paths, many of the common problems encountered in high-relief areas (layover, shadowing, phase discontinuities) can be overcome. C-band image data, coregistered with the DEM at 30 m/pixel, will also be produced.

Figure 12.4 Configuration of the Shuttle Radar Topography Mission (SRTM), showing the SIR-C/X-SAR radar in the space shuttle cargo bay, and secondary, receive-only antennas mounted at the end of a 60-m boom structure.

12.3 CASE STUDIES

The following case studies provide examples of the use of SAR data for geological interpretation in different environments and for different problems. The first example is from Sarawak in East Malaysia. Most geological mapping efforts in such tropical areas have attempted to use optical data such as *Landsat*. Heavy cloud cover at the time of satellite overpass has made acquisition of cloud-free optical imagery difficult and use of radar data ideal in this environment. Most regions in Southeast Asia and some parts of South America have similar environmental conditions and have a requirement to produce or update large-scale geological maps. The case study describes the results of the geologic interpretation of *RADARSAT* and airborne data.

Arctic arid areas present particular challenges for field logistics, site access, and difficult climatic conditions. Yet these areas and other arid areas are among the most active current mineral exploration sites. The second case study, of Bathurst Island in the Canadian Arctic, examines a variety of sedimentary rock types and provides a rich test site for determining the effect of radar imaging parameters on backscatter.

The third case study explores use of SAR as a tool to monitor geological hazards in Hawaii. SAR imaging and interferometry have have proven to be extremely useful in the study of volcanic terrains. The sensitivity of SAR backscatter to surface roughness variations near the scale of the radar wavelength has allowed discrimination and mapping of a variety of volcanic deposits, in a number of different terrestrial environments. The active volcanic Kilauea, on the Big Island of Hawaii, has been the subject of many investigations of the application of radar remote sensing to volcanic terrains, including repeat-pass interferometry for surface deformation mapping.

In the fourth section we explore the use of SAR interferometry for geologic studies. Two powerful applications are described, generation of digital topography models (DEMs) and mapping of surface deformation caused by earthquakes. Examples of topographic mapping are provided from airborne and spaceborne systems, and the use of *ERS-1* data for analysis of the Landers, California earthquake is reviewed.

12.3.1 Case Study: Tropical Environment, Sarawak, Malaysia

The Sarawak Province of East Malaysia covers much of the northwest part of the island of Borneo. Sarawak is, in large part, underlain by an arcuate belt of deformed flysch deposits and subduction melange that accumulated in an outer-arc basin. Hamilton (1978) describes this deformation as a complex history of contemporaneous deposition and deformation where basinal sediments and shelf deposits have been strongly sheared, imbricated, and deformed into broad folds during Eocene subduction of the South China Sea floor. The case study site is located in the Tubau Bukit Lumut area, which was previously mapped by Kirk (1957) and later revised by Liechti et al. (1960). The latest geological map of Sarawak published in (Yin 1992) incorporates the information of the earlier work with some further revisions particularly on the formational names.

RADARSAT Standard mode beam 6 data were acquired (Figure 12.5) in 1996 as part of an international SAR application development program (Campbell et al., 1994). The SAR image covers an area of approximately 50 km × 50 km in the

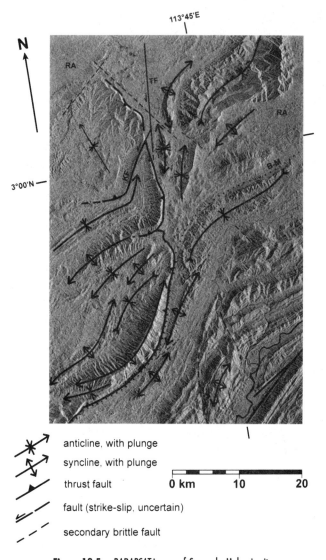

Figure 12.5   *RADARSAT* image of Sarawak, Malaysia site.

Bintulu area, Sarawak, Malaysia. Previously acquired airborne SAR data of the same area were interpreted geologically in terms of lithology and structure (Ling, 1994; D'Iorio et al., 1995). Lithological groups observed in the imagery were riverine alluvium, molasse deposits, and flysch deposits. They were differentiated on the basis of their morphological and textural appearances. Three major faults were identified: the Bukit Mersing Line, the Tubau Fault, and the newly named Kalo Fault.

The riverine alluvium is found in the northern portion of Figure 12.5 and has a relatively smooth SAR image texture. The SAR signature is that of the forest canopy and little information can be derived concerning the sediment types or variability. The riverine alluvium can only be separated from other major deposit types.

The molasse deposits are generally arenaceous and commonly contain thick beds of sandstone dipping steeply south or north and may be moderately folded. The sandstone beds stand out clearly in the SAR imagery as areas of resistant, continuous, parallel ridges. Boundaries between resistant and recessive units are frequently visible due to small erosional escarpments at their contacts, which correspond to changes or gaps in the forest canopy. High precipitation rates and warm temperatures facilitate chemical weathering of bedrock and the pattern of erosion seen from SAR data can be diagnostic of rock type. The traces of bedding elucidate complex fold structures, in particular, a large type 3 refolded fold that occurs in the central part of the study site. The SAR image texture of areas underlain by these deposits is coarser than areas with the riverine alluvium whose deposits are flat-lying and finer.

The flysch deposits are more intensely folded and are composed of more argillites than the other formations. In the SAR image, broad folds trending east to northeast and exhibiting reversals of plunge are recognized by discontinuous and subdued ridges which are somewhat rounded in appearance. The areas represented by flysch also appear to have a coarser SAR image texture than the areas occupied by molasse deposits.

The traces of three major faults were identified or inferred from the SAR image of this area. They are the Bukit Mersing Line, the Tubau Fault (Bannert et al., 1993), and a third, newly inferred fault which is situated to the west of Sungai Tubau (referred to as the Kalo Fault).

The Bukit Mersing Line separates the flysch deposits in the south from the molasse deposits in the north. It trends more or less in the NE–SW direction in the area. The trace of this fault is offset in a left-lateral sense by a minor NW trending brittle fault, as can be observed in the southern part of SAR image. The Tubau Fault, which trends almost N–S, follows the Sungai River. This fault extends farther south, outside the area of the SAR image. The Kalo Fault trends roughly in the direction of NE-SW and is interpreted to be the result of accommodation of the sinistral displacement along the Tubau Fault. Sinistral movement of the Kalo Fault is apparent in the SAR imagery by the dragging of beds on the eastern side of the fault. The sinistral movement of the Tubau Fault is interpreted to be syn-kinematic with thrusting at the eastern end of the older Bukit Mersing Line. Numerous rectilinear features of uncertain age, presumed to be brittle structures, trend mainly north to northeast and are also apparent on the SAR imagery.

12.3.2 Case Study: Arctic Arid Environment, Bathurst Island, Canadian Arctic

Bathurst Island, located in the Arctic, forms the eastern part of the Parry Islands Fold Belt and is bordered to the east by fault structures related to the Cornwallis Fold Belt. The structure and stratigraphy of the Phanerozoic rocks on Bathurst Island has been described by Kerr (1974) and published as a 1:250,000-scale geological map. Much of the island is characterized by gently west-to-southwest-plunging open synclinal folds spaced about 10 to 15 km apart and separated by thrust- and normal-faulted anticlines. The sedimentary succession consists of a series of shallowing-upward sequences from deepwater shales and siltstones to platformal carbonates and nonmarine sandstones (de Frietas et al., 1993). Thus a wide variety of sedimentary rock types occur on Bathurst Island, making the area suitable for examining the effects of lithological control on radar backscatter. Although bedrock is not always exposed, the style of cryogenic fragmentation and displacement of rock appears to be characteristic of the underlying lithology.

Studies using *Landsat* TM data (bands 4, 5, 7) and satellite *ERS-1* radar data (Budkewitsch et al., 1996a, b) reveal a strong spatial correlation between radar units and the lithological units and structures outlined on the geological map of Kerr (1974). Airborne radar data were collected by the C-band sensor onboard the Convair-580 aircraft (Livingstone et al., 1987, 1988) in April 1995.

The site presented in this case study is located near Erskine Inlet. It was chosen to illustrate the information extraction process from radar imagery. Bright radar units shown in the vicinity of Erskine Inlet (Figure 12.6) correspond to resistant ridges that are composed of quartz arenites. In the field, the blocks are large (7 to 31 cm) subequant, angular cobbles or 2- to 5-cm-thick slabs of quartzite. The ridges are consistently radar bright but include one distinct dark band (and a few minor ones) that runs parallel to the ridge for several kilometers. These bright and dark bands probably correspond to mappable members of the geological formation.

The radar unit overlying the bright, folded ridges has a distinct coarse (mottled) texture that corresponds to an upper member of another mapped formation. This texture is best observed where the unit is almost flat-lying, such as near hinge areas of synclines. The upper unit is a poorly cemented fine-grained quartzite that disintegrates quite readily, leaving sandy flats and some well-rounded (1 to 9 cm) friable pebbles. These areas have relatively low backscatter and appear dark on radar images. Resistant parts of the upper member remain exposed as small hills or hoodoos (Kerr, 1974), which accounts for the bright areas that yield the mottled texture for this unit. The mottled radar unit terminates against a second bright radar unit that appears as a resistant ridge on images. This unit is much thinner than the underlying one and marks the onset of a well-cemented quartz arenite, corresponding to a member of another mappable unit.

South of Erskine Inlet, a thrust fault (not present on Kerr's map) is inferred from the relationships of the radar units. The lateral continuity of the bright radar unit (lower Hecla member) is abruptly truncated against itself in the western part of the image. In the center and east, the truncated margin of the bright unit can be seen to cut across the mottled radar unit, thinning the unit in a manner consistent with thrust faulting along a north-dipping fault plane.

12.3.3 Case Study: Volcanic Terrain, Kilauea, Hawaii

Kilauea is a broad basaltic shield volcano built up along the southeast flank of the much larger Mauna Loa volcano, which rises nearly 10 km from its base on the seafloor. Eruptive activity at Kilauea is dominated by effusive eruptions of fluid basaltic lava, from rift zone or summit caldera vents. The volcano is nearly continuously active, and its surface contains few flow units older than several hundred years (Holcomb, 1987). Occasional explosive eruptions (1790, 1924; Decker and Christiansen, 1984) have deposited pyroclastic materials, primarily ash with scattered blocks, that remain at the surface today and affect remote sensing signatures in some areas. Many other morphologic features are prominent in image data, including vents, fissures, cones, pit craters, and fault scarps.

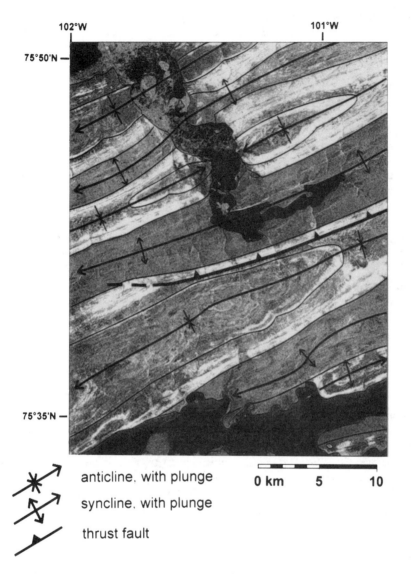

Figure 12.6 C-band CCRS airborne SAR image of a series of anticlines and synclines near Erskine Inlet, Bathurst Island, Canada. Outlined radar unit (solid lines) corresponds to quartz arenites of the Hecla Bay Formation. Area shown is about 35 km × 35 km.

The first significant study of Kilauea using radar imaging data was reported by Gaddis et al., (1989, 1990; see also Kaupp et al., 1986). Using two L-band images acquired at different incidence angles by the Shuttle Imaging Radar-B (SIR-B) in 1984, the authors were able to discriminate several textural classes of lava flows, map large-scale topographic features with radar-facing slopes, and identify pyroclastic deposits. SIR-B data were adequate for discrimination of aa lava flows, but pahoehoe flows were not easily distinguished from other smooth deposits, such as tephra. The smaller-incidence-angle scene (28°) was preferred over the larger-incidence-angle scene (48°) for sensitivity, unit discrimination, and signal/noise ratio. The study concluded that the SIR-B data allowed correct interpretation of many aspects of the eruptive history of Kilauea, including the overall shield morphology, the origin of the summit caldera by multiple collapses, identification of the major rifts and fracture systems, and the existence of a well-developed down-rift magma plumbing system.

The multiwavelength, polarimetric NASA/JPL AIRSAR system has flown several campaigns at Kilauea (Figure 12.7; see the color insert). Campbell et al. (1993) used AIRSAR data of Kilauea to explore the polarization properties of the lava flows and to investigate the scattering mechanisms responsible for the observed behavior. They found that the flow surfaces were generally too rough to be modeled by the "small perturbation" mechanism. The high correlation of HH and VV backscatter values at C- and L-band indicates that the scattering is dominated by single scattering events, either from "diffuse" randomly oriented dipoles or from "small facets" that are a significant fraction of the wavelength in size. By assuming that all HV power was due to diffuse scattering, the authors constructed maps of the fraction of backscatter due to facet scattering. Rough aa flows showed low values of facet fraction, due to the high diffuse component. At P-band, VV echoes stronger than HH were observed on many surfaces. Because lava flows do not have unusually high dielectric constants, the VV enhancement may be due to penetration and volume scattering of the P-band signal. In a related study, Campbell and Garvin (1993) analyzed the effects of sampling frequency on the calculation of roughness parameters in microtopographic profiling of lava flows. They note that the commonly used roughness parameter, RMS slope, will vary depending on the length scale at which the measurement is made. This effect was more pronounced on rougher (aa) flow surfaces than on smoother (pahoehoe) ones.

Kilauea was again a primary target for the space shuttle imaging radar system when SIR-C/X-SAR (Spaceborne Imaging Radar-C/X-band Synthetic Aperture Radar) flew twice in 1994. Data were acquired at a variety of incidence angles, in several different radar system modes (Mouginis-Mark, 1995a). One of the goals of the SIR-C/X-SAR studies at Kilauea was to extend the knowledge of the relationships between backscatter behavior and known field and eruptive characteristics of lava flows at this well-known site to remote volcanoes in other areas of the world. The L-band data were found to be most useful for discriminating lava flow roughness types as well as for distinguishing vegetated and rough lava surfaces.

A portion of the second flight of SIR-C/X-SAR (known as *SRL-2*, Space Radar Laboratory 2) was dedicated to obtaining repeat-pass interferometry data. One pair of interferometric data takes of Kilauea was acquired with a 6-month time separation, and a four-pass set of 24-hour interval data takes was acquired at the end of the *SRL-2* mission. As reported by Rosen et al. (1996) the 6-month interferometric pair contain a topographic change signal related to volcano-tectonic deformation of the Kilauea edifice. However, comparison with global positioning system benchmark deformation data indicates that an additional signal of similar magnitude but probably unrelated to the volcano contaminates the interferogram, making interpretation of the interferometric deformation signal difficult. A robust deformation signal was identified at the Pu'u O'o vent area, consistent with vertical deflation of about 14 cm. The competing phase delays are ascribed by Rosen et al. (1996) as being due to atmospheric refractivity variations, probably related to tropospheric water vapor.

This effect can severely limit the use of repeat-pass interferometry for not only surface displacement measurement, but for simple digital elevation mapping as well. The four-pass 24-hour interval data were fortuitously acquired during a period of active lava "breakouts" on Kilauea's lower flanks (Mouginis-Mark, 1995b, Zebker et al., 1996). Although the absolute interferometric phase also suffers from atmospheric delays in these data (Zebker et al., 1997), Zebker et al. (1996) were able to use the correlation between the complex-valued resolution elements of the 24-hour image pairs to identify lava flow surfaces that underwent significant changes (Figure 12.8; see the color insert). These included newly erupted flow surfaces and freshly emplaced flows that experienced inflation or deflation. This technique shows some promise for determining lava flow eruption rates and for monitoring ongoing eruptions that may not be easily accesible to other sensors or field investigation.

12.4 INTERFEROMETRY

12.4.1 Introduction

A powerful, emerging application of radar observations of the earth's surface is the technique known as SAR interferometry. SAR interferometry exploits the coherent nature of SAR echoes (i.e., the signals are recorded in both amplitude and phase) to measure differences in the phase from each patch of the surface when observed from slightly different locations and/or times. These phase differences can be ascribed to differences in path length between the two signals. The path-length differences are a function of (1) the separation of the antennas making the observations, (2) the topography of the surface, (3) change in the position of patches of the surface, and (4) differences in atmospheric or ionospheric conditions along the two paths. If the two observations are made simultaneously from a pair of antennas on a single platform, item 1 above is well known and the effects of items 3 and 4 are negligible, leaving item 2, the surface topography, as the controlling factor. Thus a fixed-baseline dual-antenna interferometric system is the preferred method for topographic map generation. Such a configuration has been implemented on several airborne radar systems (see Section 12.2) and is planned for the next flight of the SIR-C/X-SAR space shuttle radar, SRTM.

Substantial interest has been generated in the geosciences community over the potential applications of SAR interferometry for surface change detection (Dixon, 1995). In this technique, two or more observations are made from approximately the same location in space at different times, the phase differences due to topography and antenna separation are removed, and the residual phase differences are interpreted in terms of change in position of patches of the surface. The conditions for successful application of this technique are somewhat restrictive. If only two observations are used, a digital elevation model of the imaged region with a spatial resolution on the order of the SAR data is required to remove topographic effects. In addition, the surface conditions must be sufficiently preserved in any pair of images to allow correlation of the phase measurements. If the radar scattering properties of the surface have changed significantly, due to changes in vegetation for example, the measured phase differences cannot be analyzed for surface position changes. Finally, differences in atmospheric and ionospheric conditions can alter the path length and therefore the relative phase of the two observations (Goldstein, 1995; Massonnet et al., 1994). Despite these restrictions, however, a number of workers have successfully extracted maps of surface deformation from interferometric analysis of multiple SAR images. The most promising applications to date are in studies of coseismic and postseismic surface deformation, deformation of active volcanoes, response of the surface to changes in groundwater, and detection of motions of glaciers and ice sheets.

In this chapter the application of SAR interferometry for topographic mapping is treated in examples for both air- and spaceborne sensors. A review is then provided of a series of studies of surface deformation associated with the 1992 Landers, California earthquake, based on spaceborne SAR data from the *ERS-1* satellite.

12.4.2 Topographic Mapping with SAR Interferometry

Digital topographic maps, also know as digital elevation models (DEMs), can be obtained from aircraft or spacecraft SAR interferometry systems. Aircraft systems typically utilize dual antennas rather than repeated passes to obtain interferometric image pairs. Aircraft systems will generally have higher-resolution-output DEMs (5 to 10 m per pixel) than those obtained from spacecraft, with swath widths in the range 10 to 20 km. Generating DEMs for large areas with aircraft data therefore requires mosaicking of adjacent swaths of data. An example of a large-area DEM mosaic obtained by the NASA/JPL TOPSAR system is shown in Figure 12.9 (see the color insert). The image of western Los Angeles County was constructed from eight separate swaths, each approximately 10 km × 80 km. Pixel size in this DEM is 10 m at 30 to 60 looks per pixel. The 5-m-height accuracy of the DEM was achieved using a combination of onboard GPS measurements, ground control points, and an image-matching algorithm along overlapping swath segments.

During the second flight of the SIR-C/X-SAR system (SRL-2) on the space shuttle *Endeavour* in 1994, experimental repeat-pass interferometry data were acquired for several dozen sites around the world (Stofan et al., 1995). Shuttle controllers were able to repeat the orbital tracks at 24-hour intervals with baselines ranging from about 1 km to less than 100 m. Successful phase correlation and fringe formation was obtained at all three radar frequencies of the SIR-C/X-SAR system (L-, C-, and X-bands). The SIR-C system was operated primarily in a 20-MHz single-polarization mode, yielding swath widths of 20 to 40 km. Some quad-polarization data were also acquired. In an effort to demonstrate the capability to acquire and process interferometry swaths at continental scales, a number of data passes extended as much as several thousand kilometers in length. One such pass, which extended from the Oregon coast southeast through eastern California and into Mexico, was processed at JPL into a single continuous DEM, 40 km × 1600 km with a horizontal pixel spacing of 25 m (Figure 12.10 see the color insert). L-band VV polarization data passes taken approximately 48 hours apart were used to construct interferograms in 100-km segments. The short time interval and long radar wavelength provided excellent correlation between the two passes. The relatively short baseline (mostly less than 100 m) limited the statistical height accuracy of the DEM to 5 to 15 m, while additional errors up to 100 m may have been introduced by atmospheric path-length differences. The interferograms were mosaicked together, and a set of ground control points were selected from existing topographic maps to refine the baseline estimate. C-band amplitude data were ortho-rectified along with the L-band data, to allow generation of three-dimensional perspective views with backscatter image overlays (Figure 12.11; see the color insert). Many of the processing steps were accomplished automatically, including interferogram formation, correlation estimation, phase unwrapping, mosaicking, and geocoding. This experiment demonstrated the feasibility of large-scale interferometric topography mapping, which is planned to be accomplished on the next flight of the SIR-C/X-SAR system.

12.4.3 Earthquake-Related Deformation, Landers, California

The magnitude 7.3 (Mw) Landers, California earthquake of June 28, 1992 (Kanamori et al., 1992; Sieh et al., 1993) provided the first opportunity for detection and measurement of earthquake-related surface deformation using SAR interferometry. The earthquake ruptured over 85 km of the surface in the lightly populated southeastern Mojave Desert. The focal mechanism of the earthquake was dominantly strike-slip, with surface displacements as large as 6 m. The sparsely vegetated, semiarid setting of the ground deformation proved to be stable enough to allow excellent correlation of SAR measurements over a time period of a year or more.

The *ERS-1* SAR satellite obtained data of the Landers area two months before the quake, in April 1992. Additional data passes were acquired in the months following the quake. The first published interferometric analysis of the *ERS-1* data of the Landers event was by Massonnet et al. (1993). They used a "before-and-after" pair of *ERS-1* images (April 24 and August 7, 1992) to form an interferogram or image map of the phase differences. A 90-m-resolution digital elevation model (DEM) was used to remove the effects of topography on the interferometric phases. The resulting interferogram could then be interpreted as a contour map of the change in distance in the direction of the radar line-of-sight of each 90-m patch of the surface. Each "fringe," or 360° cycle of phase, corresponds to 2.8 cm of displacement (one half of the 5.6-cm radar wavelength). A preliminary comparison of the radar-derived displacements and those obtained by conventional geodetic means showed an RMS difference of 3.4 cm, while the actual displacements were on the order of tens of centimeters. As a test of the validity of their approach, Massonnet et al. (1993) generated a synthetic interferogram based on observed fault offsets, seismic constraints on the fault geometries, and a model of dislocation of an elastic half-space. Comparison of the synthetic and actual interferograms (Figure 12.12) shows remarkable agreement. Many details of the SAR-derived interferogram are not reproduced in the synthetic interferogram, especially in areas closest to the fault rupture and in blocks between overlapping fault strands that likely experienced tilting and/or rotation.

Peltzer et al. (1994) conducted a more detailed analysis of the interferograms produced by Massonnet et al. (1993). They successfully applied the synthetic interferogram modeling technique to explain complexities in the observed interferogram, using models of block rotation and tilt between subfaults, as well as distributed shear due to fault rupture at depth.

In a follow-up study, Massonnet et al. (1994) analyzed *ERS-1* interferometry data of the Landers site up to one year following the earthquake. They detected post-

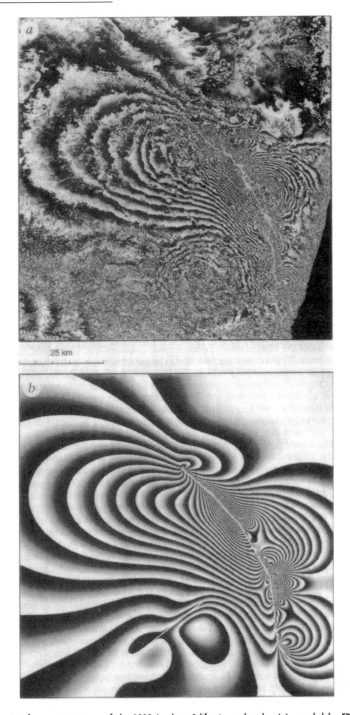

25 km

Figure 12.12 Interferometric signature of the 1992 Landers, California earthquake: (*a*) recorded by ERS-1 repeat-pass interferometry; (*b*) modeled using field and seismic observations of fault geometry. Much of the observed interferometric signature is reproduced in the model, except for areas closest to the fault. Each cycle of phase represents 2.8 cm of diplacement in the radar line of sight. (From Massonnet et al., 1993.)

seismic slip signatures on the order of 10 cm in the vicinity of the fault. Aftershocks as small as magnitude 5 produced distinct phase signatures in the later interferograms. Reanalysis of the coseismic interferograms showed triggered slip on secondary faults, which was consistent with field observations of displacements on these faults.

Zebker et al. (1994) applied the technique of differential interferometry to *ERS-1* data of the Landers area. In this technique, three passes of SAR data are used to remove the effects of topography and recover the surface deformation field simultaneously. A digital elevation model is not required. Instead, two of the three SAR images are used to generate a topographic interferogram. This interferogram is then differenced with a "before-and-after" interferogram, thus removing the effects of topography. Zebker et al. (1994) used *ERS-1* data from April, July, and August 1992 to generate a deformation map of the Landers area automatically. In bypassing the need for an existing DEM, the authors were able to improve the precision of the deformation measurements by about an order of magnitude over the earlier studies. In another significant advance over the analysis of Massonnet et al. (1994), they "unwrapped" the phase information in the differential interferogram to create a map of absolute displacements (along the radar look direction). This map was then directly compared to 18 geodetic measurements from both Global Positioning System (GPS) and electronic distance measurement (EDM) data sets. Comparison of the geodetic and radar-derived displacement measurements showed good correlation, but there was a puzzling bias in the correlation between the data sets that may have been caused by the lack of good vertical control in the geodetic measurements. The authors also documented an unusual "tiling" effect in the the deformation interferogram, in which areas of the surface several hundreds of meters in extent and not in the immediate vicinity of the surface ruptures were bounded by sharp discontinuities in phase. These phase discontinuities were attributed to surface cracking at the edge of near-surface "tiles" or coherent blocks.

A determination of the focal mechanism of an aftershock of the Landers sequence, using *ERS-1* interferometric data, was published by Feigl et al. (1995). The pattern of interferometric fringes from a pair of ERS-1 passes that bracket the December 4, 1992 M5.1 aftershock was inverted using an elastic dislocation formulation to estimate the focal mechanism and magnitude. The results were consistent with the seismologically determined source parameters. A discrepancy in magnitude between the two methods was ascribed by the authors to a possible contribution from other smaller aftershocks in the area that occurred in the interval between the radar passes.

Activity in the Landers seismic zone has continued for over three years since the mainshock. While conventional seismic techniques are being used to monitor aftershocks, SAR interferometry is being used to monitor continued postseismic deformation. Peltzer et al. (1996) used differential (three-pass) interferometry to document ongoing postseismic surface deformation in the Landers zone. Predominantly vertical deformation signatures were observed in blocks bounded by surface ruptures of Landers' subfaults. This postseismic adjustment may represent recharge of fluid pore pressure as the crust enters the next phase of the proposed earthquake cycle.

References

Bannert, D., V. Hon, and D. Johari, 1993. *Landsat MSS Interpretation: A Contribution to the Structural Evolution of Western Borneo*, Geological Survey of Malaysia, Kuala Lumpur

Bernstein, R. L., ed., 1982. *Seasat* Special Issue I, *J. Geophys. Res.*, 87(C5).

Born, G. H., J. A. Dunne, and D. B. Lame, 1979. *Seasat* mission overview, *Science*, 204, 1405–1406.

Budkewitsch, P., M. A. D'Iorio, and J. C. Harrison, 1996a. An examination of the relationship between lithology and radar signatures in arctic environments: preliminary results from Bathurst Island, N.W.T., *Curr. Res.*

Budkewitsch, P., M. A. D'Iorio, and J. C. Harrison, 1996b. SAR expressions of geology in the Canadian Arctic, in *Proceedings of the 18th Canadian Symposium on Remote Sensing*, Vancouver, British Columbia, Canada, pp. 88–91.

Campbell, B. A., and J. B. Garvin, 1993. Lava flow topographic measurements for radar data interpretation, *Geophys. Res. Lett.*, 20, 831–834.

Campbell, B. A., R. E. Arvidson, and M. K. Shepard, 1993. Radar polarization properties of volcanic and playa surfaces: applications to terrestrial remote sensing and Venus data interpretation, *J. Geophys. Res.*, 98, 17099–17113.

Campbell, F. H. A., R. J. Brown, M. E. Kirby, D. Benmouffok, and D. Lapp, 1994. The Canadian GlobeSAR Program in *Proceedings of the International Geoscience and Remote Sensing Symposium*, IGARSS '94, Pasadena, Calif., pp. 1523–1524.

Cimino, J. B., and C. Elachi, eds., 1982. *The Shuttle Imaging Radar-A (SIR-A) Experiment*, JPL Publ. 82-77, Jet Propulsion Laboratory, California Institute of Technology, Pasadena, Calif.

Cimino, J. B., B. Holt, and A. Richardson, 1987. *The SIR-B Experiment Report*, JPL Publ. 87-88-2, Jet Propulsion Laboratory, California Institute of Technology, Pasadena, Calif.

Decker, R. W., and R. L. Christiansen, 1984. Explosive eruptions of Kilauea volcano, Hawaii, in *Explosive Volcanism: Inception, Evolution and Hazards*, National Academy of Science, Washington D.C., pp. 122–132.

de Freitas, T., J. C. Harrison, and R. Thorsteinsson, 1993. New field observations on the geology of Bathurst Island, Arctic Canada: A. Stratigraphy and sedimentology of the Phanerozoic succession, GSC Pap. 93–1B, in *Current Research, Part B*, Geological Survey of Canada, Ottawa, Ontario, Canada, pp. 1–10.

D'Iorio, M. A., Ling Nan Ley, P. A. Budkewitsch, and R. Richardson, 1995. Geological map update using airborne and *RADARSAT* simulated SAR data in Sarawak, Malaysia, *Geocarto Int.* 10 (2), 43–50.

Dixon, T. H., ed., 1995. *SAR Interferometry and Surface Change Detection: Report of a Workshop Held in Boulder, Colorado, February 3–4, 1994*, RSMAS Tech. Rep. TR 95–003. University of Miami Rosenstiel School of Marine and Atmospheric Science, Miami, Fla.

Elachi, C., ed., 1986. SIR-B Special Issue, *IEEE Trans. Geosci. Remote Sensing*, 24.

Elachi, C., J. B Cimino, and M. Settle, 1986. Overview of the Shuttle Imaging Radar-B preliminary scientific results, *Science*, 232.

Evans, D. L., et al., 1993. The Shuttle Imaging Radar-C and X-band Synthetic Aperture Radar (SIR-C/X-SAR) mission, *EOS Trans. AGU*, 74(13); 145–158.

Feigl, K., A. Sergent, and D. Jacq, 1995. Estimation of an earthquake focal mechanism from a satellite radar interferogram: application to the December 4, 1992 Landers aftershock, *Geophys. Res. Lett.*, 22, 1037–1040.

Ford, J. P., ed., 1988. Special Issue on Advances in Shuttle Imaging Radar-B Research, *Int. J. Remote Sensing*, 9.

Ford, J. P., R. G. Blom, M. G. Bryan, M. L. Daily, T. H. Dixon, C. Elachi, and E. C. Xenos, 1980. *Seasat Views North America, the Carribean, and Western Europe with Imaging Radar*, JPL Publ. 80–67, Jet Propulsion Laboratory, California Institute of Technology, Pasadena, Calif.

Ford, J., J. B. Cimino, and C. Elachi, 1982. *Space Shuttle Columbla Views the World with Imaging Radar: The SIR-A Experiment*, JPL Publ. 82–95, Jet Propulsion Laboratory, California Institute of Technology, Pasadena, Calif.

Ford, J. P., J. B. Cimino, B. Holt, and M. Ruzek, 1986. *Shuttle Imaging Radar Views the Earth from Challenger: The SIR-B Experiment*, JPL Publ. 86–10, Jet Propulsion Laboratory, California Institute of Technology, Pasadena, Calif.

Gaddis, L., P. Mouginis-Mark, R. Singer, and V. Kaupp, 1989. Geologic analysis of Shuttle Imaging Radar (SIR-B) data of Kilauea volcano, Hawaii, *Geol. Soc. Am. Bull.*, 101, 317–332.

Gaddis, L. R., P. J. Mouginis-Mark, and J. N. Hayashi, 1990. Examination of lava flow surface textures: SIR-B image texture, field observations and terrain measurements, *Photogramm. Eng. Remote Sensing*, 56, 211–224.

Goldstein, R., 1995. Atmospheric limitations to repeat-track radar interferometry, *Geophys. Res. Lett.*, 22, 2517–2520.

Gonzalez, F. I., R. C. Beal, W. E. Brown, P. S. DeLeonibus, J. W. Sherman III, J. F. Gower, D. Lichy, D. B. Ross, C. L. Rufenach, and R. A. Schuchman, 1979. *Seasat* synthetic aperture radar: Ocean wave detection capabilities, *Science*, 204, 1418–1421.

Hamilton, W., 1978. *Tectonics of Indonesia*, USGS, Prof. Pap. 1078, U.S. Geological Survey, Washington, D.C.

Holcomb, R. T., 1987. *Eruptive History and Long-Term Behavior of Kilauea Volcano*, USGS Prof. Pap. 1350, U.S. Geological Survey, Washington, D.C., pp. 261–350.

Holt, B., ed., 1988. Special issue on shuttle imaging radar experiment, *J. Geophys. Res.*, 33.

Jordan, R. L., 1980. The *Seasat*-A synthetic aperture radar system, *IEEE J. Ocean. Eng.*, 5(2), 154.

Jordan, R. L., B. L. Huneycutt, and M. Werner, 1995. The SIR-C/X-SAR synthetic aperture radar system, *IEEE Trans. Geosci. Remote Sensing*, 33, 829–839.

Kanamori, H., H.-K. Thio, D. Dreger, and E. Hauksson, 1992. Initial investigation of the Landers, California, earthquake of 28 June 1992 using TERRAscope, *Geophys. Res. Lett.*, 19, 2267–2270.

Karnevi, S., E. Dean, D. J. Q. Carter, and S. S. Hartley, 1993. *Envisat*'s advanced synthetic aperture radar: ASAR, *ESA Bull.*, 76, 30–35.

Kaupp, V. H., L. R. Gaddis, P. J. Mouginis-Mark, B. A. Derryberry, H. C. MacDonald, and W. P. Waite, 1986. Preliminary analyses of SIR-B radar data for Hawaii, *Remote Sensing Environ.*, 20, 283–290.

Kerr, J. W., 1974. *Geology of Bathurst Island Group and Byam Martin Island, Arctic Canada*, GSC Mem. 378, 152p. (+ 1:250 000 scale map, #1350A, 2 sheets), Geological Survey of Canada, Ottawa, Ontario, Canada.

Kirk, H. J. C., 1957. *The Geology and Mineral Resources of the Upper Rajang and Adjacent Area*: Mem. 8, British Territory of Borneo Geological Survey Department, 181 pp.

Kirwan, A. D., T. J. Ahrens, and G. H. Bern, eds., 1983. *Seasat* special issue II, *J. Geophys. Res.*, 88(C3).

Liechti, P., F. W. Roe, and N. S. Haile, 1960. *The Geology of Sarawak, Brunei, and the Western Part of the North Borneo*, Bull. 3, British Territory of Borneo Geological Survey Department, 360 pp. + portfolio.

Ling Nan Ley, 1994. Geologic interpretation of GlobeSAR image of Tubau Bukit Lumut area, Bintulu, Sarawak, Malaysia, in *Proceedings of the GlobeSAR Southeast Asian Regional Workshop*, Bangkok, Thailand, pp. 149–154.

Livingstone, C. E., A. L. Gray, R. K. Hawkins, R. B. Olsen, J. G. Halbertsma, and R. A. Deane, 1987. CCRS C-band airborne radar: system description and test results in *Proceedings of the 11th Canadian Symposium on Remote Sensing*, June 22–25, Waterloo, Ontario, Canada, pp. 503–518.

Livingstone, C. E., A. L. Gray, R. K. Hawkins, and R. B. Olsen, 1988. CCRS X/C airborne synthetic aperture radar: an R and D tool for the *ERS-1* timeframe, *Aerosp. Electron. Syst. Mag.*, 3(10), 15–21.

Livingstone, C. E., A. L. Gray, R. K. Hawkins, P. Vachon, T. I. Lukowski, and M. Lalonde, 1995. The CCRS airborne SAR systems: radar for remote sensing research, *Can. J. Remote Sensing* 21(4), 468–491.

Lou, Y., Y. Kim, and J. van Zyl, 1996. The NASA/JPL airborne synthetic aperture radar system, *Summaries of the 6th Annual AIRSAR Earth Science Workshop*, JPL Publ. 96–4, Vol. 2, Jet Propulsion Laboratory, California Institute of Technology, Pasadena, Calif., pp. 51–56.

Luscombe, A. P., et al., 1993. The *RADARSAT* synthetic aperture radar development, *Can. J. Remote Sensing*, 19(4), 301.

Massonnet, D., M. Rossi, C. Carmona, F. Adragna, G. Peltzer, K. Feigl, and T. Rabaute, 1993. The displacement field of the Landers earthquake mapped by radar interferometry, *Nature* 364, 138–142.

Massonnet, D., K. Feigl, M. Rossi, and F. Adragna, 1994. Radar interferometric mapping of deformation in the year after the Landers earthquake, *Nature* 369, 227–230.

McCauley, J. F., G. G. Schaber, C. S. Breed, M. J. Grolier, C. U. Haynes, B. Issawi,

C. Elachi, and R. Blom, 1982. Subsurface valleys and geoarchaeology of the eastern Sahara revealed by shuttle radar, *Science*, 218, 1004–1019.

McCauley, J. F., C. S. Breed, G. G. Schaber, W. P. McHugh, B. Issawi, C. V. Haynes, M. J. Grolier, and A. El Kilani, 1986. Paleo-drainages of the eastern Sahara: the radar rivers revisited (SIR-A/B implications for a mid-Tertiary trans-African drainage system), *IEEE Trans. Geosci. Remote Sensing*, 24, 624–648.

Mouginis-Mark, P. J., 1995a. Preliminary observations of volcanoes with the SIR-C radar, *IEEE Trans. Geosci. Remote Sensing*, 33, 934–941.

Mouginis-Mark, P. J., 1995b. Analysis of volcanic hazards using radar interferometry, *Earth Observ. Q.* 47, 6–10.

Murakami, M., S. Fujiwara, M. Nemoto, and T. Saito, 1995. Application of the interferometric *JERS-1* SAR for detection of crustal deformations in the Izu Peninsula, Japan, *Eos Trans. AGU*, 76 suppl., F63.

Peltzer, G., K. Hudnut, and K. Feigl, 1994. Analysis of coseismic surface displacement gradients using radar interferometry: new insights into the Landers earthquake, *J. Geophys. Res.*, 99, 21971–21981.

Peltzer, G., P. Rosen, F. Rogez, and K. Hudnut, Postseismic rebound in fault stepovers caused by pore fluid flow, *Science*, 272, 1202–1204.

Raney, R. K., A. P. Luscombe, E. J. Langham, and S. Ahmed, 1991. *RADARSAT*, *Proc. IEEE*, 79, 839–849.

Rosen, P., S. Hensley, Y. Lou, S. Shaffer, and E. Fielding, 1995. A digital elevation model from Oregon to Mexico derived from SIR-C radar interferometry, *Eos Trans. AGU*, 76 suppl., F64.

Rosen, P. A., S. Hensley, H. A. Zebker, F. H. Webb, and E. Fielding, 1996. Surface deformation and coherence measurements of Kilauea volcano, Hawaii from SIR-C radar interferometry, *J. Geophys. Res.*, 101,23, 109–23,126.

Schaber, G. G., J. F. McCauley, C. S. Breed, and G. R. Olhoeft, 1986. Shuttle Imaging Radar: Physical controls on signal penetration and subsurface scattering in the eastern Sahara, *IEEE Trans. Geosci. Remote Sensing*, 24, 603–623.

Settle, M., and J. Taranik, 1982. Use of the space shuttle for remote sensing research: recent results and future prospects, *Science*, 218, 993–995.

Sieh, K., and 19 others, 1993. Near-field investigations of the Landers earthquake sequence, April to July, 1992, *Science*, 260, 171–176.

SIR-B Science Team, ed., 1984. *The SIR-B Science Investigations Plan*, JPL Publ. 84–3, Jet Propulsion Laboratory, California Institute of Technology, Pasadena, Calif.

Stofan, E. R., D. L. Evans, C. Schmullius, B. Holt, J. Plaut, J. van Zyl, S. D. Wall, and J. Way, 1995. Overview of results of Spaceborne Imaging Radar-C, X-band Synthetic Aperture Radar (SIR-C/X-SAR), *IEEE Trans. Geosci. Remote Sensing*, 33, 817–828.

Weissman, D. E., ed., 1980. *Seasat-1* special issue, *IEEE J. Ocean. Eng.*, 5.

Yin, E. H., 1992. *Geological Map of Sarawak*, Geological Survey of Malaysia, Kuala Lumpur.

Zebker, H. A., S. Madsen, J. Martin, K. Wheeler, T. Miller, Y. Young, G. Alberti, S. Vetrella, and A. Cucci, 1992. The TOPSAR interferometric radar topographic mapping instrument, *IEEE Trans. Geosci. Remote Sensing*, 30, 933–940.

Zebker, H. A., P. A. Rosen, R. M. Goldstein, A. Gabriel, and C. L. Werner, 1994.

On the derivation of coseismic displacement fields using differential radar interferometry: the Landers earthquake, *J. Geophys. Res.*, 99, 19617–19634.

Zebker, H., P. Rosen, S. Hensley, and P. J. Mouginis-Mark, 1996. Analysis of active lava flows on Kilauea volcano, Hawaii, using SIR-C radar correlation measurements, *Geology*, 24, 495–498.

Zebker, H., P. Rosen, and S. Hensley, 1997. Atmospheric effects in interferometric synthetic aperture radar surface deformation and topographic maps, *J. Geophys. Res.*, 102, 7547–7563.

Geophysical Methods

John Broome

Geoscience Integration Section
Geological Survey of Canada
Ottawa, Ontario, Canada

13.1 INTRODUCTION

A chapter on geophysics within a remote sensing manual may at first seem out of place; however, the distinction between remote sensing and geophysics is equivocal. In common usage, most remote sensing data are seen to originate from satellites, whereas geophysical data are collected by aircraft. In reality, the collection platform is not a useful distinction, and some so-called geophysical data are collected by satellite and some remote sensing data are collected by aircraft. Some data, for example gamma ray spectrometry data, are identical in many ways to multiband satellite data but are routinely analyzed by geophysicists rather than remote sensing specialists. I would argue that since "geophysical" data are collected by remote instrumentation, they could correctly be called remote sensing data, and since "remote sensing" data are controlled by the physical properties of the earth, they are geophysical. Any differences between geophysics and remote sensing are related more to the processing and interpretation methodology than the data themselves.

Geophysicists are often required to interpolate point or line data and grid them to produce continuous coverages, and the resulting images are similar in form to remote sensing images. Obviously many of the enhancement and analysis techniques applied to these data can also be used for analysis of remote sensing data. Until the late 1980s there was relatively little cross-pollination between the two disciplines. With the increased emphasis on interdisciplinary studies and data integration, communication between these disciplines has increased greatly, to the benefit of both

Remote Sensing for the Earth Sciences: Manual of Remote Sensing, 3 ed., Vol. 3, edited by Andrew N. Rencz.
ISBN: 0471-29405-5 © 1999 John Wiley & Sons, Inc.

groups. The general availability of sophisticated GIS and image processing software has facilitated data integration and sharing of methodology.

Increasingly, the remote sensing specialist will be involved in projects where remote sensing data are interpreted together with geophysical and other geoscience data. Geophysical data and the geophysical images that can be produced are extremely useful aids to geological mapping and resource exploration. To contribute to these projects, a basic understanding of the geophysical data is required. The intent of this chapter is to briefly describe background theory, data collection, specialized processing, and appropriate applications for several types of geophysical data commonly used for mineral exploration and geological mapping. Wherever possible, references are provided for the material covered.

13.1.1 Geophysics

Geophysics covers the study of the physics of the earth and atmosphere. This discipline traditionally encompasses studies of magnetic, gravity, and electrical fields, radioactivity, and elastic, thermal, and electrical properties. This range of topics is extremely wide and there is considerable overlap between subdisciplines. Applications of physical principles to studies of the Earth are found in the fields of meteorology, geodesy (studies of the form of the earth), oceanography, tectonophysics, resource exploration, geological mapping, geochronology, hydrology, seismology, and many others. Many of these applications utilize the geophysical data in ways fundamentally different from those used for remote sensing data. Geophysical data are often three-dimensional in nature, compared to the largely two-dimensional nature of remote sensing data. In addition, coarse sampling of the measured geophysical parameter can result in data too sparse for interpolation to produce continuous raster coverage.

There are also differences in the interpretation methods used. In addition to the qualitative analysis of features, patterns, textures, and anomalies, interpretation of geophysical data includes the use of quantitative methods. Quantitative analysis involves a coordination of theoretical understanding of the relationship between a source body and perturbations of the measured parameter to model a source with geometry, location, and properties that could produce the observed variations.

In this chapter we discuss only a few selected geophysical data sets. Gravity, aeromagnetic, and airborne gamma ray spectrometric data were selected because they are readily available for large areas, commonly used for geological mapping and mineral exploration, and collected with a sampling density appropriate to produce regional spatial imagery. Airborne electromagnetic methods are also covered briefly. Types of geophysical imagery not covered in this chapter include seismic reflection and electromagnetic imaging of vertical sections in the crust, and ground probing radar. Gamma ray spectrometric and electromagnetic detectors measure energy levels at the opposite extremes of the electromagnetic spectrum shown in Figure 13.1.

This chapter is organized as follows. An introduction covering basic theory, data collection, and specialized processing is provided for each data set, followed by some examples of qualitative interpretation for specific geological applications.

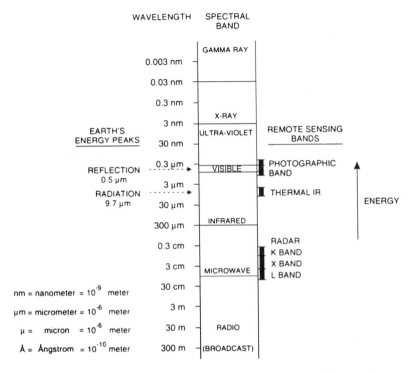

Figure 13.1   Two geophysical methods discussed in this chapter, airborne electromagnetic techniques and gamma ray spectrometry, measure energy wavelengths at the extremes of the electromagnetic spectrum.

13.2   GEOPHYSICAL DATA: THEORY, DATA COLLECTION, PROCESSING

13.2.1 Potential Fields

The gravity and magnetic methods are different from most remote sensing methods because they are based on potential field theory. A brief introduction to potential field theory as applied to the gravity method follows.

Newton's law of gravitation is the basis of the gravity method and states that the force between two particles of mass (M_1 and M_2) is directly proportional to the product of the masses and inversely proportional to the square of the distance between the centers of mass (r):

$$F = - \gamma \frac{M_1 M_2}{r^2} r_1 \tag{13.1}$$

where γ is the universal gravitational constant and r_1 is a unit vector directed from M_1 toward M_2. The universal gravitational constant is one of the fundamental constants of nature. The acceleration due to gravity (g) can be found by dividing F by M_2. If M_1 is the mass of the Earth, M_e, the accelerations of a mass M_2 at the surface of the earth is

$$g = \frac{F}{M_2} = -\gamma \frac{M_e}{r_e^2} r_1 \tag{13.2}$$

with r_e the radius of the earth and r_1 a unit vector extending outward from the center of the Earth.

Potential fields have the property that if a mass is moved from one position to another, the sum of positional (potential) energy and kinetic energy is constant in a closed system. The work required to move a unit mass from infinity to a position a distance R from the center of gravity of M following any path can be used to define the scalar potential U:

$$U(r) = \int_{\infty}^{R} g \; dr = \frac{\gamma M}{R} \tag{13.3}$$

The ability to define the gravitational potential at any point in a closed system allows the calculation of the acceleration due to gravity, g, at any point in the system. This opens many possibilities for interpretation. For the solution of gravity problems, the scalar potential (U) is often used rather than the vector (g). The Newtonian, or three-dimensional, potential can be approximated for an arbitrary body by dividing the body into discrete elements and integrating to determine the total gravity effect:

$$U = \gamma\sigma \int_x \int_y \int_z \frac{1}{r} \; dx \; dy \; dz \tag{13.4}$$

where σ is the density and $r^2 = x^2 + y^2 + z^2$.

Discrete survey measurements can be used to define the potential field throughout the working system from well-defined relationships. Other information, such as derivative fields and the potential and field strength at other datum levels, can then be determined from the field definition. Potential field theory also allows calculation of the theoretical effect of a mass at any position in the system, allowing the development of systems to model the anomalies that would result from hypothetical source bodies and compare the results to measured anomalies.

The gravity and magnetic fields can be interpreted to provide information about density and magnetization variations in the earth's crust that is useful for geological mapping and mineral exploration. Potential fields, which are intrinsic to the distribution of the causative parameter and act at a distance from the source, vary in proportion to the parameter variations. Gravity and magnetic methods are similar, as they both involve measurement of small variations in a huge force field, which varies with respect to both position and time.

Most remote sensing methods have limited penetration below the surface of the earth and provide no direct subsurface information to help interpret the three-dimensional crustal geology. Both gravity and magnetic data provide subsurface information about the Earth and are particularly useful for understanding areas where bedrock is obscured from view. Although exact definition of density and magnetization distributions in the crust is impossible due to an ambiguity problem, and knowledge of these parameters does not allow definitive identification of lithology, incorporation of other supporting information can help to constrain the interpretation.

Because of the nature of potential fields, qualitative interpretation can be used to model the three-dimensional subsurface distribution of density and magnetization. Due to the ambiguity of solutions, this process is most effective when undertaken interactively by computer and many programs are available for this purpose. In this chapter we deal with the qualitative interpretation of anomaly patterns on potential field imagery. Simple linear patterns of anomalies can easily be interpreted without training, but effective interpretation of more complex anomalies is more difficult. Experimentation with an interactive modeling tool is a excellent way to test the feasibility of a range of qualitative interpretations of more complex anomalies.

As with each type of remote sensing data, method-specific processing is required to correct for variations in the measured parameters unrelated to geological sources. These variations may be due to instrumentation drift, geoid variations, or other external or cosmic effects. Although detailed descriptions of these corrections and processing procedures are beyond the scope of this manual, it is important to know the significance of each step to determine whether the necessary corrections have been applied to a given data set.

13.2.2 Gravity

13.2.2.1 EARTH'S GRAVITY FIELD.

The Earth's gravity field is a monopole potential field explained by Newton's law of gravitation, which states that every particle of matter exerts a force of attraction on every other particle (Section 13.2.1). The Earth's field averages 980 gal, but its magnitude on the Earth's surface depends primarily on the five factors shown in Table 13.1 (in order of decreasing importance).

Gravity anomalies of interest for resource exploration and geological mapping and interpretation are due to density variations in the Earth's crust (item 4). Table 13.1 shows that the variations in the gravity due to latitude, elevation, and topography produce variations larger than those due to density variations. An image of the gravity field intensity over Canada generated from data gridded and supplied by the Canadian Geophysical Data Centre is shown in Figure 13.2a (see the color insert).

TABLE 13.1 Factors Influencing the Earth's Gravity Field in Decreasing Order of Intensity

Factor	Effect
1. Latitude (maximum at poles, minimum at equator)	5000 mgal
2. Elevation (approx. 0.3 mgal/m)	1000 mgal
3. Geometry of the surrounding terrain	Variable
4. Variations in the density of the subsurface	1–10 mgal
5. Earth tides	0.3 mgal

13.2.2.2 GRAVITY DATA COLLECTION.

Most gravity data have been collected as ground measurements made at stations at a more or less uniform spacing. Station spacing can vary from 100 m for detailed mineral exploration surveys to the more typical 5 to 15 km spacing of regional surveys. Because of the requirement of measuring changes as small as 0.1 mgal, relative rather than absolute measurements are made. A number of types of gravity meters are used, including Lacoste Romberg, Scintrex, Worden, and others.

Accurate positioning, particularly vertically, is essential for accurate data because of the strong correlation between the gravity field and elevation. Regional surveys are typically conducted at surveyed gravity stations. Regional survey stations must be positioned with an accuracy of 20 m horizontally and 3 m vertically. Detailed surveys must be positioned at approximately 10 times the regional accuracy. Station positioning has traditionally been done by surveying but is now done increasingly by global positioning system (GPS). During surveying, measurements at control stations are required periodically to allow later removal of gravity meter drift during data processing.

A recent development is the use of airborne gravity data collection for regional surveying (Hammer, 1983; Halpenny and Darbha, 1995). Accuracy is limited by aircraft positioning errors, but continuing improvements have resulted in systems that are now acceptable for some applications. When airborne gravity systems develop to the stage that accuracy is adequate for regional surveying, greater sampling densities will allow the production of higher-resolution imagery, which is more suitable for enhancement and qualitative interpretation.

13.2.2.3 GRAVITY DATA CORRECTION.

The gravity anomalies of interest for exploration or mapping of crustal density variations (item 4 in Table 13.1) are typically less than 0.001% of the total field, smaller than variations due to gravity station latitude, elevation, and terrain variation. Therefore, these effects must be carefully removed. These corrections must be applied to raw gravity station (point) data to bring them to a common datum level before interpolation onto a grid for imaging and interpretation. A brief description of each correction follows. More details can be found in geophysical references (Telford et al., 1978)

A. Latitude Correction. A correction is required to remove the increase of gravity with latitude due to (1) centrifugal acceleration from the earth's rotation and (2) the slight equatorial bulge in the Earth's shape which results in an increase in the Earth's radius from the poles to the equator.

B. Free-Air Correction. This correction removes variations in the gravity field due to the distance between the measurement point and a datum surface parallel to the geoid. Sea level is commonly chosen as the datum level. The free-air correction corrects for variations in this distance related to station elevation changes due to topography to reduce the data to the datum surface. This correction does not consider the effect of material between station and datum plane elevation.

C. Bouguer Correction. The Bouguer correction removes variations in gravity due to the density of the material between station elevation and datum level. The correction assumes a slab of rock with infinite horizontal extent and uniform density.

The density of the slab can be set based on local lithology but is often set to 2.67 g/cm³, which approximates the average density of continental crust.

D. Terrain Correction. Terrain corrections remove variations in gravity due to surface irregularity in the vicinity of the station. Without this correction, measurements made at the top of hills will be greater than measurements made in valleys, due to the gravitational attraction of material in the adjacent valley walls. This correction is essential in areas with rugged topography.

E. Earth Tide Correction. The earth tide correction is required to remove gravity variation due to movement of the sun and moon. The correction can be up to 0.3 mgal and is variable with both latitude and time.

F. Isostatic Correction. To remove regional gravity variations due to isostacy. This correction is usually not required for exploration scale data sets because the wavelength of the isostatic variations is an order of magnitude longer than wavelengths of gravity anomalies of interest for most exploration.

13.2.3 Magnetic Methods

13.2.3.1 EARTH'S GEOMAGNETIC FIELD.

The theory of the magnetic method has much in common with the gravity method. Both are potential fields; however, the theory of magnetic fields is somewhat more complicated because it is a vector dipolar field (i.e., north and south poles). Since the orientation of the dipoles that determine magnetization may lie in any direction, definition of the magnetic state of a body requires both magnitude and direction rather than simply magnitude (mass) as in the case of the gravity field. The vector defining this component of the field is defined by declination, inclination, and magnitude. Figure 13.3a is a contour map of variation in the magnitude of the earth's total field. Figure 13.3b shows the variation in the inclination of the field from near vertical near the poles to horizontal near the equator.

Units used for magnetic field measurement are somewhat complicated and a number of systems and terminologies have developed. The common systems of measurement are SI and cgs. The more modern SI system of units utilizes the *tesla* as the basic unit; the cgs system uses the *gauss*, where 1 tesla = 10^4 gauss. Most magnetic work uses smaller units [i.e., the nanotesla (nT; 10^{-9} tesla) and gamma (10^{-5} gauss) which are numerically equivalent].

Earth's magnetic field is 40,000 to 65,000 nT and has the following components:

1. *Earth's core field.* This accounts for 99% of the geomagnetic field. In geologic time context the field direction reverses periodically. Analysis of lava sequences and other selected geological environments has indicated that reversals occur on average every 10^5 years during some periods (Stacey, 1977). Contour maps of the inclination and intensity of the Earth's magnetic field are shown in Figure 13.3a and b, respectively.

2. *External field.* Most of the remaining 1% of the geomagnetic field is due to sources outside the Earth, such as electric currents in the ionized layers of the outer atmosphere related to solar diurnal variations, lunar diurnal variation,

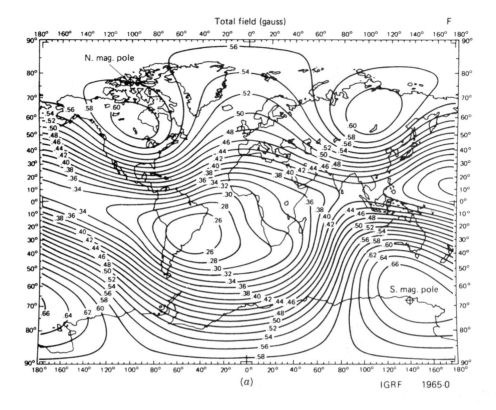

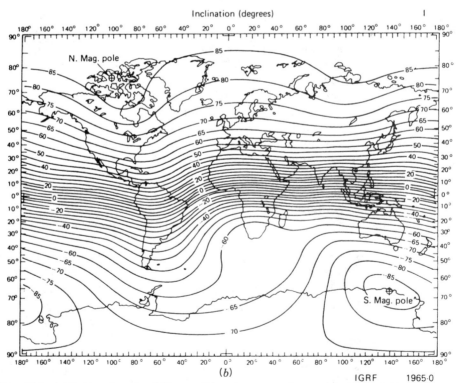

Figure 13.3 (*a*) Earth's total magnetic field computed from the international geomagnetic reference field (IGRF) at Epoch 1965; (*b*) inclination of the Earth's magnetic field, also computed from the 1965 IGRF. (After Leaton, 1971.)

and sunspot activity. This component of the field may vary significantly with time of day, latitude, and season; for example, solar diurnal variation has a range of approximately 30γ during a 24-hour period.

3. *Crustal sources.* This component of the magnetic field is the one of interest for geological applications. Concentrations of magnetic minerals in the Earth's crust produce complex regional anomalies that can occasionally be equal in size to the geomagnetic field but are usually significantly smaller. The crustal source component of the field can be due to induced magnetization by the geomagnetic field or to permanent remanent magnetization in the rock, which complicates interpretation.

An image of the total magnetic field over Canada generated from the Canadian Geophysical Data Centre aeromagnetic database is shown in Figure 13.2*b* (see the color insert).

13.2.3.2   AEROMAGNETIC DATA COLLECTION.

Aeromagnetic data are collected by flying a magnetometer-equipped aircraft along regularly spaced *traverse* lines. Perpendicular *control* lines, which are flown at a wider spacing, allow data leveling at intersections with the traverse lines to remove temporal variations. Traverse lines are typically flown at a spacing of from 5 km to 100 m, depending on the amount of detail required. Surveys are flown at *constant altitude* for fixed-wing regional surveys in mountainous terrain, whereas *drape flying* at constant terrain clearance of 80 to 300 m is used for detailed surveys. Detailed low-altitude helicopter surveys for mineral exploration or detailed geological mapping would use the closer spacing, while the 5-km spacing would be appropriate for regional constant-altitude fixed-wing surveys. The high sampling density along the flight lines results in better resolution and continuity of anomalies striking perpendicular to the flight line direction on the interpolated image. For this reason, traverse lines are best flown perpendicular to the geologic strike to enhance the representation of associated anomalies.

Magnetometers are used to measure the magnitude of the magnetic field. Types of magnetometers include, in order of increasing sensitivity, fluxgate (10 nT), proton precession (0.25 nT), and optical pumping (<0.01 nT). Actual survey accuracies of 1 nT are considered adequate for most applications. Vertical magnetic gradiometers use two magnetometers separated vertically to estimate the vertical magnetic gradient, which is more sensitive to higher wavenumber (near-surface) anomalies. One advantage of vertical gradient data is that diurnal variations and leveling problems are removed automatically, due to the differential nature of the method. The actual magnetometer sensor may be installed in a *bird* towed by a cable beneath the aircraft or installed on a *stinger*, or mounting rod, attached to the aircraft. Sophisticated compensation systems are used to cancel the magnetic fields originating in the aircraft.

Traditionally, aeromagnetic survey navigation involved manual flight path recovery using flight path photography to track the path of the aircraft on air photo mosaics. Now most navigation utilizes electronic navigation instrumentation, including GPS, Loran-C, inertial navigation system (INS), or radio navigation system.

To remove *diurnal variations* in the geomagnetic field, *ground stations* (recording magnetometers located in the survey area) collect data showing the temporal variations in the geomagnetic field which are used later to remove these variations from the survey data.

Magnetic data collected by satellite can also be used to generate images. Because of the greater distance from the sensor to the crustal sources, the resolution of satellite magnetic data is less than that of airborne data. For example, the *Magsat* satellite samples the magnetic field at 40-km intervals along the satellite orbit. In practice, resolution of crustal anomalies with wavelengths of less than 500 km is poor. This lack of resolution limits the utility of satellite magnetic data for geological applications (Taylor and Schnetzler, 1990). Downward continuation can be used to enhance the high-wavenumber components of the data, but the amount of continuation is limited by noise in the data (Whaler, 1994).

13.2.3.3 MAGNETIC DATA PROCESSING.

As with gravity data, a number of corrections must be applied before the aeromagnetic survey line data are interpolated onto a regular grid for image generation and interpretation. Since the component of the geomagnetic field due to the crustal sources of interest is small, great care must be taken to avoid creating artificial anomalies during data processing, or *artifacts*, in the data.

A. Diurnal Corrections. A fixed-location ground magnetometer may be employed during surveying to record temporal changes in the magnetic field. These variation can then be removed from the flight line data. Since diurnal variations change with location, the ground magnetometer should be located centrally in the survey area.

B. Leveling Correction. In theory, accurately reflying a traverse line should result in identical measurements. In practice this does not happen, due to variations in the magnetic field caused by a number of spatial and temporal variations including diurnal variations, instrument drift, and positioning errors. One way of eliminating these "level" errors is by flying *control lines* perpendicular to the flight lines, calculating the differences at intersections with the traverse lines and applying a level correction to the traverse line data by interpolating these differences between intersections. Since the control lines are widely separated, complete coverage of the survey area is completed rapidly, and errors in the traverse line due to temporal effects can be drastically reduced.

Positioning errors may include a directional bias due to instrument lag and position of the magnetometer sensor relative to the aircraft. If uncorrected, this directional bias combined with the opposite direction of adjacent flight lines can result in spatial oscillation of originally straight anomalies, sometimes referred to as the "herringbone" effect.

C. International Geomagnetic Reference Field (IGRF) Removal. The IGRF (Langel, 1992) is a mathematical model of the Earth's long-wavelength magnetic field that can be removed from measured data. Local features are enhanced in IGRF-corrected data, but the long-wavelength nature of the correction makes the correction useful only for regional data sets at scales smaller than 1:250,000. The data used to produce the contour maps of total magnetic field and inclination in Figure 13.3 were generated using the IGRF.

13.2.4 Electrical Methods

13.2.4.1 CONDUCTIVITY OF THE EARTH.

Electrical methods detect the effects of current flow in the Earth. Several electrical properties of rocks, including electrical potential, conductivity, and dielectric constant, can be measured to provide information about the Earth. Only conductivity is considered in this chapter, as it is the most significant electrical property. Current flow can occur as electronic, electrolytic, and dielectric conduction in rock. Electronic conduction occurs in conductive material, such as metals with free electrons. Electrolytic and dielectric conduction take place in poor conductors by ion and polarization changes, respectively.

13.2.4.2 ELECTRICAL DATA COLLECTION.

A great range of methods are available, including those that measure naturally occurring potentials, currents, and electromagnetic fields and those that are generated artificially. These include magnetotelluric, resistivity, equipotential methods, electromagnetic methods, polarization, mise-a-la-masse, and induced polarization. A comprehensive description of the many different methods is beyond the scope of this manual; only some methods used for airborne conductivity mapping are discussed.

Airborne electromagnetic surveying (AEM) has been widely used for mineral exploration and conductivity mapping for many decades (Palacky, 1987b). A wide range of systems have been designed to measure both naturally occurring and artificially generated fields. All AEM methods (Table 13.2) measure the effect of the ground on the propagation of EM fields.

Active AEM systems generate a primary EM field by passing alternating current through a coil or wire. The presence of secondary EM fields is detected by measuring alternating currents induced in receiver coils. The phase and amplitude of the secondary field can provide information about the conductivity, size, and geometry of conductive zones in the earth. Most AEM systems record a number of simultaneous channels of data, which although not independent, contain complementary information. The fundamental challenge in all AEM methods is measurement of small secondary responses related to the electrical properties of the Earth in the presence of large primary fields.

Active AEM systems can be categorized as frequency-domain or time-domain and single or multicomponent. Frequency-domain systems measure the in-phase and out-of-phase, or quadrature, components of the induced field from a continuously operating transmitter, while time-domain systems use a pulsed transmitter and measure

TABLE 13.2   Types of AEM Systems

Type	Frequency Range (Hz)	Transmitter/Receiver
AFMAG	50–500	Fixed/aircraft
Time domain	50–5000	Aircraft/aircraft or bird
Frequency domain	400–2300	Aircraft or bird/bird
VLF	10,000–25,000	Fixed/aircraft or bird

time slices of the decay of the secondary response to avoid the technical problem of measuring a weak secondary field in the presence of the much stronger transmitter field. Since the time-domain receiver is measuring when the transmitter is off, variations in the primary field due to relative motion between the transmitter and receiver are eliminated. These variations are difficult to distinguish from changes in the secondary field. Single-component systems measure only the horizontal component of the secondary field along the survey line (x component) direction, whereas multicomponent systems attempt to resolve the secondary field into both x and y or x,y, and z components (Smith and Keating, 1996). Many frequency-domain systems acquire data using multiple sensors, with each receiver having a corresponding transmitter and operating frequency. These systems are essentially multiple single-component systems rather than true multicomponent systems. The imaging and display possibilities of this wealth of data are huge, but the interpretation of the data is complex and system dependent.

Important passive systems are very low frequency (VLF), which use powerful (10 to 25 kHz) marine navigation transmitters as a field source and audio-frequency magnetic (AFMAG) systems that measure variations in the dip angle of naturally occurring primary fields due to natural phenomenon such as lightning discharges. The frequencies used by VLF, although low relative to other radio frequencies, are high relative to frequencies used by other AEM methods. This relatively high frequency limits depth penetration.

13.2.4.3 AEM DATA CORRECTION: VELOCITY.

Because of the time constant of the EM receiver and the speed of the aircraft, AEM measurements are not representative of discrete points on the earth's surface. The velocity correction corrects for signal stacking due to the receiver time constant and aircraft speed. Without this correction anomaly shape is flattened, and the apparent depth of anomaly sources will be greater than their true depth. Velocity corrections to AEM data are specific to the particular systems and may be performed by spectral analysis (Bartel and Becker, 1990).

13.2.5 Radioactivity

13.2.5.1 EARTH'S RADIOACTIVITY.

All rocks and soils are radioactive due to at least 20 naturally occurring radioactive elements. Because they are the relatively abundant, uranium (U), thorium (Th), and potassium (K) (Table 13.3) are the only important nuclides from a geophysical perspective (Bristow, 1983). Uranium and thorium are estimated by monitoring two of their respective daughter products, ^{214}Bi and ^{208}TI. Parts per million (ppm) are often used to indicate equivalent uranium and thorium concentrations and percent for more abundant potassium.

Gamma radiation is positioned at the high-frequency end of the electromagnetic spectrum. Gamma rays, which have different characteristic energy levels, are emitted from certain radioactive nuclides during nuclear decay. By measuring gamma rays, the concentration of U, Th, and K in the earth's surface can be determined. The particular U and Th daughters used are selected because their characteristic gamma

TABLE 13.3 Geophysically Important Isotopes

Isotope	Detection	Percent of Element	Energy Level (Mev[a])
^{238}U	Indirectly, via daughter ^{214}Bi	99.7	1.70
^{232}Th	Indirectly, via daughter ^{208}Tl	100	2.62
^{40}K	Directly	0.012	1.46

[a] Million electron-volts

rays are more easily discriminated from other gamma rays in the spectrum and because of their high energies.

In addition to radiation originating in the crust, there is always a background radiation level from cosmic radiation as well as radiation from radium in water and radon gas in the air. These background values vary with location and topography and must be measured and removed during data processing.

Gamma radiation is strongly attenuated by most materials, including water, rock, overburden, and vegetation. As a result of this attenuation, 90% of the radiation originates in the top 30 cm of rock, and any overlying bodies of water, overburden, thick vegetation, or water-saturated overburden severely attenuate the gamma radiation from the bedrock. The gamma radiation may also originate from overburden that is not in situ complicating interpretation, particularly in glaciated terrains.

13.2.5.2 RADIOMETRIC DATA COLLECTION.

Data collected by aircraft equipped with a gamma ray spectrometer can be used to produce multiband imagery similar to satellite remote sensing imagery. Gamma ray spectrometers count and sort electrical signals produced in a photomultiplier tube from *scintillations*, or tiny flashes of light, emitted in sodium iodide detectors when they absorb gamma radiation. The electrical pulses are sorted by their amplitude into channels and counted. Sodium iodide crystal volume is maximized to increase the number of gamma rays intercepted to improve the signal/noise ratio.

When airborne spectrometer surveys are designed, the variation in detection footprint of the gamma ray spectrometer with elevation must be considered. A single spectrometric reading indicates the averaged surface radiation over the entire detection footprint, typically composed of a variety of materials. As the sensor elevation increases, the dimension of the footprint increases proportionally and the area of the footprint increases as the square of the elevation. For example, at an elevation of 135 m, 70% of the measured gamma radiation originates from the area swept by a circle 200 m in diameter beneath the aircraft. Clearly, high-resolution gamma ray spectrometry imagery requires surveys to be conducted at low terrain clearance. If the ground footprint width is smaller than the interline spacing, radioactivity of areas between the lines will not be measured and the data may be uncorrelated from line to line. In this case interpolation of the line data to produce continuous grid coverage is not recommended, due to the lack of correlation between the lines, and color strip or profile plots are recommended for display of the data. Line spacing will typically vary from 75 m (elevation 50 m) for a detailed low-altitude helicopter survey to 5000 m for a regional fixed-wing survey. As with most airborne geophysical surveying, optimum line direction is perpendicular to strike to maximize sampling density across geological features. Navigation methods are the same as for aeromagnetic surveys. To detect variations in background radiation, upward-looking sodium iodide crystals are utilized to measure radiation from radon.

13.2.5.3 RADIOMETRIC DATA CORRECTION.

In addition to the instrumentation, related corrections such as dead-time corrections, and energy calibration, a number of specialized correction are required for airborne gamma ray spectrometry data (International Atomic Energy Agency, 1991; Grasty and Minty, 1995).

A. Removal of Background Radiation. Background radiation can originate from (1) radioactivity of the aircraft and equipment, (2) cosmic radiation, or (3) radon gas in the atmosphere. Methods such as upward-looking sodium iodide detectors and the spectral ratio technique are used to correct for the background radiation, which is then removed from the survey data.

B. Calculation of Effective Height. Since the airborne count rates depend on the density of the air between the detector and the ground, they will also depend on air temperature and pressure. This correction is particularly important for surveys flown with constant terrain elevation in rugged terrain.

C. Spectral Stripping. For a number of reasons, counts may be recorded in an energy window that is too low. For example, high-energy photons from a Th source may be counted in lower-energy U or K windows. Causes for counts incorrectly measured in low-energy windows include the low resolution of sodium iodide detectors, incomplete absorption of photons, and Compton scattering. Compton scattering occurs when gamma rays collide with electrons as they pass through material imparting some of their energy to the electron. These effects are corrected through application of spectral stripping coefficients (a, b, g, α, β, γ) which are shown in schematic form in Figure 13.4. The spectral stripping coefficients for an instrument are determined through the use of radioactive concrete calibration pads. From mea-

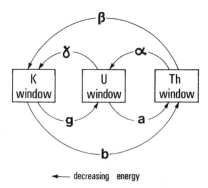

Figure 13.4 How the spectral stripping coefficients α, β, and γ are used to remove the interference between the energy windows for each radioelement. The upward stripping factors a, b, and g are generally small and are often ignored. (From Killeen, 1987.)

surements on these pads and matrix algebra, the coefficients can be determined (Grasty and Minty, 1995).

D. Altitude Correction. Corrections to the count rates are required to remove the attenuating effect of the atmosphere. A simple exponential correction determined by flights over a calibrations range is adequate.

E. Conversion of Corrected Survey Data to Ground Concentrations. Test flights of the survey system over ground with known concentrations are used to determine the sensitivity of the instrument. Survey results can then be corrected to reflect ground concentrations, allowing comparison of data between surveys collected with different instrumentation and methodology at different times.

13.3 DATA-SPECIFIC ENHANCEMENTS

Once the data are corrected, numerous enhancements are available to improve our visualization of image data. Although enhancements to improve data visualization are covered in general in this volume, it is useful to discuss the enhancements most appropriate for each type of geophysical data. In each type of geophysical data the signature of different geological sources has different spectral and spatial characteristics. Because of these different data characteristics, different methods are often required to enhance the imagery generated from those data.

13.3.1 Gravity and Aeromagnetic Data Imaging and Enhancements

For discussion of enhancements, the potential field data (gravity and magnetic) are sufficiently similar that many enhancement techniques are applicable to both. The crustal field component of the gravity and magnetic fields comprises a large number of superimposed contributions from source bodies distributed throughout the Earth's crust. This complex mix of individual anomalies is often difficult to interpret due to the overlap and the wide range of *wavenumber* (spatial equivalent of frequency, units are cycles/unit distance) and *amplitude*. Imaging this complex mix of anomalies is difficult because most display techniques emphasize a particular amplitude and wavenumber range. To facilitate interpretation, enhancement techniques are used to separate anomalies with particular amplitude and wavenumber characteristics from the remainder of the data.

It is important to remember when interpreting images generated by interpolating point or line data onto a regular grid that false anomalies, or artifacts, can be generated by the interpolation process. Since remote sensing images are usually generated directly from the data without interpolation, the remote sensing specialist may not be cognizant of these features. Interpolation artifacts result from attempting to interpolate undersampled data or data collected at a nonuniform sampling density, as in the case of aeromagnetic surveys, to produce a regular grid of data values. Where there are fewer than two samples per cycle, *aliasing* can occur, which is an ambiguity in the wavenumber represented by sampled data. A typical example of interpolation artifacts is evident in the image representation of a linear magnetic anomaly that intersects the flight line direction at 45° (Figure 13.5). If the flight lines are too far

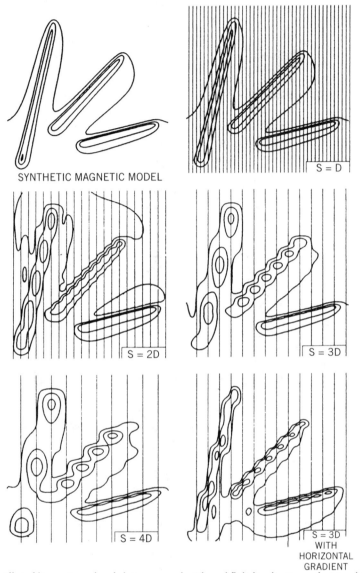

SYNTHETIC MAGNETIC MODEL

S = D

S = 2D

S = 3D

S = 4D

S = 3D
WITH
HORIZONTAL
GRADIENT

Figure 13.5 Effect of line spacing and angle between anomaly strike and flight line direction on the interpolated grid. The line spacing S as a function of height above the magnetic source. (From Hogg, 1987.)

apart, the linear feature will be represented as a stepped feature. Since interpolation artifacts are usually high-wavenumber features, they usually become more trouble-some when high-wavenumber image enhancement techniques are used. In general it is advisable to display the original measurement points or lines on geophysical images to help detect anomalies not centered on measurement locations.

13.3.1.1 GRID FILTERING METHODS.

A number of filters and transforms are used to enhance gravity and magnetic data. Some of these filters are designed from potential field theory and others are defined subjectively to enhance certain amplitude/wavenumber ranges of interest. Filters can be applied in the space domain by convolving a filter kernel with the original data or, more commonly, by converting the data to the frequency domain and applying a transfer function equivalent to the desired filter (Clement, 1973). Filtering in the frequency domain is more efficient because very large convolution kernels are required to accurately discriminate the low-wavenumber anomalies. The data are usually converted to the frequency domain using fast Fourier transform (FFT) software (Cooley and Tukey, 1965). The data must be input to the FFT in real or floating-point grid format because integer image data have inadequate resolution to represent the field accurately. A description of some of the specific filters commonly applied to both magnetic and gravity data follows. Transfer functions for some common filters can be seen in Figure 13.6.

A. Continuation.   Potential field theory allows field observations measured at one datum level to be continued upward or downward to another level. Since potential

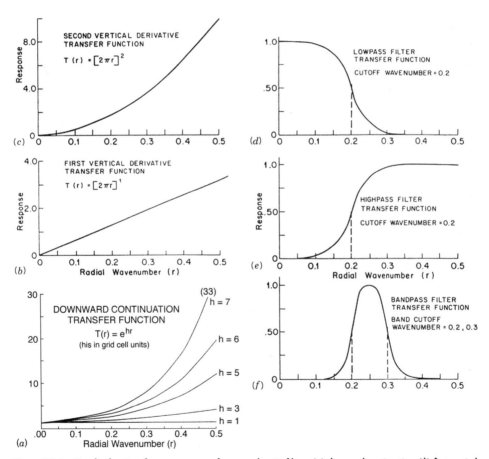

Figure 13.6   Transfer functions for some common frequency-domain filters: (*a*) downward continuation; (*b*) first vertical derivative; (*c*) second vertical derivative; (*d*) low-pass filter; (*e*) high-pass filter, (*f*) bandpass filter. Filter response is plotted against wavenumber. A wavenumber of 0 would represent an anomaly of infinitely long wavelength, while a wavenumber of 0.5 represents an anomaly with a wavelength twice the grid cell size.

fields vary with inverse distance from the source, upward-continued data contain low-amplitude and long wavelength data, whereas the reverse is true in downward continued data (Figure 13.6*a*). Caution must be taken with downward continuation since enhancement of the short-wavelength anomalies can also enhance noise in the data. The signal/noise ratio of the data ultimately limits the distance to which data can be continued downward.

B. Derivatives.   Derivative or gradient filters (Figure 13.6*b* and *c*) enhance the high-wavenumber component of the data by representing the rate of change of amplitude in a given direction. First-order derivatives in the horizontal and vertical directions are commonly used. The first vertical derivative is generally equivalent to measured vertical gradient data, which has the useful characteristic that geological contacts in areas with vertical structure can be delimited by following the zero level in the data. Second-order derivatives enhance very high wavenumber components in the data, and their utility is often limited by noise in the data. Horizontal derivatives are less often used, as they locate maxima on the flanks of anomalies in the total field data, somewhat complicating interpretation. Horizontal derivatives are used more often for gravity data than for magnetic data (Sharpton et al., 1987).

C. High-, Low-, and Bandpass Filters.   Particular ranges of wavenumber can also be extracted by empirically designed filters. These filters must be designed carefully to avoid creation of undesirable artifacts in the data. When designing filters a gradual roll-off must be used, rather than abrupt truncation, to eliminate oscillations in the filtered data. Low-and high-pass filters (Figure 13.6*d* and *e*) retain information below or above a specified cutoff frequency, whereas bandpass filters (Figure 13.6*f*) retain information in a range of wavenumbers between an upper and a lower cutoff wavenumber. *Regional/residual separation* is an adaptive form of filtering where the cutoff wavenumber(s) are based specified on the wavenumber range of the regional and residual components of the field being measured. This method is commonly used to separate the core field from the upper crustal anomalies of interest.

13.3.1.2   AEROMAGNETIC-SPECIFIC GRID FILTERING.

Due to the dipolar nature of the magnetic field, some filter and transform operators have been developed specifically for application to these data. These operators are discussed in detail in a classic paper by Bhattacharyya (1965).

A. Reduction to the Pole.   Due to the dipolar nature of the Earth's magnetic field, anomalies from identical sources change as the inclination of the field changes with latitude. For example, the 0° inclination of the field at the equator will produce a null directly over a magnetic body with flanking anomalies to the north and south rather than a single anomaly directly over the source as is observed at the magnetic poles where the inclination is 90°. The reduction to the pole operator (Baranov, 1957) transforms data to appear as it would if collected at the magnetic pole, which greatly simplifies interpretation (Figure 13.7). The result of the filtering operation is sometimes called a pseudogravity map.

B. Apparent Susceptibility Transform.   The apparent susceptibility transform applies a series of filter operators to produce an estimate of the average magnetic susceptibility of a vertical-sided prism extending to infinite depth located under each grid cell. The transform makes a number of assumptions regarding the uniformity

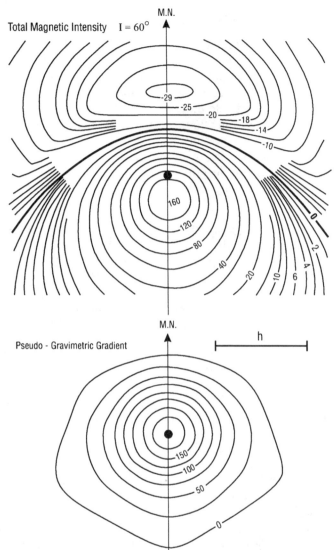

Figure 13.7 The upper figure is the magnetic intensity anomaly from a sphere (solid circle) at a depth of h below the surface magnetized by the Earth's field at an inclination of 60°. After reduction-to-the-pole, which corresponds to an inclination of 90°, the anomaly is symmetrically located over the source with the maximum directly over the body. (After Baranov, 1957.)

magnetic field, lack of remanent magnetization in the prisms, and the structural orientation of the geology. If these assumptions are correct, the results can be very useful; however, results can be misleading if the actual situation does not fit the assumptions. A number of variations of the basic transform are available to fit different geometrical scenarios.

13.3.1.3   IMAGING POTENTIAL FIELD DATA.

Although image generation and enhancement are covered in some detail in this volume, some additional comments are required for potential field data. Imaging enhancements are often used in combination with the grid-based enhancements just described; for example, a calculated vertical-derivative image may be shaded.

A. Interpolation of Potential Field Data to a Regular Grid. Before images can be generated, survey data must be corrected and interpolated onto a regular grid of floating-point values. The interpolation method typically must handle an order-of-magnitude-greater data density along lines than across lines. Interpolation methods commonly used for potential field data, which is inherently smoothly varying, are cubic spline (Evenden, 1989) and minimum curvature interpolation (Webring, 1981; Mendoca and Silva, 1995). Best results are obtained when the grid cell size is larger than 20% of the interline spacing (i.e., if the line spacing is 1 km, the cell should be larger than 200 m). In general the grid size of geophysical data is an order of magnitude larger than remote sensing data.

B. Quantization of Data into Image Form. Image generation of potential field data involves generation of an integer grid from the floating-point grid by quantization. The number of quantization levels are determined by the output devices and end-user requirements, but typically 14 to 256 levels are created. The following types of quantization are commonly used for potential field data:

1. *Linear quantization:* equal intervals between minimum and maximum
2. *Equal-area or histogram-equalized quantization:* equal number of pixels for each level
3. *Histogram specification:* customized to match data characteristics (e.g., vertical magnetic gradient where most data are concentrated about zero)

C. Color Selection for Image Display. Potential field data are most often displayed using a smoothly varying palette. Most people intuitively relate to a color scheme utilizing blue for low values through green, yellow, orange, and red for increasing intensity. Avoid palettes where the colors change erratically. Colors may be specified using any of the system definition systems described in this volume. The data are typically quantized into a minimum of 16 to the usual maximum of 256 levels. Any number over 40 is usually adequate for a smoothly varying palette.

Precise discussion of color requires that a representation scheme be specified. Common color representations, including red–green–blue (RGB), cyan–magenta–yellow (CMY), and intensity–hue–saturation (IHS), are described in detail in this volume. Since the human visual system tends to group areas with the same color hue, the same image may be perceived differently when a different color palette or quantization method is used. Gray-tone palettes eliminate this effect and are particularly useful when directional trends of anomaly patterns and textures are being studied.

13.3.1.4   IMAGE ENHANCEMENT FOR POTENTIAL FIELD DATA.

A. Shading. Shading is useful for enhancing small, high-wavenumber anomalies without the risk of creating processing artifacts associated with filtering. Shaded-relief images are intensity images of the calculated reflectance of a calculated surface defined by the original data intensity illuminated by an imaginary light source (Figure 13.8; Dods et al., 1987; Broome, 1990). The method is most useful for imagery with significant high-wavenumber content such as aeromagnetic data. The shading effect

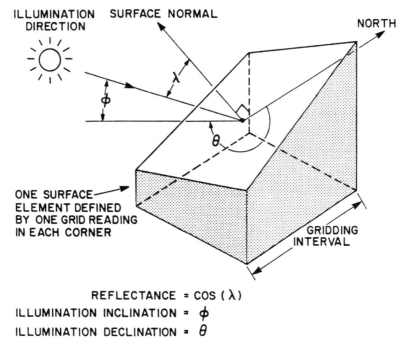

Figure 13.8 Geometry used to determine reflectance for one grid cell. The resulting grid of reflectance values is used to generate a shaded magnetic relief image.

enhances features striking perpendicular to the pseudoillumination direction, allowing selective enhancement of anomalies oriented in particular directions. Gray-tone palettes give optimum results. The disadvantage of the shading method is that regional amplitude information is lost.

B. Color-Shaded Images. Shading of color intensity images is the preferred display method for aeromagnetic data because both regional amplitude information and small high-wavenumber anomalies are visible. Hue on these images reflects the intensity, while the saturation is controlled by the reflectance to create shading. The resulting image is perceived as a three-dimensional colored surface (e.g., Figure 13.13*d*). On normal color-intensity images, anomalies with amplitudes below the quantization interval are invisible, but shaded-color images retain this information.

13.3.2 Airborne Electromagnetic Data Enhancement

Many of the enhancement methods used for potential field imagery can be applied to AEM data. Although the multichannel nature of modern AEM survey data presents many imaging possibilities, single-component imagery is generally used, and stacked-profile maps (Figure 13.19*d*; color insert) or multitrace profile strips are used for more detailed interpretation. Profiles allow analysis of the anomaly shapes in the different channels to be compared easily.

13.3.2.1 INVERSION PRODUCTS.

A number of specialized image products have been developed for most commercial AEM systems. These derivative products are typically developed through a combination of theoretical understanding of system characteristics and validation with field examples. Due to the proprietary nature of these methods, detailed descriptions are often not available. Figure 13.19*e* and *f* (color insert) show lake depth and conductive layer thickness estimates generated from a detailed helicopter-borne six-frequency AEM survey using singular-value deconvolution (Kovacs and Holladay, 1990).

13.3.3 Gamma Ray Spectrometry Data Enhancement

13.3.3.1 GAMMA RAY SPECTROMETRY RATIO IMAGES.

The ratios of Th/K, U/K, and U/Th are commonly calculated in addition to the individual radioelement concentrations. Ratios are useful for lithology discrimination and alteration detection because they minimize the effects of attenuation due to water, overburden, and so on. For example, K/Th ratio images help the interpreter to distinguish between potassium anomalies caused by variations in bedrock exposure and anomalies due to alteration. Anomalies due to alteration have a high K/Th ratio, due to potassium enrichment combined with little change in the thorium content, whereas anomalies due to good bedrock exposure or high overall radioactivity will not generate a high K/Th ratio. Ratio images tend to be noisier than single radioelement images because the ratioing sums the noise components of the two-component images.

13.3.3.2 TERNARY RADIOELEMENT PLOTS.

These color images are optimized three-component ratio images. Each radioelement is assigned a primary hue, and intermediate colors reflect mixing of the primary hues and radioelements. Areas with consistent radioelement signatures, possibly reflecting consistent lithology, tend to appear in consistent colors. Color saturation may also be controlled by the overall radioactivity level to enhance areas with high radioactivity which, due to better signal/noise ratios, have more accurate radioelement ratios (Broome et al., 1987). The ternary radioelement image shown in Figure 13.23 (see the color insert) demonstrates the use of this method to distinguish the zones with consistent radioelement signatures in a composite granitic intrusion.

13.3.3.3 HEAT GENERATION IMAGES.

Since most of the heat generated in the crust is due to radioelement decay, the radioelement concentrations can be used to calculate the radiogenic heat generation in the crust (Richardson and Killeen, 1980).

13.3.4 Composite Images

Each type of geophysical data has its strengths and weaknesses for particular applications. With a single method, the geology and structure of an area can rarely be discriminated; however, in combination, multiple data sets complement each other and can provide significantly more information (Broome, 1990). For example, regional gravity data may be useful for identifying areas of low density that are likely to be granitic, but the resolution to define the contacts of the intrusion with the denser country rock is lacking. Since aeromagnetic data can often resolve the location of the contacts, a combined plot showing both the magnetic and gravity fields is useful (Figure 13.12*d*; color insert). Gamma ray spectrometry data are also complementary, as they allow detection of subtle radioelement ratio variations that can be used to characterize different intrusive phases.

Composite images are most easily produced by assigning different component colors (e.g., red, green, and blue) to images generated from three original data sets. These simple color composite images are easily interpreted because each data set is clearly associated with a unique color. Where images that illustrate the variations of more than three data sets are required, statistical methods such as principal components analysis can be used to combine the information of more than three data sets into three or fewer images. These component images can then be displayed using the standard three-component color method, with the result that the correlations and variability of all the original data sets can be visualized. Although the principal components method is an excellent way of merging information from a large number of data sets, anomalies and the features visible in the resulting composite images can be difficult to relate to the original data sets since each component color actually represents a linear combination of the original data sets. Composite images generated by assigning each data set to one parameter of the common color models (RGB, CMY, HLS) are usually effective and easy to interpret. Reference line work overlays such as geological contacts and contour intensity maps can sometimes be used effectively to add another component to the images.

Geophysical data can also be combined with remote sensing data (Harris et al., 1994). Figure 13.9 (see the color insert) is an IHS composite image generated by combining *Landsat* TM band 5 (intensity) with total magnetic intensity (hue). Saturation was fixed at 100%. In areas where base maps are not available, remote sensing data can act as a substitute base during interpretation of the geophysical data.

13.4   RECOGNITION OF GEOLOGICAL FEATURES

Many geological features have characteristic signatures in the geophysical imagery. Characteristics such as amplitude, predominant trend direction, texture, and wavenumber distribution can all assist the geoscientist in detection of zones of consistent or contrasting lithology and structure. Typically, the image characteristics useful for detecting a particular lithology or structure are obscured by other data components. In this case selected enhancement techniques can be applied to help visualize the components of particular interest.

The remaining part of this chapter will provide examples of the expression of lithology and structure on imagery produced from the three types of geophysical data being discussed. Examples have been selected where the feature of interest is particularly clearly represented. In actual applications, the imagery will usually have less perfect expressions of these features, due to interference from overlapping features and other effects, such as metamorphism, regional structural deformation, and lack of compositional uniformity. In other cases the geophysical expression of known lithological variations and structure is simply absent, or areas mapped as having consistent lithology will have widely varying geophysical responses.

A key point to consider when contemplating the use of geophysical imagery is whether sufficient contrast is expected in the rock properties controlling the measured parameter to produce a visible effect on the imagery. For example, detection of strike-slip faulting using aeromagnetic imagery in a region of uniform lithology will probably be impossible, due to the lack of marker horizons indicating displacement. In cases where one data set is inadequate to distinguish or delineate features of interest, composite images generated using a combination of different data types may provide improved discrimination of the features.

For all methods, the sampling interval, survey elevation, and instrument resolution must be appropriate to allow the generation of images suitable for the type and scale of the investigation. However, images generated from each of the geophysical methods must also be interpreted with careful consideration of the causative mechanism. As with other remote sensing methods, ground follow-up with sampling and subsequent rock properties measurement is essential to accurate interpretation.

For each of the geophysical methods, some observations follow on:

1. Characteristics and interpretation guidelines
   a. Resolution of the data: wavenumber content
   b. Sensitivity to the depth of the source material
2. Applications
   a. Ease of interpretation
   b. Lithological discrimination
   c. Structural interpretation
   d. Direct ore body detection

13.4.1 Gravity Imagery

13.4.1.1 CHARACTERISTICS AND INTERPRETATION GUIDELINES.

The gravity anomalies of interest for geological interpretation are caused by variations in rock density at different levels in the earth. The density of different rocks and minerals vary between 1 and 20 g cm^{-3}. The density of a rock is based on its composition and, in a given area, is likely to vary from 2.5 to 3.5 g cm^{-3}. Since the composition of different rock types is variable, their densities also vary enough that there is considerable overlap in the density of different rocks, usually making it impossible to identify lithology from density alone. Density is a bulk property of the rock and relatively unaffected by tectonic, structural, and metamorphic events compared to magnetic susceptibility and conductivity.

Accurate interpretation of the gravity field requires an understanding of these density contrasts. Although type densities are published in most geophysical texts, the range of variation is large for a given rock type. Table 13.4 provides a summary of the density ranges of a number of common rocks and minerals. Although exceptions can be found, some general rules can be given:

1. Sedimentary rocks have lower average densities than those of igneous and metamorphic rocks.

TABLE 13.4   Density of Some Common Earth Materials

Material	*Density (g cm⁻³)*
Igneous rocks	
Rhyolite	2.3–2.7
Granite	2.5–2.8
Gabbro	2.7–3.5
Anorthosite	2.6–2.9
Metamorphic rocks	
Granulite	2.5–2.7
Gneiss	2.6–3.0
Marble	2.6–3.9
Amphibolite	2.9–3.0
Sedimentary rocks	
Limestone	2.6–2.7
Coal	1.2–1.5
Sandstone	2.0–2.5
Minerals	
Biotite	2.7–3.2
Quartz	2.5–2.7
Gypsum	2.2–2.6
Galena	7.5
Magnetite	5.2
Other materials	
Unconsolidated material	1.7–2.3
Rock salt	2.1–2.6
Petroleum	0.6–0.9

Source: Telford et al. (1978).

2. Intrusive igneous rocks are more dense than extrusive.
3. Basic rocks are denser than acidic rocks.
4. Density usually increases slightly with metamorphic grade.

Interpretation of gravity data can be direct or indirect. Direct interpretation involves taking advantage of significant density contrasts between rocks and minerals present in an area to locate rocks of interest. An example would be the detection of massive sulfide deposits within generally less dense host rock (Pemberton, 1987). Unfortunately, direct detection is usually impossible because the density contrasts are inadequate and/or the bodies of interest are too small.

More often, indirect interpretation is used. Indirect interpretation takes advantage of density differences that are inadequate to discriminate lithology but large enough to cause detectable gravity field variations. Combined with a field program to sample rocks and measure their density, these subtle trends aid in the refinement of geological contacts and structural mapping.

Wherever possible, a program of field sampling for density measurements should be part of the interpretation process. Density measurements from drill cores are always preferred over surface sample measurements because the subsurface samples are unaffected by weathering. Since rock samples for density measurement are rarely available from deep in the earth, interpreters must often rely on extrapolation of surface geology and textbook density values. Since overburden is less dense than bedrock, overburden thickness can significantly affect the gravity field; however, the gravity method is rarely used for surficial geology applications because of resolution limitations for mapping near the surface and because field variations from bedrock density contrasts obscure overburden contributions.

A. Resolution of the Data: Wavenumber Content. Regional gravity data are usually generated from ground measurements spaced 3 to 12 km apart, a resolution more suitable for regional studies. However, the lack of resolution is only a function of the sampling interval, and detailed surveys can produce imagery suitable for claim-scale studies. Although gravity data produce inherently smoother and less detailed data than other types of imagery, many geological features are represented, and useful information is provided on the lithology and structure at deep levels in the crust.

B. Depth of the Anomaly Sources. The gravity method has the greatest depth penetration of the three methods considered, and the source of gravity anomalies can originate deep in the crust and mantle of the earth. In general, the intensity of a given gravity anomaly will diminish inversely with distance from the source, whereas the corresponding magnetic anomaly from the same body will diminish inversely with the square of the distance. This is a simplification because the fall-off rates of anomalies are dependent on the body geometry, but it is generally true. A related effect is that anomalies from deep bodies are not obscured by anomalies from near-surface bodies to the same degree as with magnetic data (Kane and Godson, 1987). Due to the ambiguity of potential field interpretation, the shape and intensity of a given anomaly do not indicate a unique depth for the source (Figure 13.10).

13.4.1.2 APPLICATIONS OF GRAVITY IMAGES.

A. Ease of interpretation. Interpretation of gravity imagery is relatively easy compared to aeromagnetic data but more complex than interpretation of most remote sensing images because the gravity field is composed of contributions from an unknown three-dimensional distribution of bodies. The monopole field results in positive anomalies positioned directly over bodies with positive density contrasts. Density is a bulk property controlled by composition; therefore, the gravity field is not greatly affected by many geologic processes. The gravity field is composed of anomalies due to sources at a range of depths, but for a given body geometry, depth is related inversely to wavenumber. Ambiguity is a problem with all potential field interpretation. An infinite number of density–geometry–position combinations can produce a given anomaly. In Figure 13.10, three different configurations of depth, body geometry, and density are shown that produce the same gravity anomaly.

B. Lithological Discrimination. Since density is a bulk property of the rock, it is not affected by metamorphism, phase change, and recrystalization to the same degree as is magnetic susceptibility, which may change by several orders of magnitude. Unfortunately, the narrow range of density found in different rock types limits the utility of gravity for lithological discrimination. In addition, bedrock geology maps usually indicate only the surface lithology, whereas the gravity field is indicative of density variation in the entire crust. The low spatial resolution of available regional data also limits the applicability of the data to regional bedrock mapping.

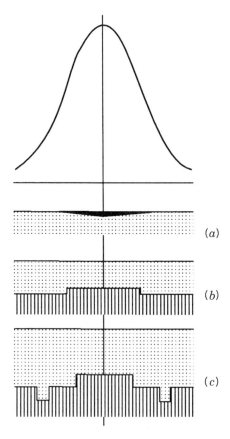

(a)

(b)

(c)   **Figure 13.10** Ambiguity in gravity interpretation. The rock units in the three geologic cross sections can be assigned densities so that the resulting anomaly profile is the same. This ambiguity greatly complicates the interpretation of gravity and magnetic data.

C. Structural Interpretation. Due to its excellent depth penetration, one of the major applications of regional gravity data is tectonic-scale structural investigation (Thomas, 1987). A skilled interpreter can relate the shape of anomalies to body geometry, dip of contacts, and the depth of source. Less skilled interpreters will find use of quantitative modeling software useful for understanding interpretation of gravity data. The resolution of available data often limits the utility of gravity data for detection of fault offsets and precise location of contacts. Density contrasts must be adequate to produce significant field variations. Granite bodies are often significantly less dense than their host rock, and their geometry can often be resolved from gravity data.

D. Direct Ore Body Detection. Direct ore body detection is usually impractical. Despite the existence of significant density contrasts for ore minerals, their quantity is often insufficient to produce an anomaly that is large enough to be discriminated from topographic effects and overburden effects. Although economic ore bodies may be quite large, the ore minerals are often so disseminated that the overall density contrast between host rock and economic ore is often quite small. For example, the No. 6 deposit in the Brunswick mining camp, New Brunswick, Canada, consisting of 12 million tons of massive sulfide ore with an average density of 4.4 g cm^{-3}, generated a Bouguer gravity anomaly of only 4 mgal (Pemberton, 1987). The key to direct detection of any ore body using gravity data is a significant density contrast between the ore and the host rock and a large-enough ore body.

13.4.1.3 INTRUSIVES.

The low density of granitic intrusives (ca. 2.6 to 2.65 g/cm^3) relative to many crustal rocks results in negative anomalies. Figure 13.11a (see the color insert) shows a color Bouguer gravity image of a large part of the Rae Province, Northwest Territories, Canada (Figure 13.12; Broome, 1990). The image is dominated by a 125 km $\times$ 200 km elliptical gravity low of about 30 mgal. This anomaly is centered on a zone mapped as granite and granodiorite with an average density of 2.64 g cm^{-3} in host rock of average density 2.73 g cm^{-3}. Three-dimensional modeling of this anomaly using the measured densities indicates that the anomaly is compatible with a body of granitic rock with a convex-upward shape extending to a depth of 8 to 10 km.

Figure 13.13 is a gray-tone image of the Bouguer gravity intensity over the sub-Paleozoic portion of the Flin Flon/Snow Lake belt (Leclair et al. 1997). The locations of granitic and gabbroic (melagabbro/diorite/peridotite) intrusions shown on the figure were determined using a combination of aeromagnetic data interpretation and drilling. A gray, shaded-relief aeromagnetic image covering the same area can be seen in Figure 13.17. Based on the relatively high average density of gabbroic rocks (see Table 13.4), one would expect a positive gravity anomaly over the gabbroic intrusion. The absence of this anomaly is probably due to an inadequate number of gravity stations, shown by white dots on Figure 13.13, which results in undersampling of the gravity field. Since there are no gravity stations within the area underlain by the gabbroic intrusive, no anomaly is indicated on the image. This example demonstrates the importance of knowing the locations of gravity stations during interpretation.

The large elongated NW/SE-striking gravity low is associated with a complex granitic intrusive, part of which is named the Cormorant Batholith. Due to its size, the relatively low density of the granitic rocks forming this intrusive produces a coincident gravity low. The diffuse character of the gravity low is partly due to the widely separated gravity stations, but is also typical of gravity imagery. Coincidence of a gravity low and the concentric aeromagnetic anomalies shown in Figure 13.17 often indicates a granitic intrusive.

13.4.1.4 GREENSTONE BELTS.

Greenstone belts are generally denser than the surrounding granitic and gneissic rocks, due largely to the high proportion of mafic volcanic rocks. Greenstone belts therefore often are associated with positive gravity anomalies. Figure 13.14 shows (a) a simplified geological map of northeastern Botswana and (b) a contoured Bouguer gravity map of the same area. The general correspondence between the known greenstone belts (A, B) and gravity highs can be seen as well as possible extensions of the belts (C, circles) interpreted from the gravity data (Reeves, 1987).

13.4.1.5 GRANULITE TERRAIN.

Although definitive classification of metamorphic grade from density is generally impossible due to the complex history and lithological variability of rocks, density variations associated with changes in metamorphic grade can result in gravity anomalies. In the Bouguer gravity intensity image shown in Figure 13.11a (color insert),

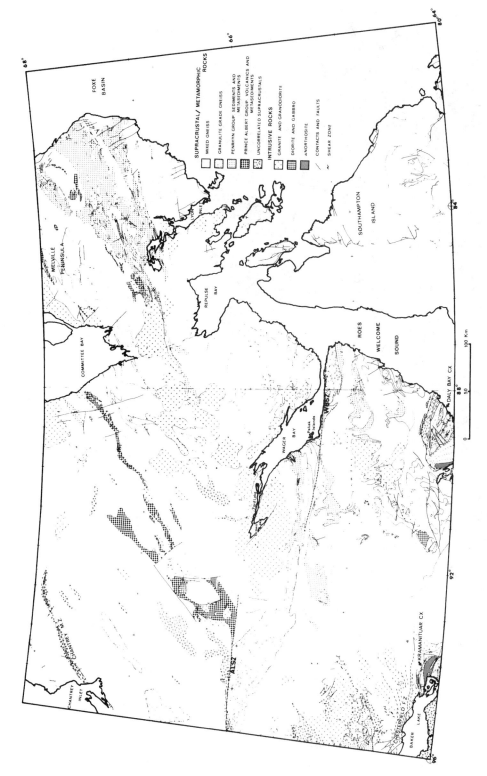

Figure 13.12 Simplified geology of the Rae Province, Northwest Territories, Canada.

671

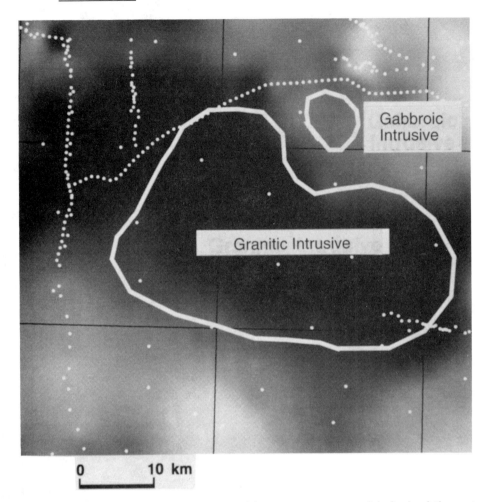

Figure 13.13 This roughly 65 km-square gray-tone image of the Bouguer gravity intensity is located in the sub-Phanerozoic portion of the Flin Flon/Snow Lake Belt, Manitoba, Canada. Gravity highs and lows are shown in white and black, respectively, and gravity station locations are marked by white dots. The locations of granitic and gabbroic (melagabbro–diorite–peridotite) intrusions were determined using a combination of aeromagnetic data interpretation (Figure 13.18) and drilling. A large diffuse gravity low is roughly coincident with the relatively low density granitic intrusive, but an expected gravity high, coincident with the gabbroic intrusive, is absent, due to inadequate sampling of the gravity field.

no obvious gravity anomaly is associated with the granulite terrain mapped south of the Amer and Wager Bay shear zones (Figure 13.11*e*; Henderson and Broome, 1990). In Figure 13.11*b* the vertical derivative of the gravity gradient has been calculated to enhance high-wavenumber features. A moderate gradient anomaly can now be seen directly over the granulite terrains. The noncontinuous nature of the gradient anomaly is due to inadequate sampling of the gravity field. The anomaly was not apparent in the Bouguer gravity intensity image because its maximum 1-mgal amplitude is lost during quantization because of a coincident, approximately 30 mgal,

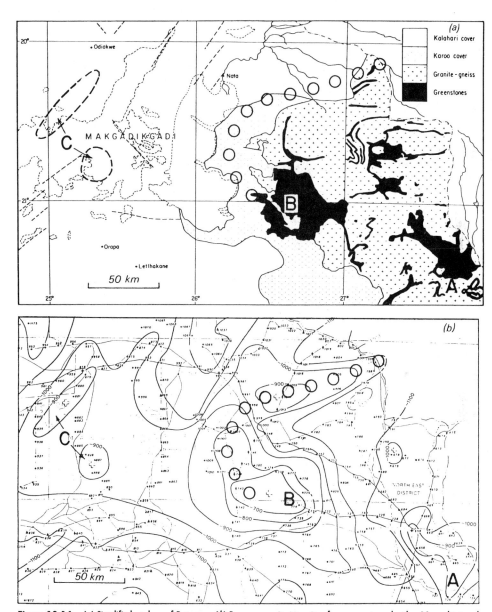

Figure 13.14   (*a*) Simplified geology of Botswana, (*b*) Bouguer gravity intensity of area *a* contoured with a 10-mgal interval. The location of the Tati Greenstone Belt (A), Matasitama Greenstone Belt (B) with its extenuation (circles) interpreted from the gravity data (B), and interpreted greenstone belts (C) are shown. (From Reeves, 1987.)

negative regional anomaly. This large regional anomaly is proposed to be caused by a large granitic intrusive (see Section 13.4.1.3). The vertical gradient image enhances high-wavenumber anomalies, such as the granulite anomaly relative to low-wavenumber anomalies such as the anomaly due to the granitic intrusive.

13.4.2 Aeromagnetic Imagery

13.4.2.1 CHARACTERISTICS AND INTERPRETATION GUIDELINES.

The crustal magnetic field is due to variations in the magnetization of crustal rock caused by relatively low concentrations of magnetic minerals. Magnetization can be *soft*, or *induced magnetization*, due to the presence of magnetic minerals with high *magnetic susceptibility* in the Earth's geomagnetic field. The magnetic susceptibility of rock typically varies from 10^{-6} to 10^{-3} SI. Common strongly magnetic minerals are, in decreasing order of magnetization and importance, magnetite, pyrrhotite, and ilmenite.

Less commonly, magnetization can also be *remanent magnetization*, which is permanent magnetization locked in place in a historical field direction different from the induced field. The presence of significant remanent magnetization complicates interpretation and modeling in particular. Most significant is *thermoremanent magnetization*, which is caused by cooling from above Curie temperature (575°C for magnetite).

The magnetic susceptibility of rocks and minerals is determined by their magnetic mineral content. All materials can be classified into three groups, *diamagnetic, paramagnetic,* and *ferromagnetic*, according to their magnetic properties. Diamagnetic substances have weak negative magnetic susceptibilities. Graphite, gypsum, marble, quartz, and salt are common diamagnetic materials. All substances with positive magnetic susceptibilities are called paramagnetic. The ferromagnetic substances are part of the paramagnetic group, which contains groups of magnetically aligned atoms, or *domains*. The ferromagnetic minerals, of which magnetite is the most important, are the cause of most magnetic anomalies. Magnetite rarely represents more than 1 % of the bulk rock composition.

Table 13.5 shows the magnetic susceptibilities of common rocks and minerals. Note the large range in the susceptibilities of individual rock types and the overlap between different rock types. Clearly, direct interpretation of lithology from the magnetic field is very difficult, due to the high degree of variability and overlap. Note also, however, the tremendously wide range of susceptibility, a factor of about 10^5, compared to a factor of 2 for density variation. Direct interpretation of some geologic structures is possible, notably iron ore bodies, massive sulfides (with coincident EM anomaly), kimberlite pipes, and mafic dikes, due to their high relative susceptibilities and distinctive geometry.

A comparison of aeromagnetic and gravity imagery for a given area will usually reveal that the magnetic imagery is more complex than the gravity. This is due to both the high degree of variability in rock magnetic susceptibility and the smaller sampling interval used for aeromagnetic surveying than gravity. Indirect interpretation is more commonly used for aeromagnetic data than for gravity data. Although magnetic mineral content is related to lithology, magnetic properties alone cannot

TABLE 13.5   Magnetic Susceptibility of Some Common Earth Materials

Material	Magnetic Susceptibility k (SI units $\times$ 10⁵)
Igneous rocks	
Rhyolite	25–375
Granite	0–5000
Gabbro	100–9000
Basalts	25–18,000
Metamorphic rocks	
Schist	30–300
Gneiss	13–2500
Slate	0–3750
Amphibolite	75
Sedimentary rocks	
Limestone	3–350
Shale	6–1600
Coal	3
Sandstone	0–2000
Minerals	
Quartz, gypsum, rock salt	−1
Chalcopyrite	40
Pyrite	5–525
Pyrrhotite	125–625,000
Ilmenite	30,000–375,000
Magnetite	125,000–2,000,000

Source: Adapted from Telford et al. (1978).

identify rock type in the conventional classification system, which is based on silicate and carbonate mineralogy (Grant, 1985).

Despite this theoretical limitation, investigation often reveals many magnetic features and textures that relate closely to the mapped structure and lithology. In other areas correlation with lithology is poor. This variability can be explained by changes in magnetic mineral content due to complex oxidation, crystallization, and remanent magnetization effects related to local and regional metamorphic and deformation processes. In general, different lithologies are characterized by different magnetic mineral content; therefore, aeromagnetic data can, to some degree, map units with significantly different magnetic mineralogy. However, ground sampling must generally be used to identify the lithology. It is important to recognize that magnetic anomalies often reflect the history of the rocks as much or more than their bulk composition (Grant, 1985).

The interpretation process should include a program of field sampling and magnetic susceptibility measurement to help understand the observed anomalies. In situ susceptibility measurements are possible using compact and inexpensive magnetic susceptibility meters. Since magnetic minerals can be destroyed by oxidation in weathered rock, measurements taken from unweathered rock or drill core are preferred. More accurate susceptibility measurements and remanent magnetization are possible using laboratory equipment. Wherever possible a number of measurements should be taken from a given rock unit and averaged because magnetic mineral content can be very variable within a given unit.

As with gravity data, anomalies due to bedrock sources typically are orders of magnitude stronger than anomalies from surficial sources, thus limiting the utility of the magnetic method for mapping surficial geology. In some cases, however, the presence of high concentrations of magnetic minerals in the overburden combined with relatively nonmagnetic bedrock allows effective use of the magnetic method. One example is the use of detailed ground magnetic surveys to map placers with high magnetite concentrations (Schwarz and Wright, 1988)

A. Resolution of the Data: Wavenumber Content. Dense sampling along flight lines results in detailed images with a strong high-wavenumber content. The high variability of rock susceptibility and its sensitivity to change caused by common geological processes also contributes to the high-wavenumber content. Among the commonly used geophysical data sets, aeromagnetic images provide the most detailed information about structure and lithology. As with gravity data, the shape of anomalies can be related to body geometry and dip of contacts.

B. Sensitivity to the Depth of the Source Material. The depth penetration of magnetic data is dependent on survey parameters but ranges from the surface down to a maximum depth determined by the depth of the Curie isotherm of the magnetic mineralization (580°C for magnetite). The Curie point temperature is usually found at 30 to 40 m depth, depending on the local geothermal gradient. The effective depth penetration of the magnetic method is usually less than gravity data for the same area because of the greater rate of anomaly amplitude attenuation with distance from the source compared to gravity. The magnetic field image is therefore more indicative of near-surface sources. This is beneficial for geological mapping and mineral exploration, which tend to focus on near-surface geology. In general, the wavelength of anomalies increases and the amplitude decreases with increasing depth to the source. The rate of decrease is dependent on the geometry of the source body; for long, wide sources in an approximately linear fashion, for long, narrow horizontal bodies as the inverse square, and for equidimensional concentrated sources as the inverse cube of the distance (Hinze and Zeitz, 1987).

13.4.2.2 APPLICATIONS OF AEROMAGNETIC IMAGES

A. Ease of interpretation. Interpretation of aeromagnetic imagery can be quite complicated due to (1) the dipole field, (2) the complex superposition of anomalies, (3) the possible presence of remanent magnetization, and (4) the degree to which rock susceptibility can be changed by geological processes. The dipole nature of the field causes anomalies to vary depending on field inclination (latitude). In the northern hemisphere positive anomalies are associated with a weaker negative anomaly to the north. The magnitude of the flanking negative anomaly varies from nonexistent at the north pole to approximately equal to the positive anomaly near the equator.

Despite these difficulties in exact interpretation, many of the patterns and lineations observed in magnetic imagery can be related to structure and lithology. The high information content and wavenumber range in magnetic data makes it ideal for application of a wide range of enhancements. Aeromagnetic data are currently the most widely used geophysical data set for geological mapping and mineral exploration.

*B. **Lithological Discrimination.*** Since the magnetic susceptibility of a rock is generally determined by less than 3 % of its bulk composition, it is generally impossible to relate magnetic character directly to lithology in the conventional classification system, which is based on silicate and carbonate mineralogy (Grant, 1985). For example, both magnetic and nonmagnetic granites are common.

Magnetite content can be created, remobilized, or destroyed by emplacement processes and metamorphic effects (Figure 13.15), resulting in clear delineation of boundaries and generation of characteristic textures and trend directions. There are four major processes that affect the distribution of magnetite: (1) the amphibolite–granulite phase transition, (2) serpentinization of ultramafic rocks, (3) oxidation in fracture zones, and (4) magnetite grain size changes during metamorphic and structural events. Magnetic imagery can therefore be very useful for identifying areas with uniform lithology.

*C. **Structural Interpretation.*** The high resolution of magnetic data often results in clear delineation of structural features. Folding, faulting, deformation, and deposition processes often affect the distribution of magnetic minerals or result in their creation or destruction. These changes often produce offsets, lineations, and trends in the magnetic imagery, which can be very useful for tracing and delineating structure (Henderson and Broome, 1990).

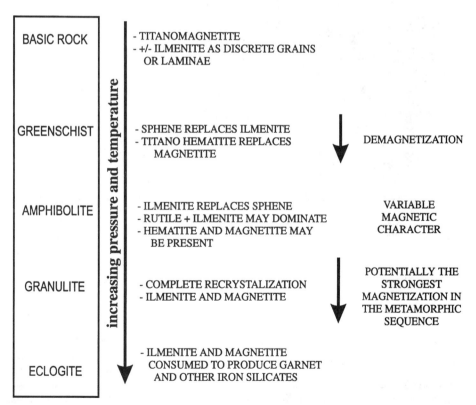

Figure 13.15 Magnetization changes with metamorphism. With increasing metamorphic grade, magnetite in a simple basalt may be destroyed and re-created producing variations in the magnetic anomaly produced by the basalt. (From Reeves, 1987.)

D. Direct Ore Body Detection. Other than the obvious example of iron ore deposits, magnetic anomalies offer only indirect detection of ore bodies, due to the association of magnetic minerals with ore, such as in the case of massive sulfide deposits or with diamonds in kimberlite pipes. The value of magnetic imagery for prospecting is usually its ability to provide structural information, which helps to define the regional and local geology to find environments favorable for the particular ore body model.

13.4.2.3 FAULT OFFSET.

The images in Figure 13.16 were generated from a high-resolution aeromagnetic survey flown with a survey altitude of 150 m and line separation of 300 m located in the Ecum Secum Area, Nova Scotia, Canada. The data were gridded to 50 m. Figure 13.16*a* is a gray-tone image of the total magnetic field. Darker areas in this image are magnetic highs. Figure 13.16 is a black-and-white image generated from

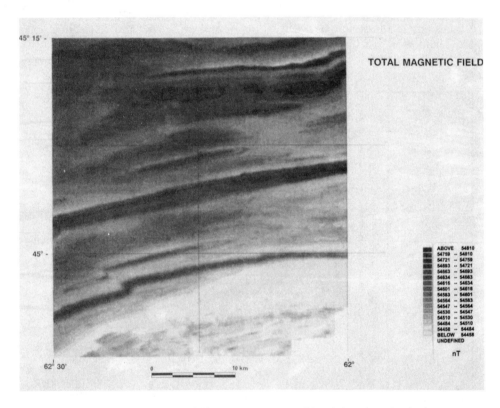

Figure 13.16 (*a*) Gray-tone total magnetic field intensity image generated from high-resolution data for the Ecum Secum Area, Nova Scotia, Canada. (*b*) This image was created by equal-area quantization (histogram equalization) of the measured vertical gradient into 16 levels. Note how the image detail is sharpened and the regional gradient is lost compared to image (*a*). Many individual anomalies in the southeastern corner of the image are now visible. A and B represent associated anomalies. (*c*) Results of total magnetic field modeling for the data over profile P3 in (*b*) demonstrate how anomaly shape and magnetic property data can be used to interpret subsurface geometry. In this figure the subsurface cross section of zones of varying pyrrhotite concentrations (indicated in %) are defined so that anomaly profiles calculated match the data measured.

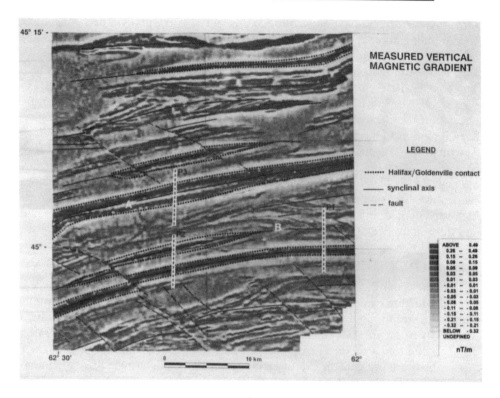

(b)

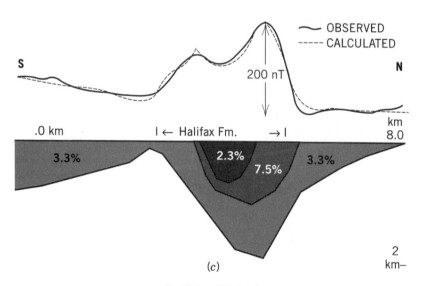

(c)

Fig 13.16 (Continued)

measured vertical gradient data collected during the same survey. The stronger east/west-trending anomalies (A) are due to pyrrhotite in the Halifax Formation slates (Schwarz and Broome, 1994; Henderson, 1984). The weaker subparallel anomalies (B) are due to low concentrations of magnetite in the Goldenville Formation beds. The contact between the Halifax and Goldenville Formations is gradational. This data set is excellent for demonstrating the magnetic expression of various features because of the clear representation of the structure and geology due to the lack of overlapping metamorphic effects and the uniformity of the structure and lithology. The offset of magnetic anomalies due to sinistral strike-slip movement on a NW-to-SE fault is clearly visible. In the total field image (Figure 13.16*a*) the main E-W anomaly (A) is clearly visible, but the fault offset is not as well defined where it crosses the Goldenville Formation. In the shaded relief image (Figure 13.16*b*) the high-wavenumber information is enhanced, resulting in better definition of the fault and the appearance of other parallel anomalies.

13.4.2.4 SURFACE TRACE OF FOLDS.

The Ecum Secum imagery can also be used to show how magnetic marker beds in layered structure can provide evidence of deformation. The surface trace of folding within the steeply dipping Goldenville structure is clearly visible in the southeast corner (B) of Figure 13.16*b*. Marker beds with elevated magnetite concentration delineate the fold structure here and elsewhere in the image. If strongly magnetic marker beds are present, the dip of bedding can also be determined from the aeromagnetic imagery by analysis of anomaly shape.

13.4.2.5 INTRUSIONS AND CONTACT AUREOLES.

Figure 13.17 is a shaded-relief image of the sub-Paleozoic portion of the Flin Flon Snow Lake Belt, Manitoba, Canada. These aeromagnetic data were collected at 150-m elevation and 300-m line spacing. Since the bedrock geology in this area is covered by several hundred meters of nonmagnetic Paleozoic rock (Leclair et al., 1997) basement mapping must be accomplished by a combination of geophysical methods and drilling. The magnetic field largely reflects variations in the magnetization of basement sources below the nonmagnetic cover rocks. Areas of uniform magnetic intensity with internal concentric anomalies cutting the regional magnetic signature are characteristic of granitic intrusions. Since the magnetite content of granitic intrusions relative to the surrounding host rock is highly variable, the intensity of the associated anomalies is correspondingly variable. The zone of concentric anomalies which disrupts the regional SW/NE anomaly trend is due to a complex granitic intrusive which contains the Cormorant Batholith, a medium-to course-grained granite–granodiorite intrusion. Bouguer gravity intensity imagery over this area, shown in Figure 13.13, can assist the interpreter in determining the density of the intrusive rock relative to the surrounding rock and thus its lithology. Internal concentric magnetic anomalies are probably due to the contact metamorphic effect of individual intrusive phases and subsequent differentiation during cooling and indicate the location of the margins of these phases. Mafic to ultramafic intrusives, such as the gabbroic intrusive marked on the figure, typically generate a magnetic high. The combination of magnetic and gravity imagery is particularly useful for mapping of intrusive rocks in Precambrian terrain.

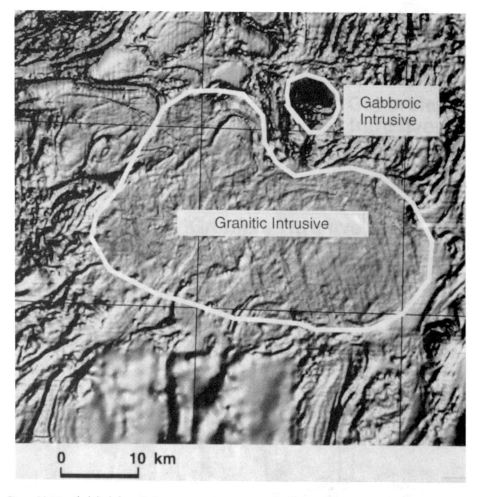

Figure 13.17 Shaded-relief image of aeromagnetic data over the sub-Paleozoic portion of the Flin Flon/Snow Lake belt. The outline of gabbroic and granitic intrusions, determined by a combination of gravity, aeromagnetic, and drill hole data, is overlain. Intrusions are typically characterized by areas of somewhat uniform texture that cross-cut the regional fabric. Concentric anomalies, often found over the intrusion, are usually due to concentration of magnetite along the margins of intrusive phases.

13.4.2.6 MAFIC DIKES.

Mafic dikes typically cause strong, distinctly linear anomalies that are clearly visible on magnetic images. Due to their directional high-wavenumber character, the anomalies are best displayed by shading. Several NW/SE-striking anomalies are clearly visible on Figure 13.11d (color insert; Broome, 1990) from the Rae Province, northern Canada. In some dikes, remanent magnetization may predominate over induced magnetite, resulting in negative anomalies. The location of dikes may be important for mineral exploration as heat engines to remobilize base metals or to map structural breaks, but in many cases dikes are not of particular economic interest and their anomalies overprint more important magnetic anomalies caused by structure and lithology.

13.4.2.7   GREENSTONE BELTS.

Within the Superior Province of the Canadian Shield there is a strong correlation between greenstone belts and regional magnetic lows, suggesting that greenstone belts are relatively depleted in magnetite (Hood et al., 1987). The magnetic effect of the granitic and volcanic rocks of a typical greenstone belt can be seen clearly in Figure 13.18, a vertical gradient image from the Abitibi Greenstone Belt, Val d'Or area, Quebec, Canada (Hood et al., 1982). The vertical gradient data used to generate this image were collected at 150-m flight elevation and 300-m line spacing. The granitic rocks generally correspond with featureless lows and the anomalies due to volcanic rocks may have a banded character, due to alternating tholeiite and komatiite (magnetic) layers in the volcanic pile. The interpreter must, however, be cautioned that elsewhere in the Rae Province in Canada, Ivory Coast, and Uganda (Reeves, 1987), anomaly patterns similar to these are found in granitic rocks. In these cases the magnetic field may be reflecting the magnetite distribution of a paleolithology derived from ancestral rock.

13.4.2.8   SHEAR ZONES.

Shear zones are often clearly represented on aeromagnetic imagery as generally linear anomalies coincident with the shear zone and cutting across the regional anomaly

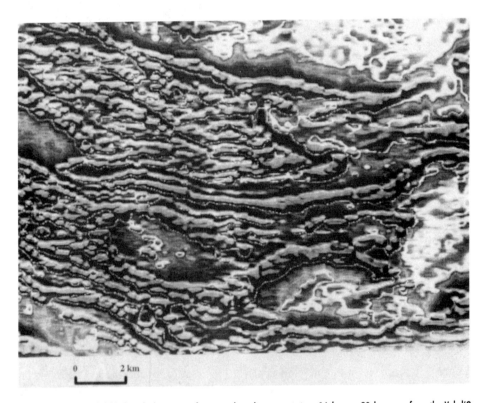

0   2 km

Figure 13.18   Detailed black-and-white copy of a vertical gradient survey in a 16 km × 20 km area from the Val d'Or area of the Abitibi Greenstone Belt, Quebec, Canada. The striped anomaly pattern is due to the alternating mafic and felsic flows in the volcanic rocks. The magnetically quiet zones are due to granitic intrusions.

trend. The anomalies have numerous causes: destruction of magnetite by oxidation, creation or destruction of magnetite by metamorphism, or enhancement of susceptibility by grain-size change. The magnetic expression of several large-scale shear zones is clearly visible in Figure 13.11 *c* and *d* (color insert; Broome, 1990). A more detailed view of the internal structure of the Wager Bay Shear Zone, Northwest Territories, Canada, is shown by the vertical magnetic gradient image (Figure 13.11*e*). This image was generated by regridding the data used to produce the total field image at 200 m and using a frequency-domain transform to calculate the vertical magnetic gradient.

13.4.2.9 KIMBERLITE PIPES.

Kimberlite pipes can be magnetic or nonmagnetic. When pipes contain magnetite, their near-vertical dip and generally circular shape can produce characteristic "bulls-eye" anomalies. Due to the small size of the pipes, they often do not appear on regional aeromagnetic images, due to the wide survey line spacing. The color total field image shown in Figure 13.19*a* (see the color insert) shows the negative anomaly (blue) over a pipe in the Lac de Gras area, Northwest Territories, Canada (square box). This anomaly is approximately 500 m in diameter and is surrounded by numerous smaller positive anomalies (red) of unknown source. The vertical gradient image shown in Figure 13.19 (color insert) shows a similar negative anomaly centered over the pipe, surrounded by a weak positive ring. The negative gradient anomaly over the pipe is surrounded by smaller positive and negative anomalies.

13.4.3 Electromagnetic Imagery

13.4.3.1 CHARACTERISTICS AND INTERPRETATION GUIDELINES.

Initially, AEM systems were developed for direct detection of massive sulfide ore bodies in glaciated terrains. Many AEM surveys have resulted in discoveries because the conductive ore bodies are surrounded by resistive host rock. Other important bedrock conductors in shield areas are due to graphite and graphite schists. Table 13.6 shows the highly variable conductivity in millisiemens per meter ($mS\ m^{-1}$) of various material (Palacky, 1987a). Values of over 500 $mS\ m^{-1}$ are considered high and are typically found in massive sulfide ores, graphite, and salt water. Sedimentary rock, glacial sediments, weathered layers, and alteration zones generally have moderate conductivities ranging from 1 to 500 mS/m.

Refinements in the method and reduced demand for direct detection surveys have resulted in an increased emphasis on conductivity mapping to support geological mapping through detection of graphite rich layers, clays in alteration zones, weathered layers, and glaciolacustrine sediments. Since the type of clay encountered in saprolite (thoroughly chemically weathered rock) depends on the lithology of the parent rock, AEM data can assist the mapping of lithology in areas with in situ weathered layers. This application is therefore more suitable for nonglaciated terrains.

A. Resolution of the Data: Wavenumber Content. AEM surveys exhibit variable resolution, due to great variation in survey specifications, instrumentation, and processing, but can show detail comparable to aeromagnetic imagery. AEM data are often

TABLE 13.6   Conductivity of Some Common Earth Materials

Material	Conductivity (mS m^{-1})
Massive sulfides	10,000–10,000,000
Graphite	500–10,000
Alteration zones	10–500
Saprolite	50–500
From mafic volcanic rocks; schist	
Felsic volcanic rocks; granite, gneiss	5–50
Mottled crust	0.5–5
Duricrust (canga)	0.03–0.5
Glaciolacustrine clays	10–200
Glacial tills	0.5–20
Gravel and sand	0.1–2
Sedimentary rocks	0.01–500
Igneous and metamorphic rocks	0.0005–1
Seawater	1000–5000
Fresh water	100–500
Sea ice	1–50
Permafrost	0.001–2

Source: Palacky (1987a).

collected together with other data sets in multiparameter surveys. The wide range of conductivity present in common materials and the degree to which conductivity varies with moisture content and metamorphic processes contribute to the high-wavenumber content. As with potential field data, high-resolution data are beneficial to interpretation, as they allow accurate definition of anomaly shape, which can provide information about source geometry, depth, and conductivity. Spectral analysis in the wavenumber domain of multichannel AEM data can be used; however, interpretation of results is complex.

B. Sensitivity to the Depth of the Source Material. The depth of penetration of AEM methods varies inversely with both the frequency of the electromagnetic field and the conductivity of the Earth. Problems with generating and measuring low-frequency signals limit the depth of penetration to about 500 m (Kearey and Brooks, 1984), and many systems have less depth penetration. In general, the depth of penetration varies inversely with instrument frequency. In surveys designed to image conductive mineralization at depth, conductive surface overburden, such as water-saturated clays, can produce strong anomalies that obscure bedrock anomalies.

Multicomponent systems, which resolve the secondary field in two or three dimensions, increase the possibility of detecting discrete conductors, and the *z*-component data can be used to ascertain the dip and depth to the conductor. In areas of layered structure, measuring the *z* component improves the signal/noise ratio, which increases the effective depth of penetration as well as providing information on lateral symmetry (Smith and Keating, 1996). A Hilbert transform can be used to calculate the *z* component from the *x* and *y* components, but the calculated *z* component provides no information on lateral symmetry.

13.4.3.2 APPLICATIONS OF ELECTROMAGNETIC IMAGES

A. *Lithological discrimination.* Lithological classification of rocks is usually possible only indirectly using conductivity variations in *saprolite*, the most conductive component of the weathered layer located immediately below the water table (Palacky, 1987b). Not all rock types weather consistently. Rocks containing high proportions of resistant mineral such as quartz weather more slowly, producing thin saprolite layers. Less resistant mafic rocks contain more minerals that are replaced by conductive clays during weathering, with the result that more conductive saprolite is produced. Figure 13.20 shows the relative quartz content and conductivity of common igneous rocks. AEM can be used to complement other geophysical methods for lithological mapping or when an in situ layer of saprolite is present, on its own. AEM techniques are also a useful tool for mapping the lithology and thickness of Quaternary sediments because of their sensitivity to near-surface changes in conductivity.

B. *Structural Interpretation.* AEM surveys can indirectly provide regional structural information. Passive techniques, and in particular VLF surveys, have been used successfully to map shear and fracture zones. VLF anomalies coincident with these features are believed to be due to current flow in water-saturated clays in the fracture zones. Active systems are less useful for regional conductive features, due to the inability of their relatively weak primary field to generate current flow in regional features.

C. *Direct Ore Body Detection.* AEM surveying was developed to detect buried volcanogenic massive sulfide (VMS) ore bodies in shield terrains. In many cases the conductive AEM anomalies associated with massive sulfide anomalies were found to be due to the presence of pyrrhotite. Before 1970, VMS ore bodies were believed always to have associated magnetic anomalies. Anomalies due to magnetite rather than VMS ore could be detected in multiparameter geophysical surveys by the associated negative in-phase EM response and positive magnetic anomaly. The use of AEM for direct ore-body detection fell out of favor because it was discovered that many VMS deposits had no associated magnetic anomaly and because of the belief that most of the VMS deposits directly detectable using AEM anomalies have been found.

13.4.3.3 KIMBERLITE DETECTION.

Kimberlite pipes, which are typically recessive and rarely outcrop, usually have anomalous geophysical properties compared to the surrounding host rock, allowing

Figure 13.20 Conductivity of saprolite from common igneous rocks. The susceptibility of rock to weathering and saprolite conductivity increases with decreasing quartz content.

their detection using multiparameter geophysical surveys (Smith et al., 1996). The kimberlites themselves are characterized by low resistivity, variable magnetization, and variable radiometric signature. Although both magnetic and AEM surveying are useful for kimberlite detection, the AEM response is most often due to weathered alteration products in near surface, while the magnetic material, if present, is located at greater depth in the unweathered pipe. Due to the small size of typical kimberlites, high-resolution low-level helicopter-borne survey techniques are usually utilized. Profile data selected from a six-frequency single-component AEM survey, shown in Figure 13.19*d* (color insert), clearly show the anomalous responses over a pipe.

Figure 13.19*c, e,* and *f* show interpretive images produced by specialized processing of the AEM data. Figure 13.19*c* is an apparent resistivity image produced from the 4175-Hz coplanar coils. The image reflects resistivity variation in the top 100 m in host rock, with an averaged resistivity of 100 Ω-m and shows anomalies over the altered pipe and elsewhere. In most areas there are a number of possible causes for conductivity anomalies. In this area, conductive responses are due to both conductive clays in lake sediments and weathered kimberlite pipes. To distinguish pipes from conductive sediments, the AEM data are processed using inversion techniques (Kovacs and Holladay, 1990) to determine lake depth (Figure 13.19*e*) and conductive layer thickness (Figure 13.19*f*). The conductive layer thickness image is particularly effective for delineating the pipe. The coincidence of a magnetic gradient anomaly, resistivity high, and conductive layer thickness anomaly is a good indicator for kimberlite pipes in this area.

13.4.3.4 OVERBURDEN MAPPING.

Due to its intermediate depth penetration range and the usually high conductivity of overburden relative to bedrock, images generated from AEM data are often useful for mapping overburden, which if in situ can provide information on the bedrock lithology. Figure 13.21*a* (see the color insert) is a Quaternary geology map from a glaciated region in the area of Kirkland Lake mining camp, Ontario, Canada. Figure 13.21*b* (color insert) is a conductivity image calculated for the area generated from time-domain single-component AEM data collected using the GEOTEM system (Annan and Lockwood, 1991). Based on the conductivity image, the most conductive overburden unit appears to be the "organics and alluvium," mapped in black. The black line on both images follows a linear conductivity high on the conductivity image that correlates well with three black polygons, representing the organics and alluvium, on the overburden map. The continuity of the linear anomaly suggests that the organics and alluvium unit is probably continuous along the black line in the subsurface. One could speculate that other linear anomalies on the image represent unexposed linear zones of this unit.

13.4.4 Gamma Ray Spectrometry Imagery

13.4.4.1 CHARACTERISTICS AND INTERPRETATION GUIDELINES.

The measured gamma radiation above the Earth originates in radioelements contained in the material on the surface of the Earth. Because of limited penetration, gamma ray spectrometry is more useful than potential field methods for surficial applications. Properly conducted and corrected surveys yield quantitative geochemical data. Interpretation is complicated by strong attenuation of the gamma radiation by any dense material, including water. Radiometric data can, of course, be used directly to detect uranium and thorium deposits and kimberlites.

Different lithologies can be characterized by different concentrations of the three principal radioelements. Radioelement ratios can also be used to map lithological variations, such as different intrusive phases in granite. This methodology is most effective in areas with little overburden. In glaciated areas with significant overburden, radiometric anomalies originating in the overburden can be displaced from the source location. Common radioactive minerals are listed for each radioelement in Table 13.7. All rock contains trace amounts of the radioactive minerals, but some, such as potassium-rich granites and some basalts, are more radioactive. Due to attenuation caused by water and overburden, radioelement ratios are often more consistent with lithology across an area and therefore more useful for discrimination of surface lithology than are individual radioelement concentrations.

When evaluating ratio images it is important to consider the radioelement intensities determining the ratio as well. Strong, but insignificant ratio anomalies can occur where the radioelement intensities are low due to poor signal/noise ratios. Display methods such as ternary radioelement plots (Broome et al., 1987) incorporate both ratio and intensity information in the image, allowing the significance of ratio anomalies to be determined.

Wherever possible, a ground follow-up component should be included in the interpretation program. A portable spectrometer, can be used in the field to determine the spectral characteristics of different rocks, greatly aiding interpretation.

A. Resolution of the Data: Wavenumber Content. The resolution of gridded spectrometry data is highly dependent on the survey altitude and line spacing. Since, like most remote sensing methods, the data originate from the surface, they lack the low-wavenumber component from depth found in potential field data and can suffer from high-wavenumber noise, due to poor counting statistics. The imagery suffers from

TABLE 13.7   Radioactive Minerals and Their Occurrence

Mineral	*Occurrence*
Potassium (K)	
Orthoclase + microcline feldspars	Acid igneous rocks and pegmatites
Muscovite	Acid igneous rocks and pegmatites
Alunite	Alteration in acid volcanic rocks
Sylvite	Saline deposits in sediments
Thorium (Th)	
Monazite	Granites, pegmatites, placers
Thorianite	Granites, pegmatites, placers
Thorite, uranothorite	Granites, pegmatites, placers
Uranium (U)	
Uraninite (pitchblende)	Granites, pegmatites with Au, Pb, Cu vein deposits
Carnotite	Sandstones
Gummite	Associated with uraninite

Source: Telford et al. (1978).

considerable noise, which limits the utility of high-wavenumber enhancement methods such as shading. Various types of two-and three-component ratio images are useful for deemphasizing attenuation effects and removing the high correlation between individual radioelement images, which complicates direct interpretation of radioelement intensities. Classification methods can be useful for identifying zones with relatively homogeneous radioelement signatures.

B. Sensitivity to the Depth of the Source Material. The top 30 cm of the rock is the source of 90% of the measured radiation. In areas with overburden thicker than 30 cm, most of the radiation originates in the overburden; however, general bedrock composition is reflected in the overburden even in heavily glaciated regions because the overburden is largely derived from the local bedrock and transport distances are typically a few kilometers or less. In the case of small buried radioactive sources such as carbonatite intrusions, airborne surveys may clearly outline the extent of glacial dispersion (Ford et al., 1988)

13.4.4.2 APPLICATIONS OF GAMMA RAY SPECTROMETRY IMAGES.

A. Ease of interpretation. The interpretation of the responses is generally straightforward, due to the surface origin of the responses and the direct indication of radioelements. Attenuation from vegetation, water and water-saturated overburden, topography, and gamma radiation from radon are the main complications. (Shives et. al., 1995)

B. Lithological Discrimination. Different lithologies often have distinctive radioelement signatures that can be detected using gamma ray spectrometry. Since individual radioelement intensities are strongly affected by attenuation, radioelement ratios are very useful for lithological discrimination. Where overburden is present, it attenuates radiation from the bedrock and is the origin of much of the gamma radiation. In glaciated terrain where much of the overburden is not in situ, gamma radiation originating in the overburden may not reflect the bedrock lithology, complicating interpretation. Gamma ray data are particularly useful as an indicator for granophile element differentiation within complex suites of granitic rocks (Broome et al., 1987).

C. Structural Interpretation. Gamma ray spectrometry can aid structural interpretation but not to the same degree as potential fields because it is a surficial method and subject to attenuation effects. Radioelement distribution data also lack the detail of aeromagnetic data, due to the averaging effect of the detection method. In relatively rare cases the presence of marker units with elevated radioelement concentrations allows the surface trace of folds to be mapped or fault offsets to be detected.

D. Direct Ore Body Detection. Gamma ray spectrometry is best known for its application to uranium exploration (Killeen, 1979). Uranium deposits are often found in regions somewhat enriched in U and are often associated with high U/Th and U/K ratios. U and Th are also pathfinder elements for Li, Cs, Be, Nb, Ta, Zr, and other rare earth elements. In general, gamma ray data can reflect geochemical alterations commonly associated with mineralization (Shives et al., 1997). For example, porphyry copper deposits are often associated with K enrichment.

13.4.4.3   REGIONAL GAMMA RAY SPECTROMETRY IMAGES.

Regional gamma ray spectrometric data were collected in the 1970s for most of the Precambrian Shield area of Canada. These data were collected at a line spacing of 5 km, which is too wide for the generation of detailed images. More useful imagery can be generated from data collected at a line spacing of 1 km. Figure 13.23 (see the color insert) is a ternary radioelement image generated from 1-km-spaced data collected from an area in southwest Newfoundland, Canada. Ternary radioelement images are a specialized form of composite imagery designed to enhance subtle variations in the ratios of the three radioelements (see Section 13.3.3). To optimize color variability in regional radiometric images, the ratio range in the image is quantized using an equal-area method. In areas of low gamma radiation the ratios tend to be very noisy, causing high variability in the color hue. To minimize this effect the color saturation in the image is controlled by the total radioactivity, causing areas with low radioactivity or strong attenuation from lakes, swamps, or other surface material to appear white on the image. Zones of consistent color hue on the image have a uniform radioelement signature, which generally indicates lithologic uniformity. Since the color hues on the image can be related to combinations of the primary colors representing each image, the actual radioelement contributions can easily be inferred from the color hue. For example, an orange area is due to the presence of K (magenta) and Th (yellow).

13.4.4.4   GLACIAL DISPERSAL TRAINS.

Gamma ray spectrometric imagery can show anomalies due to glacial transport of material eroded from radioactive sources. In Figure 13.22 (color insert) glacial erosion has produced a dispersal train of lithologically distinct till. The elevated thorium concentration of the dispersal train is apparent in the thorium intensity image shown in Figure 13.22. The source body is the Allan Lake Carbonatite (Ford et al., 1988), which is a small (about 0.4 km^2) unexposed ankeritic–sideritic carbonatite. The dispersal train covers approximately 10 km^2 and is characterized by a 10-to 20-fold increase in thorium and 5-to 10-fold increase in uranium compared to background. This anomaly was detected initially on the regional 5-km line spacing survey. Data from a subsequent detailed survey with 0.5-km line spacing was used to generate this image.

13.4.4.5   GRANITIC PHASE DIFFERENTIATION, SOUTHWEST NEWFOUNDLAND.

Rocks within a large composite intrusion created during different periods of intrusion commonly have different radioelement signatures which are clearly visible on gamma ray spectrometry images. Figure 13.23 is a (see the color insert) color radiometric image generated from gamma ray spectrometry data collected at a line spacing of 1 km (from Broome et al., 1987). Color hue variations clearly visible within the mapped granite bodies can be easily used to detect zones of granite with different composition. Notice in particular the Francois Granite in the south.

13.4.4.6   MULTI-INSTRUMENT DATA FOR MAPPING AND EXPLORATION.

Gamma ray spectrometry is often used in combination with other geophysical methods to support regional mapping and mineral exploration programs. The aeromagnetic and radioelement images shown in Figures 13.24 and 13.25 were generated from a multi-instrument survey in Ghana. This detailed survey collected data at a nominal sensor terrain clearance of 100 m along approximately northwest traverse lines separated by 200 m. The aeromagnetic data were corrected using a base station magnetometer. Since Ghana is located near the equator, the magnetic inclination is near 0°, complicating interpretation. At low inclination angles, a single source body will produce a magnetic high located south of the body and a magnetic low north of the body rather than the single anomaly centered over the body at near-vertical inclinations.

The most striking features of the geology of Ghana are the northeast-striking parallel linear Birmian volcanic belts, shown in Figure 13.24. This northeast–southwest direction is the predominant texture of the aeromagnetic image shown in Figure 13.25*a* (see the color insert) and to a lesser degree in the radiometric images shown

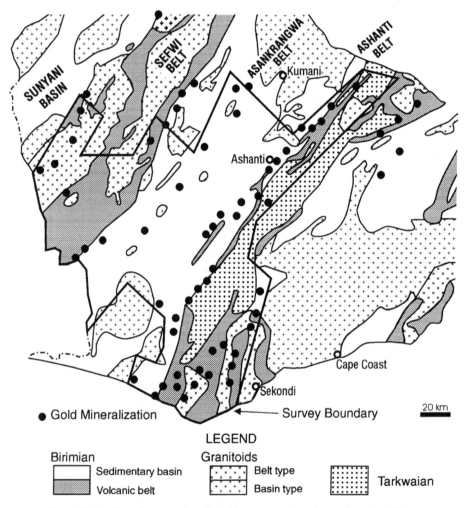

Figure 13.24   The study area geology, derived from a regional map from Leube et al. (1990).

in Figure 13.25*b* to *d* (color insert). Many mapped sections of the volcanic belts are coincident with linear magnetic anomalies. Although the thorium and potassium intensity shown in Figure 13.25*a* and *b* show some relation to the geology, the thorium/potassium ratio shown in Figure 13.25*c* reflects the mapped geology more accurately. Comparison of the aeromagnetic and gamma ray spectrometry imagery reveals that the two methods provide complementary information about the area. For example, in the southeast corner, aeromagnetic anomalies trend northeast–southwest, while in the same area, many of the thorium/potassium ratio image features trend north–south.

On a regional scale the vast majority of the Birmian gold deposits, marked by circles on the images, occur aligned along the flanks of the volcanic belts and in particular along the Ashanti Belt, which follows the eastern limit of the survey area. In this area, and many others, the combination of aeromagnetic and radiometric survey data can support mapping of volcanic belts important for gold and base metal exploration.

13.4.5 Summary of Data Characteristics and Applications

To conclude this introduction to the characteristics and applications of geophysical data and methods, an attempt has been made in Table 13.8 to summarize the relative utility of the various methods for geoscience applications. Generalizations on the utility of the geophysical methods are difficult because of the diversity of geological environments and the wide range of data collection parameters for each type of data. For each case where the utility of a method is rated "poor" for a particular application, there are specific geological environments and data collection parameters that could allow the method to work effectively, or conversely, situations where a method rated as "excellent" would not be effective.

Important characteristics for determining the utility of methods for a particular application are depth penetration, surface exposure, and the likelihood of a significant contrast in the measured parameter. Clearly, without a contrast in the measured parameter, no method will be effective. An important consideration for methods with little subsurface penetration is the degree to which the surface reflects the underlying geology. The value of a particular method is very dependent on the application. For lithological discrimination, the value of gamma ray spectrometry and *Landsat* data is clearly diminished if the surface is extensively weathered, masked with vegetation, or covered with overburden that is not in situ. These same methods may be very useful if, for example, the aim of the survey is to detect alteration or vegetation stress related to the overburden geochemistry. Potential field and electromagnetic methods are much more effective as an aid to bedrock mapping in areas with poor exposure, due to their ability to measure variations in subsurface materials.

The other problem when determining ratings is the broad range of data collection specifications compared to satellite scanner data, which generally have consistent specifications. Very detailed surveys utilizing small sampling intervals can produce data that are much more useful than regional data for the small study areas typical of mineral exploration applications. Despite this "disclaimer," the generalizations in Table 13.8 serve as a useful guide to the various methods.

TABLE 13.8 Summary of Geophysical Data Characteristics and Applications

Method		Data Characteristics			Applications			
		Sampling *1 interval	Typical grid resolution	Crustal Depth penetration	Bedrock lithology discrimination	Structural interpretation	Quaternary mapping	Direct ore detection
Gravity	regional ground	3-12 km	>1000 m	>100 km	FAIR	GOOD	POOR	POOR
	detailed ground	0.5-2 km	100-500 m	>100 km	FAIR	EXCELLENT	FAIR	FAIR
Magnetic	satellite	40 km	50 km	<40 km	POOR	POOR	POOR	POOR
	regional airborne	1-5 km	200-1000 m	<40 km	FAIR	GOOD	POOR	FAIR
	detailed airborne	0.1-0.5 km	50- 200 m	<40 km	FAIR	EXCELLENT	POOR	GOOD
Electromagnetic	detailed airborne	0.1-0.5 km	50-100 m	<1 km	FAIR	POOR	FAIR	EXCELLENT
Gamma-ray Spectrometry	regional airborne	1-5 km	200-1000 m	<1 m	GOOD	POOR	GOOD	GOOD (U)
	detailed airborne	0.1-0.5 km	50-200 m	<1 m	GOOD	POOR	GOOD	GOOD (U)
LANDSAT	TM	30 m	30 m	<1 cm	GOOD	FAIR/GOOD	GOOD	POOR

[a] For airborne geophysical surveying, the sampling interval refers to the spacing between survey lines. Data collection is much denser along the line.

692

13.5 CONCLUSION AND THE FUTURE

The airborne geophysical techniques discussed briefly in this chapter were developed to obtain systematic coverage of large areas using adaptations of instrumentation originally developed for ground use. For many decades these data were processed and interpolated onto regular grids to produce raster data sets which were subsequently used to produce contour maps. In the 1970s, gray-tone and color imagery produced by computer from geophysical data became popular and some scientists began to experiment with the application to geophysical data of image enhancement and classification techniques originally developed for multiband satellite data. Some of the remote sensing techniques proved to be very useful for improving the interpretability of airborne geophysical data. Despite the fact that remote sensing specialists and geophysicists utilize data collection methods, data, and interpretation techniques which are quite similar, the two disciplines have developed along similar but distinct paths. There are some clear differences between some geophysical data and remote sensing data: for example, the variable depth of the origin of potential field anomaly sources. Nevertheless, some geophysical methods, such as gamma ray spectrometry, produce data that are very similar to remote sensing data and can be interpreted using similar tools.

With the widespread move toward routine digital archiving, a range of geoscience data are now often readily available, facilitating their integrated interpretation using the increasingly sophisticated spatial analysis and imaging tools available in a range of image analysis and GIS software. Remote sensing and geophysical data are complementary for many geoscience applications. For example, potential field methods can provide information on subsurface structure and lithology with radar or *Landsat* data imaging the surface trace of the subsurface structures. When planning a study it is worthwhile to check for the availability of geophysical data. Regional aeromagnetic and gravity data may be available from government at a fraction of the cost of satellite data.

The general availability to end users of easily integrated digital data is creating an increased demand for interpretation tools that can be used by nonspecialists. There is also a need for improved understanding and quantification of the relationships between different data sets, such as geophysical and remote sensing data. Development of this understanding will require joint research between geophysicists and remote sensing specialists but will ultimately benefit both disciplines.

ACKNOWLEDGMENTS

This chapter benefited from discussions with many of my peers at the Geological Survey of Canada, the academic community, and the Canadian mineral exploration community. In particular, I would like to acknowledge the staff of the Geoscience Integration Section at the GSC—Jeff Harris, Cameron Bowie, David Viljoen, and Don Desnoyers—for constructive discussions; the editor of this volume, Andy Rencz; and Pierre Keating, who helped with the electromagnetic section.

References

Annan, A. P., and R. Lockwood, 1991. An application of airborne GEOTEM in Australian conditions, *Explor. Geophysi.*, 22, 5–11

Baranov, V., 1957. A new method for interpretation of aeromagnetic maps:pseudo-gravimetric anomalies, *Geophysics*, 22(2), 359–383.

Bartel, D. C., and A. Becker, 1990. Spectral analysis in airborne electromagnetics; *Geophysics*, 55(10), 1338–1346.

Bhattacharyya, B. K., 1965. Two-dimensional harmonic analysis, as a tool for magnetic interpetation, *Geophysics*, 30 (5), 829–857 [reprinted in *Geophysics*, 50(11), 1878–1906].

Bristow, Q., 1983. Airborne γ-ray spectrometry in uranium exploration: principles and current practice, *Int. J. Appl. Radiat. Isot.*, 34(1), 199–229.

Broome, J., 1990. Generation and interpretation of geophysical images with examples from the Rae Province, northwestern Canadian Shield, *Geophysics*, 55(8), 977–997.

Broome.J., Brodaric, B., Viljoen, D., and Baril, D. 1994 The NATMAP digital geoscience data management system; Computers and Geosciences, 19, 10, 1501–1516.

Broome, J., J. M. Carson, J. A. Grant, and K. L. Ford, 1987. *A Modified Ternary Radioelement Mapping Technique and Its Application to the South Coast of Newfoundland*, GSC Pap. 87–14, Geological Survey of Canada, Ottawa, Ontario, Canada, 1 map sheet.

Clement, W. G., 1973. Basic principles of two dimensional filtering, *Geophys. Prospect.*, 21, 125–145.

Cooley, J. W., and J. W. Tukey, 1965. An algorithm for the machine computation of complex Fourier series, *Math. Comput.*, 19, 297–301.

Dods, S. D., D. J. Teskey, and P. J. Hood, 1987. The new 1: 1 000 000-scale magnetic anomaly maps of the Geological Survey of Canada: compilation techniques and interpretation, in *The Utility of Regional Magnetic and Gravity Anomaly Maps*, W. J. Hinze, ed., Society of Exploration Geophysicists, p. 69–8.

Evenden, G. I., 1989. *Review of Three Cubic Spline Methods in Graphic Applications*, Open File Rep. 89-0019, United States Geological Survey, Woods Hole, Mass., 14 pp.

Ford, K. L., R. N. W. Dilabio, and A. N. Rencz, 1988. Geological, geophysical, and geochemical studies around the Allan Lake carbonatite, Algonquin Park, Ontario; *J. Geochem. Explor.*, 30, 99–121.

Grant, F. S., 1985. Aeromagnetics, geology, and ore environments: I. Magnetite in igneous, sedimentary, and metamorphic rocks: an overview, *Geoexploration*, 23, 303–333.

Grasty, R. L., and B. R. S. Minty, 1995. *A Guide to the Technical Specifications for Airborne Gamma-ray Surveys*, Rec. 1995/60, Australian Geological Survey Organization, Canberra City, New South Wales, Australia, 65 pp.

Halpenny, J. F., and D. M. Darbha, 1995. Airborne gravity tests over Lake Ontario, *Geophysics*, 60, 61–65.

Hammer, S. 1983. Airborne gravity is here! *Geophysics*, 48, 213–223.

Harris, J. R., C. Bowie, A. Rencz, and D. Graham, 1994. Computer enhancement techniques for the integration of remotely sensed, geophysical, and thematic data for the geosciences, *Can. J. Remote Sensing*, 20(3).

Henderson, J. R., 1984. *Geology, Ecum Secum Area, Nova Scotia*; Map 1648A, 1: 50 000 scale, Geological Survey of Canada, Ottawa, Ontario, Canada, 1 map sheet.

Henderson, J. R., and J. Broome, 1990. Geometry and kinematics of Wager Bay shear zone from structural fabrics and magnetic data, *Can. J. Earth Sci.*, 27, 590–604.

Hinze, W. J., and I. Zeitz, 1987. The magnetic anomaly map of the conterminous United States; in *The Utility of Regional Magnetic and Gravity Anomaly Maps*, W. J. Hinze, ed., Society of Exploration Geophysicists, pp. 1–24.

Hogg, R. L. S., 1987. Recent advances in high sensitivity and high resolution aeromagnetics, in *Proceedings of Exploration '87: 3rd Decennial International Conference on Geophysical and Geochemical Exploration for Minerals and Groundwater*, OGS Spec. Volume ISSN 0827-181X, Ontario Geological Survey, Ottawa, Ontario, Canada, pp. 153–169

Hood, P., J. Irvine, and J. Hansen, 1982. The application of the aeromagnetic gradiometer survey technique to gold exploration in the Val d'Or mining camp, Quebec, *Can. Min. J.*, Sept.

Hood, P. J., P. H. McGrath, and D. J. Teskey, 1987. Evolution of Geological Survey of Canada magnetic-anomaly maps: a Canadian perspective, in *The Utility of Regional Magnetic and Gravity Anomaly Maps*, W. J. Hinze, ed., Society of Exploration Geophysicists, pp. 62–68.

International Atomic Energy Agency, 1991. *Airborne Gamma Ray Spectrometer Surveying*, Tech. Rep. Ser. 323, International Atomic Energy Agency, Vienna, 96 pp.

Kane, M. F., and R. H. Godson, 1987. Features of a pair of long-wavelength (>250 km) and short-wavelength (<250 km) Bouguer gravity maps of the United States, in *The Utility of Regional Magnetic and Gravity Anomaly Maps*, W. J. Hinze, ed., Society of Exploration Geophysicists, pp. 46–61.

Kearey, P., and M. Brooks, 1984. *An Introduction to Applied Geophysics*, Blackwell Scientific Publications, London, 296 pp.

Killeen, P. G., 1979. Gamma ray spectrometric methods in uranium exploration: application and interpretation, in *Geophysics and Geochemistry in the Search for Metallic Ores*, Econ. Geol. Rep. 31, P. J. Hood, ed., Geological Survey of Canada, Ottawa, Ontario, Canada, pp. 163–179.

Kovacs, A., and J. S. Holladay, 1990. Sea ice measurement using a small airborne electromagnetic sounding system, *Geophysics*, 55, 1327–1337.

Langel, R. A., 1992. International Geomagnetic Reference Field, 1991 revision: International association of Geomagnetism and Aeronomy (IAGA) Division V, Working Group 8; analysis of the main field and secular variation, *Phys. Earth Planet. Inter.*, 70, 1–6.

Leaton, B. F., 1971. I.G.R.F. charts, in A. J. Zmuda, (ed.), *World Magnetic Survey, 1957–1969*, IAGA Bull. 28, International Union of Geology and Geophysics, Paris. pp. 189–203.

Leclair, A. D., S. B. Lucas, H. J. Broome, D. W. Viljoen, and W. Weber, 1997. Regional mapping of Precambrian basement beneath Phanerozoic cover in southeastern Trans-Hudson Orogen, Manitoba and Saskatchewan, *Can. J. Earth Sci.* 34, 618–634.

Leube, A., W. Hirdes, R. Mauer, and O. K. Godfried, 1990. The early Proterozoic Birmian Supergroup of Ghana and some aspects of its associated gold mineralization, *Precambrian Res.* 46, 139–165.

Mendoca, C. A., and J. B. C. Silva, 1995. Interpolation of potential field data by equivalent layer and minimum curvature, *Geophysics*, 60, 399–407.

Palacky, G. J., 1987a. Geological background to resistivity mapping, in *Airborne Resistivity Mapping*, G. J. Palacky, ed., GSC Pap. 86–22, Geological Survey of Canada, Ottawa, Ontario, Canada, pp. 19–28.

Palacky, G. J., 1987b. Advances in geological mapping with airborne electromagnetic systems, in *Proceedings of Exploration '87: 3rd decennial International Conference on Geophysical and Geochemical Exploration for Minerals and Groundwater* OGS Spec. Vol. ISSN 0827-181X, Vol. 3, Ontario Geological Survey, Ottawa, Ontario, Canada, pp. 137–152.

Pemberton, R. H., 1987. Geophysical response of some common Canadian Massive sulphide deposits, in *Proceedings of Exploration '87: 3rd Decennial International Conference on Geophysical and Geochemical Exploration for Minerals and Groundwater*, OGS Spec. vol. ISSN 0827–181X, Vol. 3, Ontario Geological Survey, Ottawa, Ontario, Canada, pp. 137–152.

Reeves, C. V., 1987. Geophysical mapping of Precambrian granite–greenstone terranes as an aid to exploration, in *Proceedings of Exploration '87: 3rd Decennial International Conference on Geophysical and Geochemical Exploration for Minerals and Groundwater* OGS Spec. Vol. ISSN 0827-181X, Vol. 3, Ontario Geological Survey, Ottawa, Ontario, Canada, pp. 254–266.

Richardson, K. A., and P. G. Killeen, 1980. Regional radiogenic heat production mapping by airborne gamma ray spectrometry, in *Current Research, Part B*, GSC Pap. 80-1B, Geological Survey of Canada, Ottawa, Ontario, Canada, pp. 227–232.

Schwartz, E. J., and J. Broome, 1994. Magnetic anomalies due to pyrrhotite in Paleozoic metasediments in Nova Scotia, Canada, *J. Appl. Geophys.*, 32, 1–10.

Schwarz, E. J., and N. Wright, 1988. The detection of buried placer deposits by ground magnetic survey, *Geophys. Prospect.* 36, 919–932.

Sharpton, V. L., R. A. F. Grieve, M. D. Thomas, and J. F. Halpenny, 1987. Horizontal gravity gradient: an aid to the definition of crustal structure in North America, *Geophys. Res. Lett.*, 8, 808–811.

Shives, R. B. K., K. L. Ford, and B. W. Charbonneau, 1995. *Applications of Gamma*

Ray Spectrometric/Magnetic/VLF-EM Surveys: Workshop Manual, GSC Open File Rep. 3061, Geological Survey of Canada, Ottawa, Ontario, Canada, 82 pp.

Shives, R. B. K., Charbonneau, B. W., and Ford, K. L. 1997 The detection of potassic alteration by gamma-ray spectrometry—recognition of alteration related to mineralization; in Proceedings of Exploration 1997, 4th Decennial Conference on Mineral Exploration, Toronto, Canada, Sept 14–18., 741–762

Smith, R. S., and P. B. Keating, 1996. The usefulness of multicomponent, time-domain airborne electromagnetic measurements, *Geophysics*, 61(1), 74–81.

Smith, R. S., P. Annan, Lemieux J. and R. N. Pederson, 1996. Application of a modified GEOTEM system to reconnaissance exploration for kimberlites in the Point Lake area, NWT, Canada, *Geophysics*, 61(1), 82–92.

Stacey, F. D., 1977. *Physics of the Earth*, Wiley, New York, 414 pp.

Taylor, P. T., and C. C. Schnetzler, 1990. Satellite magnetic data: the exploration industry rates their usefulness, *Geophys. Lead. Edge Explor.*, 9(10), 42–43.

Telford, W. M., L. P. Geldart, R. E. Sherrif, and D. A. Keys, 1978. *Appl. Geophys.*, Cambridge University Press, New York, 860 pp.

Thomas, M. D., 1987. Gravity studies of the Grenville province: significance for Precambrian plate collision and the origin of anorthosite, in *The Utility of Regional Magnetic and Gravity Anomaly Maps*, W. J. Hinze, ed., Society of Exploration Geophysicists, pp. 109–123.

Webring, M. W., 1981. *MINC: A Gridding Program Based on Minimum Curvature*, USGS Open File Rep. 81–1224, U.S. Geological Survey, Reston Va., 43 pp.

Whaler, K. A., 1994. Downward continuation of *Magsat* data lithospheric anomalies to the earth's surface, *Geophys. J. Int.*, 116, 267–278.

Index

TUCKI WASH FAN

BLACKWATER WASH FAN

TRAIL CANYON FAN

Figure 2.21   Decorrelation stretched TIMS image of Death Valley, California displaying channels 5, 3, and 1 in red, green, and blue, respectively.

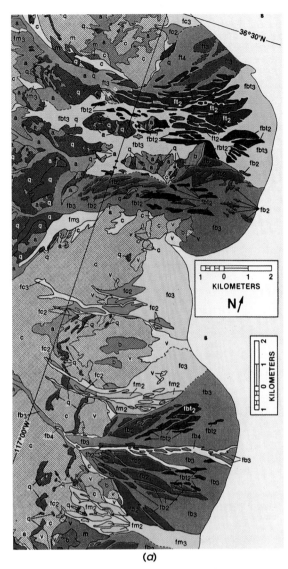

(a)

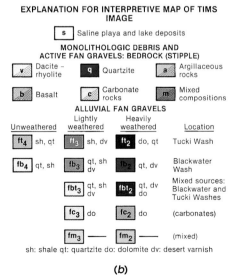

EXPLANATION FOR INTERPRETIVE MAP OF TIMS IMAGE

| s | Saline playa and lake deposits |

MONOLITHOLOGIC DEBRIS AND
ACTIVE FAN GRAVELS: BEDROCK (STIPPLE)

| v | Dacite – rhyolite | q | Quartzite | a | Argillaceous rocks |
| b | Basalt | c | Carbonate rocks | m | Mixed compositions |

ALLUVIAL FAN GRAVELS

Unweathered	Lightly weathered	Heavily weathered	Location
ft₄ sh, qt	ft₃ sh, dv	ft₂ do, qt	Tucki Wash
fb₄ qt, sh	fb₃ qt, sh dv	fb₂ qt, dv	Blackwater Wash
	fbt₃ qt, sh dv	fbt₂ qt, dv do	Mixed sources: Blackwater and Tucki Washes
	fc₃ do	fc₂ do	(carbonates)
	fm₃ ———	fm₂ ———	(mixed)

sh: shale qt: quartzite do: dolomite dv: desert varnish

(b)

Figure 2.22   Photo-interpretation map of the decorrelation stretch image shown in Figure 2.21.

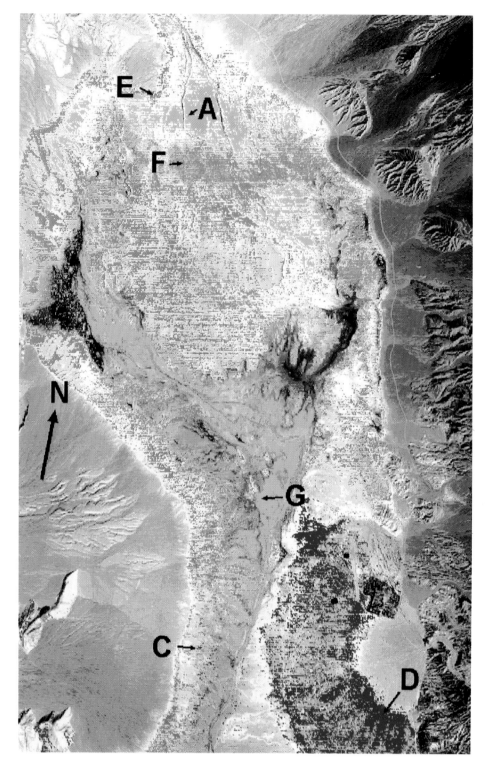

Figure 2.24   Pseudotemperature TIMS image of the Cottonball Basin study area with color overlays showing the distribution of various spectral classes. Labels A to G indicate areas used to extract the single-pixel alpha residual spectra shown in Figure 2.25. Surficial units represented by various image colors are: A, yellow pixels, thenardite-rich crusts in saline facies of sulfate zone; B, orange pixels, silty halite, smooth facies and carbonate zone, silty facies; C, red pixels, gypsum crusts; D, dark blue pixels, illite/muscovite-rich alluvial deposites; E, green pixels, quartz-rich fan gravels and mudflats; F, cyan pixels, massive halite and silty halite, rough facies; G, light green pixels, mixed silicate and evaporite mineral crusts on floodplains. The image has not been georeferenced and there is some geometric distortion. (Unit descriptions are modified from Hunt et al., 1966.)

Figure 2.27 Decorrelation stretched TIMS image of part of the north flank of Mauna Loa. Bands 5, 3, and 1 are displayed as red, green, and blue, respectively. Age and vegetation cover information are given in Figure 2.28.

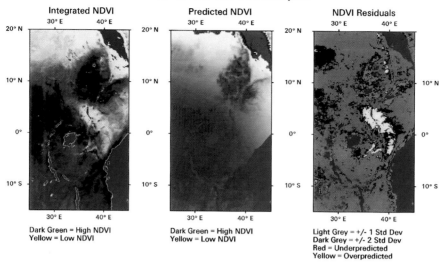

East Africa NDVI Analysis

Integrated NDVI

Dark Green = High NDVI
Yellow = Low NDVI

Predicted NDVI

Dark Green = High NDVI
Yellow = Low NDVI

NDVI Residuals

Light Grey = +/- 1 Std Dev
Dark Grey = +/- 2 Std Dev
Red = Underpredicted
Yellow = Overpredicted

Figure 4.4 Trends in NDVI values for East Africa. (Left) integrated image of the sum of the six NDVI multitemporal composites; (middle) NDVI surface predicted by the general linear model; (right) residual image when the predicted image is subtracted from the observed. Anomalous regions are indicated by red and yellow colors.

Geobotanical and Structural Interpretation Map of the Volcano Region, WV

Lineaments with strike sub-parallel to Burnings Springs Anticline

Lineaments with strike near-perpendicular to Burning Springs Anticline

Geobotanically anomalous areas

Fold axes

Figure 4.6 (Left) Vegetation classification using nPDF analysis of *Landsat* TM October image of the Volcano Oil Field, West Virginia, showing a geobotanical anomaly related to hydrocarbon seepage. Structural features, superimposed on the image, were interpreted from *Landsat* data.

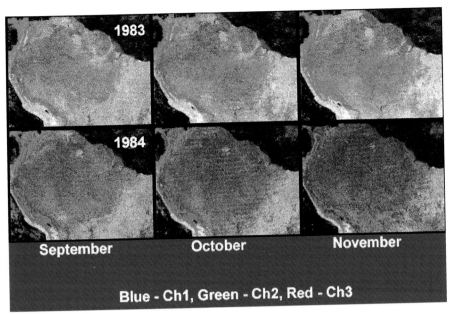

Figure 4.8   Color composite of AVHRR images for two years in the central Amazon Basin area. The 1983 El Niño is different from other years due to the widespread smoke in central and eastern Brazil, as noted by the increased blue. The smoke results from increased burning during the dryer periods of El Niño.

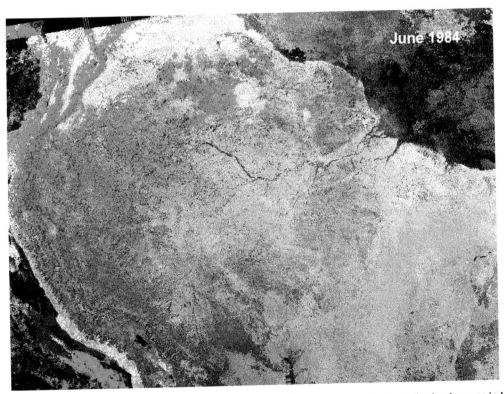

Figure 4.10   The AVHRR channel 3 provides an independent measure of subtle phenologic changes in tropical vegetation communities. RGB = AVHRR-Ch3, Ch2, and Ch1. Normally, the reddish areas designate savannas during the dry period. In June the northern savannas are beginning to green while the savannas south of the equator are senescing. Unusual for 1984 is the reddish color of the upland terra firma forests around Manous, Brazil, in the center of the image. This change is evident a year after the 1983 El Niño. To first order, we find that NPV is expressed in AVHRR-Ch3 for the 10-year AVHRR record. In the savannas, seasonal cycles of Ch3 lag temperature and NDVI by two months, also indicating that Ch3 is a consistent independent spectral measure of the surface.

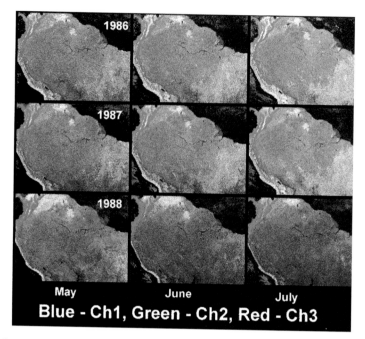

Figure 4.11   Surface change with the greatest spectral contrast corresponds to the year following the 1987 El Niño. The senescence of both the northern and southern savannas are delayed by three months in 1988 in comparison to 1986 and 1987. For both the 1983 and 1987 El Niño, no coincident spectral surface effects are detected. However, there are delayed phenologic responses of vegetation communities one year after both these periods.

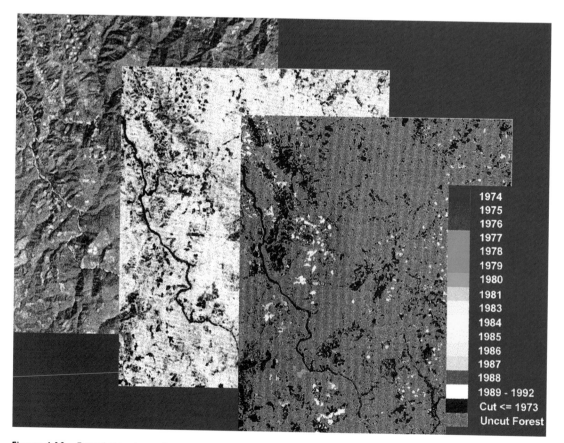

Figure 4.13   Forest stand age for an area around Bluff Creek, California, as derived from multidate *Landsat* MSS and TM imagery.

Thematic Mapper Color Composite Images

"Natural" Color: 3=R 2=G 1=B   Vis-NIR Color: 7=R 4=G 1=B

Figure 5.1 *Landsat* thematic mapper (TM) color composite images for a section of the Grand Canyon, Arizona: (top) "natural" color composite with TM band 3 in red, 2 in green, and 1 in blue; (bottom) color composite with visible and near-infrared bands, with TM band 7 in red, 4 in green, and 1 in blue.

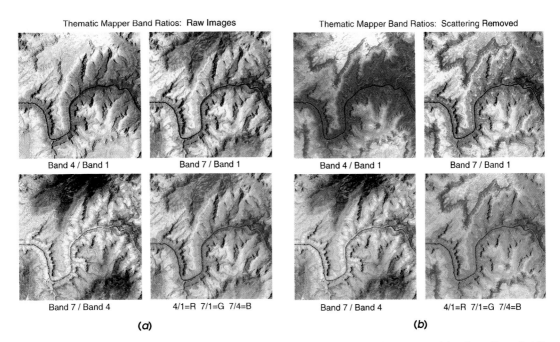

Figure 5.2 Effect of atmospheric scattering on the calculation of band-ratio images: (*a*) ratios of bands 4/1, 7/1, and 7/4 using raw data, along with a color composite of the three ratio images; (*b*) same three ratios and color composite as in (*a*), but with an estimate of the atmospheric scattering component removed.

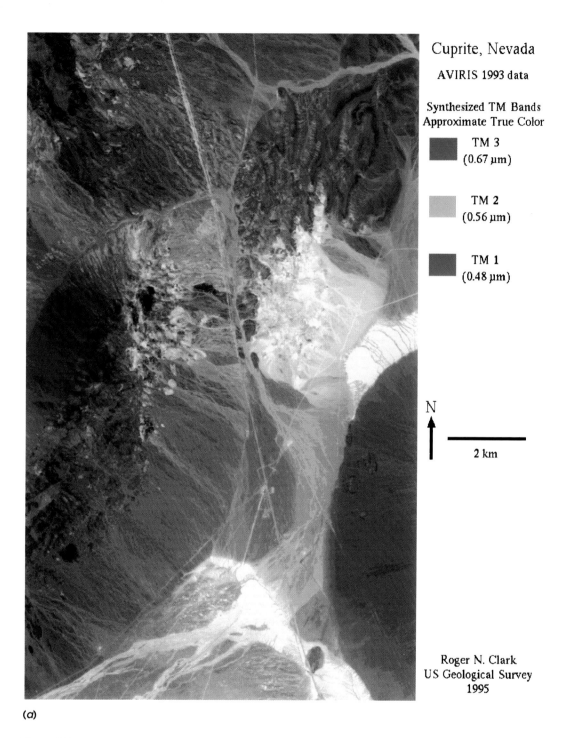

Cuprite, Nevada

AVIRIS 1993 data

Synthesized TM Bands
Approximate True Color

TM 3
(0.67 μm)

TM 2
(0.56 μm)

TM 1
(0.48 μm)

N

2 km

Roger N. Clark
US Geological Survey
1995

(a)

Figure 5.7   Example of mineral mapping using complete band shape for AVIRIS data of Cuprite, Nevada. (a) Pseudo-true-color composite of scene acquired in 1993. The scene is approximately 17 km long and 10.5 km wide. (b) Results of applying the full absorption band mapping technique, emphasizing minerals with relatively narrow, vibrational absorptions in the spectral region 2 to 2.5 μm. Regions in black indicate an absence of the minerals mapped. (c) Similar application, with emphasis instead on electronic absorption features which are broader than the vibrational absorptions and typically found in the spectral region 0.4 to 1.2 μm.

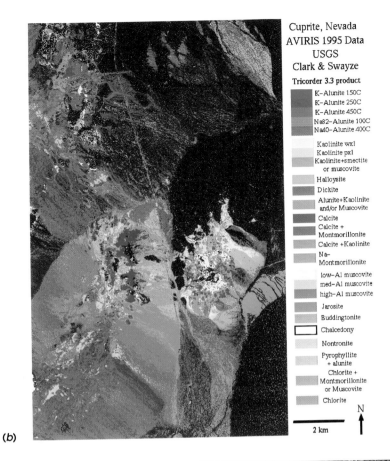

(b)

Cuprite, Nevada
AVIRIS 1995 Data
USGS
Clark & Swayze

Tricorder 3.3 product

K–Alunite 150C
K–Alunite 250C
K–Alunite 450C
Na82–Alunite 100C
Na40–Alunite 400C

Kaolinite wxl
Kaolinite pxl
Kaolinite+smectite
or muscovite
Halloysite
Dickite
Alunite+Kaolinite
and/or Muscovite
Calcite
Calcite +
Montmorillonite
Calcite +Kaolinite
Na–
Montmorillonite
low–Al muscovite
med–Al muscovite
high–Al muscovite
Jarosite
Buddingtonite
Chalcedony
Nontronite
Pyrophyllite
+ alunite
Chlorite +
Montmorillonite
or Muscovite
Chlorite

N

2 km

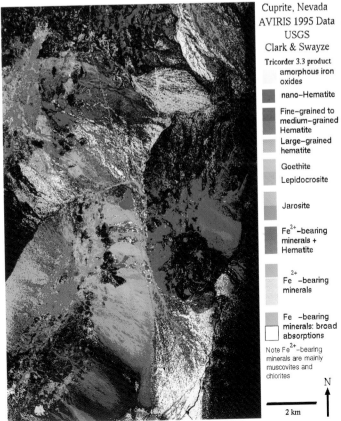

Cuprite, Nevada
AVIRIS 1995 Data
USGS
Clark & Swayze

Tricorder 3.3 product
amorphous iron
oxides
nano–Hematite

Fine–grained to
medium–grained
Hematite
Large–grained
hematite

Goethite
Lepidocrosite

Jarosite

Fe^{2+}–bearing
minerals +
Hematite

Fe^{2+}–bearing
minerals

Fe^{2+}–bearing
minerals: broad
absorptions

Note Fe^{2+}–bearing
minerals are mainly
muscovites and
chlorites

N

2 km

(c)

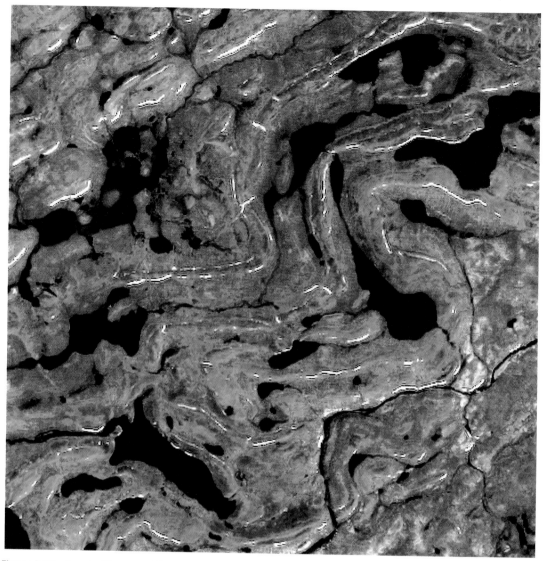

Figure 5.12 *Landsat* image of Cape Smith with TM band 2 in blue, band 3 in green, and band 4 in red. This image is approximately 15 km across, with north toward the top of the image. Areas with abundant green vegetation appear red, and well-exposed outcrops are blue. Snow banks appear white.

Figure 6.6   Color bar—perceived brightness of the gray
line is affected by brightness of surrounding color.

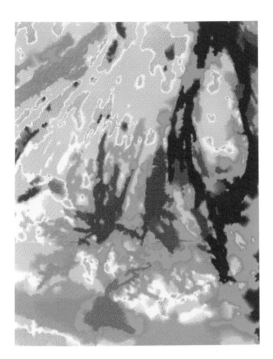

Figure 6.7 Digital elevation model (DEM) — different elevation ranges for each color.

Figure 6.8   Digital elevation model—Intensity modulated with shaded relief.

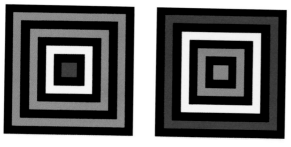

Figure 6.9   Color cubes perceived either as a hallway or pyramid.

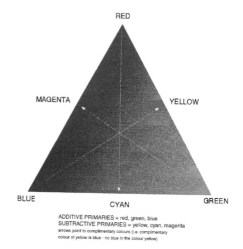

ADDITIVE PRIMARIES = red, green, blue
SUBTRACTIVE PRIMARIES = yellow, cyan, magenta
arrows point to complimentary colours (i.e. complimentary
colour of yellow is blue - no blue in the colour yellow)

Figure 6.13   RGB ternary triangle relating additive primary to subtractive primary colors.

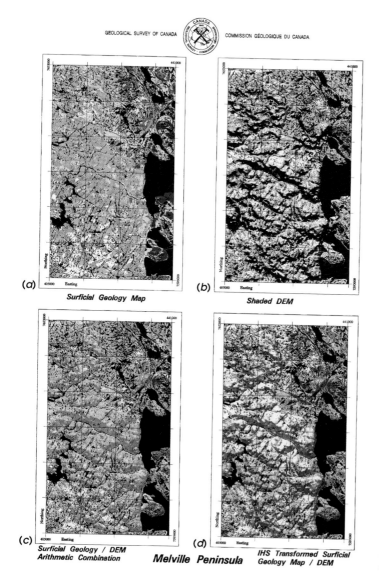

(a) Surficial Geology Map

(b) Shaded DEM

(c) Surficial Geology / DEM Arithmetic Combination

Melville Peninsula

(d) IHS Transformed Surficial Geology Map / DEM

Figure 6.17   IHS and arithmetic combination of surficial geology and digital elevation (DEM) data.

(a)   (b)

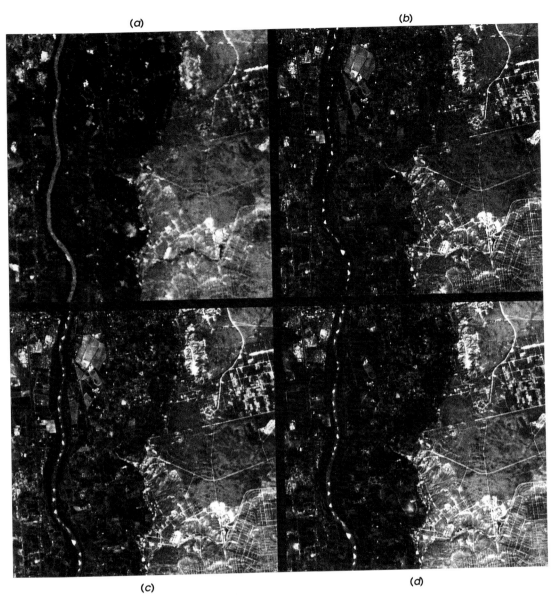

(c)   (d)

Figure 6.18   (a) *Landsat* TM image; (b) *SPOT* image; (c) IHS combined *Landsat* TM and *SPOT* images; (d) *Landsat* TM and *SPOT* images combined using the MWD technique (see the text for details). (Courtesy of David Yocky, Sandia National Laboratories, Albuquerque, New Mexico.)

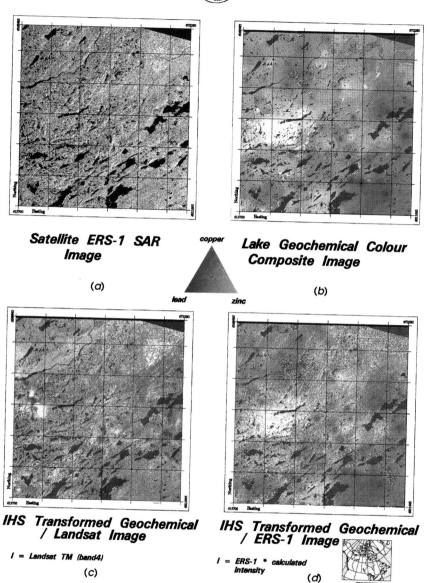

Satellite ERS-1 SAR Image

(a)

copper

lead   zinc

Lake Geochemical Colour Composite Image

(b)

IHS Transformed Geochemical / Landsat Image

I = Landsat TM (band4)

(c)

IHS Transformed Geochemical / ERS-1 Image

I = ERS-1 * calculated intensity

(d)

Figure 6.19   IHS combination of *ERS-1* radar, *Landsat* TM, and lake sediment geochemical data: (*a*) *ERS-1* radar image; (*b*) lake sediment color composite image; (*c*) IHS combined *Landsat* TM and geochemical image; (*d*) IHS combined *ERS-1* and geochemical image. (Geochemical data courtesy of the Geological Survey of Canada.)

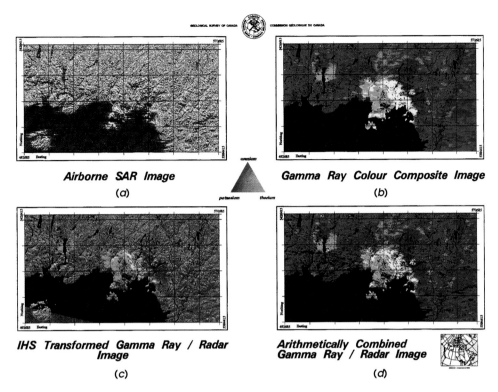

Airborne SAR Image

(a)

Gamma Ray Colour Composite Image

(b)

IHS Transformed Gamma Ray / Radar Image

(c)

Arithmetically Combined Gamma Ray / Radar Image

(d)

Figure 6.21   IHS and arithmetic combination of airborne radar and gamma ray spectrometer data: (*a*) airborne radar image; (*b*) gamma ray spectrometer color composite image (red-uranium, green-thorium, blue-potassium); (*c*) IHS combined radar and gamma ray spectrometer data; (*d*) arithmetic combination of radar and gamma ray spectrometer data. (Gamma ray spectrometer data courtesy of the Geological Survey of Canada; airborne radar courtesy of the Canada Centre for Remote Sensing.)

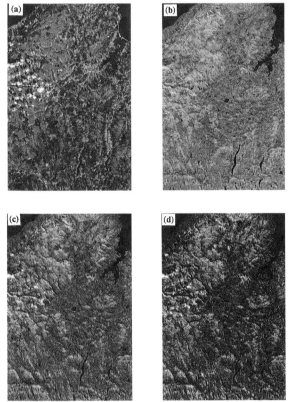

Figure 6.22   Combination of *Landsat* and airborne radar data: (*a*) *Landsat* MSS false-color composite image; (*b*) RGB *Landsat* MSS and radar color composite image; (*c*) arithmetic combination of *Landsat* MSS and radar imagery; (*d*) IHS-combined *Landsat* MSS and radar image. (Radar imagery courtesy of the Canada Center for Remote Sensing.)

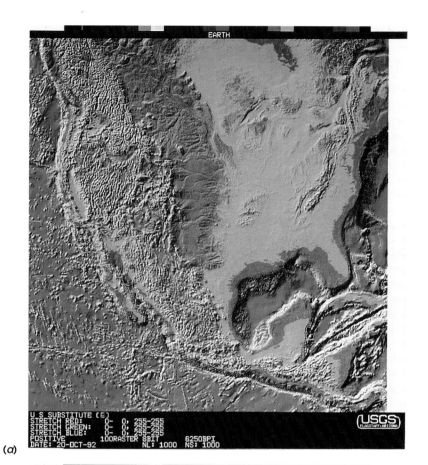

(a)

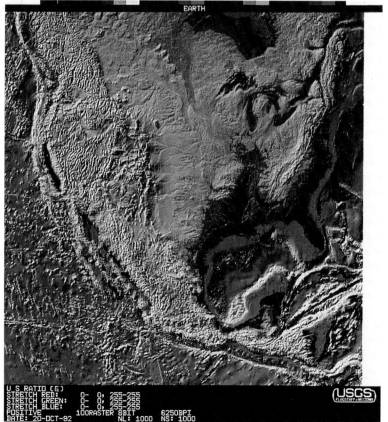

(b)

Figure 6.23 IHS combination of shaded relief topographic and color-coded topographic data: (a) intensity substitution method; (b) intensity-adjustment technique. (Imagery courtesy of K. Edwards, U.S. Geological Survey; original topographic information from NOAA color-shaded relief images provided by the U.S. Geological Survey.)

Saturation - 100%

(a)

Saturation - 25%

(c)

Figure 6.24 IHS combined airborne magnetic and gravity data: (*a*) saturation 100%; (*b*) saturation 50%; (*c*) saturation 25%. (Magnetic and gravity data courtesy of the Geological Survey of Canada.)

Saturation - 50%

(b)

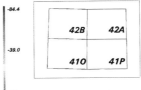

-84.4

-39.0

-6.5
units in milligals

42B	42A
41O	41P

NTS Map Sheets Covered - 1:250,000

LOCATION MAP - LOCALISATION DE LA CARTE

IHS TRANSFORMED MAGNETICS/GRAVITY IMAGE - ONTARIO

I = shadow enhanced magnetics
H = gravity data
S = gravity data

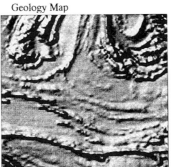

(a) Geology Map

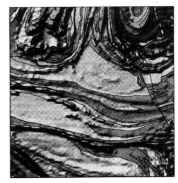

(c) IHS Combined Geology and Magnetic Data

Figure 6.26 IHS combination of geological and airborne magnetic data: (*a*) geological map; (*b*) shaded relief magnetic image; (*c*) IHS combination of geological and magnetic data. (Geological map courtesy of Ken Ashton, Saskatchewan, Energy & Mines; magnetic data from Geological Survey of Canada.)

(b) Shaded Relief Magnetics Image

LOCATION MAP - LOCALISATION DE LA CARTE

(d)

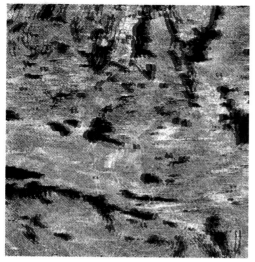 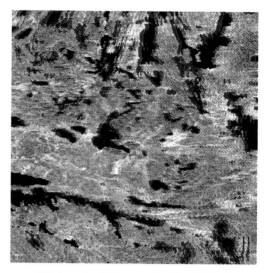

(a) Landsat/DEM Analglyph   (b) Landsat/Magnetics Analglyph

Figure 6.29   Analglyphs: (a) analglyph in which relief is provided by topographic data (DEM), (b) analglyph in which relief is provided by airborne magnetic data.

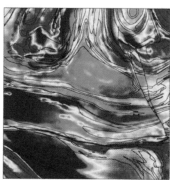 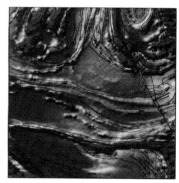

(a)   Coloured Total Field Magnetic Image   (c)   Chromosterescopic Magnetics Image

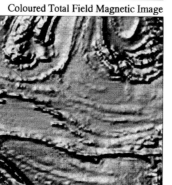 

Figure 6.30   Chromostereoscopic image: (a) color-enhanced total field magnetic image; (b) shaded relief magnetic image; (c) IHS combined chromostereoscopic image.

(b)   Shaded Relief Magnetics Image   (d) LOCATION MAP - LOCALISATION DE LA CARTE

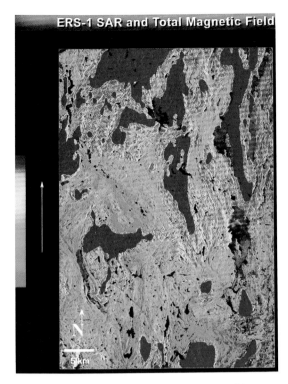

Figure 6.31 IHS combined *ERS-1* and airborne magnetic chromostereoscopic image. (Courtesy of B. Rivard (University of Alberta) and T. Toutin, Canada Centre for Remote Sensing.)

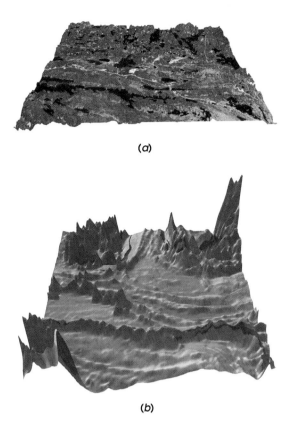

(a)

(b)

Figure 6.32 Perspective views: (*a*) *Landsat* TM, perspective view with topographic relief; (*b*) geologic map, perspective view with magnetic relief.

Figure 6.33 3D view of geologic map, seismic reflection trace and gravity profile—gravity data courtesy of Geological Survey of Canada, seismic data courtesy of Lithoprobe project, image courtesy of D. Desnoyers (Geological Survey of Canada).

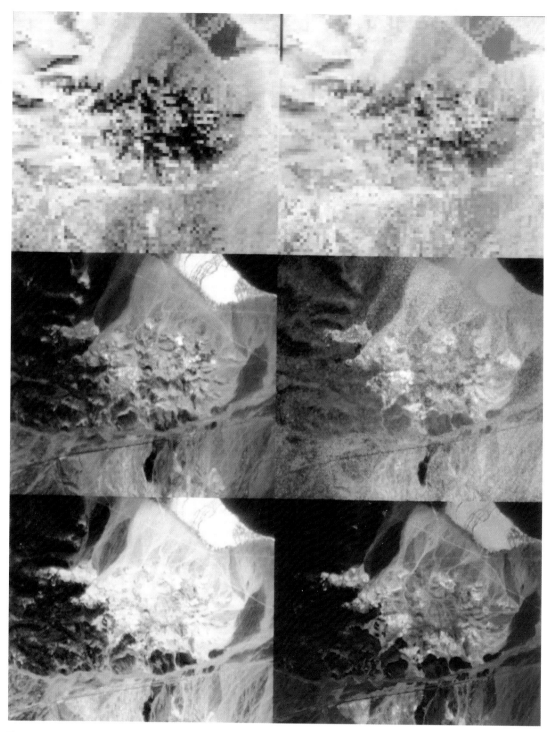

Figure 8.1   Color composite images of Cuprite, Nevada, processed from AVIRIS and TIMS data to simulate ASTER images. The top row of images are linearly stretched composites, and the bottom row are decorrelation stretched images. Images on the left depict ASTER VNIR bands 3, 2, and 1 in red, green, and blue (RGB), respectively. The center pair displays SWIR bands 4, 6, and 9 (RGB), and the pair on the right displays TIR bands 13, 12, and 10 (RGB). ASTER bandpasses are given in Table 8.1, and a geologic map of the Cuprite area appears in Chapter 11. (From Abrams and Hook, 1995.)

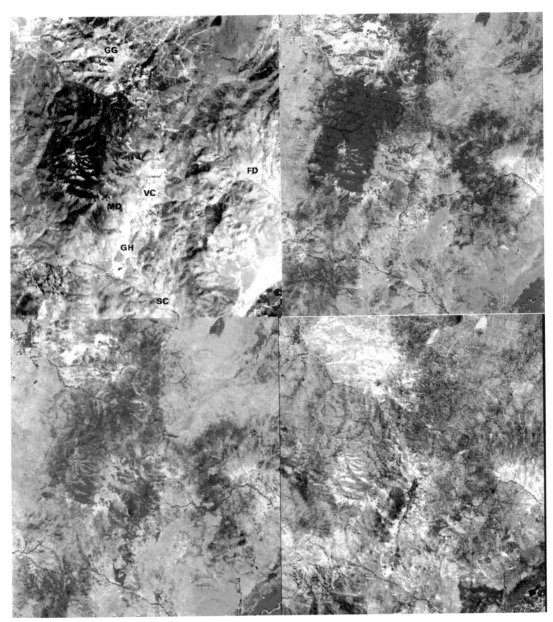

Figure 8.8 *Landsat* TM images of the Comstock mining district and surrounding area, Virginia City, Nevada: (*a*) True-color composite of bands 3, 2, and 1 (RGB); (*b*) band-ratio color composite of TM 5/7 (red), TM 5/4 (green), and 3/1 (blue); (*c*) selective principal components color composite made from second principal components of the same band pairs as (*b*) and represented in the same colors; (*d*) feature-oriented principal components image depicting pixels dominated by ferric oxides in blue and pixels dominated by hydroxyl-bearing minerals in red. Pixels with abundant ferric oxide and hydroxyl minerals appear white. Labeled localities are identified in Figure 8.9. (Courtesy of Richard Bedell, Homestake Mining Company.)

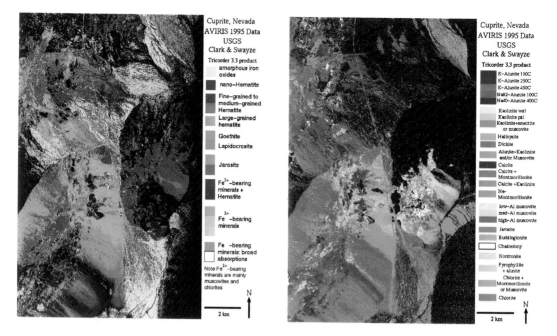

Figure 8.10 Classified mineral map of the Cuprite, Nevada, area constructed from 1995 AVIRIS images using the USGS tricorder algorithm: (*a*) mineral distributions mapped from the VNIR bands; (*b*) minerals mapped from SWIR bands. (Courtesy of Roger Clark, USGS Spectroscopy Laboratory.)

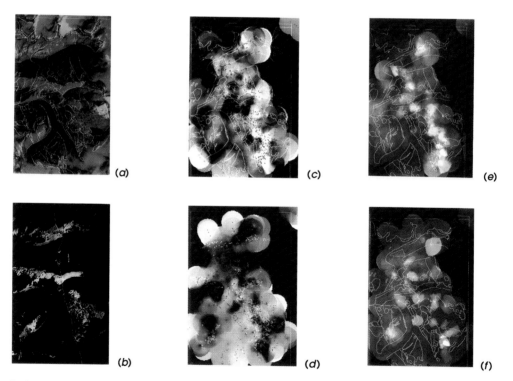

Figure 8.12 *Landsat* TM and geochemical data from the Sulphurets–Bruce Jack Lake district, British Columbia: (*a*) TM 543 (RGB) color composite image; vegetation appears in various shades of yellow, glacial ice in blue, and surficial sediments and outcrop in shades of red; (*b*) two-component band-ratio composite image with TM 5/7 in red and TM 3/1 in green; pixels dominated by ice, snow, and vegetation are masked and appear black; (*c*) continuous surface map of SiO_2 derived from lithogeochemical data; (*d*) continuous surface map of alkali index values derived from lithogeochemical data; (*e*) ternary continuous surface map of gold (red), silver (green), and antimony (blue) values derived from lithogeochemical data; (*f*) ternary continuous surface map of copper (red), zinc (green), and lead (blue) values derived from lithogeochemical data.

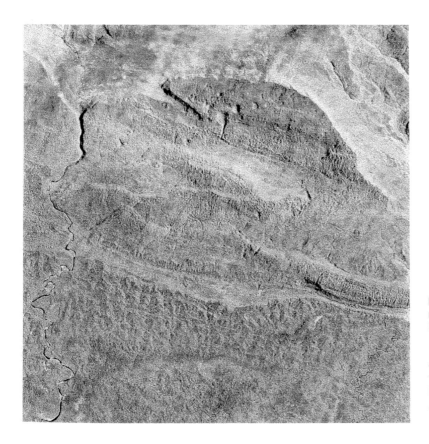

Figure 9.2 *Landsat* thematic mapper image of P'nyang area. Path/row 100/64 acquired Sept. 22, 1986. Location is shown on Figure 9.1. P'nyang Anticline is the NW–SE trending cigar-shaped feature just above image center. (From Valenti, 1993.)

Northern Grapevine Mountains, Nevada/California – Remote Sensing Data Comparisons

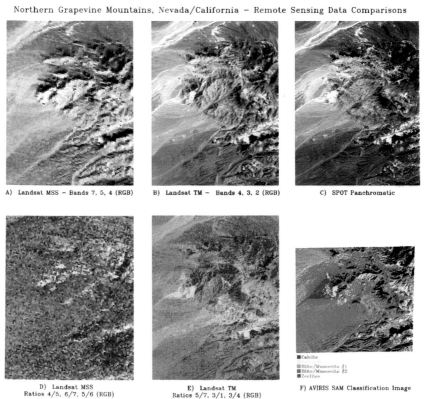

A) Landsat MSS – Bands 7, 5, 4 (RGB)

B) Landsat TM – Bands 4, 3, 2 (RGB)

C) SPOT Panchromatic

D) Landsat MSS
Ratios 4/5, 6/7, 5/6 (RGB)

E) Landsat TM
Ratios 5/7, 3/1, 3/4 (RGB)

F) AVIRIS SAM Classification Image

Figure 11.1 Comparison of images from *Landsat* MSS, *Landsat* TM, SPOT, and AVIRIS for the northern Grapevine Mountains, Nevada site.

A. AVIRIS Bands 30, 18, 8 (0.67, 0.55, 0.45 μm)
RGB, True Color Image, Goldfield, NV, USA

B. AVIRIS Unmixing Result, Kaolinite, Alunite,
Illite/Muscovite (RGB), Goldfield, NV, USA

Figure 11.16 AVIRIS true-color composite for reference (left) and overlay of color composite of Goldfield unmixing results on AVIRIS band 30 (0.67 μm) (right). Red areas are predominately kaolinite, green areas are predominantly alunite, and blue areas are predominately illite/muscovite. Mineral mixtures appear as intermediate colors (e.g. yellow = red + green = kaolinite + alunite).

Figure 12.7 AIRSAR three-frequency HV polarization image of Kilauea volcano, Hawaii. Red is P-HV, green is L-HV, and blue is C-HV. Kilauea caldera (about 3.2 km in diameter) is at the right. Lava flows of varying ages and surface roughness characteristics have different backscatter responses at each wavelength, giving the different colors and brightnesses in this image. Aa flow of 1974 is the bright meandering feature in lower center and left. Scarps of the Kaoe Fault system are seen in the lower center. Image dimensions are 9.3 km × 20.3 km. North is toward the upper right.

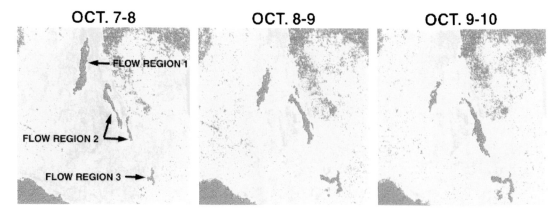

OCT. 7-8   OCT. 8-9   OCT. 9-10

Figure 12.8   Active flow field of Kilauea volcano, showing L-band decorrelation signatures of active areas observed by SIR-C in 24-hour repeat-pass interferometry, October, 1994. Area shown is 3.8 km × 3.8 km. Correlated areas are shown in yellow and green; decorrelated areas in pink and light blue. Among the three active flow regions, region 1 shows decreased activity with time, while regions 2 and 3 show increased activity. Decorrelated area in lower left is the ocean; upper right is vegetation. (From Zebker et al., 1996.)

Scale (Km)

↑N

Three-dimensional p⊘
metry data obtaine⊘
the center and the
n is applied. Color i⊘
n is CVV, and blue is

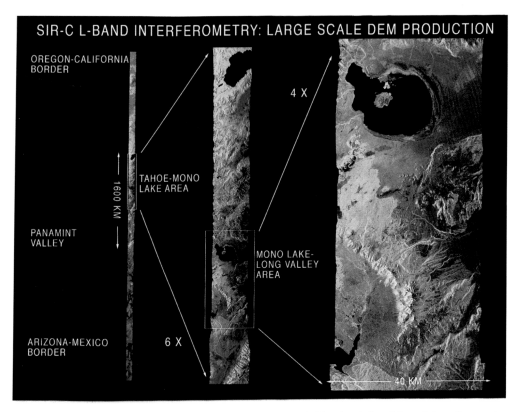

OREGON-CALIFORNIA
BORDER

1600 KM

TAHOE-MONO
LAKE AREA

PANAMINT
VALLEY

4 X

MONO LAKE-
LONG VALLEY
AREA

ARIZONA-MEXICO
BORDER

6 X

40 KM

Figure 12.10   Large-scale digital elevation model of eastern California derived from L-band spaceborne repeat-pass interferometry by SIR-C. The full swath measures 1600 km × 40 km, with 25-m pixel spacing. Colors represent elevations, brightness is derived from L-band backscatter. This experiment demonstrated the feasibility of interferometric topo map generation at continental scales, which is essential for future global topographic mapping missions. (Courtesy of P. Rosen, JPL, 1995.)

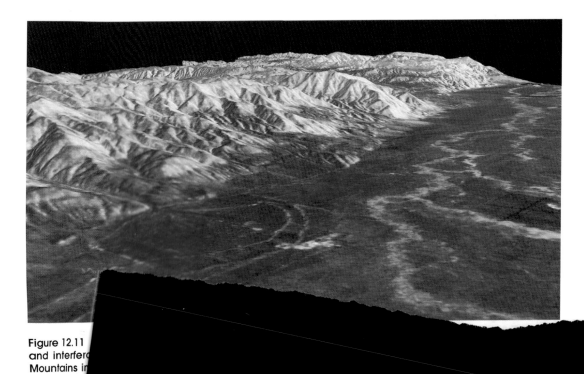

Figure 12.11
and interfer
Mountains i
exaggeratic
is LVV, gree

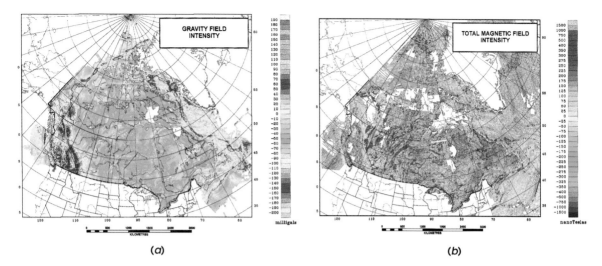

(a)   (b)

Figure 13.2   Potential field images for Canada: (a) color intensity image showing the Bouguer gravity field (the Bouguer gravity correction is explained in Section 13.2.2.3; (b) total magnetic field image. Note that 1 gauss is equivalent to 10^5 nanotesla. (Images generated by the Canadian Geophysical Data Centre.)

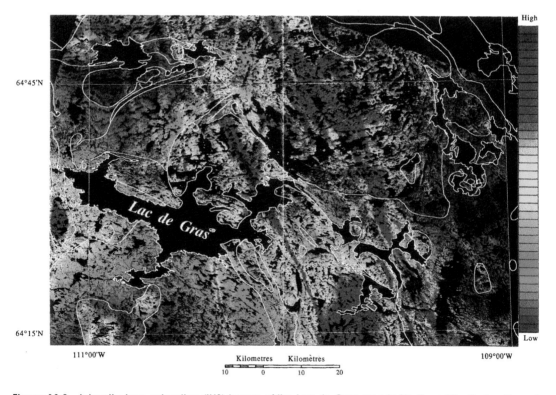

Figure 13.9   Intensity–hue–saturation (IHS) image of the Lac de Gras area in Northwest Territories, Canada, generated by combining Landsat TM band 5 with total magnetic intensity (hue). Saturation is fixed at 100%. The line overlay shows the location of geological contacts. (From Broome et al., 1993).

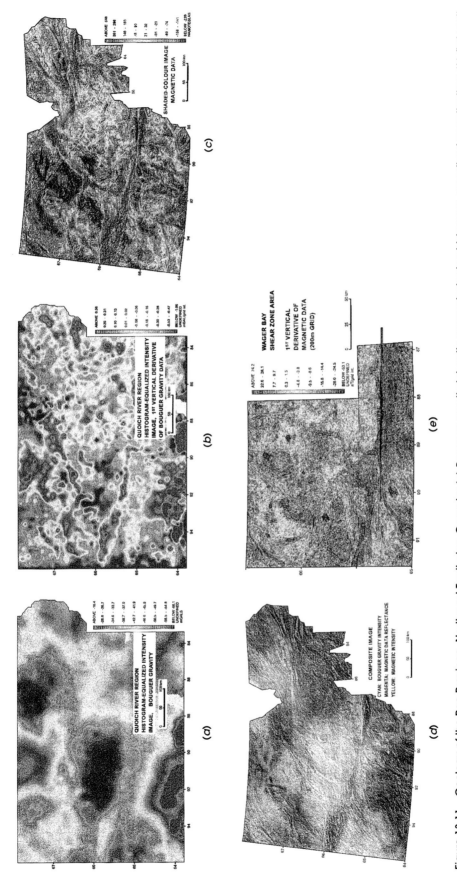

Figure 13.11 Geology of the Rae Province, Northwest Territories, Canada: (*a*) Bouguer gravity intensity generated using histogram-equalized quantization; (*b*) first vertical derivative of the Bouguer gravity data shown in Figure 13.11; (*c*) shaded-color image of magnetic intensity (hue is determined by magnetic intensity and color saturation is controlled by reflectance); (*d*) composite magnetic/gravity image created with magnetic intensity (magenta), gravity intensity (cyan), and shaded magnetic relief with illumination from the northwest (yellow); (*e*) vertical gradient calculated from magnetic data for the area outlined in (*c*). A dotted line follows the Wager Bay Shear Zone (WBSZ).

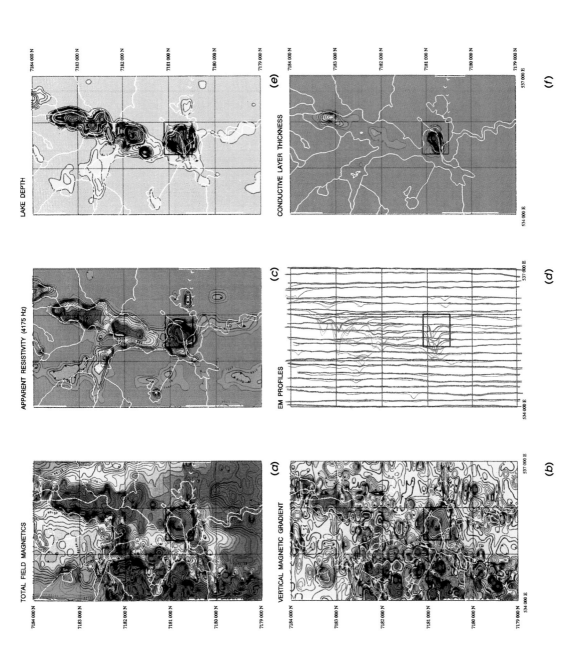

Figure 13.19 Images derived from a detailed multi-parameter helicopter survey over a Kimberlite pipe in the Lac de Gras area, Northwest Territories, Canada (courtesy of Aerodat), (a) total field, (b) vertical magnetic gradient, (c) apparent resistivity, (d) EM profiles, (e) lake depth, and (f) conductive layer thickness.

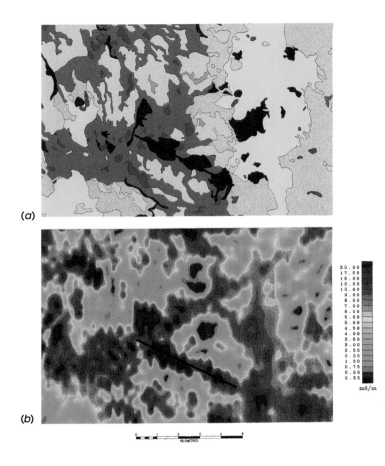

(a)

(b)

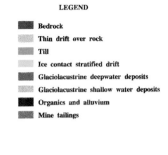

Figure 13.21 Overburden geology (a) and conductivity image (b) for a glaciated region in the Kirkland Lake area, Ontario, Canada. The black line following a linear conductivity high corresponds with three zones of a relatively conductive unit, "organics and alluvium." The continuous nature of the linear anomaly suggests that the discrete mapped areas on the geology map may be continuous in the subsurface.

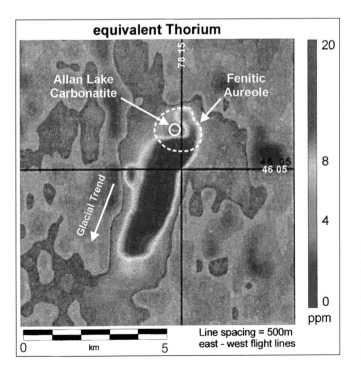

equivalent Thorium

Allan Lake
Carbonatite

Fenitic
Aureole

Glacial Trend

20

8

4

0

ppm

Line spacing = 500m
east - west flight lines

0   km   5

Figure 13.22 Image generated from a high-resolution gamma ray spectrometry survey of the Allan Lake Carbonatite: from p. 1072. Although the carbonatite and fenite alteration halo are only about 1 km² in combined area, a 10-km-long dispersal train is clearly outlined in the thorium images. (From Ford et al., 1988.)